Project 233

IUGS UNESCO IGCP

Terranes In The Circum-Atlantic Paleozoic Orogens

R.D. DALLMEYER, Athens, USA
J.D. KEPPIE, Halifax, Canada
(Directors)

Volumes already published in the series:

R.D. DALLMEYER and E. MARTINEZ GARCIA (Eds.):
Pre-Mesozoic Geology of Iberia (1990)

R.D. DALLMEYER and P.P. LECORCHE (Eds.):
The West African Orogens and Circum-Atlantic Correlatives (1991)

R.D. Dallmeyer
E. Martínez García (Eds.)

Pre-Mesozoic Geology of Iberia

With 187 Figures

Springer-Verlag
Berlin Heidelberg New York
London Paris Tokyo
Hong Kong Barcelona

Dr. R. DAVID DALLMEYER
Department of Geology
University of Georgia
Athens, Georgia 30602, USA

Prof. Dr. ENRIQUE MARTINEZ GARCIA
Departamento de Geologia
Universidad de Oviedo
33005 Oviedo, Spain

ISBN 3-540-51792-8 Springer-Verlag Berlin Heidelberg New York
ISBN 0-387-51792-8 Springer-Verlag New York Berlin Heidelberg

Library of Congress Cataloging-in-Publication Data. Pre-Mesozoic geology of Iberia / R. D. Dallmeyer, E. Martínez García (eds.). p. cm.
Papers from an international conference hosted by IGCP Project 233 (Terranes in the Circum-Atlantic Paleozoic Orogens) in Sept. 1986 in Oviedo, Spain. 1. Geology, Stratigraphic-Precambrian-Congresses. 2. Geology, Stratigraphic-Paleozoic-Congresses. 3. Geology-Spain-Congresses. 4. Geology-Portugal-Congresses. I. Dallmeyer, R. D., 1944- . II. Martínez-García, E. (Enrique), 1940- . III. IGCP Project 233-Terranes in the Circum-Atlantic Paleozoic Orogens.
QE653.P744 1990 551.7'00946-dc20 90-9772

Typesetting and printing: Druckerei Appl, Wemding
Binding: J. Schäffer, Grünstadt
32/3145-543210 - Printed on acid-free paper

Preface

The Iberian Massif constitutes the southwesternmost and largest tectonic element of the so-called Armorican plate, a continental block which appears to have rifted from Gondwana in the Early Paleozoic. It was then separated from Laurentia by a proto-Atlantic Ocean, from Baltica by the Tornquist Sea and from Gondwana by the Theic Ocean. Subsequent closing of these intervening oceans and the sequential reassembly of the various continental elements resulted in multiphase Paleozoic orogenesis within the circum-Atlantic mobile belts. The Iberian Massif therefore plays an important role in any attempt to reconstruct the Paleozoic tectonothermal evolution of the circum-Atlantic continents, because it displays the largest remnant of the western and southern margins of Armorica. Moreover, the area has largely escaped significant rectonic reworking during Alpine tectonothermal events.

In view of its critical tectonic position in the circum-Atlantic realm, Project 233 (Terranes in the Circum-Atlantic Paleozoic Orogens) of the International Geological Correlation Program hosted an international conference in September 1986, in Oviedo, Spain. The conference focused on the geologic evolution of the Iberian Massif and its regional correlatives. During the conference it became clear that the knowledge of Iberian geology had increased markedly in the last decade, during which the Spanish Geological Survey (IGME) itself had published over 300 1:50,000 map sheets. The overwhelming quantitiy of new geologic data presented during the conference prompted us to organize a volume dedicated to synthesizing the Pre-Mesozoic Geology of Iberia. This is not merely a collecting of papers presented during the Oviedo meeting, but, instead, represents a coordinated volume in which regional experts were solicited for contributions. We feel the result reflects the serious efforts of all contributors, and represents a major step forward in Iberian geology. We sincerely appreciate the work of all reviewers whose comments have greatly improved this volume.

Athens/Oviedo, Autumn 1990

R. D. Dallmeyer
E. Martinez Garcia

Contents

Part III West Asturian-Leonese Zone

Part IV Central-Iberian Zone

Part V Ossa-Morena Zone

Part VI South Portuguese Zone

Part VII Sedimentary and Faunal Domains in the Iberian Peninsula During Lower Paleozoic Times

Contributors

Addresses listed at beginning of respective contributions.

Apalategui, O.
Arbizu, M. A.
Arenas, R.
Arribas, A.
Baeza, L.
Barriga, F. J. A. S.
Bastida, F.
Carracedo, M.
Corretgé, L. G.
Dallmeyer, R. D.
de San José, M. A.
Dias, R.
Díez Balda, M.
Eguiluz, L.
Florido, P.
Gabaldón, V.
Galán, G.
García Alcalde, J. L.
García-Hidalgo, J. F.
García López, S.
Gil Ibarguchi, J. I.
González Lodeiro, F.
Gumiel, P.
Gutiérrez Marco, J. C.
Herranz, P.
Heredia, N.
Liñán, E.
Locutura, J.
Luque, C.
Marcos, A.
Martínez, F. J.
Martínez Catalán, J. R.
Martínez Chacón, M. L.
Martínez García, E.
Menéndez Alvarez, J. R.
Méndez Bedia, I.
Méndez Fernández, C.
Montesinos, J. R.
Munhá, J.
Oliveira, J. T.
Palacios, T.
Pascual, E.
Peláez, J. R.
Pereira, E.
Perejón, A.
Pérez Estaún, A.
Pieren, A. P.
Pulgar, J. A.
Quesada, C.
Ribeiro, A.
Robardet, M.
Rodríguez Fernández, L. R.
Ruíz, F.
Sánchez Carretero R.
Sánchez de Posada, L. C.
Silva, J. B.
Soto, F.
Suárez, O.
Tornos, F.
Truyóls, J.
Truyóls Massoni, M.
Ugidos, J. M.
Vegas, R.
Wagner, R. H.
Villa, E.
Vilas, L.

Part I Introduction

Introduction to the Pre-Mesozoic Geology of Iberia

R. D. DALLMEYER[1] and E. MARTINEZ GARCIA[2]

The Iberian Massif constitutes the largest expanse of pre-Permian rocks within the Iberian Peninsula (Fig. 1). Other exposures include the Sierra de la Demanda, Iberian Ranges, Basque Massifs, the axial zone of the Pyrenees, Catalonian Coastal Ranges, and several poorly known areas within the Alpine Betic Cordilleras (Fig. 1). The Iberian Massif represents the westernmost exposure of the European Hercynides, and constitutes the largest Hercynian domain to have escaped significant reworking during Alpine tectonothermal events. Only local Tertiary continental deposits within fault-bounded basins cover older sequences. Nearly all post-orogenic cover was removed following eastward regional tilting in the Holocene, which resulted in extensive erosion and entrenchment of west-flowing rivers into a pre-Premian peneplain. The Iberian Massif is bounded to the north, west, and southwest by the Cantabrian Sea and the Atlantic Ocean. Its eastern margin is covered by post-Hercynian and post-Alpine continental deposits, and the southeastern boundary is marked by the Guadalquivir Fault.

Calderón (1985) and Macpherson (1988) considered the Iberian Meseta to represent a block of ancient rocks surrounded by younger orogenic belts. Staub (1926) suggested that the Massif was composed of central exposures of Archean rocks which had been successively reworked during Caledonian and Hercynian orogenic events. By contrast, Stille (1924, 1927) interpreted the Iberian Massif to have been totally constituted during the Hercynian Orogeny. The extent of Hercynian tectonothermal activity was more clearly defined by Lotze (1929b), Carlé (1945), and Schmidt-Thomé (1945). These new views of Iberian geology led Lotze (1945) to subdivide the Iberian Massif into six different zones on the basis of contrasts in stratigraphic, structural, and/or metamorphic characteristics. From northeast to southwest these included (Fig. 1) the Cantabrian, West Asturian-Leonese, Galician-Castilian, Lusitanian-Alcudian, Ossa-Morena, and South Portuguese Zones. This subdivision was modified slightly with publication of the Tectonic

[1] Department of Geology, University of Georgia, Athens, Georgia 30602 USA

[2] Departamento de Geología, Universidad de Oviedo, 33005 Oviedo, Spain

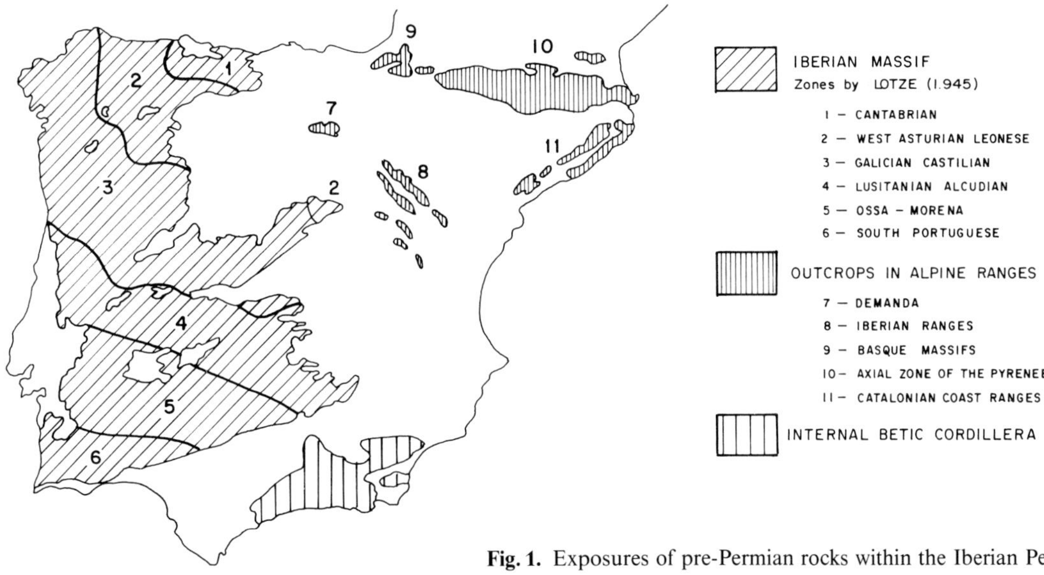

Fig. 1. Exposures of pre-Permian rocks within the Iberian Peninsula

R. D. Dallmeyer and E. Martínez García (Eds.)
Pre-Mesozoic Geology of Iberia

Map of the Iberian Peninsula (Julivert et al. 1972), where the Galician-Castilian and Lusitanian-Alcudian Zones of Lotze were combined into a Central Iberian Zone (Fig. 1) in which both allochthonous and autochthonous units were outlined. With few exceptions, the resultant fivefold division of the Iberian Massif has been generally accepted by most geologists working in Iberia. However, there has been a considerable controversy regarding the nature and exact location of some zone boundaries (e. g., Marcos et al. 1980; vs Martínez García 1986; Herranz Araujo 1984; vs Robardet 1976). In spite of these difficulties, the fivefold zonal division of the Iberian Massif is non-genetic and largely based on significant geologic contrasts. Therefore, this volume has been organized following this subdivision. Each zone is discussed in a separate chapter which begins with an introductory statement outlining its overall characteristics, location, boundaries, history of definition, and major points of controversy. Following are systematic descriptions of the stratigraphic, structural, igneous, metamorphic, and metallogenetic characteristics. The closing two chapters attempt to synthesize the preceding data and present a broader view of the Iberian Massif and its tectonic setting in the circum-Atlantic realm.

References

Calderón S (1985) Ensayo orogénico sobre la Meseta Central de España. Anal Soc Esp Hist Nat 15 (1): 131-160

Carlé W (1945) Ergebnisse geologischer Untersuchungen im Grundgebirge von Galicien (Nordspanien). Geotekt Forsch 6: 13-36

Herranz Araujo P (1984) El Precámbrico del NEE de "Ossa Morena" Planteamiento y estado de la cuestión, unidades, bases para su correlación y esquema evolutivo. Cuad Geol Iberica 9: 119-211

Julivert M, Fontboté JM, Ribeiro A, Conde LN (1972) Mapa Tectónico de la Península Ibérica y Baleares. Inst Geol Min Esp

Lotze F (1929a) Das Präkambrium Spaniens. Neues Jahrb Geol Paläontol 8: 373-380

Lotze F (1929b) Stratigraphie und Tektonik des Keltiberischen Grundgebirges (Spanien). Abh Ges Wiss Gött Math Phys Kl (N Ser) 14 (2): 1-320

Lotze F (1945) Zur Gliederung der Varisziden in der Iberischen Meseta. Geotekt Forsch 6: 78-92

Macpherson J (1988) Del carácter de las dislocaciones de la Península Ibérica. Anal Soc Esp Hist Nat 17: 331-367

Marcos A, Martínez FJ, Julivert M, Pulgar JA (1980) Explicación de la Hoja n° 50 (Cangas de Narcea) del Mapa Geológico Nacional (MAGNA) a escala 1:50.000. Inst Geol Min Esp, pp 1-31

Martínez García E (1986) The Iberian Massif. In international conference on Iberian terranes and their regional correlation, Oviedo, Spain, 1986. Guidebook Excursion A-1: Oviedo, 1-39

Robardet M (1976) L'originalite du segment hercynien sudiberique au Paleozoique inferieur, Ordovicien, Silurien et Devonien dans le nord de la province de Seville (Espagne). C R Acad Sci Paris (Serie D) 283: 999-1002

Schmidt-Thomé P (1945) Paläozoisches Grundgebirge und junges Deckegebirge im Westlichen Zentralspanien (Provinz Salamanca und Cáceres). Geotekt Forsch 6: 37-77

Staub R (1926) Gedanken zur Tektonik Spaniens, Viertels. Naturforsch Ges Zürich 71: 196-260

Stille H (1924) Grundfragen der Vergleichenden Tektonik, Borntraeger, Berlin

Stille H (1927) Über Westmediterrane Gebirgszusammenhänge. Abh Ges Wiss Gött Math Phys Kl 12: 3

Part II Cantabrian and Palentian Zones

1 Introduction

A. PEREZ ESTAUN[1]

The Cantabrian Zone constitutes the foreland thrust and fold belt of the Hercynian Mountain belt in the northwestern part of the Iberian Peninsula. It is placed in the core of the Ibero-Armorican arc, defined by the trace of the Hercynian structures in these areas. The Cantabrian Zone is a stratigraphically and structurally well established domain. The first map showing some of the characteristics of this zone was that of Schulz (1858), and the presence of faults and folds with vergence towards the central part of the arc was pointed out by Barrois (1882). The Cantabrian Zone boundaries and some of its features, based mainly on the stratigraphical differences with respect to the neighbouring zones, were characterized by Lotze (1945). Julivert presented the first comprehensive review and established the

[1] Departamento de Geología, Universidad de Oviedo, 33005 Oviedo, Spain

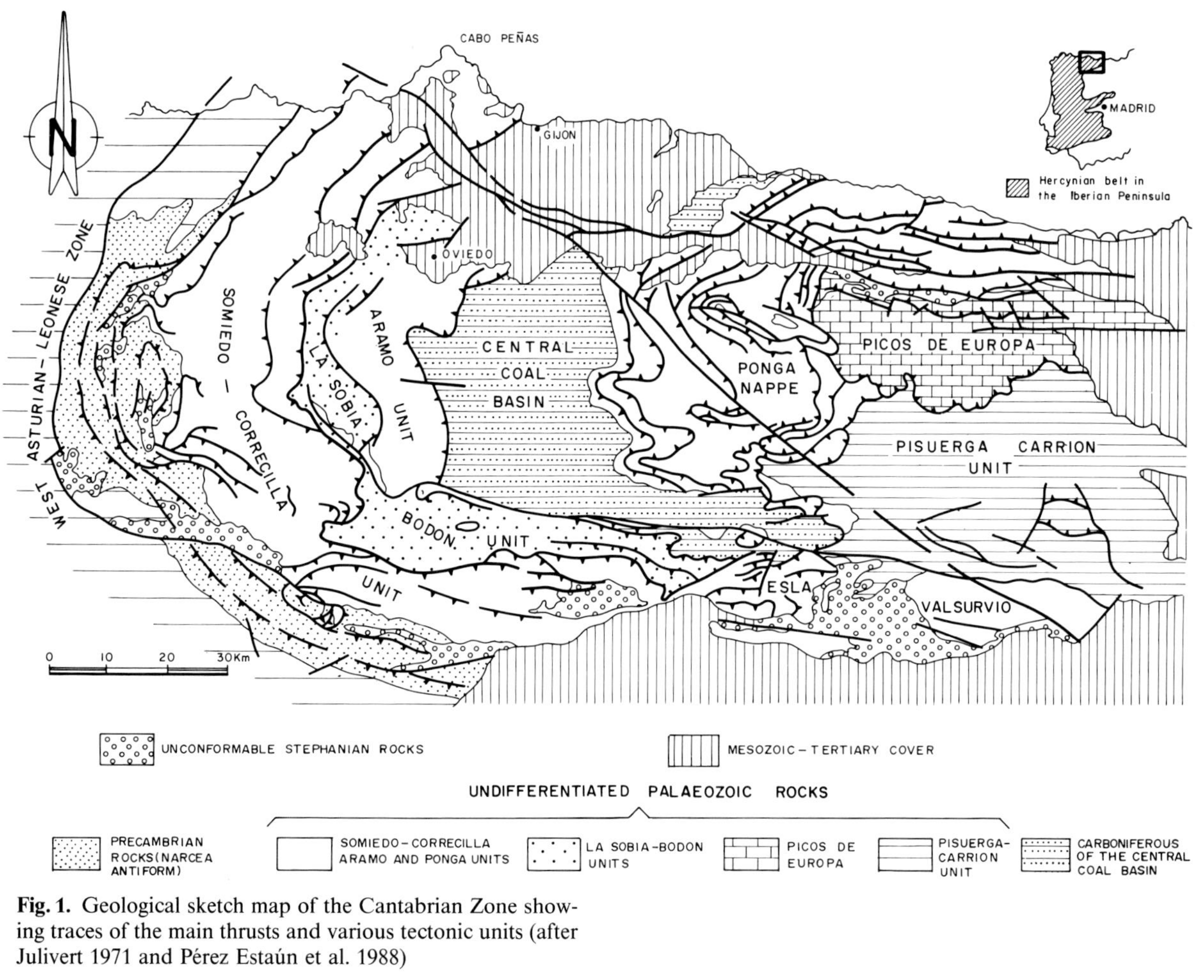

Fig. 1. Geological sketch map of the Cantabrian Zone showing traces of the main thrusts and various tectonic units (after Julivert 1971 and Pérez Estaún et al. 1988)

R. D. Dallmeyer and E. Martínez García (Eds.)
Pre-Mesozoic Geology of Iberia

main structural characteristics of this zone. There are a large number of references dealing with the geology of the Cantabrian Zone. A general synthesis of the geology of this zone can be found in the papers of Julivert (1979, 1981, 1987), Julivert and Marcos (1973), Truyóls and Marcos (1978), Marcos and Pulgar (1982), Truyóls (1983), and Pérez Estaún et al. (1988).

The structure of the Cantabrian Zone has a thin-skinned type geometry complicated by its arcuate shape. The deformation took place practically without metamorphism and penetrative cleavage is generally absent. The western boundary of the Cantabrian Zone, the boundary to the more internal zones of the orogen (West Asturian-Leonese Zone), is formed by Precambrian rocks cropping out in the core of an antiform with a complex structure (Narcea Antiform). The other limits are constituted by Mesozoic and Tertiary rocks and the Cantabrian Sea.

Stratigraphically, the Cantabrian Zone is characterized by an incomplete and comparatively thin Cambrian-Ordovician sedimentary sequence, a Siluro-Devonian succession which varies from being complete in the west to being absent in the east, and a very well developed Carboniferous sequence.

Within the Cantabrian Zone, several major allochthonous units (Fig. 1) (Julivert 1971; Pérez Estaún et al. 1988) and a unit forming part of the relatively autochthonous Ibero-Armorican arc core can be distinguished. This later unit has been called the Pisuerga-Carrión Unit by Julivert (1971) and Palentian Unit by Martínez García (1981). It displays some features differing from the overlying allochthonous units. For this reason and because of the results of recent work in this unit, the description of their characteristics has been separated from the Cantabrian Zone in the following chapters.

References

Barrois C (1882) Recherches sur les terrains anciens des Asturies et de la Galice. Mém Soc Geol Nord II 1: 1-630

Julivert M (1971) Decollement tectonics in the Hercynian Cordillera of NW Spain. Am J Sci 270 (1): 1-29

Julivert M (1979) A cross-section through the northern part of the Iberian Massif: its position within the Hercynian fold belt. Krystallinikum 14: 51-67

Julivert M (1981) A cross-section through the northern part of the Iberian Massif. Geol Mijnbouw 60 (1): 107-128

Julivert M (1987) The structure and evolution of the Hercynian fold belt in the Iberian Peninsula. In: Schaer JP, Rodgers J (eds) The anatomy of mountain belts. pp 65-102

Julivert M, Marcos A (1973) Superimposed folding under flexural conditions in the Cantabrian Zone (Hercynian Cordillera, NW Spain). Am J Sci 273: 353-375

Lotze F (1945) Zur Gliederung der Varisziden in der Iberischen Meseta. Geotekt Forsch 6: 78-92

Marcos A, Pulgar JA (1982) An approach to the tectonostratigraphic evolution of the Cantabrian foreland thrust and fold belt, Hercynian Cordillera of NW Spain. N Jahrb Geol Paläont Abh 163 (2): 256-260

Martínez García E (1981) El Paleozóico de la Zona Cantábrica Oriental (Noroeste de España). Trab Geol 11: 95-127

Pérez Estaún A, Bastida F, Alonso JL, Marquínez J, Aller J, Alvarez Marrón J, Marcos A, Pulgar JA (1988) A thin-skinned tectonics model for an arcuate fold and thrust belt: the Cantabrian Zone. Tectonics 7 (3): 517-537

Pérez Estaún A, Martínez Catalán JR, Bastida F (in press) Crustal thickening and deformation sequence in the footwall to the suture of the Hercynian Belt of NW Spain. Tectonophysics

Schulz G (1858) Descripción geológica de la Provincia de Oviedo. Jose González, Madrid, 138 pp

Truyóls J (1983) El Carbonífero del NW de España. In: Martínez Diaz (coord) Carbonifero y Pérmico de España, Inst Geol Min Esp, pp 27-30

Truyóls J, Marcos A (1978) La cartografía geológica de Asturias desde Guillermo Schulz a nuestros dias. Trab Geol 10: 5-18

2 Stratigraphy

2.1 Introduction

A. PEREZ ESTAUN[1]

The Cantabrian Zone can be divided into two areas from the stratigraphic point of view: the Asturian-Leonese domain and the Palentine domain. The Asturian-Leonese domain comprises the following tectonic units (see Introduction, Fig. 1): Somiedo, Sobia and Aramo Units, Central Coal Basin Unit, Ponga Unit and Picos de Europa Unit. The Palentian Unit corresponds to the Pisuerga-Carrión or Palentian Zone. On the other hand, the Paleozoic sequence of the Cantabrian Zone can be subdivided into preorogenic and synorogenic (Carboniferous) successions, since they show a very distinctive character. As a consequence, the description of the stratigraphy of the Cantabrian Zone is going to be subdivided into the following chapters: Pre-Carboniferous rocks of the Asturian-Leonese domain, Carboniferous rocks of the Asturian-Leonese domain, Pre-Carboniferous rocks of the Palentine domain, Carboniferous rocks of the Palentine domain, and Stephanian and Permian basins.

[1] Departamento de Geología Universidad de Oviedo 33005 Oviedo, Spain

R. D. Dallmeyer and E. Martínez García (Eds.)
Pre-Mesozoic Geology of Iberia

2.2 The Asturian-Leonese Domain (Cantabrian Zone)

J. Truyols, M. A. Arbizu, J. L. Garcia Alcalde, S. Garcia Lopez, I. Mendez Bedia, F. Soto and M. Truyols Massoni[1]

2.2.1 Introduction

The most complete Palaeozoic succession of the Iberian Massif, from Lower Cambrian to Lower Permian, is developed in the Asturian-Leonese Domain of the Cantabrian Zone.

These rocks rest unconformably on Precambrian. However, variations in the extent of the basin at different times make for gaps in the Palaeozoic succession in different places. There is sedimentary uniformity over the entire Cantabrian Zone up to the Lower Ordovician. From Late Ordovician to Late Devonian times, an Asturian uplift, located in the northeastern part of the Cantabrian Zone, limited the sedimentary basin and acted as its main source area. After a Late Devonian levelling episode, a slight eastward tilting of the marine platform virtually eliminated the Asturian borderland, which ceased to act as a source area. The more extensive Carboniferous basin received clastic input from the West Asturian-Leonese Zone. Although the deformation during the Hercynian orogeny produced nappes and thrust-sheets facing the concave part of the Asturian Arc (AA), the palaeogeographic reconstructions are not too difficult.

Precambrian. The oldest strata, the Narcea Schists of Lotze (1956) (= Mora Fm. of Comte 1959), crop out in the western part of the Cantabrian Zone, in the so-called Narcea Antiform (Fig. 1; see Pérez Estaún, this Vol., Part II, Chap. 1). This formation is a turbidite succession consisting of 1000–2000 m of greywackes and siltstones (Pérez Estaún 1973). In some places, porphyritic rocks affected by schistosity are present; these are probably older than the turbidites in which they occur. Although unfossiliferous, a Late Vendian age is assumed. The formation is unconformably overlain by Lower Palaeozoic rocks (Lotze 1956); the Precambrian-Cambrian boundary is a featureless surface, probably due to peneplanation, perhaps equivalent to the Cadomian unconformity in other places in Europe.

Cambrian. The Palaeozoic commences with a clastic succession, ca. 500–1400 m thick, consisting of pinkish feldspathic quartzites, sandstones, and conglomerates, the Herrería Fm. of Comte (1938c) (= Cándana Fm. of Lotze 1961). It frequently shows the presence of dolomitic beds near the base, and of green or violet shales at the top (i. e. the Barrios Member of Lotze 1961). These beds represent shallow platform deposits with some intertidal episodes. Trace fossils occur occasionally (Seilacher 1970; Crimes et al. 1977). In the Barrios Mb., trilobites are found: *Dolerolenus, Lunolenus*, and *Anadoxides* (Sdzuy 1961). The age of the formation is early Cambrian (Tommotian-Atdabanian; Ovetian Stage of Sdzuy, 1971), but a Vendian age cannot be excluded for the lower part.

The Herrería Fm. is overlain by ca. 40 m of shallow water limestones (Zamarreño 1972), the Láncara Fm. of Comte (1937). This formation is constituted of grey limestones and yellow stromatolitic dolomites in its lower half, and by nodular, generally red limestones ("Cambrian griotte"), in the upper half. The formation begins in the Atdabanian, reaching to the outermost part of the AA, the highest Solenopleuropsidae "Stage" (Zamarreño 1972). The Lower-Middle Cambrian boundary practically coincides with the lithological change mentioned above. The upper part of the lower Member, has yielded an archaeocyathid fauna of Elankian age (Debrenne and Zamarreño 1970), with *Archaeocyathus* cf. *laqueus*, and *Pycnoidocyathus* cf. *erbiensis*, the main localith being at Valdoré (León province). The upper Member has yielded, just above its base, brachiopods and trilobites be-

[1] Departamento de Geología, Universidad de Oviedo, c/ Jesús Arias de Velasco, s/n, 33005 Oviedo, Spain

R. D. Dallmeyer and E. Martínez García (Eds.)
Pre-Mesozoic Geology of Iberia

longing to the *Paradoxides (Acadoparadoxides)* "Stage", *Conocoryphe ovata* "Substage" (Sdzuy 1961). The very abundant fossil contents comprise mainly trilobites, but also brachiopods, echinoderms, and algae (Zamarreño 1972).

After the carbonate deposition of the Láncara Fm., siliciclastic sedimentation was restored with the Oville Fm. (Comte 1937), ca. 100–400 m of glauconitic sandstones and green shales, with many facies changes. Towards the upper part, quartzitic beds predominate. These are still shallow platform deposits and the upper part represents partly intertidal and deltaic environments. The base of this formation provides abundant trilobites and carpoids of Mid-Cambrian age (Sdzuy 1961). The diachronism between this formation and the preceding one is shown by the presence of faunas in different places, ranging from the *P. (Acadoparadoxides)* "Stage" (with *Acadolenus decorus, A. inornatus, Parabailiella matutina, Acadoparadoxides* aff. *sacheri*) to the top of the Solenopleuropsidae "Stage" (with *Badulesia juliverti, Solenopleuropsis marginata, S. simula, Conocoryphe heberti, Paradoxides brachyrhachis, P. pradoanus,* etc.). The top of the formation has yielded some ichnofossils (Baldwin 1978) and palynomorphs (Fombella 1978), which still represent the Middle Cambrian and the Tremadocian. This seems to indicate either the absence of the Upper Cambrian or an extremely condensed development of the same.

Ordovician. The Oville Fm. is overlain by the Barrios Fm. (Comte 1937), ca. 150–400 m of white quartz sandstones with some interbedded quartzitic conglomerates and, sometimes, with an intercalation in the middle of brown to greenish sandy shales. These beds were deposited in shallow water, ranging into intertidal to supratidal environments. The formation has been mainly dated by trace fossils (Crimes and Marcos 1976; Baldwin 1978). In the lower part, the occurrence of *Cruziana breadstoni* or *C. semiplicata* indicates the Tremadocian; higher up, the association of *C. furcifera, C. rugosa* and *C. goldfussi* points to the Arenigian. At the top of this unit, or at the bottom of the next one up (Gutiérrez Marco and Rodríguez 1987a, b; Aramburu 1987), the late Arenigian graptolite *Azygograptus undulatus* of the-*hirundo* Biozone, has been found.

Usually, the Barrios Fm. is paraconformably overlain by Silurian shales (Formigoso Fm.) (W and S of the Central Coal Basin) or by uppermost Famennian rocks (Ermita Fm.) (Ponga Nappe Region; Fig. 1; see Pérez Estaún, this Vol., Part II, Chap. 1). In the northeastern part of the La Sobia-Bodón Nappe, and in some places of the Ponga Nappe Region (Fig. 1; see Pérez Estaún, this Vol., Part II, Chap. 1), the formation was progressively eroded or even eliminated. Where an apparently uninterrupted succession occurs (coastal Asturian area, Peñas and Vidrias Capes, Fig. 1, see Pérez Estaún, this Vol., Part II, Chap. 1), the Barrios Fm. is overlain by a ca. 400 m thick, black or dark grey shaly unit, the Luarca Fm. (Barrois 1882). In the latter, sedimentation took place in deeper, quieter, and more poorly oxygenated conditions. Sometimes the lower part of the Luarca Fm. consists of a Transition Series (50–100 m thick), with nodular shales, quartz sandstones, and tuffs, locally with basic volcanic sills.

At the base of the Luarca Fm. Llanvirnian graptolites have been found (*Didymograptus artus, D. murchisoni,* etc.) (Pello and Philippot 1967; Julivert et al. 1968; Gutiérrez Marco 1986). Higher in the sections near the Peñas and Vidrias Capes, Llandeilian graptolites (*Orthograptus calcaratus, Gymnograptus linnarssoni,* etc.) and trilobites occur (*Neseuretus tristani, Placoparia borni, Eodalmanitina destombesi destombesi, Ectillaenus giganteus,* etc.) (Julivert et al. 1968; Julivert and Truyóls 1972; Truyóls and Julivert 1976; Montesinos 1981a, b; Gutiérrez Marco 1986). In the same area, the Luarca Fm. is overlain by the mainly Caradocian to early Ashgillian El Castro Volcano-Detritic Complex (Radig 1962), which is composed of ca. 100–500 m of sandstones, tuffites, and volcanic agglomerates, with basaltic lavas in the upper part. The Llandeilian/Caradocian boundary can be traced here in the uppermost part of the Luarca Fm. or in the basal one of the next unit, above beds where *Prionocheilus mendax, Destombesia hispanica* and *Ectillaenus giganteus* occur (Guitérrez Marco 1986).

The El Castro Complex is followed by the Sarreo Fm., a thin calcareous unit, ca. 30 m thick, with sandy shales and tuffs in the middle part. The lower part, with conodonts such as *Trichonodella* cf. *excavata* and *T.* cf. *symmetrica,* is probably Ashgillian; the upper one contains Silurian conodonts such as *Panderodus unicostatus, Neoprioniodus* cf. *planus, Hindeodella* cf. *equidentata* and *Lonchodina* cf. *walliseri.* Within the middle shaly member, *Flexicalymene* sp. and *Eospirifer* sp. occur (Julivert and Truyóls 1972).

Southeastwards of the Somiedo-Correcilla Nappe (Fig. 1; see Pérez Estaún, this Vol., Part II, Chap. 1), the Transition Series of the basal Luarca Fm. is sometimes overlain unconformably by Ashgillian limestones (8 m thick), containing brachiopods *(Nicolella actoniae, Eoanastrophia pentamera, Leptestiina prantli)*, cystoids and crinoids (*Mesopilocystites* cf. *tregarvanicus, Conspectocrinus celticus, Aonodiscus spinosus*, etc.) (Gutiérrez Marco 1986).

Silurian. The Silurian basin extended almost throughout the western and southern parts of the AA. Thereafter, up to the latest Famennian, the remainder of the Asturian-Leonese Domain belonged to the Asturian uplift.

Excepting the Upper Sarreo Fm. limestone (see above), the Cantabrian Silurian is always represented by a shaly unit, the Formigoso Fm. of Comte (1937). This formation consists of a terrigenous unit, ca. 100–300 m thick, with black or grey soft siltstones, the Bernesga Member of Kegel (1929), which become more compact towards the upper part and which alternate with fine-grained sandstones (the Villasimpliz Member of Kegel 1929). These deposits were formed in an euxinic, quiet water environment, not far from the coast, containing pelagic graptolite faunas and palynomorphs. The lower part of the Bernesga Mb. represents the Late Llandovery as follows from the presence of graptolites of de *-convolutus* and *-sedgwicki* Biozones (*Monograptus tenuis, M. sedgwicki, M. runcinatus, Petalograptus altissimus, Pseudoplegmagraptus obesus*, etc.). In the upper part of the formation, Early Wenlockian graptolites of the *-centrifugus* or *-murchisoni* Biozones occur (*Retiolites geinitzianus, Monograptus priodon, M. vomerinus*, etc.). The Llandovery-Wenlock boundary should be located within the Villasimpliz Mb. (Kegel 1929; Comte 1959; Truyóls et al. 1972). Palynomorphs confirm the age based on graptolite data (Cramer 1964, 1966, 1967, 1970).

The Formigoso Fm. is overlain by the Furada Fm. (Barrois 1882; =San Pedro Fm. of Comte 1936a), ca. 80–200 m of red, sometimes grey or white, ferruginous sandstones, with thin-bedded, green, shaly intercalations and sandy limestones lenses, the latter being especially abundant in the upper part. These sediments were deposited in an agitated to more quiet water platform environment. From the lower part of the formation, Late Wenlockian palynomorphs have been recovered (Rodríguez 1982). The middle part of the unit still provides Silurian (Ludlovian) brachiopods *(Orbiculoidea striata, Dalmanella orbicularis)*, graptolites *(Saetograptus fritschi fritschi, S. chimaera salweyi)* and palynomorphs, but the uppermost 30 m represents already the Devonian with brachiopods, trilobites, conodonts and other Early Lochkovian fossils (Comte 1959; Cramer 1967, 1970; Poll 1970; Truyóls et al. 1974; Rodríguez 1982, and new data; Figs. 1, 2).

Devonian. The Devonian developed in an overall regressive context. The most complete succession occurs in the external part of the AA and consists of ca. 2000 m of alternating terrigenous and calcareous, sometimes reefal, units. Towards the Asturian uplift, lying during the Devonian around the area where the present-day Central Coal Basin is and northeastwards, the Devonian thins and practically disappears. The sedimentation took place in the shallow marine neritic realm, sometimes in intertidal to supratidal conditions, and sporadically in deeper shelf conditions.

The Furada and San Pedro Fms. are overlain by a 400–600 m thick, predominantly calcareous, series. In the northern branch of the AA this unit is termed the Rañeces Group (Comte 1959) (cf. Arbizu 1972). It includes four distinct lithostratigraphic units: the Nieva Fm. (Barrois 1882), with limestones, dolostones and shales; the Bañugues Fm. (Zamarreño 1976), chiefly with dolostones; the Ferroñes Fm. (Barrois 1882), with argillaceous and crinoidal limestones and shales; and the Aguión Fm. (Radig 1962; cf. Arnao Fm. of Barrois 1882), with red or pink coloured, crinoidal limestones. Towards the Central Coal Basin, the Rañeces Group becomes more terrigenous and less fossiliferous. In the southern branch of the AA the unit receives the name of La Vid Group (Comte 1936). It can be divided into four informal members (García Alcalde et al. 1979): sandy and argillaceous limestones and then dolomitic marls and dolostones, sometimes with stromatolites; argillaceous limestones and shales; dark brown shales with limestone lenses; and red detrital limestones and marlstones interbedded with shales. In the Valsurbio Region (Fig. 1; see Pérez Estaún, Part II, Chap. 1), the Compuerto Fm. (Koopmans 1962) correlates rather well with the upper part of the above units. The sedimentation took place in subtidal to intertidal or supratidal conditions (cf. Vera de la Puente 1985). Fossils (rugose and tabulate corals, stromatoporoids, bryozoans, brachiopods, bivalves, gastropods, nautiloids, tentaculitoids, trilobites, ostracods, crinoids, blastoids, conodonts, etc.) are in general very abundant (Figs. 1, 2). The Lochkovian/Pragian boundary can be provisionally traced at the appearance of the brachiopod *Vandercammenina sollei*, in the Nieva Fm. The Pragian/Emsian boundary would be best placed at the middle part of the Ferroñes Fm., and within the unit 2 of the La Vid Group in the conodont Biozone *P. dehiscens*, just before an important faunal renewal (Figs. 1, 2).

The Moniello (Barrois 1882), and Santa Lucía (Comte 1936) Formations, include ca. 250 m of rather sublittoral, very fossiliferous (Arbizu et al. 1979; García Alcalde et al. 1979), dark grey argillaceous limestones. Several members can be distinguished (Coo et al. 1971; Méndez Bedia 1976; Arbizu et al. 1979) (Fig. 1). The lower part consists of argillaceous and crinoidal limestones with interbedded shaly layers; the middle, partially massive, member comprises limestones and argillaceous limestones with abundant reef-building elements; the upper part is characterized by argillaceous limestones and thin shale layers, with several reefal episodes, and near the top by limestones and crinoidal limestones. In the Valsurbio Region, the correlative Otero Fm. (Koopmans 1962), is thinner and less fossiliferous. Shallower, littoral conditions, represented by increasing content of birdseyes limestones and mudcracks, developed towards the Asturian uplift (Méndez Bedia 1976). Near the Aramo Unit thrust (Fig. 1; see Pérez Estaún, this Vol., Part II, Chap. 1) the Moniello Fm. laterally passes to a more sandy, unnamed formation. Most of the Moniello-Santa Lucía Fms. corresponds to the Emsian. The Emsian/Eifelian boundary is marked by the entry of *Icriodus retrodepressus* and *Arduspirifer intermedius* (Figs. 1, 2), within the upper part of both formations.

A major sedimentary shift is represented by the succeeding siliciclastic Naranco (Adaro and Junquera 1916), Huergas (Comte 1936) Formations. Both of these ca. 300–500 m thick units display two vertically superposed sedimentary models, the lower characterized by a coarser and greater elastic supply than the upper one (García Ramos 1978; García Ramos and Colmenero 1981). The top and bottom of both formations show crinoidal and sandy limestones and dolostones. Coarse, cross-bedded, ferruginous red sandstones abound in the lower and middle parts of the Asturian Naranco Fm. Ichnofossils, brachiopods, solitary corals, and bryozoans are the most representative fossils. The Huergas Fm. chiefly consists of euxinic, nodular black shales, sometimes associated with pelagic faunas (ammonoids, dacryoconarids, entomozoids, trilobites, etc.; García Alcalde and Arbizu 1976; Fig. 2). In the Valsurbio Region, the Hornalejo Fm. (Koopmans 1962) is very similar to the Naranco Fm. These units have been deposited in varied, more or less rough, shore-distant, and deep, sublittoral conditions. The age of the formations is Eifelian-Givetian. The boundary between both Stages can be provisionally traced in the upper part of the Huergas Fm., near the acmé of the genus *Cabrieroceras*.

A return to reefal conditions is recorded, in the northern branch of the AA, at the Candás Fm. (Barrois 1882; *sensu* García Alcalde 1985, *non* Radig 1962, Truyóls and Julivert 1976, Julivert et al. 1983), and in the southern branch at the Portilla Fm. (Comte 1936), and the Valdoré Limestone (Comte 1938a) of the Nocedo Fm. The Candás Fm. is composed of a 200 m of littoral to sublittoral, very fossiliferous limestones and argillaceous limestones, in which several members have been distinguished (Bereskin 1978; García Alcalde et al. 1979). The Portilla Fm. *sensu stricto* is a complex biostromal reef facies with ca. 50 m of fossil-rich, shallow-marine, argillaceous and crinoidal limestones, variably dismembered by different authors (Mohanti 1972; Reijers 1972; García Alcalde et al. 1979). The lower part of the Nocedo Fm. includes calcareous sandstones laterally varying to a more calcareous development with sandy and biostromal limestones (Valdoré Lms.). The Portilla Fm. *s. s.*/Valdoré Lms. boundary, is usually underlined by a more or less conspicuous siliciclastic interval. However sometimes it is not easy to draw the boundary between both units, and most authors have gathered them in a Portilla Fm. *s. lato* (García Alcalde et al. 1985). In the Valsurbio Region, the Valcovero Fm. (Koopmans 1962), correlates with the Portilla Fm. *s. l.* The above units are of Givetian age; nonetheless, the top of the Valdoré Lms. could locally reach the earliest Frasnian.

The Candás Fm. is overlain by a fairly siliciclastic series, ca. 400 m thick: the Piñeres Fm. (Radig 1962; *sensu* García Alcalde 1985). The lower part of this unit consists of ferruginous and calcareous sandstones, and very fossiliferous argillaceous and sandy limestones and marlstones, showing sporadically thin restricted reefoid growths. The upper part shows cross-bedded ferruginous, calcareous, and microconglomeratic quartz sandstones, and thin interbedded layers of siltstones and shales (Sánchez de la Torre et al. 1976). The Piñeres Fm. was deposited in shallow to very shallow neritic conditions. The occurrence of *Ancyrodella binodosa* and *Polygnathus asymmetricus* in the lower part of the unit (García López 1986) indicates the beginning of the

Figs. 1, 2. (pages 14 and 15) Devonian stratigraphy of the Asturian-Leonese Domain. Fossil information by J. L. García Alcalde (brachiopods), M. Truyóls Massoni (tentaculitoids), F. Soto (corals), J. R. Montesinos (ammonoids). Trilobite data by M. Arbizu and selected references from Z. Smeenk (1983). Conodont data by S. García López and selected references from J. G. M. Raven (1983), J. L. García Alcalde et al. (1985), and J. Grötsch (1987). *BA*. Bañugues Formation; *FERR*. Ferroñes Formation ▶

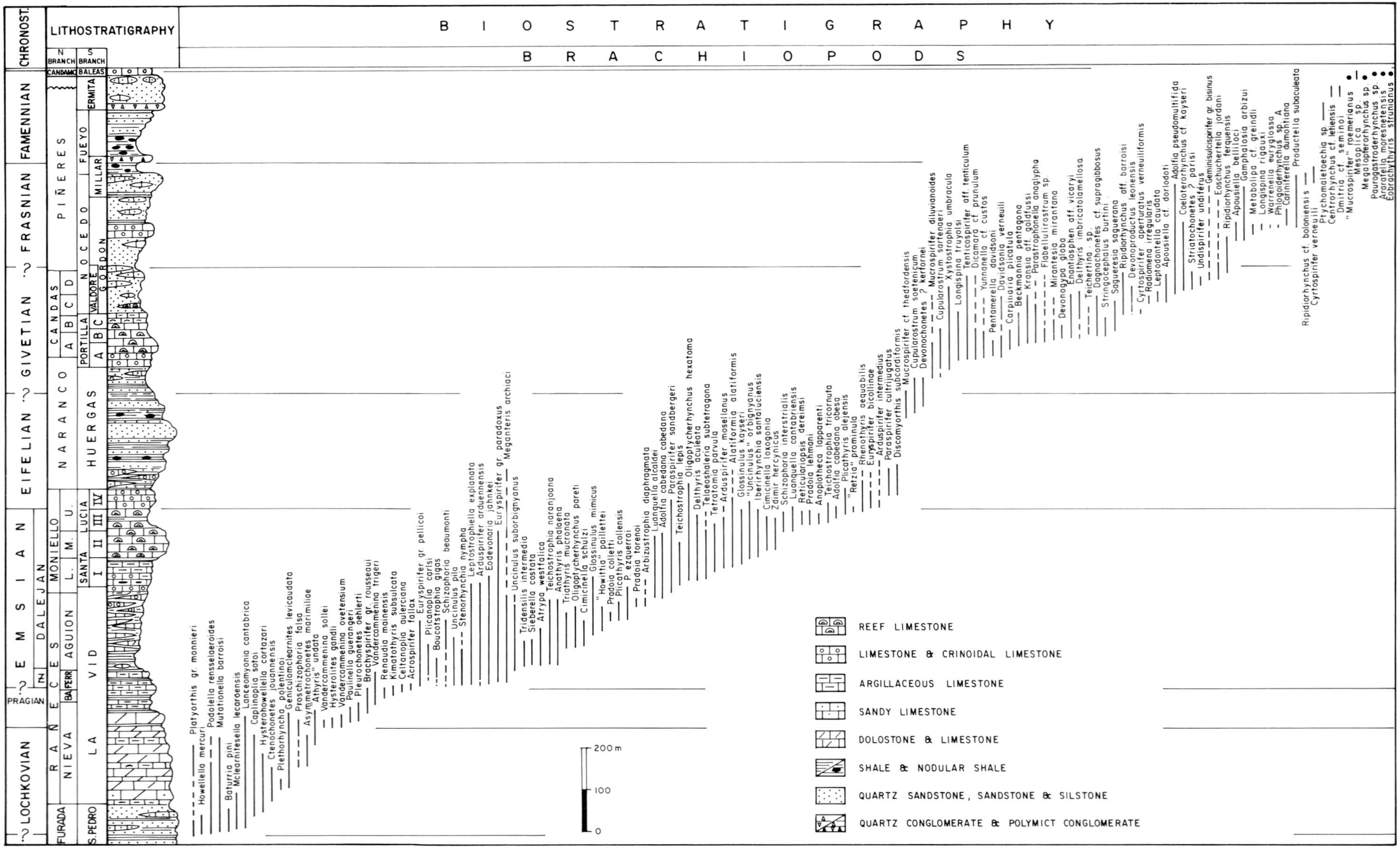
CHRONOST.
LITHOSTRATIGRAPHY
N BRANCH
S BRANCH
BIOSTRATIGRAPHY
BRACHIOPODS
LOCHKOVIAN
PRAGIAN
EMSIAN
DALEJAN
EIFELIAN
GIVETIAN
FRASNIAN
FAMENNIAN
FURADA
RAÑECES
NIEVA
BAFERR
AGUION
MONIELLO
L.
M.
U.
NARANCO
CANDAS
A
B
C
D
PIÑERES
CANDAMO
S.PEDRO
LA VID
SANTA LUCIA
I
II
III
IV
HUERGAS
PORTILLA
A
B
C
VALDORE
NOCEDO
GORDON
MILLAR
FUEYO
ERMITA
BALEAS
200 m
100
0
REEF LIMESTONE
LIMESTONE & CRINOIDAL LIMESTONE
ARGILLACEOUS LIMESTONE
SANDY LIMESTONE
DOLOSTONE & LIMESTONE
SHALE & NODULAR SHALE
QUARTZ SANDSTONE, SANDSTONE & SILSTONE
QUARTZ CONGLOMERATE & POLYMICT CONGLOMERATE

Fig. 1

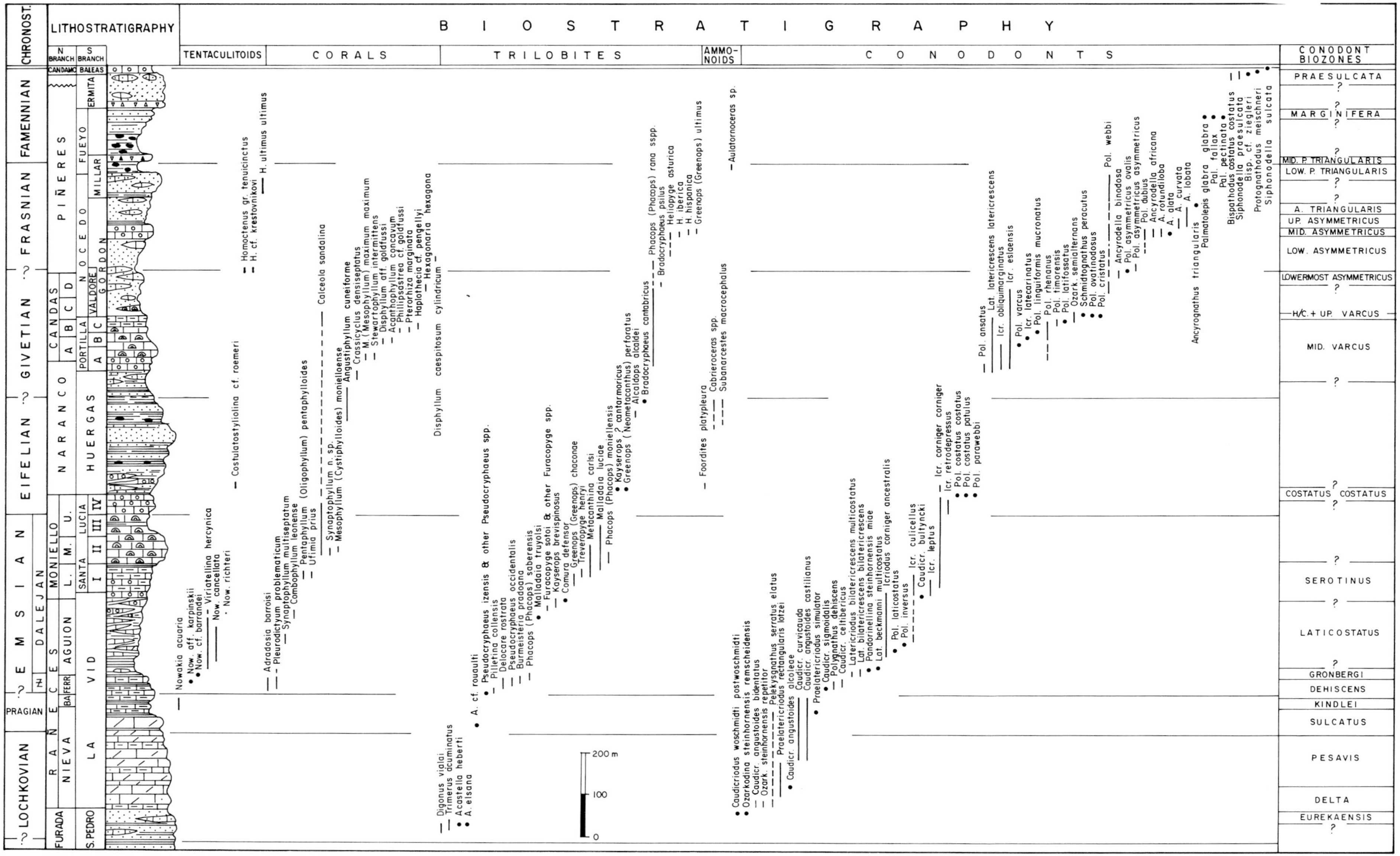

CHRONOST.
LITHOSTRATIGRAPHY
N BRANCH
S BRANCH
BIOSTRATIGRAPHY
TENTACULITOIDS
CORALS
TRILOBITES
AMMONOIDS
CONODONTS
CONODONT BIOZONES
FAMENNIAN
FRASNIAN
GIVETIAN
EIFELIAN
EMSIAN
DALEJAN
PRAGIAN
LOCHKOVIAN
CANDAMO
BALEAS
ERMITA
PIÑERES
FUEYO
MILLAR
NOCEDO
GORDON
VALDORE
CANDAS
PORTILLA
NARANCO
HUERGAS
MONIELLO
SANTA LUCIA
RANECES
AGUION
BAFERR
NIEVA
LA VID
FURADA
S. PEDRO
Nowakia acuaria
Now. aff. karpinskii
Now. cf. barrandei
Viriatellina hercynica
Now. cancellata
Now. richteri
Costulatostyliolina cf. roemeri
Homoctenus gr. tenuicinctus
H. cf. krestovnikovi
H. ultimus ultimus
Adradosia barroisi
Pleurodictyum problematicum
Synaptophyllum multiseptatum
Combophyllum leonense
Pentaphyllum (Oligophyllum) pentaphylloides
Ufimia prius
Calceola sandalina
Synaptophyllum n. sp.
Mesophyllum (Cystiphylloides) moniellоense
Angustiphyllum cuneiforme
Crassicyclus densiseptatus
M. (Mesophyllum) maximum maximum
Stewartophyllum intermittens
Disphyllum aff. goldfussi
Acanthophyllum concavum
Phillipsastrea cf. goldfussi
Pterorhiza marginata
Haplothecia cf. pengellyi
Hexagonaria hexagona
Disphyllum caespitosum cylindricum
Digonus vialai
Trimerus acuminatus
Acastella heberti
A. elsana
A. cf. rouaulti
Pseudocryphaeus izensis & other Pseudocryphaeus spp.
Pilletina collensis
Delocare rostrata
Pseudocryphaeus occidentalis
Burmeisteria pradoana
Phacops (Phacops) saberensis
Malladaia truyolsi
Furacopyge sotoi & other Furacopyge spp.
Kayserops brevispinosus
Comura defensor
Greenops (Greenops) chaconae
Treveropyge henryi
Metacanthina carlsi
Malladaia luciae
Phacops (Phacops) moniellensis
Kayserops ? cantarmoricus
Greenops (Neometacanthus) perforatus
Alcaldops alcaldei
Bradocryphaeus cantabricus
Phacops (Phacops) rana sspp.
Bradocryphaeus psilus
Heliopyge asturica
H. iberica
H. hispanica
Greenops (Greenops) ultimus
Foordites platypleura
Cabrieroceras spp.
Subanarcestes macrocephalus
Aulatornoceras sp.
Caudicriodus woschmidti postwoschmidti
Ozarkodina steinhornensis remscheidensis
Caudicr. angustoides bidentatus
Ozark. steinhornensis repetitor
Pelekysgnathus serratus elatus
Praelatericriodus rectangularis lotzei
Caudicr. angustoides alcoleae
Caudicr. curvicauda
Caudicr. angustoides castilianus
Praelatericriodus simulator
Caudicr. sigmoidalis
Polygnathus dehiscens
Caudicr. celtibericus
Latericriodus bilatericrescens multicostatus
Lat. bilatericrescens bilatericrescens
Pandorinellina steinhornensis miae
Lat. beckmanni multicostatus
Icriodus corniger ancestralis
Pol. laticostatus
Pol. inversus
Icr. culicellus
Caudicr. bultyncki
Icr. leptus
Icr. corniger corniger
Icr. retrodepressus
Pol. costatus costatus
Pol. costatus patulus
Pol. parawebbi
Pol. ansatus
Lat. latericrescens latericrescens
Icr. obliquimarginatus
Icr. eslaensis
Pol. varcus
Icr. latecarinatus
Pol. linguiformis mucronatus
Pol. rhenanus
Pol. timorensis
Pol. latifossatus
Ozark. semialternans
Schmidtognathus peracutus
Pol. ovatinodosus
Pol. cristatus
Pol. webbi
Ancyrodella binodosa
Pol. asymmetricus ovalis
Pol. asymmetricus asymmetricus
Pol. dubius
Ancyrodella africana
A. rotundiloba
A. alata
A. curvata
A. lobata
Ancyrognathus triangularis
Palmatolepis glabra glabra
Pal. fallax
Pal. pectinata
Bispathodus costatus costatus
Siphonodella praesulcata
Bisp. cf. ziegleri
Protognathodus meischneri
Siphonodella sulcata
200 m
100
0
PRAESULCATA
MARGINIFERA
MID. P. TRIANGULARIS
LOW. P. TRIANGULARIS
A. TRIANGULARIS
UP. ASYMMETRICUS
MID. ASYMMETRICUS
LOW. ASYMMETRICUS
LOWERMOST ASYMMETRICUS
H/C. + UP. VARCUS
MID. VARCUS
COSTATUS COSTATUS
SEROTINUS
LATICOSTATUS
GRONBERGI
DEHISCENS
KINDLEI
SULCATUS
PESAVIS
DELTA
EUREKAENSIS

Fig. 2

Frasnian (Fig. 2). The upper part could already belong to the Famennian, but the lack of fossils precludes an unambiguous datation.

The Nocedo Fm. (Comte 1936) overlies the Portilla Fm. It consists of a ca. 300 m thick alternation of shales, siltstones, calcareous sandstones and limestones, usually arranged in two regressive-transgressive sequences: the Gordón and the Millar Members (Loevezijn 1983, 1986a, b). Brachiopods, corals and crinoids are the most conspicuous fossil elements. In the bioclastic limestones, rich-conodont faunas are also found.

The uppermost part of the Nocedo Fm. in the Esla Region (Fig. 1.; see Pérez Estaún, this Vol., Part II, Chap. 1) is a very fossiliferous biostromal limestone: the Crémenes Lms. (Rupke 1965; cf. Loevezijn et al. 1986). The Nocedo Fm. was deposited in shallow water ranging into subtidal to supratidal environments. The Givetian/Frasnian boundary lies in the lower part of the Formation (Figs. 1, 2). The Crémenes Lms. enclose a very interesting latest Frasnian fauna (Westbroek 1964; Loevezijn et al. 1986; Loevezijn 1986a, b).

An important transgressive event is recorded in the Fueyo Fm. (Comte 1936), ca. 130 m of black nodular shales, sometimes with a thin, mostly quartzitic, polymict conglomerate in the lower part, and an alternation of shales and sandstones in the upper one (Rodríguez Fernández et al. 1985). The conglomerate and the underlying shales have been interpreted by Loevezijn (1983, 1986a, b, 1988) and by Loevezijn and Raven (1983) as the top of the Millar Mb. Sparse pelagic faunas (ammonoids, nautiloids, bivalves, ostracods), and flora debris are found. The basal layers still contain Frasnian douvillinids, whereas in the shales overlying the conglomerate, Early Famennian conodonts (Loevezijn 1986a) and brachiopods occur (Fig. 1).

The regressive phase initiated in the Upper Fueyo Fm. culminated in the succeeding Ermita Fm. (Comte 1936). This unit consists of ca. 60 m of cross-bedded sandstones and quartz sandstones, siltstones, thin shale layers, and sandy limestone lenses; microconglomeratic sandstones are rather common in the lower part of the formation. The sedimentation took place in intertidal to supratidal conditions. Late Famennian fossils (mainly brachiopods, bivalves, gastropods, bryozoans, crinoids and flora debris) are frequent in calcareous sandstones. The topmost calcareous lenses of the unit usually give conodonts of the *Siphonodella praesulcata* to *S. sulcata* Biozones (García Alcalde and Menéndez Alvarez 1988).

In the Valsurbio Region, the Camporredondo Fm. (Koopmans 1962), a rather thick unit containing quite homogeneous sandy rocks, may be the equivalent of the Piñeres Fm. and of the whole Nocedo to Ermita Fms. The scarcity of fossils, however, prevents accurate correlations.

The levelling and tilting undergone by the Cantabrian Zone during the Late Devonian, allowed the deposition of a thin transgressive sedimentary marine wedge over most of the Asturian-Leonese Domain. This cover rests disconformably on increasingly older rocks towards the internal part of the AA. It consists of ca. 12 m of microconglomeratic and calcareous sandstones, with sandy crinoidal limestone lenses that grade upwards to a pure limestone shoal sequence with bioclastic, sometimes sandy, grainstone and packstone limestones (García Alcalde and Menéndez Alvarez 1988). The siliciclastic member is commonly referred to as the Ermita Fm., whereas the limestone member is termed the Candamo (Pello 1976) or Baleas (Wagner et al. 1971) Formation. The fossil content indicates that the Devonian/Carboniferous boundary (*S. praesulcata-S. sulcata* boundary) usually lies inside these last Formations (Figs. 1, 2).

2.2.2 Devonian Reef Building

The reef development began in the Cantabrian Zone during the Devonian. Although sparse coral biostromes are found in the La Vid Group (Stel 1975), important reef buildings are only present in the Moniello-Santa Lucía (Late Emsian-Eifelian), and Candás-Portilla *s. l.* (Late Givetian) Fms. The last reefal episodes occur during the Late Devonian in a very restricted way and locally in calcareous levels within the Nocedo Fm. (Frasnian) (Soto 1986).

The reefal carbonates are of very varied nature, and it is possible to identify there patch reefs, bioherms or true reefs, and bedded structures or biostromes. Among the faunas contributing to their formation, the rugose corals (Fig. 2) are better known than the tabulates and the stromatoporoids.

The sedimentation model of the Moniello-Santa Lucía Fms. (Coo 1974; Méndez Bedia 1976) represents a wide back-reef environment with sublittoral and occasionally off-shore facies to the south and west, and with shallower water, lagoonal and perilittoral deposits towards the AA core. Biostromal masses of stromatoporoids and corals develop, mainly in the more distal carbonate facies, within the middle member of the Formations, although small patch reefs and, locally, true reefs or bioherms are also found.

The best exposed and illustrative example of the internal structure of a true reef is provided by a bioherm, 140 m thick, developed on the Asturian coast (Arnao) (Méndez Bedia 1976; Sánchez de la Torre and Manjón 1976). A detailed study of this reef (Méndez Bedia and Soto 1984) led to setting up a vertical paleoecological succession, reflecting different growth stages, reaching the high energy surf zone. The reef mainly contains large (up to 2 m in diameter) subspherical-shaped stromatoporoids, framestones and bindstones, and a smaller amount of rugose and tabulate corals. The contributing stromatoporoids are (Méndez Bedia 1984): *Actinostroma verrucosum, A.* cf. *stellulatum, Atelodictyum fallax, Anostylostroma?* sp., *Stromatoporella* sp. and *Stromatopora* sp. The frequent occurrence of the algae *Sphaerocodium* and *Girvanella,* associated with the building organisms, is also noteworthy.

The biostromal deposits, typical of back-reef areas, are formed by stromatoporoids, with morphology dependent on the mud supply, and by rugose and tabulate corals. The following stromatoporoids have been mentioned there (Sleumer 1969): *Actinostroma papillosum, A. verrucosum, Stromatoporella? granulata, S.? selwyni* and *Stromatopora hüpschii.*

The reef facies distribution of the Candás-Portilla *s. l.* Fms. (Reijers 1972, 1985; Raven 1983) indicates the occurrence, from north to south, of a back-reef zone, a reef track, and a fore-reef zone. In the lower part, an incipient, fairly biostromal, barrier reef is developed. In the upper part, a continuous organic barrier, evidenced by bioherms and biostromes rimming the carbonate platform, occurs. Good examples of these reef buildings are visible in the Esla Region (Cistierna Bioherm, Matallana Biostrome).

The reef-building community is dominated, as in the Moniello-Santa Lucía Fms., by tabulate and rugose corals and stromatoporoids. The last fossils are less frequent here than in the above-mentioned formations, especially in the biostromal argillaceous carbonates. Among them the following species have been mentioned (Sleumer 1969): *A. stellulatum, Stromatopora concentrica, S.* sp. *2, Stromatoporella? granulata* and *S.* sp. *2.*

Fry and Bereskin (1977), Bereskin (1978) and Raven (1983), describe several biostromal sequences within the Candás Fm., consisting of spherical and laminar stromatoporoid framestones with rugose and tabulate corals, at the base. Towards the top they are built by branching tabulates, associated with a greater argillaceous content. And finally, the sequences are overlain by mudstones and shales.

The last recorded Devonian reefal levels occur at the top of the Nocedo Fm. (Crémenes Lms.). Loevezijn et al. (1986) and Loevezijn (1986a, b) describe the facies of these biostromal limestones, formed by coral-stromatoporoid boundstones, in which *S. granulata, S.* sp. and *Actinostroma* sp. are found.

Acknowledgement. Financial support for this work was provided by the Comisión de Investigación Científica y Técnica (CTCYT) Spain, PB-86/0251 research project.

References

Adaro L, Junquera G (1916) Hierros de Asturias. In: Criaderos de Hierro en España, vol 2, text 1-160, vol láms, 10 láms

Aramburu C (1987) Descubrimiento de graptolitos arenigienses en la Escama de Rioseco (Zona Cantábrica; N. de España): Discusión. Cuad Lab Xeol Laxe 11: 221-223

Arbizu M (1972) El Devónico Inferior de la costa asturiana entre la Punta de Narvata y la Ensenada de Moniello. Breviora Geol Astúrica 16 (3): 33-39

Arbizu M, García Alcalde JL, García López S, Méndez Bedia I, Sánchez de Posada LC, Soto F, Truyóls M, Truyóls J, Alvarez F, Méndez C, Menéndez JR (1979) Biostratigraphical study of the Moniello Formation (Cantabrian Mountains, Asturias, NW Spain). Geol Palaeontol 13: 103-124

Baldwin CT (1978) A comparison of the stratigraphy and depositional processes in the Cambro-Ordovician rocks of the Cantabrian and West Asturian-Leonese Zones, NW Spain. In: Geología de la parte norte del Macizo Ibérico (Edición Homenaje a I Parga Pondal). Cuad Semin Estud Cerámicos Sargadelos 27: 43-78

Barrois C (1882) Recherches sur les terrains anciens des Asturies et de la Galice. Mém Soc Géol Nord 2 (1): 1-630

Bereskin SR (1978) Subdivisión estratigráfica de la caliza de Candás. Devónico del Noroeste de España. Supl Cien Bol Inst Estud Asturianos 23: 89-97

Comte P (1936) Le Dévonien Inférieur du Léon. C R Acad Sci Paris 202: 771-773

Comte P (1937) La série cambrienne et silurienne du Léon. C R Acad Sci Paris 204: 604-606

Comte P (1938a) Les faciès du Dévonien supérieur dans la Cordillère cantabrique. C R Acad Sci Paris 206: 1496-1498

Comte P (1938b) La succession lithologique des formations cambriennes du Léon (Espagne). 71 Congrès de la Societé des Savants de Nice, pp 181-183

Comte P (1959) Recherches sur les terrains anciens de la Cordillère Cantabrique. Mem Inst Geol Min Esp 60: 1-440

Coo JCM de (1974) Lithostratigraphy of the Devonian Santa Lucía Limestones in León, Spain. PhD Thesis, Leiden University, 87 p

Coo JCM, Deelman JC, Baan V van der (1971) Carbonate facies of the Santa Lucía Formation (Emsian-Couvinian) in León and Asturias, Spain. Geol Mijnbouw 50 (3): 359-366

Cramer FH (1964) Some Acritarchs from the San Pedro Formation (Gedinnian) of the Cantabrian Mountains in Spain. Bull Soc Belge Géol Paléontol Hydrol 73: 33-38

Cramer FH (1966) Additional morphographic information on some characteristic Acritarchs of the San Pedro and Furada Formations (Silurian-Devonian boundary) in León and Asturias, Spain. Notas Comun Inst Geol Min Esp 83: 27–48

Cramer FH (1967) Chitinozoans of a composite section of Upper Llandoverian to basal Lower Gedinnian sediments in northern León, Spain. A preliminary report. Bull Soc Belge Géol Paléontol Hydrol 75: 69–129

Cramer FH (1970) Distribution of selected Silurian Acritarchs. Revista Española de Micropaleontología, núm extra, 1–102

Crimes TP, Marcos A (1976) Trilobite traces and the age of the lowest part of the Ordovician reference section for NW Spain. Geol Mag 113 (2): 350–356

Crimes TP, Legg I, Marcos A, Arboleya ML (1977) Late Precambrian-low Lower Cambrian trace fossils from Spain. In: Trace Fossils. Geol J, Spec Issue, 2 (9): 91–138

Debrenne F, Zamarreño I (1970) Sur la découverte d'Archéocyathes dans le Cambrien du NW de l'Espagne. Breviora Geol Astúrica 14 (1): 1–11

Fombella MA (1978) Acritarcos de la Formación Oville, edad Cámbrico medio-Tremadoc, provincia de León, España. Palinología, núm extra, pp 245–261

Fry SA, Bereskin SR (1977) Arrecifes del Devónico Medio y Superior del Miembro de Perán (Caliza de Candás) en el Cabo de Peñas (Asturias, España). Resumen Comunicaciones VIII Congreso Nacional Sedimentologia Oviedo, Grupo III

García Alcalde JL (1985) La extensión de la Biozona de Phlogoiderhynchus (braquiópodo rinconélido, Givetiense terminal-Frasniense Inferior). Trab Geol 15: 77–86

García Alcalde JL, Arbizu M (1976) Les faunes pélagiques du Dévonien moyen de León (versant méridional des Montagnes Cantabriques, NO de l'Espagne). Ann Soc Géol Nord 96 (4): 413–417

García Alcalde JL, Arbizu M, García López S, Méndez Bedia I (eds) (1979) Cantabrian Mountains (NW Spain). Guidebook of the Field Trip, Meeting of the International Subcommission on Devonian Stratigraphy, 1–41. Serv Publ Univ Oviedo

García Alcalde JL, Menéndez Alvarez JR (1988) The Devonian-Carboniferous boundary in the Asturo-Leonese Domain (Cantabrian Mountains, NW Spain). Cour Forschungs Inst Senckenb 100: 21–37

García Alcalde JL, Menéndez Alvarez JR, García López S, Soto F (1985) El Devónico Superior y el Carbonífero Inferior del Sinclinal de Beberino (Pola de Gordón, León, NO de España). C R X Congr Int Geol Estratigr Carbonífero 2: 375–386

García López S (1986) Los Conodontos y su aplicación al estudio de las divisiones cronostratigráficas mayores del Devónico Astur-leonés (España). Publ Espec Bol Geol Min 97 (3–5): 1–112

García Ramos JC (1978) Estudio e interpretación de las principales facies sedimentarias comprendidas en las formaciones Naranco y Huergas (Devónico Medio) en la Cordillera Cantábrica. Trab Geol 10: 195–247

García Ramos JC, Colmenero JR (1981) Evolución sedimentaria y paleogeográfica durante el Devónico en la Cordillera Cantábrica. Real Academia de Ciencias Exactas, Físicas y Naturales, Progr Int Correlación Geol (PIGC) 2: 61–76

Grötsch J (1987) Conodonten und Stratigraphie der unterdevonischen La Vid Formation (Kantabrisches Gebirge, NW-Spanien). Dipl-Arbeit Erlangen, pp 1–94

Gutiérrez Marco JC (1986) Graptolitos del Ordovícico español. Thesis, Univ Complutense de Madrid, 1–701

Gutiérrez Marco JC, Rodríguez L (1987a) Descubrimiento de graptolitos arenigienses en la escama de Rioseco (Zona Cantábrica, N de España). Cuad Lab Xeol Laxe 11: 209–220

Gutiérrez Marco JC, Rodríguez L (1987b) Descubrimiento de graptolitos arenigienses en la escama de Rioseco (Zona Cantábrica, N de España): réplica a la discusión. Cuad Lab Xeol Laxe 11: 225–228

Julivert M, Marcos A, Philippot A, Henry JL (1968) Nota sobre la extensión de las pizarras ordovícicas al E de la Cuenca Carbonífera central de Asturias. Breviora Geol Astúrica 12 (4): 1–4

Julivert M, Truyóls J (1972) La coupe du Cabo Peñas, une coupe de référence pour l'Ordovicien du Nord-Ouest de l'Espagne. C R Sommaire Séances Soc Géol Fr 1972 (6): 241–243

Julivert M, Truyóls J, Verges J (1983) El Devónico en el Macizo Ibérico. In: Geología de España (Libro Jubilar JM Rios). Inst Geol Min Esp, núm extra 1: 265–311

Kegel W (1929) Das Gotlandium in den Kantabrischen Ketten Nordspaniens. Z Dtsch Geol Ges 81: 35–62

Koopmans BN (1962) The sedimentary and structural history of the Valsurvio dome, Cantabrian Mountains, Spain. Leidse Geol Meded 26: 121–232

Loevezijn GBS van (1983) Upper Devonian block movements and sedimentation in the Asturo-Leonese Basin (Cantabrian Mountains, Spain). Leidse Geol Meded 52 (2): 185–192

Loevezijn GBS van (1986a) Late Devonian basin-fill history of the southern Cantabrian Mountains, Spain. Bol Geol Min 97 (6): 704–716

Loevezijn GBS van (1986b) Stratigraphy and facies of the Nocedo, Fueyo and Ermita formations (Upper Devonian to lowermost Carboniferous) in León, N Spain. Scr Geol 81: 1–116

Loevezijn GBS van (1988) The Famennian of the Cantabrian Mountains, northwestern Spain. Neues Jahrb Geol Paläontol Monatsh 1988 (5): 278–292

Loevezijn GBS van, Raven JGM (1983) The Upper Devonian deposits in the northern part of León (Cantabrian Mountains, northwestern Spain). Leidse Geol Meded 52 (2): 179–183

Loevezijn GBS van, Raven JGM, Pol W van der (1986) The Crémenes Limestone, a Late Frasnian biostrome in the Cantabrian Mountains (northwestern Spain). Neues Jahrb Geol Paläontol Monatsh 1986 (10): 599–612

Lotze F (1956) Das Präkambrium Spaniens. Neues Jahrb Geol Paläontol Monatsh 1956 (8): 373–380

Lotze F (1961) Das Kambrium Spaniens. T 1: Stratigraphie. Abh Math-Naturwiss Kl 1961 (6): 1–498

Méndez Bedia I (1976) Biofacies y litofacies de la Formación Moniello-Santa Lucía (Devónico de la Cordillera Cantábrica, NW de España). Trab Geol 9: 1–93

Méndez Bedia I (1984) Primera nota sobre los Estromatopóridos de la Formación Moniello (Devónico de la Cordillera Cantábrica, NW de España). Trab Geol 14: 151–159

Méndez Bedia I, Soto F (1984) Paleoecological succession in a Devonian organic buildup (Moniello Fm., Cantabrian Mountains, NW Spain). Geobios, Mém spécial 8: 151–157

Mohanti M (1972) The Portilla Formation (Middle Devonian) of the Alba Syncline, Cantabrian Mountains, prov León, northwestern Spain: carbonate facies and rhynchonellid palaeontology. Leidse Geol Meded 48 (2): 135–205

Montesinos JR (1981a) El Ordovícico medio en el área de Cabo de Peñas. Correlación con Cabo Vidrias (Asturias, N de España). Cuad Lab Xeol Laxe 2: 175–185

Montesinos JR (1981b) Las Pizarras de Luarca (Ordovícico medio) en el área de Cabo Vidrias (Asturias, N de España). Cuad Lab Xeol Laxe 2: 187-193

Pello J (1976) Memoria del Mapa Geológico de España E 1: 50.000. Proaza, 52, 12-5. Inst Geol Min Esp, 1-53

Pello J, Philippot A (1967) Sur la présence du Llanvirn au Puerto Sueve (Zone orientale des Asturies, NW de l'Espagne). C R Sommaire Séances Soc Géol Fr 1967 (4): 156-157

Pérez Estaún A (1973) Datos sobre la sucesión estratigráfica del Precámbrico y la estructura del extremo Sur del Antiforme del Narcea (NW de España). Breviora Geol Astúrica 17 (1): 5-16

Poll K (1970) Stratigraphie und Tektonik an der Wende Silurium/Devon im Westlichen Kantabrischen Gebirge (Nordspanien). Sitzungsber Phys Med Soz Erlangen 83-84: 32-113

Radig F (1962) Zur Stratigraphie des Devons in Asturien (Nordspanien). Geol Rundschau 51: 249-267

Raven JGM (1983) Conodont biostratigraphy and depositional history of the Middle Devonian to Lower Carboniferous in the Cantabrian Zone (Cantabrian Mountains, Spain). Leidse Geol Meded 52 (2): 265-339

Reijers TJA (1972) Facies and diagenesis of the Devonian Portilla Limestone Formation between the river Esla and the Embalse de la Luna, Cantabrian Mountains, Spain. Leidse Geol Meded 42: 163-249

Reijers TJA (1985) Devonian basin-fill histories of the Spanish Cantabrian Mountains and the Belgian Ardennes; a comparison. Geol Mijnbouw 64: 41-62

Rodríguez RM (1982) Las palinofacies: su relación con el medio sedimentario. Aplicación al estudio de la cuenca de Furada/San Pedro (Silúrico Superior-Devónico Inferior) de la Cordillera Cantábrica, NO de España. Neues Jahrb Geol Paläontol Abh 163 (2): 176-182

Rodríguez Fernández LR, García Alcalde JL, Menéndez Alvarez JR (1985) La sucesión del Devónico Superior y Carbonífero Inferior en el Sinclinal de Alba (León, NO de España). C R 10 Congr Int Geol Estratigr Carbonífero 1: 133-144

Rupke J (1965) The Esla Nappe, Cantabrian Mountains (Spain). Leidse Geol Meded 32: 1-74

Sánchez de la Torre L, Colmenero JR, Suárez de Centi C (1976) Serie de la Arenisca del Devónico Superior en la playa de La Gargantera. Trab Geol 8: 187-201

Sánchez de la Torre L, Manjón M (1976) Facies de la Caliza de Moniello en la zona de Arnao. Trab Geol 8: 109-129

Sdzuy K (1961) Das Kambrium Spaniens. T 2: Trilobiten. Abh Math-Naturwiss Kl 1961 (7-8): 219-411

Sdzuy K (1971) Acerca de la correlación del Cámbrico inferior en la Península Ibérica. I Congr Hispano-luso-americano Geol Económica, Madrid-Lisboa 1971, 2: 753-768

Seilacher A (1970) *Cruziana* stratigraphy of "non fossiliferous" Palaeozoic sandstones. In: Trace fossils. Geol J Spec Issue (3): 447-476

Sleumer BHG (1969) Devonian stromatoporoids of the Cantabrian Mountains (Spain). Leidse Geol Meded 44: 1-136

Smeenk Z (1983) Devonian trilobites of the southern Cantabrian Mountains (northern Spain) with a systematic description of the Asteropyginae. Leidse Geol Meded 52 (2): 383-511

Soto F (1986) Asociaciones coralinas del Devónico Astur-Leonés (Cordillera Cantábrica, NO de España). Trab Geol 16: 25-35

Stel JH (1975) The influence of hurricanes on the quiet depositional conditions in the Lower Emsian La Vid shales of Colle (NW Spain). Leidse Geol Meded 49: 475-486

Truyóls J, Julivert M (1976) La sucesión paleozoica entre Cabo Peñas y Antromero. Trab Geol 8: 5-30

Truyóls J, Philippot A, Julivert M (1974) Les formations siluriennes de la Zone Cantabrique et leurs faunes. Bull Soc Géol Fr 16 (1): 23-35

Vera de la Puente C (1985) Características de los depósitos de llanura mareal del Devónico Inferior de la Zona Cantábrica (Cuenca Astur-Leonesa). Trab Geol 15: 11-17

Wagner RH, Winkler Prins CF, Riding RE (1971) Lithostratigraphic units of the lower part of the Carboniferous in Northern León, Spain. Trab Geol 4 (2): 507-531

Westbroek P (1964) Systématique et importance stratigraphique des rhynchonelles du Calcaire de Crémenes (Dévonien Supérieur, province de León, Espagne). Leidse Geol Meded 30: 243-252

Zamarreño I (1972) Las litofacies carbonatadas del Cámbrico de la Zona Cantábrica (NW España) y su distribución paleogeográfica. Trab Geol 5: 1-118

Zamarreño I (1976) Depósitos carbonatados de tipo "tidal flat" en el Devónico inferior del NW de España: las Dolomías de Bañugues. Trab Geol 8: 59-85

2.3 The Palentine Domain (Palentian Zone)

J. L. GARCIA ALCALDE[1], J. R. MONTESINOS[2], M. TRUYOLS MASSONI[1], S. GARCIA LOPEZ[1], M. A. ARBIZU[1] and F. SOTO[1]

The Palentine Domain (Brouwer 1964), belonging to the Pisuerga-Carrión Unit (Cantabrian Zone; Fig. 1, see Pérez Estaún, this Vol., Part II, Chap. 1), has recently been interpreted as a predominantly Silurian-Devonian allochthonous unit, coming from the southwestern-located West Asturian-Leonese Zone, overlying the Asturo-Leonese Domain (Fig. 1, see Pérez Estaún, this Vol., Part II, Chap. 1) (Marcos 1979; Frankenfeld 1983; Marquínez and Marcos 1984; Rodríguez Fernández et al. 1985). In the Palentine Domain various tectonic events would have produced a complex system of superposed thrust units with later refolding and faulting. This picture does not agree with the data presented by Wagner et al. (1984), which suggest that the relative position of the Palentine and Asturian-Leonese Domains is essentially unchanged, even though large-scale thrusting, mainly though not exclusively southward directed, has shortened distances eliminating transitional areas. All authors agree on the presence in the Palentine Domain of nappe structures (first proposed by Wagner 1955, 1971).

A stratigraphic differentiation between Palentine and Asturian-Leonese Domains is evident from Emsian onwards. The Palentine Domain shows a typical Hercynian "magnafacies" consisting mainly of shaly facies with carbonates of anoxic or semi-anoxic environments unsuitable for benthic faunas, and an average sedimentation rate of ca. 1.5 cm per 1000 years (according to the scale of Harland et al. 1982) (García Alcalde et al. 1988)

[1] Departamento de Geología, Universidad de Oviedo, c/ Jesús Arias de Velasco s/n, 33005 Oviedo, Spain
[2] Departamento de Biología Animal, Universidad de León, 24071 León, Spain

R. D. Dallmeyer and E. Martínez García (Eds.)
Pre-Mesozoic Geology of Ibria

The oldest unit of the Palentine Domain, the Carazo Sandstone Formation (*sensu* Binnekamp 1965; thickness not well established, ca. 500 m or more), consists predominantly of sandstones and shales with important quartz sandstones episodes (Fig. 1). The ratio of calcareous rocks increases upwards, and particularly near the top of the Formation. The Carazo Formation *sensu lato* has been subdivided into three Formations (Ambrose 1974): Robledo Sandstone and Quartzite, Arroyacas Shale, and Carazo *sensu stricto*. The relative stratigraphic position and exact age of the first two Ambrose's Formations are still subject to discussion, because of the tectonic complexity and sparse faunal record (García Alcalde et al. 1988). However, the palynomorphs, trilobites, graptolites and brachiopods found in these formations are unquestionably Silurian (Ambrose 1974; Cramer and Rodríguez 1977; Lobato et al. 1977; Jahnke et al. 1983). The Carazo Fm. *sensu* Ambrose (1974) consists predominantly of quartzites, ferruginous sandstones, and shales; the upper part of this unit (member C or Upper Member: Binnekamp 1965; Veen 1965; Lobato 1977; Jahnke et al. 1983) yields rich Lochkovian faunas (Fig. 1), but reliable palaeontological information on the position of the Silurian/Devonian boundary is lacking. A general lithological column of the Devonian formations is given in Fig. 1.

Early Lochkovian sedimentation (Fig. 1) took place in a rather stagnant, possibly anoxic basin. It was punctuated by short periods of biotic recovery (due to storms?). In the Late Lochkovian and Early Pragian times, a wide and shallow carbonate platform came into being (the Lebanza Formation, Alvarado and Sampelayo 1945; Krans et al. 1982). Later in the Pragian a rather sudden deepening took place and basinal conditions were re-established; these remained up to the end of Devonian times.

Fig. 1. Devonian stratigraphy of the Palentine Domain (After García Alcalde et al. 1988). *M* Man Member; *L* Lezna Mb.; *R* Requejada Mb. (*sensu* Montesinos and Truyóls Massoni 1987); *A, B, C, D,* and *E* Members of the Lebanza Fm., according to Krans et al. (1982) ▶

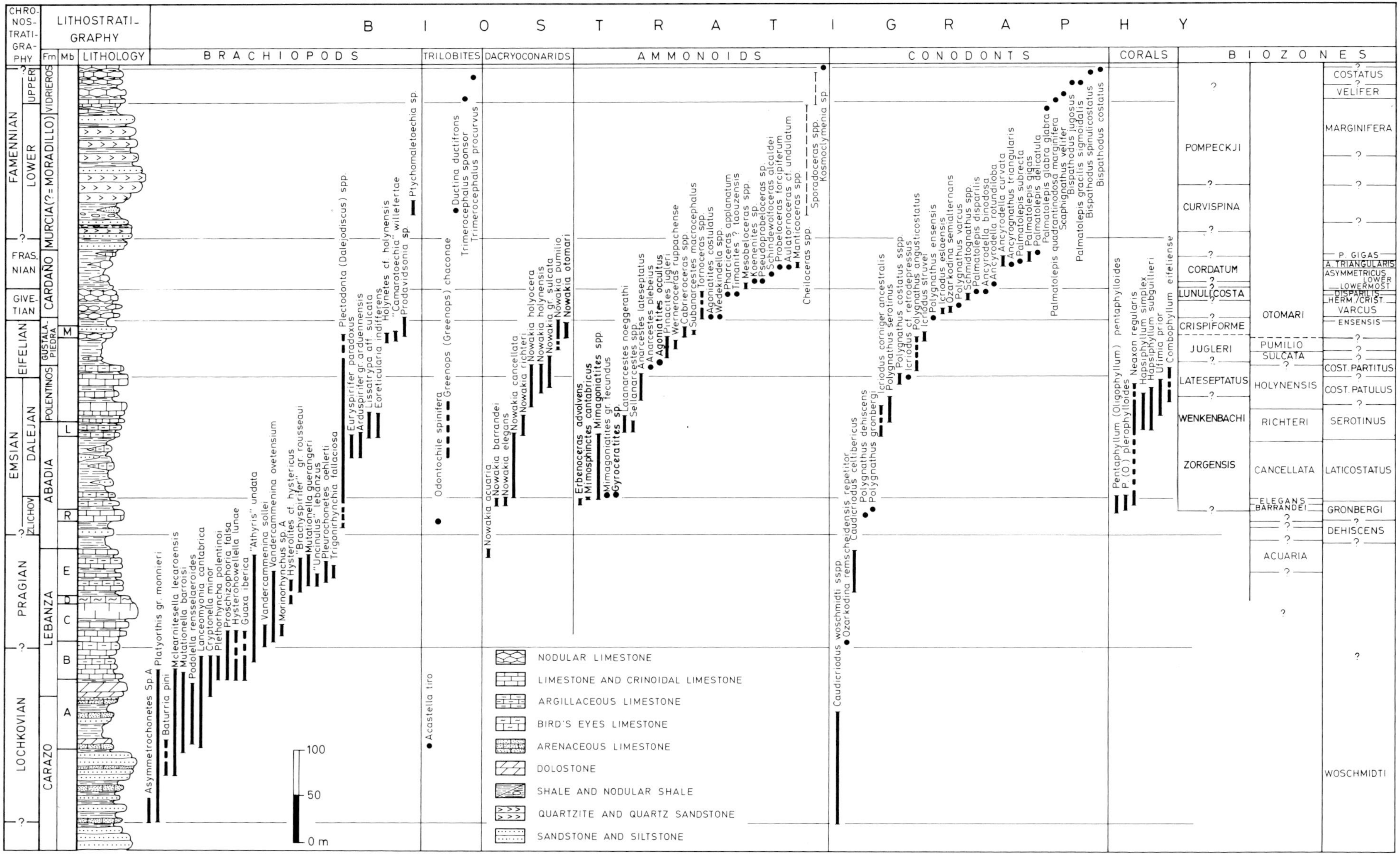
CHRONOSTRATIGRAPHY
LITHOSTRATIGRAPHY
Fm
Mb
LITHOLOGY
BIOSTRATIGRAPHY
BRACHIOPODS
TRILOBITES
DACRYOCONARIDS
AMMONOIDS
CONODONTS
CORALS
BIOZONES
FAMENNIAN
UPPER
LOWER
FRASNIAN
GIVETIAN
EIFELIAN
EMSIAN
DALEJAN
ZLICHOV
PRAGIAN
LOCHKOVIAN
VIDRIEROS
MURCIA(?=MORADILLO)
CARDAÑO
GUSTALAPIEDRA
POLENTINOS
ABADIA
LEBANZA
CARAZO
M
L
R
E
D
C
B
A
Asymmetrochonetes Sp.A
Platyorthis gr. monnieri
Baturria pini
Mclearnitesella lecaroensis
Mutationella barroisi
Podolella rensselaeroides
Lanceomyonia cantabrica
Cryptonella minor
Plethorhyncha polentinoi
Proschizophoria falsa
Hysterohowellella lunae
Guaxa iberica
"Athyris" undata
Vandercammenina sollei
Vandercammenina ovetensium
Morinorhynchus sp.A
Hysterolites cf. hystericus
"Brachyspirifer" gr. rousseaui
Mutationella guerangeri
"Uncinulus" lebanzus
Pleurochonetes oehlerti
Trigonirhynchia fallaciosa
Plectodonta (Dalejodiscus) spp.
Euryspirifer paradoxus
Arduspirifer gr. arduennensis
Lissatrypa aff. sulcata
Eoreticularia indifferens
Holynetes cf. holynensis
"Camarotoechia" willefertae
Prodavidsonia sp.
Ptychomaletoechia sp.
Acastella tiro
Odontochile spinifera
Greenops (Greenops) chaconae
Ductina ductifrons
Trimerocephalus sponsor
Trimerocephalus procurvus
Nowakia acuaria
Nowakia barrandei
Nowakia elegans
Nowakia cancellata
Nowakia richteri
Nowakia holyocera
Nowakia holynensis
Nowakia gr. sulcata
Nowakia pumilio
Nowakia otomari
Erbenoceras advolvens
Mimosphinctes cantabricus
Mimagoniatites spp.
Mimagoniatites gr. fecundus
Gyroceratites sp.
Latanarcestes noeggerathi
Sellanarcestes spp.
Anarcestes lateseptatus
Anarcestes plebeius
Agoniatites occultus
Pinacites jugleri
Werneroceras ruppachense
Cabrieroceras spp.
Subanarcestes macrocephalus
Tornoceras spp.
Agoniatites costulatus
Wedekindella spp.
Pharciceras applanatum
Timanites ? taouzensis
Mesobeloceras spp.
Koenenites sp.
Pseudoprobeloceras sp.
Schindewolfoceras alcaldei
Probeloceras forcipiferum
Aulatornoceras cf. undulatum
Manticoceras spp.
Cheiloceras spp.
Sporadoceras spp.
Kosmoclymenia sp.
Caudicriodus woschmidti sspp.
Ozarkodina remscheidensis repetitor
Caudicriodus celtibericus
Polygnathus dehiscens
Polygnathus gronbergi
Icriodus corniger ancestralis
Polygnathus serotinus
Polygnathus costatus sspp.
Icriodus cf. retrodepressus
Polygnathus angusticostatus
Icriodus struvei
Polygnathus ensensis
Icriodus eslaensis
Ozarkodina semialternans
Polygnathus varcus
Schmidtognathus spp.
Palmatolepis disparilis
Ancyrodella binodosa
Ancyrodella rotundiloba
Ancyrodella curvata
Ancyrognathus triangularis
Palmatolepis subrecta
Palmatolepis gigas
Palmatolepis delicatula
Palmatolepis glabra glabra
Palmatolepis quadrantinodosa marginifera
Scaphignathus velifer
Bispathodus jugosus
Palmatolepis gracilis sigmoidalis
Bispathodus spinulicostatus
Bispathodus costatus
Pentaphyllum (Oligophyllum) pentaphylloides
P. (O.) plerophylloides
Neaxon regularis
Hapsiphyllum simplex
Hapsiphyllum subguillieri
Ufimia prior
Combophyllum eifeliense
POMPECKJI
CURVISPINA
CORDATUM
LUNULICOSTA
CRISPIFORME
JUGLERI
LATESEPTATUS
WENKENBACHI
ZORGENSIS
OTOMARI
PUMILIO
SULCATA
HOLYNENSIS
RICHTERI
CANCELLATA
ELEGANS
BARRANDEI
ACUARIA
COSTATUS
VELIFER
MARGINIFERA
P. GIGAS
A. TRIANGULARIS
ASYMMETRICUS
LOWER
LOWERMOST
DISPARILIS
HERM./CRIST.
VARCUS
ENSENSIS
COST. PARTITUS
COST. PATULUS
SEROTINUS
LATICOSTATUS
GRONBERGI
DEHISCENS
WOSCHMIDTI
NODULAR LIMESTONE
LIMESTONE AND CRINOIDAL LIMESTONE
ARGILLACEOUS LIMESTONE
BIRD'S EYES LIMESTONE
ARENACEOUS LIMESTONE
DOLOSTONE
SHALE AND NODULAR SHALE
QUARTZITE AND QUARTZ SANDSTONE
SANDSTONE AND SILTSTONE
100
50
0 m

The Lockkovian/Pragian boundary cannot be traced in the absence of a formal definition. The entry of the brachiopod species *Vandercammenina sollei,* and a related faunal turnover in the lower part of the Lebanza Fm. (Fig. 1), can be used as a provisional boundary marker (Carls 1987; García Alcalde et al. 1988).

Tectonic complexity, a paucity of fossil remains, and a lack of formal definition of the boundary prevent the positioning of a convenient boundary between Pragian and Emsian.

The Zlichovian/Dalejan boundary lies above the Requejada Limestone Member (Veen 1965; *sensu* Montesinos and Truyóls Massoni 1987) of the Abadía Shale Formation (Binnekamp 1965; *sensu* Montesinos and Truyóls Massoni 1987) at the transition between the *Nowakia elegans* and *N. cancellata* Biozones, just overlying the level where the sudden evolution of the so-called *Anetoceras* Fauna (Chlupáč 1976) occurs (Fig. 1).

Icriodus cf. *retrodepressus* and other conodonts allow the Emsian/Eifelian boundary to be placed at the top of the Polentinos Limestone Fm. (Veen 1965; Jahnke et al. 1983; *sensu* García Alcalde et al. 1988; Fig. 1). The extinction of numerous anarcestids and the appearance of the first tornoceratids are the criteria used provisionally to recognize the Eifelian/Givetian boundary, at the top of the Gustalapiedra Shale Fm. (Veen 1965; *sensu* García Alcalde et al. 1988; Fig. 1).

In the Cardaño Shale and Limestone Fm. (Veen 1965), the pharciceratids are substituted by beloceratids as *Mesobeloceras* spp. at or near the base of the Frasnian (lower limit of the Lower *P. asymmetricus* conodont Biozone; Fig. 1).

The Lower Famennian commences at the base of the Murcia Quartzite Fm. (Veen 1965; equivalent to the Moradillo Fm. of Wagner and Wagner Gentis 1963; cf. Wagner et al. 1984) and continues into the basal part of the overlying Vidrieros Limestone Fm. (Veen 1965) with the *Cheiloceras* Biozone (Arbizu et al. 1986; García Alcalde et al. 1988; Fig. 1). The remainder of the Vidrieros Fm. contains clymenids and other ammonoids (Kullmann 1960; Wagner et al. 1984), conodonts of the *S. velifer* to *B. costatus* Biozones (Adrichem Boogaert 1967), and other Upper Famennian fossils, especially ostracods of Thuringian provenance (Becker 1981, 1982; Wagner et al. 1984; Arbizu 1985; Fig. 1).

At the top of the Vidrieros Limestone Fm., conodonts of the *S. sulcata* Biozone (Lower Tournaisian) have been recorded, i. e. *Polygnathus inornatus* (Adrichem Boogaert 1967).

In the Devonian, several euxinic black shale episodes probably represent global, mainly transgressive pulses, related to biostratigraphic turnovers such as have described by House (1985), Walliser (1985), Chlupáč and Kukal (1986), and others; these are the so-called Daleje, Jugleri, Kačak-Otomari, Taghanic-Pharciceras and Frasnes Events (García Alcalde et al. 1987; Truyóls Massoni et al. in litt.).

Acknowledgement. Financial support for this work was provided by the Comisión de Investigación Científica y Técnica (CICYT)-Spain, PB-86/0251 research project.

References

Adrichem Boogaert HA van (1967) Devonian and Lower Carboniferous conodonts of the Cantabrian Mountains (Spain) and their stratigraphic application. Leidse Geol Meded 39: 129-192

Alvarado A, Sampelayo AH (1945) Zona occidental de la cuenca del Rubagón. Bol Inst Geol Min Esp 58: 1-44

Ambrose T (1974) The Lower Palaeozoic rocks of northern Palencia. Breviora Geol Astúrica 28 (4): 49-53

Arbizu MA (1985) Trilobites Phacopinae de la Formación Vidrieros, en el área de Gildar-Montó (León, NO de España) y su distribución estratigráfica. Trab Geol 15: 67-75

Arbizu MA, García Alcalde JL, Montesinos JR (1986) La edad de la Formación Murcia en el Dominio Palentino (Cordillera Cantábrica, NO de España). Paleontol Evol 20: 87-91 (Sabadell)

Becker G (1981) Ostracoda aus Cephalopoden-führenden Oberdevon im Kantabrischen Gebirge (N. Spanien). I. Hollinacea, Primitiopsacea, Kirkbyacea, Healdiacea und Bairdiocypridacea. Palaeontographica A 173 (1-4): 1-63

Becker G (1982) Ostracoda aus Cephalopoden-führenden Oberdevon im Kantabrischen Gebirge (N. Spanien). II. Bairdiacea, Cytheracea und Entomozoacea. Palaeontogr Abt A Palaeozool Stratigr 178 (4-6): 109-182

Binnekamp JG (1965) Lower Devonian brachiopods and stratigraphy of North Palencia (Cantabrian Mountains, Spain). Leidse Geol Meded 33: 1-62

Brouwer A (1964) Deux faciès dans le Dévonien des Montagnes Cantabriques Méridionales. Breviora Geol Astúrica A8 (1-4): 3-10

Carls P (1987) Ein Vorschlag zur biostratigraphischen Redefinition der Grenze Gedinnium/Siegenium und benachbarter Unter-Stufen. Erster Teil: Stratigraphische Argumente und Korrelationen. Cour Forschungsinst Senckenb 92: 77-121

Chlupáč I (1976) The oldest goniatite faunas and their significance. Lethaia 9: 303-315

Chlupáč I, Kukal Z (1986) Reflection of possible global Devonian Events in the Barrandian Area, C.S.S.R. In: Walliser OH (ed) Global bio-events. Lect Notes Earth Sci 8: 169-179

Cramer FH, Rodríguez R (1977) Robledo and Arroyacas Formation (Arroyo de Las Arroyacas, province of Palencia, Spain) palynologically dated as Late Silurian. Breviora Geol Astúrica 21 (1): 2-4

Frankenfeld H (1983) El Manto de Montó-Arauz: interpretación estructural de la Región del Pisuerga-Carrión (Zona Cantábrica, NW de España). Trab Geol 13: 37-47

García Alcalde JL, Montesinos JR, Truyóls Massoni M, García Lopez S, Arbizu MA, Soto F (1988) El Silúrico y el Devónico del Dominio Palentino (NO de España). Rev Soc Geol Esp 1 (1-2): 7-13

Harland WB, Cox AV, Llewellyn PC, Pickton CAG, Smith AG, Walters R (1982) A geologic time scale. Cambridge Earth Science Series, Cambridge University Press, 131 pp

House MR (1985) Correlation of mid-Palaeozoic ammonoid evolutionary events with global sedimentary perturbations. Nature 313: 17-22

Jahnke H, Henn A, Mader H, Schweineberg J (1983) Silur und Devon im Arauz-Gebiet (Prov. Palencia, N-Spanien). Newsl Stratigr 13 (1): 40-66

Krans TF, Guit FA, Ofwegen LP van (1982) Facies patterns in the Lower Devonian carbonates of the Lebanza Formation (Cantabrian Mountains, Province of Palencia, NW Spain). Neues Jahrb Geol Paläontol Abh 163 (2): 192-230

Kullmann J (1960) Die Ammonoidea des Devon im Kantabrischen Gebirge (Nordspanien). Akad Wiss Lit Abh Math Naturwiss Kl Wiesbaden 7: 1-105

Lobato L (1977) Geología de los valles altos de los ríos Esla, Yuso, Carrión y Deva. Institución "Fray Bernardino de Sahagún", CSIC, León, 193 pp

Lobato L, Cramer FH, Rodríguez R (1977) Descripción de la Formación Carazo de la provincia de Palencia, España. Edad palinológica de su miembro inferior: Ludloviense Medio o Superior. Breviora Geol Astúrica 21 (2): 25-29

Marcos A (1979) Facies differentiation caused by wrench deformation along a deep-seated fault system (León Line), Cantabrian Mountains, North Spain. Discussion. Tectonophysics 60: 303-309

Marquínez J, Marcos A (1984) La estructura de la Unidad del Gildar-Montó (Cordillera Cantábrica). Trab Geol 14: 53-64

Montesinos JR, Truyóls Massoni M (1987) La Fauna de Anetoceras y el límite Zlichoviense-Dalejiense en el Dominio - Palentino (NO de España). Cuad Lab Xeol Laxe 11: 191-208

Rodríguez Fernández R, Heredia N, Lobato L, Velando F (1985) Memoria de la Hoja Geológica E. 1:50.000 n° 106 (16-7), Camporredondo de Alba. IGME, Serv Publ Minist Industr Ener (Madr) 98 pp

Truyóls Massoni M (1981) Primera aportación sobre los tentaculites dacrioconáridos dela región del Pisuerga-Carrión. Rev Esp Micropaleontol 13 (1): 69-88

Truyóls Massoni M, Montesinos R, García Alcalde JL, Leyva F (in press) The Kačak-Otomari event and its characterization in the Palentine Domain (Cantabrian Zone, NW Spain). Lecture Notes in Earth Sciences

Veen J van (1965) The tectonic and stratigraphic history of the Cardaño Area, Cantabrian Mountains, northwest Spain. Leidse Geol Meded 35: 45-104

Wagner RH (1955) Rasgos estratigráfico-tectónicos del Paleozoico Superior de Barruelo (Palencia). Estud Geol (Madr) 11: 145-202

Wagner RH (1971) Carboniferous nappe structures in northeastern Palencia (Spain). Trab Geol 4 (2): 431-459

Wagner RH, Carballeira J, Ambrose T, Martínez García E (1984) Memoria de la Hoja Geológica E. 1:50.000 n° 107 (17-7), Barruelo de Santullán. IGME. Serv Publ Minist Industr Ener (Madr) 113 pp

Wagner RH, Wagner Gentis CHT (1963) Summary of the stratigraphy of the Upper Palaeozoic rocks in NE Palencia, Spain. Proc K Ned Akad Wet (Amsterdam) 66: 149-163

Walliser OH (1985) Natural boundaries and Commission boundaries in the Devonian. Cour Forschungsinst Senckenb 75: 401-408

2.4 Carboniferous Pre-Stephanian Rocks of the Asturian-Leonese Domain (Cantabrian Zone)

L. C. Sanchez de Posada[1], M. L. Martinez Chacon[1], C. Mendez Fernandez[1], J. R. Menendez Alvarez[1], J. Truyols[1] and E. Villa[2]

2.4.1 Introduction

Carboniferous rocks ranging in age from Tournaisian to Stephanian crop out over some 15 000 km^2 in the Cantabrian Zone. Unlike Devonian rocks, they are present in every part of this zone and their maximum thickness could reach several 1000 m. The presence of abundant marine facies, or locally, total marine successions, gives the Carboniferous of NW Spain a peculiar character among the Carboniferous strata of Western Europe.

The limits of the present area occupied by Carboniferous strata are roughly coincident with the boundaries of the Cantabrian Zone. Nevertheless, the original extent of the basin must have been considerably larger, since the Carboniferous rocks have been tectonically shortened (to about 50% according Marcos and Pulgar 1982) and are covered at the borders of the Cantabrian Zone by Mesozoic and Tertiary strata or overthrusted by rocks of the West Asturian-Leonese Zone.

The reader is referred to Pérez Estaún and Bastida (Part II, Chap. 3.1) for details on the structural geology and subdivisions of the Cantabrian Zone.

[1] Departamento de Geología, Universidad de Oviedo, 33005 Oviedo, Spain
[2] Unidad de Geología, Facultad de Biología, Universidad de León, Spain

R. D. Dallmeyer and E. Martínez García (Eds.)
Pre-Mesozoic Geology of Iberia

2.4.2 The Carboniferous Succession

Carboniferous strata in the Asturian-Leonese domain (Fig. 1) comprise three conspicuously different sequences: a lower sequence (ranging in age from Tournaisian to approximately Kinderscoutian) that remains fairly constant throughout the basin, a middle sequence (Kinderscoutian to Late Westphalian D), with rocks showing sharp differences in lithology from area to area or even undergoing important lateral changes within a single area; and finally, a sequence, Stephanian in age, also rather inconstant in lithology, exposed in isolated outcrops and unconformably overlying older rocks. Only the first two sequences are dealt with in this chapter. The third sequence is described in another article (Martínez García, this Vol.).

2.4.2.1 The Lower Succession (Tournaisian - Kinderscoutian)

The base of Carboniferous rocks in the Asturian-Leonese area is composed of some 5–10 m of light grey bioclastic limestone (Baleas Fm.) overlying Upper Devonian sandstones. The most recent and reliable research on the conodont content of the Baleas Fm. was carried out by Río and Menéndez Alvarez (1978) and García Alcalde et al. (1985). The latter authors investigated the type section of this formation and concluded that it was Late Famennian (*-costatus*) - Early Tournaisian (*-sulcata*) in age. The condensation of the stratigraphic sequence makes it difficult to clearly identify the whole sequence of conodont zones established for these ages in Western Europe and North America. Nevertheless, although discrepancies exist concerning the exact zone reached at the top of the Baleas Fm., there is no doubt that it is Tournaisian in age.

The Baleas Fm. is absent in many places in the Cantabrian Zone. At these sites, the Carboniferous starts with the Alba Fm., whose base is Late Tournaisian (cf. infra) or with a succession (usually less than 5 m thick) of shales, typically black, occasionally grey, containing chert lenses, phosphatic nodules and sporadic sandstone (Vegamián Fm.). Data on the age of the Vegamián Fm. are scarce. At the Genicera locality, Wagner et al. (1971) found *Muensteroceras arkansanum*, previously known from the Kinderhookian and Lower Osagian of the USA. The formation includes at least rocks of Middle Tournaisian (*-cooperi-communis*) to Upper Tournaisian (*-anchoralis*) as documented by conodont faunas recovered in its lower and upper parts (Higgins 1974; Raven 1983). However, at least at some localities, the Vegamián Fm. includes strata of an earlier age judging from a conodont fauna reported by Higgins et al. (1964) and Higgins and Wagner Gentis (1982), at the locality of Santiago de las Villas (León).

The Alba Fm. (Genicera Fm. or Griotte) is the most widespread lithostratigraphic unit in the Carboniferous of the Cantabrian Zone. It is made up of red, nodular limestones (occasionally pink or grey) and is less than 30 m in thickness. Usually a thin chert band several metres thick exists near the base, splitting the formation into three members (the Gorgera, Canalón and Lavandera members). The age of the Alba Fm. has been established from numerous cephalopod and conodont finds. Its base is unquestionably Upper Tournaisian and its top probably Arnsbergian. At the base there exists a hard ground containing abundant reworked conodonts, the younger species being characteristic of the *-anchoralis*. With the exception that no Tournaisian cephalopods have been found in the Alba Fm., the results from the study of these fossils agree with those of the research on conodonts. On the whole, Goniatitina of these strata are species ranging in age from Early Visean (CUII β-γ) to Arnsbergian. Kullmann (1962) and Wagner Gentis (1963) reported cephalopod species typical of the Upper *Perycyclus* (II β-γ) to *Eumorphoceras* Zones. Namurian faunas are found in the uppermost levels of the Alba or even within the transitional beds to the overlying formation. In these beds there are conodonts characteristic of the upper part of *-bollandensis*.

Except for the innermost unit of the Cantabrian Zone (Somiedo-Correcilla Unit, Fig. 1), the Alba Fm. is followed by a succession of limestones, which constitute the lower part of the Caliza de Montaña and which is now generally regarded as an independent formation (Barcaliente Fm.). The Barcaliente Fm. consists of dark grey, largely nonfossiliferous micritic laminated limestones (usually some 200–300 m in thickness), and locally a few levels of breccias. In the last years, several authors (Marquínez et al. 1982; Navarro et al. 1986) working in the Picos de Europa Unit and neighbouring areas have pointed out that, at least in some sections, the distinction between the Barcaliente and Valdeteja Fms. is not as sharp as was supposed by other workers. The age of the Barcaliente Fm. is based on very limited data. Cephalopods found in the transitional Alba-Barcaliente beds permit the allocation of these strata to the *Eumorphoceras* Zone (probably E2, Kullmann 1962; Wagner Gentis 1963). Conodont faunas have been found at only a few localities. At the type section of the formation, Menéndez Alvarez (unpublished data) found conodonts characteristic of the Zones of *bilineatus bollandensis* (Arnsbergian, E2), *noduliferus-lateralis* (Chokerian and Alportian) and, in the uppermost levels, *Idiognathoides sinuatus*, which in other areas appears at the base of the *Reticuloceras* Zone (Kinderscoutian). In some sections where the Barcaliente Fm. is not overlain by the Valdeteja Fm., but by a succession of shales that replaces the Valdeteja, some cephalopods characteristic of the Kinderscoutian stage have been found that set a limit for the minimum possible age of the Barcaliente Fm. at such localities.

In the Somiedo-Correcilla Unit, the Carboniferous succession is different. Some 8 m of reddish to green shales (Olaja Beds) overlie the Alba Fm. and in turn are followed by a terrigenous succession of shales, limestones, sandstones and conglomerates with turbiditic intervals (Cuevas Fm.).

The lower part of the Cuevas Fm. was named Olleros Fm. by Wagner et al. (1971), who considered its upper limit to be the base of a thick limestone exposed at the section of Olleros de Alba, some 250 m above the base of the formation.

The Olaja Beds and the lower strata of the Olleros Fm. (Entomozoen Schiefer, Becker et al. 1975) contain cephalopods of Arnsbergian age. The same age can be assigned to the Olleros Fm. at its type section, judging from E2 conodont faunas found by Menéndez Alvarez in the limestone lenses below the upper boundary of this formation. However, the uppermost strata of this terrigenous sequence can be younger in some places, as indicated by cephalopods characteristic of the *Reticuloceras* Zone found at several localities, *R. paucicrenulatum*, *R.* cf. *nodosum* among others (Bouroz 1962; Wagner et al. 1971; Kullmann 1979).

2.4.2.2 The Middle Succession (Kinderscoutian - Westphalian D)

The top of Barcaliente Fm. (or the top of Cuevas Fm.) roughly coincides with a remarkable increase in the diversity of Carboniferous sedimentation within the Cantabrian Zone. Actually this diversity starts before this time, during the period in which the terrigenous sequence was being deposited in the Somiedo-Correcilla Unit and limestones of the Barcaliente Fm. were forming over wide areas of the basin. However, after the deposition of Barcaliente Fm., differences in sedimentation in different areas are much greater.

The top of the middle succession has been dated as Westphalian D in most of the Cantabrian Zone. Nevertheless, in the eastern part of the basin, a marine or nearly marine succession is exposed in which the youngest strata are Upper Moscovian (Myachkovsky Horizon) or exceptionally Kasimovian in age.

2.4.2.2.1 The Succession in the Fold and Nappe Province (Somiedo-Correcilla, Sobia-Bodón and Aramo Units)

In the Somiedo-Correcilla Unit, rocks belonging to the Middle Succession are mostly lacking. The succession above the Barcaliente Fm. in the Sobia-Bodón Unit includes rocks that reach, as maximum, an Early Moscovian age according to available data.

On the whole this succession includes a lower limestone portion directly above the Barcaliente Fm. and an upper part consisting of alternating calcareous and terrigenous rocks. The amount of terrigenous sediment increases towards the top, so that in the uppermost levels there are not conspicuous limestones. Coal seams, which are occasionally workable, occur in some sections.

The Valdeteja Fm. (grey, generally light coloured, bioclastic and massive limestones) constitutes the lower part of the succession. Its thickness is variable (average thickness is some 400 m, but in some localities it is very much thinner or locally even absent). The top of the Valdeteja Fm. is sharply diachronic. Studies carried out since 1965, especially research on foraminiferal and conodont faunas of these rocks (Ginkel 1965; Villa 1982; Menéndez Alvarez unpublished data), show that it extends from Marsdenian in the basal beds to, as a maximum, Lower Moscovian at the top in certain sections, such as the type section. Accordingly, noteworthy differences in the age render it difficult to make generalizations about the age of the uppermost levels of the Valdeteja Fm. Usually its top does not exceed a Bashkirian age, as indicated both by the data obtained in the upper part of this unit and by the Bashkirian age of the strata above (Ginkel 1965; Truyóls et al. 1982).

The strata succeeding the Valdeteja Fm. are alternating sandstones, shales, sandy shales, limestones and some coal seams (San Emiliano Fm.), the last being more abundant and important in the upper part of the succession. At the type area, the San Emiliano Fm. is 1800 m thick. On the basis of the different proportions of limestone, terrigenous sediment and coal seams, Bowman (1982, 1985) divided this formation into three members (Pinos, La Majúa and Candemuela). In contrast to the other two members, the middle one contains abundant limestone units (some of them of remarkable thickness). The upper member is characterized by the presence of coal seams, sometimes workable. On the northern slope of the Cantabrian Mts., the San Emiliano Fm. is exposed in the area of Teverga (Asturias) where the upper part constitutes a succession of more than 1000 m and is subject to active mining. In contrast to the south and east, the mixed platform sequence which crops out in the San Emiliano area is replaced, at a short distance, by a succession containing turbidites which are well exposed at Villafeliz, Robledo de Caldas and Villanueva de la Tercia (Carballeira et al. 1985).

Even though some problems have not yet been solved, palaeontological data allow the age of the San Emiliano Fm. to be determined with some accuracy. The lower part of this formation in the San Emiliano Syncline has yielded Yeadonian spores; limestones and shales of La Majúa Mbr. have yielded Bashkirian faunas (Foraminifera and Brachiopoda essentially). Only in the higher levels of the series were foraminifers and brachiopods of Early Moscovian age found. The Bashkirian/Moscovian boundary was not documented in detail (cf. Wagner and Bowman 1983; Carballeira et al. 1985). Floral remains found in the lower half of the Candemuela Mbr. indicate a Westphalian A age for these strata. The age of the rocks of the San Emiliano Fm. in the Teverga region is roughly the same as for the rocks of the type area. Even though the contact of the Valdeteja and San Emiliano Fms. is usually faulted, brachiopods of the calcareous units are the same as those collected in the San Emiliano area (Martínez Chacón 1979), and limestones have yielded foraminifers of Late Bashkirian age. On the other hand, coal-bearing levels contain a microflora regarded as Late Westphalian A or Early Westphalian B (Cachán 1978).

As is usual in these successions above the Caliza de Montaña, rocks exposed around Quirós (Asturias) have a lower part containing abundant limestones and an upper part essentially terrigenous, with coal seams that, as in Teverga, are being worked. The age of the productive strata is roughly the same as the age of the San Emiliano Fm. but differs from the age of the productive coals in the Asturian Central Coal Basin (cf. supra). In fact, strata immediately above the Caliza de Montaña contain Bashkirian foraminifers (Truyóls et al. 1982). Granados et al. (1985) and Leyva et al. (1985) considered the foraminiferal faunas obtained from the uppermost strata of the lower part of this succession to be characteristic of a level intermediate between the top of the Bashkirian and the base of the Moscovian (Transition Interval). This perhaps can also be recognized in other areas of the Cantabrian Zone. The upper part of this succession has yielded only a few floral remains attributed to either Upper Namurian or Lower Westphalian (Wagner 1959; Julivert 1960).

To the north, a rather different succession is known to occur below the Mesozoic cover 6 km E of Gijón (Asturias), in the La Camocha Coal Mine, where some 2300 m of a succession containing workable coal seams as old as Upper Namurian is exposed. Thus the rocks exposed at La Camocha seem to indicate that non-marine conditions were established here before they occurred in other areas of the La Sobia-Bodón Unit (such as Teverga, San Emiliano or Quirós).

2.4.2.2.2 The Succession in the Asturian Central Coal Basin

A thick succession of Carboniferous strata containing workable coal seams in its upper part is exposed in the Asturian Central Coal Basin. These rocks are folded into a complex structural syncline (Julivert and Marcos 1973; Aller 1986).

The relative abundance of limestone in the sequence allowed Barrois (1882) to distinguish a lower succession (Assise de Lena) rich in limestones from an upper succession (Assise de Sama) nearly devoid of limestone. The thick sequence of Carboniferous rocks above the Caliza de Montaña in the Asturian Central Coal Basin is usually split into several intervals, or Paquetes (Fresnedo, Levinco, Llanón, Tendeyón, Caleras, Generalas, San Antonio, María Luisa, Sotón, Entrerregueras, Sorriego, Modesta and Oscura, from bottom to top). This division is rooted in the tradition of mining and has earned geological significance since the papers by Bless (1967, 1968) and García Loygorri et al. (1971).

The lower part of the rocks above the Caliza de Montaña (often regarded as unproductive strata) consists of some 400 m of shales (Paquete Fresnedo), a limestone unit (Caliza de Peña Redonda) also some 400 m thick (sometimes comprising several limestone units with interbedded shales), 650-700 m of alternating shales, siltstones, sandstones and limestones (together with the Caliza de Peña Redonda known as Paquete Levinco), some 100 m of diverse rocks, essentially shales, sandstones and thin coal seams near the top (Paquete Llanón) and some 950 m of shales, siltstones, sandstones, quartz sandstones and coals (Paquete Tendeyón).

According to the data of García Loygorri et al. (1971), García Loygorri (1974) and Leyva et al. (1985), the earliest Moscovian faunas occur in the Paquete Levinco above the Caliza de Peña Redonda. This, in turn, contains a mixture of fossils of uppermost Bashkirian and lowermost Moscovian ages (Transition Interval of Granados et al. 1985 and Leyva et al. 1985). The Lower/Upper Moscovian boundary is located within the Paquete Tendeyón, close to its base. On the basis of the macroflora the upper part of the Paquete Llanón (containing *Linopteris oblicua*) was dated as Westphalian C and most of the Paquete Tendeyón as Westphalian D (Leyva et al. 1985; Sáenz de Santa María et al. 1985). In the extreme NE of the Asturian Central Coal Basin the sequence is wholly marine up to the upper Lower Moscovian (Kashirsky Horizon).

A noteworthy change in the stratigraphic characteristics of the succession containing workable coals takes place at the level of the Paquete Generalas. This last Paquete is in fact the boundary between the lower part of the sequence, with scarce coal seams and abundant limestones, and an upper part nearly barren of limestones, which includes the Paquetes San Antonio, María Luisa, Sotón, Entrerregueras, Sorriego, Modesta and Oscura (in total little under 2000 m thick). In general, the frequency of continental sedimentation increases from bottom to top. The Paquete María Luisa is the interval most extensively worked for coal.

The age of the upper part of the productive sequence has been disputed in the literature. Since the studies by Adaro (1914) and Jongmans (1951) several authors have dealt with the possibility of an Early Stephanian age for the uppermost strata (Paquete Oscura) of this sequence. Although this possibility has not been entirely rejected, it now does not seem likely that Lower Stephanian rocks are present in the Asturian Central Coal Basin (Wagner 1971; Wagner and Laveine, in Laveine 1977). Moreover, a recent revision of floral records in this area has shown that the top of the succession reaches a level below that of the highest Westphalian D (Wagner and Alvarez Vázquez in press).

Most of the productive sequence of the Asturian Central Coal Basin must be Westphalian D in age, despite the disagreement concerning the age of the lower Paquetes. The Paquete Caleras has been dated as Westphalian C by García Loygorri et al. (1971). According to these authors the same age might apply to the overlying Paquete Generalas and even to the lower part of the Paquete San Antonio, whereas Sáenz de Santa María et al. (1985) and Luque et al. (1985) still consider most of the Paquete Tendeyón to be Westphalian D in age. The comprehensive revision carried out by Wagner and Alvarez Vázquez (in press) suggests that the base of Westphalian D lies close to the base of Paquete Caleras.

Foraminifera in this part of the sequence are scarce. On the basis of data provided by these fossils, García Loygorri et al. (1971) and García Loygorri (1974) dated as Westphalian D the part of the sequence extending from the Paquete Caleras to the lower part of the Paquete Entrerregueras. According to Martínez Chacón (in Luque et al. 1985) the transition between the Podolsky and Myachkovsky Horizons probably takes place within the Paquete San Antonio or the Paquete María Luisa.

On the southern slopes of the Cantabrian Mts. the general sequence is broadly similar to that found in the north and can be divided into two parts: a lower part containing abundant limestones that thicken toward the E and an upper part nearly devoid of calcareous horizons. However, differences in lithology between these rocks and those exposed to the north prevent the use of the paquetes in the above subdivision. These facts, coupled with the geological continuity of both areas led Aller (1986) to split the Lena Fm. into three subdivisions (Units Lena I, Lena II and Lena III). Using another scheme, Brouwer and Ginkel (1964) proposed a different name (Lois-Ciguera Fm.) for the portion of the sequence cropping out in the southern part of the Asturian Central Coal Basin. Limestones of the Lois-Ciguera Fm. (sensu Brouwer and Ginkel 1964 and Rácz 1965) have yielded foraminifers and algae of Late Bashkirian to Late Moscovian age, providing evidence for the correlation of these horizons with the sequence in the northern part of this region.

A fault extending from the Huerna valley to N of Mieres separates a smaller area to the west (Riosa-Olloniego Unit, García Loygorri et al. 1971) that has stratigraphical characteristics different from other areas in the Asturian Central Coal Basin (see Truyóls 1983; Wagner and Alvarez Vázquez in press).

2.4.2.2.3 The Succession in the Ponga Unit

Although noticeably diachronic, the Carboniferous succession above the Barcaliente Fm. is rather constant in lithology in a transect across the structures. However, parallel to the direction of the main structures, changes in facies take place. To the NE, sections rich in carbonate are found that resemble the typical sequences of the Picos de Europa Unit more than those to the south in the Ponga Unit.

In the southern and central parts of the Ponga Unit the Valdeteja Fm. is missing. Above the Barcaliente Fm. the succession is as follows: a variable thickness (usually less than 20 m, although in some places up to 100 m) of shales, usually red, occasionally brown or green, commonly containing manganese and iron nodules and sometimes calcareous lenses (Ricacabiello Fm.); alternating shales, siltstones, sandstones and scarce limestones (300 m in thickness on average, Beleño Fm.); some 300 m of light coloured limestones (Escalada Fm. or Caliza Masiva) and a series of alternating limestones, shales, siltstones and sandstones, usually 500–800 m thick but which in some places exceed 1500 m (Fito Fm.).

Discoveries at several localities have shown the presence in the Ricacabiello Fm. of several fossil groups, most of them represented by peculiar taxa of older aspect than that corresponding to its true age. Thus, the associations of the Ricacabiello Fm. have a distinctive character with regard to other fossil associations from the Carboniferous of the Cantabrian Zone (Martínez Chacón et al. 1985). Characteristic cephalopod taxa such as *Retites* and, to a lesser extent, some finds of conodonts and flora, along with the discovery (Sjerp 1967) of a foraminiferal fauna in a limestone lens above the Ricacabiello Fm. enable a Kinderscoutian-Westphalian A age to be assumed for these rocks (Martínez Chacón et al. 1985). Few stratigraphically significant fossils are known from the lower part of the Beleño Fm. Only the above-quoted finding by Sjerp and some foraminifers (Villa and Heredia 1988) and conodonts (Menéndez Alvarez herein) found in calcareous lenses near the base of this formation are significant, pointing to an Early Moscovian (probably Vereisky Horizon). Beginning in the upper part of the Beleño Fm. the content of both limestone and fossils increases in these rocks. Foraminifers and brachiopods supply the essential data for the dating of these sequences.

Lithostratigraphical units show a high degree of diachronism if their age is considered in the context of different structural units. For instance, the Escalada Fm. ranges in age from Early Moscovian (upper part of Kashirsky Horizon) to Late Moscovian (Podolsky Horizon) in the western units and from an age equivalent to that of the Podolsky Horizon to, probably, an age equivalent to that of the Myachkovsky Horizon in the eastern units. The Fito Fm. ranges in age from a time equivalent to the Podolsky Horizon to a time equivalent to the Myachkovsky Horizon.

Research on the foraminiferal content at several localities, coupled with a review of previously published data, enabled Villa and Heredia (1988) to point out the following facts: (1) The age of each formation is younger toward the east; (2) the Fito Fm. is absent in the easternmost units; (3) the more eastern the unit under consideration is, the fewer number of formations the sequence reaches, although (as a consequence of the diachronism) the age of the highest strata is not necessarily very much older; and (4) the main stratigraphical pattern in the Ponga Unit matches that in the Asturian Central Coal Basin. It must be pointed out, however, that noteworthy differences exist between both sequences. Only in one unit (Campo de Caso Unit) is there a terrigenous series (uppermost levels of Fito Fm.) overlying the clastic and calcareous part of this formation. It can be compared with the Sama Fm., but the terrigenous series starts at a younger age and contains less coals.

As stated above, a relative decrease of terrigenous materials and an increase of calcareous rocks in northern and eastern directions results in wholly calcareous successions in the northeastern part of the Ponga Unit. This fact, illustrated by Navarro et al. (1986), means that the northeastern successions bear more resemblance to successions of the Picos de Europa Unit than to those of the southern part of the Ponga Unit.

2.4.2.2.4 The Succession in the Picos de Europa Unit

The Carboniferous succession in the Picos de Europa Unit is fairly different from that exposed in other areas, only resembling the sequence exposed in the northern part of the Ponga Unit. A succession more than 1000 m thick, made only of limestones ranging in age from Tournaisian to Late Moscovian (at an age equivalent to the Myachkovsky Horizon) or even in some places to the Kasimovian, crops out throughout the Picos de Europa Unit.

The lowermost beds in this succession are composed of rocks which can be assigned to the Baleas, Alba, Barcaliente and Valdeteja Fms., as previously described for other areas of the Cantabrian Zone, even though Martínez García (1981) proposed the name Portillas Fm. for a series of limestones exposed in the Picos de Europa that underlie limestones of the Alba Fm. Overlying the Valdeteja Fm. there exists a sequence of limestones (Picos de Europa Fm.), bedded and containing chert and intercalations of calcareous shales in the lower part (Bedded Lms. Member of Maas 1974; 200 m in thickness on average), and massive, usually light colored with red intercalations in the upper part (Massive Lms. Member, Maas 1974; 350 m in thickness on average).

As in other areas of the Cantabrian Zone, the top of the Valdeteja Fm. is noticeably diachronic. Maas (1974) considered its age to be Bashkirian (-*Millerella* and lower part of Subzone A of -*Profusulinella*). Although this is the case for some sections, Truyóls et al. (1984) described some localities where a younger age (Early Moscovian, Vereisky Horizon) might be attained. Accordingly, foraminifers reveal some variations in the age of the Picos de Europa Fm. At some localities (as in the section of Las Llacerias in the northern part of the area), the base of the Picos de Europa Fm. is Bashkirian in age. Only several metres above the base do the lowermost Moscovian fossils (*Eofusulina triangula fusiformis*) appear, followed in the section by fusulinids characteristic of the Horizon of Kashira (even in the Bedded Mbr.), and finally by species indicating a presumably Late Moscovian age. In contrast, in other sequences (such as the succession of Río Nevandi, in the southern part of the area) some 150–200 m above the base, fusulinids (*Hemifusulina ex* gr. *dutkevitchi*) pointing to an age close to the boundary of the Middle/Upper Carboniferous appear, and higher in the section species indicating the subdivision B3 of the -*Fusulinella* are found.

Rocks immediately above the Picos de Europa Fm. show a great variation in lithology ranging from platform limestones in the north (Puentellés Fm.) to turbiditic sequences, like those exposed at Lebeña and Aliva (Lebeña or Aliva Fm.). Fusulinid faunas found in the Puentellés Fm. contain taxa such as *Protriticites* and *Triticites* characteristic of the Kasimovian. This seems to be the minimum age of the Lebeña Fm., judging from Kasimovian fusulinids found in calcareous blocks included within clastic sediments. These rocks and the paralic mate-

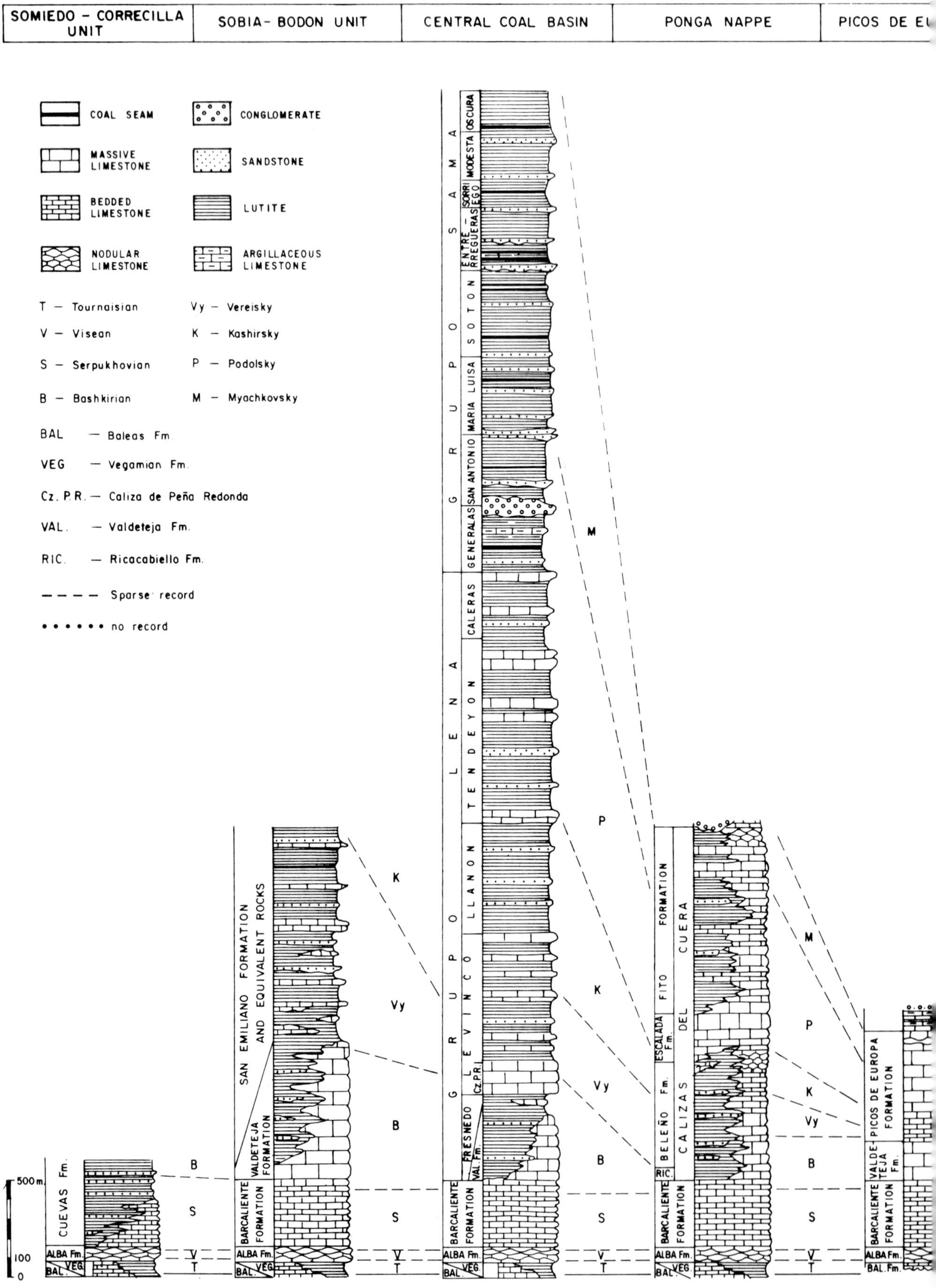

Fig. 1. Generalized Carboniferous succession in the structural units of the Asturian-Leonese area and distribution of selected fossil taxa

rials of post-Westphalian age are dealt with in another chapter of this volume.

2.4.3 Conclusion

A rather complete succession of pre-Stephanian Carboniferous rocks is exposed in the Cantabrian Zone. These rocks reflect a wide spectrum of environments ranging from continental to open marine conditions.

The first episode of Carboniferous sedimentation is characterized by rocks of quite uniform character throughout the basin, containing a poor diverse benthic fauna. Cephalopods and conodonts are the key fossil groups for dating these strata. In Namurian times, differentiation in the basin produced different stratigraphical sequences in the sundry units of the Cantabrian Zone. A variegated benthic fauna developed in shallow marine environments and a rather prolific flora inhabited Cantabrian lands. Several benthic fossil groups, essentially foraminifers (which can be directly compared with Soviet faunas) and to a lesser extent brachiopods, are specially useful in dating the sequences in terms of the marine Soviet stratigraphical scale, whereas plants allow us to date the rocks according to the continental West European scale.

In general, and probably as a consequence of a deformation spreading from the west and south (convexity of the Asturian Arc), the sequences cropping out in the eastern or northern structural units include younger strata and more marine rocks than do those sequences exposed in the western or southern structural units.

Acknowledgements. We wish to thank Prof. C. Hepburn for assisting in the revision of the English manuscript and Dr. R. Wagner and Miss Carmen Alvarez Vázquez for their comments and data on the distribution of flora in the Cantabrian Zone.

References

Adaro L de (1914) Emplazamiento de sondeos para investigar la probable prolongación de los senos hulleros por bajo de los terrenos mesozoicos. Bol Inst Geol Min Esp 34: 9-79

Aller J (1986) La Estructura del Sector Meridional de las Unidades del Aramo y Cuenca Carbonífera Central. Principado de Asturias. Consejería de Industria y Comercio, 180 pp

Barrois C (1882) Recherches sur les terrains anciens des Asturies et de la Galice. Mém Soc Géol Nord 2 (1): 1-630

Becker G, Bless MJM, Kullmann J (1975) Oberkarbonische Entomozoen-Schiefer im Kantabrischen Gebirge (Nordspanien). Neues Jahrb Geol Paläontol Abh 150 (1): 92-110

Bless MJM (1967) On the marine beds of some cyclothems in the Central Carboniferous Basin of Asturias with special reference to their ostracode fauna. Notas Comun Inst Geol Min Esp 99: 91-134

Bless MJM (1968) On two hollinid ostracode genera from the Upper Carboniferous of northwestern Spain. Leidse Geol Meded 43: 157-212

Bouroz A (1962) Sur la présence de *Reticuloceras paucicrenulatum* Bisat et Hudson dans le Carbonifère du NW de la Province des Asturies. CR Acad Sci Paris 255 (D): 1968-1970

Bowman MBJ (1982) The stratigraphy of the San Emiliano Formation and its relationship to the other Namurian/Westphalian sequences in the Cantabrian Mts., NW Spain. Trab Geol 12: 23-35

Bowman MBJ (1985) The sedimentology and palaeogeographic setting of Late Namurian-Westphalian A basin-fill succession in the San Emiliano and Cármenes areas of NW León, Cantabrian Mts., NW Spain. In: Lemos de Sousa MJ, Wagner RH (eds) Papers on the Carboniferous of the Iberian Peninsula (Sedimentology, Stratigraphy, Palaeontology, Tectonics and Geochronology). Ann Fac Ciênc Porto 64 (Suppl): 117-169

Brouwer A, Ginkel AC van (1964) La succession carbonifère dans la partie des Montagnes Cantabriques. CR V Congr Int Strat Géol Carbonifère Paris 1963 (1): 307-319

Cachán CJ (1978) Palinoflora del Westfaliense A superior y B inferior de la cuenca hullera de Teverga (Oviedo, España). Palinología 1: 103-113

Carballeira J, Corrales I, Valladares I et al. (1985) Aportaciones al conocimiento de la estratigrafía de la Formación San Emiliano (Carbonífero, Cordillera Cantábrica) en su área tipo. CR X Congr Int Strat Géol Carbonifère Madrid 1983 (1): 345-362

García Alcalde JL, Menéndez Alvarez JR, García López S, Soto F (1985) El Devónico Superior y el Carbonífero Inferior del Sinclinal de Beberino (Pola de Gordón, León, NO de España). CR X Congr Int Strat Géol Carbonifère Madrid 1983 (2): 375-386

García Loygorri A (1974) Caracteres generaux du Bassin Central des Asturies (Espagne). CR VII Congr Int Strat Géol Carbonifère Krefeld 1971 (3): 114-128

García Loygorri A, Ortuño G, Caride de Liñán C, Gervilla M, Greber C, Feys R (1971) El Carbonífero de la Cuenca Central Asturiana. Trab Geol 3: 101-150

Ginkel AC van (1965) Carboniferous fusulinids from the Cantabrian Mountains (Spain). Leidse Geol Meded 34: 1-225

Granados L, Solovieva MN, Reitlinger EA, Martínez Díaz C (1985) The Bashkirian - Moscovian boundary problem in the Asturias (northwest Spain). CR X Congr Int Strat Géol Carbonifère Madrid 1983 (1): 27-35

Higgins A (1974) Conodont zonation of the Lower Carboniferous of Spain and Portugal. Int Symp Belgian Micropal limits from Emsian to Visean Namur 1974. Publ no 4, 17 p

Higgins A, Wagner Gentis CHT (1982) Conodonts, goniatites and biostratigraphy of the earlier Carboniferous from the Cantabrian Mountains. Palaeontology (Lond) 25 (2): 313-350

Higgins A, Wagner Gentis CHT, Wagner RH (1964) Basal Carboniferous strata in part of northern León, NW Spain. Bull Soc Belge Geol Paleontol Hydrol 72 (2): 205-248

Jongmans WI (1951) Las floras carboníferas de España. Estud Geol 14 (7): 281-330

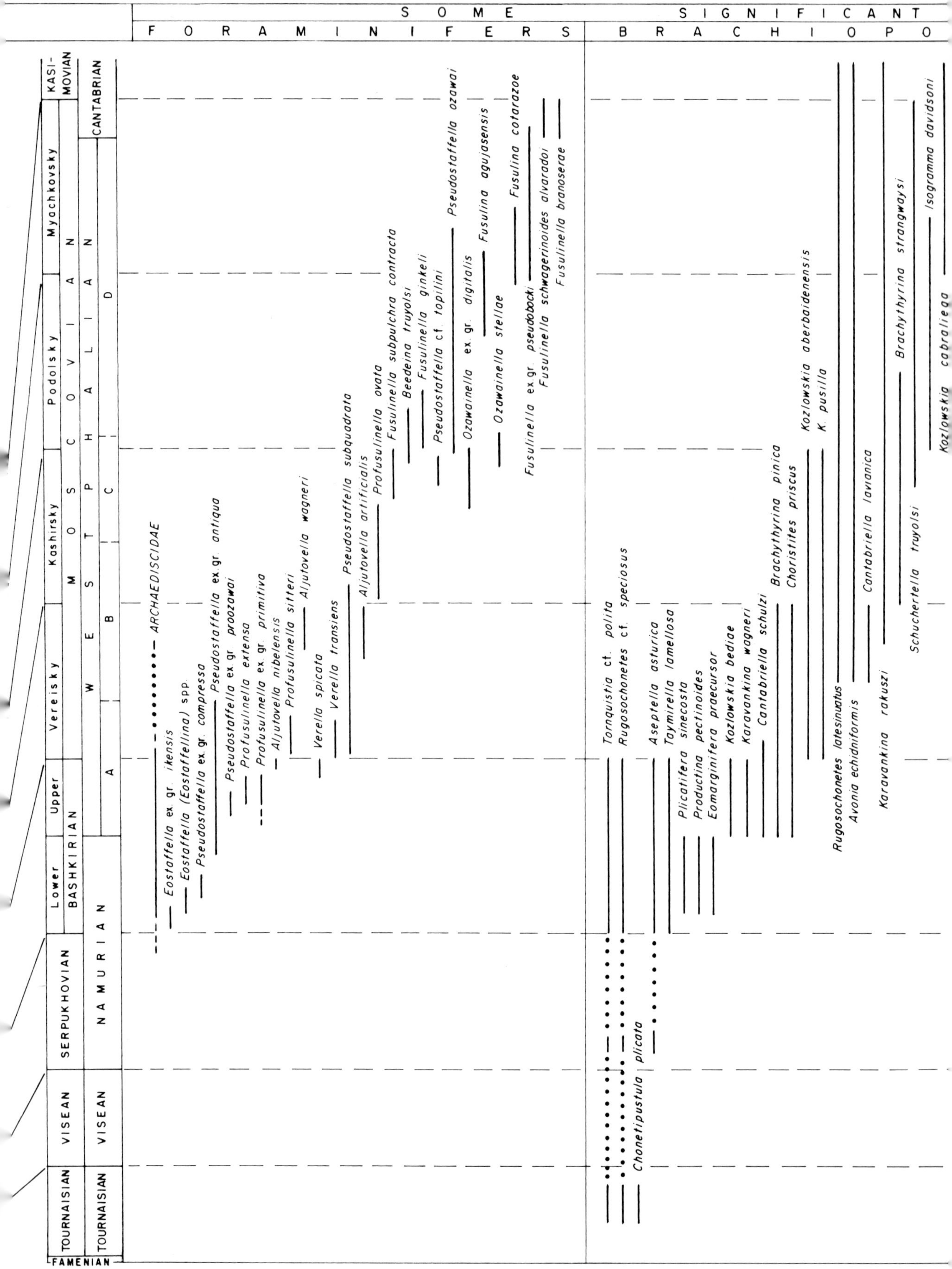

SOME SIGNIFICANT
FORAMINIFERS
BRACHIOPO
KASIMOVIAN
Myachkovsky
Podolsky
Kashirsky
Vereisky
Upper
Lower
BASHKIRIAN
MOSCOVIAN
SERPUKHOVIAN
VISEAN
TOURNAISIAN
FAMENIAN
CANTABRIAN
WESTPHALIAN
A
B
C
D
NAMURIAN
ARCHAEDISCIDAE
Eostaffella ex gr. ikensis
Eostaffella (Eostaffellina) spp.
Pseudostaffella ex gr. compressa
Pseudostaffella ex gr. antiqua
Pseudostaffella ex gr. proozawai
Profusulinella extensa
Profusulinella ex gr. primitiva
Aljutovella nibelensis
Profusulinella sitteri
Aljutovella wagneri
Verella spicata
Verella transiens
Pseudostaffella subquadrata
Aljutovella artificialis
Profusulinella ovata
Fusulinella subpulchra contracta
Beedeina truyolsi
Fusulinella ginkeli
Pseudostaffella cf. topilini
Pseudostaffella ozawai
Ozawainella ex gr. digitalis
Ozawainella stellae
Fusulina agujasensis
Fusulina cotarazoe
Fusulinella ex gr. pseudobocki
Fusulinella schwagerinoides alvaradoi
Fusulinella branoserae
Tornquistia cf. polita
Rugosochonetes cf. speciosus
Chonetipustula plicata
Aseptella asturica
Taymirella lamellosa
Plicatifera sinecosta
Productina pectinoides
Eomarginifera praecursor
Kozlowskia bediae
Karavankina wagneri
Cantabriella schulzi
Brachythyrina pinica
Choristites priscus
Kozlowskia aberbaidenensis
K. pusilla
Rugosochonetes latesinuatus
Avonia echidniformis
Cantabriella lavianica
Karavankina rakuszi
Brachythyrina strangwaysi
Schuchertella truyolsi
Isogramma davidsoni
Kozlowskia cabraliega

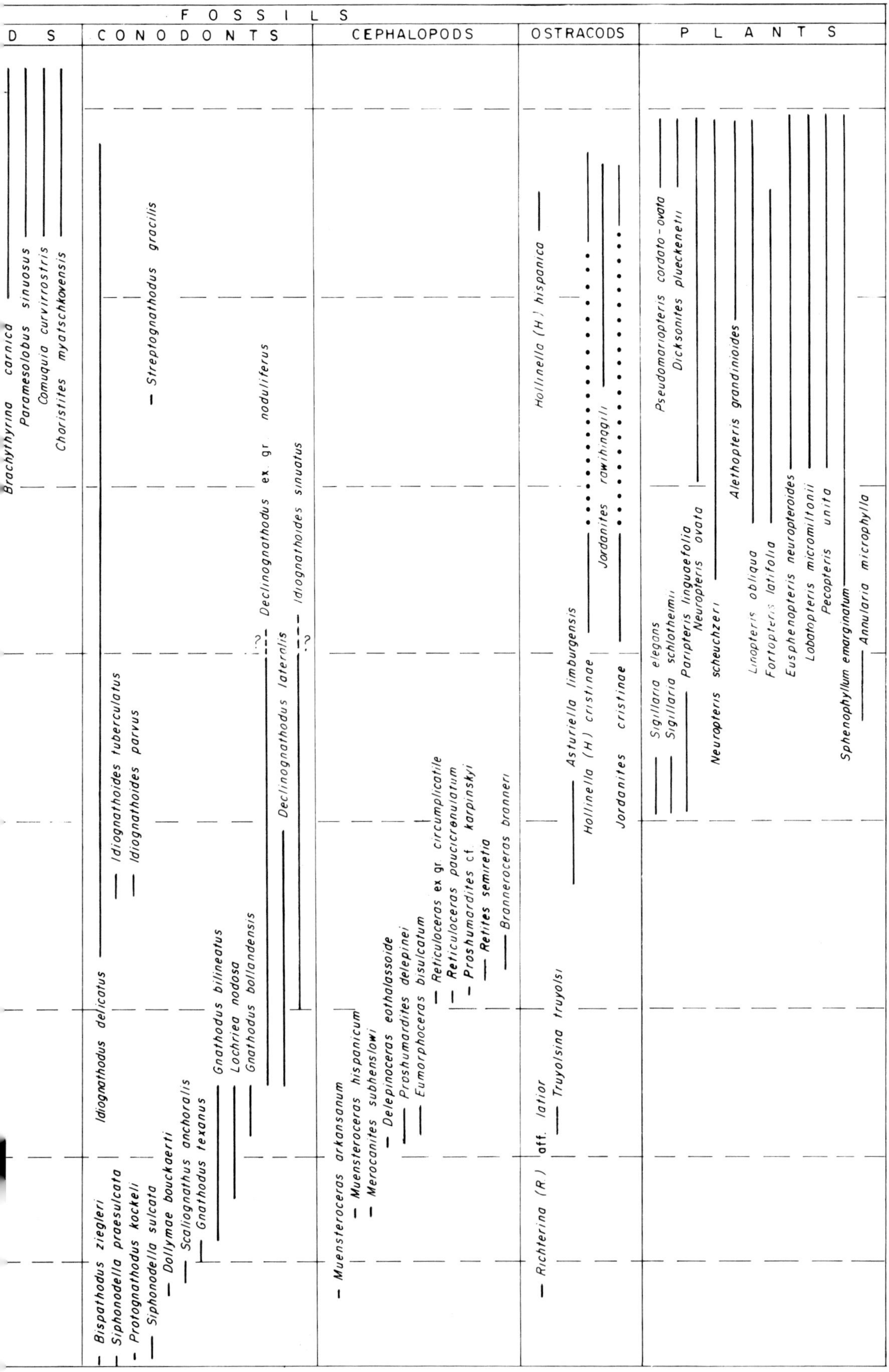
FOSSILS
DS
CONODONTS
CEPHALOPODS
OSTRACODS
PLANTS
Brachythyrina carnica
Paramesolobus sinuosus
Comuquia curvirostris
Choristites myatschkovensis
Bispathodus ziegleri
Siphonodella praesulcata
Protognathodus kockeli
Siphonodella sulcata
Dollymae bouckaerti
Scaliognathus anchoralis
Gnathodus texanus
Idiognathodus delicatus
Idiognathoides tuberculatus
Idiognathoides parvus
Streptognathodus gracilis
Gnathodus bilineatus
Lochriea nodosa
Gnathodus bollandensis
Declinognathodus lateralis
Declinognathodus ex gr. noduliferus
Idiognathoides sinuatus
Muensteroceras arkansanum
Muensteroceras hispanicum
Merocanites subhenslowi
Delepinoceras eothalassoide
Proshumardites delepinei
Eumorphoceras bisulcatum
Reticuloceras ex gr. circumplicatile
Reticuloceras paucicrenulatum
Proshumardites cf. karpinskyi
Retites semiretia
Branneroceras branneri
Richterina (R.) aff. latior
Truyolsina truyolsi
Hollinella (H.) cristinae
Asturiella limburgensis
Hollinella (H.) hispanica
Jordanites cristinae
Jordanites rawihinggili
Sigillaria elegans
Sigillaria schlotheimii
Paripteris linguaefolia
Neuropteris ovata
Neuropteris scheuchzeri
Pseudomariopteris cordato-ovata
Dicksonites plueckenetii
Alethopteris grandinioides
Linopteris obliqua
Fortopteris latifolia
Eusphenopteris neuropteroides
Lobatopteris micromiltonii
Pecopteris unita
Sphenophyllum emarginatum
Annularia microphylla

Julivert M (1960) La terminación septentrional de la Cuenca de Quirós (Cordillera Cantábrica). Breviora Geol Astúrica 4 (1-2): 19-34

Julivert M, Marcos A (1973) Superimposed folding under flexural conditions. Am J Sci 273: 353-375

Kullmann J (1962) Die Goniatiten der Namur-Stufe (Oberkarbon) im Kantabrischen Gebirge, Nordspanien. Abh math Naturwiss Kl Akad Wiss Lit 1962 (6): 260-377

Kullmann J (1979) Die "Reticuloceras-Zeit" (Oberkarbon) in der Entwicklung des kantabrischen Variszikums. Clausth Geol 30: 34-44

Laveine JP (1977) Report of the Westphalian D. In: Reports by the Working Groups of the IUGS Subcommission on Carboniferous Stratigraphy. Symp Carb Strat Praha 1973: 71-83

Leyva F, Granados L, Solovieva MN et al. (1985) La estratigrafía del Carbonífero Medio en el Sector Central de la Unidad Estructural de La Sobia-Bodón (Cuencas de Quirós-Teverga). CR X Congr Int Strat Géol Carbonifère Madrid 1983 (1): 213-231

Luque C, Gervilla M, Sáenz de Santa María JA, Leyva F, Laveine JP, Loboziak S, Martínez Chacón ML (1985) Características sedimentológicas y paleontológicas de los paquetes productivos en el corte de La Inverniza-El Cabo (Cuenca Central Asturias). CR X Congr Int Strat Géol Carbonifère Madrid 1983 (1): 281-303

Maas K (1974) The geology of Liébana, Cantabrian Mountains, Spain; deposition and deformation in a flysch area. Leidse Geol Meded 49: 379-465

Marcos A, Pulgar JA (1982) An approach to the tectonostratigraphic evolution of the Cantabrian Foreland thrust and fold belt, Hercynian Cordillera of NW Spain. Neues Jahrb Paläontol Abh 163 (2): 256-260

Marquínez J, Méndez C, Menéndez Alvarez JR, Sánchez de Posada LC, Villa E (1982) Datos biostratigráficos de la sucesión carbonífera (Turnesiense-Kasimoviense) de Las Llacerias, Picos de Europa, N de España. Trab Geol 12: 187-193

Martínez Chacón ML (1979) Braquiópodos carboníferos de la Cordillera Cantábrica (Orthida, Strophomenida y Rhynchonellida). Mem Inst Geol Min Esp 96: 1-291

Martínez Chacón ML, Menéndez Alvarez JR, Sánchez de Posada LC, Truyóls J (1985) Aportaciones al conocimiento de la formación Ricacabiello (Carbonífero de la Zona Cantábrica, N de España) y su contenido paleontológico. Trab Geol 15: 53-65

Martínez García E (1981) El Paleozoico de la Zona Cantábrica Oriental (Noroeste de España). Trab Geol 11: 95-127

Navarro D, Leyva F, Villa E (1986) Cambios laterales de facies en el Carbonífero del Oriente de Asturias (Cordillera Cantábrica, Norte de España). Trab Geol 16: 87-102

Rácz L (1965) Carboniferous calcareous algae and their associations in the San Emiliano and Lois-Ciguera Formations (Prov. León, NW Spain). Leidse Geol Meded 31: 1-112

Raven JGM (1983) Conodont biostratigraphy and depositional history of the Middle Devonian to Lower Carboniferous in the Cantabrian Zone (Cantabrian Mountains, Spain). Leidse Geol Meded 52: 265-339

Río P del, Menéndez Alvarez JR (1978) Estudio lito y bioestratigráfico de la Caliza de Candamo (Loredo, zona central de Asturias). Trab Geol 10: 379-387

Sáenz de Santa María JA, Luque C, Gervilla M et al. (1985) Aportaciones al conocimiento estratigráfico y sedimentológico del Carbonífero Productivo de la Cuenca Central Asturiana. CR X Congr Int Strat Géol Carbonifère Madrid 1983, Inst Geol Min Esp (1): 303-327

Sjerp N (1967) The geology of the San Isidro-Porma area (Cantabrian Mountains, Spain). Leidse Geol Meded 39: 55-128

Truyóls J (1983) La Cuenca Carbonífera Central. In: Martínez Díaz C (coord) Carbonífero y Pérmico de España. Inst Geol Min Esp: 60-81

Truyóls J, Arbizu M, García Alcalde JL et al. (1982) Mapa Geológico de España, E. 1:50 000. Hoja núm. 74 (La Plaza, Teverga). Memoria explicativa. Inst Geol Min Esp 64 p

Truyòls J, González Lastra J, Marquínez J, Martínez Díaz C, Méndez C, Menéndez Alvarez JR, Sánchez de Posada LC (1984) Preliminary note on two marine sections (Tournaisian-Kasimovian) in the Picos de Europa area (Cantabrian Mountains, NW Spain). CR IX Congr Int Strat Géol Carbonifère Washington/Champaign-Urbana 1979 (2): 148-156

Villa E (1982) Observaciones sobre la edad de la Formación Valdeteja (Carbonífero de la Cordillera Cantábrica) en su área-tipo. Rev Esp Micropaleontol 14: 63-72

Villa E, Heredia N (1988) Aportaciones al conocimiento del Carbonífero de la Región de Mantos y de la Cuenca Carbonéfera Central (Cordillera Cantábrica, NO de España). Bol Geol Min 99 (5): 757-769

Wagner RH (1959) Flora fósil y estratigrafía del Carbonífero de España NW y Portugal N. Estud Geol 15 (41-44): 393-420

Wagner RH (1971) The Westphalian D floras of the Olloniego and Esperanza formations in the Central Asturian coalfield. Trab Geol 4: 461-505

Wagner RH, Alvarez Vázquez C (in press) Floral characterisation and biozones of the Westphalian D stage in NW Spain. Neues Jahrb Geol Paläontol Abh

Wagner RH, Bowman MBJ (1983) The position of the Bashkirian/Moscovian boundary in the West European chronostratigraphy. News Stratigr 12 (3): 132-161

Wagner RH, Winkler Prins CF, Riding RE (1971) Lithostratigraphic units of the lower part of the Carboniferous in northern León, Spain. Trab Geol 4: 603-663

Wagner Gentis CHT (1963) Lower Namurian goniatites from the Griotte Limestone of the Cantabrian Mountains Chain. Notas Comun Inst Geol Min Esp 69: 5-23

2.5 Carboniferous of the Palentian Zone

N. HEREDIA[1], L. R. RODRIGUEZ FERNANDEZ[1]
and R. H. WAGNER[2]

2.5.1 Introduction

This area, which is also called the Pisuerga-Carrión Unit (Julivert 1971,[3] Fig. 1 in Part II, Chap. 1, p. 7) or Palentian Region (Martínez García 1981), shows a different stratigraphic development from the Asturian-Leonese domain. The significance of the observed differences can be explained in various ways and there is no consensus at present between the different authors. There is evidence of considerable synsedimentary tectonism, with recurrent turbidite and submarine fan facies, and the reported presence of olistoliths or alluvial and fluviodeltaic facies related to strong subaerial reliefs. The rate of sedimentation is high from the lower Westphalian onwards, and values of 1000 m per million years are obtained for some parts of the succession. The continuous synorogenic conditions from Namurian onwards into the Stephanian stage and the presence of several unconformities and lateral sedimentary changes constitute some of the characteristics of this area. Four successions have been described, corresponding to the Tournaisian-Namurian, Bashkirian-Westphalian A, Westphalian B-D (excepting the highest Westphalian D) and Stephanian B(?)-C. The latter is restricted to a few unconformable outliers which will be dealt with in a separate chapter.

[1] Instituto Tecnológico Geo-Minero de España, 23, Rios Rosas, 28003 Madrid, Spain
[2] Jardín Botánico de Córdoba, Apartado 3048, 14080 Córdoba, Spain
[3] For references, see Part II, Chap. 3.1.

R. D. Dallmeyer and E. Martínez García (Eds.)
Pre-Mesozoic Geology of Iberia

2.5.2 Tournaisian to Namurian Deposits: the Preorogenic Successions

There is apparent continuity with the Devonian, and the nodular limestones of the Vidrieros Formation (Famennian to V–VI) reach locally into the Lower Tournaisian (Adrichem Boogaert 1967). Overlying dark grey siliceous shales and siltstones are usually recorded as Vegamián Formation, in comparison with the Asturian-Leonese area, but proper dating is absent and the facies is comparable but not wholly identical. This unit is only a few tens of metres thick. There is an apparent break accompanied by a marked facies change at the base of grey, thinly bedded limestones, with occasional reddish mottled limestone bands. Although this unit appears on the published maps as "caliza de montaña", it is a lateral equivalent of the red and grey nodular limestones of the Genicera (Villabellaco or Alba) Formation of Late Tournaisian, Viséan and Early Namurian ages. The latter does, in fact, occur in the southernmost part of the Palentian Zone, and there may be a lateral transition. However, the better - known occurrences of nodular limestone corresponding to the Genicera Formation are in allochthonous units (nappes) and do not belong to the Palentian Zone proper. The thinly bedded, bluish grey limestones contain Viséan goniatites (C. H. T. Wagner Gentis, pers. comm.) and, at perhaps 50 to 100 m from the base, a coquina with basal Namurian goniatites has been recorded by Wagner Gentis (1980). In the northern part of the Palentian Zone (Fig. 2), i. e. in the Liébana, an olistolith with thinly bedded fetid limestones yielded Viséan conodonts (Maas 1974). This is the Lon Formation of Martínez García (1981). The depositional environment of all these formations is a marine platform with a very low rate of sedimentation, with a restricted or euxinic area for the Vegamián Formation and a shallow and oxygenated environment for the Genicera Formation (Sánchez de la Torre et al. 1983).

This palaeogeographic situation suggests a tectonically quiet environment accompanied by a slow subsidence rate in the whole Palentian Zone.

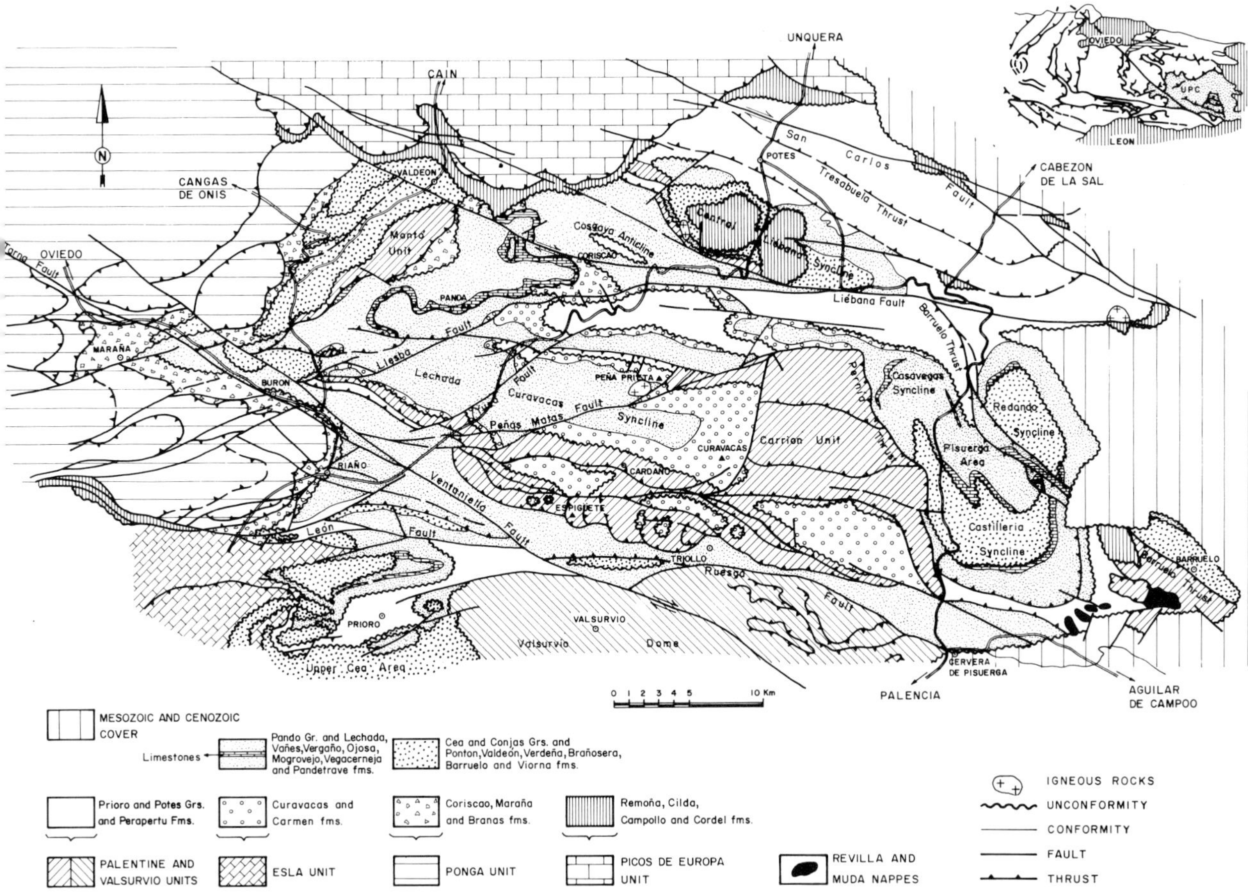

Fig. 2. Geological map of the Palentian area (or Pisuerga-Carrión Unit), showing the main tectonic units and related synorogenic successions

2.5.3 The Synorogenic Succession: Bashkirian to Westphalian A Deposits

Overlying the basal Tournaisian-Namurian strata predominantly siliciclastic sediments appear. These consist of lutites, sandstones, calcareous breccias and/or polymict conglomerates. These are the shales, sandstones and limestones of the Cosgaya Formation which have been dated as (Late?) Bashkirian. Overlying deposits of the Vejo formation have also been dated as Late Bashkirian (Martínez García 1981). These are currently incorporated in the informal Potes "Group" (Rodríguez Fernández et al. 1986, Fig. 3), a ca. 1500–2000 m thick succession of (1) tidal flat deposits, (2) deltaic and subtidal flat strata and (3) turbidites with arenaceous-pelitic lobes and proximal conglomerates. The Potes "Group" is overlain by the Piedrasluengas Limestone (part of the Perapertú Formation of Wagner and Wagner Gentis 1963), which has been dated as Early Moscovian, Vereisky (Van Ginkel 1965) or Atokan (Wagner Gentis 1985). Southwards, the Perapertú Formation (Fig. 4) shows small limestone platforms with intervening basinal areas characterized by mudstones with turbidites, occasional debris-flow conglomerates and calcareous debris-flow deposits (Wagner et al. 1984). There is a clear stratigraphic break between the Perapertú Formation and the overlying siliciclastics (characteristically turbidites with conglomerate debris-flow deposits) of the Carmen Formation (Wagner and Wagner. Gentis 1963; Wagner et al. 1984; Fig. 3). Where the Carmen Formation rests on Perapertú (Piedrasluengas) limestones, a palaeokarst has been developed. Lateral facies changes in the Carmen Formation show important developments of conglomerate (often confused with the later Curavacas conglomerates), and some near-shore terrestrial deposits include rootlet beds and drifted plant remains which are not later than Westphalian A (Langsettian) in

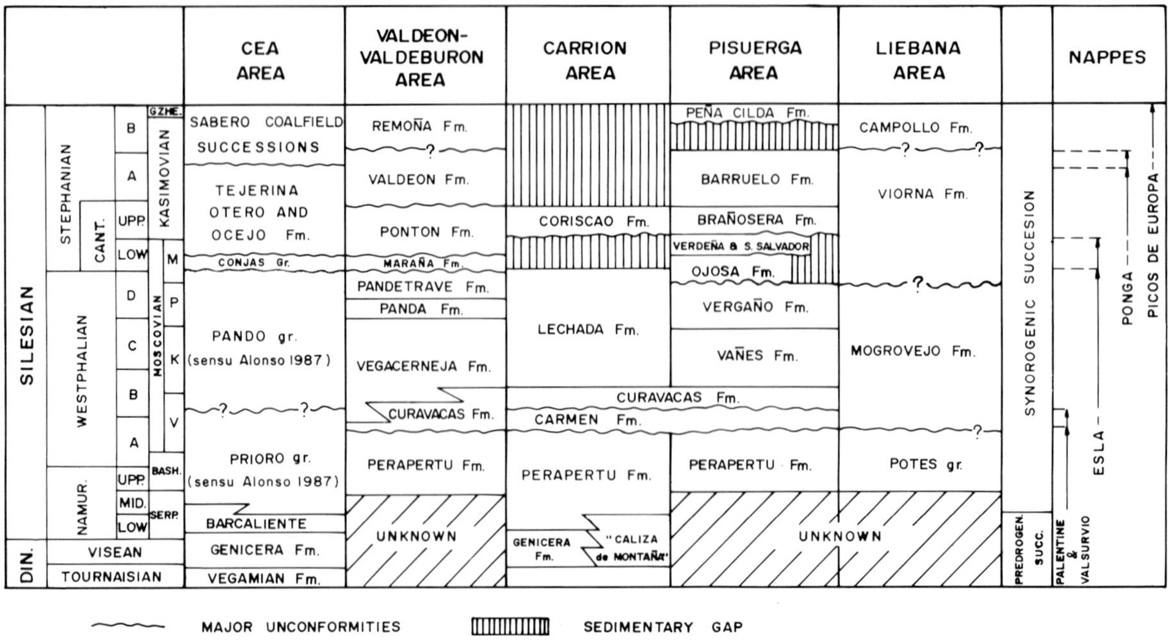

Fig. 3. Lithostratigraphic units in different parts of the Palentian Zone (*Gr.* informal group; *Fm.* Formation). Cea Area, after Van Loon (1971), Wagner and Winkler Prins (1989) and Alonso (1987). Valdeón-Valdeburón, Carrión and Liébana areas modified after Rodríguez Fernández and Heredia (1987 a). Pisuerga area, after Wagner and Winkler Prins (1985). The correlation by the lithostratigraphic units and tectonic nappes after Rodríguez Fernández and Heredia (1987 a)

age (Wagner 1959; Wagner and Bowman 1983). It has not always been easy to distinguish between the Perapertú and Carmen Formations where both are in siliciclastic facies, and no clear discontinuity as shown by a palaeokarst can be demonstrated. The combined succession has therefore been marked as the Cervera Formation (Brouwer and Van Ginkel 1964) on some maps. Martínez García (1981) distinguished a Cucayo Formation overlying the Piedrasluengas Limestone. This is probably an equivalent of the Carmen Formation.

2.5.4 Westphalian B, C, D Deposits

A widespread angular unconformity below the Curavacas Conglomerate Formation (Fig. 3) of Westphalian B age (Wagner 1960; Wagner and Bowman 1983) marks the end of the first tectonic deformation that occurred in the Palentian Zone [a single record of Westphalian A at the base of the Curavacas Formation (Stockmans and Willière 1965) cannot be safely attributed to this formation and may well refer to the Carmen Formation (see comments by Wagner and Bowman (1983)]. The facies of the Curavacas Formation changes both upwards and laterally (northwards) from fluvial and fluviodeltaic to submarine fan. The changes are accompanied by a diminishing thickness northwards and westwards, and it is also noted that the quartzite pebbles and boulders show a degree of recrystallization not found in the Cantabrian Zone. The source area must therefore lie in a more metamorphic area to the SSE. The Curavacas conglomerates are at least 500 m thick. This formation is followed in conformable succession by a turbidite formation which is called the Vañes Formation (ca. 700 m) in the southeastern part (Brouwer and Van Ginkel 1964; Van de Graaff 1971 b) and the Lechada Formation in the northwestern part (Savage 1967). Westwards, a similar though more generally fine-grained succession following upon only partly exposed Curavacas Conglomerates has been subdivided into the Prioro and Pando Formations by Van Loon (1972). These are dated as Westphalian C (Van Loon 1971). Alonso (1987 a) combined these as the Pando "group" (Fig. 3). The Lechada Formation contains limestones in the upper part, viz. the El Ves, Vallines and Panda limestones (Van Veen 1965; Savage 1967; Lobato 1974). The lateral continuity of the Panda Limestone allows the Pandetrave (Savage 1967) and Vegacerneja Formations to be separated from the Lechada Formation in the northern part of the area.

The next and highest unit of this succession is the Vergaño Formation (*sensu* Van de Graaff 1971 b; Wagner et al. 1984; Fig. 3), which corresponds to a high-destructive delta and which contains a number of fairly continuous limestone units, the most important of which is the Sierra Coriza Limestone at the top of the formation. The Vergaño Formation, which is ca. 1500 m thick, has been dated as Early-to-Mid-Westphalian D (Highest Podolsky and Lower Myachkovsky of the Late Moscovian). A lateral passage to turbidites has been described as the Covarres Formation (Van de Graaff 1971 b).

In the northern area (Liébana) there is little palaeontological control, but Alonso and Rodríguez Fernández (1983); Rodríguez Fernández et al. (1986) and Rodríguez Fernández and Heredia (1987) distinguish three different unconformable successions that are genetically related to the synsedimentary reactivation of the Central Liébana Syncline (Figs. 2, 3); these are called the Mogrovejo, Viorna and Campollo Fms. Sedimentary facies are associated with submarine fans and all contain olistoliths.

According to these authors, the Mogrovejo Fm. is equivalent to the Lechada Formation, and the basal conglomerate Cuvo or Bárcena Conglomerate (Maas 1974) is equivalent to the Curavacas Formation.

2.5.5 Upper Westphalian D, Cantabrian and Lower Barruelian Succession

The geographical distribution of the various types of sediment suggest a subdivision in sub-basins of the Palentian Zone at this time. In the eastern part of the Palentian Zone (Pisuerga area, Fig. 2), the succession, up to ca. 6000 m thick, is one of the best studied in the Cantabrian Mountains (see the recent summary by Wagner and Winkler Prins 1985). Uplift accompanied by faulting on a major scale, and folding produced a new basin configuration in Late Westphalian D times (Leonian movements). Foraminiferal dating indicates a Late Moscovian age (upper part of the Myachkovsky Horizon) (Lys in Wagner et al. 1977). This equates to Upper Westphalian D and Lower Cantabrian. Westwards, at the Guardo Coalfield (Cea area; Figs. 2, 3), the basin becomes less marine and shows alternating shallow marine and terrestrial formations laid down at a rate of 1000 m per million years and showing clear evidence of tectonic control, as the result of which widespread transgressions swept the basin. Western and northwesterly onlap is accompanied by evidence of faulting on the basin margin and synsedimentary tectonism is further emphasized by the presence of valley fill deposits on the WNW basin margin; preservation of the valley fills is due to the reactivation of faults in the underlying structure (Alonso 1982), rejuvenating topographic relief and depressing the valley floors sufficiently to preserve up to 450 m of local sediment, which includes terrestrial debris flows of a coarse conglomeratic nature (Iwaniw 1985). These valleys were apparently straight-sided and orientated at approximate right angles to the basin margin (op. cit.). In Late Cantabrian times (earliest Kasimovian) a widespread transgression corresponding to the base of the Brañosera Formation (Fig. 3), accompanied a major expansion of the basin which, once again, was linked to synsedimentary faulting. This gave rise to olistostromal deposits (carbonate debris flows and turbidites).

To the N and W of the Palentian Zone (Valdeón valley and surrounding areas, Fig. 2), sedimentary successions, 2500 m thick, lie unconformably over Westphalian and Devonian outcrops. The lower part consists of olistostromal units with large carbonate olistoliths, with ages ranging from Famennian to Kasimovian (Maas and Van Ginkel 1983). They are called Maraña, Brañas and Coriscao Formations (Rodríguez Fernández and Heredia 1987). The upper part (Pontón and Valdeón Formations) are characterized by a predominant occurrence of large quartzitic conglomerate lenses. Maas and Van Ginkel (op. cit.) suggested a turbiditic origin for these deposits. In more recent papers, Rodríguez Fernández and Heredia (1987) propose a deltaic and fandeltaic origin.

The Brañosera Formation, which is almost 900 m thick in its stratotype in the Pisuerga area, is followed in conformable succession by the coal-bearing Barruelo Formation (ca. 1100 m) of deltaic and shallow marine (mudflat) facies. This formation is partly Late Cantabrian, but mainly Early Barruelian (ex-Stephanian A) in age (Wagner and Winkler Prins 1985), and contains the Barruelian boundary stratotype.

The latest synorogenic sediments are called the Lebeña Formation or Aliva Formation in the Picos de Europe Unit (Maas 1974) or Bedoya "Group" (Martínez García and Rodríguez Fernández 1984) to the north of Liébana. Maas and Van Ginkel (op. cit.) referred to it in the Valdeón area as Remoña Olistostrome Member, since they took it as the basal part of the "Valdeón Formation". More recently, Rodríguez Fernández and Heredia (op. cit.) recognized its unconformable portion with regard to the other formations. For this reason the name of Remoña Fm is suggested (Fig. 3) for all of them except the units located in the Picos de Europa area. The Remoña Fm. consists of olistostromal successions with 600 m of calcareous breccias, quartzitic conglomerates and limestone olistoliths of the Picos de Europa units.

The age of the Maraña, Brañas, Coriscao, Pontón and Valdeón groups is Cantabrian-Barruelian (ex-Stephanian A), and for the Remoña group a minimum age of Stephanian B is considered.

2.6 Stephanian and Permian Basins

E. MARTINEZ GARCIA[1]

2.6.1 Introduction

The Stephanian and Permian basins in the northeastern Iberian Massif have been studied in some detail only in the last few years, the conclusion being that they are of the maximum interest to unravel the tectonic development of this region and the relationships of this part of the Iberian Massif to Europe and Russia. Here, the only outcrops of marine deposits of Stephanian and Permian age in Western Europe, with the exception of the Eastern Alps and Sicily, have been described (Martínez García and Wagner 1971; Ginkel 1971; Martínez García 1983a; Wagner and Martínez García 1982). Stephanian basins in this area show different relationships with the structural scenery, since they were deposited in the paroxysm of the Hercynian Orogeny and either continental or marine sediments are found everywhere (Fig. 1). Compressional tectonics characterize the Stephanian basins, whereas the Permian (and Upper Stephanian C) shows a tensional environment that reveals plate-tectonic conditions suggestive of aborted rifting and associated widespread mineralization (Martínez García 1981b, 1983a, b; Martínez García and Tejerina 1985; Luque and Martínez García 1983, 1985; Gutiérrez Claverol et al. 1987, 1988, 1989; Luque et al. this Vol.).

With regard to the Stephanian basins, we shall consider them as occurring in three separate areas: The *western Cantabrian Area,* including the Arnao, Ferroñes, Tineo, Cangas, Carballo, Villablino, La Magdalena, Ciñera-Matallana, Puerto Ventana, Canseco, Rucayo and Salamón areas, the *eastern Cantabrian Area,* which includes the Sebarga and Picos de Europa areas, and the *Palentian Zone,* with the Guardo, Pernía, Redondo, Barruelo, Valdeón, and Bedoya areas.

The Permian basins crop out mainly in the central Cantabrian Zone as well as in the eastern part of the Cantabrian and Palentian Zones, although the two successions that constitute it were probably covering completely both general areas.

[1] Departamento de Geología, Univ. Oviedo, 33005 Oviedo, Spain

2.6.2 Stephanian

Stephanian deposits in the Cantabrian Mountains have been considered by Wagner (1971a) to be two separate basins that are best defined in the Palentian Zone and eastern end of the western Cantabrian Area. These have been termed respectively post-Leonese and post-Asturian basins, for they were deposited respectively after the Leonese and Asturian phases of the Hercynian Orogeny.

2.6.2.1 The Post-Leonese Basin

The post-Leonese succession starts in the Upper Westphalian D, following tectonic movements developed mainly in northern León (eastern part of the western Cantabrian Area); (Leonese phase of Wagner 1959), being followed by sediments belonging to the Cantabrian stage (Wagner 1966a, b) which is now considered as the basal Stephanian. Cantabrian rocks have been described mainly at the Guardo-Tejerina (Wagner et al. 1971; Wagner and Fernández García 1983), La Pernía (Wagner and Varker 1971; George and Wagner 1972; Wagner et al. 1977) Barruelo (Wagner and Winkler Prins 1985), and Picos de Europa areas (Wagner 1967; Marcos 1967; Martínez García and Wagner 1971, 1982, 1984; Ginkel 1971; Martínez García 1981a; Martínez García and Rodríguez-Fernández 1984; Wagner and Martínez García 1982; Martínez García et al. 1985; Martínez García and Villa 1986; Villa and Martínez García in press). The unit stratotype of the Cantabrian stage lies in the post-

R. D. Dallmeyer and E. Martínez García (Eds.)
Pre-Mesozoic Geology of Iberia

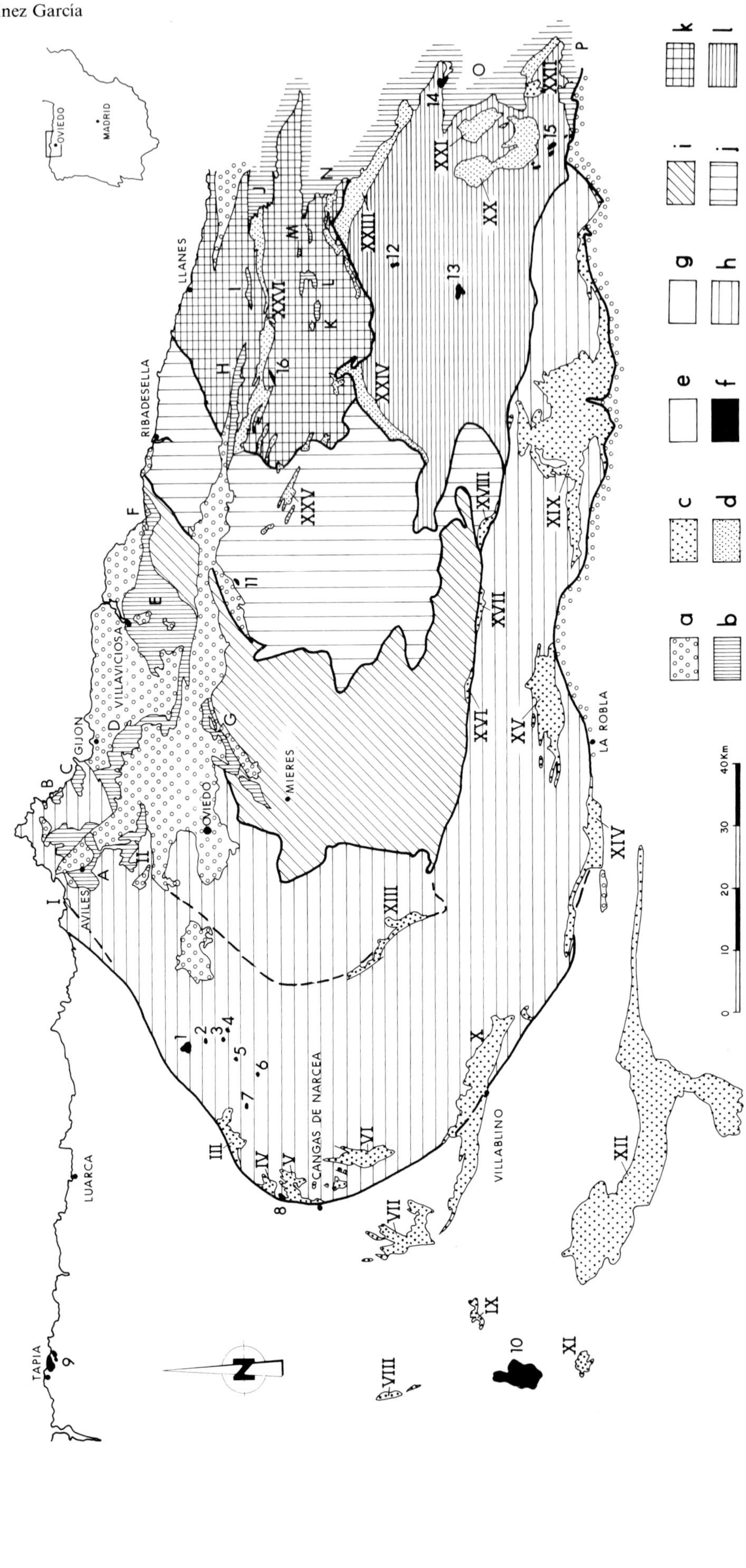
OVIEDO
MADRID
TAPIA
LUARCA
AVILES
GIJON
VILLAVICIOSA
RIBADESELLA
LLANES
OVIEDO
MIERES
CANGAS DE NARCEA
VILLABLINO
LA ROBLA
0 10 20 30 40Km

Leonese sedimentary basin in northern Palencia and northeastern León.

2.6.2.1.1 La Pernía Area

The earliest deposits of the post-Leonese succession correspond to a small area near Casavegas in La Pernía, where the Maldrigo Limestone of the Sierra Coriza Member of the Vergaño Formation flexed downwards to form a small basin sag. The post-Leonese strata clearly abut laterally onto the sides of the basin sag which represents the initial post-Leonese basin. This basin was bound in the east by the synsedimentary Los Llazos Fault, which functioned as the margin to the siliciclastic basin, separating it from a carbonate platform to the east in Late Westphalian D and earliest Cantabrian times (Castillería Limestone, Ginkel 1972). The Castillería Limestone is 90 m thick and is the lateral equivalent of 2500 m of siliciclastic sediments with calcareous intercalations (Rozo and Ojosa Formations). West of the Los Llazos Fault, the succession commences with the Rozo Fm., which is made up of a large number of small fining-upward sequences ranging from siltstones to calcareous mudstones and local lenses of limestone conglomerates, considered to have been laid down near the source area. Above it there is a 2300-m-thick succession (Ojosa Fm.) of coarsening-upward deltaic sequences, commencing with marine mudstones and ending with cross-bedded sandstones with channeling. One of the most important calcareous beds is the lenticular Lores Limestone, which contains fusulinid foraminifera and other benthic remains. It was selected as the boundary-stratotype of the Cantabrian stage (George and Wagner 1972), although later on this limit was moved some 1000 m above it (Wagner 1983). Above them are the Rosa María Coals which constitute a single terrestrial interval in an almost exclusively marine succession. The Urbaneja Limestone, a carbonate debris flow deposit high in the Ojosa Formation, is correlated with the Villanueva marine formation of the Guardo Area in the western part of the post-Leonese Basin, which has been proposed as the boundary-stratotype of the Cantabrian Stage (Wagner and Winkler Prins 1985).

◀ **Fig. 1.** Stephanian and post-Hercynian rocks in the northeastern Iberian Massif (NW Spain). Geological subdivisions: **a** Jurassic to Tertiary cover; **b** Permian; Stephanian; **c** continental; **d** marine; **e** Paleozoic and Precambrian; **f** igneous rocks of Permian age. Regional background: **g** West Asturian Leonese Zone; Cantabrian Zone: **h** Western Cantabrian Zone; Central Cantabrian Zone: **i** Central Coal Basin, **j** Beleño-Ribadesella Region, **k** Picos de Europa Region; **l** Palentian Zone.
Igneous rocks: *1* Godán; *2, 3* Carlés; *4* Courio; *5* Pontigos; *6* Villaverde; *7* Boinás; *8* Linares; *9* Salave; *10* Campo del Agua; *11* Lozana; *12* Pico Jano; *13* Peña Prieta; *14* Pico Cordel; *15* Cervera de Pisuerga.
Stephanian: *I* Arnao; San Juán de Nieva; *II* Ferroñes; *III* Tineo; *IV, V* Cangas de Narcea; *VI* Carballo; *VII* Rengos; *VIII* Ibias; *IX* Tormaleo; *X* Villablino; *XI, XII* Ponferrada; *XIII* Puerto Ventana; *XIV* La Magdalena; *XV* Ciñera-Matallana; *XVI* Canseco; *XVII* Rucayo; *XVIII* Salamón; *XIX* Guardo; *XX* La Pernía; *XXI* Redondo; *XXII* Barruelo; *XXIII* Bedoya; *XXIV* Valdeón; *XXV* Sebarga; *XXVI* Gamonedo-Cabrales.
Permian: *A, B, C* Avilés; *D* Gijón-La Collada; *E* Villaviciosa; *F* Caravia; *G* Siero; *H* Onís; *I, J* Panes; *K, L, M* Picos de Europa; *N* La Hermida; *O, P* Peña Labra

2.6.2.1.2 Guardo Area

In the Guardo Area, representing the western part of the post-Leonese basin, the time-equivalent succession to the Ojosa Formation of La Pernía comprises ten alternating terrestrial and marine formations. The succession there is still largely marine, particularly in the eastern part. Westwards, the proportion of terrestrial strata increases, but the marine formations are seen to be laterally extensive and sufficiently continuous to serve as markers for geological mapping (Wagner and Fernández García 1983). An entirely terrestrial succession of strata of Early Cantabrian age is found on the western edge of the post-Leonese basin, in the area of Ocejo de la Peña. This is accompanied by evidence of a rising and north-northwestward shifting basin margin and a marked paleotopography (Iwaniw 1984, 1985).

The major marine transgression of the Otero Fm. in the Guardo Coalfield correlates with the Verdeña Limestone of the La Pernía-Castillería area further east. The Verdeña Limestone is one of the first deposits laid down in the eastern part of the post-Leonese basin after the synsedimentary Los Llazos Fault had ceased to function and an eastward expansion of the basin was taking place.

2.6.2.1.3 Barruelo and Redondo Areas

A very major event, presumably affecting both basin margins but most strikingly on the eastern side of the basin, is represented by the major transgression associated with the base of the Brañosera Formation. This is correlated with the base of the Taranilla Marine Formation in the western Guardo Area. A substantial rejuvenation of the eastern basin margin is apparent from the nature of the sediments in the lower part of the Brañosera Formation. The major marine transgression associated with the base of the Brañosera Formation has led to a substantial withdrawal of the sediment source area and the consequent deposition of fine-grained sediments in the area available for observation. The Brañosera Formation in its type section near Brañosera (Rubagón Valley) commences with a limestone which is followed by mudstones and massive-bedded sandstones. Subsequently, mudstones were laid down with minor amounts of carbonate debris flow deposits. This interval correlates with the more spectacular debris flows and megabreccias in the Redondo Valley Area. Shallow water turbidites follow in succession in both sections. They are overlain by mudstones with cross-bedded channel sandstones of fluviatile influence. In the upper part, several limestones appear in the mudstones, representing probably a quiet embayment with distributary channels. Further shallowing is apparent where the coal seams of the Peñacorba Member of the overlying Barruelo Formation come in at a slightly higher level of the succession.

The Barruelo Formation only occurs in the Barruelo Syncline which occupies the Barruelo Coalfield and the area of the Redondo Valley, and starts with the coals of Peñacorba in which a limestone lens shows the marine influence. The coal-bearing unit is followed by a fully marine succession which has been described as the Loma Member, in which mudrocks with ostracodes show a shallow marine condition with lenticular sandstone bodies. Overlying mudstones contain abundant floral remains which form an assemblage comparable to the Lower Stephanian of Rive-de-Gier and Carmaux in the Massif Central of France. A thin coal-bearing unit (Carboneros Member) is the boundary stratotype of the Barruelian (ex Stephanian A) Stage (George and Wagner 1972; Wagner and Winkler Prins 1985). Deltaic coarsening-upward sequences present in the upper part of the Carboneros Member, continue in the fully marine Polvorín Member, which is the overlying unit. In keeping with the general tendency towards more terrestrial facies in a southeastward succession, the overlying Calero Member in the Redondo Syncline is seen to be almost entirely marine, while the Barruelo Area contains the majority of workable coals. Within the Barruelo Coalfield a gradual increase in terrestrial facies is noted from the area northwest of Barruelo de Santullán to Vallejo de Orbó, at the southeastern end of this coalfield.

The overturned succession of Upper Cantabrian and Lower Barruelian strata is overlain with an angular unconformity by Upper Stephanian B or Lower Stephanian C strata of the Peña Cildá Formation. This unconformity is due to the Asturian tectonic movements.

2.6.2.2 *The Post-Asturian Basin*

The tectonic movements of the Asturian Phase which took place after the Lower Barruelian strata were laid down, extended across the entire Cantabrian Zone. Fairly intensive folding and thrusting characterized the Asturian movements in most of the zone, and this includes gravity structures with major decollements in post-Palentian and post-Leonese strata of northern Palencia. Only the general region of the Picos de Europa, in the concavity of the Asturian arcuate fold belt, experienced merely uplift as a result of the Asturian movements.

The configuration of the post-Asturian basin of sedimentation was quite different to that of the preceding post-Leonese basin. In general outline, it consisted of an extensive floodplain area bordering a rising mountain chain which coincided more or less with the West Asturian-Leonese Zone of Lotze (1945). This floodplain area constituted the terrestrial part of a costal basin which sloped north and eastwards into a fully marine area which coincided more or less with the Region of the Picos de Europa, where upper Kasimovian carbonate deposits of the Puentellés Limestone (Martínez García 1981 a) are followed by the turbiditic Cavandi Formation (Martínez García and Wagner 1982, 1984). The coastal floodplain area accumulated coal-bearing deposits of mainly Stephanian B and Early Stephanian C ages, but which also show the presence of Upper Barruelian strata at the base of the post-Asturian succession in the Saberno Coalfield of northeastern León. It is not quite clear whether sub-basins existed in the coastal floodplain area, but there was certainly communication between the various parts of the general coastal basin. Well documented onlap relationships show that the post-Asturian coastal basin was initiated in the area of the Sabero Coalfield, immediately west of the post-Leonese outcrops of the Guardo-Valderrueda Coalfield, and that the basin of sedimentation extended gradually west-northwestwards from Late Barruelian through Stephanian B into Early Stephanian C times.

2.6.2.2.1 *Northern Palentian Area*

In the northern part of the Palentian Zone, there are some Stephanian deposits that have been known since Kutterink (1966), who named the Valdeón Formation in the western part. A flora found by this author gave a Stephanian age (Amerom in Kutterink 1966). It was considered by Maas (1974) as equivalent to the Lebeña Formation. Later on, Maas and Ginkel (1983) distinguished several separate outcrops of similar sediments, describing them as olistostromic and naming them the Remoña, Brañas, and Coriscao olistostromes. They considered these rocks to have been deposited in front of the advancing nappes coming from the north, the Picos de Europa, Gildar-Montó, and Panda Nappes respectively. The Picos de Europa Nappe (First Nappe) is composed of mainly limestones of Carboniferous age and Cantabrian affinity, while the Gildar-Montó Nappe is a thrust empilement of Devonian and Carboniferous rocks of Palentian character, and the Panda Nappe is made up of Carboniferous limestones belonging to the Palentian Zone.

The wedge-shaped olistostromic units have a faint layering and consist of angular to subrounded clasts from microscopic to boulder size. They are made up of several debris-flow units but they differ from one another with regard to the matrix and the exotic blocks. Their thickness is around 500 m. In the Coriscao Olistostrome, below the Panda Nappe, the youngest exotic blocks found have been dated by means of fusulined foraminifera (Maas and Ginkel 1983), as Uppermost Myachkovian (limestone beds of the N_2 interval in the Donetz Basin), which probably correlate with the earliest Middle Cantabrian. Since the age of the overlying Panda Limestone has been established as probably Middle to Upper Podolskian (Ginkel 1965), there is no doubt that the contact must be of tectonic character. The age of the Remoña Olistostrome, below the Picos de Europa Nappe, has been established by the existence of limestone blocks with fusulinids of Upper Myachkovsky age (Lower Cantabrian).

In the eastern part of the northern Palentian Zone, and below the Picos de Europa Nappes, there are also olistostromic sediments that can be compared with the Valdeón Formation, although no detailed study of these rocks has been made.

2.6.2.2.2 *Picos de Europa Area*

In the Picos de Europa Area of the Cantabrian Zone, two Stephanian basins have been described, the lower of Cantabrian age and the upper of Stephanian B age. Sediments of Cantabrian age have only been described to date in the Gamonedo Area, east of Arenas de Cabrales (Marcos 1967, 1968a; Wagner 1967).

The Lebeña Formation was proposed by Maas (1974) in the southern part of this area for a turbidite sequence made up of shales, conglomerates, and debris flows unconformable on the Picos de Europa Formation. Although the fossil content of the Lebeña Formation corresponds to reworked faunas, the presence of *Protriticites* in calcareous clasts allowed the attribution to these rocks of a minimum Kasimovian age (Ginkel in Martínez García and Rodríguez Fernández 1984).

The stratigraphic succession established by Maas cannot be extended to the whole region since in the northern area there are many differences, as is the appearance of important carbonate beds of Kasimovian age (Ginkel 1971) being defined by Martínez García (1981a) as Puentellés Formation, overlain in turn by turbidites of the Cavandi Formation (Martínez García and Wagner 1982, 1984), also of Kasimovian age. On the other hand, Martínez García and Wagner (1971) described a terrestrial sequence with a flora of Stephanian B age, below limestone beds similar to the Puentellés Limestone at Arenas de Cabrales. Towards the south, these two formations are substituted by the Lebeña Formation, which itself resembles the Cavandi Formation.

In the same Gamonedo Area, Villa and Martínez García (in press) have already described a section named the Dobros Section, in which disconformable sandy limestones and conglomerates rest on older Carboniferous formations. Fusulinids in the upper part of these limestones have yielded *Triticites (Montiparus) priscus* sp. nov., *Triticites (Montiparus)* spp., *Fusiella* ex gr. *lancetiformis* Putrya, *F.* ex gr. *rawi* (Lee) and *Ozawainella* sp. The presence of *Triticites (Montiparus)* indicate that these beds can be correlated with the Khamovnichesky Horizon of the Kasimovian.

Above them, there are apparently disconformable sandstones and limestones, in which *Quasifusulinoides* aff. *parafusiformis* Bensh and *Triticites (Triticites)* ex gr. *ferganensis* Miklukho-Maklay have been found, indicating that they can be correlated with the *Triticites (Triticites)* Zone and therefore with the Upper Kasimovian, although the primitive character of these seem to point to the lower part of the Dorogomilovsky Horizon.

These data indicate that in the northern part of the Picos de Europa Area there are rocks of Cantabrian age (Llacerias Section, Marquínez et al. 1984), of Stephanian A age (Dobros Limestones at the Dobros section, Villa and Martínez García in press) and of Stephanian A-B age (Puentellés Limestones at Puentellés and Dobros sections, Martínez García 1981a; Villa and Martínez García in press), all of them of marine character. Interbedded terrestrial sandstones with plant remains confirm the age of the Puentellés Formation, since they yielded a Stephanian A or B age. In the southern part, only rocks of probably Lower Kasimovian age have been recorded (upper part of the Picos de Europa Formation, Lebeña Formation). These are the youngest marine formations of Carboniferous age recorded to date and point to a physical relation between the Carboniferous Basin in the Cantabrian and Palentian Zones and the Donetz Basin in the USSR, based on the exact correlation of the fusulinid faunas as expressed by Ginkel (1971) and Villa (1989). This relation could be expressed by eastward transport of the Iberian Peninsula along the Pyrenean Fault (Martínez García 1983b).

2.6.2.2.3 Sebarga Area

In the Sebarga Area, west of the Picos de Europa Area, plants of Stephanian A, B or C age were already described by Wagner (1965, 1970) within an unconformable marine succession with terrestrial intercalations, later studied in detail by Colmenero Navarro and Bahamonde Rionda (1986) as a fan-delta system. Martínez García et al. (1985) cited a fusulinid fauna in limestones at the base of the succession, which was dated as basal Kasimovian, determining the existence of a hyatus from Kashirian to Kasimovian below the limestones, unconformable on the Escalada Formation. These rocks can be correlated with the top limestones in the Llacerias Section of the Picos de Europa Area (Marquínez et al. 1982).

2.6.2.3 Correlations with the USSR

The fusulinid faunas from the eastern part of the post-Leonese basin in León and Palencia, which occur in upper Westphalian D and Cantabrian strata, combine with the fusulinids found in probable Stephanian B deposits of eastern Asturias (Ginkel 1971; Martínez García and Rodríguez Fernández 1984; Martínez García and Wagner 1982, 1984; Villa 1989) to provide correlations with the Soviet Union.

The Lores Limestone (Ojosa Fm.), originally accepted as the base of the Cantabrian stage (George and Wagner 1972) but presently referred to as the Upper Westphalian D (Wagner 1983; Wagner and Winkler Prins 1985), contains a definite Myachkovsky fusulinid fauna (i.e. the *Fusulinella* B_3 subdivision of Ginkel 1965). The Urbaneja Limestone, near the top of the Ojosa Fm., and currently regarded as forming part of the lowermost Cantabrian, has been dated rather tentatively by Ginkel (1971) as the top Myachkovsky Horizon of the upper Moscovian.

A limestone lens in the San Salvador coal-bearing Formation (Highest Lower Cantabrian) has yielded a rich fusulinid fauna including *Pseudotriticites* cf. *P. Lebedevi* (Putrya) indicating the very highest Myachkovsky Horizon, virtually on the Moscovian/Kasimovian boundary (Lys in Wagner et al. 1977). No foraminiferal faunas have been recorded from the Verdeña Limestone, in between the Ojosa and San Salvador Formations.

The basal part of the Brañosera Formation which has been assigned to the Upper Cantabrian, has yielded a definite *Protriticites* fauna which both Ginkel (1971) and Lys in Wagner et al. (1977) have attributed to the Lower Kasimovian, i.e. the Kreviakinsky Horizon of the Moscow Syneclise. Higher in the Brañosera Formation, similar faunas occur.

Fusulinids from probable Stephanian B deposits at Arenas de Cabrales and elsewhere in eastern Asturias (Martínez García and Wagner 1984) belong to the *Triticites* zone and have been dated as Early Late Kasimovian, comparable to the Dorogomilovsky Horizon and the O_2-O_3 limestones of the Donbass succession (Ginkel 1971; Villa 1989).

Plant megafossils, which occur throughout the Cantabrian and Barruelian stratotypes and correlative sections in the same area in NE León and NW Palencia, can only be compared in some respects with those of the Donetz Basin. In the latter area the records from time-equivalent beds are apparently sparser than in northwest Spain. Fissunenko and Laveine (1984), in a recent analysis of the Donbass floras, mention that more work needs to be done in the Donetz Basin before the Westphalian-Stephanian boundary can be defined properly in this area. It is currently placed at the N_2 Limestone level but may have to be shifted to a higher level (that of the O_1 Limestone). Wagner and Higgins (1979) noted a floral change in the Donbass at the level of the n_1 coal seam (below the N_2 Limestone) which they identified as corresponding to the Mid-Westphalian D floral change. This floral change was analyzed in detail from Lorraine, France, by Laveine (1977) and from South Wales by Cleal (1978) and is therefore rather well known. This corresponds to the lower boundary of the *vestita* Zone as described by Wagner (1984). Fissunenko and Laveine (1984) also suggested that the N_2 Limestone level might represent the Mid-Westphalian D, mentioning that the change in floral assemblages at this level relates to the disappearance of a number of Westphalian species rather than the appearance of Stephanian elements.

2.6.2.4 Correlations with the USA

Plant fossils have been used to suggest the time equivalence of Lower Cantabrian with the Lower Desmoinesian of Illinois. This is based on an examination of the floral assemblage from No. 6 coal of the Carbondale Formation recorded originally by Gastaldo (1977), and which has been referred to the *cantabrica* Zone (Wagner 1984). It thus seems likely that the Desmoinesian/Missourian boundary falls within the Upper Cantabrian.

Correlations between the American and USSR successions on the basis of fusulinid faunas suggest that the Desmoinesian/Missourian boundary coincides approximately with the Moscovian/Kasimovian boundary (Aisenverg et al. 1979; Rauser-Chernussova et al. 1979; Wilde 1984). This does not contradict the above mentioned correlation on plant megafossils.

2.6.3 Continental Stephanian Deposits in the Central and Western Cantabrian Zone

From east to west, there are numerous outcrops of Stephanian rocks, most of them fault-related and following the Asturian Arc, with extension varying from a few km^2 to several hundreds (Fig. 1). In the eastern part, south of the León Line (Sitter 1962), the main outcrops are the Sabero and Ciñera-Matallana areas, near the southern border of the Cantabrian Zone, while along the León Line, there are aligned the Salamón, Rucayo, Canseco and Puerto Ventana small outliers. In the western part, there are numerous small to very large outcrops (San Juán de Nieva, Arnao, Ferroñes, Tineo, Cangas de Narcea, Carballo, Villablino, and La Magdalena in the Cantabrian Zone, and Rengos, Ibias, Tormaleo, and Ponferrada in the West Asturian-Leonese Zone.

2.6.3.1 Central Cantabrian Zone

The Sabero and Ciñera-Matallana outcrops are related to the Sabero-Gordón Line (Rupke 1965), important strike-slip fault (Reijers 1985) parallel to the León Line. The Sabero Area had already been studied by Ezquerra del Bayo (1844) and Prado and de Verneuil (1850). It is composed of ca. 2500 m of sediments (Helmig 1965; Knight 1971), in which Knight (1974) separated seven formations (Raposa, Gonzalo, Sucesiva, Quemadas, Herrera, Unica, and Perla Formations), comparing them with the ones described by Wagner (1971 b) in the Ciñera-Matallana Area, although the lowermost beds (Raposa, Gonzalo, and Sucesiva Formations) in the Sabero Area seem to be somewhat older (Upper Stephanian A). The succession starts with some 100 m of limestone conglomerates at the base, unconformable also on the Upper Stephanian A Alejico beds (Knight 1975, 1983 a), which are the oldest post-Asturian sediments in the Cantabrian Zone, and followed by an alternation of shales and sandstones with coal seams.

Plant impressions occur throughout the coal-bearing intervals at Sabero, and these provide the means of biostratigraphic correlation with the Stephanian A/B succession at Carmaux, on the southern margin of the French Massif Central.

The Ciñera-Matallana Area is the best known from the stratigraphic and palaeontologic point of view. It was studied by Mallada (1886, 1898), but the knowledge has increased greatly in the last 40 years, thanks to the contributions of Gómez de Llarena (1950), Amerom and Dillewijn (1963); Helmig (1965); Evers (1967); Staalduinen (1973); Wagner and Artieda (1970); Wagner (1971 b), etc. The succession starts with paleovalley-filling limestone conglomerates 250 m thick, followed by marsh facies with important coal seams and fluvial beds with sandstones, shales and coal seams. Heward (1978) interpreted the Ciñera-Matallana Basin, as related to the La Magdalena Area, as more distal alluvial fan facies that the latter. Other authors considered that the Ciñera-Matallana succession was similar to the one at Sabero and described both as belonging to a single formation (Cea Formation of Helmig 1965; Sabero Formation of Evers 1967; Prado Formation of van Staalduinen 1973). Wagner (1971 b), on the contrary, established the existence of seven formations or paquetes, as is usually done in other areas. These formations were named San Francisco, Pastora, Cascajo, Roguera, San José, Bienvenidas, and Matallana. This basin is at present a reference mark in the Upper Carboniferous of northern Spain, thanks to the detailed paleobotanical study made by Wagner (1963, 1964 a, 1965). The flora is of Stephanian B age.

The Salamón, Rucayo, and Canseco areas, which are located along the León Line, are narrow strips up to 1 km wide on either side of the fault, and probably represent isolated outcrops of a once continuous exposure. Evers (1967) called the stratigraphic succession Rucayo Formation and described it to be made up of ca. 500 m of siliceous conglomerates at the base, followed by sandstones, shales, and coal seams that have been exploited elsewhere. Stephanian B floras have been cited from the Rucayo Area (Wagner 1963), and Canseco and Salamón Areas (Amerom 1965).

The Puerto Ventana Area is a somewhat larger outcrop of Upper Carboniferous sediments more to the west, along the same León Line. This is an important fault that has been considered as a fracture controlling the development of the Upper Paleozoic in the Cantabrian Zone (Sitter 1962), or a Late Carboniferous strike-slip fault (Marcos 1968a). In the Puerto Ventana Area there are 900 m of a succession composed mainly of well-rounded siliceous conglomerates in the north, while in the south these are mainly at the base and as intercalations in the upper levels of sandstones and shales with coal beds. Precise dating and study of this Stephanian B succession was accomplished by Madariaga (1928); Gómez de Llarena and Rodríguez Arango (1948); García Fuente (1959); Wagner (1965); etc.

2.6.3.2 *Western Cantabrian Zone*

In the north, there are some very small outcrops near the coast, related to faults whose importance has not yet been established. These are the Ferroñes, San Juan de Nieva, and Arnao outcrops. The Ferroñes exposure was one of the first to be known in the Cantabrian Zone (Paillette et al. 1845; Schulz 1858), Zeiller (1882) had already placed these rocks in the Upper Carboniferous. The succession is made up of siliceous conglomerates, sandstones, and shales with coal seams, unconformable on Devonian rocks, and its flora has been determined as being Stephanian B or C in age (Wagner 1965, 1970).

The Arnao outcrop, west of the town of Avilés, is thrusted over by Devonian rocks, as was observed by Schulz (1858). Its flora was studied by Geinitz (1867), and Grand'Eury (1877), and attributed to the Upper Carboniferous. A modern revision of the flora showed that it must be Stephanian B-C or C (Wagner 1965; Lorenzo 1977).

More to the east are, from north to south, different outcrops related to the Cangas de Narcea Fault, probably active in Late Carboniferous times as a strike-slip left-lateral accident. Most of them are either on top of the Cambrian and Precambrian rocks of the West Asturian-Leonese Zone. These are the Tineo, Cangas de Narcea, Carballo, Villablino, and La Magdalena areas. The northernmost outcrops were studied by Barrois (1882) and served to establish his Tineo Subdivision of the Upper Carboniferous.

The Tineo Area had already been studied by Schulz (1858); Barrois (1882), and dated by Grand'Eury (1877) and Zeiller (1882). The succession is made up of a few meters of local reddish breccias with overlying siliceous conglomerates, followed by alternating coal-bearing shales and sandstones within the Gera-La Rasa Formation. In the upper part are conglomerate beds with cycles including sandstones, shales, and coal seams, with a total thickness of ca. 800 m (Alvarez Ramis 1966; Wagner 1964b; Julivert et al. 1977). The studied floras have been dated as Stephanian B-C (Doubinger and Alvarez Ramis 1964; Wagner 1966b).

In the Cangas de Narcea Area, there are 200 m of breccias and conglomerates followed by fluvial sandstones, shales, and coal beds (Virgili and Corrales 1966; Corrales 1970, 1971). The floras have been dated as Stephanian B or C (Vega Rollán 1959, 1964, 1968; Wagner 1970).

The Carballo Area presents a succession 800 m thick, with basal breccias followed by 400 to 500 m of conglomerates with sandstones, shales, and coal seams. On top are several cycles of sandstones, shales, and coal seams (Crespo 1982). The floras found have been determined as Stephanian B-C (Vega Rollán 1964; Crespo 1982).

The Villablino Area is one of the biggest outcrops and reaches up to 3000 m of sediments. Although there are older data on this basin (Rubio 1876; Monreal 1878; Revilla 1906; Durán and Arango 1918), the main studies have been accomplished in the last 40 years (Vidal Box 1943; Alvarado 1952; Alvarez Ramis 1966; Corrales and Peláez 1966; Bosch 1969; Corrales 1970, 1971). The basal succession is made up of breccias and siliceous conglomerates followed by fluvial sandstones and shales in cycles, as in the Cangas de Narcea outcrop, although conglomerates are more rare here. At the top of the succession, there is a constant level of quartz conglomerates, which has been considered by Wagner (1965) as equivalent to the basal conglomerate of the Tineo Area. Several formations (Orallo and Calderón below this conglomerate, and Paulina and María Bolsada above it) have been described by Corrales (1970). The age of the succession is considered as Stephanian B-C by Wagner (1965, 1970). Some intrusive prophyry dykes of granitic affinity have been described in the Villablino as well as in the Tineo Areas.

The La Magdalena Area is located in the southern part, close to the Ciñera-Matallana Area, and is considered a lateral equivalent of it (Heward 1978). The succession there is ca. 1500 m thick, starting with a few meters of reddish polygenic breccias and abundant siliceous conglomerates in the basal and middle parts with interbedded coal seams. Heward

(1978) considered this succession as being formed in alluvial fans with several superposed megasequences. It has been dated as Stephanian B (Wagner 1959, 1963, 1964a, 1965; Alvarez Ramis 1966; Stockmans and Williere 1965; Lorenzo 1978).

More to the west are several outcrops that follow the arcuate trend and overlie the rocks of the West Asturian-Leonese Zone. These are, from north to south, the Rengos, Ibias, Tormaleo, and Ponferrada Areas. The Rengos Area had already been mentioned by Barrois (1882) and by Durán and Arango (1918), and more recently studied by Corrales (1970, 1971). The succession is up to 1500 m thick and commences with local breccias, followed by sandstones, shales, and coal seams in cyclic sequences, similar to the other areas already described. The age is comparable to them and has been attributed to the Stephanian B (Vega Rollán 1959, 1964; Wagner 1965; Alvarez Ramis 1966), or to the Stephanian B-C (Wagner 1970). The Ibias outcrop is very small and is composed of breccias and conglomerates in the basal part, followed by sandstones, shales, and coal seams in the upper part (Marcos et al. 1979).

The Tormaleo Area is also very reduced and starts with basal conglomerates, followed by shales and alternating sandstone-shale cycles with topping coal seams. Its thickness is up to 350 m (Pulgar et al. 1981). The studied floras have been dated as Stephanian B-C (Wagner 1970), although Lorenzo (1978) attributed the basal succession to the Stephanian B and the upper part probably to Stephanian C. The Ponferrada or Bierzo Area is the most important of all these outcrops, well over 300 km^2, and has only been studied recently (Alvarado and Sobrino 1946; Hernández Sampelayo and Hernández Sampelayo 1952; Alvarado 1952; Almela and del Valle 1963; Velando and Martínez Díaz 1973; Pérez Estaún et al. 1982; Matas et al. 1982). The stratigraphic succession starts with quartzitic conglomerates 100 to 150 m thick, followed by sandstones, conglomerates, shales, and coal seams. Several formations or paquetes were described by Alvarado (1952; Cuervo, Congosta, Llamazares, La Posada, Modroño, and Valdebraña, from bottom to top) with a total thickness of 1500 m. The floras found there have been dated as Stephanian B or B-C (Alvarez Ramis 1966; Wagner 1965, 1970).

2.6.4 The Permian Basins

2.6.4.1 Introduction

Barrois (1882) was the first to attribute a Permian age to rocks in the Cantabrian Mountains near Villaviciosa and Langreo, including various volcanic formations. Adaro and Junquera (1916) showed different sections of the same rocks as being of Triassic age, due to their strong red color. The first author to date the existence of Permian sediments was Patac (1920), by means of plant remains found in a mine shaft near Pola de Siero (Asturias). Later on, Meléndez (1950) also described different sections of supposedly Permian rocks near Colunga in Asturias, although lacking paleontological proofs. Since then there have been only general references to the existence of a probable Permian near Villaviciosa, whose stratigraphy and petrology were described in detail for the first time by Prado (1972), all reddish formations unconformable on the rocks affected by the Hercynian orogeny being generally described as Permo-Triassic. Jong (1971) and Maas (1974) described the presence of Permian plants below the thick Triassic formations of the eastern cover of the Iberian Massif, and Wagner (in Wagner and Martínez García 1982) studied a flora of Stephanian or Permian age found by Gervilla et al. (1973b) in the Central Coal Basin. Martínez García (1981b, 1983a) attributed to the Permian all the previously considered Permo-Triassic and Triassic rocks in the Cantabrian Zone, with the exception of the eastern border. Wagner and Martínez García (1982) finally demonstrated the Permian age of diverse successions, like the ones in Villaviciosa and Sotres, by means of floral findings.

Two units, corresponding to different basins, have been distinguished in the Permian succession of the Cantabrian and Palentian Zones by Martínez García (1983a). The lower one or Viñón Beds, is made up of basal, grey, well-rounded calcareous conglomerates, followed by alternating sandstones, black shales, tuffs, volcanic agglomerates, lava flows, and thick limestone intercalations. These limestone bands have a marine intertidal character and bear algae that strongly suggest the presence of a shallow sea in which the Viñón Beds were deposited. Above them are volcanic tuffs and ashbeds of increasingly more acid character, which make up more than 900 m of succession in some areas, with rapid lateral thickness changes. The presence of thick conglomeratic volcanic formations that laterally wedge out indicates intense fault activity with associated volcanic effusion. The upper succession or Villaviciosa Beds is quite different in character and unconformable, either on the preceding one or directly on Carboniferous or pre-Carboniferous rocks. It is composed of red sandstones and shales with interbedded siliceous conglomerates, followed by red calcareous breccias and caliche limestones. Above them are dolomite beds and marls topped by alternating shales and sandstones with gypsum intercalations.

2.6.4.2 *Geographic Extent of the Permian Outcrops*

Outcrops of Permian rocks are widespread in the Cantabrian and Palentian Zones, indicating the existence of a once regional cover that was later dismantled. They can be seen from east of Avilés, at the Salinas Beach to the Santander Basin, and down south to the vicinity of Cervera de Pisuerga in the Palentian Zone. Many other indications, such as the strong dolomitization of the underlying calcareous basement, point to the existence of this Permian cover in areas to the west where no remains are seen at present. The main outcrops can be defined geographically by the names of nearby localities, as the Avilés, Gijón-La Collada, Siero, Villaviciosa, Caravia, Onís, Sotres, Panes, La Hermida, and Peña Labra Areas. The most abundant rocks in these outcrops belong to the Villaviciosa Beds, the Viñón Beds having been described only in a few of the outcrops, namely at Villaviciosa, Sotres, Siero, Panes, and Peña Sagra, although they could also be concealed in some instances beneath the Villaviciosa Beds, as in the case of Villabona in the Avilés Area, where similar rocks to the Viñón Beds have been reported from a drill core.

2.6.4.2.1 Avilés Area

Near the town of Avilés (Asturias), there are several isolated outcrops of the Villaviciosa Beds, unconformably resting on older, already folded and faulted Paleozoic rocks. The most important outcrops are near the locality of Villabona, between Avilés and Oviedo, but there are also some smaller ones near Salinas, Luanco, Candás and Xivares. The first ones have received more attention for their relation to economic fluorspar deposits. On the Paleozoic basement the succession starts with a few meters of red clay causing regional rubefaction of the underlying basement, followed by siliceous conglomerates with sandy matrix and fluorspar as cement. Above them are some 70 m of alternating shales and sandstones with rare conglomerate beds, followed by a limestone bed 3 to 6 m thick of caliche character, which is the generally mineralized horizon in this once important mining district. It can laterally pass to a red-colored limestone breccia. Above it are shales and sandstones with centimetric dolomite lenses and finally red and grey shales with gypsum intercalations, some 120 to 200 m thick and usually found below the Liassic limestones.

2.6.4.2.2 Gijón-La Collada Area

South of Gijón, there is a continuous outcrop of rocks belonging to the Villaviciosa Beds, extending south to the La Collada Area, which appear below the Liassic limestones and dolomites, and are unconformable on Carboniferous limestones. The succession usually starts with siliceous and calcareous conglomerates (5 to 50 m) followed by red sandstones and shales (20–60 m) and marly limestones that laterally pass into limestone conglomerates. Above them are again sandstones and predominant shales with some gypsum (60 to 150 m).

2.6.4.2.3 Siero Area

In the vicinity of Pola de Siero, there are diverse outcrops of Permian rocks that are not very well known, such as the Bergueres, Aramil, San Tirso, and Nava outcrops. In the Bergueres outcrop, Patac (1920) found the first Permian flora in the Cantabrian Mountains, which was dated as Autunian. The succession here starts with a red calcareous breccia, followed by alternating sandstones and shales with gypsum intercalations in the upper part. In drillings described by Almela and Rios (1962) there is a similar succession, with marls and black limestones unconformable on the basement, followed by sandy limestones, limestone conglomerates, red and grey shales and sandstones, marly limestones, and finally red and grey marls with gypsum. Thickness changes in the limestone conglomerates are quite strong, as expressed by the 361 m cut in a nearby drill-hole. In the area south of the Careses Fault, Permian sediments are covered by Cretaceous rocks, while north of it they always lie below Liassic sediments.

In the San Tirso outcrop, Gervilla et al. (1973b) found plant remains considered by Wagner (in Wagner and Martínez García 1982) as Stephanian C or Permian in a coal-bearing succession below volcanic rocks with calcareous basal conglomerates similar to the Viñón Beds. This unit could represent a transition with the Stephanian.

2.6.4.2.4 *Villaviciosa Area*

The succession in this area had already been considered as Permian by Barrois (1882) and later by Patac (1956), and partly by Gervilla et al. (1973 a), although other authors who studied the successions there, such as Adaro and Junquera (1916); Prado (1972); and Sánchez de la Torre et al. (1977), described them as Permo-Triassic or Triassic.

Several units have been distinguished in the Viñón Beds by Suárez Rodríguez (1988; calcareous conglomerates, Niao, Valbuena, Torazo, El Cayo, Arboleya, and Villaescusa). In the volcanoclastic rocks of the El Cayo Unit, a fossil flora was found by Wagner and Martínez García (1982) and dated as Autunian. The Villaviciosa Beds are made up of some 800 m of red shales, conglomerates, sandstones, and gypsum intercalations, and several units have been distinguished in them by Suárez Rodríguez (1988; Viacaba, Lugás, and Fuentes).

2.6.4.2.5 *Caravia Area*

Stratigraphic studies in this area were made mainly by mining geologists interested in the relationships of the fluorspar mineralization and the Permian rocks, such as Forster (1974); Jullien (1974); Kuchelka (1974); Schäfer (1974); Huber (1976); etc. Forster (1974) attributed these rocks to the Permian on erroneous paleontological grounds. The described successions belong to the Villaviciosa Beds and start with red limestone conglomerates (20 m), followed by alternating sandstones, marls, carbonate beds, and conglomerates. The succession ends with alternating marls, clays and sandstones with gypsum (150 m).

On the eastern Sueve Nassif, there is also a small outcrop at the Cofiño locality, where a few meters of red limestone conglomerates, sandstones, and shales have been studied by Huber (1976). Gervilla et al. (1973 a) also describe some small outcrops of Permian rocks of the Villaviciosa Beds below Liassic rocks north of Borines in the southern Sueve Massif.

2.6.4.2.6 *Onís Area*

A discontinuous strip of probably Permian light sandstones and grey shales with plant fragments is found along the Cangas de Onís-Robellada Fault below Cretaceous rocks. No evidence of its age has been obtained to date, but they could also be of Stephanian age.

2.6.4.2.7 *Sotres Area*

Within the Picos de Europa Massif, there are several isolated outcrops of Permian rocks, as are the Sotres, Pandébano, Alles, Arangas, Mier, Beges, Cueto Moro, and Tresviso outliers. In the Sotres outcrop, an Autunian flora has been found (Wagner and Martínez García 1982) below some 70 m thick probably marine limestones also dated as Autunian by means of palinology (Neves in Martínez García 1981 a). On top of them disconformably lies a succession similar to the Villaviciosa Beds, with red calcareous conglomerates, sandstones, and shales (Martínez García 1981 a). Most of the remaining outliers only show the presence of the Villaviciosa Beds, with perhaps the exception of the Mier outcrop, where some light sandstones are found at the base.

2.6.4.2.8 *Panes Area*

In this area are several outcrops where Permian rocks are seen. Sometimes, as happens near Suarías and Merodio, these are mainly grey limestone conglomerates with associated rocks of volcanic origin. In other places, the base of the succession is made up of sandstones and conglomerates with barite veins (Merodio, Ortiguero). More to the east, near Carmona (Cantabria), the succession is similar to the Villaviciosa Beds, with red limestone breccias and caliches, followed by sandstones and shales that are unconformably overlain by the thick Buntsandstein deposits of the Santander Basin. To the east, on the Carboniferous limestones of Las Caldas de Besaya, there are also light sandstones and shales of Permian age, unconformably overlain by Triassic rocks.

2.6.4.2.9 *La Hermida Area*

Permian and Mesozoic rocks are well developed in this area, and two sectors can be distinguished. To the north, the Permian sediments are reduced to red limestone breccias, sandstones, and shales, with gypsum that are unconformably overlain by the thick Buntsandstein clastic rocks of the Santander Basin (Burkhardt 1976). South of Lebeña, Permian sediments are thicker than 1000 m and composed of alternating sandstones and shales of grey color, with some carbonates interbedded at Peña Sagra, which could be equivalent to the Viñón and Villaviciosa Beds.

2.6.4.2.10 Peña Labra Area

The presence of Permian sediments in this area was noted by Maas (1968, 1974) and Jong (1971), with the finding of plant remains of this age. Maas described there the Labra Formation, made up of lahar volcanic rocks (70 m), volcanic agglomerates (130 m), volcanic tuffs (130 m), and alternating sandstones, shales and conglomerates with nodules (225 m), below the Buntsandstein clastic rocks. Some acid volcanic rocks also exist below this formation, which could be either of Lower Permian or Stephanian age.

2.6.5 Development of Permian Sedimentation in the Cantabrian and Palentian Zones

After the Hercynian orogeny took place in the Cantabrian and Palentian Zones, a shallow marine basin was installed mainly in the eastern and central part, with extension (probably transtension) producing NE-SW oriented fault-bound troughs and extrusion of bimodal volcanic rocks. This has been considered as a continental rifting by Martínez García (1981 a, 1983 a) and Martínez García and Tejerina (1985). After the deposit of the Viñón Beds, of predominant volcanic and volcanoclastic character and nearly 1000 m thick in some places, a new tensional or transtensional episode started, with the main activity along NW-SE directed faults. The Villaviciosa Beds were then deposited in a continental semi-arid environment of endorrheic basins with development of caliche in the center and breccias near the boundary faults. New faulting episodes ended this period of sedimentation, erosion took place and, later on, Triassic rocks were deposited mainly in the eastern part, in the Santander Basin. Fault activity also took place after the deposition of the Triassic succession and erosion of most of it was accomplished in the Cantabrian and Palentian Zones, where Liassic rocks unconformably overlie the Permian Villaviciosa Beds in most instances.

References

Adaro L de, Junquera G (1916) Criaderos de hierro de España. T II. Criaderos de Asturias. Mem Inst Geol Min Esp, pp 1-676

Aisenverg DE, Brazhnikova NE, Nassilyuk NP, Vdovenko MV, Gorak SV, Dunaeva NN, Zernetskaya NV, Poletaev VI, Potievskaya PD, Rotai AP, Sergeeva MT (1979) The Carboniferous sequence of the Donetz Basin: a standard section for the Carboniferous System. In: Wagner RH, Higgins AC, Meyen SV (eds) The Carboniferous of the USSR. Yorkshire Geol Soc Occas Publ 4: 197-224

Almela A, Rios JM (1962) Investigación del Hullero bajo los terrenos de la costa cantábrica (zona de Oviedo-Gijón-Villaviciosa-Infiesto). Empr Nac Adaro Inv Min, pp 1-171

Almela A, Valle de Lersundi J del (1963) Estudio geológico de la zona Ponferrada-Tremor de Abajo. Breviora Geol Astúrica VIII, 1-4: 3-26

Alvarado A de (1952) Limites stratigraphiques du Carbonifére du NW de León (structure et caractéristiques des couches exploitées). In: C R III Congr Int Strat Géol Carbonifère Heerlen 1951, I: 5-12

Alvarado A de, Sobrino M (1946) Mancha carbonífera de El Bierzo. Not Comun Inst Geol Min Esp 16: 5-53

Alvarez Ramis C (1966) Primera aportación al estudio de los "Sphenopteris" del Carbonífero español (a Estefaniense: 1ª parte). Estud Geol 21: 1-77

Amerom HW van (1965) Note préliminaire sur quelques flores stéphaniennes de la bordure nord des Léonides dans les Montagnes Cantabriques (Espagne du nord-ouest). Leidse Geol Meded 32: 151-156

Amerom HW van, Dillewijn J van (1963) Note sur le bassin houillier de Ciñera-Matallana. Leidse Geol Meded 29: 303-312

Barrois C (1882) Recherches sur les terrains anciens des Asturies et de la Galice. Mém Soc Géol Nord 2, 1: 1-630

Bosch WJ van den (1969) Geology of the Luna-Sil region. Cantabrian Mountains (NW Spain). Leidse Geol Meded 44: 137-225

Burkhardt R (1976) Geologie und Lithologie der Permotriassischen Schichtabfolgen und deren Paläozoischer Rahmengesteine im Bereich östlich der Picos de Europa im Kantabrischen Gebirge, Nordspanien. PhD Thesis, Technische Universität München, pp 1-119 (unpubl)

Cleal CJ (1978) Floral biostratigraphy of the upper Silesian Pennant Measures of South Wales. Geol J 13, 2: 165-194

Colmenero JR, Bahamonde JR (1986) Análisis estratigráfico y sedimentológico de la cuenca estefaniense de Sebarga (Región de Mantos, Zona Cantábrica). Trab Geol 16: 103-119

Corrales I (1970) Estratigrafía y sedimentología del Estefaniense de Cangas de Narcea, Rengos y Villablino. PhD Thesis, Univ Oviedo (unpubl)

Corrales I (1971) La sedimentación durante el Estefaniense B-C en Cangas de Narcea, Rengos y Villablino (NW de España). In: Wagner RH (ed) The Carboniferous of northwest Spain, Oviedo. Trab Geol 3: 69-73

Corrales I, Peláez JR (1966) La sedimentación marginal en la cuenca estefaniense de Villablino (León). Breviora Geol Ast X, 1-4: 33-54

Crespo A (1982) Explicación de la Hoja nº 76 (Pola de Somiedo) del Mapa Geológico de España (MAGNA) a escala 1/50.000. Inst Geol Min Esp, pp 1-51

Doubinger J, Alvarez Ramis C (1964) Sur la flore du bassin houillier de Tineo (Asturies). C R Acad Sci Paris 259: 1885-1887

Durán M, Arango G (1918) Estudio industrial de los manchones carboníferos de Gillón, Gedrez, Monasterio de Hermo y Cerredo (Oviedo). Bol Of Min Met 14: 1-32

Evers HJ (1967) Geology of the Leonides between the Bernesga and Porma rivers. Cantabrian Mountains, NW Spain. Leidse Geol Meded 41: 83-151

Ezquerra del Bayo J (1844) Descripción geognóstica y minera de la provincia de Palencia. Bol Of Minas 14: 160-163

Fissunenko OP (1974) Analog Cantabriya v razrezye karbona Donetskogo basseina. (Analogue of the Cantabrian in the Carboniferous section of the Donetz Basin). Izvest Akad Nauk SSSR (Ser Geol) 7: 152-156

Fissunenko OP, Laveine JP (1984) Comparaison entre la distribution des principales espèces-guides végétales du Carbonifère moyen dans le bassin du Donetz (URSS) et les bassins du Nord-Pas-de Calais et de Lorraine (France). In: Gordon M Jr (ed) C R IX Congr Int Strat Géol Carbonifère, Washington DC, Champaign/Urbana, 1979, Southern Illinois University Press, Carbondale and Edwardsville 1: 95-100

Forster A (1974) Die Fluss-Spatlagerstätten Asturiens, Nordspaniens und deren Genese. Geol Rundsch 63: 212-263

García Fuente S (1959) Explicación de la Hoja nº 77 (La Plaza) del Mapa Geológico de España (1ª Serie) a escala 1:50.000. Inst Geol Min Esp, pp 1-69

Gastaldo RA (1977) A Middle Pennsylvanian nodule flora from Carterville, Illinois. In: Romans RC (ed) Geobotany. Plenum, New York, pp 133-155

Geinitz HB (1867) Über organische Überreste aus dem Steinkohlengraben Arnao bei Avilés in Asturien. Neues Jahrb Min, pp 283-286

George TN, Wagner RH (1972) Subcommission on Carboniferous Stratigraphy. Proceedings and report of the General Assembly at Krefeld, August, 21-22. C R VII Congr Int Strat Géol Carbonifère Krefeld 1971, 1: 139-147

Gervilla M, Beroiz C, Barón A, Ramírez del Pozo J, Giannini G (1973a) Explicación de la Hoja nº 30 (Villaviciosa) del Mapa Geológico de España (MAGNA) a escala 1:50.000. Inst Geol Min Esp, pp 1-54

Gervilla M, Beroiz C, Barón A, Ramírez del Pozo J, Giannini G (1978) Explicación de la Hoja nº 29 (Oviedo) del Mapa Geológico de España (MAGNA) a escala 1:50.000. Inst Geol Min Esp, pp 1-64

Ginkel AC van (1965) Spanish Carboniferous fusulinids and their significance for correlation purposes. Leidse Geol Meded 34: 137-225

Ginkel AC van (1971) Fusulinids from uppermost Myachkovian and Kasimovian strata of northwestern Spain. Leidse Geol Meded 47: 115-161

Ginkel AC van (1972) Correlation of the Myachkovian and Kasimovian in the USSR with the West European subdivision. Leidse Geol Meded 49: 1-7

Gómez de Llarena J (1950) Nuevos datos geológicos y paleontológicos sobre la Cuenca carbonífera de Santa Lucía-Matallana (León). Estud Geol 11: 51-98

Gómez de Llarena J, Rodríguez Arango C (1948) Datos para el estudio geológico de la Babia Baja (León). Bol Inst Geol Min Esp 61: 79-206

Grand'Eury C (1877) Determination des etapes houillères a l'aide de la flore fossile. Resumé des travaux de M..., par Mr Zeiller. Ann Mines 12, 6: 341-391

Gutiérrez Claverol M, Luque C, Suárez V (1987) El lineamiento tectónico Nazaré (W de Portugal)-Luarca (NW de España) y su implicación metalogenética (Macizo Hespérico). In: Geología de los granitoides y rocas asociadas del Macizo Hespérico. Libro Homenaje a LC García de Figuerola, Editorial Rueda, Madrid, pp 447-455

Gutiérrez Claverol M, Luque C, Martínez García E (1988) Importancia metalogenética de los lineamientos en Asturias. Com Congr Geol Esp, Granada, 1988, 2: 253-256

Gutiérrez Claverol M, Luque C, Martínez García E, Ruíz F, Suárez V (1988) Los lineamientos deducidos de imágenes de satélite de las zonas Cantábrica y Astur Occidental Leonesa (NW de España). Estud Geol 44: 263-270

Helmig HM (1965) The geology of the Valderrueda, Tejerina and Sabero coal basins (Cantabrian Mountains, Spain). Leidse Geol Meded 32: 75-149

Hernández Sampelayo P, Hernández Sampelayo A (1952) Explicación de la Hoja nº 159 (Bembibre) del Mapa Geológico de España (1ª Serie) a escala 1:50.000. Inst Geol Min Esp, pp 1-96

Heward AP (1978) Alluvial fan and lacustrine sediments from the Stephanian A and B (La Magdalena, Ciñera-Matallana, Sabero) coalfields, northern Spain. Sedimentology 25: 451-488

Heward AP, Reading HG (1980) Deposits associated with a Hercynian to late-Hercynian continental strike-slip system, Cantabrian Mountains, northern Spain. Spec Publ Int Assoc Sediment 4: 105-125

Hüber G (1976) Geologisch-lagerstättenkundliche Untersuchungen horizont-gebundener Fluss-Spatlagerstätten in Ostasturien und Möglichkeiten zu ihrer Darstellung. PhD Thesis, Clausthal-Zellerfeld, Technische Universität Clausthal, pp 1-206 (unpubl)

Iwaniw E (1984) Lower Cantabrian basin margin deposits in NE León, Spain - a model for valley-fill sedimentation in a tectonically active, humid climatic setting. Sedimentology 31: 91-110

Iwaniw E (1985) The sedimentology of lower Cantabrian basin margin deposits in NE León, Spain. In: Sousa MJL, Wagner RH (eds) Papers on the Carboniferous of the Iberian Peninsula (sedimentology, stratigraphy, palaeontology, tectonics and geochronology), An Fac Cienc Porto (Suppl) 64: 49-115

Jong JD de (1971) Molasse and clastic wedge sediments of the southern Cantabrian Mountains (NW Spain) as geomorphological and environmental indicators. Geol Mijnbouw 50, 3: 399-416

Julivert M, Marcos A, Pulgar JA (1977) Explicación de la Hoja nº 27 (Tineo) del Mapa Geológico de España (MAGNA) a escala 1:50.000. Inst Geol Min Esp, pp 1-28

Jullien J (1974) Recherches sur les minéralisations fluorées liées a l'évolution du bassin Permo-Triassique des Asturies (Espagne). PhD Thesis, Univ Paris IV, pp 1-298

Knight J (1971) The sequence and stratigraphy of the eastern end of the Sabero coalfield. In: Wagner RH (ed) The Carboniferous of northwest Spain. Trab Geol 3: 193-229

Knight J (1974) The Stephanian A-B flora and stratigraphy of the Sabero Coalfield (León, NW Spain). In: C R VII Congr Int Stratigr Géol Carbonifère Krefeld 1971, 3: 283-316

Knight J (1975) The systematics and stratigraphic aspects of the Stephanian flora of the Sabero Coalfield. Pt II. The systematic palaeobotany of the Sabero Coalfield. PhD Thesis, Sheffield, Univ Sheffield, pp 1-676 (unpubl)

Knight J (1983a) The stratigraphy of the Stephanian rocks of the Sabero Coalfield, León (NW Spain) and an investigation of the fossil flora. Part I. The stratigraphy and general geology of the Sabero Coalfield. Palaeontographica, Abt B, 187, 1/3: 1-88

Knight J (1983b) The stratigraphy of the Stephanian rocks of the Sabero Coalfield, León (NW Spain) and an investigation of the fossil flora. Part II. Systematic palaeobotany: introduction; Pteridospermae. Palaeontographica, Abt B, 187, 4/6: 155-248

Kuchelka R (1974) Die Erkundung und Darstellung der Fluss-Spätlagerstätten Cueto del Aspa/Asturien sowie ein Versuch zur genetischen Deutung. PhD Thesis, Clausthal-Zellerfeld, Technische Universität Clausthal, pp 1-108 (unpubl)

Kutterink JA (1966) Geologie van het Valdeón gebied. Int Rept Geol Inst Leiden

Laveine JP (1977) Report on the Westphalian D. Reports by the working Group of the IUGS Subcommission on Carboniferous Stratigraphy. Symp Carb Strat Prague, 1973, pp 71-87

Lorenzo P (1977) La edad de las cuencas de San Juán de Nieva y Arnao (NW de España). Breviora Geol Astúrica XXI, 2: 29-32

Lorenzo P (1978) Flora fósil del Carbonífero de La Magdalena (León). Thesis, Univ Oviedo (unpubl)

Lotze F (1945) Zur Gliederung der Variszíden der Iberischen Meseta. Geotekt Forsch 6: 78-92

Luque C, Martínez García E (1983) Depósitos minerales en el Carbonífero de la Cordillera Cantábrica. In: Martínez Díaz C (coord) Carbonífero y Pérmico de España, Inst Geol Min Esp, Madrid, pp 163-188

Luque C, Martínez García E (1985) Análisis tectometalogenético de las mineralizaciones en el Carbonífero de la Zona Cantábrica (NW de España). In: C R X Congr Int Stratigr Geol Carbonifère Madrid 1983. Inst Geol Min Esp 3: 95-104

Maas K (1968) Geologie van het Polaciones-Comunidad gebied, Sedimentologie. Int Rept Geol Inst Leiden

Maas K (1974) The geology of Liébana, Cantabrian Mountains, Spain. Deposition and deformation in a flysch area. Leidse Geol Meded 49: 379-465

Maas K, Ginkel AC van (1983) Variscan olistostrome deposition and synsedimentary nappe emplacement, Valdeón Area, Cantabrian Mountains, Spain. Leidse Geol Meded 52: 341-381

Madariaga R de (1928) Introducción a un ensayo de sincronización de cuencas carboníferas españolas. Not Comun Inst Geol Min Esp 1: 51-59

Mallada L (1886) Datos para el estudio geológico de la cuenca hullera de Ciñera y Matallana. Bol Com Mapa Geol Esp XIV: 173-207

Mallada L (1898) Explicación del Mapa Geológico de España. T III. Sistemas Devoniano y Carbonífero. Mem Com Mapa Geol Esp, pp 1-405

Marcos A (1967) Estudio geológico del reborde NW de los Picos de Europa (región de Onís-Cabrales, Cordillera Cantábrica). Trab Geol 1: 39-46

Marcos A (1968a) Nota sobre el significado de la "León Line". Breviora Geol Astúrica XII: 1-5

Marcos A (1968b) Sobre la existencia de niveles de tránsito entre el Westfaliense D y el Estefaniense A en el Oriente de Asturias (NW de España). Acta Geol Hisp 3, 1: 7-14

Marcos A, Pérez Estaún A, Pulgar JA, Bastida F (1979) Explicación de la Hoja nº 74 (Fonsagrada) del Mapa Geológico de España (MAGNA) a escala 1:50.000. Inst Geol Min Esp, pp 1-28

Marquínez J, Méndez CA, Menéndez Alvarez JR, Sánchez de Posada LC, Villa E (1982) Datos biostratigráficos de la sucesión carbonífera (Turnesiense-Kasimoviense) de Las Llacerias, Picos de Europa, Norte de España. Trab Geol 12: 187-193

Martínez García E (1971) The age of the Caliza de Montaña in the eastern Cantabrian Mountains. In: Wagner RH (ed) The Carboniferous of northwest Spain. Trab Geol 3: 267-276

Martínez García E (1981a) El Paleozóico de la Zona Cantábrica oriental (Noroeste de España). Trab Geol 11: 95-127

Martínez García E (1981b) Tectónica y mineralizaciones pérmicas en la Cordillera Cantábrica oriental (Noroeste de España). Cuad Lab Xeol Laxe 2, II: 263-270

Martínez García E (1983a) El Pérmico de la Región Cantábrica. In: Martínez Díaz C (coord) Carbonífero y Pérmico de España. Inst Geol Min Esp, Madrid, pp 389-402

Martínez García E (1983b) Permian mineralizations in the Cantabrian Mountains (NW Spain). In: Schneider HJ (ed) Mineral deposits of the Alps and of the Alpine Epoch in Europe. Springer, Berlin Heidelberg New York, pp 259-274

Martínez García E, Rodríguez Fernández LR (1984) Explicación de la Hoja nº 56 (Carreña-Cabrales) del Mapa Geológico de España (MAGNA) a escala 1/50.000. Inst Geol Min Esp, pp 1-45

Martínez García E, Tejerina L (1985) Fluorspar deposits associated with Carboniferous and Permian rocks in Asturias and León (NW Spain). In: Cross AT (ed) CR IX Congr Int Stratigr Géol Carbonifère, Washington DC Urbana/Champaign 1979, Southern Illinois University Press, Carbondale and Edwardsville 4: 467-478

Martínez García E, Villa E (1986) Una discordancia importante en el Carbonífero de los Picos de Europa (Asturias, NW de España). In: Actas IX Congr Nac Sedim, Salamanca 1980, Acta Salmanticensia, Salamanca 50, II: 719-726

Martínez García E, Wagner RH (1971) Marine and continental deposits of Stephanian age in eastern Asturias (N Spain). In: Wagner RH (ed) The Carboniferous of northwest Spain. Trab Geol 1: 285-305

Martínez García E, Wagner RH (1982) Una cuenca marina del Estefaniense superior en el Noroeste de España. Trab Geol 12: 119-124

Martínez García E, Wagner RH (1984) The post-Asturian marine basin of late Stephanian age in northwest Spain. In: Belt ES, MacQueen RW (eds) CR IX Congr Int Strat Géol Carbonifère, Washington DC, Champaign/Urbana 1979, Southern Illinois University Press, Carbondale and Edwardsville 3: 508-516

Martínez García E, Corrales I, Carballeira JR (1971) El flysch carbonífero de Pendueles (Asturias). In: Wagner RH (ed) The Carboniferous of northwest Spain. Trab Geol 3: 277-283

Martínez García E, Rodríguez Arango R, Villa E (1985) The unconformable marine Kasimovian strata of the Sellaño Basin (Cantabrian Zone, NW Spain). In: Lemos de Sousa ML, Wagner RH (eds) Papers on the Carboniferous of the Iberian Peninsula (sedimentology, stratigraphy, paleontology, tectonics and geochronology), An Fac Cienc (Suppl) 64: 35-40

Martínez García E, Wagner RH, Lobato L, Fernández L, Alonso JL (1983) El Carbonífero de la Región Oriental (Pisuerga-Carrión). In: Martínez Díaz C (coord) Carbonífero y Pérmico de España, Inst Geol Min Esp Madrid, pp 116-132

Matas J, Abejaro V, Fernández L, Fernández Pello J (1983) Explicación de la Hoja nº 127 (Noceda) del Mapa Geológico de España (MAGNA) a escala 1:50.000. Inst Geol Min Esp, pp 1-63

Meléndez B (1950) Nota previa sobre los terrenos pérmicos de Colunga y Caravia (Asturias). Bol R Soc Esp Hist Nat 48, 2: 141-154

Monreal LN (1878) Datos geológicos acerca de la provincia de León recogidos durante la campaña de 1877 a 1878. Bol Com Mapa Geol Esp V: 201-207

Paillette A, Verneuil E de, D'Archiac A (1845) Recherches sur quelques unes des roches qui constituent la province des Asturies. Note sur les fossiles du terrain paléozoique des Asturies. Bull Soc Géol Fr (2e Ser) 2: 439-482

Patac I (1920) La formación uraliense asturiana. Estudios de cuencas carboníferas. Cia Asturiana Artes Gráficas, Gijón, pp 1-50

Patac I (1956) La geología del litoral asturiano y su importancia industrial. Bol Inst Nac Carbón, Mayo-Junio, pp 107-123

Pérez Estaún A, Pulgar JA, Bastida F, Marcos A, Sánchez de la Torre L, Galán J, Vargas I, Ruíz F (1982) Explicación de la Hoja nº 126 (Vega de Espinareda) del Mapa Geológico de España (MAGNA) a escala 1:50.000. Inst Geol Min Esp, pp 1-35

Prado C de, Verneuil E de (1850) Sur les terrains de Sabero et des environs dans les montagnes de León (Espagne). Bull Soc Géol Fr (2e Ser) VII: 137-155

Prado JG (1972) Nota sobre la petrografía de la zona de Viñón (Asturias). Stud Geol III: 7-32

Pulgar JA, Bastida F, Marcos A, Pérez Estaún A, Galán J, Vargas I (1981) Explicación de la hoja nº 100 (Degaña) del Mapa Geológico de España (MAGNA) a escala 1:50.000. Inst Geol Min Esp, pp 1-35

Rauser-Chernousova DM, Ivanova EA, Grozdilova LP, Makhlina MK, Alksne AE, Kireeva GD, Konovalova MB, Meyen SV, Morozova IP, Rosovskaya SE, Faddeeva IZ, Shamov DF, Schegolev AK, Shcherbakova MV (1979) The Upper Carboniferous Series. In: Wagner RH, Higgins AC, Meyen SV (eds) The Carboniferous of the USSR. Yorkshire Geol Soc, Occas Publ 4: 147-174

Reijers T (1985) Evidence for strike-slip movement along the Sabero-Gordón and associated faults in the Luna-Porma Area, León, NW Spain. Trab Geol 15: 177-187

Revilla J (1906) Riqueza minera de la provincia de León. Su descripción industrial y estudio de soluciones para explotarla. Impr Alemana, Madrid, pp 1-311

Rubio A (1876) Reseña físico-geológica del valle de Laceana (provincia de León). Bol Com Mapa Geol Esp III: 333-345

Rupke J (1965) The Esla Nappe (Cantabrian Mountains, Spain). Leidse Geol Meded 32: 1-74

Sánchez de la Torre L, Agueda Villar JA, Colmenero JR, Manjón M (1977) La serie permotriásica en la región de Villaviciosa (Asturias). Cuad Geol Ibérica 4: 329-338

Schäfer D (1974) Zur Geologie einiger Permotrias-Vorkommen in Asturien/Spanien. PhD Thesis, Clausthal-Zellerfeld, Technische Universität Clausthal, pp 1-75 (unpubl)

Schulz (1858) Descripción geológica de la provincia de Oviedo. Impr José González, Madrid, pp 1-138

Sitter LU de (1962) The structure of the southern slope of the Cantabrian Mountains. Explanation of a geological map with sections, scale 1:100.000. Leidse Geol Meded 26: 255-264

Staalduinen CJ van (1973) Geology of the area between the Luna and Torío rivers, southern Cantabrian Mountains, NW Spain. Leidse Geol Meded 49: 167-205

Stockmans F, Willière Y (1965) Documents paléobotaniques pour l'étude du houiller dans le nord-ouest de l'Espagne. Mém Int Roy Sci Nat Belg (2), 79, pls I-XXXVIII

Suárez Rodríguez A (1988) Estructura del área de Villaviciosa-Libardón (Asturias, Cordillera Cantábrica). Trab Geol 17: 87-98

Truyóls J, González Lastra J, Marquínez García J, Martínez Díaz C, Méndez Férnandez C, Méndez Alvarez JR, Sánchez de Posada LC (1984) Preliminary note on two marine sections (Tournaisian-Kasimovian) in the Picos de Europa area (Cantabrian Mountains, NW Spain). In: C R IX Congr Int Stratigr Géol Carbonifère, Washington DC, Urbana/Champaign, 1979, Southern Illinois University Press, Carbondale and Edwardsville 2: 148-156

Vega Rollán C de la (1959) Flora carbonífera de Asturias y su distribución estratigráfica. Breviora Geol Astúrica III, 1-2: 3-70

Vega Rollán C de la (1964) Contribución al estudio de la flora carbonífera del occidente de Asturias. Breviora Geol Astúrica VIII, 1-4: 37-52

Vega Rollán C de la (1968) Flora de las series molásicas estefanienses del occidente de Asturias. Acta Geol Hisp 3, 1: 7-14

Velando F, Martínez Díaz M (1973) Explicación de la Hoja nº 159 (Bembibre) del Mapa Geológico de España (MAGNA) a escala 1:50.000. Inst Geol Min Esp, pp 1-34

Vidal Box C (1943) Notas previas a un estudio morfológico y geológico de la alta cuenca del rio Sil, cuenca de Laceana y Babia alta (prov de León). Rev R Acad Ci Ex Fis Nat 38: 95-117

Villa E (1989) Fusulináceos carboníferos del Este de Asturias (N de España). PhD Thesis, Univ Oviedo, pp 1-378 (unpubl)

Villa E, Martínez García E (in press) El Carbonífero superior marino de Dobros (Picos de Europa, Asturias, NW de España). Trab Geol 18

Virgili C, Corrales I (1966) Las series molásicas estefanienses del occidente de Asturias. Acta Geol Hisp I, 4: 17-21

Wagner RH (1959) Sur la présence d'une nouvelle phase tectonique "léonienne" d'âge Westphalien D dans le nord-ouest de l'Espagne. C R Acad Sci Paris 249: 2804-2806

Wagner RH (1963) Stephanian B flora from the Ciñera-Matallana coalfield (León) and neighboring outliers, I: introduction, Neuropteris. Not Comun Inst Geol Min Esp 72: 5-70

Wagner RH (1964a) Stephanian B flora from the Ciñera-Matallana coalfield (León) and neighboring outliers, II: Mixoneura, Reticulopteris, Linopteris and Odontopteris. Not Comun Inst Geol Min Esp 75: 5-56

Wagner RH (1964b) Stephanian floras in NW Spain, with special reference to the Westphalian D-Stephanian A boundary. C R V Congr Strat Géol Carbonifère Paris 1963, II: 835-851

Wagner RH (1965) Stephanian B flora from the Ciñera-Matallana coalfield (León) and neighboring outliers, III: Callipteridium and Alethopteris. Not Comun Inst Geol Min Esp 78: 5-70

Wagner RH (1966a) Notes on the geology of Paleozoic rocks in the northeastern part of the province of Palencia, NW Spain. Not Comun Inst Geol Min Esp 86: 31-40

Wagner RH (1966b) Paleobotanical dating of Upper Carboniferous folding phases in NW Spain. Mem Inst Geol Min Esp 66: 1-169

Wagner RH (1967) Apuntes sobre las floras de la zona Gamonedo-Cabrales en la parte oriental de Asturias. Trab Geol 1: 47-58

Wagner RH (1970) An outline of the Carboniferous stratigraphy of north-western Spain. In: Streel M, Wagner RH (eds) Colloque sur la stratigraphie du Carbonifère, Liege, 1969, Congr Coll Univ Liege 55: 429-463

Wagner RH (1971a) Proposal for the recognition of a new "Cantabrian" stage at the base of the Stephanian series. C R VI Congr Int Strat Géol Carbonifère Sheffield, 1967 1: 140-154

Wagner RH (1971b) The stratigraphy and structure of the Ciñera-Matallana coalfield (prov. of León, NW Spain). In: Wagner RH (ed) The Carboniferous of northwest Spain. Trab Geol 4: 385-429

Wagner RH (1971c) Carboniferous nappe structures in northwestern Palencia (Spain). In: Wagner RH (ed) The Carboniferous of northwest Spain. Trab Geol 4: 431-459

Wagner RH (1983) Late Westphalian D and Early Cantabrian floras of the Guardo Coalfield. In: Geology and palaeontology of the Guardo Coalfield (NE León-NW Palencia), Cantabrian Mts. Inst Geol Min Esp, Madrid, pp 57-92

Wagner RH (1984) Megafloral zones in the Carboniferous. In: Sutherland PK, Manger WL (eds) CR IX Congr Int Stratigr Géol Carbonifère, Washington DC, Urbana/Champaign, 1979, Southern Illinois University Press, Carbondale and Edwardsville 2: 109-134

Wagner RH, Artieda JI (1970) La Cuenca Minera Ciñera-Matallana, León, Soc. Hullera Vasco-Leonesa, León, pp 1-288

Wagner RH, Fernández García L (1971) The lower Carboniferous and Namurian rocks north of La Robla (León). In: Wagner RH (ed) The Carboniferous of northwest Spain. Trab Geol 4: 507-531

Wagner RH, Higgins AC (1979) The Carboniferous of the USSR: its stratigraphic significance and outstanding problems of worldwide correlation. In: Wagner RH, Higgins AC, Meyen SV (eds) The Carboniferous of the USSR. Yorkshire Geol Soc, Occas Publ 4: 5-22

Wagner RH, Martínez García E (1982) Description of an early Permian flora from Asturias and comments on similar occurrences in the Iberian Peninsula. Trab Geol 12: 273-287

Wagner RH, Varker WJ (1971) The distribution and development of post-Leonian strata (upper Westphalian D, Cantabrian, Stephanian A) in northern Palencia, Spain. In: Wagner RH (ed) The Carboniferous of northwest Spain. Trab Geol 4: 533-601

Wagner RH, Winkler Prins CF (1985) The Cantabrian and Barruelian stratotypes: a summary of basin development and biostratigraphic information. In: Lemos de Sousa ML, Wagner RH (eds) Papers on the Carboniferous of the Iberian Peninsula (sedimentology, stratigraphy, paleontology, tectonics and geochronology), An Fac Cienc (Suppl) 64: 359-410

Wagner RH, Winkler Prins CF, Riding RE (1971) Lithostratigraphic units of the lower part of the Carboniferous in northern León, Spain. In: Wagner RH (ed) The Carboniferous of northwest Spain. Trab Geol 4: 603-663

Wagner RH, Villegas FJ, Fonollá F (1971) Description of the lower Cantabrian stratotype near Tejerina (León, NW Spain). In: CR VI Congr Int Strat Géol Carbonifère, Sheffield 1967, 1: 115-139

Wagner RH, Jones DG, Spinner E, Wagner Gentis CHT (1970) The upper Cantabrian rocks near Inguanzo, eastern Asturias, Spain. In: Streel M, Wagner RH (eds) Colloque sur la stratigraphie du Carbonifère, Liege, 1969, Congr Coll Univ Liege 55: 465-486

Wagner RH, Park RK, Winkler Prins CF, Lys M (1977) The post-Leonian basin in Palencia: a report on the stratotype of the Cantabrian Stage. In: Symposium on Carboniferous Stratigraphy, Prague, Spec Publ Geol Surv Prague, Prague, pp 89-146

Wilde GL (1984) Systematics and the Carboniferous-Permian Boundary. In: Sutherland PK, Manger WL (eds) CR IX Congr Int Strat Géol Carbonifère, Washington DC, Urbana/Champaign, 1979, Southern Illinois University Press, Carbondale and Edwardsville 2: 543-558

Zeiller MR (1882) Note sur la flore houillère des Asturies. Mém Soc Géol Nord 1, 3: 1-22

3 Structure

3.1 Cantabrian Zone

A. Perez Estaun and F. Bastida[1]

3.1.1 Introduction

The foreland thrust and fold belt of the Variscan collisional orogen in the northwestern Iberian Península, Cantabrian Zone (CZ), is located in the inner part of an arc described by the structures of the orogen (Asturian arc or Iberoarmorican arc; Fig. 1 in Part II, Chap. 1, p. 7). The structure of this zone has a thin-skinned type of geometry, complicated by the existence of more than one level of decollement and by a wide range of different thrust systems. These combine to produce complex and irregular map outcrop patterns of the thrust units, as shown in Fig. 1. The deformation took place under shallow crustal conditions, without metamorphism and with only the local development of cleavage.

There are a large number of references dealing with the geology of the CZ. General syntheses on the structure can be found in the papers of Julivert (1971, 1979, 1981, 1983), Julivert and Marcos (1973), Julivert et al. (1977), Pérez Estaún et al. (1988).

3.1.2 General Structure

To the west, the CZ is separated from the more internal areas of the orogen, the West Asturian-Leonese Zone, by the Narcea Antiform (Fig. 1), which shows Precambrian rocks exposed in its core. A thrust situated in the central part of this antiform forms the boundary between the Cantabrian and West Asturian-Leonese zones. This boundary divides paleogeographically different areas and represents approximately the upper front of cleavage and metamorphism, which are generally developed to the west. To the east and south, the Paleozoic rocks of the CZ are covered by Mesozoic and Tertiary sediments.

The general structure of the CZ is shown in the geological sketch map in Fig. 1 and in the cross-section through the central part of the Asturian arc (Fig. 2). The thrust surfaces mostly diverge from a decollement located within the limestones and dolomites of the Lancara Formation (Lower-Middle Cambrian), although in the western part of the cross-section it is possible to observe that the thrusts affect Precambrian rocks. Geometrically there are two major thrust systems in the section. The first to develop was in the rear or westernmost units: Somiedo-Correcilla, Sobia-Bodón and Aramo units (Fig. 3, *1*). The second thrust system started its development beneath this earlier system and caused out-of-sequence thrusts, a thickening of the orogenic wedge and the progression of deformation towards the foreland producing successive thrusts. The out-of-sequence thrusts, superimposed over the first thrust system, were responsible for the geometry of the Narcea Antiform and the thickening of the orogenic wedge (Fig. 3, *2*). Further along, the successive thrusts generated as a consequence of the orogenic belt thickening, and gave rise progressively from west to east, to the Central Coal Basin Unit, Ponga Unit and Picos de Europa Unit.

An important feature of the CZ is its structural heterogeneity. This can be observed by comparing cross-sections along the CZ from different orientations and by observing the mapped disposition of the main thrust units. These do not form a concentric set, but are arranged like the leaves of a photographic camera iris (Fig. 4).

The Pisuerga-Carrión Unit (Julivert 1971) or Palentine Unit (Martínez García 1981) formed as a relatively autochthonous part of the Asturian Arc in the southwestern sector of the CZ (Fig. 1). It has been overthrusted by three separate nappes coming from the south, west and north (Esla thrust Unit, Ponga Unit, and Picos de Europa Unit, respectively). Since the structure of this unit displays some

[1] Departamento de Geología, Universidad de Oviedo, 33005 Oviedo, Spain

R. D. Dallmeyer and E. Martínez García (Eds.)
Pre-Mesozoic Geology of Iberia

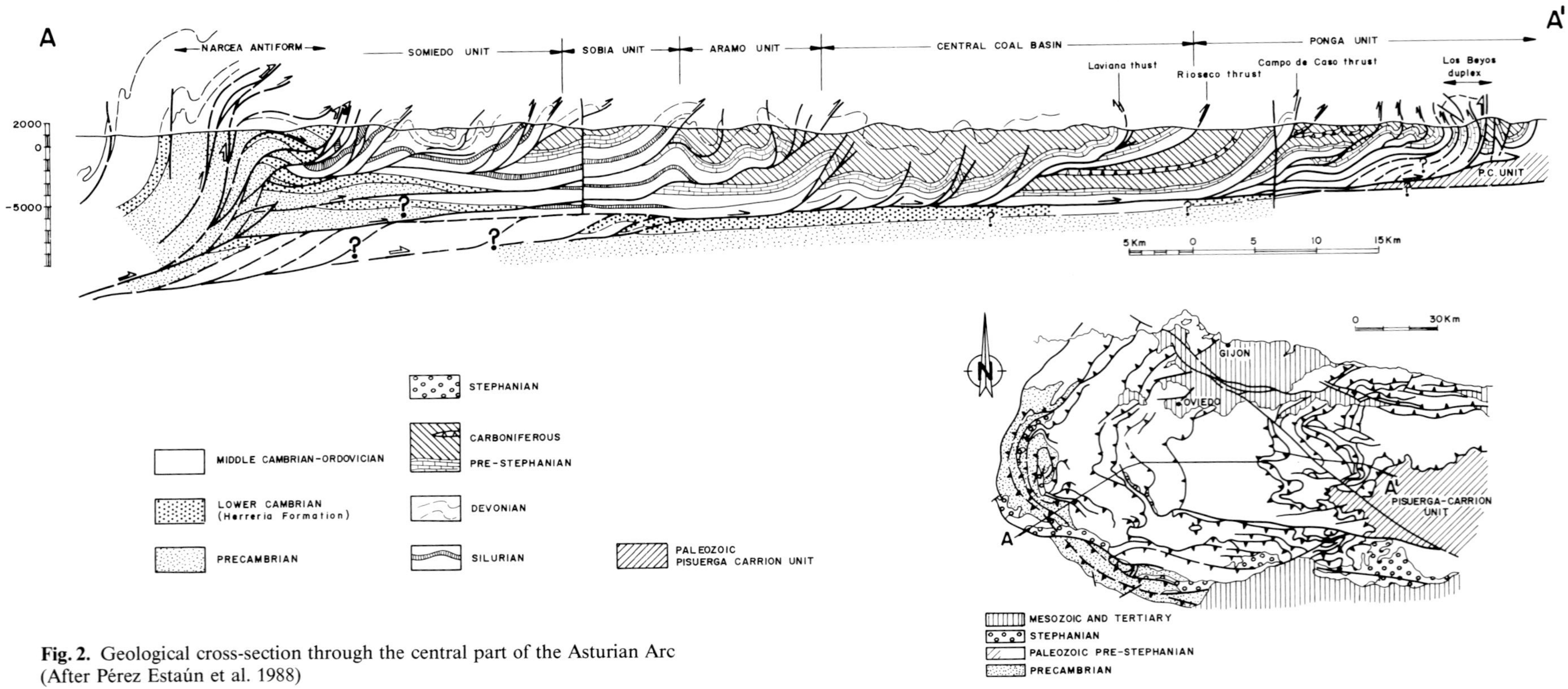

Fig. 2. Geological cross-section through the central part of the Asturian Arc (After Pérez Estaún et al. 1988)

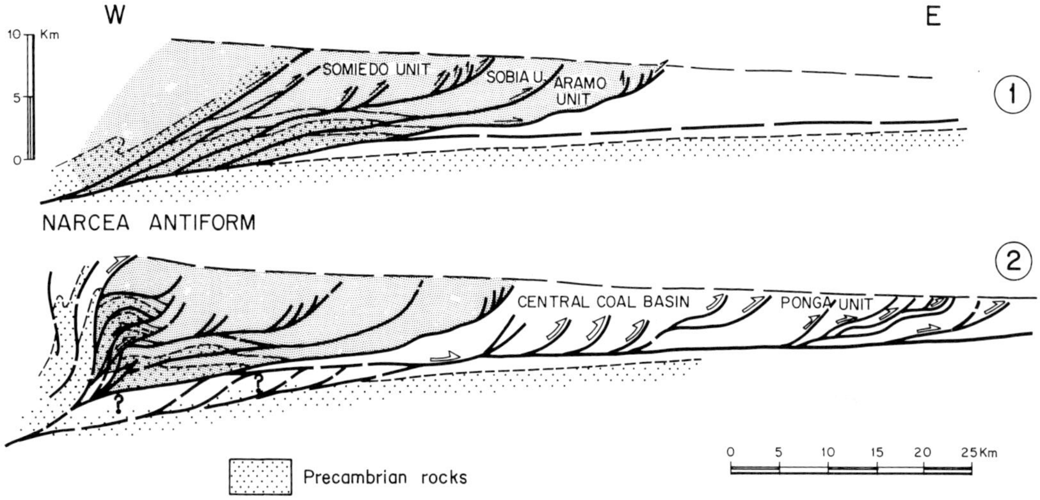

Fig. 3. Cross-sections showing two structural evolution stages of Fig. 2 cross-section

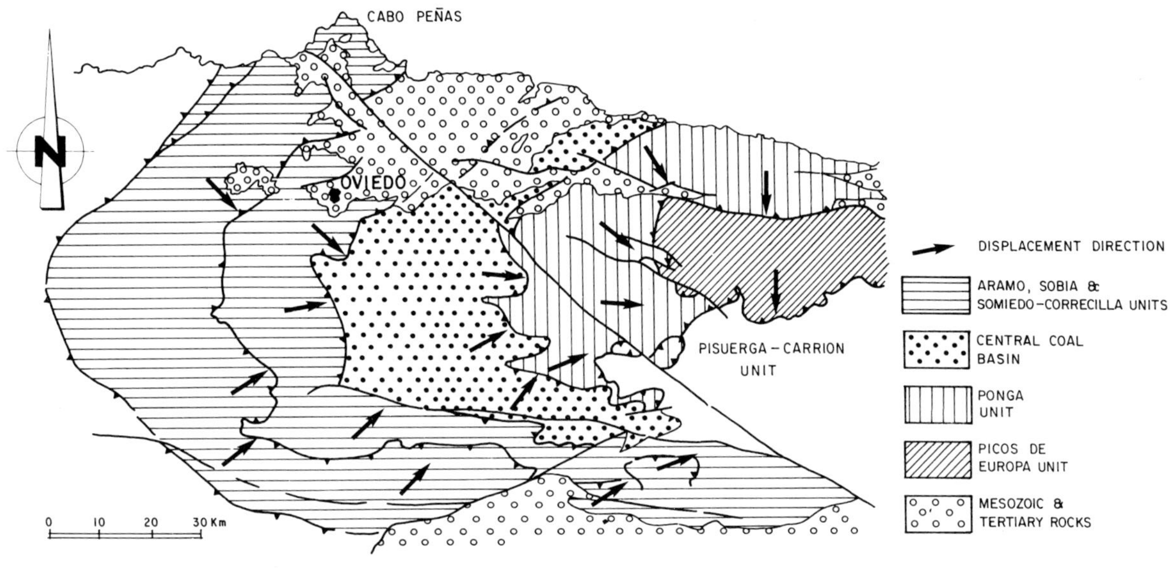

Fig. 4. Sketch map showing a disposition of major thrust units similar to that of the leaves in a photographic camera iris. *Arrows* indicate local transport directions of thrusts

features that make it different from overlying allochthonous units it will be described in a separate chapter in this volume.

There are two sets of folds in the CZ, named transverse (or radial) and arched because of their orientation in relation to the Asturian Arc (Julivert and Marcos 1973). The arched set runs parallel to the thrust traces, whereas the radial set ist transverse to them. The existence of transverse folds allows analysis of the deep geometry of the thrust sheets, since, due to thrust surfaces tilting, a map view of the cross-sections can be observed in the eroded core of the antiforms.

The folds of the arched set are related to the thrust sheet geometry and can be interpreted as leading edge folds (Boyer 1986), dorsal and frontal culmination walls, etc. (Butler 1982). The folds of the radial set can be related to lateral structures of the thrust. Both fold sets underwent an important tightening after the emplacement of the unit or units to which they are related. The tightening of the radial folds was particularly intense in the Ponga Unit clearly visible through the sinuous trace of the thrust surfaces (Pérez Estaún et al. 1988; Alvarez Marrón and Pérez-Estaún 1988; Fig. 1).

3.1.3 Geometry of the Cantabrian Thrust Sheets

The thrust sheets in the CZ show the characteristic features of thin-skinned tectonics including a general ramp and flat staircase geometry as well as associated structures, especially folds. However, there is considerable geometrical variability among the different thrust associations that can be differentiated in detail in the CZ. Representative examples of these associations follow.

Esla Nappe Region. In the Esla nappe region (Fig. 5; Rupke 1965; Alonso 1987a, b), the development of three superimposed major thrust sheets and three duplexes have led to a large total shortening and complex geometry of tectonic superposition. An accumulated horizontal displacement of about 90 km has been inferred, as well as replication of part of the stratigraphic sequence to three times its original thickness. An important characteristic of this region is the presence of out-of-sequence thrusts which have complicated the geometry and kinematics of the duplexes. To accommodate the deformation produced by the development of out-of-sequence thrusts, an antiformal stack originated in the back part of the structure of the Esla nappe region. Its geometry can be compared, although on a smaller scale, to that of the Narcea antiform. This deformation also caused the formation of a synform (Aguasalio synform) between the antiformal stack and the frontal ramps of the nappes. Although not shown on the cross-sections presented here, a large number of folds originated from the accommodation of the thrust sheets to the hanging-wall ramps. These are commonly modified by simple shear induced by the displacement of the units.

Somiedo, Sobia and Aramo Thrust Sheets. These thrust sheets do not display a large number of stratigraphic repetitions (Evers 1967; Julivert et al. 1968; Marcos 1968a; van den Bosch 1969; Basti-

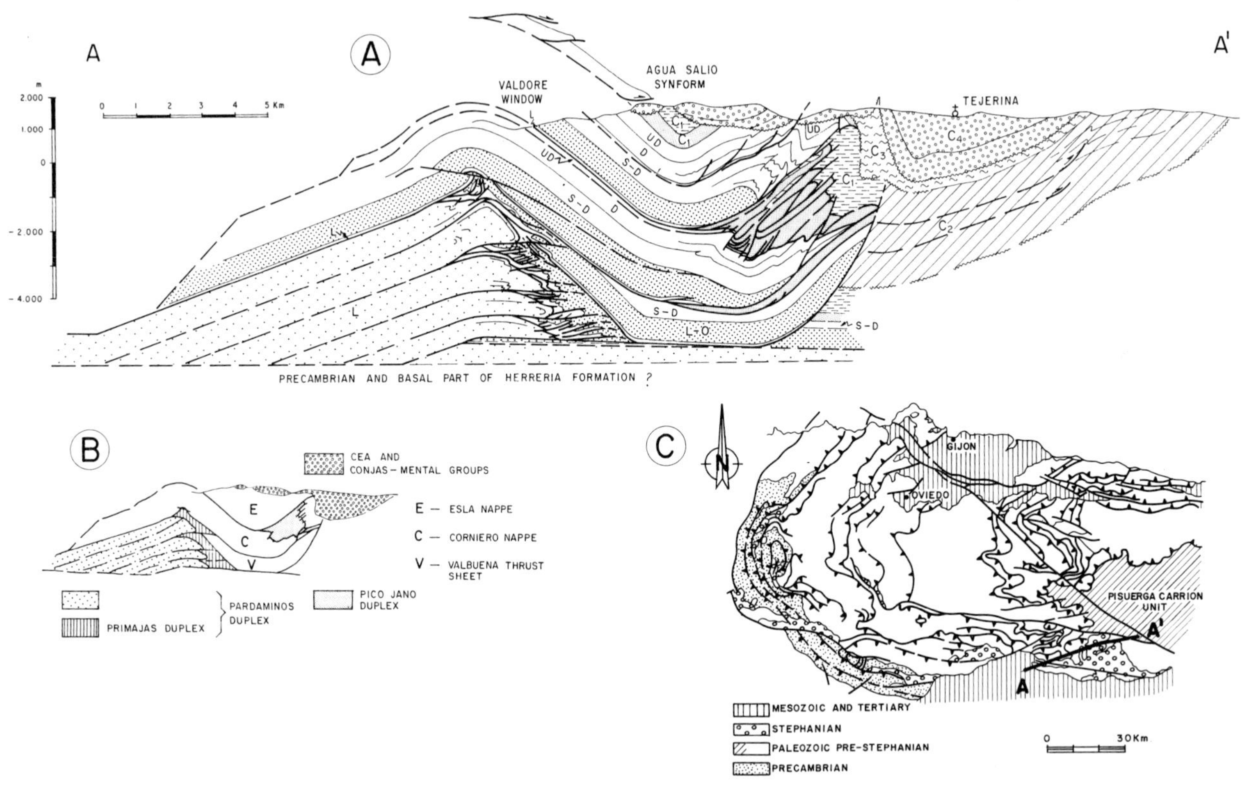

Fig. 5. A Geological cross-section through, the Esla Unit. **L** Cambrian; **L-O** Cambrian-Ordovician; **S-D** Silurian-Devonian; **D** Devonian; **UD** Upper Devonian; $\mathbf{C_1}$ Lower Carboniferous + Namurian; $\mathbf{C_2}$ Upper Carboniferous (Pando Group); $\mathbf{C_3}$ Upper Carboniferous (Conjas-Mental Group); $\mathbf{C_4}$ Upper Carboniferous (Cea Group). **B** Structural units destinguished in section **A. C** Cross-section location (After Alonso 1987a)

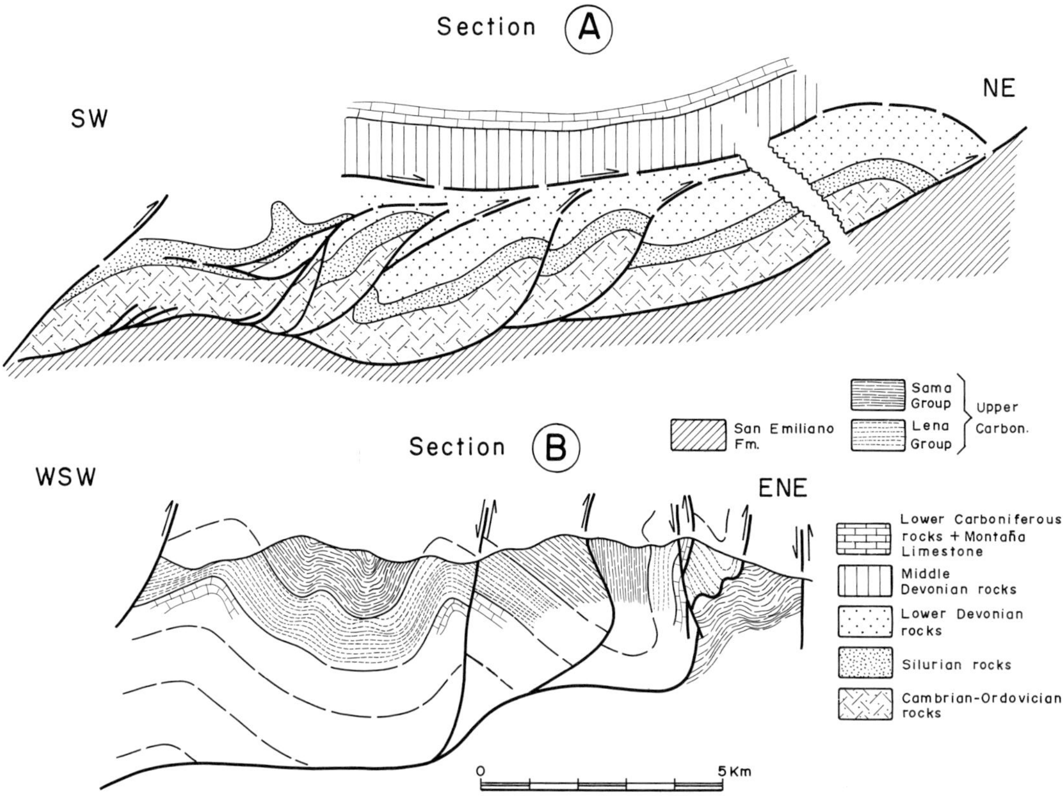

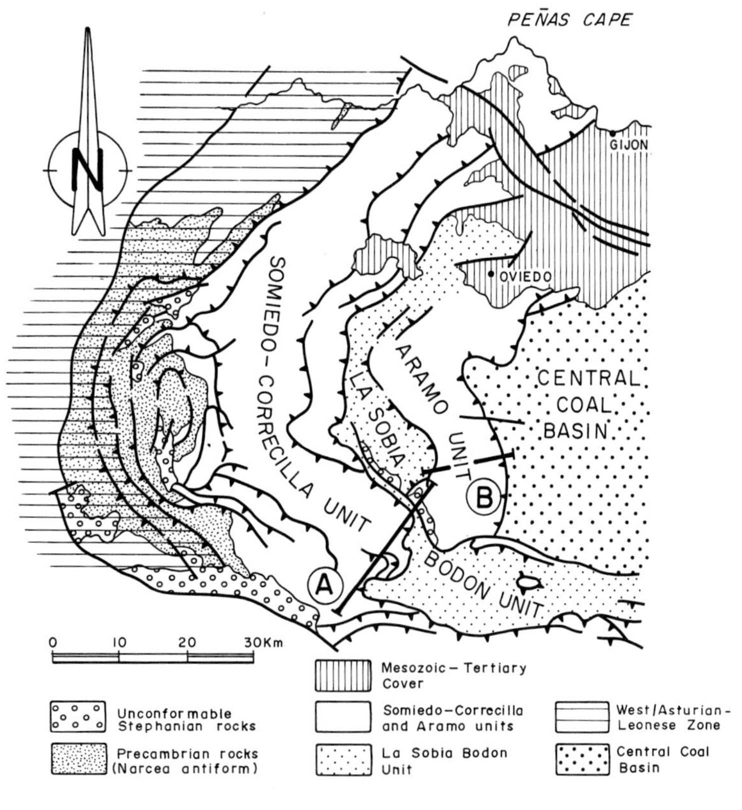

Fig. 6. **A** Geological cross-section through the Somiedo-Correcilla Unit (After Bastida et al. 1984) and **B** Aramo Unit (After Aller 1986)

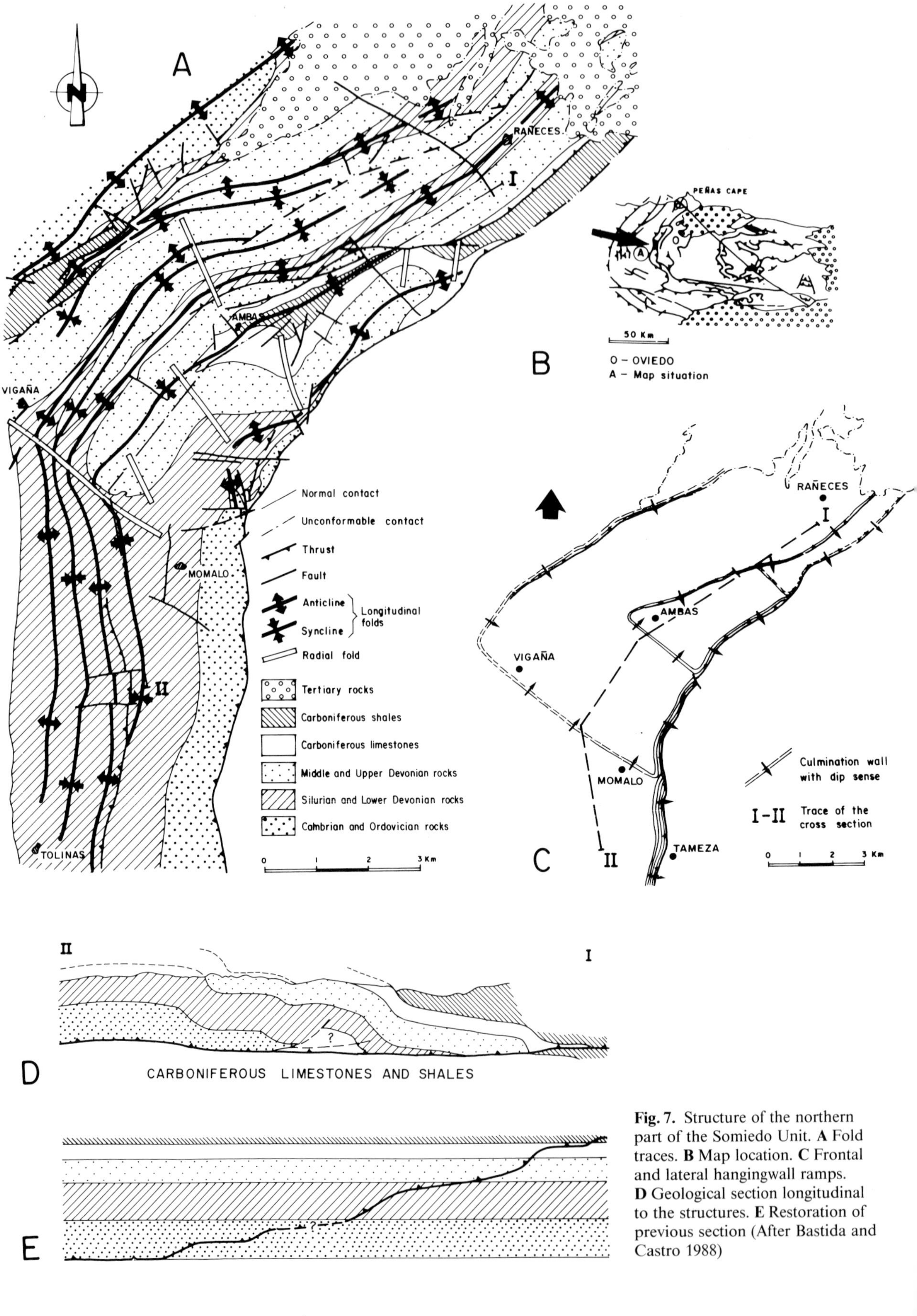

Fig. 7. Structure of the northern part of the Somiedo Unit. **A** Fold traces. **B** Map location. **C** Frontal and lateral hangingwall ramps. **D** Geological section longitudinal to the structures. **E** Restoration of previous section (After Bastida and Castro 1988)

da et al. 1984; Aller 1986). Generally, the structure is less complex than that of the Esla nappe region. In the Somiedo nappe a duplex develops within the lower part of its hangingwall (Fig. 6 A). Also important in this nappe is the great length and height of the footwall frontal ramp of the basal thrust. The position of the major folds in the Somiedo, Sobia and Aramo thrust units is controlled by the position of the ramps, while the size and geometry of the units is controlled by the distance between the ramps (Fig. 6 B). These rearmost (westernmost) thrust sheets show good examples of footwall and hangingwall lateral ramps, such as the footwall lateral ramps in the southern part of the Somiedo and Sobia nappes. The most remarkable effects of these ramps are the sudden structural truncation of the structures in the sector, in particular that of the Lagos syncline (Somiedo nappe, Fig. 6), and the dramatic change in the orientation shown by the map pattern of the thrust surface. Additional lateral structures also occur in the northern part of the Somiedo nappe (Bastida and Castro 1988) (Fig. 7), where hangingwall lateral ramps are separated by small flats that describe a staircase trajectory.

Central Coal Basin. This is a thrust sheet with a thick synorogenic Carboniferous sequence (about 5 km thick). Its map outcrop shows a cross-folding pattern which developed domes and basins (Fig. 8; Julivert and Marcos 1973; Aller 1986). The absence of remarkable thrusts within this unit makes the setting up of the relationships between thrusts and folds difficult. Often, the folds in the Central Coal Basin, especially in its eastern part, are nothing but a continuation of those already present in the easternmost unit (Ponga Unit). In some cases, they are the result of amplification and tightening suffered by the transverse fold related to the lower thrust sheets (Ponga Unit). In other cases, they may be folds overlying blind thrusts.

Ponga Unit. The Ponga Unit consists of a large number of individual thrust sheets (Julivert 1967a, b). Many of these sheets have accumulated signifi-

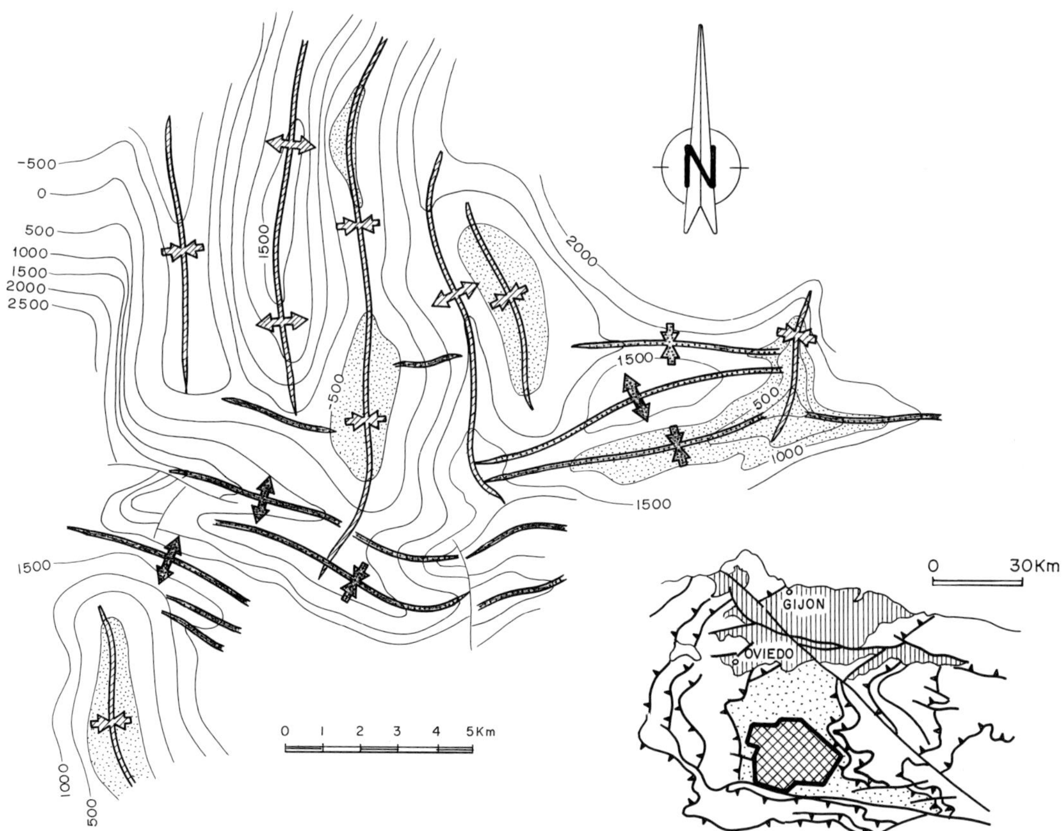

Fig. 8. Structural contours showing the interference figures in domes and basins formed by the superposition of arched and transverse folds (After Aller 1986)

cant displacements (Fig. 9). The accumulated displacement of this unit is about 62 km (Alvarez Marrón and Pérez Estaún 1988). In this unit, there are frequent out-of-sequence thrusts with a dominant southward transport direction, some extending into the adjacent unit (Picos de Europa Unit). Transverse folds are well developed in the Ponga Unit. These folds are mainly associated with lateral structures and significantly amplified after their initiation. The most outstanding examples of the lateral structures in the unit are in Río Monasterio and Río Color antiforms (Alvarez Marrón and Pérez Estaún 1988; Fig. 9) where they constitute ramps of the east-verging thrust system reactivated by later south-verging, out-of-sequence thrusts. In the same way, in the Mampodre region, an important tear fault, also reactivated, separates blocks having very different structures (Sjerp 1967; Pérez Estaún et al. 1988).

Picos de Europa Unit. This unit (Fig. 10; Maas 1974; Marquínez 1978; Farias 1982) is a typical imbricated thrust system in which the individual thrusts display a listric fault geometry without flat or intermediate ramps. The number of individual thrust sheets is large, but the shortening is less significant than in the above units. The accumulated displacement within this unit is of the order of 35 km. It is important to emphasize that in both the Ponga and Picos de Europa units, the deformation took place in a stratigraphic succession that is much thinner than in the other allochthonous units. The succession there is composed almost entirely of monotonous and massive limestones, as opposed to well-layered sequences with alternating lithologies in the other units. This probably accounts for the distinctive structure of the internal faults in this unit.

3.1.4 Faults

In addition to the faults that form the inner structure of each thrust unit, there are several faults that have very long traces and affect several units. Those in the southern part of the CZ have an E-W trace (e. g. the León and Sabero-Gordón faults) and affect Stephanian rocks. These faults have a long history of movements which implied initial strike-slip

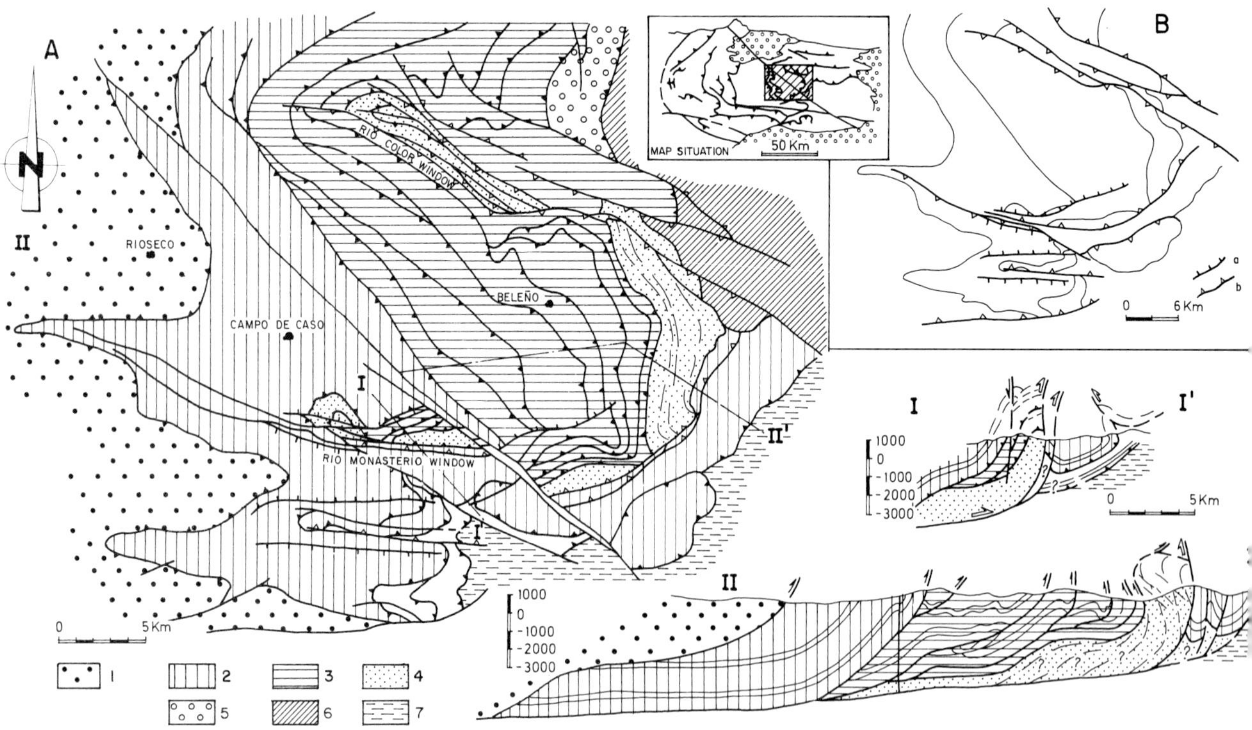

Fig. 9. The Ponga Unit. **A** Main thrust sheets: **1** Laviana and Rioseco thrust nappes; *2* Campo de Caso thrust nappe; *3* Espinaredo thrust nappe, Beleño imbricated system and Sebarga thrust nappe; *4* Los Beyos duplex; *5* Tornin Unit; *6* Picos de Europa Unit; *7* Pisuerga-Carrión or Palentine Unit. *I–I'* Geological cross-section through Río Monasterio window. *II–II'* Balanced cross-section. **B** Cartographic sketch of the main later transverse faults: *a* normal faults; *b* reverse faults (After Alvarez Marrón and Pérez Estaún 1988)

movements and later reactivations as reverse faults during the successive emplacements of the different nappes. Strike-slip components in the León and Sabero-Gordón faults were recognized by Marcos (1968b, 1979); Julivert et al. (1971); Bastida et al. (1976); Aller (1986) and Alonso (1987a), and dip-slip components are evident from the disposition of rocks and tectonic units in the maps.

In the northern part of the CZ, there are also a number of faults with an E-W direction, which rep-

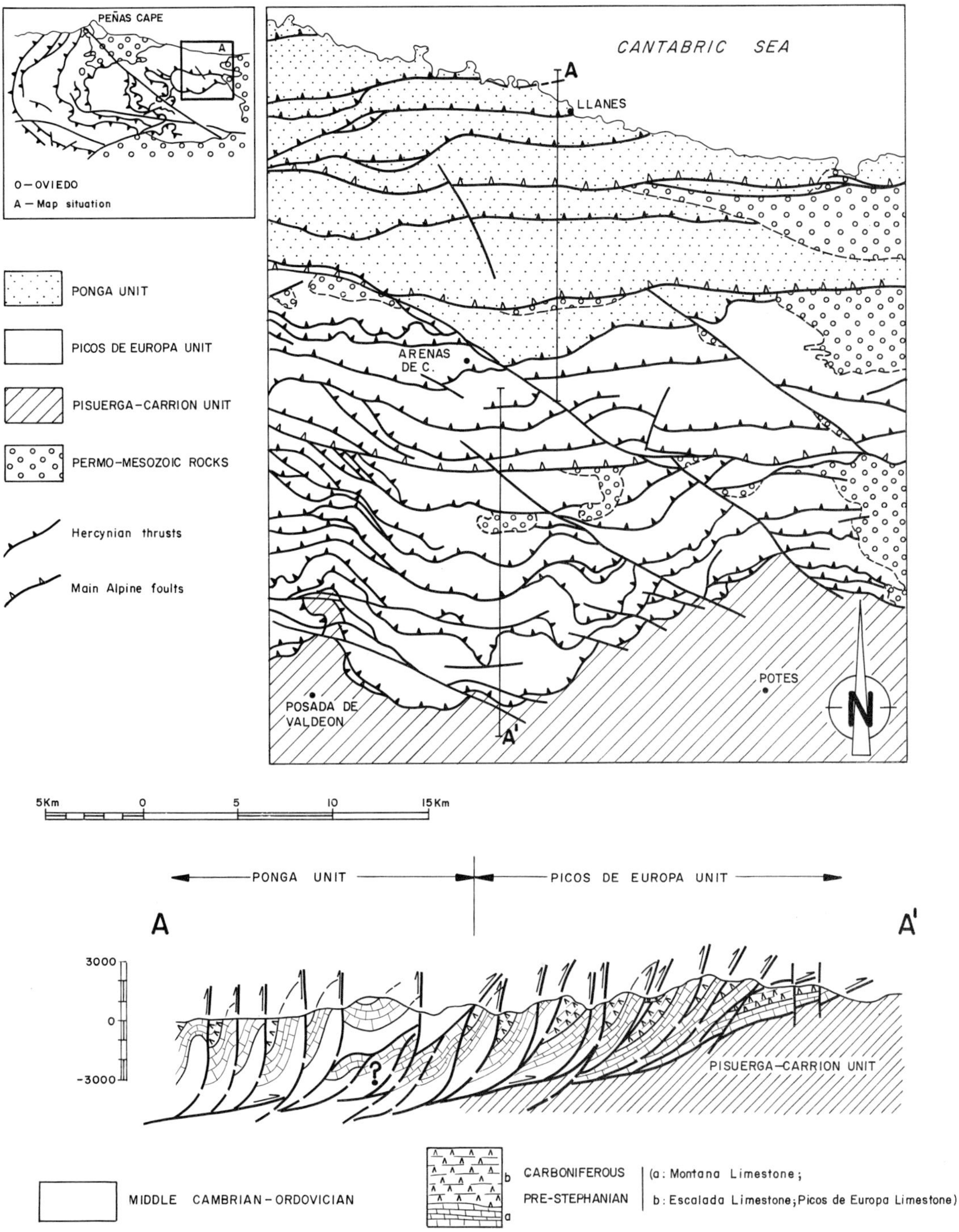

Fig. 10. Geological sketch map and cross-section of the Picos de Europa Unit and northeastern part of Ponga Unit (After Marquínez in press)

resent alpine reactivations of previous thrusts (Tosal 1968).

A particularly prominent fault is the Ventaniella fault, which crosses the CZ in a NW-SE direction (Fig. 1) and has been interpreted as a post-Variscan wrench fault (Julivert et al. 1971).

3.1.5 Cleavages

Although cleavage is not common in the CZ, some zones can be found wherein cleavage has been developed along with anchizone-epizone metamorphism (van der Pluijm and Kaars-Sijpesteijn 1984; Raven and van der Pluijm 1986; Aller et al. 1987; Fig. 11). One of these zones is the southern part of the Central Coal Basin, where cleavage and metamorphism were produced later than the major structures of the CZ. Here, they are probably related to a local thermal gradient. Another area where cleavage has been developed is the Pisuerga-Carrión Unit. The most widespread cleavage here seems to be coeval to that of the Central Coal Basin.

3.1.6 Structural Evolution of the Cantabrian Zone

In order to determine the structural evolution of the CZ, it is necessary to know the emplacement sequences and the emplacement direction of the different thrust sheets. The thrust faulting migrates, in a general way, from the hinterland towards the foreland, in a forward-type sequence. This can be demonstrated with a palinspastic restoration of the synorogenic sequence in an E-W section (Marcos and Pulgar 1982). Indeed, the presence of several synorogenic clastic wedges that relate to the formation of successive nappes progressively younging towards the east shows that the first units to be emplaced were the westernmost ones (Somiedo, Sobia and Aramo Units). The Central Coal Basin Unit followed and then the Ponga Unit. This sequence agrees with the data obtained from a study of the age of olistostrome sediments related to the emplacement of the nappes. The deposits associated with the rearmost units give Westphalian B ages (Arboleya 1981), Upper Moscovian ages in the Ponga Unit (Rodríguez Fernández and Heredia in press) and, in the Picos de Europa Unit, the olistostrome rocks are of Kassimovian age (Maas 1974; Marquínez 1978). The same conclusion is reached when analyzing the distribution of the orogenic fa-

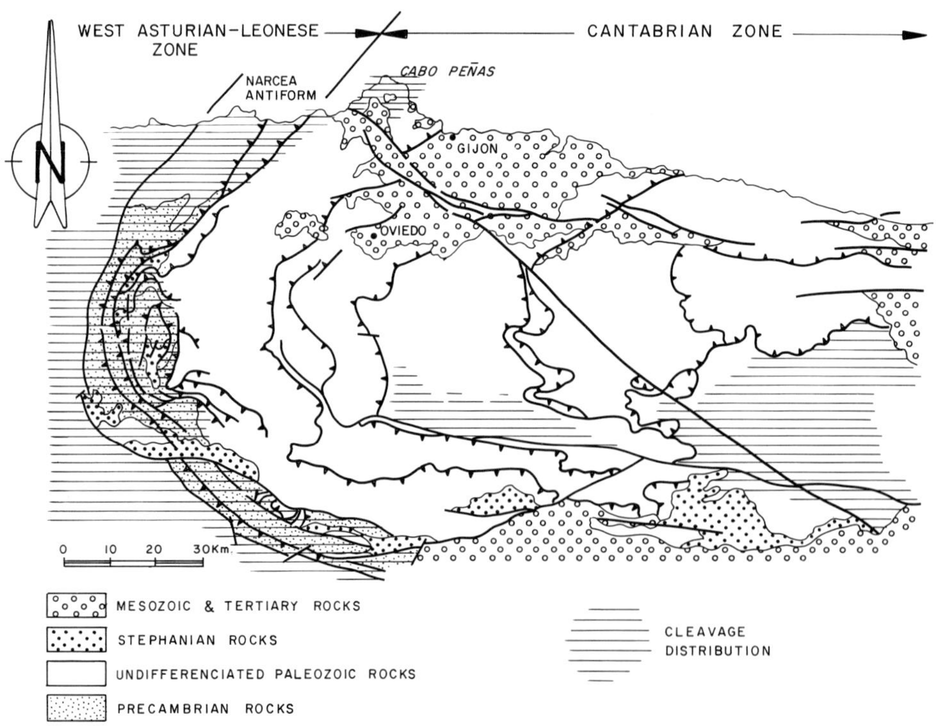

Fig. 11. Cleavage distribution in the Cantabrian Zone (After Raven and van der Pluijm 1986; Aller et al. 1987)

cies. Hence, for instance, rocks of Cantabrian age (Lowermost Stephanian) show continental facies unconformably over the thrust sheets already emplaced in the Esla Region, while they show synorogenic marine facies in the Picos de Europa Unit (Marquínez 1978). Similar relations can also be seen in the earlier Westphalian rocks.

Although the deformation sequence can be said to be a forward type, there are several anomalous structures that do not fit this pattern, such as faults that affect previously emplaced thrust sheets. Also, some thrust sheets were apparently folded by lower imbricated thrusts or duplexes, as in the Esla Region and the Narcea antiform.

Emplacement directions of the thrust sheets in the CZ have been deduced from kinematic indicators such as frontal and lateral elements of the nappes and, less often, from rocks and structures associated with the thrust surfaces (Arboleya 1981; Farias 1982; Bastida et al. 1984; Julivert and Arboleya 1986; Alonso 1987b; Bastida and Castro 1988; Alvarez Marrón and Pérez Estaún 1988). The displacement pattern, in its current configuration, presents the following features (Fig. 4):

1. The displacement vectors of all of the thrust sheets in the CZ show a distribution that verges towards the core of the Asturian Arc (centripetal orocline).
2. Small changes in the direction of the displacement vectors may occur between adjacent units and gradational variations may occur within the very same unit.

The disposition of the major thrust units is another feature that needs consideration in stating the structural evolution of the CZ. The thrust units display curving wedge shapes, which widen in a clockwise direction around the Asturian Arc and have a distribution not unlike that of the iris of a photographic camera.

To agree with these considerations, the emplacement of the nappes would have had to occur from

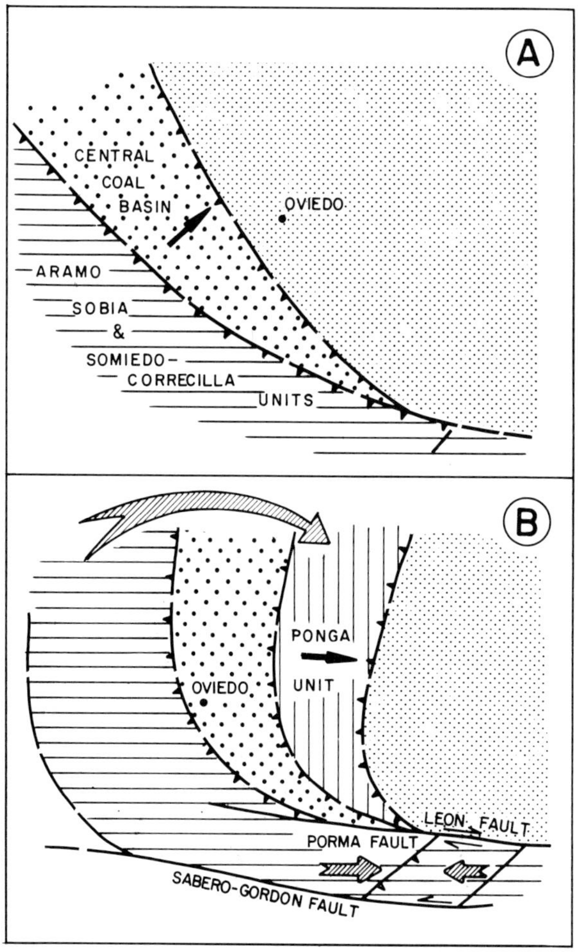

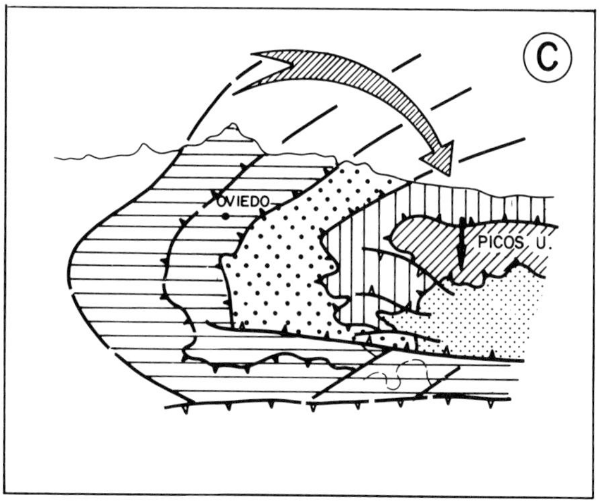

Fig. 12. Idealized scheme of the Asturian Arc development

different directions. Thus, the rearmost units would have displaced with a considerable northern component of movement. During the displacement of the Central Coal Basin and the Ponga Unit, an eastern component prevailed and, eventually, during the emplacement of the Picos de Europa Unit, a southern component predominated. This evolution is shown in Fig. 12, according to Pérez Estaún et al. (1988).

According to this displacement pattern, the emplacement of each thrust unit implies a shortening direction different from that of the previous nappes. An amplification of the previous folds and the reactivation and development of previous faults took place. This reactivation process was particularly strong when the Picos de Europa Unit was emplaced, dominantly towards the south. This caused the reactivation of the lateral structures of the Ponga Nappe as frontal structures in relation to the emplacement of the Picos de Europa (Alvarez Marrón and Pérez Estaún 1988). In the same way, at this stage in the evolution, there was a tightening and axial reorientation and overturning of the axial surfaces of frontal and lateral structures that already existed in the units of the southern branch of the CZ (Alonso 1987a). Furthermore, the large E-W faults in this same area were reactivated as reverse faults (León and Sabero-Gordón Faults). This kinematic model gave rise to a progressive rotation of the tectonic units and to the formation of the Asturian Arc, and is in agreement with the paleomagnetic data (Ries et al. 1980; Perroud 1982).

The evolution model outlined here solves the space problems raised by the development of a centripetal arc. Indeed, this model accounts for the whole structure of the CZ in terms of thin-skinned tectonics. An alternative model of buckling or bending, involving the basement, could not solve the space problems mentioned. Also, it would be difficult to explain the way in which the basement would fit the deformation taking into account the fact that the CZ is part of the foreland of the Variscan Orogeny.

Acknowledgements. This paper ist the result of continuous studies of the Cantabrian Zone by the structural geology group of Oviedo. We would like to thank colleagues at Oviedo. We also thank Ana Ojanguren and Chris Hepburn for the English version of the manuscript.

Bibliography and References for Part II, Chap. 2.5, Part II, Chap. 3.1 and Part II, Chap. 3.2

Adrichem Boogaert HA van (1965) Conodont-bearing formations of Devonian and Lower Carboniferous age in northern León and Palencia (Spain). Leidse Geol Meded 31: 165-178

Adrichem Boogaert HA van (1967) Devonian and Lower Carboniferous conodonts of the Cantabrian Mountains (Spain) and their stratigraphic application. Leidse Geol Meded 39: 129-192

Aller J (1986) La estructura del sector meridional de las unidades del Aramo y Cuenca Carbonífera Central. Principado de Asturias, Consejería de Industria y Comercio, 180 pp

Aller J, Bastida F, Brime C, Pérez Estaún A (1987) Cleavage and its relation with metamorphic grade in the Cantabrian Zone (Hercynian of north-west Spain). Sci Geol Bull 40: 255-272

Alonso JL (1982a) Las discordancias progresivas de cobertera carbonífera de Ocejo de la Peña: testigos de la reactivación de un pliegue del basamento por "flexural slip" (Cordillera Cantábrica). Bol Geol Min XCIII: 214-225

Alonso JL (1987a) Estructura y evolución tectonoestratigráfica de la región del Manto del Esla, (Zona Cantábrica, NO de España). Diputación Provincial de León. Inst Fray Bernardino de Sahagún, 275 pp

Alonso JL (1987b) Sequences of thrust and displacement in the superposed duplexes of the Esla Nappe Region (Cantabrian Zone, NW Spain). J Struct Geol 9 (8): 969-983

Alonso JL, Rodríguez Fernández LR (1983) Las discordancias carboníferas de la Región del Pisuerga-Carrión (Cordillera Cantábrica, NO de España). Significado orogénico. C R X Congr Int Strat Géol Carbonifère Madrid 1983, Inst Geol Min Esp, 3: 533-540

Alvarez Marrón J, Pérez Estaún A (1988) Thin skinned tectonics in the Ponga Region (Cantabrian Zone, NW Spain). Geol Rundschau 77 (2): 539-550

Ambrose T (1974) The Lower Paleozoic rocks of northern Palencia. Breviora Geol Astúrica XVIII (4): 49-53

Arboleya ML (1981) La estructura del Manto del Esla (Cordillera Cantábrica, León). Bol Geol Min XCII-(I): 19-40

Bastida F, Castro S (1988) Estructura de la prolongación septentrional de la Escama de Tameza (Zona Cantábrica, NW España). Trab Geol 17: 67-85

Bastida F, Marcos A, Arboleya ML, Mendez I (1976) La Unidad de Peña Corada y su relación con el Manto del Esla (Zona Cantábrica, NW de España). Breviora Geol Astúrica XX (4): 49-55

Bastida F, Marcos A, Pérez Estaún A, Pulgar JA (1984) Geometría y evolución estructural del Manto de Somiedo (Zona Cantábrica, NO España). Bol Geol Min XCV-(VI): 517-53

Boch W van den (1969) Geology of the Luna-Sil region. Cantabrian Mountains (NW Spain). Leidse Geol Meded 44: 137-225

Boyer SE (1986) Styles of folding within thrust sheets: examples from the Appalachian and Rocky Mountains of the U. S. A. and Canada. J Struct Geol 8 (3/4): 325-339

Brouwer A (1964) Deux facies dans le Devonien des Montagnes Cantabriques meridionales. Breviora Geol Astúrica, VIII (1/4): 3-10

Brouwer A, Ginkel AC van (1964) La succesion carbonifére dans la partie méridionale des Montagnes Cantabriques (Espagne du Nord-Ouest). CR V Congr Strat Géol Carbonifére Paris, 1963: 307-319

Budinger P, Kullman J (1984) Zur Frage von Sedimentationsunterbrechungen im Goniatiten und Conodonten führenden Oberdevon und Karbon des Kantabrischen Gebirges (N Spanien). Neues Jahrb Geol Paläontol Abh 7: 414-429

Butler RWH (1982) The terminology of structures in thrust belts. J Struct Geol 4: 239-245

Comte P (1959) Recherches sur les terrains anciens de la Cordillère Cantabrique. Mem Inst Geol Min Esp 60: 1-440

Evers HJ (1967) Geology of the Leonides between the rivers Bernesga and Porma (Cantabrian Mountains, NW Spain). Leidse Geol Meded 41: 83-151

Farias P (1982) La estructura del sector central de los Picos de Europa. Trab Geol 12: 63-72

Frankenfeld H (1983) El manto del Monto-Arauz: interpretación estructural de la Región del Pisuerga-Carrión (Zona Cantábrica, España): Trab Geol 13: 37-47

Ginkel AC van (1965) Carboniferous fusulinids from the Cantabrian Mountains (Spain). Leidse Geol Meded 34: 1-225

Graaf WJE van de (1971a) The upper Carboniferous, limestone rich, high destructive, delta systems with submarine fan deposits. Cantabrian Mountains, Spain. Leidse Geol Meded 46: 157-215

Graaf WJE van de (1971b) The Piedrasluengas limestone, a possible model of limestone facies distribution in the Carboniferous of the Cantabrian Mountains. Trab Geol 3: 151-159

Graaf WJE van de (1971c) Facies distribution and basic configuration in the Pisuerga area before the Leonian phase. Trab Geol 3: 161-177

Heward AP, Reading HG (1980) Deposits associated with a Hercynian to late Hercynian continental strike-slip system, Cantabrian Mountains, northern Spain. Spec Publ Int Assoc Sedimentol 4: 105-125

Higgins AC (1971) Conodont biostratigraphy of the late Devonian - early Carboniferous rocks of the south-central Cantabrian Cordillera. Trab Geol 3: 179-192

Higgins AC (1974) Conodont zonation of the lower Carboniferous of Spain and Portugal. Int Symp Belgian Micropaleontological limits. Namur 1974: 4, 14

Higgins AC, Wagner Gentis CHT, Wagner RH (1964) Basal Carboniferous strata in part of northern León, NW Spain. Stratigraphy, conodont and goniatite faunas. Bull Soc Belge Geol 72 (2): 205-248

Ivaniw E (1985) The sedimentology of lower Cantabrian basin margin deposits in NE León, Spain. In: Sousa MJL, Wagner RH (eds) Papers on the Carboniferous of the Iberian Península (sedimentology, stratigraphy, paleontology, tectonics and geochronology). An Fac Cienc Porto (Suppl) 64 (1983): 49-115

Jordan H, Bless MJM (1970) Nota preliminar sobre los ostrácodos de la Formación Vegamián. Breviora Geol Astúrica LIV (4): 37-44

Julivert M (1967a) La ventana tectónica del Rio Color y la prolongación spetentrional del Manto del Ponga. Trab Geol 1: 1-26

Julivert M (1967b) La ventana de Río Monasterio y la terminación meridional del Manto del Ponga. Trab Geol 1: 59-76

Julivert M (1971) Decollement tectonics in the Hercynian Cordillera of NW Spain. Am J Sci 270: 1-29

Julivert M (1979) A cross-section through the northern part of the Iberian Massif: its position within the Hercynian fold belt. Krystallinikum 14: 51-67

Julivert M (1981) A cross-section through the northern part of the Iberian Massif. Geol Minjbouw 60: 107-128

Julivert M (1983) La estructura de la Zona Cantábrica. In: Libro Jubilar JM Ríos, Geología de España. Inst Geol Min Esp 1: 339-380

Julivert M, Arboleya ML (1986) Areal balancing and estimate of areal reduction in a thin skinned fold and thrust belt (Cantabrian Zone, NW Spain): constraints on its emplacement mechanism. J Struct Geol 8: 407-414

Julivert M, Marcos A (1973) Superimposed folding under flexural conditions in the Cantabrian Zone (Hercynian Cordillera, NW Spain). Am J Sci 273: 353-375

Julivert M, Pello J, Fernández García L (1968) La estructura del Manto de Somiedo (Cordillera Cantábrica). Trab Geol 2: 1-44

Julivert M, Ramírez del Pozo J, Truyóls J (1971) Le réseau de failles et la couberture post-hercynienne dans les Asturies. In: Histoire structurale du Golfe de Gascogne. Publications de l'Institute Francais du Pétrole. Editions Tech 2: V.3.1-V.3.34

Julivert M, Marcos A, Pérez Estaún A (1977) La estructure de la chaîne hercynienne dans le secteur ibérique et l'arc Ibero-Armoricain. In: La Chaine Varisque d'Europe Moyenne et Occidentale. Coll Int Cent Nat Réch Sci Rennes 1974 (243): 429-440

Kanis J (1956)) Geology of the eastern zone of the Sierra del Brezo (Palencia, Spain). Leidse Geol Meded 21: 377-446

Koopmans BN (1962) The sedimentary and structural history of the Valsurvio Dome (Cantabrian Mountains, Spain). Leidse Geol Meded 26: 131-232

Kullman J (1961) Die goniatiten des Unterkarbons im Kantabrischen Gebirge (Nordspanien): I. Stratigraphie und Paläontologie der U. O. Goniatitina HYATT. Neues Jahrb Geol Paläontol Abh 113 (3): 219-326

Kullman J (1963) Las series Devónicas y del Carbonífero inferior con ammonoideos de la Cordillera Cantábrica. Estud Geol XIX: 161-169

Kullman J, Schönenberg R (1975) Geodynamische und Paläokologische Entwicklung im Kantabrischen Variszikum (Nordspanien). Ein Interdisziplinares Arbeitskonzept. Neues Jahrb Geol Paläontol Abh 13: 151-166

Lobato L (1977) Geología de los valles altos de los ríos Esla, Yuso, Carrión y Deva (Cordillera Cantábrica). Diputación Provincial de León. Instituto Fray Bernardino de Sahagún (CSIC), 200 pp

Loon AJ van (1970) Grading of matrix and pebble characteristic in syntectonic pebbly mudstone and associated conglomerates, with examples from the Carboniferous of northern Spain. Geol Mijnbouw 49 (1): 41-56

Loon AJ van (1971) The stratigraphy of the Westphalian C around Prioro (prov. Leon, Spain). Trab Geol 3 (I): 231-266

Loon AJ van (1972) A prograding deltaic complex in the Upper Carboniferous of the Cantabrian Mountains (Spain): the Prioro-Tejerina basin. Leidse Geol Meded 48: 1-81

Lotze F (1945) Zur Gliederung der varisziden der Iberischen Mesetas. Geotekt Forsch 6: 78-92 (Translation in: Publicaciones Extrajeras sobre Geología de España V: 149-166

Maas K (1974) The geology of Liébana, Cantabrian Mountains. Depositions and deformation in a flysch area. Leidse Geol Meded 49: 379-465

Maas K, Ginkel AC van (1983) Variscan olistotrome deposition and synsedimentary nappe emplacement, Valdeón Area, Cantabrian Mountains Spain. Leidse Geol Meded 52 (2): 341-381

Marcos A (1968a) La tectónica de la unidad de la Sobia-Bodón. Trab Geol 2: 59–84
Marcos A (1968b) Nota sobre el significado de la "León Line". Breviora Geol Astúrica 7: 1–5
Marcos A (1979) Facies differentiation caused by wrench deformation along a deep-seated fault system (León line, Cantabrian Mountains, north Spain). Discussion. Tectonophysics 60: 303–308
Marcos A, Marquínez J (1984) La estructura de la Unidad Gildar-Montó (Cordillera Cantábrica) Trab Geol 14: 53–64
Marcos A, Pulgar JA (1982) An approach to the tectonostratigraphic evolution of the Cantabrian Foreland thrust and fold belt, Hercynian Cordillera of NW Spain. Neues Jahrb Geol Paläontol Abh 163 (2): 256–260
Marquínez J (1978) Estudio geológico del sector SE de los Picos de Europa (Cordillera Cantábrica, NW de España). Trab Geol 10: 295–315
Marquinez J (in press) La estructura geológica en una sección transversal a los Picos de Europa y sector NE de la Unidad del Ponga. Trab Geol 18:
Marqínez J, Marcos A (1984) La estructura de la Unidad de Gildar-Montó (Cordillera Cantábrica). Trab Geol 14: 53–64
Martínez García E (1981) El Paleozoico de la Zona Cantábrica Oriental (Noroeste de España). Trab Geol 11: 95–127
Martínez García E, Rodríguez Fernández LR (1984) Mapa Geológico de España E. 1:50000, 2. Serie (MAGNA). Hoja 56 (Carreña-Cabrales). Inst Geol Min Esp, Madrid
Pérez Estaún A, Bastida F, Alonso JL et al. (1988) A thin-skinned tectonics model for an arcuate fold and thrust belt: the Cantabrian Zone (Variscan Ibero-Armorican Arc). Tectonics 7 (3): 517–537
Perroud H (1982) Contribution a l'étude paleomagnetique de l'arc Ibero-Armoricain. Bull Soc Geol Mineral Bretagne Ser C 15: 1–1114
Pluijm BA van der, Kaars-Sijpestijn CH (1984) Chlorite-mica aggregates: morphology, orientation, development and bearing on cleavage formation in very low-grade rocks. J Struct Geol 6: 399–407
Raven JGM, Pluijm BA van der (1986) Metamorphic fluids and transtension in the Cantabrian Mountains of northern Spain: an application of the conodont colour alteration index. Geol Mag 123 (6): 673–681
Reuther CD (1977) Das Namur im südlichen Kantabrischen Gebirge (Nordspanien). Krustenbewegungen und Faziesdifferenzierung im Übergang Geosynklinale-Orogen. Clausth Geol Abh 28: 122 pp
Reuther CD (1979) Tektonik und Oberkarbonische Faziesentwicklung des westlichen Valderrueda-Beckens und seiner Umrandung südliches Kantabrisches Gebirge, Nordspanien. Clausth Geol Abh 30: 45–61
Reuther CD (1982) The Lower Carboniferous facies levelling and the first Upper Carboniferous tectonic events in the Cantabrian Mountains and the Pyrenees (Spain). A comparison. Neues Jahrb Geol Paläontol Abh 163 (2): 244–249
Ries AC, Richardson A, Shackleton (1980) Rotation of the Iberian arc: palaeomagnetic results from north Spain. Earth Planet Sci Lett 70: 301–210
Rodríiguez Fernández LR (1983) Evolución estructural de la Zona Cantábrica durante el Carbonífero. In: Martinez Diaz C (coord) Carbonífero y Pérmico de España. Inst Geol Min Esp, pp 151–162
Rodríguez Fernández LR, Heredia N, Lobato L, Velando F (1985) Memoria de la Hoja del Mapa Geológico Nacional a escala 1:50000, 2. Serie (MAGNA) nº 106 (Camporredondo de Alba). Inst Geol Min Esp, Madrid
Rodríguez Fernández LR, Heredia N, Navarro D et al. (1986) Mapa Geológico Nacional E. 1:50000, 2. Serie (MAGNA) Hoja 82 (Tudanca). Inst Geol Min Esp, Madrid
Rodríguez Fernández LR, Heredia N (1987) La estratigrafía del Carbonífero y la estructura de la Región del Pisuerga-Carrión. Cuad Lab Xeol Laxe 12: 207–229
Rupke J (1965) The Esla nappe, Cantabrian Mountains (Spain). Leidse Geol Meded 32: 1–74
Sánchez de la Torre L, Agueda-Villar JA, Colmenero Navarro JR, García-Ramos JC, González Lastra J (1983) Evolución sedimentaria y paleogeográfica del Carbonífero de la Cordillera Cantábrica. In: Martinez Diaz (coord) Carbonífero y Pérmico de España. Inst Geol Min Esp, pp 133–150
Savage JF (1967) Tectonic analysis of Lechada and Curavacas Synclines, Yuso Basin, León, NW Spain. Leidse Geol Meded 39: 193–246
Savage JF (1979) The Hercynian orogeny in the Cantabrian Mountains. N Spain. Kristallinikum 14: 91–108
Savage JF (1980) Geotectonic cross sections through the Cantabrian Mountains, northern Spain. Geol Mijnbouw 81: 35
Schöler S (1982) Le Namurien et le Westphalien. A au Sud-est de la Cordillére Cantabrique. Neues Jahrb Geol Paläontol Abh 163 (2): 250–255
Sitter LU de (1961) Establecimiento de las épocas de los movimientos tectónicos durante el Paleozoico en el cinturón meridional del Orógeno Cántabro-astur. Not Comun Inst Geol Min Esp 61: 51–61
Sjerp N (1967) The geology of the San Isidro-Porma area (Cantabrian Mountains, Spain). Leidse Geol Meded 39: 55–128
Stockmans F, Williere Y (1965) Documents paléobotaniques pour l'etude du houiller dans le nord-ouest de l'Espagne. Mém Inst R Sci Natur Bélgique 2: 79
Tosal JM (1968) Relaciones zócalo-cobertera en el límite de las provincias de Oviedo y Santander. Breviora Geol Astúrica A 12: 9–14
Truyóls J (1983) El Carbonífero de la Región de Pliegues y Mantos. En Carbonífero y Pérmico de España. Inst Geol Min Esp, pp 39–59
Veen J van (1965) The tectonic and stratigraphic history of the Cardaño area, Cantabrian Mountains, northwest Spain. Leidse Geol Meded 35: 45–104
Wagner RH (1959) Flora fósil y estratigrafía del Carbonífero en España NW y Portugal N. Estud Geol XV: 398–420
Wagner RH (1960) Middle Westphalian floras from northern Palencia (Spain) (in relation with the Curavacas phase of folding). Estud Geol XVI (2): 55–92
Wagner RH (1964) Stephanian floras in NW Spain, with special reference to the Westphalian D – Stephanian A boundary. C R V Int Congr Strat Géol Carbonifère París 1963: 835–851
Wagner RH (1965) Palaeobotanical dating of Upper Carboniferous folding phases in NW Spain. Mem Inst Geol Min Esp 66: 1–169
Wagner RH (1970) An outline of the Carboniferous stratigraphy of north-western Spain. In: Streel M, Wagner RH (eds) Colloque sur la stratigraphie du Carbonifére). Congr Coll Univ Liége 55: 429–463
Wagner RH (1971) Carboniferous nappe structures in northwestern Palencia (Spain). Trab Geol 4: 431–459
Wagner RH, Bowman MBJ (1983) The position of the Bashkirian-Moscovian boundary in West European chronostratigraphy. Newsl Stratigr 12 (3): 132–161
Wagner RH, Martínez García E (1974) The relation between geosynclinal folding phases and foreland movements in northwest Spain. Stud Geol VII: 131–158

Wagner RH, Varker WJ (1971) The distribution and development of post-Leonian strata (Upper Westphalian D. Cantabrian, Stephanian A) in northern Palencia, Spain. The Carboniferous of northwest Spain. Trab Geol 4: 533-601

Wagner RH, Wagner Gentis CHT (1963) Summary of the stratigraphy of Upper Paleozoic rocks in NE Palencia, Spain. Prod K Ned Akad Wetensch Amsterdam LXVI (3): 149-163

Wagner RH, Winkler Prins CF (1979) The Lower Stephanian of Western Europe. CR VIII Congr Strat Géol Carbonifère Moscow 1975 (3): 11-140

Wagner RH, Winkler Prins CF (1985) Stratotypes of the two lower Stephanian stages, Cantabrian and Barruelian. C R X Congr Int Strat Géol Carbonifère, Madrid, 1983, vol 4, Working Group No. 2, IUGS Subcommission on Carboniferous Stratigraphy, pp 473-483

Wagner RH, Carballeira J, Aambrose T, Martínez García E (1984) Mapa Geológico Nacional E. 1:50000, Hoja 107 (Barruelo de Santullán). Inst Geol Min Esp, Madrid, 113 pp

Wagner RH, Park RK, Winkler Prins CF, Lys M (1977) The post-Leonian basin in Palencia: a report on the stratotype of the Cantabrian Stage. In: Holub VM, Wagner RH (eds) Symp Carboniferous stratigraphy. Geol Surv, Prague, pp 89-146

Wagner Gentis CHT (1962) Visean and Lower Namurian faunas in NW Spain (resumé). Breviora Geol Astúrica VI (1-4): 83-84

Wagner Gentis CHT (1980) Goniatites from the Visean-Namurian junction beds in Palencia, NW Spain. Scr Geol 55: 1-43

Wagner Gentis CHT (1985) A possible Atokan goniatite fauna from basal Moscovian (Westphalian A) deposits at Piedrasluengas, NW Spain. Senckenb Lethaea 65 (4/6): 337-345

3.2 Palentian Zone

L. R. Rodriguez Fernandez and N. Heredia[1]

3.2.1 Introduction

The Palentian Zone (Martínez García 1981) or Pisuerga-Carrión Unit (Julivert 1971)[2] forms the most external part (foreland basin) of the Hercynian-Iberian Massif, and is located in the inner part of the Asturian Arc (Fig. 1 in Part II, Chap. 1, p. 7).

Both the Palentian Zone (PZ) and the CZ are characterized by the predominance of thrust tectonics where deformation occurred under shallow crustal conditions.

However, the PZ and the CZ are different in some other aspects: First, the presence of allochthonous units containing Silurian-Devonian rocks with deeper facies than those of the CZ (Brouwer's Palentian facies 1964). These units have often been regarded as gravitational nappes (Wagner 1971), which would have either rooted in the PZ (Ambrose 1974; Savage 1979) or entered it as olistoplates originated in the south or southeast (Frankenfeld 1983; Alonso and Rodríguez Fernández 1983; Rodríguez Fernández et al. 1985, 1986; Marcos and Marquínez 1984); second, the abundance of synorogenic sediments and of unconformities, as described in the chapter on stratigraphy; and finally, in the PZ the metamorphic gradient is higher than in the CZ, although it never surpasses the lowest level of the greenschists facies. The presence of one or two penetrative cleavages, though not generalized, and the existence of a large amount of small Late Hercynian granites, usually located all along large faults, also characterize the PZ.

3.2.2 Tectonic Events of the PZ: Discussion and Suggested Patterns for its Structural Evolution

The relationship between tectonic and sedimentary events in the PZ has been widely discussed, especially the meaning of the unconformities. Some authors have considered the existence of several deformation phases with a regional character, marked by major unconformities.

The first structural evolution pattern was that of Sitter (1961), who, apart from the Bretonic deformation, took into account a series of stages: the Sudetic

[1] Instituto Tecnológico GeoMinero de España, Rios Rosas 23, Madrid, Spain

[2] For references, see Part II, Chap. 3.1

R. D. Dallmeyer and E. Martínez García (Eds.)
Pre-Mesozoic Geology of Iberia

stage, subdivided into two other stages (post-Namurian and post-Westphalian A), the Asturian (Upper Westphalian-Stephanian in the northern part of León and Stephanian A-B in the north of Palencia) and the Saalic stage (post-Stephanian and pre-Triassic). Wagner (1959, 1965) recognized three deformation phases called: Palentian (pre-Westphalian B), Leonian (pre-Upper Westphalian D) and Asturian (pre-Lower Stephanian B), which would be marked by three unconformed stratigraphic successions. Later works have correlated these unconformities with tectonic events and structures all through the northwest of the Iberian Peninsula (Wagner and Martínez García 1974; Wagner 1971; Ambrose 1972, 1974; Lobato 1977; Martínez García 1981).

The authors of the Dutch school (Savage 1967, 1979, 1980; Maas 1974; Maas and Van Ginkel 1983) interpret the heterogeneous structure of the PZ and CZ as formed under shallow crustal conditions with deformation usually caused by gravity. The development of folds and cleavage with a great spatial and temporal dispersion and the emplacement of allochthonous units, sometimes with divergent directions, were explained by the action of large fractures with vertical and wrench movements that affect the basement of the PZ (León Line, Cardaño Line, Peñas Matas Fault, etc; Fig. 2 in Part II, Chap. 2.5, p. 35). After Heward and Reading (1980) the localization of these structures is uncertain (except for the León Line) because of later deformation. For these authors, the structural history of the PZ (and of the CZ) is a succession of transpressive stages, with predominant strike-slip movements which would have occurred during the periods of maximum tectonic activity (Westphalian B, Upper Cantabrian) and transtensive with extensional vertical movements and initiation of quick subsidence basins. Kullman and Schönenberg (1975) also suggested patterns where the vertical tectonics would play a very important role. All these patterns are based on the current disposition of the Paleozoic sediment outcrops and on the faults that limit them without considering the important tectonic shortening caused by the thrust tectonics, which is of about 50% for the whole CZ (Marcos and Pulgar 1982).

More recent papers (Frankenfeld 1983; Rodríguez Fernández et al. 1985, 1986) consider the important role played by thrust tectonics and suggest models with the emplacement of large nappes with translations of many kilometers. These papers also take into account the influence of such thrust emplacement in the space and time distribution of the unconformities, synorogenic sedimentary facies and cleavage. These authors agree in pointing out the great importance of gravitational mechanisms for the emplacement of the nappes of the PZ. They also consider these nappes to have a southern origin, from an area situated to the south of the Folds and Nappes Region.

Recently, Rodríguez Fernández and Heredia (1987) have suggested a tectonosedimentary evolution model for the PZ which would be directly linked to the structural evolution of the CZ. For these authors, the centripetal character of the vergences in the CZ produced structures in the PZ that have an equally centripetal vergence. In the same way the emplacement of each allochthonous unit, or contemporary units, originated a synorogenic clastic wedge whose basis is usually unconformed. These unconformities weaken as they recede from the allochthonous units and eventually vanish. The centripetal character, stronger in the PZ, would account for the fact that one unit can overthrust, in time, with different directions and generate equally different clastic wedges, with depocenters located in several positions.

According to these authors, the first structural event was the emplacement of the Palentian Nappes (these units contain Devonian sediments with Palentine facies) with associated cleavage. The roots of these nappes are placed in the south, to the south of the Valsurvio Dome, and they verge towards the north or northeast. The basal part of the Potes Group and of the Perapertú Formation would be formed by the synorogenic sediment related to this structural event, which could dated back to the Namurian (Figs. 2, 3).

During the Namurian C and the Westphalian A, the Palentian Nappes are emplaced to the N or NW as gravity nappes in any case (Montó, Liébana, Fig. 2). The synorogenic sedimentation does not show any remarkable unconformities, except for the base of the Triollo Conglomerate (Fig. 3 in Part II, Chap. 2.5, p. 36).

The emplacement of the Esla and Valsurvio units to the NE (Alonso 1987a) and the last stage of the emplacement of the Palentian Nappes as individual units (Carrión, Montó and Liébana units; Fig. 2) is evidenced by thick unconformable synorogenic series (Carmen and Curavacas Fms.), from Upper Westphalian A or B, over the allochthonous units, or over the former synorogenic sediments. The erosion of the southern area of the PZ (Valsurvio and Esla) and of the very same PZ (Carrión unit) took place during the rest of the Westphalian at the same time as the deposit of the synorogenic units of this age (Curavacas, Lechada, Vegacerneja, Pandetrave, Vañes, Vergaño, Pando, Covarres, Mogrovejo, etc. Formations; Fig. 3).

A displacement towards the east of several units of the PZ occurred during the Cantabrian and Stephanian A ages in relation to the emplacement of the Ponga Nappe (Montó, Panda, Carrión, etc; see Figs. 2, 3). At the same time, some synorogenic series deposited, sometimes with a basal unconformity (Leonian unconformity).

During the Lower Stephanian B, the Picos de Europa Unit and some thrusts with S vergence and asymmetrical E-W fold associated were generated in the PZ (Fig. 4). Locally, cleavage developed associated with these folds. The Remoña Group (Fig. 3) is the synorogenic sediment of this age, and the Campollo Group would be its equivalent in the Liébana area related to the synsedimentary tightening of the Central Syncline at Liébana (Fig. 2). The late structural events in the PZ are represented by the genesis of folds of kilometric scale, with upright axial plane, E-W axis and associated crenulation cleavage with irregular distribution.

The intrusion of granodioritic and granitic igneous rocks (Peña Prieta, Iján, Jano, etc.) were related to transtensive and transpressive episodes included in the tardihercynian evolution of the Iberian Massif.

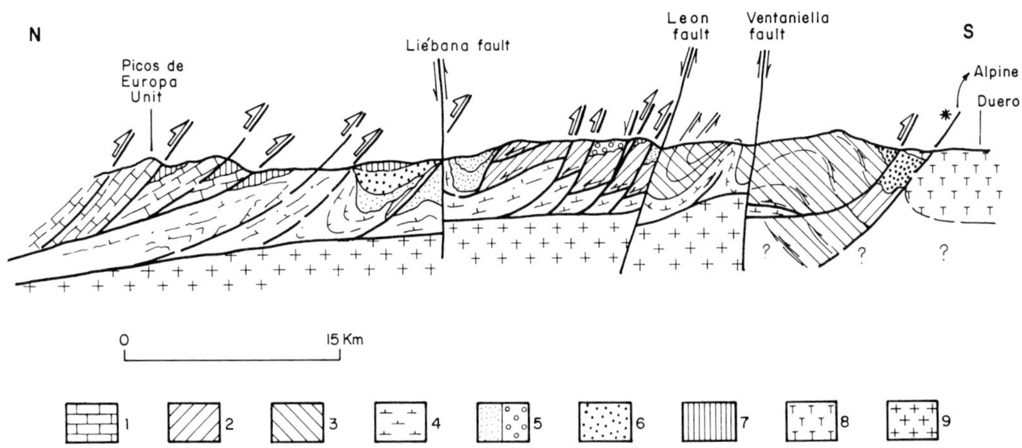

Fig. 4. Geological cross-section through the Palentian Zone (or Pisuerga-Carrión Unit). *1* Picos de Europa Unit; *2* Carrión Unit; *3* Valsurvio Unit; *4* Potes Group and Perapertú Fm. (Bashkirian); *5* Carmen, Curavacas, Lechada and Morgrovejo Fms. (Moscovian); *6* Cantabrian and Stephanian A (Barruelian); *7* Remoña Fm. (Stephanian B); *8* Tertiary deposits of Duero Basian; *9* Basement

4 Igneous Rocks

L. G. Corretge and O. Suarez[1]

The Cantabrian Zone constitutes a geological domain with a qualitatively important igneous activity. In this region plutonic bodies are scarce but volcanics are abundant in Cambrian-Upper Devonian and Permian horizons. Volcanic rocks are also found within the Narcea (Precambrian) Formation, near its limit to the West Asturian-Leonese Zone. In the Cantabrian Zone, numerous small postkinematic stocks also occur which show a variety of characteristics. The most important area for these stocks is the Palentian Zone, where more than 200 stocks and plugs have been catalogued (Corretgé et al. 1987). Research on the igneous acitivity in this area is still scarce in spite of the considerable number of articles which refer to partial aspects of the igneous and regional rocks (Frets 1965; Rupke 1965; de Sitter and Boschma 1966; van Veen 1966; Savage 1967; Suárez and Marcos 1967; Boschma 1968; Savage and Boschma 1980; Suárez 1970, unpublished work; Suárez and García 1974; Loeschke 1982, 1983; Loeschke and Zeidler 1982; Martínez García 1981, 1984; Heinz et al. 1985; Corretgé et al. 1987; Suárez and Corretgé 1987).

4.1 Volcanic Rocks

Paleozoic volcanic rocks of the Cantabrian Zone belong to rather differentiated petrographical and geochemical types (Parga 1969 and 1969, unpublished work). There is an alkaline association ranging in composition from olivine-basalts to trachybasalts-trachytes, which occurs as sills, lava flows, volcanic necks and tuffaceous layers mainly between the inner part of the Narcea Precambrian rocks and the Central Coal Basin. There is also a tholeiitic association composed of basalts, dacites and latites occurring either as tuffs or lavas of identical composition. The occurrence of the tholeiitic association is restricted to the west margin of this zone (Fig. 1). Lava flows, sills and volcanoclastic rocks of the alkaline association are particularly abundant in Mid- to Upper Cambrian and Lower Ordovician, whereas alkaline volcanic necks and pyroclastic rocks are mainly found in Ordovician and Silurian formations. The kaolin tonstein, derived from volcanic ash falls, occurs in Lower Ordovician strata (García Ramos et al. 1984) and proves the existence of pyroclastic volcanism within the alkaline association. According to the data provided by Loeschke (1983, 1986), alkaline volcanic activity extended into Upper Silurian times, since a Rb/Sr age obtained for the Cerrecedo sill is 399 ± 5 Ma.

Tholeiitic volcanism started during the Middle Ordovician, since some andesitic to basaltic lava flows have been found in the lower part of the Middle Ordovician. A period without volcanic activity subsequently followed during most of the Middle Ordovician. Finally, there was a new pulse of important igneous activity during Late Ordovician times; in this interval a thick volcanoclastic sequence was deposited mainly in the Cabo Peñas and the close area of Cabo Vidrias.

The latest Pre-Permian volcanic episodes could have taken place during the Early Carboniferous since kaolinite-sudoite rich clay layers in these rocks are interpreted as having been derived from volcanic ash of basaltic composition. However, volcanic events which could have produced these layers are unknown in the Cantabrian Mountains (Loeschke 1983).

Finally, igneous activity has been present along the Permian. Despite alteration processes, which produce important petrographical and geochemical transformation, the alkaline character of this vulcanism has been suggested by some authors (Martínez García and Tejerina 1985).

4.1.1 Pre-Permian Alkaline Association

This association shows a considerable geographical dispersion in the Cantabrian Zone. It is mainly pre-

[1] Departamento de Geologia, Universidad de Oviedo, 33080 Oviedo, Spain

R. D. Dallmeyer and E. Martínez García (Eds.)
Pre-Mesozoic Geology of Iberia

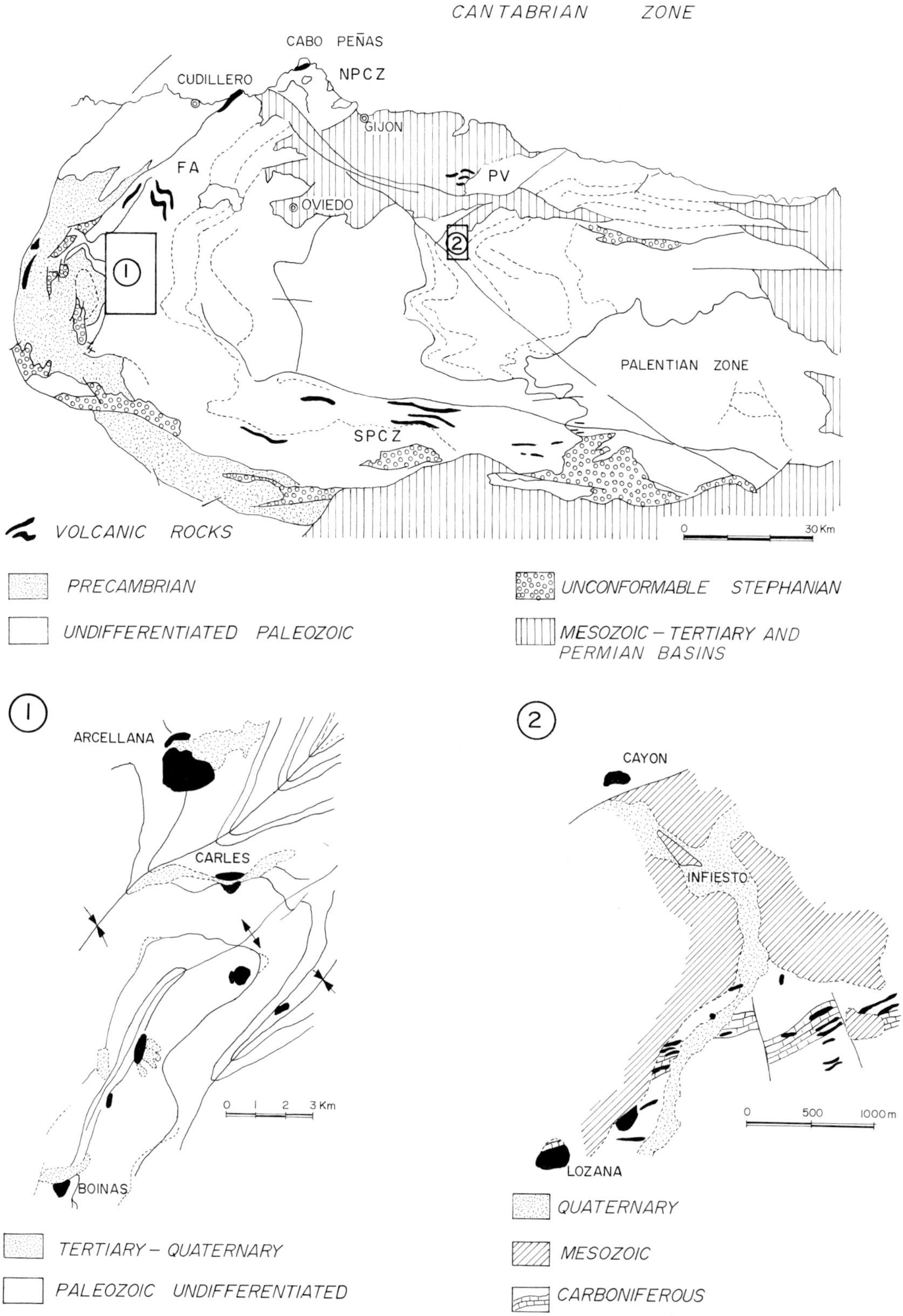

Fig. 1. Distribution of the main volcanic rocks in the Cantabrian and Palentian Zones. *1* Detail of the intrusive rocks of Salas-Belmonte; *2* detail of the intrusive and subvolcanic rocks of the Infiesto group

sent in the southern part of the Cantabrian Zone (SPCZ, Fig. 1) where volcanic rocks have been studied more thoroughly (Loeschke and Zeidler 1982; Loeschke 1983; Heinz et al. 1985). Volcanic rocks belonging to the alkaline series show a wide compositional range from more abundant basic to scarce acidic rocks. They are mostly alkali-basalts, trachybasalts, trachytes-andesites, trachytes and abundant tuffs related to phreatomagmatic volcanic activity (Heinz et al. op. cit.). These rocks are characterized by intersertal and ophitic textures, and contain basic plagioclase, clinopyroxene (diopsidic augite and titanaugite), titanobiotite, olivine, nepheline and, more rarely, potash feldspar and albite as primary minerals. The most frequent accessory minerals are titanomagnetite, apatite and zircon. They have undergone intense secondary alteration processes, which include saussuritization and the formation of important amounts of sericite, chlorite, carbonate and albite with leucoxene, limonite and large apatite crystals. Most commonly, their average composition is similar to trachybasalts but they are especially rich in K, Ti, P and Zr. This Lower Ordovician volcanism may be related to a rifting process in the Early Paleozoic (Loeschke and Zeidler 1982; Heinz et al. 1985).

In the northwestern Cantabrain Zone volcanism is less abundant (Farandón area FA; Fig. 1) and not very well known. Nevertheless, olivine-basalts and trachytes have been reported in this area (García de Figuerola and Parga Pondal 1964; Parga 1969 and 1969, unpublished work).

4.1.2 Pre-Permian Tholeiitic Association

The best-known volcanic rocks of this association are situated in the northern part of Cantabrian Zone (NPCZ) in the Cabo Peñas area (Fernández Secades 1975). They constitute a volcanoclastic sequence characterized by the association of andesites (andesite-basalts), dacites, quartz-latites and tuffs. They have diabasic and intersertal textures, often with glass and/or variolitic masses of chlorite. These rocks have suffered hydrothermal alteration to such an extent that their primary mineralogy, consisting of plagioclase, pyroxene, biotite and titanomagnetite, is seldom observable. Chlorite (as vesicles and varioles), carbonate and sericite are common as secondary minerals; quartz, serpentine, leucoxene and limonite are also present. On the basis of their chemistry, these rocks correspond to tholeiitic basalts (Fernández Secades 1975).

Some basaltic sills (of Devonian age) in the southern part of the Cantabrian Zone have P, Ti and K_2O contents that show a stronger affinity with tholeiitic basalts than with alkaline basalts. In Loeschke's (1983) point of view "the occurrence of these sills is seen in connection with a more advanced rifting which affected the continental crust of the Cantabrian Mountains, probably during Devonian time".

4.1.3 Permian Volcanic Rocks

Volcanoclastic rocks are predominant in the Permian (PV, Fig. 1) but in some areas (Prado 1972; Martínez García and Tejerina 1985; Suárez Rodríguez 1988) lava flows are also common. These rocks have microphyric textures and are commonly hydrothermally altered with plagioclase (andesine-labradorite), clinopyroxene, olivine and biotite phenocrysts pseudomorphized by secondary minerals. Plagioclase and potash feldspar microlites are the main components of the groundmass. These rocks can be classified as andesites, trachyandesites, leucobasalts, olivine-basalts and picrite-basalts on the basis of petrography, but blurring of primary igneous textures and mineralogy by hydrothermal processes has made a precise classification difficult in many cases. On the basis of geochemistry, they also vary widely (Martínez García and Tejerina 1985). Using the Ritmann norm, they can be classified as mugearites, rhyodacites, quartz-trachytes, quartz-hawaiites and quartz latite-andesites.

The nature of Permian volcanism is not well known, although its alkaline character (which has been increased by alteration processes) is suggested by some of the Miyashiro (1975) diagrams. This argument, based on geochemical aspects, is very weak, because quartz and hypersthene normative rocks are abundant in this series and also because the petrography of these rocks indicates a subalkaline (calc-alkaline or tholeiitic) character. In fact, other Permian rocks elsewhere, for instance, the Permian-Carboniferous rocks from the Iberian Chain, showing the same petrography, are K-rich calc-alkaline, and have been interpreted as compressional (Muñoz et al. 1985). In contrast, Martínez García and Tejerina (1985) suggest that the alkaline character is possibly related to a continental rifting process. This double possibility of interpreting the series (subalkaline-alkaline) could be explained by its possible dual character showing well-defined "Coomb trend" features. Nevertheless, considering the absence of fresh rock analyses and data of immobile components, the possible existence of a K-rich calc-alkaline potasium series should not be dismissed.

4.2 Plutonic Rocks

Plutonic rocks may be grouped as follows: Palentian granitoids, Infiesto igneous rocks and Salas-Belmonte granitoids (Fig. 1). The term granitoid is used here for a wide group of rocks ranging in composition from leucogabbros to granites s. s.

4.2.1 Palentian Granitoids

From a structural point of view, the granitoids of this zone seem to be related to faults or geological lineations of considerable importance (Fig. 2). Four groups have been established in more recent studies (Corretgé et al. 1987; Suárez and Corretgé 1987):

1. Granitic rocks of the Pernía Group (G-1)
2. Igneous rocks related to the León Fault (G-2)
3. Intrusive rocks of the La Reina-Riaño area (G-3)
4. Stocks of Pico Iján, Pico Jano and Peña Prieta.

Stocks of group four are not spatially related to the others, but from a structural point of view they seem to be related to the other groups, particularly with the G-1 granitoids, with which they have the strongest petrographical affinities.

4.2.1.1 Granitic Rocks of the La Pernia Group

These granitic rocks occur as stocks and minor intrusions and also as porphyritic dykes of dioritic to monzogabbroic composition. Most of these rocks have hypidiomorphic and microphyric textures. Granodiorites are true cumulates of plagioclase (An_{58-35}) and femic minerals: biotite ($Fe/Fe+Mg=0.3-0.6$), clinopyroxenes (salite and augite), hypersthene and amphibole. Quartz and potash feldspar are always intercumulus phases.

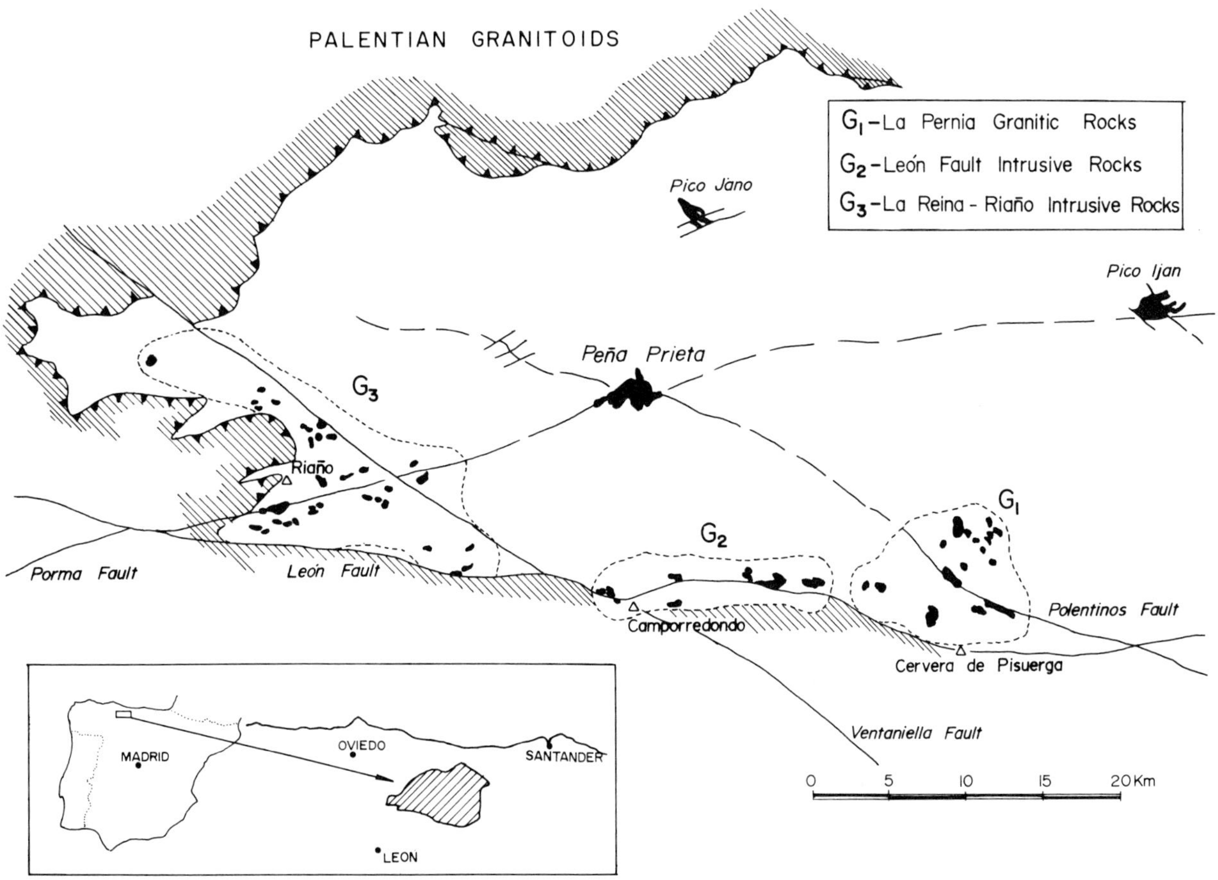

Fig. 2. Distribution of the Palentian granitoids (intrusive and subvolcanic rocks)

Among the different biotite types, late biotite is very idiomorphic and may replace pyroxenes. Common accessory minerals are: apatite, monazite, zircon, sphene, allanite, magnetite and sulfides. Among the secondary minerals, the most important are chlorite, sericite, calcite, clinozoisite-epidote and prehnite and pumpellyite as by-products of biotite alteration. Garnet, spinel, corundum, cordierite, xenoliths (of millimeter to centimeter size) of sedimentary origin either hornfelsic (biotite-rich) or restitic (with green spinel ± corundum) and fine-grained enclaves of dioritic composition can also be present.

4.2.1.2 Igneous Rocks Related to the León Fault

Sills of gabbrodioritic to nearly gabbroic composition are related to the León Fault. However, their original composition has been modified by tectonic and hydrothermal processes. The least altered rocks are hornblende-rich gabbros (although they plot near the compositional field of diorites); their texture is hypidiomorphic, sometimes subophitic or porphyritic. Plagioclase ranges in composition from An_{54} to An_{48} and amphiboles plot in the magnesiohastingsite and pargasite fields. Pyroxenes are present as a relict phase, enclosed in and replaced by amphibole. These rocks show, in most of their outcrops, a rather high content of secondary minerals: calcite, chlorite, sericite, muscovite, sulfides, limonite, goethite, rutile, leucoxene, epidote and talc, all generated as a consequence of propylitic and sericitic alteration processes (Meyer and Hemley 1967).

4.2.1.3 Intrusive Rocks of the La Reina-Riaño Area

Most of these rocks are dykes or sills of variable thickness (ranging from a few centimetres to hundreds of metres thick), and are oriented along the trends of the main fracture systems. These rocks are petrographically variable, ranging from gabbros (essentially of hornblende type) to quartz-diorites. Like the rocks of the former group, they are affected by hydrothermal alteration processes. The gabbros from this group are similar to those of group G-2, both in their mineralogy and alteration processes; their amphibole is commonly magnesio-hastingsite, although the composition of some crystals plots in the tschermakitic-hornblende field. It is important to mention the presence of biotite, as a distinguishing feature, which shows some similarities to the late biotites present in group G-1.

4.2.1.4 Stocks of Pico Ijan, Pico Jano and Peña Prieta

These outcrops form the largest plutons of the Palentian Zone and intrude Carboniferous and sometimes Permian (Pico Ijan) formations with the development of important contact metamorphism. The aureole of the Peña Prieta stock is characterized by an abundance of biotite and large chiastolite porphyroblasts and chloritoid, although the latter is not abundant. The rocks are granodioritic or dioritic in composition (Fig. 3). Suárez and García (op. cit.) distinguish two facies in the Peña Prieta stock (an intrusion with a great structural complexity, frequently featuring associated dykes). The common facies is a medium grained biotite-amphibole granodiorite. Plagioclase, in this granodiorite, ranges in composition from An_{28} to An_{36} and shows strong irregular zonation; the other two main phases are quartz and potash feldspar. The second facies is a border facies (similar to some associated dykes) and is richer in plagioclase and amphibole. Its composition is dioritic with microphyric texture.

Palentian Granitoids and Related Mineralization. Palentian granitoids frequently develop important Cu sulfides in their contact zones (Martín Izard et al. 1986). In the first group are As-Cu and Bi-Au mineralizations associated with amphibole-garnet skarns (González Montero 1986, unpublished work; Corretgé et al. 1988) or sometimes with fault breccias. In the second group, mineralizations are scarce and not important, except for a small skarn with Cu-As mineralizations. Finally, there are As-Sb with minor Au mineralizations (Paniagua et al.

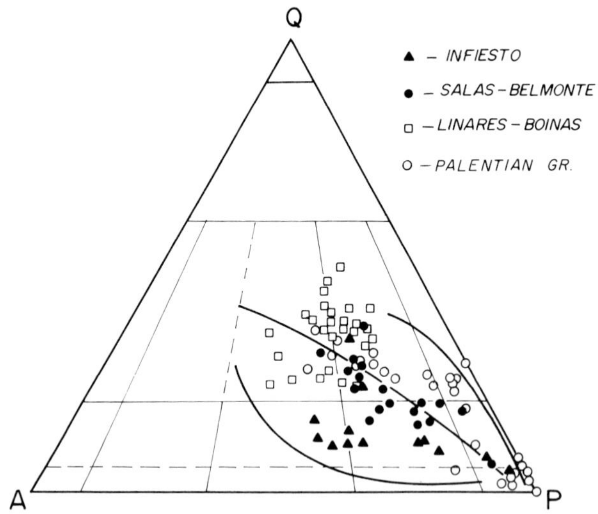

Fig. 3. Q-A-P diagram of the main four groups of granitoids of the Cantabrian and Palentian Zones

1988) associated with the G-3 stocks. Here they are related to the granodioritic or gabbroic facies or to small shear zones in areas affected by hydrothermal alteration.

4.2.2 Infiesto Group

The Infiesto group (Fig. 1) is composed of 20 outcrops which occur as dykes or sills of small thickness, and small stocks (Suárez and Marcos 1967). These rocks have developed a wide zone of contact metamorphism, which contrasts with the small size of the intrusive bodies. García Iglesias et al. (1979) have suggested that these intrusive bodies might be the apophysis of a large, deeper intrusive complex.

These rocks have been petrographically classified as porphyries, granodiorites and different kinds of gabbros (Fig. 3). Gabbros are the most abundant group and show a remarkable textural and mineralogical diversity. Some of these gabbros are quartz-monzogabbros with a hypidiomorphic texture and are composed of plagioclase (An_{42}-An_{60}), biotite, amphibole, and smaller amounts of quartz and potash feldspar. The microphyric varieties have microphenocrysts of plagioclase (up to An_{85}), brown hornblende and pyroxenes; they also contain olivine, frequently with embayments, as inclusions in clinopyroxene, amphibole or biotite.

The contact metamorphism associated with these igneous intrusives reaches the lower limit of the pyroxene-hornfels facies (580–610 °C). Skarn rocks are often present, being characterized by the abundance of clinopyroxene, amphibole, garnet, vesubianite and wollastonite. Complex sulfide-rich mineralizations showing Ag anomalies and traces of Au occur in association with the skarn paragenesis. These mineralizations have been formed at temperatures ranging from 150° to 300 °C (García Iglesias et al. 1979).

4.2.3 Salas-Belmonte Group

This group is composed of nine stocks with the total exposure surface not exceeding 5 km². They have intruded rocks ranging in age from Early Cambrian to Early Devonian. It has been observed that these intrusions couple with the anticline and syncline structures of the zone (Fig. 1).

These rocks show wide mineralogical and textural variety, ranging from granodiorites to gabbros (Fig. 3). Their texture is hypidiomorphic and they are composed of plagioclase (37–60%), orthopyroxene, clinopyroxene, biotite, amphibole, quartz and potash feldspar. Plagioclases have idiomorphic forms and are characterized by the existence of synneusis and strong oscillatory or patchy zoning. Orthopyroxene is more abundant than clinopyroxene and occurs either in gabbros or granodiorites (Corretgé et al. 1970; Suárez 1970, unpublished work), being enclosed and partially replaced by biotite and amphibole.

In the same region a quite different intrusive body is mapped: the Boinás Granite, which together with the Linares Granite (intruded into the Precambrian rocks of the Narcea Formation) are typical porphyritic biotite-granites (Suárez 1971). They are characterized by the presence of potash feldspar

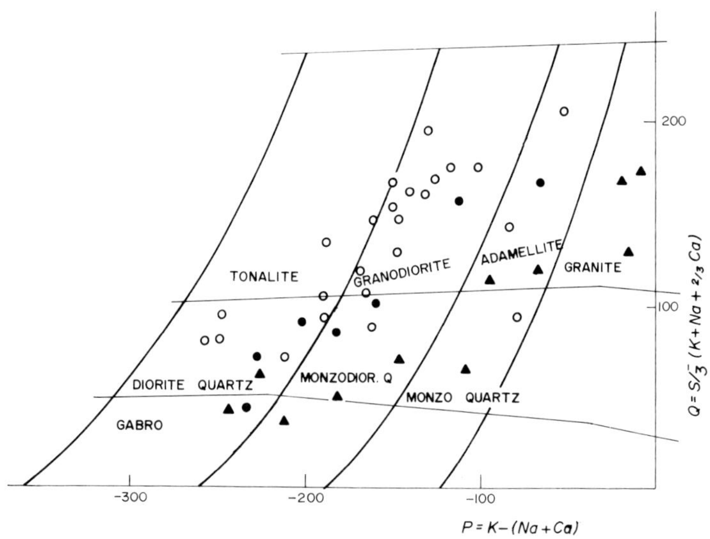

Fig. 4. P-Q diagram (La Roche 1972) of the granitoids of the Cantabrian and Palentian Zones. (symbols as in Fig. 3)

megacrysts and rounded quartz phenocrysts as well as zoned zircons and scarce molybdenite mineralization.

Within the metamorphic aureole generated by the intrusion of these rocks (which reaches the hornblende-hornfels facies) some skarns occur, characterized by parageneses especially rich in garnet, pyroxene, amphibole and Cu-sulfide. Bi-Au mineralizations are also present (Loredo and García Iglesias 1988).

4.2.4 Some Geochemical and Petrogenetic Aspects of Granitoids

Plotting of rock modal analyses of considered intrusive igneous rocks shows a wide range of compositions (Fig. 3). From the geochemical point of view (Q-P diagram of La Roche 1972, Fig. 4) there are granodioritic and quartz-dioritic rocks, and rocks richer in potash feldspar; the adamellites and quartz-monzonites are less abundant. The absence of true gabbroic rocks in the Palentian area is due to alteration processes that have affected almost all the rocks of this zone. In fact, all the rocks of the Palentian Zone show strong hydrothermal alteration, and, as a consequence, any inference on the basis of the major chemical components is strongly biased (Loeschke 1982).

Trace elements Rb, Y, Nb are an important source of information on the geotectonic setting of the granites (Pearce et al. 1984). As Rb can be mobilized during the hydrothermal alteration, any generalization should be made with the utmost care, but despite their alteration, the rocks of the Palentian Zone plot together with the (mineralogical and petrographically) equivalent fresh rocks of the westernmost part of the Cantabrian Zone, in the same Rb/(Y + Nb) field. In this case, trace element analysis shows that all the granitoids of the zone plot in the VAG field in a position characteristic of post-collision granitoids (Fig. 5). The age and structural emplacement of these granitoids are incompatible with an island arc origin. However, these rocks have been generated in late stage magmatic activity which took place in the last episodes of the Hercynian cycle. One of the problems (Corretgé et al. 1987; Suárez and Corretgé 1987) is the geochemistry of these rocks and also the associated mineralizations which are more characteristic of island arc granitoids. Thus, we cannot reject the possibility of the presence of mantle material, very close in composition to that of island arc, underlying the crustal rocks of the Cantabrian/Palentian Zone.

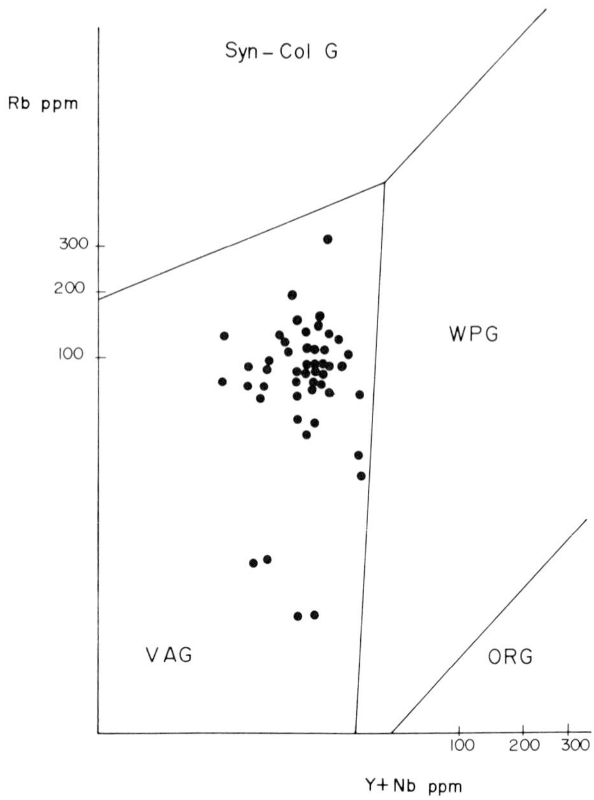

Fig. 5. Rb/(Y + Nb) diagram (Pearce et al. 1984) of the four groups of granitoids of the Cantabrian/Palentian Zones

Acknowledgements. We very much appreciate the comments and criticism of M. M. Godinho, J. M. Ugidos and an anonymous reviewer. We also wish to thank Ch. Hepburn and J. Fernández Suárez for their help with the English text. This research was supported by CAICYT, Project N° 2736/83.

References

Boschma D (1968) Provisional geological map of the southern mountains (Spain). Leidse Geol Meded 43: 217–220

Corretgé LG, Luque C, Suárez O (1970) Los stocks de la zona de Salas-Belmonte (Asturias). Bol Geol Min Esp LXXXI: 143–145

Corretgé LG, Cienfuegos I, Cuesta A et al. (1987) Granitoides de la Región Palentina (Cordillera Cantábrica, España). Publ IX Reun Geol Oeste Peninsular (1985). Mus Lab Geol Min Mem Univ Porto (1): 469–498

Corretgé LG, González Montero P, Suárez O (1988) Las trayectorias de las variables termodinámicas en la sucesion paragénetica del skarn de Carracedo (Palencia, España) Bol Soc Esp Mineral 11-1: 81–88

Fernández Secades ML (1975) Petrología y geoquímica de las manifestaciones efusivas de la región de Cabo Peñas (N. de Asturias). Breviora Geol Astúrica XIX (N° 1): 9–16

Frets DC (1965) The geology of the southern part of the Pisuerga Basin and the adjacent area of Santibañez de Resoba, Palencia, Spain. Leidse Geol Meded 31: 113–162

García de Figuerola L, Parga Pondal I (1964) Las rocas del Cámbrico. IV. Una traquita alcalina estratificada en el Cámbrico de Farandón (Valle del Narcea). Not Comun Inst Geol Min Esp 76: 79-94

García Iglesias J, Claverol M, Orueta J, Suárez O (1979) Mineralizaciones asociadas al metamorfismo de contacto del complejo igneo de Infiesto. Publ Mus Lab Min Geol Fac Cienc Porto CXI: 155-181

García Ramos JC, Aramburu C, Brime C (1984) Kaolin tonstein of volcanic ash origin in the Lower Ordovician of the Cantabrian Mountains (NW Spain). Trab Geol 14: 27-33

Heinz W, Loeschke J, Vavra J (1985) Phreatomagmatic volcanism during the Ordovician of the Cantabrian Mountains (NW Spain). Geol Rundschau 74-3: 623-639

La Roche H de (1972) Revue sommaire de quelques diagrammes chimico-mineralogiques pour l'etude des associations ignées ou sédimentaires et de leurs dérivés métamorphiques. Sci Terre 17: 31-46

Loeschke J (1982) Late Hercynian igneous rocks of the southeastern Cantabrian Mountains (NW Spain). Neues Jahrb Geol Paläontol Abh 163 (2): 260-271

Loeschke J (1983) Igneous and pyroclastic rocks in Devonian and Lower Carboniferous strata of the Cantabrian Mountains (NW Spain). Neues Jahrb Geol Paläontol Monatsh 8: 495-504

Loeschke J (in press) Magmatismus und Orogenese im Kantabrischen Gebirge Spaniens. 76 Jahrestagung der Geologischen Vereinigung, Giessen 1986

Loeschke J, Zeidler N (1982) Early Paleozoic sills in the Cantabrian Mountains (Spain) and their geotectonic environment. Neues Jahrb Geol Paläontol Monatsh 7: 419-439

Loredo J, García Iglesias J (1988) El yacimiento aurífero de Carlés (Asturias). Bol Soc Esp Min 11 (1): 47-53

Martín Izard A, Palero Fernández FJ, Reguilon Bragado R, Vindel Catena E (1986) El skarn de Carracedo (San Salvador de Cantamuda). Un ejemplo de mineralización pirometasomática en el N. de la provincia de Palencia. Stud Geol Salmant XXIII: 171-192

Martínez García E (1981) El paleozoico de la zona Cantábrica Oriental, NW de España. Trab Geol 11: 95-127

Martínez García E (1984) Outline of Paleozoic stratigraphy and structure of the eastern Cantabrian Mountains (northwest Spain). In: Nassichuk WW (ed) C R IX Congr Int Strat Géol Carbonifère Washington-Urbana 1979 (3): 329-344

Martínez García E, Tejerina L (1985) Fluorspar deposits associated with Carboniferous and Permian rocks in Asturias and León, northwest Spain. In: Cross AT (ed) CR IX Congr Int Strat Géol Carboníferère Washington-Urbana 1979 (4): 467-478

Meyer C, Hemley J (1967) Wall rock alteration. In: Barnes HL (ed) Geochemistry of Hidrotermal ore deposites. Holt, Rinehart & Winston, New York, pp 166-235

Miyashiro A (1975) Classification, characteristics, and origin of ophiolites. J Geol 83: 249-281

Muñoz M, Ancochea E, Sagredo J, Peña JA de la, Hernan F, Brandle JL, Marfil R (1985) Vulcanismo Permo-Carbonífero de la Cordillera Ibérica. CR X Congr Int Strat Geol Carboniferère, Madrid 3: 27-52

Paniagua A, Rodríguez LS, Gutiérrez JL (1988) Mineralizaciones As-Sb-Au asociadas a rocas ígneas filonianas del NE de León: Las Minas de Burón. Bol Soc Esp Min 11 (1): 35-46

Parga JR (1969) Sobre la dstribución de manifestaciones efusivas en el Cámbrico de Asturias y León. Com Serv Geol Portugal LIII: 43-56

Pearce JA, Harris NBW, Tindle AG (1984) Trace element discrimination diagrams for the tectonic interpretation of granitic rocks. J Petrol 25: 956-983

Prado JG (1972) Nota sobre la petrografía de la zona de Viñón (Asturias). Stud Geol Salmant 3: 7-32

Rupkej (1965) The Esla Nappe, Cantabrian Mountains, Spain. Leidse Geol Meded 32: 1-74

Savage JF (1967) Tectonic analysis of Lechada and Curavacas synclines, Yuso Basin, León NW Spain. Leidse Geol Meded 39: 193-247

Savage JF, Boschma D (1980) Geological maps of the southern Cantabrian Mountains (Spain). Leidse Geol Meded 50 (2): 75-114

Sitter LU de, Boschma D (1966) Explanation of a geological map of the Paleozoic of the southern Cantabrian Mountains 1:50000 Sheet Pisuerga. Leidse Geol Meded 31: 191-238

Suárez O (1971) Petrología de los afloramientos graníticos de Linares (Pola de Allande, Asturias). Est Geol II: 27-43

Suárez O, Corretgé LG (1987) Plutonismo y metamorfismo en las zonas Cantábrica y Asturoccidental-Leonesa. In: Geología de los granitoides y rocas asociadas del Macizo Hespérica Rueda, Madrid, pp 13-25

Suárez O, García A (1974) Petrología de la Granodiorita de Peña Prieta (León, Santander, Palencia). Acta Geol Hisp 9 (5): 154-158

Suárez O, Marcos A (1967) Sobre las rocas ígneas de la región de Infiesto (Zona oriental de Asturias, Cordillera Cantábrica). Trab Geol 1: 1-27

Veen J van (1966) The tectonic and stratigraphic history of the Cardaño area. Cantabrian Mountains, NW Spain. Leidse Geol Meded 35: 45-104

5 Metallogenesis

C. LUQUE[1], E. MARTINEZ GARCIA[2] and F. RUIZ[3]

5.1 The Cantabrian Zone: Introduction

In the Cantabrian Zone, there are frequent, mineralogically rich mineralizations of small size. Metallic substances include copper, cobalt, nickel, lead, zinc, mercury, arsenic, antimony, gold, iron and manganese. Among the main non-metallic substances found are fluorite, barite, talc, magnesite, kaolin and gypsum. Nearly 75% of a total of 300 localities (Figs. 1, 2) are hosted by Carboniferous rocks (Luque and Martínez García 1983). To date, only restricted review papers on single deposit types have been published on: iron, (Adaro and Junquera 1916); fluorite (García Iglesias 1972, unpublished work); barite (Fernández Fernández 1985); antimony (Gumiel Martínez 1983); mercury (Luque 1985); etc.

[1] Unidad Central de Geología de HUNOSA, Ujo (Asturias) Spain
[2] Departamento de Geología, Universidad de Oviedo, 33005 Oviedo, Spain
[3] V. Masip 7, 33013 Oviedo, Spain

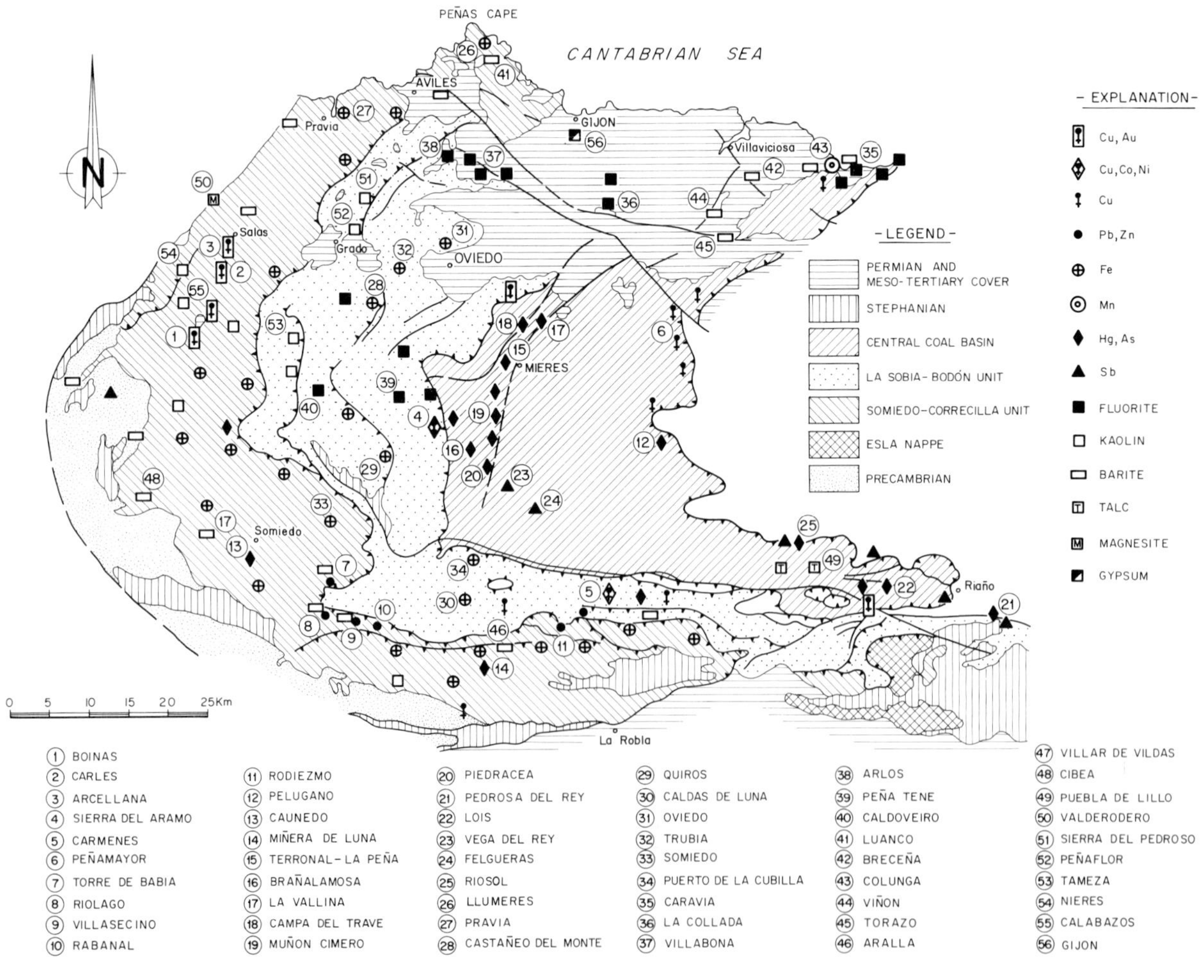

Fig. 1. Schematic geological map of the western Cantabrian Zone with location of main mineral occurrences

R. D. Dallmeyer and E. Martínez García (Eds.)
Pre-Mesozoic Geology of Iberia

5.2 Mineral Parageneses

Cu-Co-Ni. Usually hosted by Carboniferous carbonate rocks with a few exceptions in terrigenous sediments of that age near granodioritic or gabbroic igneous rocks (Corretgé et al. 1970). The usual paragenesis is arsenopyrite-chalcopyrite-pyrite, with occasional tetrahedrite, sphalerite, galena, molybdenite and gold (García de Figuerola and Peña 1964; Loredo and García Iglesias 1988). When the intrusions cut carbonate rocks of Carboniferous age (Infiesto, Fig. 2, *1*), Cambrian age (Boinás, Fig. 1, *1*) or Devonian age (Carlés, Fig. 1, *2*; Arcellana, Fig. 1, *3*), mineralizations are associated with skarn-type parageneses.

In the western Cantabrian Zone these mineralizations are hosted by Carboniferous carbonates, and a complex association of Cu, Co and Ni minerals is then found, together with intense dolomitization of the host rock. There is an extensive development of supergenic minerals, as in the Sierra del Aramo deposits (Fig. 1, *4*), with chalcopyrite, bornite, digenite, djurleite, chalcocite, azurite, malachite, cobaltite, smaltite, erytrite, annabergite, marcasite, etc. (Gómez Landeta and Solans 1981). Analysis of fluid inclusions in gangue minerals (quartz and calcite) of this mine has shown homogeneization temperatures between 80 °C and 170 °C with a maximum around 125 °C (Paniagua et al. 1988 a), which means that the main phase of ore deposition was due to low to moderate temperature (epithermal) hydrothermal solutions. In the Cármenes Area (Fig. 1, *5*) are also villamaninite, copper bravoite, linneite (Ypma et al. 1968), as well as Cu, Co and Ni disulphides (carmenesite, vaesite, cattierite), which are generally metastable (Fernández Fernández et al. 1985).

In the same area, Paniagua and Rodriguez Pevida (1988) describe an initial mineralizing episode with villamaninite, bravoite, pyrite, copper siegenite and marcasite, followed by a second one of cobaltite-gersdorffite and tennantite, a third one of chal-

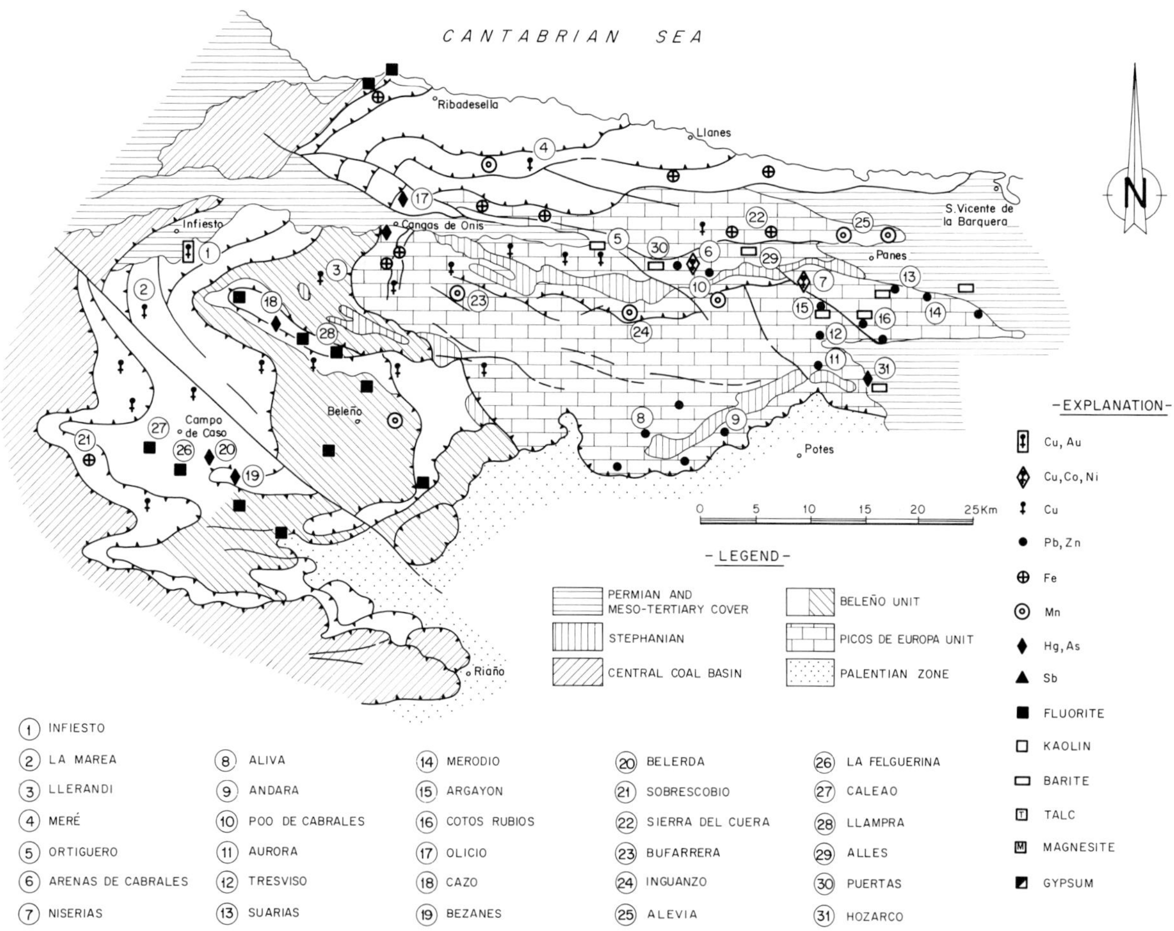

Fig. 2. Schematic geological map of the eastern Cantabrian Zone with location of main mineral occurrences

copyrite, talnakhite, bornite, X-bornite, uraninite, renierite, colusite and tucholite, and a fourth and last of supergenic alteration that produced copper sulphides, native copper and whitneyite, Co, Cu and Ni oxides and arsenates, Cu, U, Ca and Mg carbonates and arsenates, and the uranium silicate cuprosklodowskite. Temperatures of mineralization, at least in the third episode, based on alternative replacement between the different sulphide phases, seem to be around 125 °C.

In the eastern part of the Cantabrian Zone, mineralizations are mainly hosted by Carboniferous carbonates (Luque and Martínez García 1983); (Peñamayor, Fig. 1, *6*; La Marea, Fig. 2, *2*; Llerandi, Fig. 2, *3*; Meré, Fig. 2, *4*; etc.), with parageneses of tetrahedrite, copper bravoite, enargite and chalcopyrite, as fracture and karstic fillings with widespread dolomitization and silicification. In the Picos de Europa Area, there are also Cu, Co and Ni associations with erythrite, annabergite, heterogenite, tirolite and accidental schwatzite (Luque 1985; Ortiguero, Fig. 2, *5*; Arenas de Cabrales, Fig. 2, *6*; Niserias, Fig. 2, *7*; etc.).

Pb-Zn. Pb-Zn mineralizations are mainly concentrated at the eastern end of the Cantabrian Zone, within the Picos de Europa Area. The most widespread parageneses include sphalerite (caramel variety common), galena, pyrite, greenockite and chalcopyrite. There is also some gray sphalerite with Hg partially substituted for Zn in the structure (Barbanson et al. 1985; Luque 1985). Among the supergenic minerals are smithsonite, hydrocinzite, hemimorphite, cerusite, anglesite, azurite, malachite, cinnabar and goethite. Gangue minerals are calcite, ankerite, dolomite, aragonite, barite, quartz and chert. They usually occur as fillings of fractures and irregular karst cavities, usually at or near tectonic bands along faults or thrusts (Aliva, Fig. 2, *8*; Andara, Fig. 2, *9*; Poo de Cabrales, Fig. 2, *10*; Aurora, Fig. 2, *11*; etc.). Locally they are related to faults also cutting Permian sediments nearby (Tresviso, Fig. 2, *12*; Suarias, Fig. 2, *13*; Merodio, Fig. 2, *14*; Argayon, Fig. 2, *15*; Cotos Rubios, Fig. 2, *16*; etc.).

There are also small showings of Pb-Zn associated with Middle Cambrian limestones in the western part of the Cantabrian Zone (Torre de Babia, Fig. 1, *7*; Riolago, Fig. 1, *8*; Villasecino, Fig. 1, *9*; Rabanal, Fig. 1, *10*; Rodiezmo, Fig. 1, *11*). These include sphalerite, galena and barite, with some chalcopyrite and pyrite, and are probably related to volcanic rocks. Metallic minerals initially disperse in the carbonate sediments of a restricted platform were later remobilized towards intensely fractured strata in the Hercynian Orogeny (Fernández Fernández and Solans Huguet 1975).

Hg-As. Most of the Hg-As mineralizations are hosted by limestone rocks, either Carboniferous (Pelúgano, Fig. 1, *12*; Olicio, Fig. 2, *17*; Cazo, Fig. 2, *18*; Bezanes, Fig. 2, *19*; etc.), Middle Cambrian (Caunedo, Fig. 1, *13*) or Devonian (Miñera de Luna, Fig. 1, *14*), and also associated with siliceous breccias or conglomerates (Terronal-La Peña, Fig. 1, *15*), quartzites or sandstones (Olicio, Fig. 2, *17*); Brañalamosa, Fig. 1, *16*; Belerda, Fig. 2, *20*) and even with lutites or coal seams (La Vallina, Fig. 1, *17*).

The most common parageneses (Luque 1974, 1985), contain: (1) cinnabar, pyrite, melnikovite and marcasite with some sphalerite, galena and stibnite (Terronal-La Peña, Fig. 1, *15*; Campa del Trave, Fig. 1, *18*), (2) cinnabar, realgar and orpiment (Muñon Cimero, Fig. 1, *19*; Piedracea, Fig. 1, *20*; Caunedo, Fig. 1, *13*) and (3) cinnabar, stibnite, fluorite and, sometimes, livingstonnite (Riosol, Fig. 1, *25*; Pedrosa del Rey, Fig. 1, *21*; Lois, Fig. 1, *22*). In general, the gangue minerals are calcite, quartz and kaolinite. Cinnabar is also associated as supergenic mineral with some copper deposits with schwatzite (Luque 1985), as well as with others with lead-zinc-barite (Argayón, Fig. 2, *15*; Suarías, Fig. 2, *13*; Hozarco, Fig. 2, *31*) and of manganese (Bufarrera, Fig. 2, *23*), all of them in the Picos de Europa Area.

These mineralizations are almost generally fracture and cavity fillings, with association to organic or carbon content in the rocks (Luque 1985).

Sb. This is usually in Carboniferous carbonates in the neighborhood of Hg-As mineralizations, as impregnations, disseminations or small veins and irregular masses. Paragenetic minerals are arsenopyrite, stibnite, berthierite, marcasite, melnikovite, pyrite, and occasionally chalcopyrite and sphalerite (Gumiel et al. 1985), and calcite, quartz, kaolinite and fluorite as gangue minerals. They are mainly hosted by calcareous sediments (Vega del Rey, Fig. 1, *23*; Pedrosa del Rey, Fig. 1, *21*; etc.), although there are others in conglomerates (Maraña) or clastics at the contact of small intrusives (Burón, Boca de Huérgano) in the Palentine Domain.

Fe. There are two main iron deposit types in the Cantabrian Zone. One occurs as siliceous ironstones with local oolitic facies of hematite-chlorite. These are located in the Upper Silurian-Lower Devonian Furada (San Pedro) Formation of Asturias and northern León, with moderate thickness and contents between 25%–45% FeO (Adaro and Junquera 1916; Llumeres, Fig. 1, *26*; Pravia, Fig. 1, *27*; Castañeo del Monte, Fig. 1, *28*; Quirós, Fig. 1, *29*; Caldas de Luna, Fig. 1, *30*; etc.), as well as in the Naranco Formation of Couvinian-Lower Givetian age, where the iron rich concentrations are scarcer and of more limited extent (Oviedo, Fig. 1, *31*; Trubia, Fig. 1, *32*; etc.). They have formed as clastic accumulations of ferruginous sandstones and oolitic beds in epicontinental platforms after successive stages of sedimentation, weathering with Fe liberation and ooid formation, and erosion with redistribution of the iron deposits. The iron source seems to be volcanic by subaerial weathering of Lower Paleozoic basic rocks (García Ramos 1980; García Ramos et al. 1987). Less widespread are some oolitic ironstones in the Sueve Formation of Llanvirn age in Central Asturias (Hernández Sampelayo 1942; Pello and Philippot 1967).

The second type of iron deposits includes irregular karstic cavity fillings in calcareous rocks of the Lower Carboniferous (Somiedo, Fig. 1, *33*; Puerto de La Cubilla, Fig. 1, *34*; Sobrescobio, Fig. 2, *21*; Sierra de Cuera, Fig. 2, *22*; etc.), with the association of hematite, goethite and limonite, also with some content in Mn minerals, dolomite, ankerite, and calcite (Martínez Alvarez and Díaz González 1975).

Mn. In the eastern part of the Cantabrian Zone (Beleño and Picos de Europa Areas), there are frequent Mn deposits (Bufarrera, Fig. 2, *23*; Inguanzo, Fig. 2, *24*; Alevia, Fig. 2, *25*; etc.), of two main types: (1) a syngenetic synsedimentary type within the upper part of the caliza de montaña (Barcaliente Formation) interbedded with radiolarites and deep water deposits, accompanied by veins formed by remobilization in the Herycnian orogenesis and (2) a secondary type formed by erosion of the former and redeposition in karstic cavities in the Permian and Cretaceous. The manganese minerals are pirolusite, manganite, psilomelane, wad, etc. (Llopis Lladó 1958).

F. Three main types of fluorspar deposits can be distinguished in the Cantabrian Zone (Martínez García and Tejerina 1985): (1) fluorite closely related with Permian rocks unconformably overlying older Paleozoic strata. The fluorspar deposits are either strata-bound or in veins cutting the Permian and Carboniferous rocks below the unconformity, (2) mostly found in Carboniferous limestones as veins, limestone replacements, or cavity fillings, and is not directly related to Permian deposits and (3) deposits associated with folds and fractures in Carboniferous rocks. Deposits of type 1 are restricted to the northern part of the Cantabrian Zone (Caravia, Fig. 1, *35*; La Collada, Fig. 1, *36*; Villabona, Fig. 1, *37*; Arlós, Fig. 1, *38*). Types 2 and 3 are found more to the south and west, where most Permian deposits have been eroded away (Peña Tene, Fig. 1, *39*; Llampra, Fig. 2, *28*; Caldoveiro, Fig. 1, *40*; Caleao, Fig. 2, *27*; Felguerina, Fig. 2, *26*). The most common parageneses are made up of fluorite, barite, pyrite, chalcopirite, calcite, quartz and kaolinite. They are accompanied by intense silicification of the host rock (García Iglesias 1972, unpublished work).

Ba. There are three main types of deposits. These are (1) lens-shaped concentrations in Permian rocks (Luanco, Fig. 1, *41*; Breceña, Fig. 1, *42*; Colunga, Fig. 1, *43*), also with fluorite association (Caravia, Fig. 1, *35*). They are frequently related to volcanic rocks (tuffs or lavas; Viñón, Fig. 1, *44*; Torazo, Fig. 1, *45*), (2) irregular veins, either in Cambrian limestones with sphalerite, galena, pyrite and cinnabar (Torre de Babia, Fig. 1, *7*; Villasecino, Fig. 1, *9*; Riolago, Fig. 1, *8*; Aralla, Fig. 1, *46*; Villar de Vildas, Fig. 1, *47*; Cibea, Fig. 1, *48*, etc.), or in Devonian limestones (Luanco), or exceptionally, in Ordovician quartzites (Salas) and (3) karstic cavity fillings near unconformable Permian deposits. Mainly hosted by carbonate rocks of Carboniferous age in the Picos de Europa Area (Alles, Fig. 2, *29*; Suarías, Fig. 2, *13*; Argayón, Fig. 2, *15*; Ortiguero, Fig. 2, *5*; Cotos Rubios, Fig. 2, *16*; Puertas, Fig. 2, *30*; etc.). The main parageneses (Fernández Fernández 1985) are of barite, barite-cinnabar, and barite-Pb-Zn sulphides and carbonates-chalcopyrite-cinnabar.

The fluorspar and barite deposits in Permian sediments of Central Asturias are related to hydrothermal activity and Permian vulcanism that produced substitution in the carbonate beds of caliche type in the Permian sediments (Forster 1974; Endlicher 1976; García Iglesias and Touray 1977; Martínez García and Tejerina 1985).

Talc. The main deposits are located near Puebla de Lillo (León) (Fig. 1, *49*), and are related to faults cutting the Ordovician Barrios quartzites and Carboniferous limestones. The main paragenesis (Galán Huertos and Rodas 1973) is talc-chlorite-pyrite-dolomite, with silica and magnesia diffusion by hydrothermal solutions towards fracture zones that produced irregular limestone replacement.

Magnesite. The only known magnesite deposit in the Cantabrian Zone, at Valderrodero (Salas, Fig. 1, *50*), is of strata-bound character in an intensely fractured carbonate bed of the Herrería Formation (Lower Cambrian). There Gutiérrez Claverol and García Iglesias (1982) have described magnesite, dolomite, talc, chlorite and quartz with small amounts of pyrite and chalcopyrite. They are due to metasomatism and chemical diffusion probably related to basic igneous rocks nearby.

Kaolin. There is a widespread (over 1800 km^2) kaolin bed 0.4 to 1 m thick interbedded in the Ordovician Barrios Formation (García Ramos et al. 1984), having been interpreted as a kaolin tonstein originated by in situ alteration of a volcanic ash tuff. It appears mainly in the western Cantabrian Zone, with kaolinite and quartz (Sierra del Pedroso, Fig. 1, *51*; Peñaflor, Fig. 1, *52*; Tameza, Fig. 1, *53*; Nieres, Fig. 1, *54*; Calabazos, Fig. 1, *55*; etc.).

Gypsum. There are gypsum beds al the base of Liassic dolomites and limestones in Central Asturias (Gijon, Fig. 1, *56*). These locally reach 2 or 3 m with an alternation of nodular anhydrite. They are probably of "sabkha" type (Gutiérrez Claverol 1984).

5.3 Origin of Mineralizations

Most mineralizations of Cu, Au, Co, Ni, Pb, Zn, Hg, Sb, As, F, Ba and talc in the Cantabrian Zone are of *hydrothermal* origin, which evidences a common origin. Most are in carbonate rocks in intensely faulted and folded zones, showing strong control by the main fracture directions of the Late Hercynian episodes and their intersections (E-W, NE-SW and NW-SE) (Luque and Martínez García 1985).

Sometimes, they are near post-Hercynian intrusive rocks (probably of Stephanian or Permian age) or Early Permian volcanics, suggesting a relationship with the ascending or descending hydrothermal solutions (Martínez García 1981). Their age would then be Permian, as no younger strata are mineralized in this zone (Martínez García 1983a, b). Their deposition is also related to the rock porosity and reducing environment (Fernández Fernández 1985) and organic matter content (Luque and Martínez García 1983), coinciding with alteration phenomena in the host rocks (dolomitization, silicification, etc). The temperatures reached do not seem to be higher than 200 °C (Loredo and García Iglesias 1981; Luque and Martínez García 1985), so the deposits can be considered epithermal. Most of these were formed by fluids (1) of high S concentration increasing as temperature decreased and giving rise to a variety of paragenetic sulphide minerals (Paniagua and Rodríguez Pevida 1988), (2) of moderate to high salinity (10%-23% eq. NaCl) and (3) at relatively low pressures (200-600 bar; García Iglesias and Touray 1977; Loredo et al. 1988; Paniagua et al. 1987). Near-surface mixing with carbonate-rich meteoric waters circulating through karstic systems helped mineral precipitation. Weathering processes later produced supergenic minerals which increased the complexity of mineral content.

Some *metasomatic* deposits are found in carbonate rocks around small Late Hercynian granodioritic stocks in the Cantabrian Zone (Salas, Carlés, Boinás, Infiesto, etc.) with complex skarn type mineralizations including diopside, hornblende, sphene, grossularite, vesubianite, wollastonite, axinite, etc. Temperatures close to 600 °C have been established for the intrusive rocks from the study of parageneses, and from 150°-400 °C for the metallic minerals from fluid inclusion studies (García Iglesias et al. 1979; Rua Figueroa et al. 1987). The magnesite deposits in fractured Lower Cambrian carbonate rocks at Valderrodero have also been considered to be metasomatic, related to chemical diffusion of silica and magnesia through recrystallized dolomites at temperatures near 250 °C (Gutiérrez Claverol and García Iglesias 1982).

5.4 The Palentian Zone

The Palentian Zone, unlike the Cantabrian Zone, is characterized by the absence of synsedimentary or syngenetic mineralizations, and all the reported findings correspond to epigenetic deposits in relation generally with the presence of diverse Late Hercynian intrusive rocks or with tectonic features of the same age with associated hydrothermal alterations.

In a wide area between Burón, Salamón, Riaño, Peña Prieta, Peña Labra, Camporredondo, and Carracedo, is a mining district of Sb-As-Cu-Pb-Zn-Au-F-Hg, determined by the fracture systems: N 120°-140° E (Ventaniella and Polentinos Faults), N 60°-70° E (Peñas Matas Fault), and N 90° E (León Fault). In this area there are also numerous intrusive stocks associated with fractures cutting the Carboniferous rocks (Loeschke 1982; Wagner et al. 1984; Corretgé et al. 1988). The intrusives are accompanied by hydrothermal alterations, generally propilitizations, with skarn development in carbonates of the sequence. Associated mineralizations are mainly of stibnite, arsenopyrite, chalcopyrite, pyrrhotite, and pyrite in a gangue of calcite, ankerite, quartz and talc. Secondary minerals are cinnabar, fluorite, sphalerite, galena, marcasite, magnetite, ilmenite, hematite and cubanite, with traces of bismutinite and native bismuth. This mineral association is very irregularly distributed in the different deposits, the proportions of the minerals present in each case being highly variable. They are usually affected by supergenic oxidation, with a zonation of antimony ochers (if there is stibnite), malachite, and hematite.

At Burón (Fig. 3, *1*), mineralization is mainly of antimony associated with a granodiorite porphyry (N 120° E) intruded as a sill in the Carboniferous sequence. It appears as small veins 20–50 cm thick, of quartz with minute gold-bearing arsenopyrite, bismutinite, stibinite, berthierite, realgar, orpiment, sphalerite, etc., and also with pyrite, chalcopirite and fluorite in isolated cristals (Paniagua et al. 1988 b), where four mineralizing stages can be separated. There are also fluorspar occurrences near Buron (Fig. 3 *2*) with accessory stibnite, sphalerite and strontium-calcite.

The Peña Prieta copper deposit (Fig. 3, *3*) is at the contact of a Late Hercynian granodiorite stock (Suárez and García 1974) with Carboniferous lutites and sandstones, and at the crossing of the Peñas Matas and Polentinos Faults with production of a contact metamorphic aureole and hydrothermal alterations (Luque and Martínez García 1983). A vein-like talc mass is found there with disperse chalcopyrite and, more rarely, of pyrite crystals, malachite and azurite. In the Carboniferous limestones, near the granodiorite intrusive (Fig. 3, *4*), there are also SW-NE vein-like and N 120° E strata-bound mineralizations of sphalerite, galena and chalcopyrite with a calcitic dolomitic gangue. Quartz veins with chalcopyrite are also found associated with the Pico Jano granodiorite (Fig. 3, *10*).

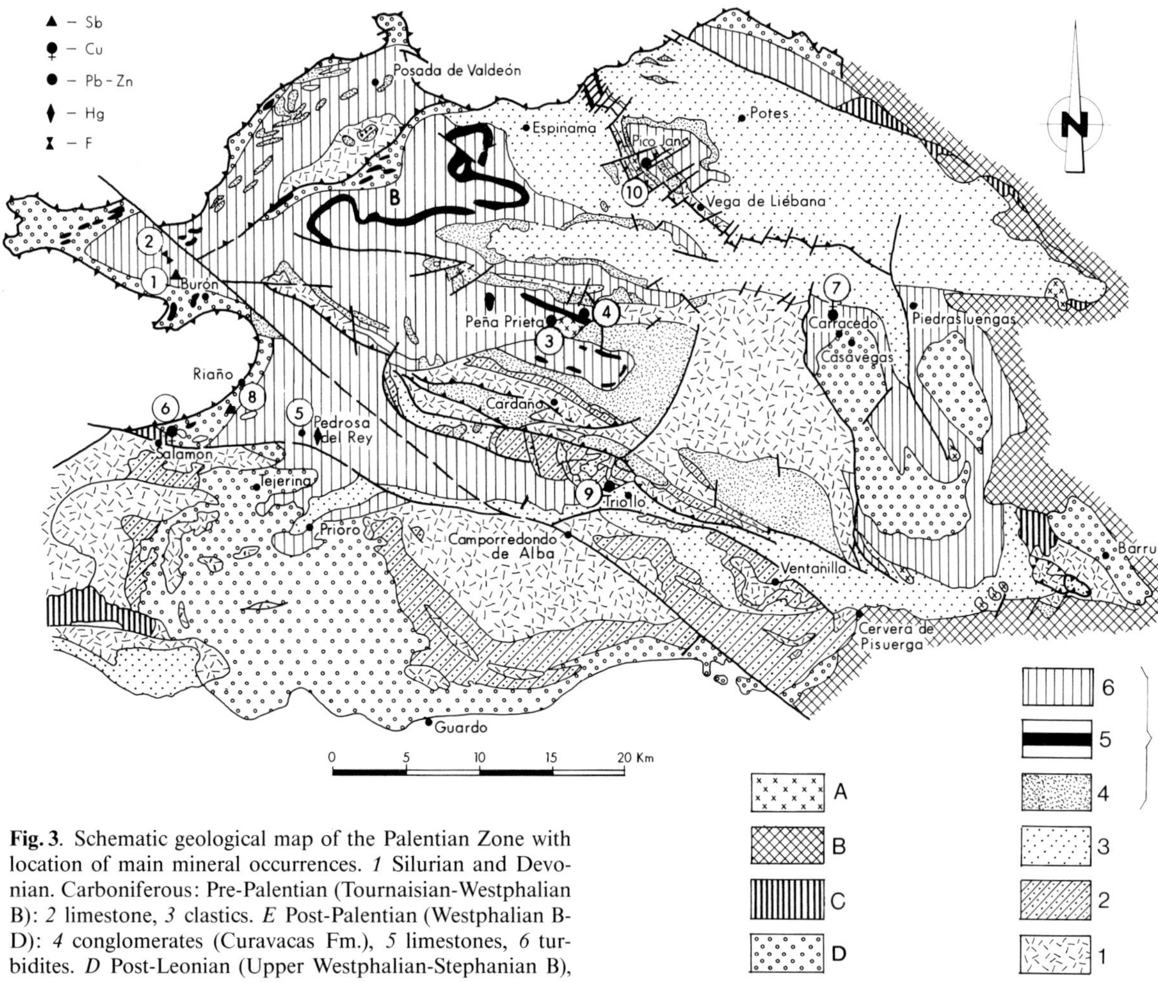

Fig. 3. Schematic geological map of the Palentian Zone with location of main mineral occurrences. *1* Silurian and Devonian. Carboniferous: Pre-Palentian (Tournaisian-Westphalian B): *2* limestone, *3* clastics. *E* Post-Palentian (Westphalian B-D): *4* conglomerates (Curavacas Fm.), *5* limestones, *6* turbidites. *D* Post-Leonian (Upper Westphalian-Stephanian B), *C* Post-Asturian (Stephanian B-C), *B* Permian and Mesozoic. *A* Granodiorite

At Pedrosa del Rey (Fig. 3, *5* and Fig. 1, *21*) mercury and antimony predominate with regard to other components. Cinnabar is found in N 90° E veins, disseminated in the limestone or as impregnations in a calcareous breccia. Stibnite appears in irregular pockets at the contact of the limestones with graphite schists. Accompanying minerals are pyrite, livingstonite, fluorite, calcite, kaolinite, goethite, realgar, stibiconite, cervantite and antimony ochers.

The Salamón mineralization (Fig. 3, *6*) is in an E-W fracture with strong accompanying dolomitization of the Carboniferous limestones. The parageneses differ within the deposit, showing the presence of chalcopyrite, gold-bearing arsenopyrite, sphalerite and galena, with supergenic alteration to malachite, azurite, smithsonite, realgar, orpiment and limonite and hematite. Sulphides are found as small crystal aggregates, while the alteration minerals form as irregular incrustations.

At Carracedo (Fig. 3, *7*), igneous rocks are found as small quartz-dioritic to granodioritic stocks and dikes of porphyries with a strike of N 120°-130° E. The intrusives form skarns (exoskarns) and contact metamorphism. The mineralization is disperse and irregularly concentrated, being constituted of pirrotite, chalcopyrite, gold-bearing arsenopyrite and pyrite as the main constituents, and marcasite, magnetite, ilmenite, hematite, cubanite, bismuthinite, chalcopyrrotite, sphalerite, löllingite, tetrahedrite and bismuth as accessories (Martín Izard et al. 1986). In the neighborhood of Riaño (Fig. 3, *8*), there are some irregular N 120° E vein-like and karst-fills mineralizations in the Carboniferous limestones (Gumiel et al. 1985). The paragenesis is gold-bearing arsenopyrite, stibnite and pyrite, with some cinnabar in the calcite or quartz gangue, together with supergenic antimony ochers, realgar, goethite, stibniconite and cervantite. At Triollo (Fig. 3, *9*), the Carboniferous limestones are also mineralized with veins and karst-fillings of galena and sphalerite, with intense supergenic alteration (Cueto and Rui Díaz 1928; Alvarado 1929).

As a whole, the Palentian terrane shows a restricted number of mineralizations that can be described as hydrothermal (mesothermal to epithermal), related to Late Hercynian fracture systems, and skarn-type associated to Late Hercynian intrusive rocks.

References

Adaro L de, Junquera G (1916) Criaderos de hierro de España. II. Criaderos de Asturias. Mem Inst Geol Min Esp pp 1-676

Alvarado A de (1929) Notas referentes a yacimientos españoles de Plomo, Cinc y metales afines. Bol Inst Geol Min Esp 51: 109-132

Barbanson L, Saulas D, Touray JC (1985) Les blendes mercuriféres de la région de Cabezón de la Sal (Santander, Espagne). Bull Min 108: 483-486

Corretgé LG, Luque C, Suárez O (1970) Los stocks de la zona de Salas-Belmonte (Asturias). Bol Geol Min 80 (2): 257-270

Corretgé LG, González MP, Suárez O (1988) Las trayectorias de las variables termodinámicas en la sucesión paragenética del skarn de Carracedo (Palencia, España). Bol Soc Esp Min 11 (1): 81-88

Cueto, Rui-Díaz P (1928) Estudio geológico-industrial de los criaderos de cinc de la provincia de Palencia. Bol Of Min Met XII (137): 821-843: (138): 871-890

Endlicher G (1976) Die Syngenetischen Flußß- und Schwerspat mineralisationen von Arlos und Villabona in West-asturien/Nordspanien. Miner Dep 11: 329-351

Fernández Fernández CJ (1985) Las mineralizaciones (CU)-Pb-Zn-Ba-(Hg) en rocas dominantemente calcáreas del Carbonífero, sector Norte de la Región de Picos de Europa, Asturias. Características metalogenéticas en relación al tipo plomo-cinc-caliza. CR X Congr Int Strat Géol Carbonifère Madrid 1983 3: 105-119

Fernández Fernández CJ, Solans Huguet J (1975) Estudio de las mineralizaciones de baritina de la Babia Baja (León). Breviora Geol Astúrica XIX (4): 54-59

Fernández Fernández CJ, Moreiras D, Paniagua A (1985) Mineralogía y paragénesis mineral de la Mina Providencia (Cármenes, León), Trab Geol 15: 239-247

Forster A (1974) Die Flußspaltlagerstätten Asturiens /Nordspanien und deren Genese. Geol Rundschau 62: 212-263

Galán Huertos E, Rodas M (1973) Contribución al estudio mineralógico de los depósitos de talco de Puebla de Lillo (León, España). Bol Geol Min 84: 347-365

García de Figuerola LC, Peña D de la (1964) El afloramiento cuarzo-diorítico de Carlés (Asturias) y la prospección geoquímica del Cu en el mismo. Bol R Soc Esp Hist Nat 62: 91-106

García Iglesias J, Touray JC (1977) A fluorite-calcite-quartz paragenesis with liquid and gaseous organic inclusions at La Cabaña, Asturias, Fluorspar District, Spain. Econ Geol 72: 298-303

García Iglesias J, Gutiérrez Claverol M, Orueta J, Suárez O (1979) Mineralizaciones asociadas al metamorfismo de contacto del complejo ígneo de Infiesto (Zona Oriental de Asturias, España). Publ Mus Lab Min Geol Fac Ciênc Porto 91: 155-181

García Ramos JC (1980) Evolución diagenética de compuestos de hierro sedimentario y su papel en la coloración de sedimentos del Devónico de la Cordillera Cantábrica. Rev Inst Invest Geol Dip Prov, Univ Barcelona 34: 281-292

García Ramos JC, Aramburu C, Brime C (1984) Kaolin tonstein of volcanic ash origin in the Lower Ordovician of the Cantabrian Mountains (NW Spain) Trab Geol 14: 27-33

García Ramos JC, Suárez De Centi C, Paniagua A, Valenzuela M (1987) Los depósitos de hierro oolítico del Paleozóico de Asturias y N de León: ambiente de depósito y relación con el vulcanismo. Geogaceta 2: 38-40

Gómez Landeta F, Solans J (1981) Procesos supergénicos en la mina de cobre del Aramo, Asturias, España. Bol Geol Min 92: 429-436

Gumiel Martínez P (1983) Metalogenia de los yacimientos de antimonio de la Península Ibérica. Tecniterrae 54: 6-120

Gumiel P, Arribas A, Vindel E (1985) Los yacimientos estratoides de Antimonio del Carbonífero de la Cordillera Cantábrica. C R X Congr Int Strat Géol Carbonifère Madrid 1983 3: 121-143

Gutiérrez Claverol M (1984) Los depósitos evaporíticos del tránsito Permotrias-Lias en Asturias (Mina Felisa). Rev Minas, Univ Oviedo 4: 37-49

Gutiérrez Claverol M, García Iglesias J (1982) El yacimiento de magnesita de Valderrodero (Asturias, España). Bol Geol Min 93: 233-243

Hernández Sampelayo P (1942) El Sistema Siluriano. Expl Nuevo Mapa Geol Esp 2: 1-592, Inst Geol Min Esp

Llopis Lladó N (1958) Sobre el karst actual y fósil de la terminación oriental de la Sierra de Cuera y sus yacimientos de hierro y manganeso. Monogr Geol X, Inst Geol Aplicada Univ Oviedo, pp 1-59

Loeschke J (1982) Late Hercynian igneous rocks of the southeastern Cantabrian Mountains (NW Spain). Neues Jahrb Geol Paläontol Abh 163 (2): 260-275

Loredo J, García Iglesias J (1981) Análisis microtermométrico del yacimiento de Fluorita de "Pié de Potro" (Caravia, Asturias). Rev Minas 2: 117-125

Loredo J, García Iglesias J (1988) El yacimiento aurífero de Carlés (Asturias). Bol Soc Esp Miner 11 (1): 47-54

Loredo J, Luque C, García Iglesias J (1988) Conditions of formation of mercury deposits from the Cantabrian Zone (Spain). Bull Min 111: 393-400

Luque C (1974) Los yacimientos de Hg Astur-Leoneses. Bol Inst Est Astur 19: 3-11

Luque C (1985) Las mineralizaciones de mercurio de la Cordillera Cantábrica PhD Thesis (unpublished), Univ Oviedo, (coord)

Luque C, Martínez García E (1983) Depósitos minerales en el Carbonífero de la Cordillera Cantábrica. In: Martinez Diaz C coord, Carbonífero y Pérmico de España. Inst Geol Min Esp, pp 163-177

Luque C, Martínez García E (1985) Análisis tectometalogenético de las mineralizaciones en el Carbonífero de la Zona Cantábrica (NW de España). C R X Congr Int Strat Géol Carbonifère Madrid 1983 3: 95-104

Martín Izard A, Palero Fernández FJ, Reguilón Bragado R, Vindel Catena E (1986) El skarn de Carracedo (San Salvador de Cantamuda). Un ejemplo de mineralización pirometasomática en el N de la Provincia de Palencia. Stud Geol Salmant XXIII: 171-192

Martínez Alvarez JA, Díaz González S (1975) Estudio de las mineralizaciones de hierro de las inmediaciones del lago La Cueva en la región de los lagos de Saliencia (Somiedo, Oviedo). Bol Geol Min LXXXVI: 498-504

Martínez García E (1981) Tectónica y mineralizaciones pérmicas en la Cordillera Cantábrica Oriental (Noroeste de España). Cuad Lab Xeol Laxe 2: 263-271

Martínez García E (1983 a) El Pérmico de la Cordillera Cantábrica. In: Martinez Diaz C (coord), Carbonífero y Pérmico de España. Inst Geol Min Esp, pp 389-400

Martínez García E (1983 b) Permian mineralizations in the Cantabrian Mountains (North-West Spain). In: Schneider H-J (ed), Mineral deposits of the Alps and of the Alpine Epoch in Europe. Springer Berlin Heidelberg New York, pp 259-274

Martínez García E, Tejerina L (1985) Fluorspar deposits associated with Carboniferous and Permian rocks in Asturias and León, northwest Spain. In: Cross AT (ed) C R IX Int Congr Strat Géol Carbonifère Washington DC-Urbana/Champaign 1979 Southern Illinois University Press, Carbondale and Edwardsville 4: 467-478

Paniagua A, Rodriguez Pevida LS (1988) Genesis y evolución de las mineralizaciones de Cu-Co-Ni-U-Pb-Zn-Au-Ag ligadas a estructuras tardihercínicas en el área de Pajares-Villamanin-Boñar (N de León). Bol Soc Esp Min 11 (2): 118-119

Paniagua A, Rodriguez Pevida LS, Garzón L, Pérez JM Quintana A (1987) Presencia de una paragénesis Cu-Ni-Co-U-As-S en la Zona Cantábrica: la Mina Profunda (Cármenes-León). Geogaceta 2: 22-24

Paniagua A, Loredo J, García Iglesias J (1988 a) Epithermal (Cu-Co-Ni) mineralization in the Aramo Mine (Cantabrian Mountains, Spain). Correlation between paragenetic and fluid inclusion data. Bull Min 111: 383-391

Paniagua A, Rodriguez Pevida LS, Gutiérrez JL (1988 b) Mineralizaciones As-Sb-Au asociadas a rocas ígneas filonianas del NE de León: Las Minas de Burón. Bol Soc Esp Min 11 (1): 35-46

Pello J, Philippot A (1967) Sur la presence du Llanvirn au Puerto Sueve (zone orientale des Asturies, NW de l'Espagne). CR Somm Soc Géol Fr 1967 (4): 156-157

Rua Figueroa A, Llavona M, Loredo J, García Iglesias J (1987) Fluid inclusions in quartz from a gold-mineralized granodioritic intrusion at Carlés, Asturias, Spain. Chem Geol 61: 217-224

Suárez O, García A (1974) Petrología de la granodiorita de Peña Prieta (León, Santander, Palencia). Acta Geol Hisp IX (5): 154-158

Wagner RH, Carballeira J, Ambrose T, Martínez García E (1984) Explicacion de la Hoja nº 107 (Barruelo de Santullán) del Mapa Geologico de Españia (MAGNA) a escala 1:50000. Inst Geol Min Esp, 113 p

Ypma PJM, Evers HJ, Woensdregt CF (1968) Mineralogy and geology of the Providencia Mine (León, Spain), type locality of villamaninite. Neues Jahrb Min Monatsh 6: 174-191

Part III West Asturian-Leonese Zone

1 Introduction

J. R. Martinez Catalan[1]

Within the Variscan orogenic belt, in the northwestern part of the Iberian Peninsula, the West Asturian-Leonese Zone (WALZ) is the transition zone between the foreland areas, represented by the Cantabrian Zone, and the more internal zones placed westwards (Central Iberian Zone). The zone was defined by Lotze (1945) as an arcuate band with the following main characteristics: existence of a very thick and nearly continuous succession ranging from the Upper Proterozoic to the Devonian, presence of a very regular train of folds with E and NE vergence, and reduced extension of the granitic massifs.

Matte (1968) restricted the zone to the eastern and northeastern parts of that defined by Lotze, and proposed a new one in its western part, Eastern Galicia, based on stratigraphic differences; notwithstanding, these are often apparent and in many cases attributable to the structure, which accounts for the outcropping of different stratigraphic ensembles in different domains. The two zones established by Matte have, however, a tectonic meaning, because the first coincides with the domain in which the D_1 folds have a steep axial surface, while the second is characterized by recumbent D_1 folds. Nevertheless, this is mainly a consequence of the later deformations.

Julivert et al. (1972) grouped again the ensemble in a single zone, and the stratigraphic differences in it allowed Marcos (1973) to propose a subdivision in three domains: Navia and Alto Sil, Mondoñedo and Caurel-Truchas. By that time, the westernmost part of the zone, which had been included by Lotze in the Galaico-Castillian Zone due to its higher grade of metamorphism and the abundance of granitic massifs, was recognized as an integral part of the WALZ from the point of view of the stratigraphic sequence.

Julivert et al. proposed two arcuate antiforms as the limits of the zone: that of Narcea, to the east, and the Ollo de Sapo Antiform, to the west. The eastern boundary can be more precisely established in an important thrust fault within the Narcea Antiform, which separates two areas with a different intensity of deformation. The Ollo de Sapo Antiform ist no longer considered the western boundary, which has been shifted towards the east and northeast to the Vivero Fault and the Sil-Truchas Syncline.

So defined, the West Asturian-Leonese Zone in the NW of the Iberian Peninsula depicts an arcuate form in a continuous outcrop that goes from the Cantabrian coast in the north, towards the south and southeast, across western Asturias, eastern Galicia and the western part of the León province, where it is covered by Tertiary deposits. It consists of an Upper Proterozoic monotonous flyschoid series and a thick Paleozoic succession, mainly of shallow water facies, folded and thrusted in Lower Carboniferous time. The metamorphism ranges from green schist to amphibolite facies and increases progressively towards the western part of the zone, where granitic rocks become abundant.

Further to the east, the WALZ outcrops again in the Sierra de la Demanda (Burgos) and in the Cordillera Ibérica (Aragón), surrounded by Mesozoic and Tertiary deposits. The next chapters on different geological aspects of this zone will deal exclusively with the outcrop of the northwest of the Iberian Peninsula.

Relevant contributions of regional importance to the knowledge of the zone are those by Riemer (1963, 1966), Walter (1966, 1968), Matte (1968), Nollau (1968), Capdevila (1969), Marcos (1973), Pérez-Estaún (1978), Bastida (1980), Pulgar (1980) and Martínez-Catalán (1985), for the main outcrop, in the northwest of Iberia, and that of Colchen (1974) for the Sierra de la Demanda (for references, see Part III, Chap. 3).

[1] Departamento de Geologia, Universidad de Salamanca, 37008 Salamanca, Spain

R. D. Dallmeyer and E. Martínez García (Eds.)
Pre-Mesozoic Geology of Iberia

2 Stratigraphy

A. Perez Estaun[1], F. Bastida[1], J. R. Martinez Catalan[2], J. C. Gutierrez Marco[3], A. Marcos[1] and J. A. Pulgar[1]

2.1. Introduction

The West Asturian-Leonese zone (WALZ; Fig. 1) is characterized by the presence of a thick series of Cambrian-Ordovician siliciclastic rocks (about 11000 m), which underwent a great deal of strain, by an overall structure formed by folds (largely recumbent) and thrusts and by the presence of cleavages and regional metamorphism.

These features differ from those of the bordering areas. For instance, in the Cantabrian Zone, which is the outermost part of the orogen, there is a nearly complete Paleozoic succession, but the thickness of the Cambrian-Ordovician rocks found there is of the order of 3000 m. In addition, in this area, the amount of strain is very small, and likewise the regional metamorphism (see, among others, Aller et al. 1987). On the other hand, in the Central Iberian and Galicia-Tras os Montes zone, the innermost parts of the Variscan orogen, the age of the metasedimentary rocks is not well established, the grade of metamorphism is quite often high and plutonic rocks are widespread.

The limits of the WALZ determined by Lotze (1945) and Julivert et al. (1972) present some problems. The western limit does not show such outstanding changes in the stratigraphy of the Paleozoic sequence. Therefore, in this chapter, the western limit has been shifted to the Vivero Fault (Fig. 1), as suggested by Martínez-Catalán (1985) and the Sil-Truchas syncline is considered to be the SW limit. Both structures separate one region to the E and NE, where a complete Cambrian sequence lies on Precambrian rocks from another one, to the W and SW, where the Cambrian is not found and where Lower Ordovician rocks rest upon a porphyritic formation known as Ollo de Sapo. The eastern limit of the WALZ is located within the Narcea Antiform, a Precambrian outcrop with a complex structure formed by the stacking of thrust sheets (antiformal stack). However, the most precise limit can be established by a thrust (Fig. 1) that separates two areas with different deformation intensity; of these two areas, the western one has undergone a stronger deformation and thus can be included in the WALZ.

On the basis of the stratigraphic variations of the Cambrian-Ordovician rocks, which will be mentioned later, the WALZ has been divided into three different domains (Fig. 1): Navia and Alto Sil Domain (NASD), Mondoñedo nappe Domain (MD) and Caurel-Truchas Domain (CTD) (Marcos 1973). The stratigraphic columns of the different domains are represented in Fig. 2. The main difference between the MD and the CTD lies in the reduced thickness of the Cambrian rocks in the Caurel-Truchas Domain, and in the absence of some of the formations of this age. Also outstanding is the presence of Lower Devonian rocks in this domain occupying the core of a recumbent fold (Caurel syncline, Fig. 1). The NASD, by contrast, represents the relatively autochthonous area of the Mondoñedo thrust sheet and crops out to the east of its frontal trace and under it in two tectonic windows (Gistral and Monte Carballosa Tectonic windows; Fig. 1). The stratigraphic sequence of the NASD is characterized by the great development of the Middle and Upper Ordovician, while in the MD, the thickness of the rocks of this age is very small.

2.2 The Precambrian Rocks in the Narcea Antiform

The Precambrian rocks in this area have been named Narcea slates (Lotze 1956) or Mora Formation (De Sitter 1961). The stratigraphic and sedimentological characteristics of these rocks can be better studied in the eastern part of the Narcea anti-

[1] Departamento de Geología, Universidad de Oviedo, 33005 Oviedo, Spain
[2] Departamento de Geología, Universidad de Salamanca, 37008 Salamanca, Spain
[3] Departamento de Paleontología, Universidad Complutense de Madrid, 28040 Madrid, Spain

R. D. Dallmeyer and E. Martínez García (Eds.)
Pre-Mesozoic Geology of Iberia

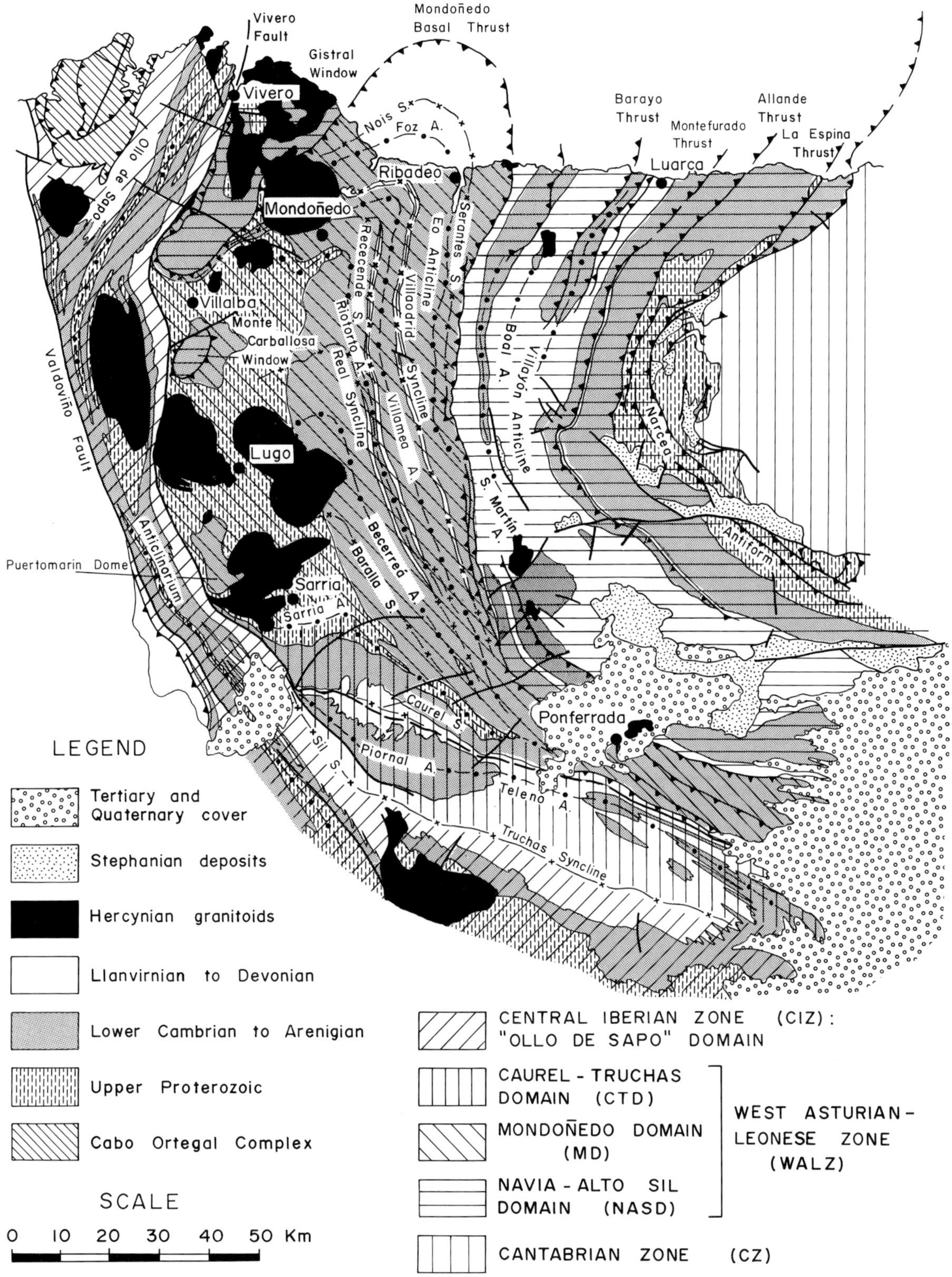

Fig. 1. Simplified geological map of the West Asturian-Leonese Zone with the names of the main structures and the subdivision in domains

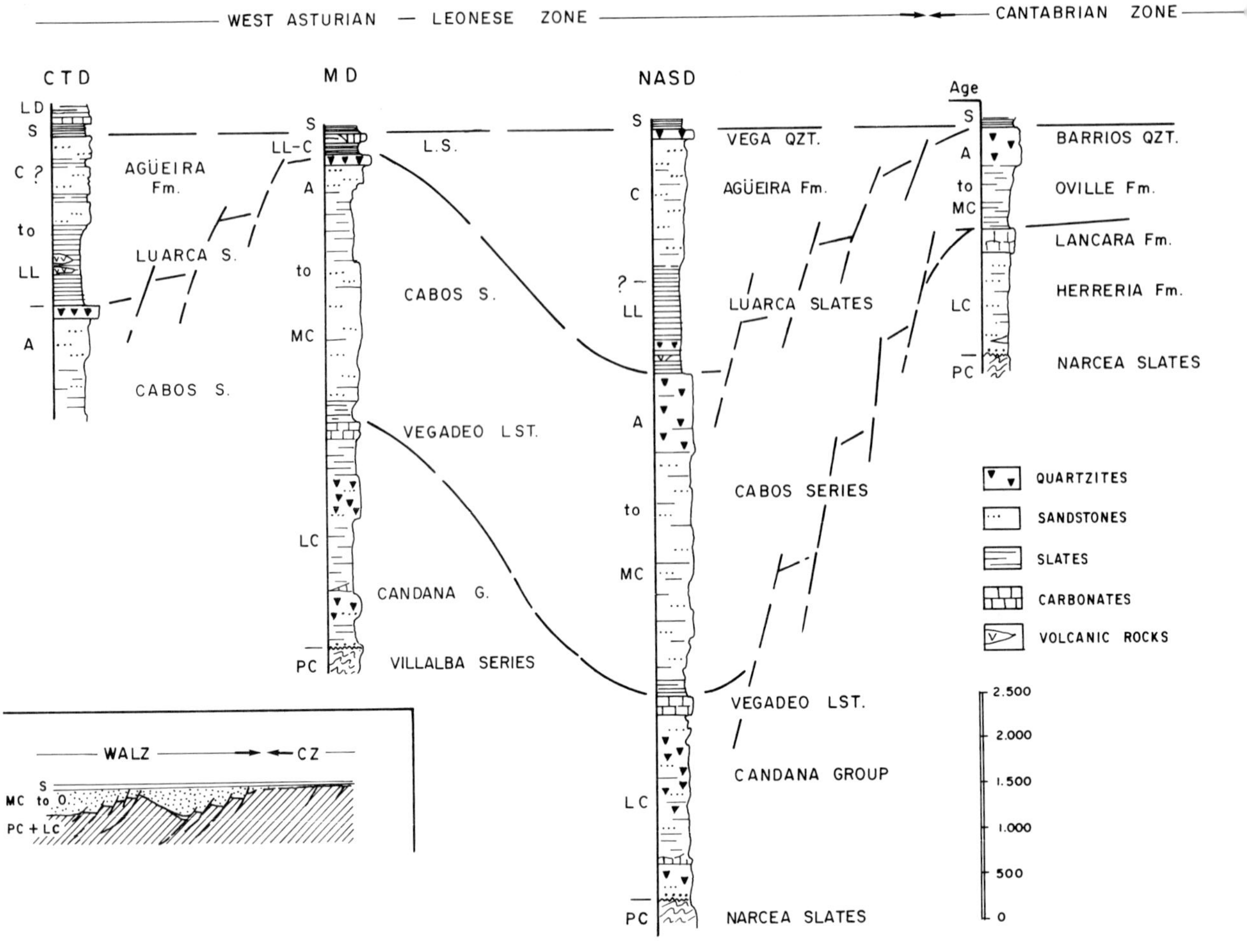

Fig. 2. Stratigraphic summary of the West Asturian-Leonese Zone. *CTD* Caurel-Truchas Domain; *MD* Mondoñedo Domain; *NASD* Navia-Alto Sil Domain. Ages: *PC* Precambrian; *LC* Lower Cambrian; *MC* Middle Cambrian; *O* Ordovician; *A* Arenig; *Ll* Llanvirn and Llandeilo; *C* Caradoc; *S* Silurian; *LD* Lower Devonian

form. There, the Precambrian succession is formed by alternating slates and sandstones, displaying turbiditic facies, and some muddy diamictites (Bosch 1969; Pérez Estaún 1973, 1978). The unconformity with Cambrian rocks is clearly visible (De Sitter 1961; Julivert and Martínez García 1967; Pérez Estaún 1978). In the western part of the antiform, because of the deformation, it is more difficult to determine the nature of the sediments. Nevertheless, they seem to share the characteristics of those in the eastern part, and the unconformity can also be seen here (Matte 1968a, b; Pérez Estaún 1978). Furthermore, there are concordant, igneous bodies in this area showing a gradual transition with the country rocks, which have been described as porphyritic gneisses and amphibolites with volcanic or volcanoclastic protolith (Lotze 1956; Färber and Jaritz 1964; Corretgé and Carpio 1968; Matte 1968b; Pérez Estaún and Martínez Fernández 1978; Marcos et al. 1980). The total thickness of the Precambrian series is unknown due to the presence of several thrusts.

2.3. The Precambrian Rocks of the Mondoñedo Domain

In the western part of the WALZ, there is an important stacking of recumbent folds (Mondoñedo Unit) in the core of which another important outcrop of Precambrian rocks appears. These Precambrian rocks, named Villalba Series (Capdevila 1969), can be divided into two parts: an upper part, consisting of schists with fine sandstone intercalations, and a lower part with a greater proportion of sandstones and some porphyritic levels. The unconformity between these units and the Cambrian series can be mapped (Martínez Catalán 1985). Most of the authors who have studied this area have emphasized the similarities between these rocks and those in the Narcea Antiform (Matte 1967, 1968a, b; Martínez Catalán 1985).

2.4 The Paleozoic Succession

The Paleozoic succession in the WALZ consists of a thick sequence of preorogenic rocks that represent the whole of the Cambrian and Ordovician, a great part of the Silurian and, in some places, the Lower Devonian. All the postorogenic Stephanian rocks lie unconformably over this succession and outcrop mainly in the eastern part of this zone.

The stratigraphy of the NASD will be used as reference in the description that follows, since it is the most complete one, and the differences from the two other domains of the zone will be indicated. The possible correlations between the different lithostratigraphic units of the WALZ domains can be seen in Fig. 2.

2.5 The Cándana Group

The Cándana Group (Lotze 1957) lies unconformably on Precambrian rocks. It has a maximum thickness of up to 2500 m in the NAD. It consists of feldspathic sandstones and slates. Conglomerates can be found not only at its base, but also as frequent microconglomerates towards the upper part of the group. These rocks were deposited in shallow-water marine facies and in continental environments (Farber and Jaritz 1964; Jaritz and Walter 1970; Arboleya 1973; Crimes et al. 1977). This group grades into the Vegadeo Limestone through the Transition Beds. The group has similar characteristics in the three domains, although it is more schistose in the MD and CTD. Within these two domains, four formations can be mapped: Lower Cándana Quartzites, Cándana Slates, Upper Cándana Quartzites and Transition Beds (Walter 1966, 1968). In the Gistral tectonic window (Fig. 1) a more than 2000-m-thick quartzite, called the Gistral Quartzite (Parga Pondal and Aleixandre 1966; Martínez Catalán 1985), can be correlated with the upper part of the Candana Group.

Lower Atdabanian (Early-Lower Cambrian) trilobites and archaeocyathids have been found near the top of the Candana Group (Sdzuy 1961, 1971a; Debrenne and Lotze 1963; Debrenne and Zamarreño 1975; Zamarreño and Perejón 1976; Perejón 1984). The lower part of this group would be Lower Cambrian (Tommotian) or even Precambrian (Vendian; Crimes et al. 1977; Crimes 1987). Diverse ichnofauna data have been referred to in this unit (Färber and Jaritz 1964; Arboleya 1973; Crimes et al. 1977), among which it is worthy to mention the ichnotaxa *Astropolithon* also common in other Lower Cambrian rocks of the Iberian Massif.

2.6 The Vegadeo Limestones

This formation, defined by Barrois (1877, 1882), consists of dolomites and limestones and its thickness ranges from 50 to 500 m. It is characterized by tidal flat facies or other shallow-water facies (Zamarreño et al. 1975; Zamarreño and Perejón 1976).

Although the Vegadeo Formation itself does not contain index fossils, the age of the formation has been determined from fossils collected near the lower and upper limits of the formation. The Transitional Beds of the underlying Candana Group present Early-Lower Cambrian faunas, whereas the overlying beds of the Vegadeo Formation show trilobite-rich Middle Cambrian faunas. This fact suggests an essentially Lower Cambrian age for the Vegadeo Formation, although it may be Middle Cambrian in its upper part.

2.7 The Cabos Series

The Cabos Series (Lotze 1958) is a very thick and extensive detrital succession. In the NASD (Fig. 3), its thickness reaches up to 4000 m (Marcos 1973; Baldwin 1975). It is made up of quartzites, sandstones and slates. In addition, a quartzite level located on its upper part can be mapped. In the other domains, the series is less thick (1000–2000 m) with a larger proportion of slates.

The organic and sedimentary structures of the Cabos Series point to shallow marine environments, although there is, however, a large variety of facies (Färber and Jaritz 1964; Jaritz and Walter 1970; Marcos 1973; Baldwin 1975, 1977; Marcos and Pérez Estaún 1981).

The first and oldest fossils of the Cabos Series recorded are located at the basal 50–100 m, consisting mainly of Middle Cambrian trilobites that represent the *Acadolenus* and the *Solenopleuropsis* sub-

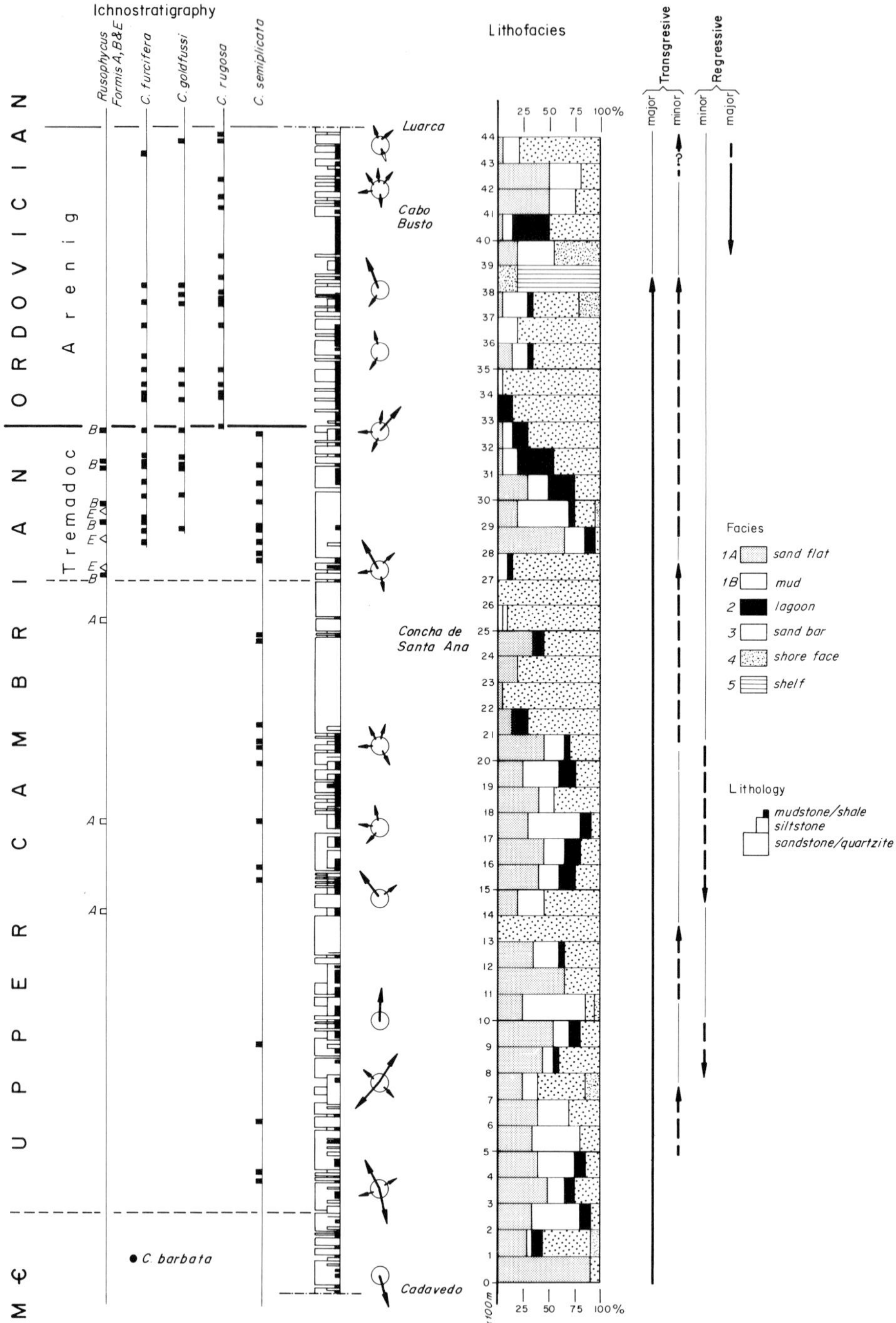

Fig. 3. Ichnostratigraphic subdivision of the Cabos Series. The boundaries between the Middle and Upper Cambrian and the boundary ot the base of Tremadoc are equivocal. The Cambrian-Ordovicien boundary is relatively certain. The principal paleocurrents are shown by *arrows* (After Baldwin 1975)

stages of Sdzuy (1968, 1971b), and on the whole comprise up to 30 different species. The remaining part of the succession contains trilobite traces (Cruziana and Rusophycus) of Upper Cambrian, Tremadoc and Arenig ages (Baldwin 1975, 1977, 1978; Marcos 1973; Marcos et al. 1980; Bastida et al. 1980). In Fig. 3, a column of NASD can be seen showing the stratigraphic position of the main ichnofossils found there. Arenig trilobites and graptolites (Extensus Zone) occur in two localities of the CTD, pertaining to the uppermost part of the Cabos Series, which here is a thick transitional sequence (Rubiana or Campañana beds) with the overlying Luarca Slates (Pérez Estaún 1974b; Hammann 1983; Gutiérrez Marco 1986). Arenig graptolites are also present in the same levels of the MD, which in this case are considered as the basal part of the Luarca Slates (Gutiérrez Marco 1983).

2.8 Luarca Slates

This formation, defined by Barrois (1882), is made up of homogeneous black slates, pyrite-rich, with varying thicknesses and oolitic ironstones in its middle part. In the NASD, the formation reaches its maximum thickness, about 1200 m (Marcos 1973). In the eastern part of this domain, a quartzite level 50-100 m thick (Sabugo Quartzite) can be found at about 200 m from the base of the formation (Marcos 1973; Marcos and Pulgar 1980). In the MD, the thickness of the Luarca Slates is very small (about 200 m), partly due to the unconformity below the Silurian. In the CTD, the thickness of the formation is greater than in the MD and several volcanic levels occur in the Truchas syncline (Matte 1964; Pérez Estaún 1974b).

The main paleontological data on the Luarca Slates in the WALZ have been summarized by Marcos (1973) and Pérez Estaún (1978), and recently revised by Gutiérrez Marco (1986). Most fossil localities belong to the lower part of the formation, which yields Llanvirn graptolites, trilobites, molluscs and echinoderms. In localities that are very close to the base of the formation, Lower Llanvirn graptolites have been found. Nevertheless, Arenigian Cruziana have been reported from the Sabugo Quartzite (Marcos 1973); this suggests that, in the eastern part of the NASD, the lower member was deposited before Llanvirn. In spite of the presence, within the formation, of three localities with Llandeilian trilobites and graptolites (*teretiusculus* Zone), there are not enough arguments to deduce the age of its upper limit, which in the MD and CTD seems to have reached the Caradoc.

2.9 The Agüeira Formation

The Agüeira Formation (Marcos 1970) is formed by metasandstones and slates that frequently display the features of a turbiditic sequence, mainly in the upper part of the formation (Marcos 1970; Jaritz and Walter 1970; Crimes et al. 1974; Pérez Estaún and Marcos 1981). Its thickness in the NASD ranges from 1500 to over 3000 m (from south to north), and the formation is absent from the MD. In the CTD, slates and sandstones are found above the Luarca Slates. The sandstones share some of the characteristics of the turbidites (Pérez Estaún 1978; Pérez Estaún and Marcos 1981). Following the lithological distribution, the variations in thickness and the sedimentary structures, a submarine fan facies has been recognized in the NASD (Pérez Estaún and Marcos 1981; Fig. 4). Some other facies of the Agüeira Formation, in this very same domain, can be interpreted as originating from the external fan or the surrounding basin plain. In the CTD, it is possible to find slate levels with limestones, quartzite or quartz pebbles. These have been interpreted as deposits of glaciomarine origin (Iglesias and Robardet 1980) or as mass flow deposits (Pérez Estaún et al. 1980; Hammann et al. 1982).

In the southern part of the NASD, the Agüeira Formation is limited at its top by a quartzite horizon, 50-200 m thick, called the Vega de Espinareda Quartzite (Matte 1968b; Pérez Estaún 1978).

After data of the three localities with brachiopods and trilobites (Marcos 1970; Pérez Estaún 1974c; Pulgar et al. 1981), a Caradocian age has been inferred for the middle part of the formation. A new locality (Rio Cua, north of Vega de Espinareda) situated in the middle part of the Agüeira Formation (Gutiérrez Marco, unpublished data) has yielded *Onnia malladai* (Oehlert), *Dalmanitina* (*D.*) cf. *acuta* Hammann, *Aegiromena meneghiniana* (Vinassa), *Svobodaina* cf. *havliceki* Villas, *Rafinesquina* sp. and undetermined bryozoans, echiroderms and gastropods which point to an Upper Caradocian-Lower Ashgillian age. Some ichnogenera have also been reported in the Agüeira Formation (Crimes et al. 1974; Pérez Estaún and Marcos 1981).

2.10 The Aquiana Limestones

This formation only appears in the CTD (Nollau 1966, 1968; Pérez Estaún 1978). It outcrops as thin, lenticular bodies with a maximum thickness of about 300 m.

The Aquiana limestones are not dated by means of fossils and they are considered as Asghill due to their correlation with the Ordovician limestones of the Iberian Ranges (Villas 1985). They lie on the Luarca Slates and, locally, even on older units sharing their slightly unconformable character, similar to certain calcareous facies of the Ashgill of scat-

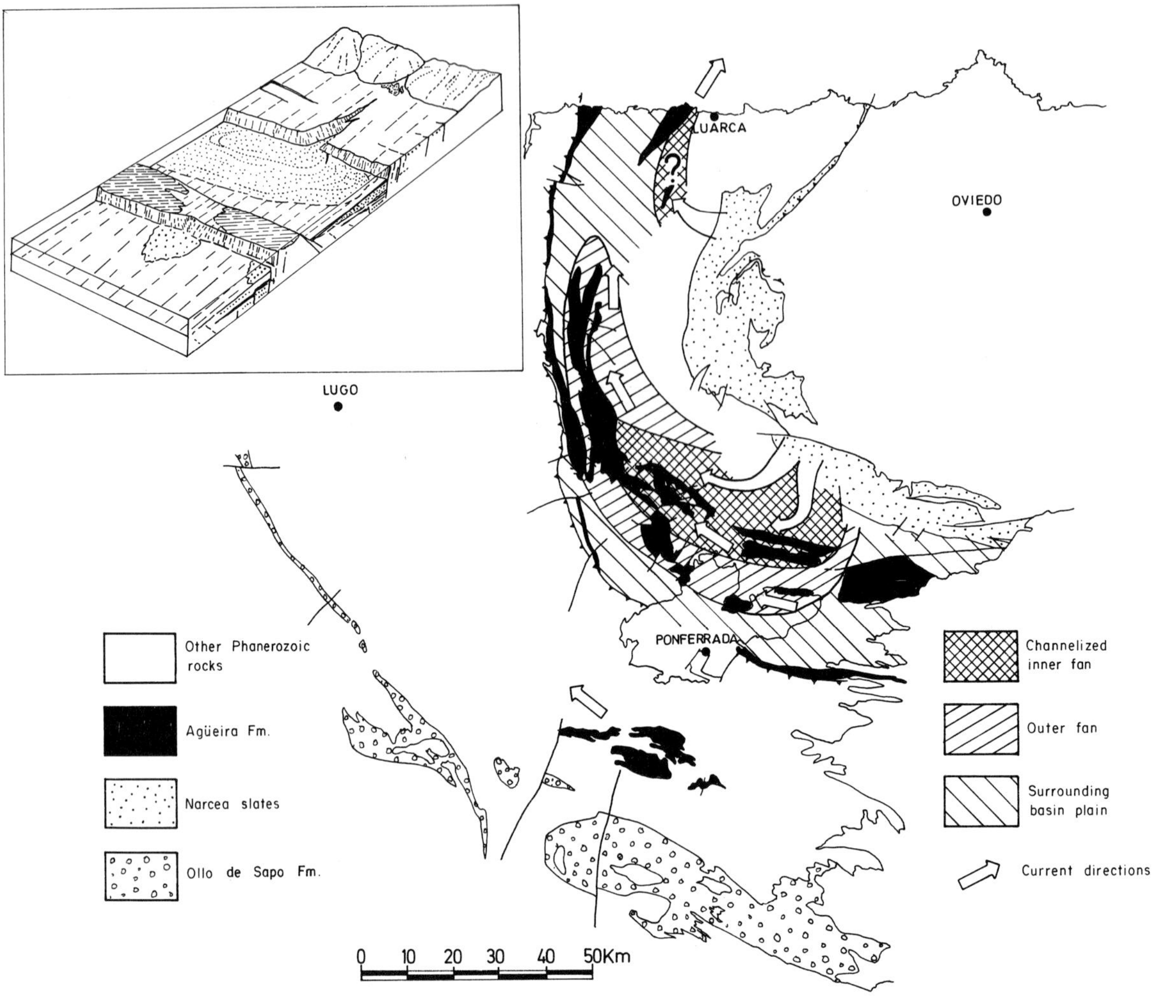

Fig. 4. Outcrops of the Agüeira Formation and facies distribution in the West Asturian-Leonese Zone. *Inset* shows an idealized reconstruction of the basin in the southern branch of the Zone during the Upper Ordovician (After Pérez Estaún and Marcos 1987)

tered areas of the Cantabrian, Central-Iberian and Ossa-Morena Zones (Gutiérrez-Marco 1986).

2.11 The Silurian Rocks

They are composed of black, chloritoid-rich metapelites and scarce lydites with abundant graptolites. Metasandstone and quartzite levels occur in the upper part of the formation. Their thickness reaches 700 m in the NASD and the CTD, but are thinner in the MD. These rocks lie upon different levels. A slight unconformity can be deduced from their outcrop pattern.

As for their age, the oldest graptolite faunas belong to the Lowermost Llandovery and the youngest to the Pridoli (Hernández Sampelayo 1915; Quintero 1962; Walter 1966; Riemer 1966; Matte 1968b; Romariz 1969; Marcos and Philipot 1972; Marcos 1973; Pérez Estaún 1978). Graptolites are the only fossils known in the formation, except for two localities which yielded a few Upper Silurian trilobites and bivalves (Riemer 1966; Nollau 1968).

2.12 Devonian Rocks

In the CTD, over the Silurian rocks, there are shales with limestone levels that are very rich in brachiopods, bryozoans, echiroderms, corals, trilobites, dacryoconarids and vertebrate remains (fish) (Drot and Matte 1967; Pérez Estaún 1978; Truyóls Massoni 1986). Their age comprises several levels within the Lower Devonian, the lowermost ones suggesting a stratigraphical continuity with the Silurian sequence, although the Lockovian have not been paleontologically proven.

2.13 Synorogenic Rocks

To the south of the WALZ, along the boundary to the Centroiberian Zone, some unconformable sediments crop out above the Silurian and Ordovician rocks in a narrow strip, bounded by faults, named San Clodio Carboniferous by several authors (Riemer 1963; Matte 1968b; Pérez Estaún 1974a). The sequence is formed by dark grey slates and sandstones, with some conglomerates (containing pebbles of metamorphic and igneous rocks). The sandstones display the characteristics of the turbidites and a NW trending paleocurrent direction can be inferred (Pérez Estaún 1974a).

The age of this sequence is Upper Devonian or most probably Lower Carboniferous (Riemer 1963; Matte 1968b; Pérez Estaún 1974a). This sequence is synorogenic since, although it is unconformable on Ordovician and Silurian rocks, it is affected by the main deformation phases of the WALZ (Matte 1968b; González Lodeiro and Iglesias Ponce 1981; Martínez Catalán 1985; Barrera et al. in press).

2.14 Stephanian Basins

Several Stephanian basins are located in the WALZ, mainly on the western limb of the Narcea Antiform and in the NASD. Among these, the Bierzo coalfield is the most important one. The thickness of the stratigraphic succession in this basin is of the order of 1800 m.

The Stephanian rocks lie unconformably over other Paleozoic and Precambrian rocks of molassic character. These rocks consist of quartzitic conglomerates, sandstones, lutites and coal seams deposited on continental basins generally related to faults (Almela and del Valle 1963; Pérez Estaún et al. 1982; Matas et al. 1982; Truyols 1983). Floras are found throughout the whole of the succession and have been dated as Stephanian B-C (Alvarez Ramis 1965; Wagner 1965, 1970).

The structure of the Stephanian basins is formed by open synclines and anticlines with steeply dipping axial planes, flanked by faults.

2.15 Paleogeographic Evolution

With the exception of the continental Stephanian deposits and some fluvial deposits of the Candana Group, the whole Paleozoic sequence in the WALZ is formed from the accumulation of marine deposits with intertidal to open shelf facies, including turbiditic levels. This fact, and the existence of igneous rocks with oceanic affinities in the northwestern part of the Iberian Peninsula (see, among others, Arenas et al. 1986), allows one to state that the WALZ was part of a continental margin in which a thick sequence of preorogenic Paleozoic rocks of mainly shallow-water facies were deposited unconformable upon an Upper Proterozoic sequence. The differences in facies and thickness in the Lower Paleozoic succession can be related to an extensional tectonic activity affecting a continental margin. It was this extensional regime that gave rise to some volcanic activity. Some of the zones with greater subsidence created troughs in which turbiditic sediments were deposited (Agüeira Formation, Upper Ordovician). The main subsiding area coincides with the NASD, where about 11000 m of Paleozoic siliciclastic sediments were deposited. The thickness of this clastic sequence decreases both towards the CZ (towards the E), and the MD and, particularly, towards the CIZ (i.e. towards the W and SW). The extensional regime occurred before the deposition of the Silurian sediments.

The stratigraphy of the WALZ differs in some aspects from that of the adjacent zones. In the CZ, the Middle and Upper Cambrian is very poorly represented and the same can be said about the Middle and Upper Ordovician. On the whole, the CZ presents a thinner preorogenic Paleozoic sequence than the WALZ, more important preserved Devonian deposits and contains a very thick synorogenic sequence (Julivert 1971; Marcos and Pulgar 1982). In the CIZ, the most important characteristic is the presence of a major unconformity separating the Lower Ordovician from its substrate, which often consists of a thick and monotonous sequence of Upper Proterozoic age (although Cambrian deposits have been preserved in some places). In the zone that is closest to the WALZ, the pre-Ordovician substrate is formed by the Ollo de Sapo Formation, a volcanic or volcanoclastic sequence probably of Upper Proterozoic age.

In spite of the differences mentioned, there is evidence to suggest that the three contiguous zones (CZ, WALZ, CIZ) were part of the same continental margin. Formations, such as the Armorican

Quartzite (or upper Cabos Series Quartzite), the Vegadeo Limestone and its equivalents, or the Silurian sediments, are well represented in the three zones and, moreover, the monotonous Upper Proterozoic sediments that crop out at both edges of the WALZ probably continue below the Paleozoic deposits of the CZ and are very similar to most of the Upper Proterozoic sediments of the CIZ.

References

Aller J, Bastida F, Brime C, Pérez Estaún A (1987) Cleavage and its relation with metamorphic grade in the Cantabrian Zone (Hercynian of north-west Spain). Sci Geol Bull 40: 255–272

Almela A, Valle J del (1963) Estudio geológico de la zona Ponferrada-Tremor de Abajo. Breviora Geol Astúrica 7: 2–26

Alvarez Ramis C (1965) Primera aportación al estudio de los "Sphenopteris" del Carbonífero español. a) Estefaniense: primera parte. Estud Geol 21: 1–141

Arboleya ML (1973) Nota sobre la icnofauna del Cámbrico de la costa asturiana entre Cudillero y Ballota (NW de España). Breviora Geol Astúrica XVII (3): 37–42

Arenas R, Gil Ibarguchi JI, González Lodeiro F et al. (1986) Tectonostratigraphic units in the complexes with mafic and related rocks of the NW of the Iberian massif. Hercynia II (2): 87–110

Baldwin CT (1975) The stratigraphy of the Cabos Series in the section between Cadavedo and Luarca, province of Oviedo, north-west Spain. Breviora Geol Astúrica 19: 4–9

Baldwin CT (1977) The stratigraphy and facies associations of trace fossils in some Cambrian and Ordovician rocks of northwestern Spain. In: Crimes TP, Harper JC (eds) Trace fossils 2. Geol J Spec Issue 9: 9–40

Baldwin CT (1978) A comparison of the stratigraphy and depositional processes in the Cambro-Ordovician rocks of the Cantabrian and West-Asturian-Leonese Zones, NW Spain. In: Julivert M, Capdevila R, Tex E den, Matte P, Ribeiro A (eds) Geología de la parte N del Macizo Ibérico (Tomo homenaje I. Parga Pondal). Ediciones do Castro, Sada, La Coruña, pp 43–70

Barrera JL, Farias P, González-Lodeiro F et al. (in press) Mapa Geológico Nacional E. 1:200000 Orense-Verín. Inst Geol Min Esp, pp 17–27

Barrois C (1877) Relations d'un voyage gèologique en Espagne. Ann Soc Géol Nord 4: 292–303

Barrois C (1882) Recherches sur les terrains anciens des Asturies et de la Galice. Mem Soc Géol Nord, t 2, nº 1, Lille

Bastida F, Marcos A, Pérez Estaún A, Pulgar JA, Galán J, Vargas I (1980) Mapa Geológico de España a escala 1:50000, Hoja de Naviego (75), 2ª serie (MAGNA). Inst Geol Min Esp, pp 1–35

Bosch WJ van den (1969) Geology of the Luna-Sil region. Cantabrian Mountains (NW Spain). Leidse Geol Meded 44: 137–225

Capdevila R (1969) Le métamorphisme régional progressif et les granites dans le segment hercynien de Galice nord oriental (NW de l'Espagne). Thesis, Univ Montpellier

Corretgé LG, Carpio V (1968) Los ortoneises básicos de Pola de Allande (Asturias). Breviora Geol Astúrica 12: 14–16

Crimes TP (1987) Trace fossils and correlation of Late Precambrian and Early Cambrian strata. Geol Mag 124: 97–189

Crimes TP, Marcos A, Pérez Estaún A (1974) Upper Ordovician turbidites in western Asturias. A facies analysis with particular reference to vertical and lateral variations. Palaeogeogr Palaeoclimatol Palaeoecol 15: 169–184

Crimes TP, Legg I, Marcos A, Arboleya ML (1977) Late Precambrian-Lower Cambrian trace fossils from Spain. In: Crimes TP, Harper JC (eds) Trace fossils 2. Geol J Spec Issue 9: 91–138

Debrenne F, Lotze F (1963) Die Archaeocyatha des spanischen Kambriums. Akad Wiss Lit Abh Math Naturwiss Kl 2: 1–38

Debrenne F, Zamarreño I (1975) Sur la faune d'archeocyathes de la Formation Vegadeo et leur rapor avec la distribution des facies carbonatés dans le NW de l'Espagne. Breviora Geol Astúrica 19: 17–27

Drot J, Matte P (1967) Sobre la presencia de capas del Devoniano en el límite de Galicia y León (NW de España). Notas Comun Inst Geol Min Esp 93: 87–92

Färber A, Jaritz W (1964) Die Geologie des westasturischen Küstengebietes. Geol Jahrb 81: 679–783

González Lodeiro F, Iglesias Ponce M (1981) Mapa Geológico Nacional E. 1:50000, 156 Monforte de Lemos. Inst Geol Min Esp, pp 1–28

Gutiérrez Marco JC (1983) Nuevos datos para el conocimiento bioestratigráfico de las Pizarras de Luarca en la Zona Asturoccidental-leonesa (NW de España). Cuad Lab Xeol Laxe 6: 71–74

Gutiérrez Marco JC (1986) Graptolitos del Ordovícico español. Thesis, Univ Madrid

Hammann W (1983) Calymenea (Trilobita) aus dem Ordovizium von Spanien; ihre Biostratigraphie, Ökologie und Systematik. Abh Senckenb Naturforsch Ges 542

Hammann W, Robardet M, Romano M (1982) The Ordovician System in southwestern Europe (France, Spain and Portugal). Int Union Geol Sci 11: 1–47

Hernández Sampelayo P (1915) Fósiles de Galicia. Nota sobre la fauna paleozoica de la provincia de Lugo. Bol Inst Geol Min Esp 36: 277–303

Iglesias M, Robardet M (1980) El Silúrico de Galicia media (Central), su importancia en la paleogeografía varisca. Cuad Lab Xeol Laxe 1: 99–115

Jaritz W, Walter R (1970) Faciesuntersuchungen in Altpalözoikum Nordwest-Spaniens/Asturien und Prov. Lugo. Geol Jahrb 88: 509–552

Julivert M (1971) Decollement tectonics in the Hercynian Cordillera of NW Spain. Am J Sci 270: 1–29

Julivert M, Martínez García E (1967) Sobre el contacto entre el Cámbrico y el Precámbrico en la parte meridional de la Cordillera Cantábrica y el papel del Precámbrico en la orogénesis herciniana. Acta Geol Hisp 2: 107–110

Julivert M, Fontboté JM, Ribeiro A, Conde LE (1972) Mapa Tectónico de la Península Ibérica y Baleares E. 1:1000000. Inst Geol Min Esp, pp 1–113

Lotze F (1945) Zur Gliederung der Varisziden der Iberischen Meseta. Geotek Forsch 6: 78–92 (Trad. en: Publicaciones Extranjeras sobre Geol Esp 5: 149–166)

Lotze F (1956) Das Präkambrium Spaniens. Neues Jahrb Geol-Paläontol Monatsh 8: 373–380 (Trad. en: Notas Comun Inst Geol Min Esp 60: 227–240)

Lotze F (1957) Zum Alter nordwestspanischer Quartzit-Sandstein-Folgen. Neues Jahrb Geol Paläontol 10: 128–139

Lotze F (1958) Zur Stratigraphie des spanischen Kambriums.

Geologie 7: 727–750 (Trad. en: Not Comun Inst Geol Min Esp 61: 131–164)

Marcos A (1970) Sobre la presencia de un flysch del Ordovícico Superior en el Occidente de Asturias. Breviora Geol Astúrica 14: 13–28

Marcos A (1973) Las Series del Paleozoico inferior y la estructura herciniana del occidente de Asturias (NW de España). Trab Geol 6: 1–113

Marcos A, Pérez Estaún A (1981) La estratigrafía de la Serie de los Cabos en la Zona de Vegadeo (Zona Asturoccidental-Leonesa, NW de España). Trab Geol 11: 89–94

Marcos A, Philippot A (1972) Nota sobre el Silurico del occidente de Asturias (NW de España). Breviora Geol Astúrica XVI (3): 39–41

Marcos A, Pulgar JA (1980) Mapa Geológico de España a escala 1:50000, Hoja de Luarca (11) 2ª serie (MAGNA). Inst Geol Min Esp, pp 1–24

Marcos A, Pulgar JA (1982) An approach to the tectonostratigraphic evolution of the Cantabrian foreland thrust and fold belt, Hercynian Cordillera of NW Spain. Neues Jahrb Geol-Paläontol Abh 163: 256–260

Marcos A, Martínez FJ, Julivert M, Pulgar JA (1980) Mapa Geológico de España a escala 1:50000 Hoja de Cangas del Narcea (50) 2ª serie (MAGNA). Inst Geol Min Esp, pp 1–31

Martínez Catalán JR (1985) Estratigrafía y estructura del domo de Lugo (Sector Oeste de la zona Asturoccidental-leonesa). Corpus Geol Gallaeciae (2ª Serie) 2, pp 1–291

Matas J, Abejero V, Fernández L, Fernández Pello J (1982) Mapa Geológico Nacional E. 1:50000, 127 Noceda. Inst Geol Min Esp, pp 1–32

Matte P (1964) Sur le volcanisme silurien du synclinal de Truchas (NW de l'Espagne). CR Somm Soc Geol Fr 2: 57–58

Matte P (1967) Le Précambrien supérieur schisto-gréseux de l'Ouest des Asturies (Nord-Ouest de l'Espagne) et ses relations avec les séries précambriennes plus internes de l'arc galicien. CR Acad Sci Paris 264: 1769–1772

Matte P (1968a) Précisions sur le Précambrien supérieur schisto-gréseux de l'Ouest des Asturies. Comparaison avec les autres affleurements précambriens du Nord-Ouest de l'Espagne. Rev Géogr Phys Géol Dynam (2ª serie) 10: 205–211

Matte P (1968b) La structure de la virgation hercynienne de Galice (Espagne). Rev Géol Alpine 44: 1–128

Nollau G (1966) El desarrollo estratigráfico des Paleozoico en el Oeste de la Provincia de León (España) Notas Comun Inst Geol Min Esp 88: 31–48

Nollau G (1968) Stratigraphie, Magmatismus und Tektonik der Montes de León zwischen Astorga und Ponferrada und Nordwest-Spanien. Geotekt Forsch 27: 71–146

Parga Pondal I, Aleixandre T (1966) La arenisca ortocuarcítica del Gistral (Lugo). Not Comun Inst Geol Min Esp 87: 59–90

Perejón A (1984) Bioestratigrafía de los arqueociatos en España. Cuad Geol Ibérica 9: 213–265

Pérez Estaún A (1973) Datos sobre la sucesión estratigráfica del Precámbrico y la estructura del extremo Sur del antiforme del Narcea (NW de España). Breviora Geol Astúrica XVII (1): 5–16

Pérez Estaún A (1974a) Aportaciones al conocimiento del Carbonífero de San Clodio (Prov. de Lugo). Breviora Geol Astúrica 18: 3–8

Pérez Estaún A (1974b) Algunas precisiones sobre la sucesión ordovícica y silúrica en la región de Truchas. Breviora Geol Astúrica 18: 23–25

Pérez Estaún A (1974c) La sucesión ordovícica en el dominio del Alto Sil (Zona Asturoccidental-leonesa). Breviora Geol Astúrica 18: 53–57

Pérez Estaún A (1978) Estratigrafía y estructura de la rama Sur de la Zona Asturoccidental-leonesa. Mem Inst Geol Min Esp 92: 1–150

Pérez Estaún A, Marcos A (1981) La Formación Agüeira en el sinclinorio de Vega de Espinareda: aproximación al modelo de sedimentación durante el Ordovícico Superior en la Zona Asturoccidental-leonesa (NW de España). Trab Geol 11: 135–145

Pérez Estaún A, Martínez-Fernández FJ (1978) El Precámbrico del antiforme del Narcea en el sector de Tineo-Cangas de Narcea (NW de España). Trab Geol 10: 367–375

Pérez Estaún A, Marquínez J, Ortega E (1980) La sucesión ordovícica y la estructura de la región de Silván (La Cabrera, León). Breviora Geol Astúrica 24: 17–24

Pérez Estaún A, Pulgar JA, Bastida F, Marcos A, Sánchez de la Torre L, Galán J, Vargas I, Ruiz F (1982) Mapa Geológico Nacional E. 1:50000, 126 Vega de Espinareda. Inst Geol Min Esp, pp 1–35

Pulgar JA, Bastida F, Marcos A, Pérez Estaún A, Vargas I, Ruiz F (1981) Mapa Geológico Nacional E. 1:50000, 100 Degaña. Inst Geol Min Esp, pp 1–35

Quintero I (1962) Graptolites de la provincia de Lugo. Notas Comun Inst Geol Min Esp 65: 61–82

Riemer W (1963) Entwicklung des Paläozoikums in der südlichen Provinz Lugo (Spanien). Neues Jahrb Geol-Paläontol Abh 117: 273–285

Riemer W (1966) Datos para el conocimiento de la estratigrafía de Galicia. Notas Comun Inst Geol Min Esp 81: 7–20

Romariz C (1969) Graptolitos silúricos del Noroeste Peninsular. Comun Serv Geol Port 53: 107–156

Sdzuy K (1961) Das Kambrium Spaniens. Teil II. Trilobiten. Akad Wiss Lit Abh Math-Naturwiss 7–8: 217–408 (499–690)

Sdzuy K (1968) Bioestratigrafía de la griotte cámbrica de los Barrios de Luna (León) y otras sucesiones comparables. Trab Geol 2: 45–67

Sdzuy K (1971a) Acerca de la correlación del Cámbrico inferior en la Península Ibérica. Publ Primer Congr Hispano-Luso-Americano Geol Econ 2: 753–768

Sdzuy K (1971b) La subdivisión bioestratigráfica y la correlación del Cámbrico Medio de España. Publ Primer Congr Hispano-Luso-Americano Geol Econ 2: 769–782

Sitter LU de (1961) Le Précambrien dans la Chaîne Cantabrique. C R Soc Geol Fr 9: 253

Truyóls J (1983) El Carbonífero superior al Oeste de la región de los Picos de Europa. In: Martínez Díaz C (ed) Carbonífero y Pérmico de España. CR X Congr Int Strat Géol Carbonifère, Madrid 1983, pp 82–94

Truyóls-Massoni M (1986) Nowakia acuaria (Praguiense, Devónico inferior) de la Zona Asturoccidental-leonesa (NO de España). Breviora Geol Astúrica 27: 12–16

Villas E (1985) Braquiópodos del Ordovícico Medio y Superior de las Cadenas Ibéricas Orientales. Mem Mus Paleontol Univ Zaragoza 1: 1–223

Wagner RH (1965) Stephanian B flora from the Ciñera-Matallana coalfield (León) and neighbouring outliers. III: Callipteridium and Alethopteris. Not Comun Inst Geol Min Esp 78: 5–70

Wagner RH (1970) An outline of the Carboniferous stratigraphy of north-western Spain. In: Streel M, Wagner RH (eds) Colloque sur la stratigraphie du Carbonifère, Congrés-Colloque Univ Liege 55: 429–463

Walter R (1966) Resultados de investigaciones geológicas en

el Noreste de la provincia de Lugo (NO-España). Notas Comun Inst Geol Min Esp 89: 7-16
Walter R (1968) Die Geologie in der nordöstlichen Provinz Lugo (nordwest Spanien). Geotekt Forsch 27: 3-70
Zamarreño I, Perejón A (1976) El nivel carbonatado del Cámbrico de Piedrafita (Zona Asturoccidental-leonesa, NW de España): tipos de facies y faunas de arqueociatos. Breviora Geol Astúrica 20: 17-32
Zamarreño I, Hermosa JL, Bellamy J, Rabu D (1975) Litofacies del nivel carbonatado del Cámbrico de la región de Ponferrada (Zona Asturoccidental-leonesa, NW de España). Breviora Geol Astúrica 19: 40-48

3 Structure

J. R. Martinez Catalan[1], A. Perez Estaun[2], F. Bastida[2], J. A. Pulgar[2] and A. Marcos[2]

3.1 General Description

The structural configuration of the West Asturian-Leonese Zone (WALZ) may be appreciated from the cross-sections of Fig. 1, the location of the main structures are depicted in Fig. 1 of the previous chapter dealing with Stratigraphy. Three major types of structure, originally recumbent folds with E vergence, thrust faults and large open folds with steep axial surfaces, are superimposed in that order. These structures determine, together with an important normal fault to the W known as the Vivero Fault, the structural pattern of the zone. Of special interest among the recumbent folds are the Mondoñedo-Lugo-Sarria anticline (Matte 1968b) and the Villaodrid syncline (Walter 1966), both of which are in the Mondoñedo Domain (MD). The reverse limb common to both folds attains 20 km in cross-sectional length. Among the thrusts, the one which bounds the Mondoñedo thrust sheet (Marcos 1973; Martínez Catalán 1980, 1985; Bastida et al. 1986) is of interest: this unit is an allochthonous structure that occupies most of the western part of the WALZ. The largest upright folds, the Lugo Antiform, and the Bretoña Synform, are found in the western part of the WALZ. Besides these, another system of folds with axes perpendicular to the system described above is well developed in the western part of the WALZ; these will be referred to as the radial set. The two systems of large open folds with steep axial surfaces gave rise to a type 1 interference pattern (Ramsay 1967), leading to the outcropping of the footwall unit of the Mondoñedo thrust sheet in two tectonic windows corresponding to domes: the Gistral and Monte Carballosa tectonic windows (see Fig. 1 of the previous chapter). The sedimentary succession in these windows corresponds to the Navia-Alto Sil Domain (NASD) and consist of an Upper Proterozoic and Lower Cambrian (Cándana) sequence. The Cándana Group includes here a thick quartzitic formation, the Gistral Quartzite, which attains a thickness of 2500 m and allows the structure to be determined in the tectonic windows. Around them, the base of the Mondoñedo thrust sheet is strongly deformed in a band whose width can be estimated at some 3 km.

The eastern part of the WALZ, i. e. what essentially constitutes the NASD, also exhibits recumbent folds, although they are smaller than those of the western part, and thrusts whose displacement is difficult to calculate. The thrust sheets are most abundant and important in the easternmost part (Fig. 1), representing the limit with the Cantabrian Zone (CZ). Open folds with steep axial surfaces are abundant in this eastern part, although they are much smaller than those of Lugo and Bretoña. The interference pattern between recumbent and steep folds are of type 3 (Ramsay 1967), as in the MD. However, whereas in the MD both the normal and the reverse limbs of the the first folds appear folded, in the NASD only the normal limbs have been folded, together with the axial surfaces and the axial plane cleavage (Figs. 1, 2). This is a result of a smaller degree of tightening of the first folds of the eastern region (Pulgar 1980), which leads to a greater interlimb angle and a greater original dip of the reverse limbs.

3.2 Phases of Deformation

The structures related to the Variscan orogeny are attributable to three main deformational phases that were developed under metamorphic conditions. The phases, which gave rise to their corresponding cleavages, can be considered as having a Lower Carboniferous age in the WALZ: on one hand, there is the synorogenic flysch of San Clodio, whose age is probably Lower Carboniferous (Pérez Estaún 1974a) and, on the other, in the Sierra de la Demanda, which is the continuation towards the E

[1] Departamento de Geología, Facultad de Ciencias, Universidad, 37008 Salamanca, Spain
[2] Departamento de Geología, Facultad de Geología, Universidad, 33005 Oviedo, Spain

R. D. Dallmeyer and E. Martínez García (Eds.)
Pre-Mesozoic Geology of Iberia

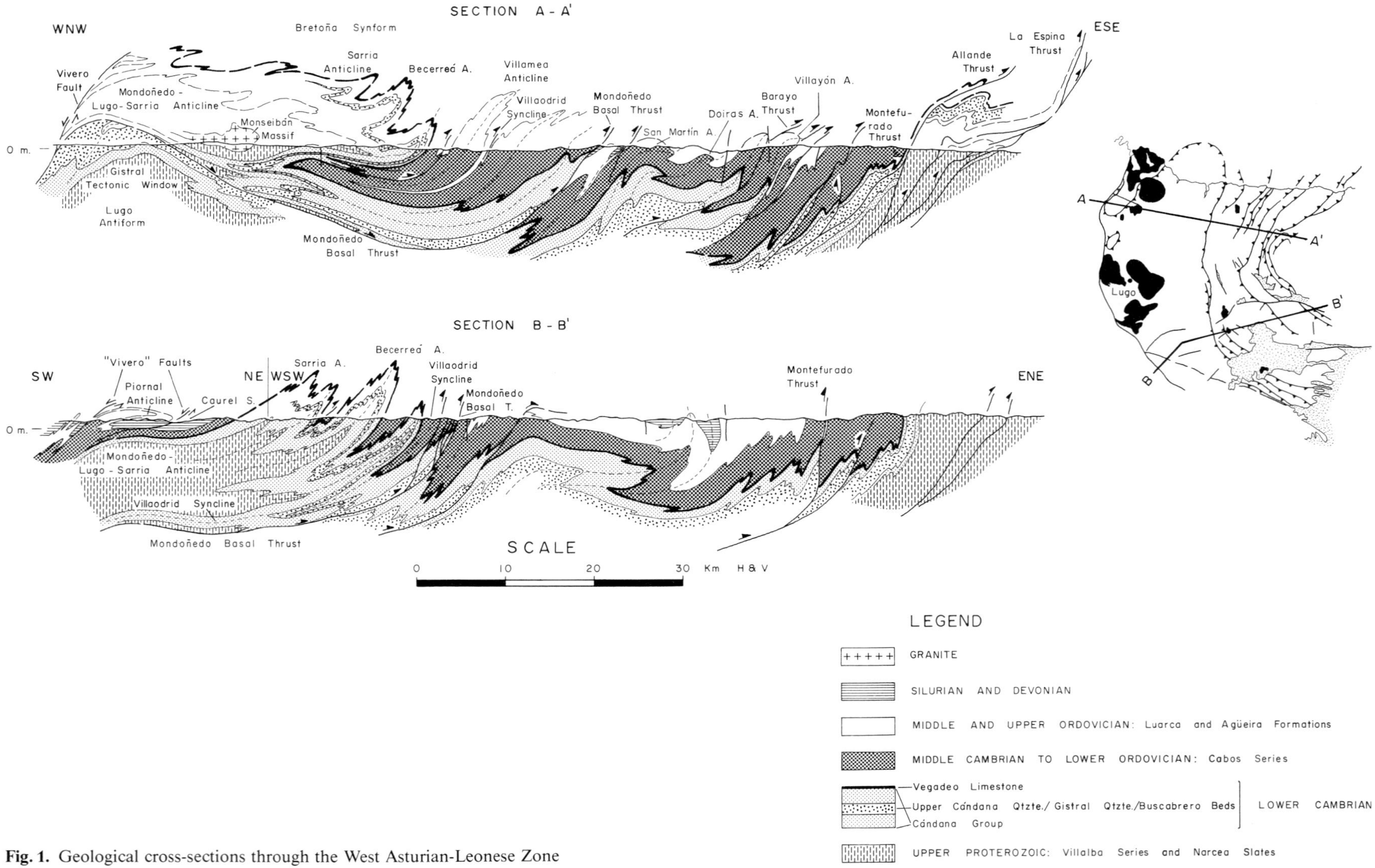

Fig. 1. Geological cross-sections through the West Asturian-Leonese Zone

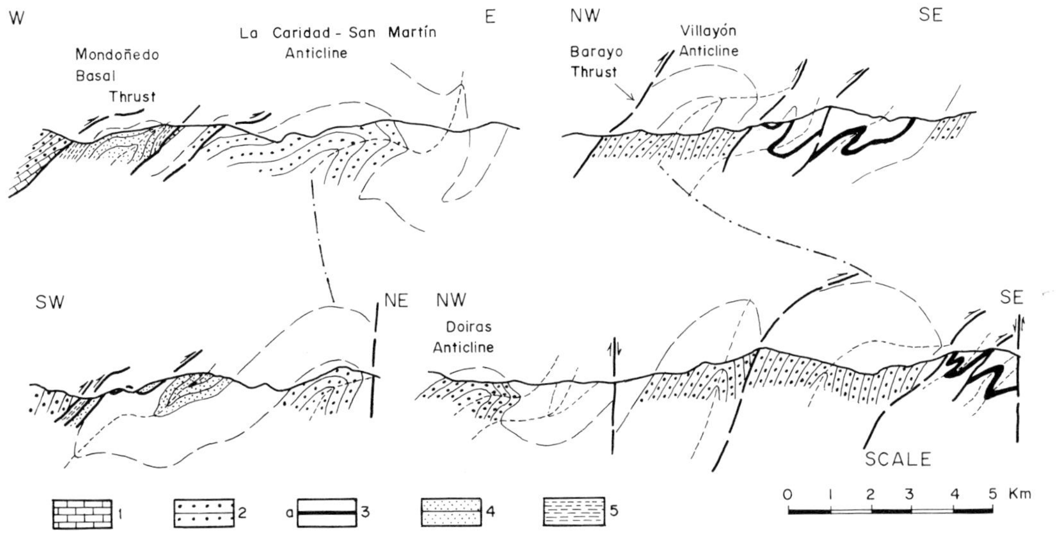

Fig. 2. Interference patterns of Ramsay's type 3 between D_1 and D_3 folds, in the Navia-Alto Sil Domain (After Pulgar 1980). *1* Vegadeo Limestone; *2* Cabos Series; *3* Luarca Slates (a-Sabugo Quartzite); *4* Agüeira Formation; *5* Silurian black slates

of the WALZ, Upper Westphalian B deposits postdate the third phase structures (Colchen 1974). The deformation probably began in the western, internal parts, and then migrated progressively towards the CZ.

The first phase, D_1, produced the recumbent folds, with a vergence towards the external part of the orogene (E, NE). Related to these folds, a primary cleavage, S_1, was developed, together with numerous minor folds. Two different classes of major folds can be identified: folds that are not very large, with fairly constant wavelengths, such as the largest that can be found in the NASD and that on the normal limb of the Mondoñedo-Lugo-Sarria anticline, and other much larger folds such as the above-mentioned anticline and the Villaodrid syncline (Fig. 1). The characteristics of the folds change from E to W. The change, which is indicated by the variation in the size and interlimb angle of the major folds, as has been pointed out above, is also evident in the minor folds (Fig. 3). Analysis of the latter in the E-W coastal cross-section through the WALZ shows the following characteristics in folds developed in quartzitic lithologies (Bastida 1980):

1. The fold profile varies between rounded and chevron-like.
2. The frequent development of fold trains with a dominant wavelength.
3. The interlimb angle decreases, in general, towards the W.
4. The amplitude (Hudleston 1973) increases towards the western part.
5. The superimposed flattening, calculated with the method of Hudleston (1973) in flattened parallel folds, increases towards the western part.

The above characteristics show that the minor folds are structures generated by buckling. The geometry of the folded layers, together with the generalized development of cleavage, reflect the existence of a superimposed flattening whose intensity seems to increase towards the west, in agreement with what may be inferred from the geometry of the major folds.

The characteristics of the associated cleavage also change from E to W. In pelitic rocks, the S_1 cleavage is a fine-grained slaty cleavage in the eastern part of the WALZ and a coarser schistosity in the western part. In the competent lithologies, it is a rough or spaced cleavage, except in the deep parts of the Mondoñedo nappe, where these rocks also show a well-developed schistosity.

A vertical deformation gradient for the first phase of deformation may also be seen in the Mondoñedo-Lugo-Sarria anticline: the major D_1 folds are more isoclinal in its lower and upper parts than in its frontal part (Fig. 1). Of special interest for the upper part of this structure are the isoclinal recumbent folds of Caurel and Piornal (Matte 1986b), in the Caurel-Truchas Domain (CTD), with a common reverse limb of 8 or 9 km in length (Fig. 1 B–B′). A greater interlimb angle and a lesser superimposed flattening can be also measured in the minor folds of the hinge zone of the Mondoñedo-Lugo-Sarria anticline in relation to those of its normal and reverse limbs.

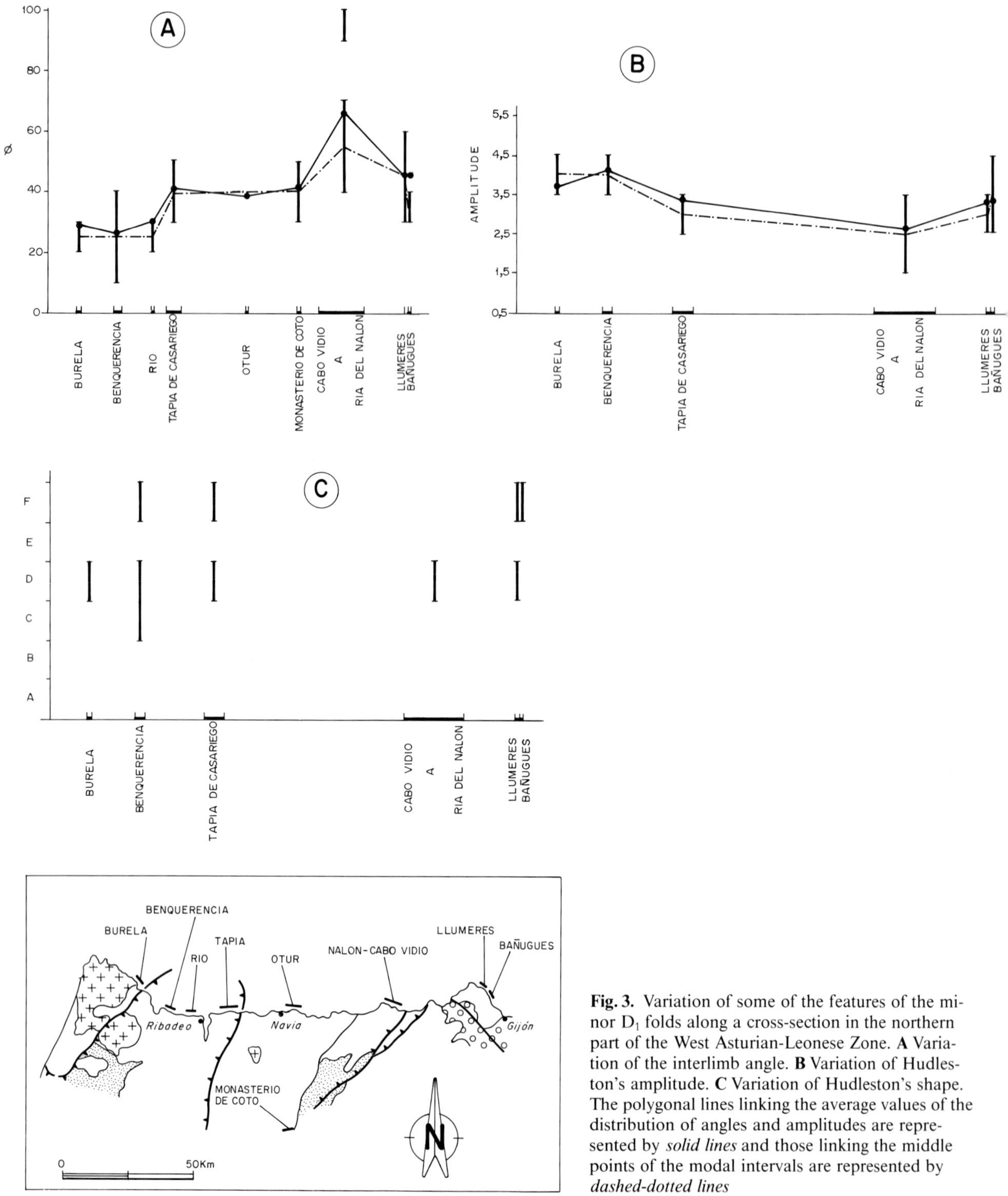

Fig. 3. Variation of some of the features of the minor D_1 folds along a cross-section in the northern part of the West Asturian-Leonese Zone. **A** Variation of the interlimb angle. **B** Variation of Hudleston's amplitude. **C** Variation of Hudleston's shape. The polygonal lines linking the average values of the distribution of angles and amplitudes are represented by *solid lines* and those linking the middle points of the modal intervals are represented by *dashed-dotted lines*

Turning our attention to the major folds, the main characteristics of those of second order, i. e. the largest in the NASD and those on the normal limb and in the hinge zone of the Mondoñedo-Lugo-Sarria anticline, are the constant amplitudes and wavelengths (Fig. 1). This suggests an origin by means of a buckling mechanism acting on a thick multilayer. However, the magnitude of the first order folds, the Mondoñedo-Lugo-Sarria anticline and the Villaodrid syncline, and the particularly intense deformation, deduced from the thinning of the formations and from the geometry of the minor folds on the reverse limb common to both folds, suggest a generating mechanism such as that proposed by Ramsay et al. (1983) for the Morcles nappe in the Helvetic Alps: a generalized subhorizontal ductile shear that affects the whole sedimentary sequence but which acquires a specially strong intensity in some bands which are the continuation of shear zones in the basement. In the case of the WALZ, these bands correspond to the reverse limb of the Mondoñedo-Lugo-Sarria anticline and, probably, to the normal limb too. Bastida et al. (1986) consider both bands as the upwards continuation of two shear zones of crustal importance that would have affected the basement. The arrangement of the second order major folds and the generation of the S_1 cleavage can also be explained by such a model (Fig. 4), which has received considerable attention from Casey and Huggenberger (1985).

The second deformation phase, D_2, was responsible for the appearance of thrust-type structures and associated subhorizontal shear zones. The most important of these structures is the basal fault and shear zone of the Mondoñedo thrust sheet, particularly in view of the fact that its internal zones crop out in the tectonic windows of the Lugo Antiform. Fault rocks of different types, together with minor structures, were developed in these deep zones and also in the more superficial zones outcropping in the frontal region of the thrust sheet. There, a narrow shear zone (200 m) was developed locally. This exhibits small D_2 folds, tension gashes and breccias in quartzitic rocks, and a crenulation cleavage in metapelites. In the internal zones, essentially in those surrounding the two tectonic windows, a broad ductile shear zone (up to 3 km thick) may be identified above the thrust fault. This shear zone also appears in the Puertomarín Dome, to the S of Lugo (see Fig. 1 of the previous chapter), an interference structure in which erosion has not reached the footwall unit. The shear zone shows abundant minor folds with a pronounced dispersion of their axial directions, curvilinear hinges and a general vergence towards the E, together with a S_2 crenulation cleavage or schistosity, transverse mineral lineations and mylonitic or phyllonitic rocks (Bastida and Pulgar 1978; Pulgar 1980; Martínez Catalán 1985). An inspection of the Upper Cándana Quartzite in Fig. 1 A–A′ suggests a minimum displacement of the Mondoñedo nappe over its basal thrust in the order of 40 km (Bastida et al. 1986).

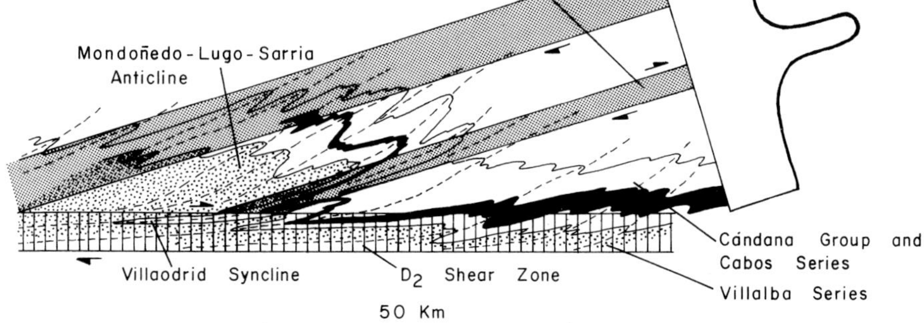

Fig. 4. Proposed relationship between the D_1 first-order folds and the development of bands with a greater intensity of simple shear during the first phase. The obliquity between the D_2 ductile shear zone *(ruled)* basal to the Mondoñedo nappe and the previous ones *(grey)* can also be appreciated. *Dots* Upper Proterozoic; *black* Upper Cándana/Gistral Quartzites

Other thrust faults exist further to the E, the most important of which are those of Barayo, Montefurado, Allande and La Espina (Marcos 1973, see Figs. 1 of this and the previous chapter). Their displacements are difficult to evaluate since they only outcrop at relatively shallow levels. However, the stratigraphic throw and the differences in facies and intensities of the regional metamorphism between the hangingwall and footwall units suggests important displacements in the order of 5 to 15 km for some of them.

The second deformation phase represents the continuation of the subhorizontal shearing, initiated during the first phase, by the development of localized shear zones that appear oblique to the more strongly deformed bands of the first phase, as may be seen by comparing the position of the reverse limb of the Mondoñedo-Lugo-Sarria anticline with the D_2 shear zone and the Mondoñedo basal thrust (Figs. 1, 4).

The third deformation phase, D_3, produced large open folds approximately homoaxial with the first folds, together with minor folds and the local development of an S_3 crenulation cleavage. The refolding during this phase gave rise to type 3 interference patterns with certain peculiarities already mentioned in the general description of the structure. The interference in the western part can be seen in Fig. 1, whereas a typical interference of the eastern domain is shown in Fig. 2 (Pulgar 1980). In the eastern part of the WALZ, the minor structures, such as folds and crenulation cleavages, were developed in bands that frequently coincide with the subhorizontal limbs of D_3 folds, whereas in the upright limbs of these folds the previous reference surfaces were rotated to vertical without the development of new fabrics or structures (Pulgar 1980). The minor folds have subhorizontal axes; steep axial surfaces, locally with dips towards the E; crenulation cleavages parallel to the axial surface; interlimb angles ranging between 10° and 150°; type D to F forms (Hudleston 1973), and amplitudes ranging between 2 and 4 (Fig. 5). A crenulation cleavage was developed in the pelitic rocks and a rough or spaced cleavage in the psammitic rocks, by mechanisms among which pressure solution seems to have played an essential role. The crenulation cleavages often gave rise to the development of a tectonic banding.

Other structures of interest are the radial folds and normal faults and, in particular, the Vivero Fault. The first are open transverse steep folds, with some associated minor structures, such as minor folds and crenulation cleavages. They are particularly well-developed in the western sector, where they refold the D_1 and D_2 structures and create, together with the D_3 folds, a dome and basin interference pattern. The Vivero Fault is a N-S striking normal fault, downthrown to the W, which cuts the Mondoñedo basal thrust and appears to be related to the third phase. Its maximum throw, comprising that of the fault itself and that of an associated ductile shear zone, is about 10 to 12 km in the N part, as inferred from stratigraphic, structural and petrological data (Martínez Catalán 1985). Towards the S its throw decreases and in the Caurel Domain it seems to bifurcate into two normal faults (see Figs. 1 of this and the previous chapter), with small displacements (Barrera et al. 1989).

The D_3 Lugo Antiform and the Vivero Fault may be structures related to the development of new deeper shear zones below the Mondoñedo thrust sheet. Shears of this type may be reflected on the surface by antiforms (Coward 1983) and it is probable that some adjustments, as normal faults, may have taken place on its limbs.

Other normal faults can be observed in the zones close to the front of the Mondoñedo thrust sheet (Marcos 1976; Marcos et al. 1976), and the Allande thrust also seems to have acted as a normal fault. These structures, as well as subhorizontal kink bands, are post-D_3 structures which were probably related to extensional movements (Matte 1969).

3.3 Structural Evolution in Relation to Metamorphism and Magmatism

Even though separate chapters concerning metamorphism and igneous rocks are included in this book, some considerations of these topics will be dealt with here in relation to the deformation phases. The metamorphism of the WALZ is essentially synkinematic (Capdevila 1969), although certain events of thermal metamorphism may be found linked to the intrusion of postkinematic granitoids. The distribution of the zones of dynamothermal metamorphism (Fig. 6) shows that this became progressively more intense towards the W, i. e. towards the internal parts of the Mondoñedo thrust sheet and its autochthon. It is a plurifacial, epi- to catazonal metamorphism in which three episodes may be distinguished in some areas. The first paragenesis includes, in the internal zones, almandine, staurolite and, locally, kyanite, indicative of a Barrovian-type metamorphism, with a medium-pressure geothermal gradient. A second paragenesis, with andalusite and sillimanite, is indicative of a meta-

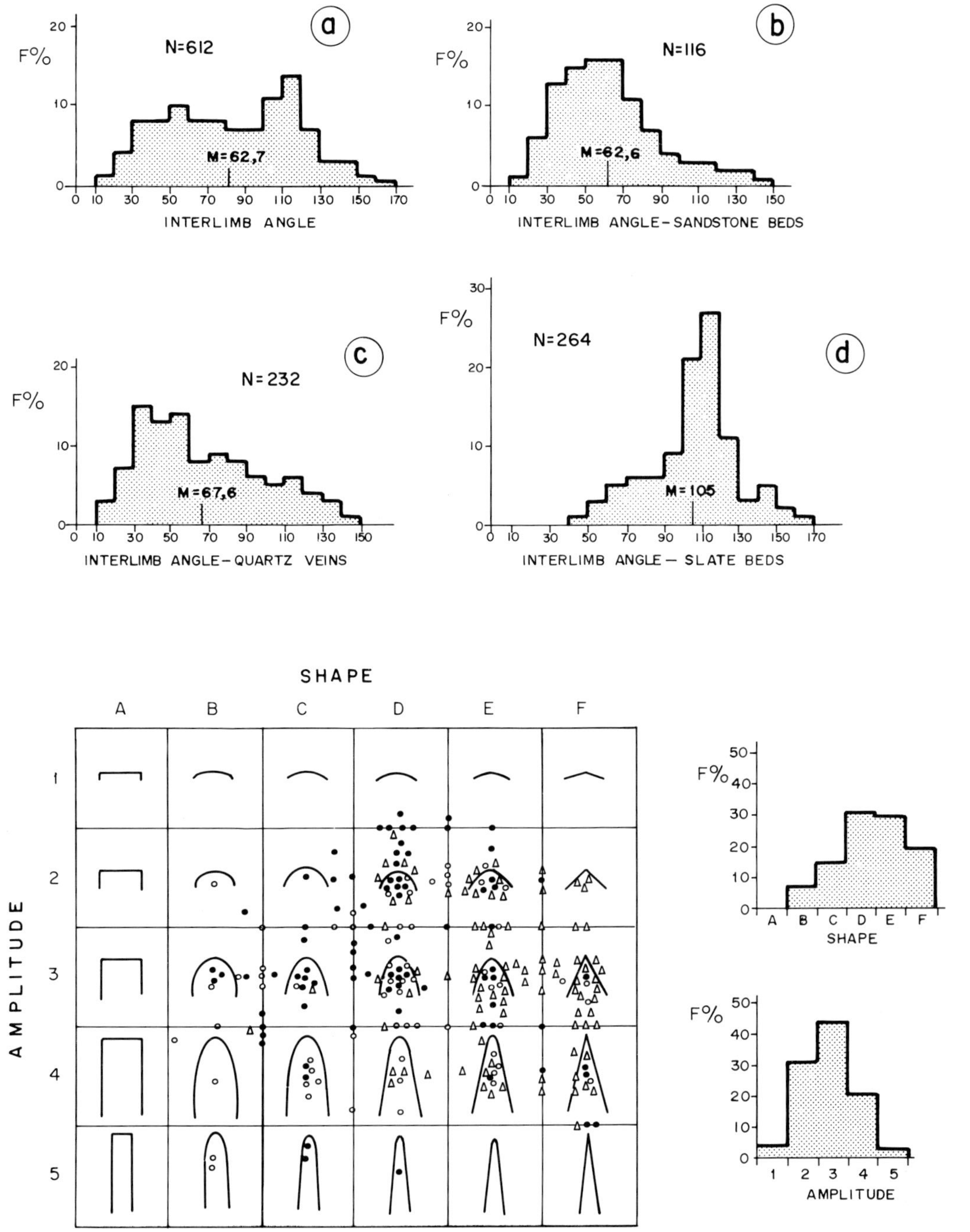

Fig. 5. Features of the D_3 minor folds in the Navia-Alto Sil Domain. **A** Frequency histograms of interlimb angles. **B** Hudleston's classification of the same folds after Pulgar (1980). *Black dots* folds in sandstone beds. *White dots* folds in quartz veins, 1 cm or more thick. *Triangles* folds in quartz veins less than 1 cm thick

morphic stage with a low-pressure geothermal gradient. Finally, a third episode, with a retrograde character, can be distinguished.

The isograds were distorted by the third deformation phase and those of biotite, almandine and staurolite of the first paragenesis cut across the D_1 folds. The sillimanite isograd reflects the interference produced by the D_3 and radial folds, because the sillimanite zone coincides with the Gistral and Puertomarín domes. However, in the Monte Carballosa dome, neither sillimanite nor andalusite appear, which suggests an initial relationship of both

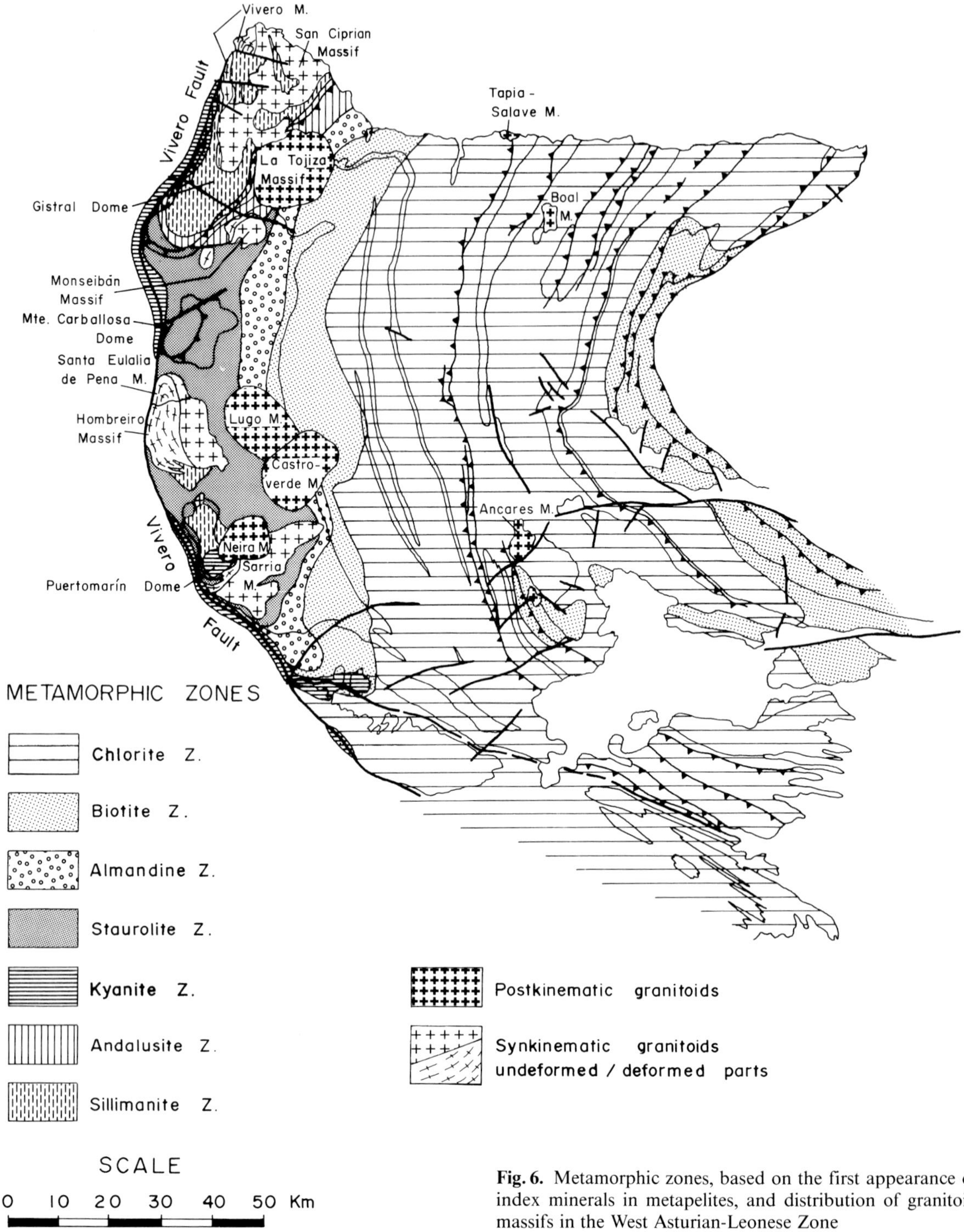

Fig. 6. Metamorphic zones, based on the first appearance of index minerals in metapelites, and distribution of granitoid massifs in the West Asturian-Leonese Zone

minerals with deeper seated and localized thermal upwarps.

The relationship between mineral crystallization and cleavage development changes from one zone to another. The first, medium-pressure paragenesis, reached its climax during D_1 in most of the chlorite and biotite zones and between D_1 and D_2 in the higher grade zones. The second, low-pressure paragenesis, did the same during the D_2 phase in the basal parts of the Mondoñedo thrust sheet, where sometimes the andalusite porphyroblasts appear with a markedly preferred orientation, owing to the effect of the deformation in the basal shear zone. However, in its autochthon, in the the Gistral tectonic window, this second paragenesis reached its climax during D_3, related to the intrusion of two-mica granites and the development of injection-type migmatites. The most significant episode of retrograde metamorphism was developed in the basal shear zone of the Mondoñedo thrust sheet, coinciding with the final stages of the D_2 deformation, but no evidence of coeval retrograde metamorphism has been found in the footwall unit. In the base of the thrust sheet, medium-grade parageneses were transformed into epizonal ones. Other retrograde processes appear locally, related to D_3 structures. Besides the above-described episodes, a localized medium-pressure dynamothermal metamorphism, frequently related to the growth of kyanite, developed in the ductile shear zone associated to the Vivero Fault (Fig. 6).

Granites are abundant in the Mondoñedo Domain and in the Gistral tectonic window but very scarce in the rest of the WALZ. They may be grouped in three sets: synkinematic biotite granitoids, two-mica granites and postkinematic granitoids. The synkinematic biotite granitoids form part of the calc-alkaline series described by Capdevila (1969) and Capdevila and Floor (1970), although they probably arose from the melting of deep contintental crust and do not seem to show any relationship with the partial melting of a subducted oceanic lithosphere. The two-mica granites are peraluminous leucogranites, frequently related to the regional metamorphism, and seem to derive from the melting of parts of the continental crust shallower than those of the former group. Massifs of this group and of the previous one can be found together, and both show the same or similar relationships with the deformation. Finally, the postkinematic granitoids have varied compositions ranging from granitic to granodioritic and gabbroic (Capdevila 1969; Suárez 1970). Their emplacement took place after the third Hercynian phase (D_3). Although there are no reliable radiometric data concerning the synkinematic granitoids of the WALZ, similar massifs in the adjacent Ollo de Sapo Domain, in the Central Iberian Zone (CIZ, see Fig. 1 of the previous chapter), have provided ages of 315–320 Ma (Visean-Namurian). The postkinematic granitoids of the WALZ give considerably younger ages: 286 ± 10 Ma (Capdevila and Vialette 1965, revised by Ries 1979).

The granitoids of the first two groups are generally contemporary with the third deformation phase in the NW of the Iberian Peninsula, but the WALZ represents an exception, since several of the synkinematic massifs found in it were intruded during the D_2 phase. This is particularly evident in the massifs of Sarria, Santa Eulalia de Pena and Hombreiro (Fig. 6), because they are partly affected by the basal shear zone of the Mondoñedo thrust sheet. In the shear zone, the massifs were intensely deformed and exhibit strongly developed plano-linear fabrics, congruent with the D_2 structures of the metamorphic host rock (Martínez Catalán 1983, 1985; Bastida et al. 1986). Other synkinematic massifs of the Mondoñedo thrust sheet also seem to show D_2 deformations, although, either because their emplacement took place at the margins of the basal shear zone or because they were late with respect to D_2, this relationship is not so clear. The granites that intruded into the Gistral tectonic window seem to be later than D_2, although a late-kinematic character for some of them cannot be ruled out (Galán 1984). The San Ciprian massif was intruded into the Mondoñedo nappe, and cuts its basal thrust in a zone in which, to the east of the Gistral window, it does not appear deformed. However, this granite, like that of Vivero (Fig. 6), shows an intense deformation to the W of the window, related to the ductile shear strain created by movement on the Vivero Fault.

3.4 Conclusion

Because the geometry of the largest D_1 folds indicates a relationship with shear zones affecting the underlying basement (Bastida et al. 1986), it is possible to visualize the tectonic evolution for the WALZ as being induced by the development of a few shears of crustal magnitude and E vergence, which would have given rise to a relatively complex structure in the sedimentary cover. Therefore, we are dealing with a case of thick-skinned tectonics (Coward 1983), with the involvement of at least part of the basement, and with a concomitant crustal thickening. The enormous thickness of the epicontinental sediments and the consistent existence of a thinned underlying continental basement seem to have created a situation uncommon in similar fold and thrust belts: the absence of outcropping basement. The structural characteristics and the tectonic evolution of the WALZ seem to be similar to those of the adjacent part of the Central Iberian Zone (CIZ). In particular, the Ollo de Sapo anticline is a structure comparable in magnitude and geometry to the Mondoñedo nappe (Matte 1968 b). By contrast, the Cantabrian Zone (CZ) represents a typical foreland thrust belt, characterized by thin-skinned tectonics, in which the basement was not involved in the deformation, and which corresponds to the external zone of the Iberian branch of the Hercynian orogen, of which both the WALZ and the CIZ constitute internal zones.

The metamorphic history fits in well with this model: the first paragenesis, with a medium-pressure gradient, would have coincided with the start of the crustal thickening, which involved a very rapid increase in pressure at deep levels but without a corresponding increase in temperature. The later increase in temperature, related to the low-pressure paragenesis, may be accounted for in terms of a reequilibration of temperature with time, by conduction. Nevertheless, the rapidity with which the increase in temperature seems to have occurred in the Mondoñedo nappe, still during the thrusting process, together with the early presence of granitoids in its deep zones, points to a very rapid heating, and the possibility that mantle rocks contributed to this increase can not be ruled out (Bastida et al. 1986).

References

Aller J, Bastida F, Brime C, Pérez Estaún A (1987) Cleavage and its relation with metamorphic grade in the Cantabrian Zone (Hercynian of North-West Spain). Sci Géol Bull (Strasbourg) 40: 1–18

Almela A, Valle J del (1963) Estudio geológico de la zona Ponferrada-Tremor de Abajo. Breviora Geol Astúrica 7: 2–26

Alvarez Ramis C (1965) Primera aportación al estudio de los "Sphenopteris" del Carbonífero español. a) Estefaniense: primera parte. Estud Geol 21: 1–141

Baldwin CT (1975) The stratigraphy of the Cabos Series in the section between Cadavedo and Luarca, province of Oviedo, North-West Spain. Breviora Geol Astúrica 19: 4–9

Baldwin CT (1977) The stratigraphy and facies associations of trace fossils in some Cambrian and Ordovician rocks of northwestern Spain. In: Crimes TP, Harper JC (eds) Trace fossils 2. Geol J Spec Issue 9: 9–40

Baldwin CT (1978) A comparison of the stratigraphy and depositional processes in the Cambro-Ordovician rocks of the Cantabrian and West-Asturian-Leonese Zones, NW Spain. In: Julivert M, Capdevila R, Tex E Den, Matte P, Ribeiro A (eds) Geologia de la parte N del Macizo Ibérico (Tomo homenaje I. Parga Pondal). Ediciones do Castro, Sada, La Coruña, pp 43–70

Barrera JL, Farias P, González Lodeiro F et al. (1989) Mapa Geológico Nacional E. 1:200000. Orense-Verín. Inst Geol Min Esp Madrid, pp 17–27

Barrois C (1877) Rélation d'un voyage géologique en Espagne. Ann Soc Géol Nord 4: 292–303

Barrois C (1882) Recherches sur les terrains anciens des Asturies et de la Galice. Mem Soc Geol Nord 2 (1): 1–630

Bastida F (1980) Las estructuras de la primera fase de deformación herciniana en la zona Asturoccidental-leonesa (Costa Cantábrica, N. W. de España). Thesis, Univ Oviedo

Bastida F, Pulgar JA (1978) La estructura del Manto de Mondoñedo entre Burela y Tapia de Casariego (Costa Cantábrica, NW de España). Trab Geol 10: 75–124

Bastida F, Martínez Catalán JR, Pulgar JA (1986) Structural, metamorphic and magmatic history of the Mondoñedo nappe (Hercynian belt, NW Spain). J Struct Geol 8: 415–430

Capdevila R (1969) Le métamorphisme régional progressif et les granites dans le segment hercynien de Galice nord oriental (NW de l'Espagne). Thesis, Univ Montpellier

Capdevilla R, Floor P (1970) Les differents types de granites hercyniens et leur distribution dans le Nord-Ouest de l'Espagne. Bol Geol Min Inst Geol Min Esp 81: 215–225

Capdevila R, Vialette Y (1970) Estimation radiométrique de l'âge de la deuxième phase tectonique hercynienne en Galice Moyenne (Nord-Ouest de l'Espagne). CR Acad Sci Paris 270: 2527–2530

Casey M, Huggenberger P (1985) Numerical modelling of finite-amplitude similar folds developing under general deformation histories. J Struct Geol 7: 103–114

Colchen M (1974) Géologie de la Sierra de la Demanda, Burgos-Logroño (Espagne). Mem Inst Geol Min Esp 85: 1–436

Corretgé LG, Carpio V (1968) Los ortoneises básicos de Pola de Allande (Asturias). Breviora Geol Astúrica 12: 14–16

Coward MP (1983) Thrust tectonics, thin skinned or thick skinned, and the continuation of thrusts to deep in the crust. J Struct Geol 5: 113–123

Crimes TP, Marcos A, Pérez Estaún A (1974) Upper Ordovician turbidites in Western Asturias. A facies analysis with particular reference to vertical and lateral variations. Palaeogeogr Palaeoclimatol Palaeoecol 15: 169–184

Crimes TP, Legg I, Marcos A, Arboleya ML (1977) Late Precambrian-low Lower Cambrian trace fossils from Spain. In: Crimes TP, Harper JC (eds) Trace fossils 2. Geol J Spec Issue 9: 91–138

Debrenne F, Lotze F (1963) Die Archaeocyatha des spanischen Kambriums. Akad Wiss Lit Abh Math Naturwiss Kl 2: 1–38

Debrenne F, Zamarreño I (1975) Sur la faune d'archeocyathes de la Formation Vegadeo et leur raport avec la distribution des facies carbonatés dans le NW de l'Espagne. Breviora Geol Astúrica 19: 17–27

Drot J, Matte P, Sobre la presencia de capas del Devoniano en el límite de Galicia y León (NW de España). Notas Comun Inst Geol Min Esp 93: 87–92

Färber A, Jaritz W (1964) Die Geologie des westasturischen Küstengebietes. Geol Jahrb 81: 679–783

Galán G (1984) Las rocas graníticas del Macizo de Vivero en el sector Norte (Lugo, NO de España). Thesis, Univ Oviedo

González Lodeiro F, Iglesias Ponce M (1981) Mapa Geológico Nacional E. 1:50000, Monforte de Lemos. Inst Geol Min Esp Madr 156

Gutiérrez Marco JC (1983) Nuevos datos para el conocimiento bioestratigráfico de las Pizarras de Luarca en la Zona Asturoccidental-Leonesa (NW de España). Cuad Lab Xeol Laxe 6: 71–74

Gutiérrez Marco JC (1986) Graptolitos del Ordovícico español. Thesis, Univ Madrid

Hammann W (1983) Calymenea (Trilobita) aus dem Ordovizium von Spanien; ihre Biostratigraphie, Ökologie und Systematik. Abh Senckenb Naturforsch Ges 542: 1–177

Hammann W, Robardet M, Romano M (1982) The Ordovician System in southwestern Europe (France, Spain and Portugal). Int Union Geol Sci Publ No 11: 1-47

Hernández Sampelayo P (1915) Fósiles de Galicia. Nota sobre la fauna paleozoica de la provincia de Lugo. Bol Inst Geol Min Esp 36: 277-303

Hudleston PJ (1973) Fold morphology and some geometrical implications of theories of fold development. Tectonophysics 16: 1-46

Jaritz W, Walter R (1970) Faciesuntersuchungen in Altpaläozoikum Nordwest-Spaniens/Asturien und Prov. Lugo. Geol Jahrb 88: 509-552

Julivert M, Martínez García E (1967) Sobre el contacto entre el Cámbrico y el Precámbrico en la parte meridional de la Cordillera Cantábrica y el papel del Precámbrico en la orogénesis herciniana. Acta Geol Hisp 2: 107-110

Julivert M, Fontboté JM, Ribeiro A, Nabais Conde LE (1972) Mapa Tectónico de la Península Ibérica y Baleares E. 1:1000000. Inst Geol Min Esp Madr

Liñán E, Sdzuy K (1978) A trilobite from the Lower Cambrian of Córdoba (Spain) and its stratigraphical significance. Senckenb Lethaea 59: 387-399

Lotze F (1945) Zur Gliederung der Varisziden der Iberischen Meseta. Geotekt Forsch 6: 78-92 (Trad. en: Publicaciones Extranjeras sobre Geol Esp 5: 149-166)

Lotze F (1956) Das Präkambrium Spaniens. Neues Jahrb Geol Paläontol Monatsh 8: 373-380 (Trad. en: Notas Comun Inst Geol Min Esp 60: 227-240)

Lotze F (1957) Zum Alter nordwestspanischer Quartzit-Sandstein-Folgen. Neues Jahrb Geol Paläontol 10: 464-471

Lotze F (1958) Zur Stratigraphie des spanischen Kambriums. Geologie 7: 727-750 (Trad. en: Notas Comun Inst Geol Min Esp 61: 131-164)

Marcos A (1970) Sobre la presencia de un flysch del Ordovícico Superior en el Occidente de Asturias. Breviora Geol Astúrica 14: 13-28

Marcos A (1973) Las Series del Paleozoico Inferior y la estructura herciniana del occidente de Asturias (NW de España). Trab Geol 6: 1-113

Marcos A (1976) Mapa Geológico Nacional E. 1:50000, San Martín de Oscos. Inst Geol Min Esp Madr 49

Marcos A, Pérez Estaún A (1981) La estratigrafía de la Serie de los Cabos en la zona de Vegadeo (Zona Asturoccidental-Leonesa, NW de España). Trab Geol 11: 89-94

Marcos A, Pulgar JA (1982) An approach to the tectonostratigraphic evolution of the Cantabrian Foreland thrust and fold belt, Hercynian Cordillera of NW Spain. Neues Jahrb Geol Paläontol Abh 163: 256-260

Marcos A, Pérez Estaún A, Pulgar JA, Bastida F (1976) Mapa Geológico Nacional E. 1:50000, Fonsagrada. Inst Geol Min Esp Madr 74

Martínez Catalán JR (1980) L'apparition du chevauchement basal de la nappe de Mondoñedo dans le dôme de Lugo (Galice, Espagne). C R Acad Sci Paris 290: 179-182

Martínez Catalán JR (1983) Deformación heterogénea en los macizos graníticos de Sarria y Santa Eulalia de Pena (provincia de Lugo). Stud Geol Salmant 18: 39-63

Martínez Catalán JR (1985) Estratigrafía y estructura del domo de Lugo (Sector Oeste de la zona Asturoccidental-leonesa). Corpus Geol Gallaeciae (2° Serie) 2: 1-291

Matas J, Abejero V, Fernández L, Fernández Pello J (1982) Mapa Geológico Nacional E. 1:50000, Noceda Inst Geol Min Esp Madr 127

Matte P (1967) Le Précambrien supérieur schisto-gréseux de l'Ouest des Asturies (Nord-Ouest de l'Espagne) et ses relations avec les séries précambriennes plus internes de l'arc galicien. C R Acad Sci Paris 264: 1769-1772

Matte P (1968a) Précisions sur le Précambrien supérieur schisto-gréseux de l'Ouest des Asturies. Comparaison avec les autres affleurements précambriens du Nord-Ouest de l'Espagne. Rev Géogr Phys Géol Dynam (2° serie) 10: 205-211

Matte P (1968b) La structure de la virgation hercynienne de Galice (Espagne). Rev Géol Alpine 44: 1-128

Matte P (1969) Les kink-bands - Exemple de déformation tardive dans l'Hercynien du Nord-Ouest de l'Espagne. Tectonophysics 7: 309-322

Nollau G (1968) Stratigraphie, Magmatismus und Tektonik der Montes de León zwischen Astorga und Ponferrada und Nordwest-Spanien. Geotekt Forsch 27: 71-146

Parga Pondal I, Aleixandre T (1966) La arenisca ortocuarcítica del Gistral (Lugo). Notas Comun Inst Geol Min Esp 87: 59-90

Perejón A (1984) Bioestratigrafía de los arqueociatos en España. Cuad Geol Ibéric 9: 213-265

Pérez Estaún A (1974a) Aportaciones al conocimiento del Carbonífero de San Clodio (Prov. de Lugo). Breviora Geol Astúrica 18: 3-8

Pérez Estaún A (1974b) Algunas precisiones sobre la sucesión ordovícica y silúrica en la región de Truchas. Breviora Geol Astúrica 18: 23-25

Pérez Estaún A (1974c) La sucesión ordovícica en el dominio del Alto Sil (Zona Asturoccidental-Leonesa). Breviora Geol Astúrica 18: 53-57

Pérez Estaún A (1978) Estratigrafía y estructura de la rama Sur de la Zona Asturoccidental-Leonesa. Mem Inst Geol Min Esp 92: 1-151

Pérez Estaún A, Marcos A (1981) La Formación Agüeira en el sinclinorio de Vega de Espinareda: aproximación al modelo de sedimentación durante el Ordovícico Superior en la Zona Asturoccidental-Leonesa (NW de España). Trab Geol 11: 135-145

Pérez Estaún A, Martínez Fernández FJ (1978) El Precámbrico del antiforme del Narcea en el sector de Tineo-Cangas de Narcea (NW de España). Trab Geol 10: 367-375

Pérez Estaún A, Marquínez J, Ortega E (1980) La sucesión ordovícica y la estructura de la región de Silván (La Cabrera, León). Breviora Geol Astúrica 24: 17-24

Pérez Estaún A, Pulgar JA, Bastida F et al. (1982) Mapa Geológico Nacional E. 1:50000. Vega de Espinareda. Inst Geol Min Esp Madr 126

Pulgar JA (1980) Análisis e interpretación de las estructuras originadas durante las fases de replegamiento en la zona Asturoccidental-leonesa (Cordillera Herciniana, NW de España). Thesis, Univ Oviedo

Pulgar JA, Bastida F, Marcos A, Pérez Estaún A, Vargas I, Ruiz F (1981) Mapa Geológico Nacional E. 1:50000. Degaña. Inst Geol Min Esp Madr 100

Quintero I (1962) Graptolites de la provincia de Lugo. Notas Comun Inst Geol Min Esp 65: 61-82

Ramsay JG (1967) Folding and fracturing of rocks. McGraw-Hill, New York

Ramsay JG, Casey M, Kligfield R (1983) Role of shear in development of the Helvetic fold-thrust belt of Switzerland. Geology (Boulder) 11: 439-442

Riemer W (1963) Entwicklung des Paläozoikums in der südlichen Provinz Lugo (Spanien). Neues Jahrb Geol Paläontol Abh 117: 273-285

Riemer W (1966) Datos para el conocimiento de la estratigrafía de Galicia. Notas Comun Inst Geol Min Esp 81: 7-20

Ries AC (1979) Variscan metamorphism and K-Ar dates in

the Variscan fold belt of S Brittany and NW Spain. J Geol Soc (Lond) 136: 89-103

Romariz C (1969) Graptolitos silúricos del Noroeste peninsular. Comun Serv Geol Port 53: 107-156

Sdzuy K (1961) Das Kambrium Spaniens. Teil II: Trilobiten. Akad Wiss Lit Abh Math Naturwiss 7-8: 217-408 (499-690)

Sdzuy K (1968) Bioestratigrafía de la griotte cámbrica de los Barrios de Luna (León) y de otras sucesiones comparables. Trab Geol 2: 45-67

Sdzuy K (1971a) Acerca de la correlación del Cámbrico Inferior en la Península Ibérica. Publ Primer Congreso Hispano-Luso-Americano Geol Econ 2: 753-768

Sdzuy K (1971b) La subdivisión bioestratigráfica y la correlación del Cámbrico Medio de España. Publ Primer Congreso Hispano-Luso-Americano Geol Econ 2: 769-782

Sitter LU de (1961) Le Précambrien dans la Chaîne Cantabrique. CR Soc Géol Fr 9: 253

Suárez O (1970) Estudio petrológico de los afloramientos plutónicos de la zona occidental Astur-leonesa (NW de España). Thesis, Univ Salamanca

Truyóls-Massoni M (1986) Nowakia acuaria (Praguiense, Devónico Inferior) de la Zona Asturoccidental-Leonesa (NO de España). Breviora Geol Astúrica 27: 12-16

Villas E (1985) Braquiópodos del Ordovícico Medio y Superior de las Cadenas Ibéricas Orientales. Mem Mus Paleontol Univ Zaragoza 1: 1-156

Wagner RH (1965) Stephanian B flora from the Ciñera-Matallana coalfield (León) and neighbouring outliers. III: Callipteridium and Alethopteris. Notas Comun Inst Geol Min Esp 78: 5-70

Wagner RH (1970) An outline of the Carboniferous stratigraphy of north-western Spain. In: Streel M, Wagner RH (eds) Colloque sur la stratigraphie du Carbonifere. Congrés-Colloque Univ Liege 55: 429-463

Walter R (1966) Resultados de investigaciones geológicas en el Noreste de la provincia de Lugo (NO-España). Notas Comun Inst Geol Min Esp 89: 7-16

Walter R (1968) Die Geologie in der Nordöstlichen Provinz Lugo (Nordwest Spanien). Geotekt Forsch 27: 3-70

Zamarreño I, Perejón A (1976) El nivel carbonatado del Cámbrico de Piedrafita (Zona Asturoccidental-Leonesa, NW de España): tipos de facies y faunas de arqueociatos. Breviora Geol Astúrica 20: 17-32

Zamarreño I, Hermosa JL, Bellamy J, Rabu D (1975) Litofacies del nivel carbonatado del Cámbrico de la región de Ponferrada (Zona Asturoccidental-Leonesa, NW de España). Breviora Geol Astúrica 19: 40-48

4 Igneous Rocks

L. G. Corretge[1], O. Suarez[1] and G. Galan[2]

With the exception of the westernmost part (Lugo Dome), where granitic plutonism becomes comparable to that in the Central Iberian Zone, magmatic activity in the West Asturian Leonese Zone (WALZ) is very scarce. For the rest, it is only represented by small plutons and stocks almost restricted to the axis of the Boal-Los Ancares metamorphic belt (Fig. 1), and by some scattered volcanic rocks of Precambrian and Lower Paleozoic age. Nevertheless, the importance of the WALZ in regard to granitic plutonism lies in the occurrence within this restricted area of almost all granite types present in the Hesperian massif.

4.1 Volcanic and Subvolcanic Rocks

Volcanic occurrences within the WALZ are rare. There are some volcanic rocks interlayered within the Precambrian metasediments of the Narcea and Lugo Dome antiforms, and also within the Cambrian and Lower Ordovician series.

In the Narcea antiform several types of porphyroid and orthogneiss levels are found in the Cudillero and Tineo areas. These rocks are derived from interlayered volcanics and tuffs within the slates and greywackes of turbiditic character of the Narcea Precambrian sequence. The main outcrops (Tineo Area) correspond to metavolcanics formed by quartz, albite, microcline, oligoclase and amphibole (tremolite-actinolite, green hornblende) chlorite, epidote, Fe oxides, leucoxene, etc. Petrographically, they form a highly varied series which includes rock types ranging from quartz-keratophyres to amphibolitic gneisses, porphyroids being the most common rocks. They have been interpreted as derived, during the regional metamorphism, from sills, shallow intrusions or extrusions of different nature, usually rhyolitic or dacitic, although there are also some more basic rocks of andesitic or gabroic type (Corretgé 1969; Pérez Estaún and Martínez 1978). In the northernmost outcrops of the Cudillero area, the porphyroids have been described as keratophyres and quartz-keratophyres and are also interpreted as derived from rhyolitic or dacitic tuffs (Suárez and Suárez del Rio 1976). On the other hand, it is important to emphasize the presence in this area of rhyolite pebbles in the basal levels of the Lower Cambrian sediments which point to the existence of a pre-Paleozoic acid vulcanism. Some metavolcanic beds also appear, interlayered within the Precambrian Villalba schists of the Lugo Dome, more to the west. They are mainly metamorphosed porphyritic paleovolcanics with quartz, microcline, plagioclase, biotite, muscovite (Bastida and Pulgar 1978), which could be compared with the porphyroids in the Narcea Precambrian sequence.

Volcanic igneous activity of Paleozoic age is scarce, but there are ultrabasic, basic and acid rocks. Most basic and ultrabasic rocks are sills, affected by Hercynian metamorphism, interlayered with Lower Cambrian quartzites near Villalba in the Lugo Dome. Among the basic rocks, metaperidotites which contain olivine crystals in a tremolite-chlorite matrix, plagioclase amphibolites, amphibolites with actinolite and chlorite-actinolite schists can be separated. All these basic-ultrabasic rocks may be considered as evidence for preorogenic Hercynian igneous activity (Capdevila 1966, 1969). Interlayered within the Lower Cambrian metasediments of the Lugo Dome, there are also some beds of acid metavolcanics of rhyolitic and rhyodacitic type. They are affected by the first Hercynian tectonic phase (P_1) (Martínez Catalán 1981; Galán 1984). These rocks are made up of quartz, microcline and muscovite or of quartz and muscovite. A larger number of volcanic rocks can be observed interbedded with metasediments of Cambro-Ordovician age, specially towards the top. According to Parga (1969, unpublished work), this vulcanism is of tholeiitic type as opposed to the alkaline vulcanism of the same age in the Cantabrian zone. The

[1] Departamento de Geología, Universidad de Oviedo, 33080 Oviedo, Spain
[2] G. P. P. G. Fac. de Geología, Universidad de Barcelona, Zona Universitaria Pedralbes, 08028 Barcelona, Spain

R. D. Dallmeyer and E. Martínez García (Eds.)
Pre-Mesozoic Geology of Iberia

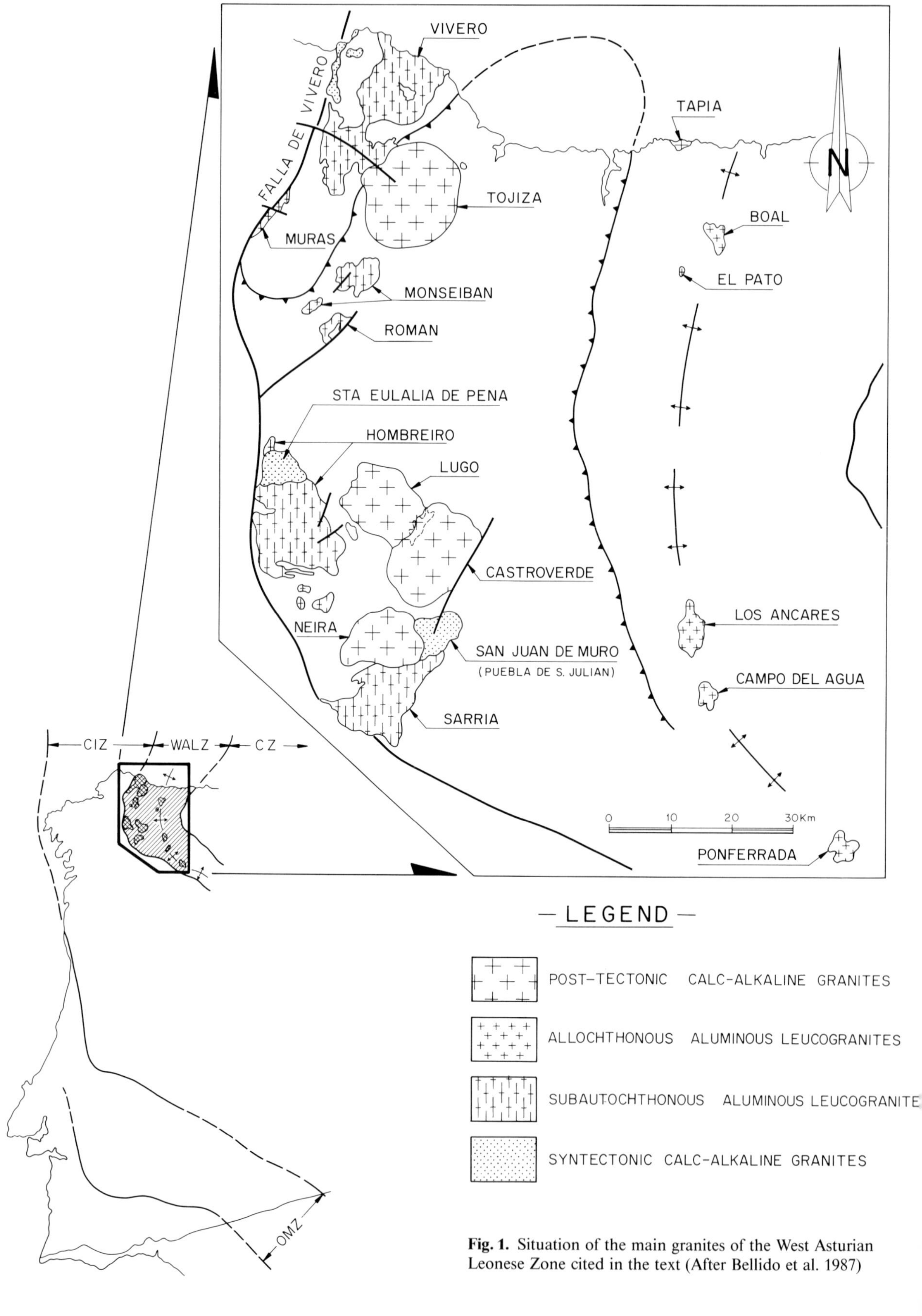

Fig. 1. Situation of the main granites of the West Asturian Leonese Zone cited in the text (After Bellido et al. 1987)

volcanic rocks of the same age, described Marcos (1973); Walter (1965); Nollau (1968), are sills of diabasic nature. These sills range in thickness from 10 to 50 m. These rocks are affected by retrograde processes brought about by conditions of the regional Hercynian metamorphism and by different alteration processes displaying a clear cleavage in some areas but either intergranular or subophitic texture can be recognized. Their primary mineralogy, preserved in very few cases, consists of plagioclase, brown amphibole, clinopyroxene, biotite, phlogopite, ilmenite and quartz. The low grade metamorphism and alteration processes produced actinolitic amphibole, chlorite, epidote, leucoxene, Fe sulphides and albitic plagioclase (An_{5-25}). The more acid vulcanism of the Truchas area (Matte 1968) is mainly represented by albitic tuffs and lavas with corroded quartz and orthoclase phenocrysts. It could represent some more differentiated magmas of the tholeiitic series (Parga 1969, unpublished work). In addition to the previously mentioned metavolcanic rocks within the WALZ, there are also some dikes of felsite and porphyries (Martínez Alvarez et al. 1975) which postdate the P_1 Hercynian structures (Bastida 1980 unpublished work).

4.2 The Hercynian Granites

The Hercynian granitic rocks in the WALZ are located along the thermal metamorphic belts of Boal-Los Ancares and Vivero-Lugo-Sarria (Suárez et al., this Vol.). These belts correspond to two antiformal structures formed during the third Hercynian deformation phase (P_3); The Boal-Los Ancares Dome and the Lugo Dome. In both domes, the two main types of orogenic granites, which characterize other Hercynian areas, are present (Didier and Lameyre 1969); calc-alkaline granites and aluminous leucogranites, both types do not necessarily agree with the I and S types proposed by Chappell and White (1974, 1984) but they are broadly similar.

4.2.1 Granites of the Boal - Los Ancares Belt

Following the Boal-Los Ancares belt axis, there are some small plutons: Tapia, Boal, El Pato, Los Ancares, Campo del Agua and Ponferrada, from north to south. All of these plutons are S-type except the Tapia stocks which are I-type, with abundant associated gabbroic and dioritic rocks, most similar to the ones in the Cantabrian Zone and Palentian Zone (Corretgé et al. 1987), particularly to the Salas Belmonte Igneous Group (see Corretgé and Suárez, this Vol.). Both groups of granites postdate the main Hercynian deformation phases.

4.2.1.1 The Allochthonous Aluminous Leucogranites

The term leucogranite is used here in Lameyre's sense (1966). The allochthonous plutons of Boal, El Pato, Los Ancares, Campo del Agua and Ponferrada belong to this group, and consist of small discrete igneous bodies which intrude along the axis of the Boal-Los Ancares belt and inside an arc-shaped band parallel to the Hercynian structure (Fig. 1). The intrusion of the Los Ancares and Campo del Agua plutons has been related to an important NE-SW lineament: the Vila Real strike-slip sinistral fault, and particularly with the intersection of this lineament with cross-cutting fractures (Gutiérez Claverol et al. 1987). However, part of these granites are elongated and follow the previously mentioned arc, being also located in the same epizonal and lithostratigraphic domain of Middle Cambrian to Middle Ordovician age. These granites clearly cut across the Hercynian structures ($P_1 + P_3$) and develop a contact metamorphism which overlaps the second episode of regional metamorphism (see Suárez et al., this Vol.). The only age determination is that of the El Pato pluton (284 ± 8 Ma), which was obtained by the Rb-Sr method (Suárez et al. 1978), and is similar to that of other post-tectonic granitoids situated more to the west of the WALZ (Cocherie 1978).

Petrographically, there are porphyritic, two-mica monzonitic granites, biotite granites and rare granodiorites, medium- to coarse-grained which often exhibit K-feldspar megacrysts (from 1 to 4 cm and occasionally up to 10 cm) (Suárez 1970, 1971, 1972, 1974, 1976). The microgranular and metamorphic enclaves are rare in all these massifs. The mineralogy is basically uniform with quartz, microcline, plagioclases, prevailing muscovite and biotite. Apatite, zircon, monazite, tourmaline, beryl, topaz and, less commonly, garnet appear as accessory minerals. Finally, in the Ponferrada pluton aluminous silicates such as andalusite, sillimanite, cordierite and, very rarely, durmortierite are also found (Suárez 1970).

There are some significant differences between the considered plutons of which the most outstanding one is the higher calcium content in the Boal pluton, whereas Los Ancares and Campo del Agua granites have more albitic plagioclases such that some of the facies are true alkali feldspar granites.

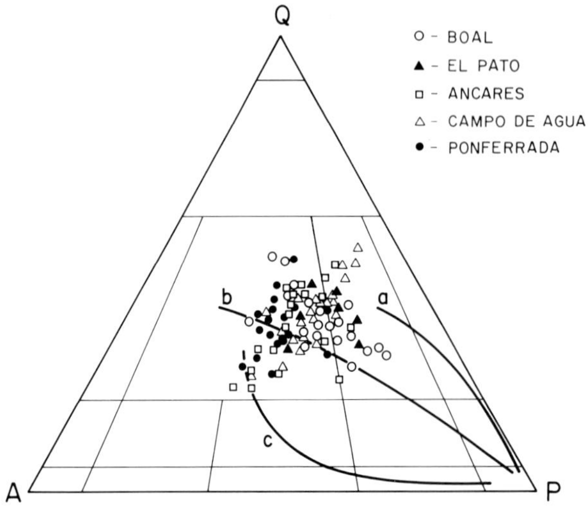

Fig. 2. *QAP* diagram for the allochthonous aluminous granites of the Boal-Los Ancares belt; *a* Calc-alkaline-trondhjemitic (low K) trend, *b* calc-alkaline granodioritic (medium K) trend; *c* calc-alkaline-monzonitic (high K) trend (After Lameyre and Bowden 1982)

The Los Ancares pluton, partly formed by leucogranites s. s., bears a stronger leucocratic character. The larger plutons of Boal (Uhling 1984) and Los Ancares present a zoned structure. In Los Ancares, the leucocratic and/or aplitic granites at the pluton border are differentiation facies (Suárez 1974). A common feature to all these granites is the presence of abundant late differentiates: aplites, some pegmatites and granitic porphyries as well as quartz dikes associated with important mineralizations mainly of W-Sn (scheelite and wolframite in Boal and Ponferrada).

The plot of these granites in the QAP diagram can be seen in Fig. 2. The lack of data on the less differentiated terms of this group, represented by the scarce microgranular enclaves (Didier 1973), prevents their positioning within the evolutionary trends of the calc-alkaline series. Generally speaking, these granites plot at the end of the calc-alkaline trends in the diagrams, but in some cases, like the Los Ancares pluton, the different facies are close to the monzonitic, calc-alkaline evolutionary trends (Lameyre and Bowden 1982).

On the whole, this granitic group forms an aluminous association, leucocratic in character and with vertical evolution (Fig. 3; Debon and Le Fort 1983). They are also characterized by low total alkali content and variable K_2O/Na_2O ratios. Besides in the Boal and Ponferrada plutons, there are some granitic rocks with low Na contents, which have been interpreted by Bea et al. (1987) as having originated during postmagmatic hydrothermal alteration.

The Rb and Sr content of these granites (Fig. 4) places them between the calc-alkaline granites and other Hercynian leucogranites. They are characterized by moderate REE contents and by a considerable REE fractionation with chondrite-normalized La/Yb ratios [La N/Yb N] between 10 and 6, as well as by large Eu negative anomalies.

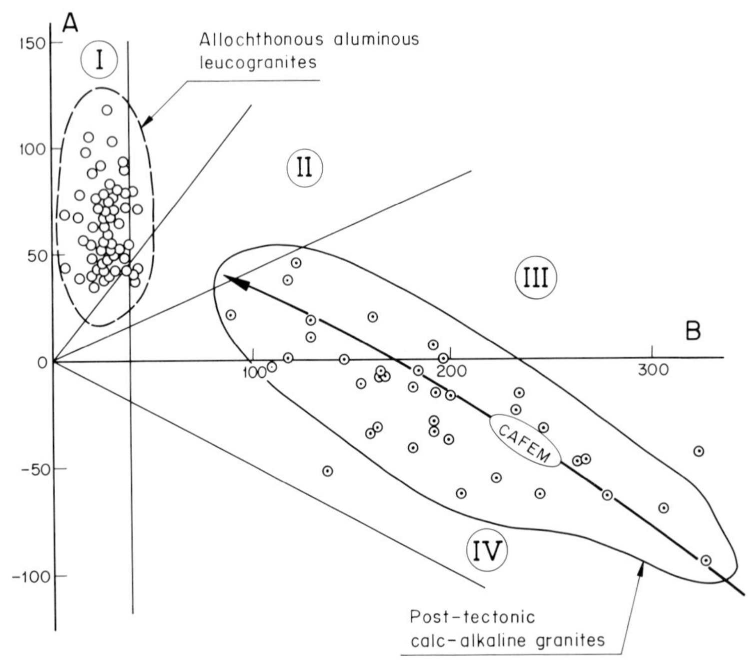

Fig. 3. Plot of allochthonous aluminous leucogranites, post-tectonic calc-alkaline granites and associated basic rocks of the Boal-Los Ancares belt in the *A*-*B* diagram (Debon and Le Fort 1983) [*A* Al-(K + Na + 2Ca); *B* Fe + Mg + Ti]

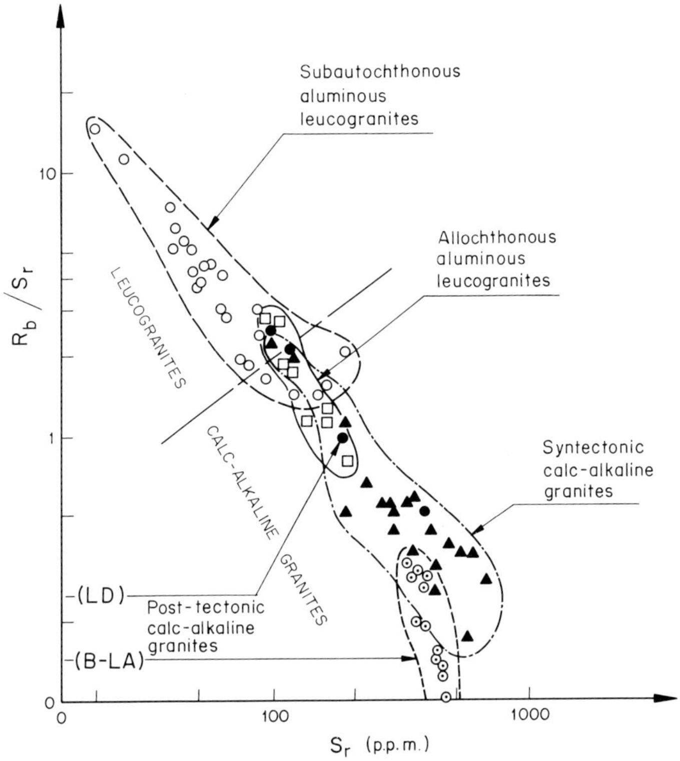

Fig. 4. Rb/Sr - Sr diagram showing the different types of granites in WALZ. The post-tectonic calc-alkaline granites of Boal-Los Ancares belt *(B-LA)* and Lugo Dome *(LD, filled circles)* plot in two different fields

The possible palingenic origin of these rocks was considered by Capdevila et al. (1973) from field, mineralogical and petrographic data. More recently, Ortega (1987, unpublished work) based on REE data attributes the origin of these granites to the fusion of pelitic and psammitic rocks similar to the country rocks and later fractional crystallization. However, this model implicates the existence of mineralogically complex residues whose existence has never been proved in the field.

4.2.1.2 Post-Tectonic Calc-Alkaline Granites and Associated Basic Rocks

Only the Tapia intrusive bodies belong to this group. They are formed by three small plutons (Porcia, Salave and Represas) which intrude epimetamorphic psammo-pelitic rocks of Cambrian-Ordovician age. They produce a metamorphic contact aureole with several zones (Suárez and Suárez 1970) and typical assemblages of pyroxene hornfels facies near the contact. Both, the intrusions and the metamorphic contact aureole, postdate the regional Hercynian structures. The age of the Salave intrusion has been established at 287±8 Ma (Rb-Sr whole rock + minerals).

These intrusions consist of basic-intermediate rocks with compositions ranging from gabbro and gabbrodiorite to granodiorite (Fig. 5). The gabbrodioritic rocks often show cumulate textures and are very heterogeneous. Four main types may be distinguished according to Suárez (1970, unpublished work): (1) biotite-amphibole gabbrodiorites; (2)

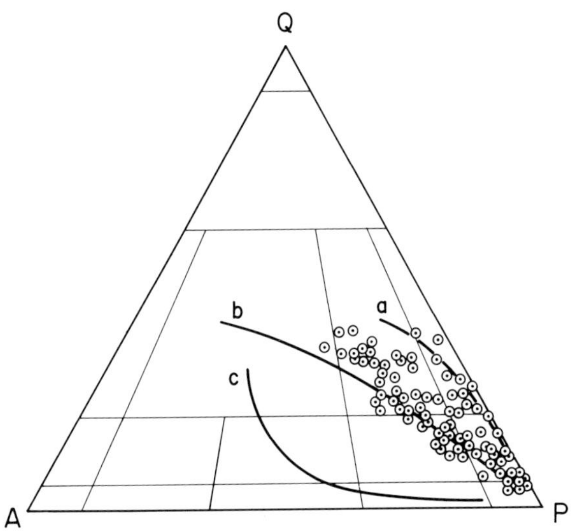

Fig. 5. *QAP* diagram of post-tectonic calc-alkaline granites and associated basic rocks of the Tapia intrusive (Boal-Los Ancares belt) showing the trend of the series (*a*, *b* and *c* as in Fig. 2)

biotite-hypersthene gabbros and diorites; (3) hypersthene granogabbros; and (4) leucogabbros. Other minerals present are clinopyroxene, always subordinated to orthopyroxene, quartz and orthoclase often as graphic intergrowths, ilmenite, magnetite and apatite. Plagioclase is the most abundant mineral (40–70%) with a composition ranging from An_{45} to An_{80}. The more differentiated rocks are biotite-amphibole quartz-diorites, tonalites and granodiorites with apatite, zircon and allanite as the most common accessory minerals. Some of these rocks were affected by important hydrothermal alterations to which mineralizations of pyrite, arsenopyrite, molibdenite, stibnite, rarer sphalerite and dispersed gold are related (Harris 1980 a, b).

The gabbrodiorites and granodiorites of Tapia enclose abundant enclaves of a different nature (Arias 1984 unpublished work): (1) country rocks more or less transformed; (2) microgranular enclaves with mineralogical and geochemical composition quite similar to the host igneous rocks; and (3) cortlandtites, norites, anorthosites and appinites which seem to be related to the same magmatic suite (Galán and Suárez 1989).

The rocks from Tapia-Salave intrusions follow a calc-alkaline series of granodioritic type (Fig. 5). In the diagram of Fig. 3, they are seen to form a cafemic association of calc-alkaline subtype (Debon and Le Fort 1983), with an evolutionary trend comparable to that of some associations from the Pyrenees. Their Rb and Sr content corresponds to the typical calc-alkaline field in Fig. 4. However, it is important to note that their Rb/Sr ratio is much lower than that of other post-tectonic calc-alkaline granites of the WALZ (as we shall see later). In this diagram, the gabbrodiorites and granites from Tapia are more similar to the syntectonic calc-alkaline granites of the Lugo Dome. Nevertheless, the composition of the biotites (Fig. 6) is very different, the biotites from Tapia corresponding to the subalkaline granite field. With regard to the other minerals, amphiboles are mainly Mg hornblendes and subordinate Fe- and edenitic hornblendes. Orthopyroxenes vary from bronzite to ferrohypersthene with

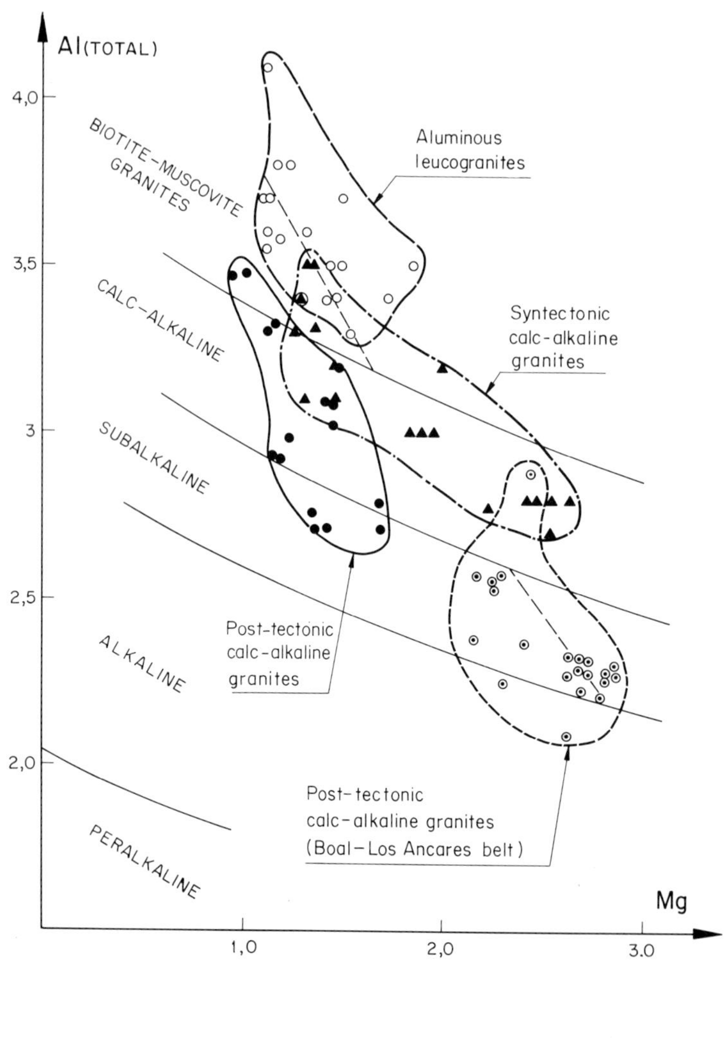

Fig. 6. $Al_{(total)}$ – *Mg* diagram (Nachit et al. 1985) for the biotites of rocks from Tapia (post-tectonic calc-alkaline granites in Boal-Los Ancares belt) and biotites from the three groups of the Lugo Dome granites

predominance of hypersthene (45–80 En), while clinopyroxenes have a more restricted composition of calcic augite or salite.

The petrographic and geochemical characteristics of the Tapia rocks point to a hybrid character with participation in their genesis of a high alumina basaltic magma (Gil Ibarguchi and Suárez 1982; Galán and Suárez, in litt.), strongly contaminated with crustal rocks (Suárez and Corretgé 1987).

4.3 Lugo Dome Granites

Granitic intrusions in the the Lugo Dome area become important as metamorphism increases in the Vivero-Lugo-Sarria belt (Suárez et al., this Vol.). Capdevila (1969) carried out the first detailed studies on the granites of the region, providing the basis for a classification which was later extended to other Hercynian granites in the Iberian Peninsula (Capdevila and Floor 1970; Capdevila et al. 1973). Later studies on granites focused on specific geochemical problems and on regional mapping (Lopez Garcia et al. 1976; Cocherie 1978; Martínez Catalán 1981; Galán 1984, 1987; Bellido et al. 1987).

Two generations of calc-alkaline granites can be differentiated in the Lugo Dome, with emplacement of a generation of leucogranites in between:

1. Syntectonic calc-alkaline granites;
2. Aluminous leucogranites;
3. Post-tectonic calc-alkaline granites.

The syntectonic calc-alkaline granites and the aluminous leucogranites appear spatially associated, forming complex massifs like those of Vivero and Santa Eulalia de Pena-Hombreiro (Fig. 1). It is often difficult to establish a clear limit between some differentiated facies of the syntectonic calc-alkaline granites and some aluminous leucogranites. This has led to some classification problems with some massifs, as happens in Puebla de San Julian (San Juan de Muro), considered by Capdevila (1969) as a syntectonic calc-alkaline granite and as belonging to the aluminous leucogranites group by others (Bellido et al. 1987). On the basis of previous studies, the main structural, petrographic, geochemical and mineralogical features of these three granitic groups are described below.

4.3.1 Syntectonic Calc-Alkaline Granites

They are rare and appear in small massifs, parallel to the regional trends. Some of the granites in the NW boundary of the Vivero massif, Santa Eulalia de Pena and Puebla de San Julian massifs belong to this group (Fig. 1). According to Martínez Catalán (1981) and Galán (1984, 1987), these granites are intruded as layers in shear zones related to the basal thrust of the Mondoñedo nappe (Hercynian P_2) and can be found both in the autochthonous (Vivero) and in the allochthonous (Santa Eulalia de Pena and Puebla de San Julian). All massifs are deformed by the Hercynian P_2 showing a clear foliation. In the westernmost part of the Vivero and Santa Eulalia de Pena massifs, they are also deformed

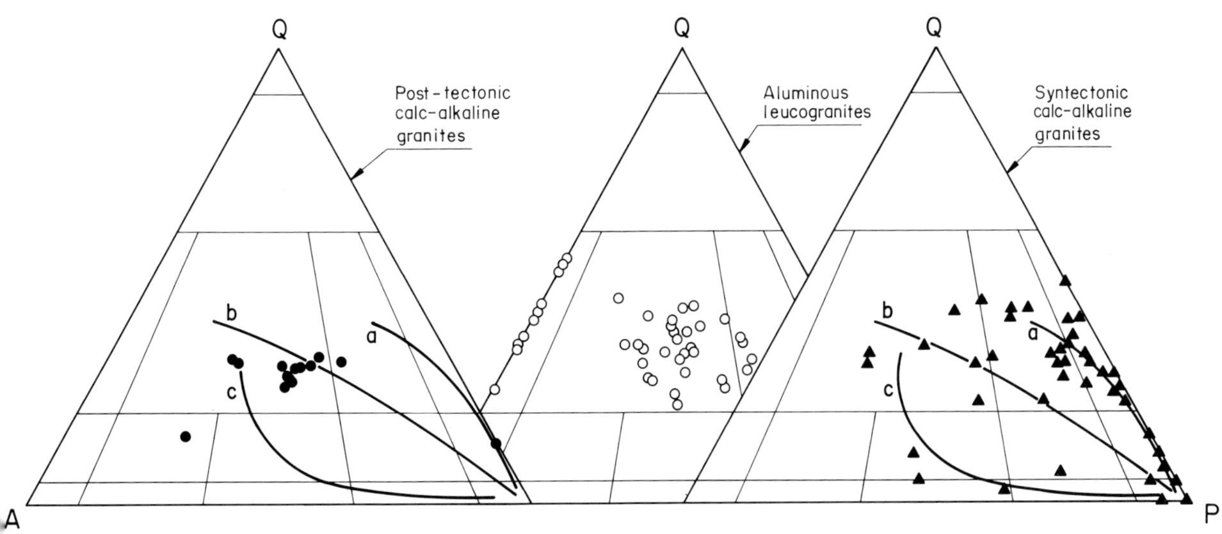

Fig. 7. Plots of modal composition for the three generations of Lugo Dome granites on *QAP* diagram. Data from Capdevila (1969) and (Galán) 1984 (*a, b* and *c* as in Fig. 2)

by the late Vivero fault (inter P_2–P_3) which caused strong mylonitization.

The country rocks for these granites are Precambrian and Lower Cambrian metasediments. Their contact is discordant with the Hercynian P_1 structures and concordant with the regional schistosity S_2 (P_2). The intrusion of these granites took place soon after the climax of regional metamorphism and their emplacement level does not surpass the staurolite isograd. Although data from Capdevila and Vialette (in Capdevila 1969) indicate an age of about 307 Ma (using the new decay constant, on whole rock + minerals isochron for the Santa Eulalia de Pena massif), so far no reliable whole rock Rb-Sr isochron has been established.

In the syntectonic calc-alkaline granites there are different facies (Fig. 7): (1) medium-grained amphibole-biotite tonalites, with relict pyroxenes and sphene, apatite, zircon, allanite, ilmenite and sulphides as accessory minerals; (2) medium-grained biotite granodiorites with apatite, zircon, allanite, ilmenite, garnet, muscovite and some sphene as accessory minerals; (3) monzonitic, biotite and biotite-muscovite granites with megacrysts and the same accessories as the granodiorites; and (4) dikes of granites, leucogranites and aplopegmatites in smaller proportion. Frequent microgranular enclaves exist also, mainly of diorites, Q-diorites, tonalites and more rarely monzonites with amphibole and biotite. Finally, there are also more basic rocks, some of them ultramafic, of cortlandtitic type, whose petrographic, mineralogical and geochemical characteristics are described elsewhere by Galán (1984, 1987).

These granitoids and microgranular enclaves, as can be seen in the diagram of Fig. 7, follow a calc-alkaline evolution of trondhjemitic to granodioritic type (Lameyre and Bowden 1982), partly conditioned by the biotite content in some of these granitoids. The calc-alkaline evolution is also observed on the AFM diagram (Fig. 8), for both granites and associated basic rocks, and on the diagram A–B (Debon and Le Fort 1983; Fig. 9) where they define a cafemic association with an alumino-cafemic tendency in the most differentiated facies of calc-alkaline subtype (Fig. 10). The Rb and Sr content of these granites can be compared to that in similar Hercynian granites of other areas (Fig. 4). On the other hand, they have a moderate REE content, presenting spectra with a variable La/Yb rato, depending on the facies, and with or without small negative Eu anomalies (Galán et al., in prep.).

The composition of the main minerals in these granites is as follows: the scarce relictic pyroxene is diopside. Amphibole is Mg-hornblende to actinolitic hornblende. Biotite is of variable composition, with decreasing Mg and increasing Al towards the most differentiated terms, thus being typical biotites of biotite-muscovite granites (Fig. 6). Finally, plagioclase ranges from andesine to oligoclase towards the most differentiated rocks.

Capdevila (1969) and Capdevila et al. (1973) suggested that these granites originated by lower crust melting with participation of deeper basic magmas. This hypothesis seems likely for the Lugo Dome, where initial isotope ratios of Sr are low: 0.702 for the granites of Santa Eulalia de Pena massif (Capdevila 1969) and slightly higher for those of

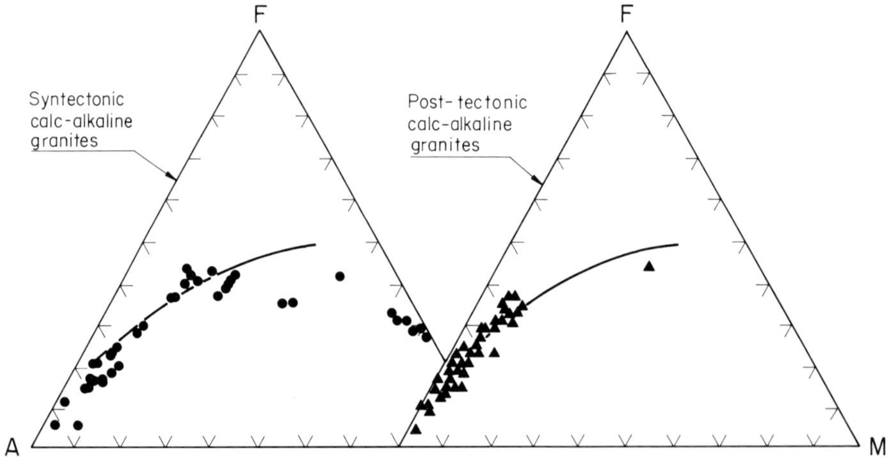

Fig. 8. *AFM* plot for the two generations of calc-alkaline granites of the Lugo Dome. Data from Capdevila (1969); Galán (1984) and IGME (1987, unpublished work). Calc-alkaline trend of Lower California Batholith (Larsen 1948) is also represented

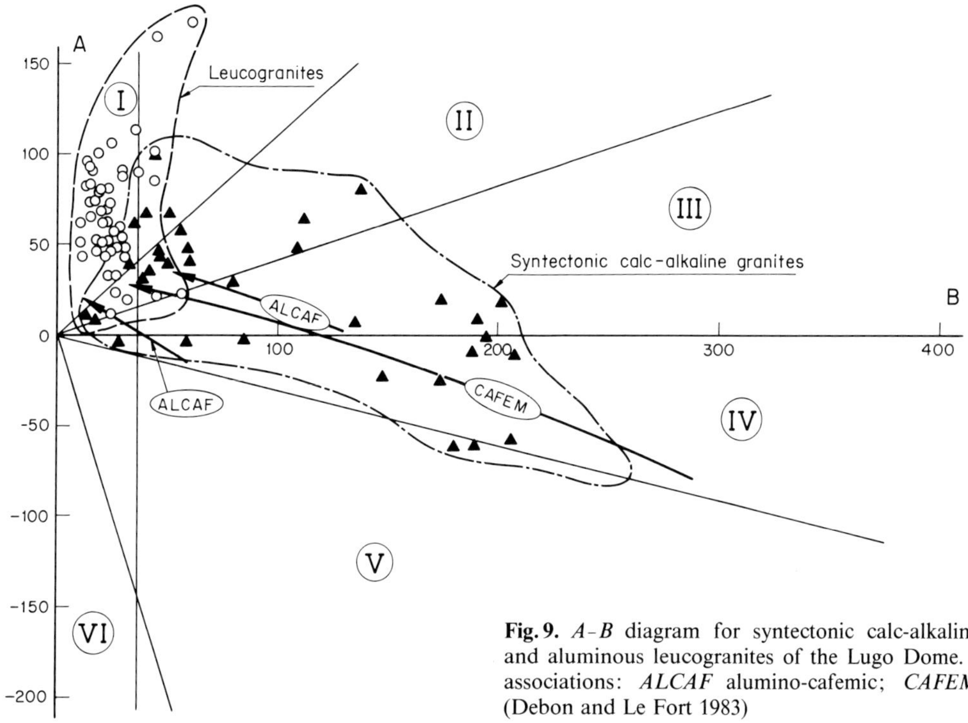

Fig. 9. *A–B* diagram for syntectonic calc-alkaline granites and aluminous leucogranites of the Lugo Dome. Magmatic associations: *ALCAF* alumino-cafemic; *CAFEM* cafemic (Debon and Le Fort 1983)

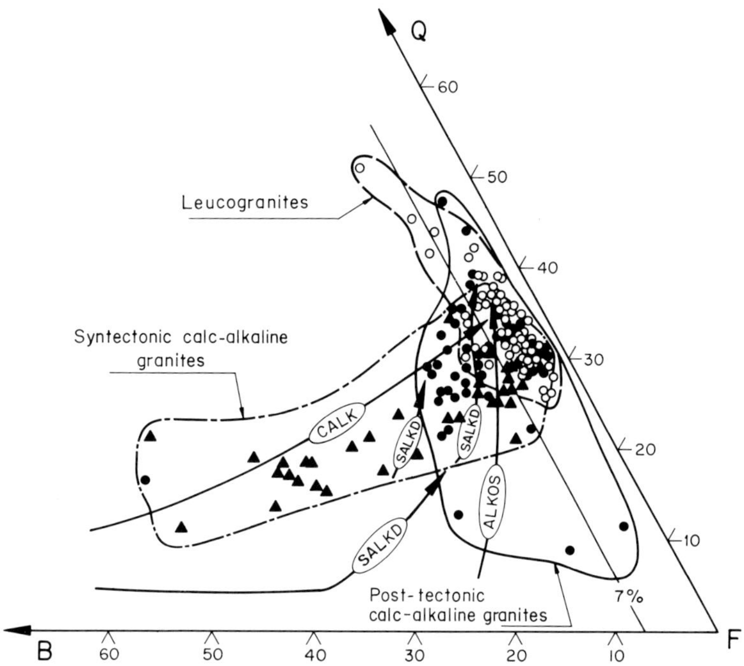

Fig. 10. Plot of the Lugo Dome granites in *Q-B-F* diagram (Debon and Le Fort 1983). See text; *Q*=Si/3-(K+Na+2Ca/3); *B*=Fe+Mg+Ti and F=55-(Q+B). *SALKD* Light-colored subalkaline trend; *ALKOS* light-colored alkaline oversaturated trend

Vivero (Galan et all., in prep.). Participation of a basaltic magma, contaminated with granitic magmas, would also explain the presence of ultramafic rocks (cortlandtites) associated with these granitoids (Galán 1987).

4.3.2 Aluminous Granites

Migmatites and granites in subautochthonous massifs with field, petrographic, mineralogical and geochemical characteristics comparable to those described by Didier and Lameyre (1969) for the Hercynian leucogranites of the French Central Massif, are included in this group.

4.3.2.1 The Migmatites

Migmatitic formations in the Lugo Dome are at the west of the Sarria massif, around the Hombreiro massif and mainly around the Vivero massif (Fig. 1). Capdevila (1969) mentioned two types of leucosomes: (1) leucosomes related to Precambrian and Lower Cambrian gneisses and diatexitic micaschists, to the west of the Sarria massif. They are leucocratic plagiogranites with quartz, acid oligoclase and secondary muscovite as accessory mineral. The associated melanosomes consist of biotite and often of garnet, sillimanite, muscovite and coarse apatites. And (2) leucosomes containing quartz, oligoclase, potassium feldspar and muscovite. The second type of leucosomes with K feldspar is less frequent and does not show associated melanosomes.

4.3.2.2 Aluminous Leucogranites in Subautochthonous Massifs

They are the most common and complex in the Lugo Dome.They are placed either along the dome axis or at the western limb. The following massifs can be differentiated from N to S (Fig. 1): Vivero, Murás, Monseiban, Román, Hombreiro and Sarria. Some authors also include the San Juan de Muro massif (Puebla de San Julian) within this group. In this chapter this massif is considered within the group of the syntectonic calc-alkaline granites with which it seems to share geochemical and mineralogical characteristics. Most of these granites are multiple intrusion laccoliths with different lithologies. They cut across P_1 folds but are emplaced from W to E following the direction of these structures and that of the basal thrust of the Mondoñedo nappe (P_2), which deforms them. They are therefore syntectonic-P_2 leucogranites which can be found both in the allochthonous (Monseibán, Román, Hombreiro and Sarria) and in the autochthonous (Muras and Vivero). The last massif shows its NW end deformed by a P_2 shear zone, while the SE boundary, cuts across the basal shear zone of the Mondoñedo Nappe, which means that emplacement of this massif extended through a longer period of time. Around the Vivero fault, some of these leucogranites (Vivero, Murás, Hombreiro and Sarria) were also affected by this later feature. No good whole rock isochrons are available on these granites, but Cocherie (1978) proposed an age of 305 Ma for the whole group.

Country rocks to these granites are Precambrian and Lower Cambrian metasediments and calc-alkaline granites from the previous group, which preceded them closely. They intrude simultaneously with the climax of the regional metamorphism and never surpass the staurolite isograd. They cut across the metamorphic isograds although sometimes they are roughly arranged around some of these granites (Hombreiro, Sarria and especially Vivero).

The least evolved facies of this group monzonitic granites and, less frequently, granodiorites (Fig. 7). Small variations occur from one massif to another but, as a whole, they are medium- to coarse-grained, equigranular and homogeneous, with quartz, plagioclase (An_{30}–An_7), perthitic microcline, biotite and muscovite as essential minerals. There are also apatite, zircon and ilmenite as accessory minerals. Muscovite often predominates over biotite. The most differentiated facies intrude either the previous ones, at the massif top or rim or as independent stocks. They are fine- to coarse-grained monzonitic leucogranites, leucogranites and alkali-feldspar granites (alaskites). They are made up of variable amounts of quartz, microcline, plagioclase (An_{16}–An_0), muscovite and subordinated biotite as essential minerals. Most characteristic of these facies are accessory minerals, with a generalized presence of tourmaline and/or garnet and, less abundant beryl, topaz, apatite, zircon and ilmenite. Postmagmatic transformations in these differentiated facies are important (albitization, muscovitization, tourmalinization), with widespread greisen zones in some massifs. Abundant aplites, pegmatites, tourmalinites and quartz veins are associated with these granites. A petrographic overlapping between the aluminous leucogranites and some facies of syntectonic calc-alkaline granites can be observed in Fig. 7.

The leucocratic and aluminous character of this group is also shown in Debon and Le Fort diagrams (Fig. 9) where they define an aluminous association with extremely variable values of parameters Q and A, with sodic to sodic-potassic character and chemical evolution following a vertical trend. The superposition of these leucogranites with some of the peraluminous facies from the syntectonic calc-alkaline granites is also obvious in Fig. 9. This superposition is less evident when comparing the Rb and Sr content (Fig. 4). The Sr content of these granites is lower than in the previous calc-alkaline granites, being in the area defined by other Hercynian leucogranites. In addition, the Rb and Sr content suggests an origin from orthogneisses unlike other Hercynian leucogranites (Fig. 4). Cocherie (1978) also deduced this origin from the rare earth element abundance and condrite normalized data from granites. These leucogranites are characterized by a low REE content, with very small Eu anomalies and high La/Yb ratios.

Mineralogically, they can be distinguished by the range of Mg concentration in biotites which is lower than in biotites of the syntectonic calc-alkaline granites and also because the Al content is higher than in the latter. In Fig. 6 a partial overlapping between the two biotite groups can be seen but both groups are clearly individualized.

All previously mentioned authors suggest that these leucogranites originated by the fusion of crustal materials during regional metamorphism. This fusion could also have been favoured by the intrusion of basic magmas which appear related to the calc-alkaline granites which precede the leucogranites closely. The initial isotope ratio Sr presented by these granites (0.709–0.718) would agree with this crustal origin.

4.3.3 Post-Tectonic Calc-Alkaline Granites

The second generation of granites with calc-alkaline affinities is better represented than the first one in the Lugo Dome. The granites of this group are mainly located along the E limb of the Dome, and from N to S the Tojiza, Lugo, Castroverde and Neira massifs are distinguished (Fig. 1). They are roughly circular or elliptical and occasionally with concentric zonation. They have vertical to subvertical sharp contacts with the country rocks which are Precambrian to Middle Ordovician metasediments and Hercynian granitoids. Weak anisotropic fabrics appear just at contact zones. Their intrusion took place after the third regional deformation phase (P_3) and they are only affected by late fractures. They also cut across the regional metamorphic isograds, reaching more superficial levels than the two previous groups of granites. Among those emplaced in more superficial levels is the Castroverde massif, which cuts across the biotite isograd. Contact metamorphism aureoles with paragenesis characteristics of the pyroxene and hornblende hornfels facies may occur in the metasedimentary country rocks. Cocherie (1978) attributed an age of 287–280 Ma to these granites based on whole rock and biotite isochrons.

The petrographic composition is less variable than that of the syntectonic calc-alkaline granites. The predominant facies is formed by monzonitic granites and, in a smaller proportion (Fig. 7), medium- to coarse-grained granodiorites with potassium feldspar megacrysts, often pink coloured. Dark minerals are rare, mainly biotite and occasionally biotite and muscovite. They also have apatite, zircon, allanite, garnet, amphibole, fluorite and ilmenite as accessory minerals. Accumulations of potash feldspar megacrysts appear gradually in some zones. These accumulation zones correspond to quartz-syenites (Fig. 7). Facies with orbicular textures may also occur. The alkali feldspar megacrysts become rarer towards the borders of the intrusions, leading to a gradation to granodiorites and monzonitic granites either more or less leucocratic than the central facies. The most differentiated facies appear in stocks and veins cutting through the granite and the country rocks. They are porphyrytic microgranites and leucogranites with biotite, muscovite, garnet and/or tourmaline. There are also scarce aplites and pegmatites and quartz veins. The most basic rocks appear in very few stocks surrounding the granites or as microgranular enclaves. These are Q-diorites and amphibole-biotite tonalites with ilmenite, sphene, zircon and apatite as accessory minerals. The relative importance of these more basic facies is smaller than in the syntectonic calk-alkaline granites. The tardi to postmagmatic transformations which affect these granites are of minor importance and restricted to fracture zones.

In the QAP diagram (Fig. 7) the different types of rocks, with the exception of the cumulative quartz-syenites, define a granodioritic calc-alkaline evolution trend (nevertheless, the few available data on microgranular enclaves should be taken into account). This tendency is also observed in the AFM diagram (Fig. 8). However, in the Debon and Le Fort diagram (1983), these massifs define an aluminous-cafemic association of light-coloured subalkaline type (Fig. 10), and at times an oversaturated light-coloured type (Tojiza massif). The evolutionary trend of the syntectonic calc-alkaline granites has also been represented in the diagrams (Figs. 10 and 11). The main differences between both groups derive from the minor importance of the rocks with basic to intermediate character and a lower melanocratic content in the peraluminous facies of the late calc-alkaline granites. These differences are also observed in the trace elements. For example, the Rb and Sr content can be compared with that in other Hercynian granites of calc-alkaline tendency, but they are generally poorer in Sr than most syn-

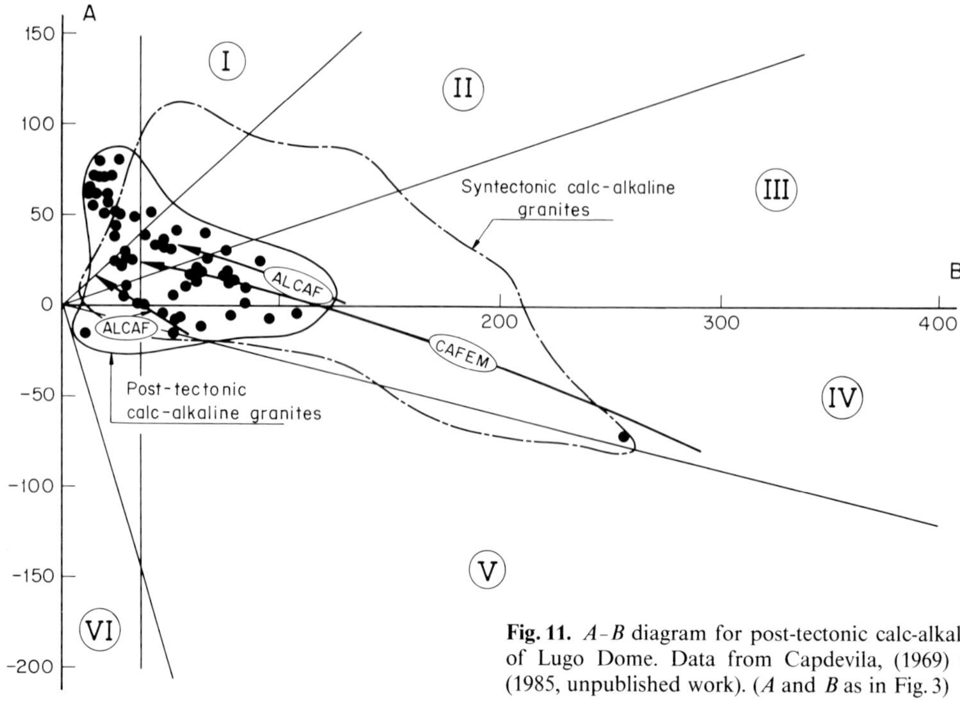

Fig. 11. *A–B* diagram for post-tectonic calc-alkaline granites of Lugo Dome. Data from Capdevila, (1969) and IGME (1985, unpublished work). (*A* and *B* as in Fig. 3)

tectonic calc-alkaline granite facies of the Lugo Dome (Fig. 4). REE elements differentiate them even better. According to Cocherie (1978) and Ortega's data (1987, unpublished work), the REE content of these granites is moderate, with slight variations from one massif to another, but in most cases they have an REE content higher than that of the syntectonic calc-alkaline granites. Their spectra are also characterized by small to moderate negative Eu anomalies and a variable La/Yb ratio depending on the facies.

Finally, the composition of the most characteristic minerals is also slightly different. The amphiboles of the most basic rocks have a Ti-richer composition of kaersutite type. Biotites are generally poorer in Mg than those of the syntectonic calc-alkaline granites and have a variable A1 content. Most of them are placed in the domain of biotites of calc-alkaline granites, but some plot in the subalkaline granite zone and others reach the aluminous granites field. Their evolutionary trend is clearly different from that of biotites in the syntectonic calc-alkaline granites (Fig. 6). Plagioclase, with the exception of the more basic rocks, is also less calcic than in the previous calc-alkaline granites, their composition ranging from oligoclase to albite.

Capdevila (1969) and Capdevila et al. (1973) suggested for these granites an origin similar to that of the first generation of calc-alkaline granites. This origin would imply the fusion of an intracrustal source with the participation of a basic magma of mantle-derived origin. Later studies, (Cocherie 1978; Ortega 1987, unpublished work), on the basis of the scarce associated basic rocks and the REE data available, attribute only a crustal genesis to most of these massifs. This crustal origin seems to be confirmed by the high initial $^{87}Sr/^{86}Sr$ ratio obtained for these granites (0.710–0.712). However, some of the authors do not dismiss a mantle participation in some of these massifs, like the Castroverde one for which a lower initial 87 Sr/86 Sr ratio is obtained (0.708 recalculated from Cocherie data).

4.4 Conclusions

The Hercynian magmatic activity in the WALZ gave rise to several generations of calc-alkaline granites (I Type) and aluminous leucogranites (S Type) in a period ranging from 307 to 284 Ma. Hercynian magmatic activity starts and ceases with the intrusion of calc-alkaline granites; aluminous leucogranites intrude between.

1. The first group of syntectonic calc-alkaline granites, with associated basic rocks, can only be found in the Vivero-Lugo-Sarria metamorphic belt where the regional metamorphism reaches

high grade conditions. Because of some of their structural and geochemical characteristics they can be compared to Cordilleran I-type (Pitcher 1982) or to I-type pre-collision granites from other orogenic chains (Debon et al. 1981).

2. Following them, a group of aluminous leucogranites is also emplaced in the Vivero-Lugo-Sarria metamorphic belt. These are spatially related to the previous group and are associated with high grade rocks. They are syntectonic granites, mostly comparable to S-type "Hercynotype" granites (Pitcher 1982).
3. A second group of S-type granites occurs in the WALZ. These differ from the previous ones in that they are post-tectonic and emplaced in zones of low grade regional metamorphism. Their petrographic and chemical characteristics are similar to those of the syntectonic leucogranites. Both groups are comparable to the associations occurring in the High Himalaya (Debon and Le Fort 1983). However, there are differences between these post-tectonic leucogranites and the syntectonic ones: (1) post-tectonic leucogranites sometimes contain A1 silicates (andalusite, sillimanite or cordierite); (2) they may enclose microgranular enclaves; (3) their Rb/Sr ratios straddle the leucogranites and calc-alkaline granite fields, and generally this ratio is lower than that of the syntectonic leucogranites; and (4) their REE content is higher. Because of these features these leucogranites constitute a group similar to those in the so-called mixed series (Capdevila et al. 1973; Corretge et al. 1974), widely represented in some other areas of the Iberian Massif. These granites are mainly emplaced in Paleozoic rocks overlying very thick Precambrian, psammopelitic series, as in the case of the S granites of the Boal - Los Ancares Belt.
4. Finally, a second group of calc-alkaline granites appears throughout the WALZ. These are I-type post-tectonic granites, with very different characteristics from those of the syntectonic calc-alkaline granites. They define associations with subalkaline trends and their structural and geochemical features are comparable to Pitcher's (1982) Caledonian I-type granites, or post-tectonic calc-alkaline granites occurring in other Hercynian areas (Barriere 1977; Orsini 1980) and collision belts. The most evolved rocks of the different associations always plot in the leucogranite field and show an overlapping of initial Sr isotope ratios, as happens with Hercynian granites elsewhere (Didier et al. 1982).

Agreement has been reached on the crustal genesis of the aluminous leucogranites with important contributions of orthogneisses leucogranites and of metasediments for the genesis of post-tectonic leucogranites. This double genetic origin leads us to agree with Didier et al. (1982), in that not all Hercynian leucogranites could be strictly comparable to the S granites defined by Chappell and White (1974).

A basic magma mantle-derived and contaminated by crustal materials is proposed as one of the contributions for the origin of the syntectonic calc-alkaline granites. The ultramafic rocks would have formed from this basic magma. The mantle contribution is not so clear for post-tectonic I-type granites because of the scarcity of associated basic rocks. However, the mantle-derived contribution cannot be rejected since the characteristics of the associated basic intrusions are typical of subalkaline series and also contain cortlandtitic enclaves.

Acknowledgements. We wish to thank Prof. E. Martínez and J. Fernández Suárez for improving the English. The manuscript has been considerably improved as a result of reviews by Dr. J. A. Whitney and Dr. M. Roden. Financial support for this work was provided by DGICYT, Project PB 86-0158.

References

Barriere M (1977) Le complexe de Ploumanach' (Massif Armoricain). Essai sur la mise en place et l'evolution pétrologique d'une association plutonique subalcaline tardi orogénique. PhD Thesis, Univ Brest, 220 pp

Bastida F, Plugar JA (1978) La estructura del manto de Mondoñedo entre Burela y Tapia de Caseriego (Costa Cantábrica, NW de España, NW de España). Trab Geol 10: 75-124

Bea F, Sánchez González del Herrero JG, Serrano Pinto M (1987) Una compilación geoquímica (elementos mayores) para los granitoides del Macizo Hespérico. In: Bea F, Carnicero A, Gonzalo IC, Lopez Placa M, Rodriguez Alonso MD (eds) Geología de los granitoides y rocas asociadas del Macizo Hespérico. Rueda, Madrid, pp 87-144

Bellido F, González Lodeiro F, Klein E, Martínez Catalán JR, Pablo Macia JG (1987) Las Rocas graníticas hercínicas del Norte de Galicia y Occidente de Asturias. Mem Inst Geol Min Esp 101: 157 pp

Capdevila R (1966) Sur la présence de sills basiques et ultrabasiques métamorphisés dans la region de Villalba (Lugo, Espagne). CR Acad Sci Paris D 262: 2193-2196

Capdevila R (1969) Le métamorphisme régional progressif et les granites dans le segment hercynien de Galice Nord Oriental (NW de l'Espagne). PhD Thesis, Univ Montpellier, 430 pp

Capdevila R, Floor P (1970) Les différents types de granites hercyniens et leur distribution dans le Nord-Ouest de L'Espagne. Bol Geol Min Esp Madrid LXXXI: 215-225

Capdevila R, Corretgé LG, Floor P (1973) Les granitoides varisques de la Meseta Iberique. Bull Soc Géol Fr (7ª Ser) 15 (3-4): 209-228

Chappell BW, White AJR (1974) Two contrasting granites types. Pac Geol 8: 173-174

Chappell BW, White AJR (1984) I- and S-type granites in the Lachlan Fold Belt, southeastern Australia. In: Xu Keqin, Tu Guanchi (eds) Geology of granites and their metallogenic relations. Science Press, Beijing, pp 87–101
Cocherie A (1978) Géochemie des terres rares dans les granitoides. Thesis, Univ Rennes, 207 pp
Corretgé LG (1969) El complejo ortoneísico de Pola de Allande (Asturias). Bol Geol Min Esp XXX-IV: 289–306
Corretgé LG, Ugidos JM, Martínez FJ (1974) Les series granitiques varisques du secteur Centre-Occidental espagnol. In: La Chaine Varisque d'Europe Moyenne et Occidentale. Coll Int CNRS Rennes 243: 453–461
Corretgé LG, Cienfuegos I, Cuesta A et al. (1987) Granitoides de la Región Palentina (Cordillera Cntábrica, España). Publ IX Reun Geol Oeste Peninsular (1985). Mus Lab Min Geol Mem Univ Porto 1: 469–498
Debon F, Le Fort P (1983) A chemical-mineralogical classification of common plutonic rocks and associations. Trans R Soc Edinb Earth Sci 73: 135–139
Debon F, Le Fort P, Sonet J (1981) Granitoid belts west and south of Tibet. About their geochemical trends and Rb-Sr isotopic studies. In: Liu DS (ed) Proceedings of Symposium on Qinghai-Xizang (Tibet) Plateau, (Beijing, China). Geological and ecological studies of Qinghai-Xizang Plateau 1: 395–405
Didier J (1973) Granites and their enclaves. Elsevier, Amsterdam, 393 pp
Didier J, Lameyre J (1969) Les granites du Massif Central Français. Etude comparée des leucogranites et granodiorites. Contrib Min Petrol 24: 219–238
Didier J, Duthou JL, Lameyre J (1982) Mantle and crustal granites: genetic classification of orogenic granites and the nature of their enclaves. J Volcanol Geotherm Res 14: 125–132
Galán G (1984) Las rocas graniticas del Macizo de Vivero en el sector Norte (Lugo NO de España) PhD Thesis, Univ Oviedo, 375 pp. (Publish. in: Corpus Geologicum Gallaeciae, Lab Geol Lage La Coruña), 2ª Serie, III (1987)
Galán G (1987) Le Massif Hercynien de Vivero (No de l'Espagne): Une association de roches ultramafiques et de granites. Comparaison avec d'autres exemples des orogenes hercynien et Caledonien. PhD Thesis, Univ Paris VI, 406 pp
Galán G, Suárez O (1989) Cortlandtitic enclaves associated with calc-alkaline granites from Tapia-Asturias (Hercynian belt, NW of Spain). Lithos 23, 4 (in press)
Gil Ibarguchi JI, Suárez O (1982) Estudio comparativo de las rocas igneas básicas de Finisterre (Galicia Occidental) y Tapia (Asturias). Cuad Lab Geol Lage (La Corunña) 2: 75–95.
Gutiérrez Claverol M, Luque C, Suárez MV (1987) El lineamiento tectónico Nazaré (W de Portugal)-Luarca (NW de España) y su implicación metalogenética (Macizo Hespérico). In: Bea F, Carnicero A, Gonzalo JC, López Plaza M, Rodriguez Alonso MD (eds) Geología de los granitoides y rocas asociadas del Macizo Hespérico. Rueda, Madrid, pp 447–456
Harris M (1980a) Gold mineralization at the Salave gold prospect, northwest Spain. Inst Mining Metall B: 1–4
Harris M (1980b) Hydrothermal alteration at the Salave gold prospect, northwest Spain. Inst Mining Metall B: 5–15
Lameyre J (1966) Leucogranites et muscovitisation dans le Massif Central Français. Ann Fac Sci Univ Clermont-Ferrand 29: 264
Lameyre J, Bowden P (1982) Plutonic rock types series: discrimination of various granitoids series and related rocks. J Volcanol Geotherm Res 14: 169–186
Larsen ES (1948) Batholith of southern California. Geol Soc Am Mem 29
López García MJ, Arce Duarte M, Fernández Tomas J, Montserin López V (1976) Estudio petrológico del Macizo de la Tojiza (provincia de Lugo). Tecniterrae 11: 7–15
Marcos A (1973) Las series del Paleozoico Inferior y la estructura herciniana del occidente de Asturias (NW de España). Trab Geol 6: 1–113
Martínez Alvarez JA, Torres Alonso M, Gutiérrez Claverol M (1975) Mapa Geológico de España. E. 1:50000. Foz. IGME. Serv Publ Minist Industr Madrid 44 pp
Martínez Catalán JR (1981) Estratigrafia y estructura del domo de Lugo (sector Oeste de la zona Asturoccidental-Leonesa) PhD Thesis, Univ Salamanca (Publish. in: Corpus Geologicum Gallaeciae, Labor Geol Lage La Coruña) (2ª Serie) II: 291 (1985)
Matte P (1968) La structure de la virgation hercynienne de Galice (Espagne). Géol Alpine 44: 1–127
Nachit M, Razafimahefa N, Stussi JM, Carron JP (1985) Composition chimique des biotites et typologie magmatique des granitoides. CR Acad Sci Paris 301 (Serie II) 11: 813–818
Nollau G (1968) Stratigraphie, Magmatismus und Tektonik der Montes de León zwischen Astorga und Ponferrada in Nordwest-Spanien. Geotekt Forsch 27: 71–146
Orsini JB (1980) Le batholite corso-sarde. Anatomie d'un batholite hercynien. Composition. Structure. Organisation d'emsemble. Sa place dans la chaine varisque française. PhD Thesis, St Jerome Univ Marseille, 390 pp
Pérez-Estaún A, Martínez FJ (1978) El Precámbrico del antiforme del Narcea, en el Sector de Tineo-Cangas del Narcea (NW de España). Trab Geol 10: 75–159
Pitcher WS (1982) Granite type and tectonic environment. In: Hsu K (ed) Mountain building processes. Academic Press, London, pp 19–40
Suárez O (1970) Los granitos de Ponferrada (León). Bol Geol Min Esp LXXXI-II-III: 185–191
Suárez O (1971) Estudio del plutón de Boal (Asturias, NW de España). Stud Geol Salmant 2: 93–113
Suárez O (1972) Petrología del stock granitico de "El Pato". (Asturias, NW de España). Stud Geol. Salmant 3: 119–129
Suárez O (1974) Estudio petrológico y geoquímico del granito de Ancares (Asturias-Lugo, España). Estud Geol 30: 167–178
Suárez O (1976) Petrogénesis del granito de Campo de Agua (zona Asturoccidental-Leonesa, España. Estud Geol 32: 425–432
Suárez O, Corretgé LG (1987) Plutonismo y metamorfismo en las zonas Cantábrica y Asturoccidental-leonesa 13–25. In: Bea F, Carnicero A, Gonzalo JC, López Plaza M, Rodríguez Alonso MD (eds) Geología de los granitoides y rocas asociadas del Macizo Hespérico. Rueda, Madrid, pp 13–25
Suárez O, Suárez V (1970) Sobre las rocas plutónicas de la zona de Tapia (Asturias, España). Bol Geol Min Esp LXXXI-II-III: 157–163
Suárez O, Suaréz Del Rio LM (1976) Estudio petrológico de los porfiroides precámbricos en la zona de Cudillero (Asturias). Estud Geol 32: 53–59
Suárez O, Ruiz F, Galán J, Vargas I (1978) Edades Rb-Sr de Granitoides del Occidente de Asturias (NW de España). Trab Geol 10: 437–442
Uhlig S (1984) Der Pluton von Boal (W-Asturien, Spanien) und seine Scheelitlagerstatte Penouta. Z Dtsch Geol Ges 135: 345–371
Walter R (1968) Die Geologie in der nordöstlichen Provinz Lugo (NW Spanien). Geotekt Forsch 27: 3–70

5 Distribution and Characteristics of the Hercynian Metamorphism

O. SUAREZ[1], L. G. CORRETGE[1] and F. J. MARTINEZ[2]

5.1 Introduction

Among the three zones into which the northwestern sector of the Hesperian Massif has been divided (Lotze and 1945; Julivert et al. 1972), the Cantabrian Zone (CZ) and the West Asturian-Leonese Zone (WALZ) clearly differ from the Central Iberian Zone (CIZ) in the scarcity of plutonism and in the absence or low grade of metamorphism, although the westernmost part of the WALZ, in the Dome of Lugo area, shows some similitudes with the CIZ in metamorphic and plutonic aspects.

The lack of metamorphism in the Cantabrian Zone (CZ) has been pointed out by some authors (Julivert 1983) but there really exists, mainly in the Palentian Zone, a low metamorphic area characterized by the presence of syntectonic quartz-albite-moscovite-chloritoid and quartz-illite-moscovite-chlorite assemblages (Lobato 1977; Suárez and Corretgé 1987). The epimetamorphic conditions in the Cantabrian Zone are clearly related to tectonic events which give rise to schistosity development (Brime 1985). Likewise changes in the illite crystallinity occur (Pérez Estaún 1978; Brime and Pérez Estaún 1980; Brime 1981, 1985; Aller and Brime 1985) have been observed in some areas in which there is a thermal gradient increase related to nappe emplacement (Brime 1981).

An increase in metamorphism and plutonism can be observed in an east-west cross section along the WALZ and CIZ, but this is not a continuous and gradual increase, since there are metamorphic bands of a higher grade, to which the greater part of granitic intrusions are associated.

[1] Departamento de Geología, Universidad de Oviedo, 33080 Oviedo, Spain
[2] Departamento de Geologia, Facultad de Ciencias, Universidad Autonoma de Barcelona, 08193 Bellaterra, Barcelona, Spain

5.2 Distribution of the Metamorphic Areas

The distribution of metamorphic isograds in both the West Asturian-Leonese Zone (WALZ) and Central Iberian Zone (CIZ) may be considered as: (1) defining elongated and, on occasions, broad belts or bands in pelite-greywacke sequences, whose age ranges from Upper Proterozoic to Low-Mid Paleozoic. These belts, almost without exception, are connected with I- and S-type granitic plutonism and (2) as round shaped synform complexes (klippen) consisting of a mixture of ultramafics, mafics and pelite-psammites that record a first metamorphic episode in high P-T (granulite, eclogite facies) These terrains are tectonically lying on top of parautochthonous units of Siluro-Devonian greenschist facies.

It is in the Northern Iberia and particularly in the West Asturian-Leonese Zone (WALZ) where the distribution in belts is outstanding. These belts in WALZ are arcuate parallel between them, following exactly the main structural trend which increases bending approaching the more external part of the chain (Cantabrian Zone, Fig. 1). The first belt in WALZ appears in the Upper Proterozoic outcrop of the Narcea antiform (the Novellana-Pola de Allande-Degaña metamorphic band; Fig. 1, 1) and the second belt, more to the west, affects Cambro-Ordovician materials, it is named the Boal-Los Ancares belt (Fig. 1, 2). These belts are roughly centred on Late Hercynian antiformal structures related to a P3 (phase 3) refolding phase that on some occasions gives rise to dome-like structures, as is the case in the already quoted Narcea antiform and in the Lugo structural dome (Fig. 1, 3). It is precisely in these antiforms where the oldest Upper Proterozoic or Lower Paleozoic materials crop out. These belts shows a broadly symmetrical metamorphic zoning; outside of their thermal axis and in opposite directions the metamorphism rapidly decreases, reaching the chlorite zone in a few kilometres.

R. D. Dallmeyer and E. Martínez García (Eds.)
Pre-Mesozoic Geology of Iberia

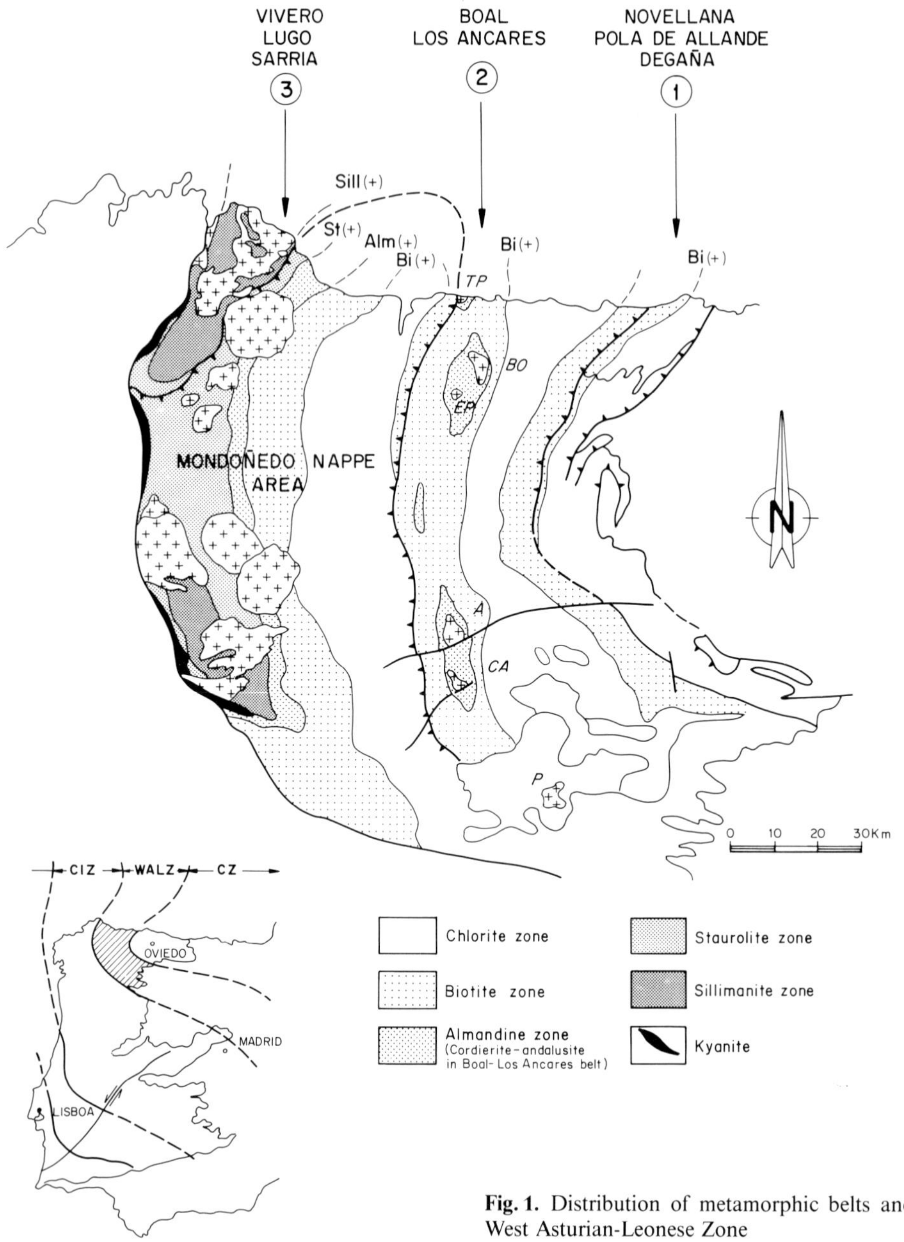

Fig. 1. Distribution of metamorphic belts and zones in the West Asturian-Leonese Zone

5.3 Characteristics of the Metamorphism

Concerning the metamorphism, two main units may be differentiated in WALZ, corresponding respectively to the Navia Domain in the east and the Mondoñedo nappe domain in the west. The first one is mostly located within the chlorite zone, with the two important exceptions earlier mentioned that constitute the Novellana-Pola de Allande-Degaña and the Boal-Los Ancares metamorphic belts (Martínez and Gil Ibarguchi 1983).

The most characteristic minerals within the largely widespread chlorite zone in WALZ are muscovite, chlorite and chloritoid, which are orientated defining the S_1 cleavage. Chlorite porphyroblasts with their longer dimension lying parallel to the original bedding are perpendicular or obliquely arranged to S_1 cleavage and show large pressure shadows within S_1 cleavage in the places where bedding and S_1 intersect at a high angle. These porphyroblasts must have grown mimetically with S_0 (bedding) and are pretectonic with respect to P_1 (S_1). Chlorite can also be found as pressure fringes around pretectonic rutile crystals (García de Figuerola and Martín Calvo 1969), which are especially abundant in Silurian slates. Chloritoid often occurs as tabular porphyroblasts and shows more complicated crystallization-deformation relationships, which have been described by Marcos (1973). This author differentiates syn-phase 1 crystals, surrounded by S_1 cleavage, post-phase 1 crystals enclosing it, and finally phase 2 related crystals arranged either along microfractures or parallel to the axial planes of second phase minor folds. As pointed out above, the most important episode of chlorite and moscovite crystallization is syntectonic with phase 1, although sometimes it might be pre-tectonic or early syntectonic. Chloritoid is most often post P_1 and covers a long crystallization period that reaches phase 2 (S_2 cleavage).

The *Novellana-Pola de Allande-Degaña* belt partially overlaps the western limb of the Narcea antiform, at the limit with the Cantabrian Zone. Biotite is the most significant mineral within the pelitic Precambrian rocks of this area; it occurs both randomly arranged or mimetically with respect to S_1 cleavage. Chlorite, moscovite and recrystallized quartz are also well developed. Biotite is abundant in psammitic Paleozoic layers placed outside of the antiform. In some pelitic strata within the Lower Cambrian limestones, biotite can also be found in carbonates with scapolite, the latter occurring as porphyroblasts post-tectonic to the main cleavage (see Fig. 2).

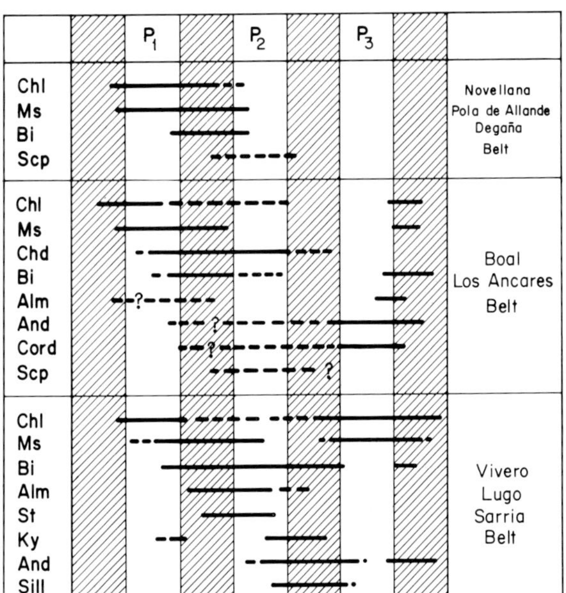

Fig. 2. Blastesis-deformation relationships in the belts shown in Fig. 1

The *Boal-Los Ancares* belt coincides with a large third phase anticline structure and appears (like other belts in inner areas of the Hercynian massif) associated to S-type granitoid intrusions represented by small plutons and stocks (Boal, El Pato, Ancares, Campo del Agua and Ponferrada Fig. 1); nevertheless, I-type affinity granitoids (like the Tapia stocks) are also associated with this belt (Corretgé et al., see this Vol.). The width and extension of this metamorphic belt seriously questions the hypothesis of an origin by simple contact metamorphism (Suárez 1970, unpublished work, 1971). The greater part of this belt is located within the biotite zone. For this zone many of the observations made in the Novellana-Pola de Allande-Degaña belt are still valid. It is remarkable that just as in Novellana-Pola de Allande-Degaña, and in the same pelitic strata above the Cambrian limestones, meionite-rich scapolite porphyroblasts can be found in the Boal-Los Ancares belt (Suárez and Arias 1981).

The second zone within the Boal-Los Ancares belt is the andalusite-cordierite isograd. It is spatially related to granitoid intrusions, although in some areas there is no evidence of such a relationship. Mineral assemblages are characterized by the presence of andalusite, cordierite and, less often, garnet as well as biotite, chlorite and muscovite. Andalusite and cordierite enclose S_3, but they also show pressure shadows in some mineral phases like garnet. It is also possible to observe the coupling of S_1 around andalusite, biotite and cordierite crystals and even rotated internal cleavages (Marcos et al. 1980; see Fig. 2).

According to the widespread occurrence of postkinematic andalusite on cordierite nodular schists, this belt seems to be generated by contact metamorphism, probably of late-syntectonic character with P_3. However it is very unlikely that such a large metamorphic belt could be generated only by the thermal effect of the above mentioned plutons, which only represent small outcrops inside the belt. The thermal anomaly responsible for the whole belt could be explained if this is considered as the result of a contact effect related to a large underlying igneous body. It seems that the thermal anomaly in this area lasted from P_1 until P_3. This can be deducted from the existence inside this belt of biotite spots, rotational garnets and chloritoid that grew synkinematic with P_1 together with the post-P_3 growth of andalusite and cordierite porphyroblasts. Suárez (1970, unpublished work, 1971), by taking into account the shape and distribution of the andalusite and cordierite zones, pointed out the possibility of a thermal anomaly onset preceding granitoid intrusion in this Boal-Los Ancares area.

The facts mentioned above indicate that the temporal evolution of the metamorphism is not just a simple late thermal event but has a more complex, longer lasting history, and also shows that the thermal history of the belt seems, at least in part, to be linked with granitoid intrusion.

5.4 Vivero-Lugo-Sarria Metamorphic Belt

This belt is also named Vivero-Lugo-Sarria thermal dome by Matínez and Gil Ibarguchi (1983). It constitutes the higher grade metamorphic belt in the WALZ and may be considered as an example of metamorphic evolution in the WALZ and CIZ. The nature of Hercynian metamorphism and the crystallization-deformation relationship have been established in the Mondoñedo nappe area (Capdevila 1969; Martínez Catalán 1985; Fig. 1). Here a plurifacial and polyphase medium or low-medium pressure type metamorphism develops. This metamorphism shows a first stage characterized by the following zone sequence: chlorite-biotite-garnet-staurolite (kyanite), and a second stage related to granitoid rise and intrusion that can be named thermal doming. This second stage is characterized by a metamorphic gradient increase and by andalusite, sillimanite prograde parageneses. Kyanite mainly occurs along a narrow band parallel to the Vivero fault and seems related to this tectonic feature. This restricted character has also been noted in other areas, as for instance, the Armorican Massif whose kyanite appears closely related to shear zones which are independent from the regional isograds (Martínez and Rolet 1988). Metamorphic trajectories reaching their top temperature with gradient of 27°C/km and 32°C/km (Martínez Catalán 1985), for assemblages respectively, with or without kyanite, can be proposed for the first metamorphic episode.

In the Vivero-Lugo-Sarria belt, andalusite occurs near the onset of the sillimanite zone and partially overlaps the staurolite zones, in such a way that the andalusite and sillimanite isograds are proximal and nearly subparallel to the contacts of some granitoid plutons. The late development of andalusite, in relationship with thermal domes and magmatic intrusions, has been pointed out by Bastida and Pulgar (1978), who describe sillimanite prisms included in andalusite. This is an essential aspect for the explanation of the Hercynian metamorphism in WALZ and CIZ, since it constitutes further evidence of a second episode with higher thermal gradients which affects part of the earlier metamorphic minerals; for instance, the staurolite porphyroblasts belonging to the first paragenesis are entirely or partially replaced by andalusite as a consequence of this second metamorphism related to the ascent and intrusion of granitic rocks which produce thermal domes.

High-grade metamorphic conditions with migmatite development are locally reached during the second episode (Capdevila 1969; Bastida et al. 1986; Galán 1987; Corretgé et al., see this Vol.). Nevertheless, high-grade conditions above the ksf-sillimanite isograd are rather uncommon in situ; although different sized, metamorphic xenoliths of gneissic character with this two minerals occur in the S-type Vivero granite (Galán 1987).

A less significant third metamorphic episode is recorded along the basal thrust of the Mondoñedo Nappe. During this episode, biotite has broken down to chlorite; garnet is resorbed to chlorite or chlorite-biotite pseudomorphs and andalusite and staurolite are pseudomorphosed by muscovite.

Blastesis-deformation relationships in different domains in the Vivero-Lugo-Sarria belt may be specified from the work by Matte (1968); Capdevila (1969); Martínez Catalán (1985); Bastida and Pulgar (1978), and Bastida et al. (1986). According to these authors' observations, the biotite, garnet and staurolite isograds intersect P_1 folds and are deformed by P_3 structures. The biotite and, partially, the almandine isograds appear well preserved in the hinge of P_3 synforms, showing no reactional changes. The andalusite and sillimanite isograds cut accross both P_1 and P_2 structure and are not parallel to the former isograds; this strongly suggests a close relationship of these later isograds with underlying heat suppliers. The general picture between metamorphic episodes and deformation phases is shown in Fig. 2. The first paragenesis, indicative of a Barrovian-tye metamorphism, reached its climax at the end of P_1 or between P_1 and P_2. The second metamorphic episode with andalusite bearing assemblages started during P_2, reaching its climax between P_2 and P_3 or at the beginning of P_3. Related with the thermal increase of the second metamorphism two mica granitic rocks were intruded into the Mondoñedo nappe area.

5.5 Age of the Metamorphism

In the Navia domain, the Upper Westphalian-Stephanian Rb-Sr whole rock age of some plutons in the Boal-Los Ancares belt (285 M a, Suárez et al. 1978) should be that of the metamorphism and this should also be the presumed age of the last thermal imprint represented by post-kinematic biotite and moscovite growth in the Narcea antiform. With regard to syn P_3 (?) garnet and biotite growth in Boal - Los Ancares, it seems sensible to consider it just a little older than Upper Westphalian, given its location inside the metamorphic area and its absence outside.

In the Vivero-Lugo-Sarria belt (Mondoñedo nappe) an age estimate for the older metamorphic event based upon the S_1-S_2 cleavage relationship with mineral growth may be made. Minerals mainly developed between P_1 (S_1) and P_2(S_2), although some late- or post-P_2 growth also occurred (see Fig. 2). An absolute post-Emsian age for the metamorphism may also be deduced, since in this Mondoñedo nappe Emsian (Low-, Mid-Devonian) materials (Drot and Matte 1967) are folded and S_1 cleaved. Therefore metamorphism should be post-Mid-Devonian and, if compared with the Boal-Los Ancares or Novellana-Pola de Allande belts, it may reasonably be considered as intracarboniferous.

Acknowledgements. Financial support for this study was provided by DGICYT, Project PB86-0158.

References

Aller J, Brime C (1985) Deformación y metamorfismo en la parte Sur de la Cuenca Carbonífera Central (NO de España). CR X Congr Int Strat Geol Carbonifère, Madrid 3: 541-548

Bastida F, Pulgar JA (1978) La estructura del manto de Mondoñedo entre Burela y Tapia de Casariego (Costa Cantábrica, NW de España, NW de España). Trab Geol 10: 75-124

Bastida F, Martínez Catalán JR, Pulgar JA (1986) Structural metamorphic and magmatic history of the Mondoñedo nappe (Hercynian belt, NW Spain). J Struct Geol 8: 415-430

Brime C (1981) Postdepositional transformation of clays in Paleozoic rocks of northwest Spain. Clay Min 16: 421-444

Brime C (1985) A diagenesis to metamorphism transition in the Hercynian of northwest Spain. Min Mag 49: 481-484

Brime C, Pérez Estaún A (1980) La transición diagénesis-metamorfismo en la région de Cabo Peñas. Cuad Lab Geol Lage. La Coruña 1: 85-97

Capdevila R (1969) Le métamorphisme régional progressif et les granites dans le segment Hercynien de Galice Nord Oriental (NW de l'Espagne). PhD Thesis, Univ Montpellier, 430 pp

Drot J, Matte P (1967) Sobre la presencia de capas del Devoniano en el límite de Galicia y León (NW de España). Notas Com Inst Geol Min Esp Madrid 93: 87-92

Galán G (1987) Las rocas graníticas del Macizo de Vivero en el sector Norte (Lugo NO de España) PhD Thesis, Univ Oviedo. (Publish. in: Corpus Geol Gallaeciae Lab Geol Lage La Corunña) (2ª Serie) III: 375 pp

García de Figuerola LC, Martín Calvo M (1969) Filitas con rutilo autígeno y cloritoide en el pantano de Navia (Asturias). Breviora Geol Astúrica XIII(1): 13-16

Julivert M, Fontboté JM, Ribeiro A, Conde L (1972) Mapa tectónico de la Península Ibérica y Baleares E 1:1000000. Inst Geol Min Esp Madrid, 113 pp

Julivert M (1983) Los tiempos precámbricos y paleozoicos. El ciclo hercínico. In: Comba JA (ed) Libro Jubilar JM Ríos, Geologia de España T: 1. IGME, Madrid, pp 59-71

Lobato L (1977) Geología de los Valles altos de los ríos Esla, Yuso, Carrión y Deva (NE. León, NW. Palencia y SW. de Santander). Inst Fr Bernardino Sahagún. CSIC León, 133 pp

Lotze F (1945) Zur Gliederung der Varisziden der Iberischen Meseta. Geotekt Forsch 6: 78-92

Marcos A (1973) Las series del Paleozoico Inferior y la estructura herciniana del occidente de Asturias (NW de España). Trab Geol 6: 1-113

Marcos A, Pulgar JA, Martínez FJ (1980) Memoria del Mapa Geológico de España E. 1:50000 n° 26, Boal Inst Geol Min Esp Madrid, pp 1-28

Martínez FJ, Gil Ibarguchi JI (1983) El metamorfismo en el macizo Ibérico. In: Comba JA (ed) Libro Jubilar JM Ríos. Geologia de España T: 1. IGME, Madrid, pp 555-569

Martínez FJ, Rolet J (1988) Late Paleozoic metamorphism in the northwestern Iberian peninsula, Brittany and related areas in southwest Europe. In: Harris AL, Fettes DJ (eds) The Caledonian-Appalachian Orogen. Geol Soc Publ 38: 611-620

Martínez Catalán JR (1985) Estratigrafía y estructura del domo de Lugo (Sector Oeste de la zona Asturoccidental-Leonesa). PhD Thesis, Univ Salamanca. (Publish. in: Corpus Geol Gallaeciae Lab Geol Lage, La Coruña) (2ª Serie) II: 291 pp

Matte Ph (1968) La structure de la virgation hercynienne de Galice (Espagne). Géol Alpine 44: 1-127

Pérez Estaún A (1978) Estratigrafía y estructura de la rama S de la zona Asturoccidental-Leonesa. Mem Inst Geol Min Esp Madrid: 92: 1-151

Suárez O (1971) Estudio del plutón de Boal (Asturias, NW de España). Stud Geol Salamant 2: 93-113

Suárez O, Arias E (1981) Origen de la escapolita en metapelitas y rocas anfibólicas de Series del Cámbrico Inferior del occidente de asturias (NW de España). Leidse Geol Meded 52: 101-107

Suárez O, Corretgé LG (1987) Plutonismo y metamorfismo en las zonas Cantábrica y Astur-occidental Leonesa. In: Bea F Carnicero A, Gonzalo JC, Lopez Plaza M, Rodriguez Alonso MD (eds) Geologia de los granitoides y Rocas asociadas del Macizo Hespérico Rueda, Madrid, pp 13-25

Suárez O, Ruiz F, Galán J, Vargas I (1978) Edades Rb/Sr de granitoides del Occidente de Asturias (NW de España). Trab Geol 10: 437-442

6 Metallogenetic Characteristics

C. Luque[1] and F. Ruiz[2]

6.1 Introduction

The West-Asturian-Leonian Zone offers an ample variety of mineralizations, although frequently of small size. The metal elements present are: iron, lead, zinc, copper, silver, gold, molybdenum, antimony, manganese, tungsten, uranium, mercury, arsenic and cobalt. The main nonmetallic minerals are: barite, kaolin, magnesite and andalucite. Their spatial distribution is compiled in Figs. 1 and 2.

6.2 Iron Mineralizations

These correspond to three types of settings:

1. *Beds*, which are organized in ferruginous levels interbedded in detrital sequences of shallow platforms, euxinic type rich in organic matter (Lunar 1977). They occupy different positions in the stratigraphic sequence, being found in the transitional series (Lower-Middle Ordovician; Valledor, No. 1), as well as in the Luarca Formation (Middle Ordovician); Villaodrid, No. 2; San Jose, No. 3; Sotelo, No. 4; Siena, No. 5; Balantex, No. 6; San Miguel de las Dueñas, No. 7; Astorga, No. 8; etc.) or in the Agüeira Formation (Upper Ordovician; Porcia, No. 9). In general, they are beds of very variable thickness (0–50 m) containing hematite, magnetite, siderite, chlorite and goethite with frequent oolitic textures, occasionally accompanied by secondary sulphides (pyrite, marcasite, chalcopyrite and arsenopyrite) as well as apatite, rutile and ilmenite. The iron content ranges from 30 to 35%;
2. *Veins*, within which there are irregular iron veins in fractured zones, composed of hematite, goethite and limonite, frequently with breccia-like structures (Villa Tascon, No. 10; La Arruñada, No. 11);
3. *Gossans*, which are derived from alteration of sulphide mineralization of complex mineralogy, consisting of limonite, goethite and hematite, mainly with boxwork to massive texture (Monteserin, No. 12).

Only the first type appears to be economically interesting, with moderate reserves, although its content in P is high (0.3 to 2%).

6.3 Lead-Zinc-Copper-Iron-Silver Mineralizations

These associations are found in three different groups with slight variations between different localities:

1. In the southern part of the West Asturian-Leonese Zone Pb-Zn deposits appear in Lower- and Middle-Cambrian limestones and dolomite levels, related to silicification, chloritization and ankeritization. The stratabound type of mineralization consists of lens-shaped deposits as well as fracture fillings, mainly with sphalerite (frequently with Hg contents) and galena, accompanied by pyrite, chalcopyrite and pyrrotite, with massive or disseminate textures, in a gangue of quartz with sericite, calcite, dolomite, siderite, chlorite and barite. Its spatial control is determined by the intersection of transversal faults and shear zones and the lithologic control is established by carbonate levels. The sulphides are distributed in a distinct zonal deposition (Merayo et al. 1984). This type of mineralization is found in Rubiales, No. 13; Toral de los Vados, No. 14; Sobrado, No. 15; Cabeza de Campo, No. 16; Castropete, No. 17; etc.
2. In the Oscos Region there are Pb-Zn deposits associated with the basal Luarca Formation with conformable levels of amphibolites and garnet metapelites derived from the regional metamor-

[1] Unidad Central de Geología de Hunosa, Ujo, Asturias, Spain
[2] Av. V. Masip, 7, 33013 Oviedo, Spain

R. D. Dallmeyer and E. Martínez García (Eds.)
Pre-Mesozoic Geolgoy of Iberia

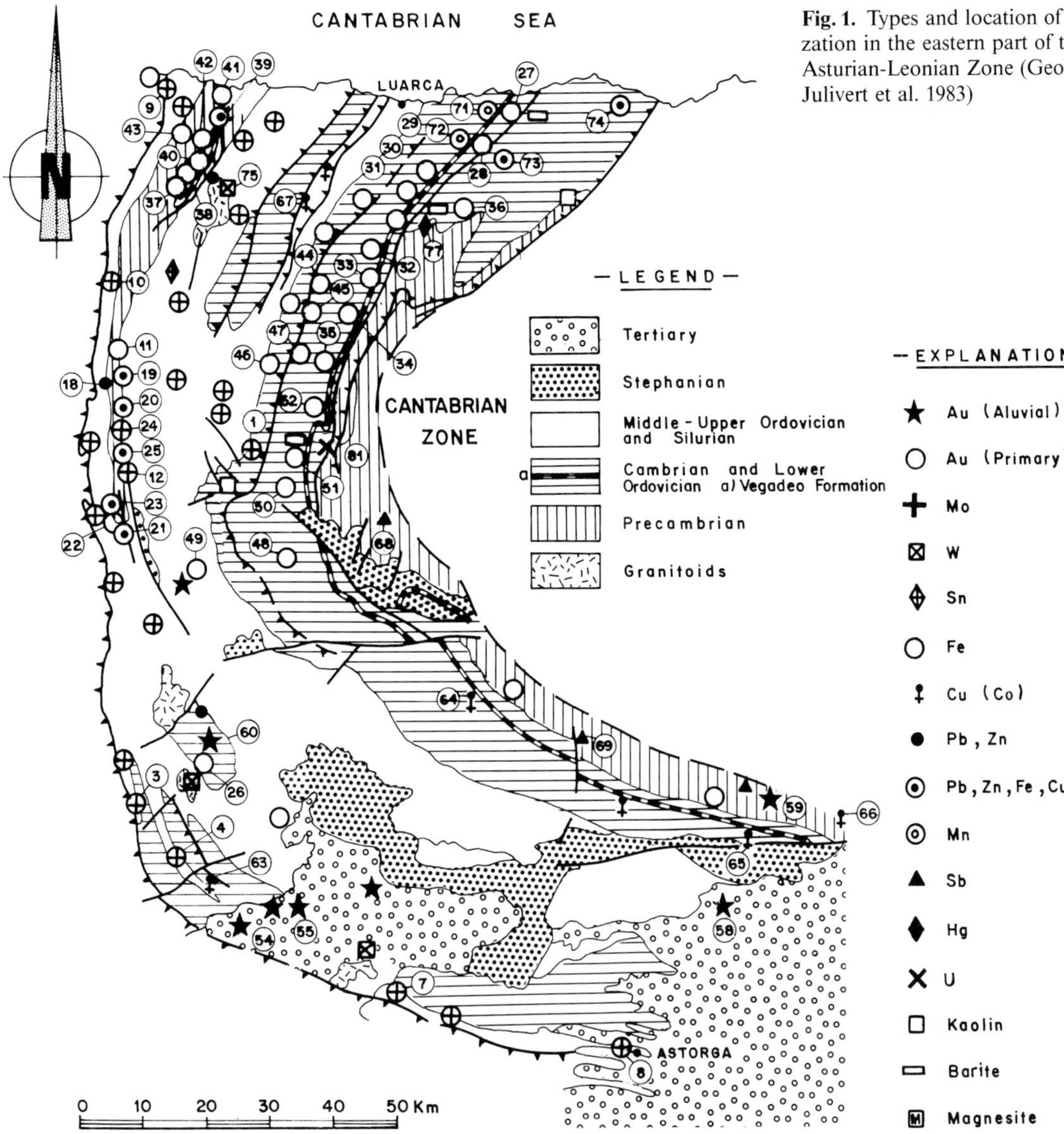

Fig. 1. Types and location of mineralization in the eastern part of the West Asturian-Leonian Zone (Geology from Julivert et al. 1983)

(1) VALLEDOR
(3) SAN JOSÉ
(4) SOTELO
(7) S. MIGUEL DE LAS DUEÑAS
(8) ASTORGA
(9) PORCIA
(10) VILLA TASCON
(11) LA ARRUÑADA
(12) MONTESERIN
(18) ASCUITA
(19) PIORNO
(20) S. MARTIN DE OSCOS
(21) PENEDA
(22) PENEDELA
(23) RIO DE PORCOS
(24) LA COBA
(25) MONTESERIN
(26) CANDIN
(27) CADAVEDO
(28) TREVIAS
(29) MERAS
(30) NARAVAL
(31) NAVELGAS
(32) RIOCASTIELLO
(33) CERREDO
(34) LAVADOIRA
(35) BACHICON
(36) LOS FORNONES
(37) BRAÑALIVEL
(38) VEGA DE OURIA
(39) ARANCEDO
(40) RAMAELLE
(41) MIUDES
(42) LA BRAÑA
(43) LA VEGUINA
(44) BUSTANTIGO
(45) PUERTO DE EL PALO
(46) VALLEDOR
(47) CARCABAS
(48) VILLALAIN
(49) SAN SALVADOR
(50) MONTERROSO
(51) SAN FELIX DE LAS MONTAÑAS
(52) IBOYO
(54) PARADASECA
(55) VALLE DE FINOLLEDO
(58) OMAÑAS
(59) RIELLO
(60) ANCARES
(63) VILLAR DE ACERO
(64) CUEVAS DEL SIL
(65) ANDARRASO
(66) CURUEÑO
(67) VILLAYON
(68) BIMEDA
(69) MURIAS DE PAREDES
(71) TREVIAS
(72) CADAVEDO
(73) MUÑAS
(74) LUIÑA
(75) BOAL
(76) LOS BARRIOS DE SALAS
(77) MUÑALEN
(81) BESULLO

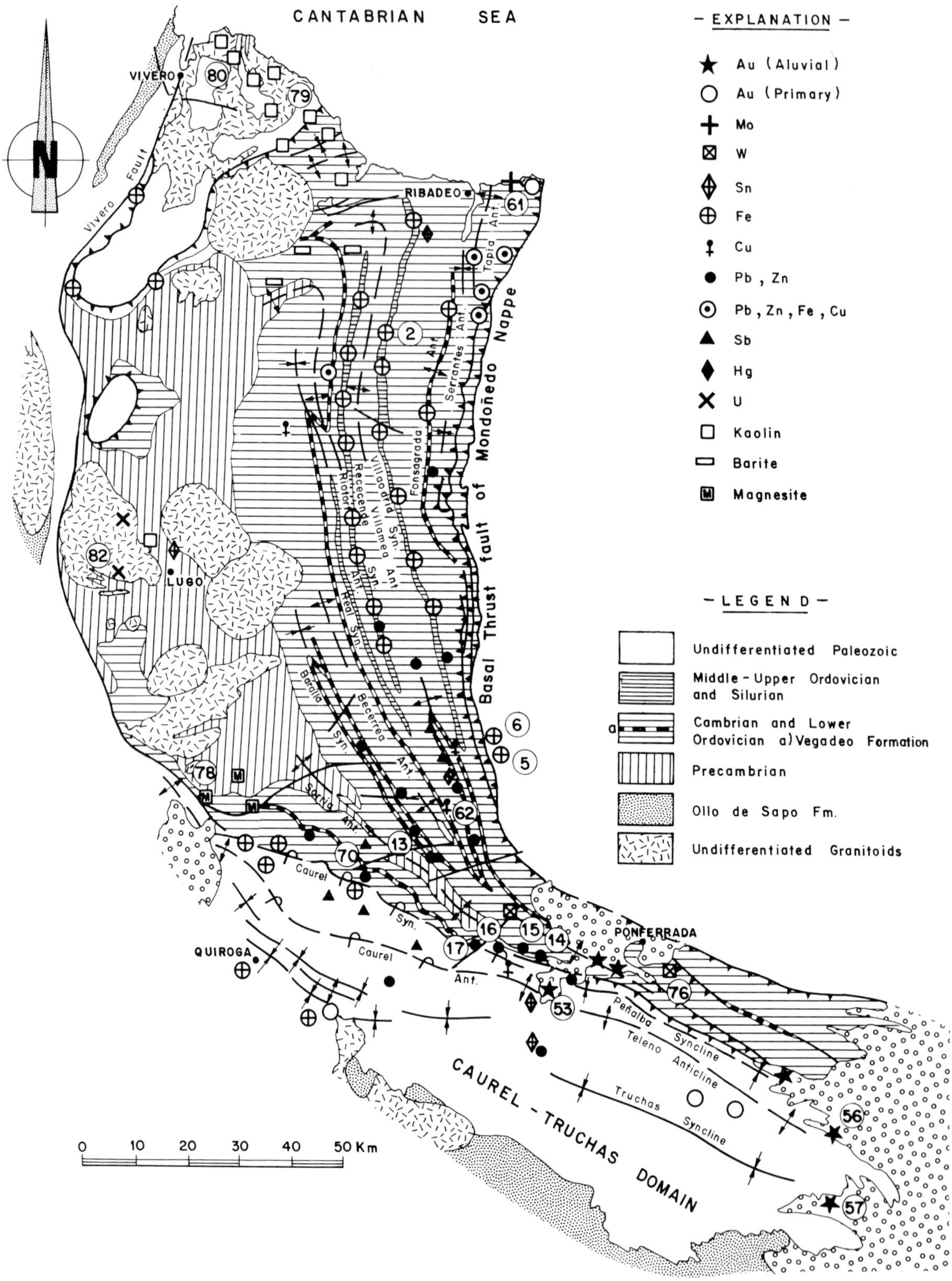

(2) VILLAODRID
(5) SIENA
(6) BALANTEX
(13) RUBIALES
(14) TORAL DE LOS VALDOS
(15) SOBRADO
(16) CABEZA DE CAMPO
(17) CASTROPETE
(53) LAS MÉDULAS
(56) VALDUERNA
(57) CASTROCONTRIGO
(61) SALAVE
(62) PIEDRAFITA DEL CEBRERO
(70) VILLARBACÚ
(76) LOS BARRIOS DE SALAS
(78) RUBIAN
(79) BURELA
(80) VIVERO
(82) LUGO

phism of volcanic and volcanoclastic rocks (Garcia Iglesias et al. 1985). They present a banded texture and highly varied mineralogical composition, with magnetite, arsenopyrite, pyrrotite, sphalerite, chalcopyrite and galena as essentials, along with tetrahedrite-freibergite, jamesonite and tennantite. An intense supergenic alteration causes the appearance of hematite, limonite, goethite, anglesite, hemimorphite, pyromorphite, pyrolusite, jarosite, coronadite, mimetite, plumbojarosite and plumbogumite (Ascuita, No. 18; Piorno, No. 19; San Martin de Oscos, No. 20; Peneda, No. 21; Penedela, No. 22; Rio de Porcos, No. 23; La Coba, No. 24; Monteserin, No. 25). Textures, structures and parageneses are similar to those found in the Hercynian volcanosedimentary districts.

3. In the northernmost sector there are stratiform sulphide mineralizations with carbonate gangue between the "Serie de los Cabos" and Luarca Formations. They are composed of siderite and associated pyrite, marcasite, galena and chalcopyrite, along with sphalerite, tetrahedrite and covellite. Exceptionally native silver appears, as well as arsenopyrite with native gold, showing also an intense supergenic alteration. Its origin would be similar to that of the Oscos mineralization.

Gold

In the West Asturian-Leonese Zone there is a large number of primary gold mineralizations at different lithostratigraphic levels (Narcea Schists, Candana-Herrería Formation, Vegadeo Limestone, Serie de los Cabos, Luarca Formation), occasionally relatively close to granitic intrusions.

A good number of them are associated with 20° N-40° E fractures, corresponding to the Nazarè-Luarca lineation (Gutiérrez Claverol et al. 1987) in relation to carbonate levels intensely silicified, as well as to quartz veins (Candin, No. 26; Caurel area) with arsenopyrite, stibnite, galena and scheelite traces. Many concentrations are found in zones of intersection with 110° N/120° E fractures.

The gold is either free or included in arsenopyrite or pyrite, in relation to hydrothermal alteration processes (chloritization, sericitization, silicification, albitization).

In the carbonate levels (Vegadeo Limestones) there is sometimes a superficial karst with sandy-clay fillings with free gold grains (Cadavedo, No. 27; Trevias, No. 28; Merás, No. 29; Naraval, No. 30; Navelgas, No. 31; Riocastiello, No. 32; Cerredo, No. 33; Lavadoira, No. 34; Bachicón, No. 35; Los Fornones, No. 36; Brañalivel, No. 37; Vega de Ouria, No. 38; Arancedo, No. 39; Ramaelle, No. 40; Miudes, No. 41; La Braña, No. 42; La Veguina, No. 43). Such fillings coexist with the primary mineralization, generally consisting of quartz stockworks and irregular pyritized aggregates with gold in the limestones.

◀ **Fig. 2.** Types and location of mineralization in the western part of the West Asturian-Leonian Zone (Geology from Julivert et al. 1983)

Along the Allande Fault, of N 40° E direction, there are several intrusive stocks, mainly albitic, subvolcanic character, mineralized and with intense hydrothermal alteration (Monterroso, No. 50, San Felix de las Montañas, No. 51). In others they develop skarn deposits in the carbonate levels of the Vegadeo Formation, in relation with small stocks (gabbros and granodiorites), as well as with associated mineralization within the same (Iboyo, No. 52).

In quartzitic formations, auriferous quartz veins usually fill E-W fractures. The erosion processes could also produce secondary concentrations of colluvium (Bustantigo, No. 44; Puerto del Palo, No. 45; Valledor, No. 46; Cárcabas, No. 47; Villalain, No. 48; San Salvador, No. 49).

Some concentrations in Tertiary sediments derived from the primary deposits are widely extended in the Bierzo (Las Médulas, No. 53; Paradaseca, No. 54; Valle de Finolledo, No. 55; etc.), Astorga (Valduerna, No. 56; Castrocontrigo, No. 57) and the Orbigo River basin (Omañas, No. 58; Riello, No. 59). Gold is also found in fluvioglacial deposits (Ancares, No. 60) and Quaternary alluvial terraces, as well as in recent sediments (Sil, Omañas, Duerna, Eria, Burbia, Navelgas, etc. rivers).

Gold-Molybdenum-Antimony

This association is related to the stock of Salave (No. 61) in which different zones composed of acid rocks (granodiorites, quartz diorites, aplitic granites) of Upper Stephanian-Lower Permian age (Suárez et al. 1978) are seen with complex alteration (sericitization, potassic alteration, silicification, chloritization, propylitization) (Harris 1980).

The mineralization is found as fissure fillings in the stock margins. The paragenesis is made up of pyrite and arsenopyrite along with molybdenite, stibnite, bornite and native gold, all accompanied by siliceous gangue. Two overlapping associations are seen, the first composed of Au-Mo-As, formed from hydrotermal solutions in the proximity of 600° C, while the second, with Au-As-Sb, is of lower temperatures.

The gold shows up as native metal, free or included in arsenopyrite, which releases it superficially when weathered. Its grain size is very fine, generally with values lower than 100 μ.

Copper
The presence of copper minerals is very limited (Piedrafita del Cebrero, No. 62; Villar de Acero, No. 63), with chalcopyrite and gray coppers associated with quartz veins in the limestones of the Vegadeo Formation. Sometimes they contain gold traces (Cuevas del Sil, No. 64; Andarraso, No. 65), are accompanied by galena (Curueño, No. 66), or associated with Co minerals, in relation to amphibolites and garnet pyroxenites interbedded in the Serie de los Cabos (Villayón, No. 67).

Antimony
It shows up in Precambrian slate materials as irregular veins (Bimeda, No. 68; Murias de Paredes, No. 69) or as stratabound occurrences (Guillou 1969; Gumiel 1983) in marly limestones, frequently remobilized towards tectonized zones (Villarbacú, No. 70) and consisting of stibnite, berthierite, pyrite and arsenopyrite, with quartz and carbonate gangue.

Manganese
In the carbonate levels of the Vegadeo Formation, lens-shaped iron deposits occur with a moderate content of Mn (20–40%). The mineralization is composed of pyrolusite, manganite and goethite, as nodules, kidney-ore aggregates and grains wrapped in a hematitic matrix. It is also possible to find jacobsite and micaceous hematite (Trevias, No. 71; Cadavedo, No. 72; Muñàs, No. 73).

Tungsten
The tungsten deposits are related to the tardi-Hercynian granodiorites of Boal, El Pato, Ponferrada, Cadafresnas and Campo del Agua, frequently of adamellitic composition (Suárez 1970).

They occur in slates of the Luarca Formation or in the Serie de los Cabos, in fractures with a 20° N/ 40° E direction, or inside the granite in greisen zones and in pegmatites (Boal, No. 75), with scheelite-wolframite, along with cassiterite, arsenopyrite, pyrite and barite or as quartz veins in the quartzitic or slate exocontacts. In the latter, the main mineral is wolframite (Boal) or scheelite (Los Barrios de Salas, No. 76), it is also possible for bismutite and galena to be present. They correspond to pneumatolythic-hydrothermal solutions.

Uranium
In the quartz-dolomitic levels of the Lower Cambrian near the Allande Fault, and in relation to intensely silicified aplitic-pegmatitic rocks, disseminated traces of pitchblende and uraninite occur, associated with pyrite. Secondary minerals (gummite), limonite and limonitic jasper filling microfractures (Besullo, No. 81) also appear. Traces of radioactive minerals exist also in relation to granitic rocks (Lugo, No. 82).

Barite, Barite-Mercury
Barite is occasionally present in the West Asturian-Leonese Zone, appearing as fillings of irregular fractures in Cambrian limestones (Luiña, No. 74). The paragenesis is composed of barite and quartz, with pyrite as individual crystals and, occasionally, associated with cinnabar (Muñalèn, No. 77).

Magnesite
This appears mainly in the carbonate levels of the Lower Cambrian. In the Rubián deposits (No. 78) lens-shaped beds occur conformably with the limestones, its mean thickness being 14 m. They were formed by deposition on restricted basins of magnesium hydrated carbonates, which, after diagenesis, became magnesite, with later remobilization related to metamorphism in the Hercynian Orogenesis.

Kaolin
The most significant kaolin deposits come from alternating acid volcanic rocks and Cambrian quartzites. Alteration of these rocks seems to have been brought about by autometamorphism, although the main process is related to meteorization in situ of the vulcanite (Burela, No. 79; Vivero, No. 80).

By erosion of these deposits, kaolinitic accumulations in Tertiary and Quaternary sediments have also been formed.

Andalucite
Abundant crystals of andalucite (chiastolite) of metamorphic origin, with centimetrical to decimetrical dimensions, occur in the exocontact of the Boal granite (No. 75), where it affects slates of the Luarca Formation.

References

García Iglesias J, Ruiz Arias F, Suárez O (1985) Mineralizaciones de afinidad volcánica en el área de San Martin de Oscos, Asturias. Trab Geol 15: 249-266

Guillou J (1969) Contribution a l'etude des mineralisations ordoviciennes en antimoine de la Sierra del Caurel (provinces de Lugo et d'Orense, Espagne). Sci Terre 14 (1): 5-26

Gumiel P (1983) Metalogenia de los yacimientos de antimonio de la Penìnsula Ibèrica. Tecniterrae 54: 6-120

Gutiérez Claverol M, Luque C, Suárez V (1987) El lineamiento tectónico Nazaré (W de Portugal)-Luarca (NW de España) y su implicación metalogenética (Mazizo Hespérico). In: Bea F, Carnicero A, Gonzalo JC, López Plaza M, Rodríguez Alonso MD (Eds) Geología de los Granitoides y rocas asociadas del Mazizo Hespérico, Libro Homenaje a L.C. García de Figuerola, Madrid, Editorial Rueda, pp 447-455

Harris M (1980) Hydrothermal alteration at Salave gold prospect, northwest Spain, Transactions of the Institution of Mining and Metallurgy. Earth Sci 89: B5-B15

Lunar R (1977) Mineralogénesis de los yacimientos de hierro del Noroeste de la Península Ibérica. Mem Inst Geol Min Esp 90: 1-211

Merayo Vega C, Arias Prieto D, Perez Cerdán F, Fernández Martínez JM (1984) Paragènesis y zonado en el yacimiento de Pb-Zn de Rubiales (Lugo). I Congreso Español de Geologìa, Segovia, tomo II, pp 599-611

Suárez O (1970) Estudio Petrológico de los afloramientos plutònicos de la zona occidental Astur Leonesa (NW de España). PhD Thesis, Univ Salamanca, 284 pp

Suárez O, Ruiz F, Galán J, Vargas I (1978) Edades Rb-Sr de granitoides del Occidente de Asturias (NW de España). Trab Geol 10: 437-442

Part IV Central-Iberian Zone

1 Introduction

A. Ribeiro[1]

The Central-Iberian Zone was defined by Julivert et al. (1972) by grouping together the Galician-Castillan and Luso Oriental (eastern Portuguese)-Alcudian Zones of Lotze (1945). In fact, the paleogeographic and tectonic evolution of these two zones of Lotze is quite similar, and the differences between them concern mainly the presence of abundant synorogenic granitoid bodies and the higher grade of regional metamorphism in the Galician-Castillan Zone when compared with the Luso Oriental-Alcandian Zone. Most authors have agreed with the subdivision proposed by Julivert et al., so the boundary between the Galician-Castillan and the Luso Oriental Zones of Lotze is today regarded as perfectly transitional, corresponding to a gradient of metamorphic rank and extent of granitoid intrusion.

The paleogeographic features that characterize the Central-Iberian Zone when compared with the surrounding zones concern mainly the pre-Ordovician sequences. In fact, the widespread Armorican Quartzite (Arenig) unconformably overlies the "slate and greywacke complex" of Cambrian und Upper Precambrian age, putting in evidence the effects of the Sardic deformation phase (Lotze 1956). The slate and greywacke complex differs from the Cambrian of both the West Asturian-Leonese and Ossa-Morena Zones; the platform facies component (arkose/quartzite/limestone) is thinner and more localized in the Central-Iberian Zone and the flyschoid facies thicker and more extended in area; this points to the presence of a deeper intracratonic trough in this zone during Cambrian times. Another difference lies in the nature of the tectonic movements in the transition Cambrian-Precambrian; a clear unconformity exists at that level, both in the West Asturian-Leonese and Ossa-Morena Zones, but not in the major part of the Central-Iberian Zone, making the distinction between Cambrian and Precambrian sedimentary sequences difficult to trace inside this Zone. A Precambrian deformation producing weak folding without cleavage development and regional metamorphism has been described only in some domains inside the Central-Iberian Zone (Ortega and González - Lodeiro 1986). The slate and greywacke complex rests on a Precambrian gneissic basement (Bischoff et al. 1986; Lancelot et al. 1985), but the inferred unconformity at its base has generally been obliterated by intense Variscan deformation, regional metamorphism and granitoid intrusion. The Paleozoic succession from Arenig to Middle Devonian in the Central-Iberian Zone is not fundamentally different from adjacent zones and can be ascribed to facies transitions between the three zones.

The age of the Variscan deformation phases is also transitional in these zones; deformation starts earlier in the Ossa-Morena Zone and migrates to the Central-Iberian Zones and later to the West Asturian-Leonese Zone, defining a large-scale tectonic polarity from SW to NE.

From the tectonic point of view the Central-Iberian Zone is regarded as the Axial Domain in the Iberian Variscan Fold Belt, separating a NE branch with predominant vergences of thrusts and folds to the NE (West Asturian-Leonese and Cantabrian Zones) and a SW branch with predominant vergences of thrusts and folds to the SW (Ossa-Morena and south Portuguese Zones). But this general picture is valid only in very general terms, because there are regional anomalies in the vergence distribution and many variations in tectonic style; upright folds dominate in the SW and recumbent folds dominate to the NE. This contrast in tectonic style, which is also marked in metamorphic grade, was the base of subdivision by Lotze in a Galican-Castillan Zone in the NE and a Luso oriental-Alcudian Zone to the SW.

A clear distinction must also be made between Autochthonous and Allochthonous sequences, on the basis of tectonostratigraphic relationships, as follows:

1. The major part of the Central-Iberian Zone is made up of *Autochthonous sequences:* where the boundary with Precambrian basement is ex-

[1] Departamento de Geologia, Faculdade de Ciências, Universidade de Lisboa, Lisboa, Portugal

R. D. Dallmeyer and E. Martínez García (Eds.)
Pre-Mesozoic Geology of Iberia

posed, no major decollement exists between it and the Paleozoic-Upper Precambrian cover, although the tight folding in the cover implies some kind of shortening mechanism inside the basement. In upper structural levels of low metamorphic grade the most probable mechanism is major thusting inside the basement, where stretching is downdip or strike-slip, combined with compression where stretching is subhorizontal;

2. In the NW segment of the Central-Iberian Zone, *Allochthonous sequences* with variable amounts of displacement are superposed in the Autochthon, from Paurautochthon that moved tens of kilometers to completely exotic terranes that moved more than 200 km. These were variably labelled: Subzone of Middle-Galicia-Trás-os-Montes (Julivert et al. 1972) to Galicia-Trás-os-Montes Zone (Farias et al., in press). The main basal thrust plane of the Allochthonous sequences is mapped in eastern Galicia, Trás-os-Montes and Minho but has not yet been followed in western Galicia, where the exact boundary between Allochthonous and Autochthonous sequences is difficult to map precisely, because it was obscured by granitoid intrusions and regional metamorphism of higher grade than elsewhere;
3. The boundaries of the Central-Iberian Zone are considered as follows: To the NE there is a transitional boundary in the western flank of the Ollo de Sapo Anticlinorium, where the "slate and greywacke complex" appears for the first time, with increasing thickness to the SW. To the SW the boundary is clearly in the Portalegre-Ferreira do Zêzere thrust (Julivert et al. 1972), which brings the typical sequences of the Ossa-Morena Zone towards the NE, on top of the Central-Iberian sequences; the displacement on this thrust zone diminishes towards the SE, in Spain, and major imbrication in the same sense is still visible around the Pedroches Batholith. To the W, the N-S Porto-Tomar dextral shear zone (Ribeiro et al. 1980) cuts the structural grain of the Central-Iberian Zone and exposes typical Ossa-Morena sequences to the west.

References

All references in the text can be found in Díez Balda et al. (this Vol.).

2 Autochthonous Sequences

2.1 Stratigraphy

2.1.1 Introduction

L. Vilas and M.A. de San Jose[1]

The Central-Iberian Zone of the Hesperian Massif was defined by Julivert et al. (1972)[2] as a composite of Galaico-Castilian and East Lusitanian-Alcudian Zones of Lotze (1945), because, in spite of the obvious differences that both have from the structural and metamorphic-magmatic point of view, common stratigraphic characteristics (particularly Lower Ordovician unconformity) show that both regions had a close sedimentary history and a jointly liable palaeogeographic evolution, but with very significant peculiarities in each of them.

This Zone is located in the central part of the Massif, between the West Asturian-Leonese Zone to the north, and the Ossa-Morena Zone to the south. The most generally accepted boundaries are tectonic or tectono-magmatic occurrences; in the north and northeast, the Vivero Fault and the northeastern limb of the antiform where the Ollo de Sapo stratigraphic unit occurs, in the south the Guadalquivir Flexure-Faulting Band and Los Pedroches-Vegas Altas-Alburquerque-Nisa Tectonomagmatic Lineament, and in the west the Ferreira do Zézere Thrust and the Porto-Tomar Shear Zone.

From a stratigraphic point of view, and notwithstanding peculiarities of each subzone or domain considered, the Central-Iberian Zone is characterized by having a record comprising four large sequences of rocks, separated by three important discontinuities: (1) Lower Alcudian; (2) Upper Alcudian-Pusian (sensu novo),[3] which includes several units separated by lower rank discontinuities; (3) Lower Ordovician-Lower Carboniferous, which can also be divided into subunits separated by discontinuities, and sealed by unconformities related to the main diastrophic Hercynian episodes; and (4) Middle and Upper Carboniferous, late and posthumotectonic.

The main characteristics of each of these sequences are as follows:

1. Lower Alcudian. Having apparent great lithological uniformity, it corresponds to a predominantly turbiditic filling of a basin whose supposed extension should largely exceed present limits of the Central-Iberian Zone. Its lower limit remains still unknown in this zone, but in northern Ossa-Morena, a likely equivalent rests with an angular unconformity on the previously deformed, metamorphosed and granitized Boeturian socle.
2. Upper Alcudian-Pusian. Having a larger lithologic variety, it ranges from deep basin turbidites to prograding shore deposits, even with emersion episodes. Siliciclastic and carbonatic shelves predominate, the last ones being peculiar to the Upper Alcudian of this zone. This sequence includes the Lower Cambrian.
3. Lower Ordovician-Lower Carboniferous. It includes the thickest part of post-Cambrian Palaeozoic, and starts with fluvial or fluvio-tidal deposits, followed by consecutive siliciclastic shore and shelf deposits with very sparse carbonatic sediments, repeated along whole Lower and Middle Palaeozoic and which culminate in Upper Devonian-Carboniferous pre- and synorogenic turbidites. This sequence presents several characteristics in common with its equivalents from other zones of the Hesperian Massif, and also with those from the neighbouring, at that time, Armorican Massif.
4. Middle and Upper Carboniferous. It comprises continental cyclothematic deposits, conserved in late or posthumotectonic intramontane limnic basins, preserved along shear bands, or forming half-grabens.

[1] Departamento de Estratigrafía, Universidad Complutense, 28040 Madrid, Spain

[2] See Part IV, Chap. 2.1.2 for references

[3] The original proposal of the Pusian (San José 1983, 1984) has a litho-chronostratigraphic purpose, whereas the new sense has an exclusively sequential meaning

R. D. Dallmeyer and E. Martínez García (Eds.)
Pre-Mesozoic Geology of Iberia

Sequence 1 and part of sequence 2 form the previously called Schist and Graywackes ante-Ordovician Complex (Carrington da Costa 1950; Teixeira 1954, 1955) or Hispanian (Teixeira 1979), while the other two sequences, 3 and 4, form a distinct sedimentary hypercycle, with successive and subordinate transgressive-regressive cycles, and the corresponding gaps and unconformities, reflecting eustatic variations and diastrophic episodes related to the Caledono-Hercynian plate motion.

Because of these characteristcs, the following stratigraphic description has been divided into two parts: the first one includes the two lower sequences, 1 and 2, comprising pre-Cambrian and ante-Ordovician Lower Palaeozoic rocks, whereas the second part groups all post-Cambrian Palaeozoic rocks (3–4), during which sedimentation homogenizes in the major part of the Hesperian massif, as well as in other perigondwanic shelf areas closely related with it.

2.1.2 Ante-Ordovician Stratigraphy

M. A. DE SAN JOSE, A. P. PIEREN, J. F. GARCIA-HIDALGO, L. VILAS, P. HERRANZ, J. R. PELAEZ and A. PEREJON[1]

2.1.2.1 Introduction

In the Central-Iberian Zone, lowermost Palaeozoic and pre-Palaeozoic rocks with diverse metamorphic grade crop out, forming the core of large antiforms flanked by Ordovician and/or younger Palaeozoic sequences. The basal Ordovician transgressive deposits rest upon the "ante-Ordovician" rocks by means of a first-rank stratigraphic discontinuity, reflected in one or several successive unconformities. This discontinuity, identified in the middle of the last century, was for many years related to the formerly called Sardic movements, even if in its original area the unconformity to which this name was applied presents a different age.

This tectonic event nevertheless constitutes the limit between two sedimentary hypercycles, which are themselves related to two important global cycles of fragmentation, separation, drift and continental regroupment. The ante-Ordovician Hypercycle, with which this subchapter deals, is characterized by the predominance of shales and graywackes, and with sparse intercalations of several other lithologies.

Stratigraphic studies carried out since the 1960s in the southern part of the Central-Iberian Zone have permitted within this hypercycle the separation of units containing Cambrian fauna from those of prior episodes. The pre-Palaeozoic rocks have been divided again in different stratigraphic units, grouped together in megasequences, separated by discontinuities which are related to eustatic changes or fini-Proterozoic diastrophic events.

Finally, in some specific areas, generally located in the northeastern part of the Central-Iberian Zone and corresponding to Lotze's (1945) Galaico-Castilian Zone, the metamorphic grade reached by ante-Ordovician rocks makes difficult the comparison with any other of the units defined in the rest of the Zone. This, combined with the existence of only scattered stratigraphic data above large areas of the Central-Iberian Zone, makes the correlation difficult, and as a result, in the following description, the megasequences and units defined in the central and southern parts of the zone are used as the standard ones (Fig. 1). The latter areas are approximately equivalent to Lotze's East Lusitanian-Alcudian Zone which, as mentioned above, have been systematic and intensively studied from the stratigraphic point of view in recent years.

Regarding the actual level of knowledge concerning the remaining areas, it is far from being homogeneous. Consequently, the main features of their units are mentioned, as well as their supposed correlation with the central and southern area, when this is possible, specifying then the problems which arise from the comparison.

2.1.2.2 Stratigraphy of the Ante-Ordovician Sequences of the Southern Part of the Central-Iberian Zone

These rocks, identified for the first time in 1834 by Le Play, were the subject of several tectonic, palaeontologic and stratigraphic works, dealing especially with the central and eastern zones of the southern part of the Central-Iberian Zone. As mentioned in the Introduction, two megaunits have been differentiated in the ante-Ordovician rocks of this area: The Lower Alcudian und the Upper Alcudian-Pusian. These megaunits are separated by first-rank discontinuities, and subsequently subdivided in tectonostratigraphic episodes and sedimentary cycles, differentiated by minor rank discontinuities. They have the lithostratigraphic significance of a Supergroup. The individuality of these megaunits and the lying relationships among them and the

[1] Instituto de Geología Económica, C. S. I. C.-U. C. M. Facultad de Ciencias Geológicas, 28040 Madrid, Spain

R. D. Dallmeyer and E. Martínez García (Eds.)
Pre-Mesozoic Geology of Iberia

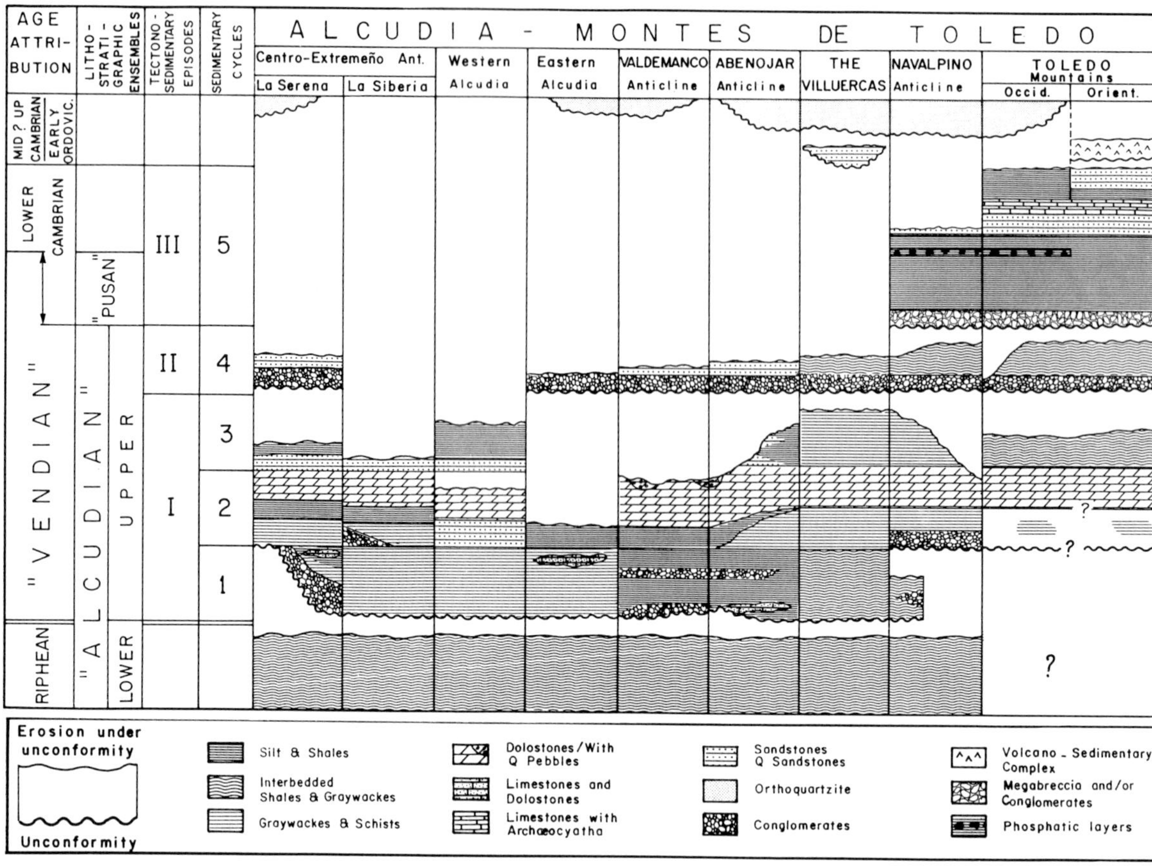

Fig. 1. Lithofacies distribution in distinctive subsectors of the Alcudia-Montes de Toledo region of the Central-Iberian Zone. Criteria used for the age distribution, division into lithostratigraphic episodes and sedimentary cycles are largely discussed throughout the text. This chart is a synthesis of several widely spaced, partial sections of the basin. The chart shows a geographic distribution where southwestern sections are represented on the *left*, and the northeastern extreme is on the *right*

Palaeozoic have been controversial, since Bouyx's (1970) initial proposal, even if nowadays there is a quite generalized consensus concerning the acceptance of these Supergroups. However, there is still some disagreement regarding the episodes, cycles and units into which they are divided, the precise stratigraphical location of some of them, and the magnitude and significance of the discontinuities dividing them. With regard to the nomenclature of these megaunits, the ancient denominations of Complexo Xisto-Grauváquico Ante-Ordoviciano or "C. X. G." (Carrington da Costa 1950; Teixeira 1954, 1955) and Transition Series (Lotze 1956) were replaced by Tamain's proposal (in Ovtracht and Tamain 1970) to formalize the name Alcudiense (Alcudian), used before colloquially with Complex rank, in the Alcudia Valley Mineral Investigation Project (Crespo et al. 1971; Crespo and Rey 1971, etc.) and later supported by Herranz et al. (1977), Vilas et al. (1981), San José (1983), etc. During the last few years, the term Alcudian has become increasingly accepted, whereas the term Hispaniense (Hispanian; Teixeira 1979) means a written-up of the former C. X. G. As for the Pusiense (Pusian; San José 1983, 1984), first proposed as a pre-Cambrian-Cambrian boundary Group name and later as a Stage-rank time equivalent of this, it is used in this work in a sequential sense, corresponding to a wider pre-Cambrian-Lower Cambrian comprehensive tectonostratigraphic Episode. Both the Alcudian and the Pusian (the latter in its original sense) constitute the Spanish counterparts of other foreign units such as the Charnian, the Zedian or the Sinian, of a similar age, and with the same meaning. Some other denominations, having no stratigraphic criterion, are not considered here. (Note: There is no unanimity among the authors; M. A. de San José disagrees on the suitability of bringing out in the present work a new significance for the term Pusian that differs from the original.)

2.1.2.2.1 Lower Alcudian

Due to the strong deformation prior to the deposition of its unconformable cover, complicated later by successive Caledono-Hercynian episodes, there is no detailed work on the Lower Alcudian stratigraphy. This unit was identified as forming part of the ante-Ordovician substratum of the North-Extremenian Great Anticlinory, of the Guadalupe-Ibor Anticline and its prolongation towards the SE, Fuenlabrada Anticline, of the Anticlines of Navalpino, Badenes and Zumajo, of the Central-Extremenian Great Anticlinory, of the Agudo-Valdemanco and Abenójar Anticlinories and of the Alcudia Anticline. The presence in the Valdelacasa Anticline is discussed (Fig. 1).

The base of the Lower Alcudian has not been identified as yet in the Central-Iberian Zone, if we consider the definition of this zone as given in Vilas and de San José, (this Vol.) (Introduction). Notwithstanding, in the northern fringe of the neighboring Ossa-Morena Zone, a Lower Alcudian counterpart appears clearly unconformable over the Boeturian metamorphic basement, which was previously folded and cleaved, granitized and locally altered (Boeturian Unconformity of Herranz et al. 1977). These outcrops are formed by volcanogenic arkoses and quartzites, with a conglomeratic base which includes pebbles of quartz, metaquartzite and lydites of Boeturian origin, and remobilized volcanodetritic or alteritic matrix. It grades upwards to interbedded schists and graywackes, with some arkosic layers, similar in facies to the typical Lower Alcudian of the adjoining Central-Iberian Zone *s. str.*, where the whole sequence is megascopically monotonous and devoid of distinct marker beds. It is formed by shales and graywackes with rare interbedded feldspar-sandstones or protoquartzites, ortho- and paraconglomerates, and sparse spilites, rhyodacites and just anecdotic lydites. The presence in the conglomerates of diverse kinds of pebbles must be emphasized: lydite, quartzite, graywacke, shale, micaschists, gneiss, aplitic granite, pegmatite and some clasts of microsparitic limestone, but most of them are vein-quartz and metaquartzite. On the other hand, fragments of diverse rocks can be found in the graywackes; among them corroded and burst quartz grains of granitic and rhyolitic origin, K-feldspar and albitic plagioclase, as well as microlithic vitreous fragments, which strongly suggest a mixed volcano-terrigenous origin for most of them. Its thickness has been estimated to be at least 4000 m, but could reach or exceed 6000 to 7000 m.

The assemblage of sedimentary structures in the areas studied is mainly characteristic of submarine fan deposits. The most common associations are: sand-mud decimeter couplets lacking any distinct sedimentary structures and "classical" turbidites, presenting complete or truncated Bouma sequences, which are thought to be lobe deposits. Several other lithofacies can also be distinguished, mainly silt to sand-mud millimeter couplets interpreted as interlobe deposits, and disorganized muddy gravels, pebbly sands, conglomerates and pebbly mudstones with some slumped levels which have been deposited in inner fan to slope environments. Wave and storm-dominated offshore facies, with hummocky cross-stratification, have been described, as well as mud-dominated outer shelf facies; both are possibly related to distal deltaic facies, which appear generally close to the unconformable basal contact of Upper Alcudian. Frequent fracture-associated silicifications, quartz veins and dolerites accompany these deposits at present.

In this region, Lower Alcudian lies unconformably below the Upper Alcudian or post-Cambrian Palaeozoic cover, generally presenting a strong deformation. It is affected by at least two main folding events (intra-Proterozoic and Hercynian), as a result of which refolded folds with a vertical axis are commonly found, whose wriggling cartographic outline confers to this unit a characteristic structural shape. In some locations, however, it seems to have a deformation grade similar to that in the paraconformable cover, so when the facies are very similar, identification is difficult. Lower Alcudian intersection lineations S_0/S_1 are strongly folded as a result of the addition of pre-Palaeozoic and Palaeozoic diastrophic episodes, while in the Upper Alcudian or Palaeozoic rocks, the intersection lineations are subhorizontal (Ortega and González Lodeiro 1986; Pieren and Herranz 1988) or are also folded by later Hercynian folding phases (García Hidalgo 1988) which also makes any tectonic study difficult. Finally, we also find in these materials large bands of microbreccias, caused by Hercynian shear zones, which can be misinterpreted as disorganized layers of sedimentary origin. Metamorphism in this area generally does not exceed middle or upper anchizone (very low grade) and normally it does not affect the cover, even the basal Palaeozoic rocks. It could correspond to a hot-spot heating, taking place in Upper Ordovician, and whose cooling age is basal Silurian, 421-423 Ma, dated by Ar40/Ar39 in whole-rock samples of Agudo-Valdemanco and Alcudia anticlines where no Hercynian metamorphic story is overprinted (Dallmeyer and Pieren 1987). Moreover, in the controversial Central-Iberian/Ossa-Morena boundary band, a pre-Hercynian, probably Pan-African tectonothermal event is conspicuous, as proposed by Llopis et al. (1970) and demonstrated by Blatrix and Burg (1981).

An accurate age for the Lower Alcudian is difficult to establish because of the nearly complete absence of microfossils. Only the local presence (one single record) of the Acritarchs *Orygmatosphaeridium* sp. and *Protosphaeridium* sp. (Mitrofanov and Timofeiev 1977) permits its attribution to a Riphean age in a broad sense (San José 1983, 1984), without discarding the possibility that the upper levels could reach a Lower Vendian age due to the wide distribution range of both taxons.

In spite of the scarcity of data regarding the Lower Alcudian, different interpretations have been suggested concerning its tectonostratigraphic setting, based on the nature of clasts, the composition of interbedded vulcanites and the sedimentary characteristics of rocks, even if most of these interpretations consider Lower Alcudian with regard to its relations with Upper Alcudian and Palaeozoic but not as a unit with an original entity. Thus, it has first been considered deposited in basins generated under a tensional regime at the beginning of a Wilson's cycle of a possible global extent: the initial miogeocline of a passive margin associated with an acid volcanic ridge (Capote et al. 1977; Vegas 1978; Herranz 1984). On the other hand, a compressive or mixed regime can also satisfy these premises: synorogenic flyschs (Dallmeyer and Pieren 1987), or turbiditic back-arc basins. A last conciliatory option would be a large foreland basin related to an emerged continental area formed by metasedimentary, metamorphic and granitic rocks; the basin would be located behind an arising collisional belt, involving the basement and including synchronous bimodal vulcanism; its deformation events are reflected and mitigated in the basin.

2.1.2.2.2 Upper Alcudian - Pusian

This sequence of rocks is typified by a wider lithological diversity and a relatively weaker structural deformation than Lower Alcudian. Internal discontinuities have been used for its division into smaller units (cycles or sedimentary episodes, Vilas et al. 1987). The higher of these units has distinctive litho- and biostratigraphic features; the latter indicating that this upper unit (original Pusian meaning) includes the lower limit of the Cambrian. Finally, in the upper part of this unit, formed by inner shore and peritidal facies, Lower Cambrian fauna is found. Unconformably overlying this episode (and remaining subensembles and Lower Alcudian also), the fluviotidal sediments of basal Lower Ordovician generally rest.

The unconformity in the base of this sequence is a stratigraphic high-range discontinuity, which implies the existence of a compressive deformation phase (locally with associated cleavage), palaeoalteration and erosion of the substratum, prior to the deposition of Upper Alcudian sediments. The existence of a palaeorelief of variable magnitude below the unconformity is proven by the fact that different units of the cover discontinuously overlie the mentioned substratum, and some of them may even disappear. This unconformity, and the diastrophic episode that causes it (Oretanian Phase, San José 1984; Nozal et al. 1988a) seems to affect a great part of the Central-Iberian Zone (San José 1983; Ortega and González Lodeiro 1986; Pieren et al. 1987). It was first identified in the Alcudia Valley (Redlin 1955; Crespo and Rey 1971) and Valdemanco Anticline (Bouyx 1970) but assigned to different ages.

Overlying this unconformity, a succession of terrigenous-carbonatic rocks appear, which present a wide facies diversity and distinctive development according to the regions. Sedimentary environments change from possibly continental (alluvial fans, fluvial), fluviotidal, deltaic, siliciclastic and carbonatic shorelines, to slope and submarine fan, with proximal or distal turbidites. These deposits are organized vertically in cycles and sedimentary episodes, described by Vilas et al. (1986a, b, 1987) and Peláez et al. (1989). These are separated by different rank discontinuities, whose importance augments upwards, reflecting the increasing instability of the basin corresponding to Late Pan-African episodes (Cadomian events). The uppermost episode (Vilas et al. 1986b), whose evolution concludes with the progradation of Lower Cambrian shallow marine facies, is separated from the two preceding ones by an important discontinuity of regional transcendency. The two former episodes are known by the global denomination of Upper Alcudian.

Upper Alcudian

In contrast to Lower Alcudian, it has a relatively low deformation pattern, and larger lithological diversity reflected by the presence of sedimentary units which include conglomerates, quartzites, limestones and dolostones, and associated minerals of economic interest. These facts have propitiated the development of stratigraphic studies over the last few years. The successive finds of fossils (macro- and microfossils, and ichnofossils) made possible the correlation of units established in distant areas, as well as the separation of distinctive sedimentary cycles. Notwithstanding, problems inciting discrepancies between different interpretations still exist. In most cases, this is due to the difficulty of identification and discrimination concerning these units from Lower Alcudian when fossils are lacking and both sequences show similar facies.

Rocks assigned to Upper Alcudian crop out, overlying unconformably Lower Alcudian in the Hercynian anticlines mentioned above. Its extent, thickness and facies distality increase east- and northwards, thus the individuality of the successive cycles or sedimentary episodes is lost or cannot be observed in this direction. Furthermore, the superposition of discontinuities of increasing importance generates the disappearance of several units, and even of complete cycles, which are not then recognized in some regions.

Bearing this in mind, and from the examination of the best and most complete sections, the Upper Alcudian sediments can be divided into the following tectonostratigraphic episodes and sedimentary cycles:

Episode I: It overlies unconformably the Lower Alcudian and comprises three sedimentary cycles. The first is not always represented, depending on the pre-Upper Alcudian palaeotopography; the second cycle is the most extensive, but the typical carbonatic facies are not always present; the upper cycle is occasionally absent due to the erosion prior to the following episode.

Cycle 1. Its thickness ranges from between 60–200 m in the southern part of the considered area to approximately 800 m in the north, where it can also be absent due to palaeogeographic reasons. It comprises continental and deltaic facies, inner siliciclastic shoreline facies, inner mixed terrigenous-carbonatic shore facies and, in the northern area, slope and deep-sea turbidites. The carbonatic sediments present discontinuous lenticular-shaped outcrops in a narrow band. The conglomeratic facies in the northern area (Navalpino) are characteristic, as well as the algal-mound limestones and shallowing-upwards sequences in Abenójar. In La Siberia (southeast) it is formed by fluviatile and shallow marine sandstones and conglomerates, topped by algal and oolitic lenticular dolostones.

When entirely marine, it is a shallowing-upwards cycle, tending to smooth the initial palaeorelief originated by the tensional episode following the diastrophic Oretanian Phase.

Cycle 2. It can reach a thickness of 500 m, but the average is 300–400 m. The base is strongly transgressive or discontinuous over Cycle 1. It can also locally rest upon the Lower Alcudian. It consists of different siliciclastic and carbonatic facies in a shallowing-upwards cycle; usually the facies grade upwards from outer to inner siliciclastic shelf, to very shallow siliciclastic and carbonatic deposits which comprise a very wide range of facies and subenvironments: thus, there are large and small mounds, flat-bedded stromatolites, dolomitic prograding bars, sandy-dolomitic channels with bimodal palaeocurrent pattern, sand to silt-mud couplets, thin terrigenous beds with hummocky cross-stratification, etc., which have been interpreted as different subtidal to intertidal subenvironments arranged in a complex, detailed pattern (García-Hidalgo 1984, 1988). It passes upwards to open shelf (below the wave-base level) carbonatic facies. In these dolostones, Calvet and Salas (1988) have distinguished a grainstone and packstone carbonated ramp and protected shelf deposits, with algal mounds and marginal oolitic shoals, and sediments of a shelf with a stromatolitic reef barrier and finally proximal slope. Locally in the top, mixed facies and siliciclastic open shelf lutites appear again below the base of the next cycle.

The punctual existence in several localities of carbonated matrix conglomerates, ankeritizations, karstic breccias and very locally erosive processes, affecting the top of the carbonatic unit, indicates late subaerial weathering.

This cycle is largely extensive over the first one, both towards the north and the south. Very likely, in this last direction, it surpasses the admitted limits of the Central-Iberian Zone.

Cycle 3. It has a maximum preserved thickness of 300 m (due to the erosion prior to Episode II). It consists of siliciclastic deltaic, shoreface and inner shelf facies and passes laterally towards the N to outer shelf facies and probably slope deposits and turbidites. The most typical facies are coarse-grained, channelled deposits, and large-scale, cross-stratified sandy bars. It is strongly transgressive on the second cycle, and also represents a shallowing-upwards cycle, which is more restricted and discontinuous than the former.

Both the second and third cycles present a clear transgressive base and a shallowing-upwards tendency, the result of which is a progressive basin homogeneity, probably due to eustatic changes as well as to tectonic substratum reactivation in a distensive post-Oretanian regime. The most extensive is the second cycle, whereas the third one represents a relative restraint in the sedimentation area, combined with a drastic change in the source area conditions (Peláez et al. 1989). These seem to be a prelude of the events which are used for the differentiation of these cycles and the following episode.

Episode II: It is separated from the previous episode by an important sedimentary discontinuity. This appears as a paraconformity or slight angular unconformity, which is responsible for the local erosive disappearance of Cycle 3 and the upper part of Cycle 2. This tectonosedimentary episode is formed by a single sedimentary cycle, whose tectonic complexity and thickness presume a further subdivision.

Cycle 4. This cycle, in the northeast, is at least 2000 m thick, decreasing southwestward to a thickness ranging between 80 to 700 m. Here, it is formed by alluvial fan-like, fluviatile and deltaic deposits grading upwards to siliciclastic shelf sediments where successively more distal facies appear towards the top of the sequence. On the other hand, in the northeast it grades vertically from massive conglomeratic facies to slope deposits, proximal and distal turbidites, upon which storm deposits and distal deltaic facies have been described in an outer shelf environment. The basal levels contain clasts originating from the removal of several previous units, while the middle part represents a transgressive pulse with more distal facies. Finally, and just in the NE, the upper part seems to form a regressive pulse onlapped by more proximal facies. These have a tendency to complete the basin, but the final evolution of this cycle is unknown due to erosion prior to the deposition of the next episode.

Episode II represents a strong deepening period with a high sedimentation rate in the NE, and reactivation of the local substratum in the SW, where a continental-transitional domain (with fluviotorrential and deltaic deposits) and the source area of these sediments are located. This tectonic instability produced a reactivation of the former fractures in a distensive environment. These movements could be the reflection of a late Pan-African event, preluding a more intense phase, generated at the limit of the next cycle. The new structuration of the basin, with a possible tilt towards NNW, causes the emersion of an area in the south exposed to erosive processes, the development of a trench in the north, and the migration of the depocenter towards the NNE, with a progressive areal restriction of the basin in the same direction, within a distensive tectonic regime, but with a strong vertical component.

To summarize the above-mentioned features, the Upper Alcudian evolved within a post-Oretanian distensive regime. It begins with the compartment of substratum, followed by the progressive replenishment of the previous palaeorelief by means of fluviatile to turbidite deposits. The usual evolution of the basin in shallowing-upwards cycles, starting with siliciclastic deposits passing upwards to carbonatic shelfs, is repeated several times by tectonic and/or eustatic pulses, which progressively restrict the extension of the basin towards the NNE. This evolution is truncated by the unconformity at the base of the following episode.

The Pusian

To date, it has just been identified in the NE part of the southern Central-Iberian Zone, where it unconformably overlies the Upper Alcudian and exhibits a weaker tectonic deformation. There is a larger lithological variety, among which characteristic units formed by megabreccias and conglomerates occur, as well as quartzites, peritidal sandstones and limestones, fine shelf silts and anoxic sediments in which phosphatic-siliceous conglomerates are interbedded. The abundance of Palaeozoic fauna especially supports the dating of the upper part of the sequence as Lower Cambrian. The unique fauna found in the lower part of this episode (Pusian as in San José 1984) display pre-Cambrian resemblances. It is mixed, however, with Cambrian-like ichnofossils and microfossils hitherto considered pre-Cambrian and widely ranging also in the previous units. Thus, it can be suggested that the Vendian-Cambrian limit should be placed in this lower part (Brasier et al. 1979; Perejón 1981). The whole comprises a single tectonostratigraphic episode, and as well as the previous one (Episode II), it has not been possible to distinguish more than one sedimentary cycle, but its great complexity and thickness preludes a further subdivision.

Episode III. An important sedimentary discontinuity separates this episode from the former one. The discontinuity has a regional extent and is represented as either angular and/or erosive unconformities, which cause at least the disappearance of several cartographic units of the preceding episodes. The absence of this unconformity was postulated by Lotze (1956); it was later identified and described for the first time by Herranz et al. (1977)

who assigned the unconformity to the Cadomian or Assyntic Phase.

Even if there is no consensus with respect to the relevance of this unconformity (the authors who assign a larger erosive range are those who minimize its tectonic significance), it is undoubtedly related to a diastrophic pulse (San José 1980, 1983; Pardo and Robles 1988; Peláez et al. 1989) followed by erosion of the substratum. This episode, which at least produces a sudden deepening of the basin, seems to be the result of a tectonic reactivation of the basement, related to the preceding event at the base of the Episode II. And although there is general agreement, concerning the mainly distensive character of this event, the presence in upper units of the fourth cycle of folds, with associated cleavage not transgressing the unconformity, was described by Alvarez-Nava and Robles (1988). These folds, below the base of Episode III and not appearing in the cover, indicate the existence of a more or less important compressive phase, prior to the normal faulting phase associated with the sudden deepening of the pre-Pusian substratum.

Cycle 5. The thickness ranges from 1500 m in the NW to more than 4500 to the E. The base is formed by calcareous-dolomitic megabreccias, including clasts proceeding from the erosive removal of underlying units, which laterally pass into quartz conglomerates and microconglomerates and/or siltstones with or without pebbles. Overlying these basal beds, massive shales and siltstones predominate, sometimes with high organic matter content (black shales or sapropelites) which are also interbedded with sandy/silty channels and partially disorganized beds (slumps and poorly developed pseudobreccias). These grade vertically into channel-like bodies formed by conglomerates and microconglomerates (polygenic, Q-K feldspar, or quartz conglomerates) with chips and escape structures, and passing into arkosic sandstones and microlaminated shales; some of the channels, mainly upwards, are filled by siliceous-phosphatic conglomerates and microconglomerates whose base is erosive, related with phosphatic laminated shales. Towards the top, there is again a broad predominance of siltstones and mudstones which grade upwards into a thick sandy-quartzitic-shaly unit, which contains Cambrian trace fossils; above are mudstones with carbonatic lenses where the first trilobites appear, succeeded by a mainly calcareous unit formed by biogenic, nodulous, oolitic and algal laminated dolomitic limestones, with archaeocyaths and trilobites and some limolitic intercalations. The sequence is topped by alternating sandstones and siltstones whose trilobite fauna has been known since the middle of the 19th Century.

The sedimentary evolution of Cycle 5 corresponds to the filling of a basin bounded at least by a tectonic escarpment, in the neighborhood of which megabreccias and slope turbidite facies are developed. These pass upwards into outer shelf deposits with anoxic facies. Consecutive sea-level changes enable the incision of gullies in the shelf (Picart 1988; Santamaría 1988; Santamaría et al. 1987a, b), in which, proceeding from the source area, down-slope coarse sediments are accumulated and redistributed. These have concentrations of organogenic deposits affixed in the anoxic lutites of the surrounding shelf. Above this prograde inner shelf, high-energy terrigenous facies and, at last, partially bioconstructed peritidal carbonates with progressive, shallowing-upwards signs. The end of the basin infill is represented by shoreface terrigenous deposits, which form the last ante-Ordovician unit conserved in the southern part of the Central-Iberian Zone, whereas in the western part of the "Montes de Toledo", terrigenous fluviotidal facies, with an important acid volcanic contribution, appear unconformably overlying this cycle below the Ordovician, i.e. the "Volcano-Sedimentary Complex" of Martín Escorza (1976, 1977). This complex is undoubtedly related to an episode apparently linked to the Ordovician evolution and not with the sequences described here, and therefore it is not presented here.

In summary, it can be said that Episode III represents a period of sudden deepening and later infilling of the basin, with a progressive reduction of the sedimentation area towards the NE. The basin repletion represents the end of the ante-Ordovician sedimentary hypercycle. The existence of compressional strains prior to the distensive initial deepening event, probably due to the tectonic reactivation of the basement, confirms the importance of this event, which must also be considered as related to the event at the base of Episode II. Both must be envisaged as different stages within a global period of fini-Proterozoic (Cadomian) deformation.

The Pusian presents peculiarities contrasting with the remaining episodes of the hypercycle in which it is inscribed; for instance, the basal megabreccias, the distinctive organic-rich anoxic facies and the profusion of palaeontological content, particularly in the upper part. Regarding the relation between the Pusian Episode and the Upper Alcudian Episodes, the progressive restriction of the sedimentary basin towards the NE accords also with the gradual emersion of sediments from the preceding cycles on areas located presently in the SW, which implies the erosion and resedimentation of these rocks as clasts in the uppermost Proterozoic-Lower Cambrian deposits. In this sense, while the sediments of Upper Alcudian indicate the presence of an adjoining emerged area, formed by dominating lutitic metasedimentary, metamorphic and granitic rocks, with a bimodal or acid volcanic participation, Pusian sediments show the same origin, with the addition of clasts from the different Upper Alcudian units, which can be perfectly identified in the gravels and blocks included in megabreccias and basal conglomerates.

All these characteristics imply at this time the existence towards the SW (as supposed by Herranz et al. 1977) of a large platform or emerged area, formed mainly by Alcudian rocks; further south, a belt formed by Boeturian-like metamorphic rocks should be simultaneously outcropping. The successive tectonic pulses during the formation of this belt had a strong repercussion both on the sedimentary evolution and the deformations of the deposits of a probable foreland basin, located successively further north. This basin is progressively restricted in the same direction, and its repletion completed during Lower Cambrian times.

Regarding the age of Upper Alcudian-Pusian sediments, it is supported by successive fossil finds, formerly in the upper part (Cambrian) of this hypercycle, and later in the pre-Cambrian, where fossils were found in progressively older rocks. The first discoveries were trace fossils and algal structures, and later, isolated or associated to them, micro- and macrofossils. All of them provide the establishment of the Faunal Associations Correlation Chart (Fig. 2). From this chart a mainly Upper Vendian age can be inferred for Upper Alcudian, without discarding the possibility that its base could reach the Middle or even the uppermost Lower Vendian. In the Pusian, the existence of hitherto

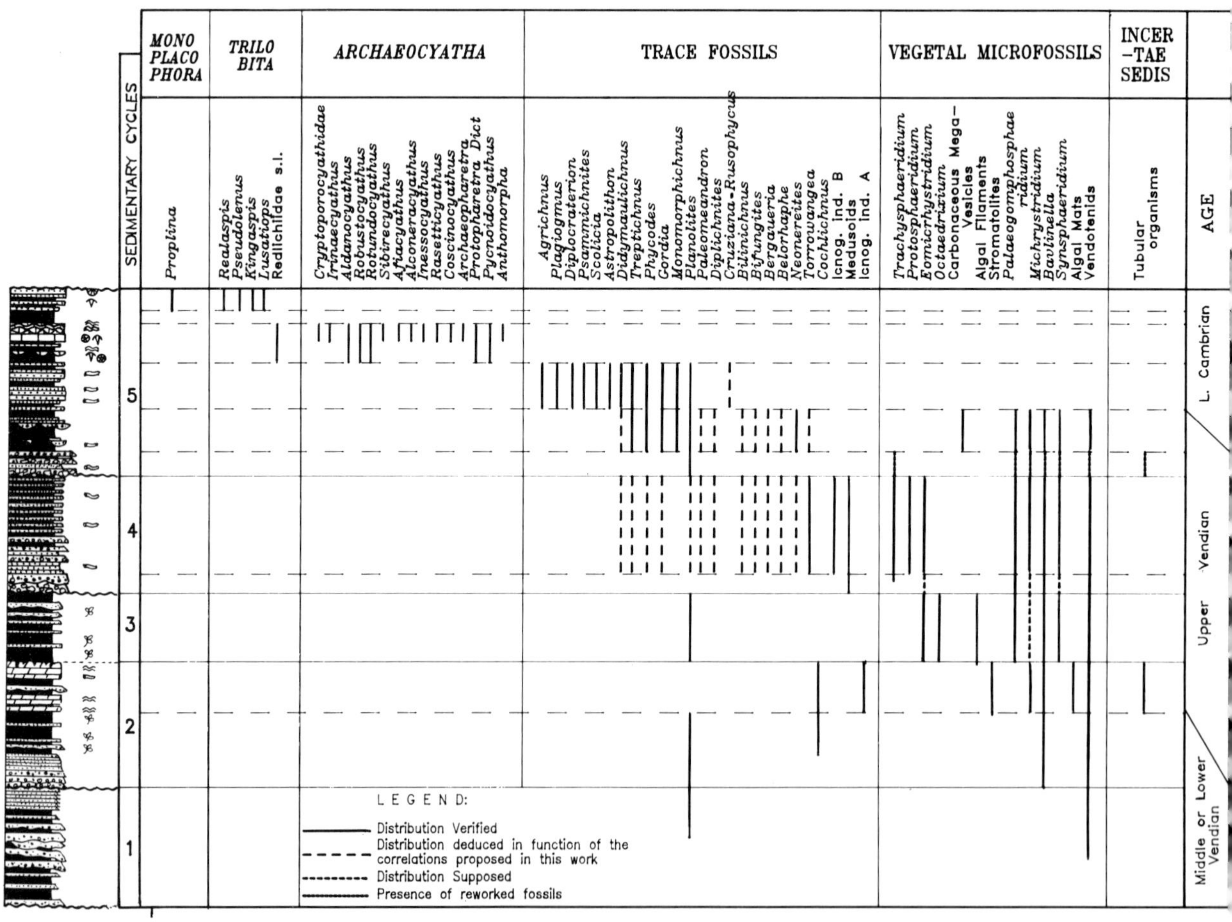

Fig. 2. Upper Alcudian-Pusian palaeontological content distribution in the Alcudia-M. de Toledo region. The log is composite. The fossil distribution of some lithological units (as stated in the legend) is based on correlations

considered "Cambrian" and "pre-Cambrian" faunas in its lower part indicates that somewhere in the mentioned unit is the Vendian-Cambrian limit (or pre-Cambrian-Cambrian) as stated by Brasier et al. (1979) and Perejón (1981). The possibility that the time equivalent of this lower part of the Pusian Episode (and perhaps Episode II also) could become in the future either a Cambrian or Vendian boundary Stage is proposed by San José (1984). With regards to the remaining units of Pusian Episode, the terrigenous shelf facies present abundant typical Lower-Cambrian trace fossils; this, combined with the Upper Ovetian age (Ovetiense Superior; Perejón 1984; Gil-Cid 1988), established from the trilobites and archaeocyaths of the overlying carbonatic formation, suggest that the terrigenous facies should be assigned to a Lower Ovetian or Cordubian age (Perejón 1984). Finally, the uppermost terrigenous coastal facies contain trilobites known since the 19th Century (*Fauna de Los Cortijos de Malagón;* Prado 1855; Sdzuy 1961; Gil Cid and Jago 1989) of an uppermost Lower Cambrian age, close to the limit of the Middle Cambrian.

Finally, the more likely ages of the discontinuities, in the light of the minimal and maximal age of the underlying and overlying rocks, are as follows:

1. The Oretanian Unconformity seems to correspond to an event located between the terminal Riphean-Lower Vendian, and the Lower?-Middle-Upper Vendian. Thus, we can suppose that the gap comprises most of the Lower Vendian and perhaps part of the Middle Vendian;
2. The Cadomian Unconformity represents a brief catastrophic event, because it just separates Upper Vendian from terminal Vendian-Lower Cambrian;
3. Among the remaining discontinuities, the one existing at the base of Episode II is the most significant, although it is extremely difficult to precisely determine the time concerned. It is intra-Upper Vendian.

2.1.2.2.3 Correlations with the Remaining Central-Iberian Zone. Synthesis Essay

Ante-Ordovician sediments similar to those described here have continuous outcrops extending towards the W, from the southern part of the Central-Iberian Zone, through Extremadura and Portugal. In this country they have large outcrops in the Beiras, Douro and Trás-os-Montes regions. Likewise, series similar to those of the Pusian Episode outcrop immediately to the S and W of Salamanca, as a prolongation of those of the northwestern Toledo Mountains. Northwards, in the part of the Central-Iberian Zone corresponding to Lotze's Galaico-Castilian Zone, the metamorphic grade is higher as well as in the Central System, the north of Zamora and along the so-called Ollo de Sapo Antiform. Therefore the stratigraphic successions are difficult to compare with those described here (Fig. 3).

A first comparison is attempted with the rocks outcropping along the Alagón river gorge, south of Coria in the High Extremadura. In this section, Liñán and Palacios (1987) separate a lower schist and graywackes unit, followed by a conglomerate unit divided into two members; the upper one grades upwards into shales. Here, the presence of two discontinuities can be suggested; the first at the base of the conglomeratic beds, and the second dividing the two conglomerate members. When this is considered, in addition to the fossil content, a Riphean-Vendian age could be deduced for the lower unit, a Vendian *s. l.* age for the lower member of the conglomeratic beds, and a higher Vendian age, even with Cambrian taxa *(Phycodes),* for the upper member. The microfossils appearing in the lower schist and graywackes unit are not determinative, but there is bioturbation (unusual in the Lower Alcudian). It seems that the correlation of this unit is possible with the Upper Alcudian cycle 1, while cycles 2–3 and 4–5 (as a whole, or just some of them) should correspond to the two members of the conglomerate beds. The lowest member is possibly related to the carbonatic lenses interbedded in sandstones mentioned by Corretgé and García de Figuerola (1971), close to the basal conglomerates of the Upper Section of their Infra-Ordovician Complex distinguished in a fringing eastward area.

The area between Salamanca, Ciudad Rodrigo, and the Las Hurdes region is another classic area for the study of the ante-Ordovician rocks. Uncomformably covered by Ordovician sequences in the "Sierras" of Tamames, Peña de Francia and Dios Padre, schists and graywackes with interbedded conglomerates and carbonatic rocks in a doubtful relationship with quartz-feldspar schists, arkoses and porphyroids, and diverse types of gneisses crop out. Above, inner shelf sandstones appear with Cambrian trace fossils (Rölz 1975), and peritidal limestones with algal mounds, archaeocyaths and trilobites (García de Figuerola and Martínez García 1972; Perejón 1972) very similar to those of the Montes de Toledo. The lower part of the succession is divided into two unconformable units (Rodríguez Alonso 1979, 1984, 1985). The lower unit consists of schists, graywackes and some conglomerates, being

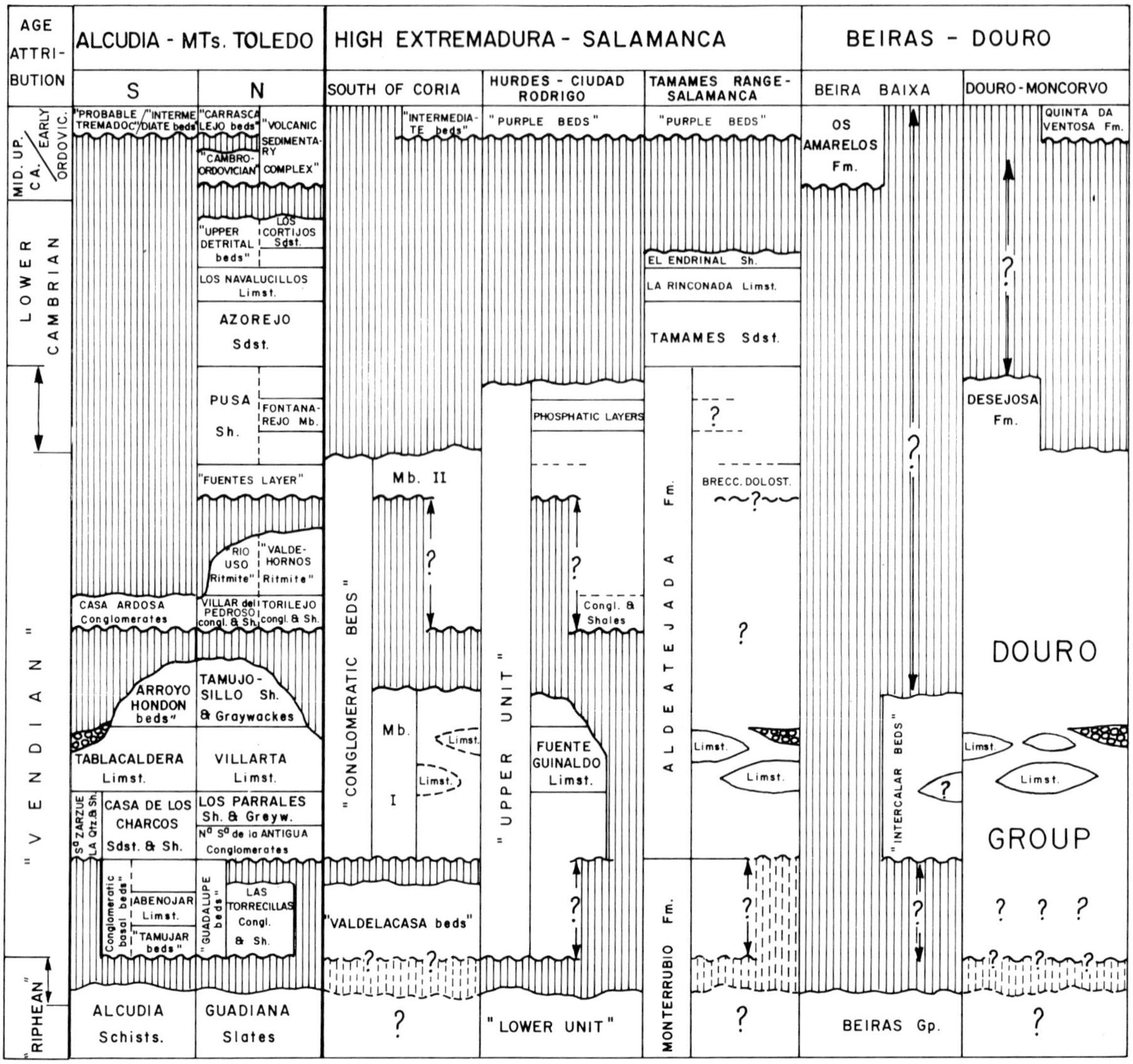

Fig. 3. Synthesis of the main ante-Ordovician lithostratigraphic units of the southern Central-Iberian Zone. A summary of the units established by the authors of this work in the region Alcudia-Mt. de Toledo is used as a pattern (*left side* of the chart). In the remaining sectors, lithostratigraphic schemes of diverse authors are used; their chronostratigraphic attributions are reinterpreted in some cases

interpreted as mainly turbiditic. In the upper unit the black shales (sometimes phosphatic) are common, as well as carbonatic rocks and also eventually conglomerates (sometimes with a carbonatic matrix) and sandstones. There are turbidite and slope facies, but shelf facies are more frequent. These are succeeded conformably by dated Cambrian sediments. This upper unit was later subdivided by Robles and Alvarez-Nava (1988) into two "series"; both are unconformable among themselves and on the lower unit. Every single unit or "series" has a distinctive peculiar deformation (Alvarez-Nava and Robles 1988) and can be compared respectively to Lower Alcudian, Upper Alcudian *(s. l.)* and the lower part of Pusian of the Toledo Mountains.

On the other hand, in the surroundings of Salamanca, the comparison is not so easy: The outer shelf lutites and conglomerates of the Monterrubio Fm. (Díez Balda 1986) or Morille Series (Martínez García and Nicolau 1973) have alternatively been considered equivalent to either the lowest part of the Pusian of the Toledo Mounts (or cycle 4?) or Lower Alcudian, whereas the upper part of the overlying shelf lutites of the Aldeatejada Fm. (Díez Balda 1986), Serie de Aldeatejada (Martínez García and Nicolau 1973) is undoubtedly related to the middle part of Pusian, close to the pre-Cambrian-Cambrian limit, because they grade upwards conformably into Lower Cambrian sediments. The problem arises with the brecciated or lenticular carbonates and the carbonate-matrix conglomerates, appearing in the middle-upper part of the mentioned unit and which have no equivalents in the south. A possible solution is that both the Monterrubio and Aldeatejada Fms. could represent all the successive units of the Upper Alcudian-Lower Pusian (and perhaps even some of the Lower Alcudian?) but forming their distal counterpart; this should render the recognition of the discontinuities difficult.

In neighboring Portugal, two areas must be distinguished: northwards, the Vale do Douro and Moncorvo (Trás-os-Montes). In this area the sandy-lutitic units, the schists with magnetite, carbonates and conglomerates forming the Douro Group (Grupo do Douro described by Bernardo de Sousa 1981, 1982, 1984) recall indeterminately the Aldeatejada and Monterrubio Fms. of southern Salamanca, in facies even more distal. This group has dating problems because poorly preserved unclassifiable trilobites were found in their upper units (Rebelo and Romano 1986). This discovery provides a double option: (1) these trilobites are the oldest found to date in the Central-Iberian Zone, and the rocks correspond to the middle part of Pusian, or (2) the unit would be immediately prior to the Ordovician, lying paraconformably on the substratum. The *Lingulellae,* found by Teixeira (1964) in outcrops of the same unit in a different locality, are not very helpful in this matter, but such large-sized unarticulate brachiopods appear in other places, especially in the Upper Cambrian or Tremadocian (in the formerly denominated Lingula flags). This is also the presumed age for the Os Amarelos Fm. of central Portugal (Gama Pereira 1983). Below this formation, or presenting a doubtful relation with it, the Serie Intercalar appears (Conde 1971), formed by arkoses and conglomerates, which unconformably lies on the Beiras Group, certainly equivalent to Lower Alcudian. The radiometric age of acid vulcanites interbedded in the lower part of the Serie Intercalar (563 Ma Rb/Sr) in Sardoal-Mação suggests its possible correlation with the Upper Alcudian as stated by Conde (see Fig. 3).

As can be seen, the correlations (depending on the varying detail of the papers discussed) can only concern the units established in the southern part of the Central-Iberian Zone, at the level of Episodes if possible, and very rarely with Cycles. The correlation that should be proposed with the NE of the Central System Ranges, Zamora or northwestern Galicia would be even more dubious. In these areas the controverted lying relationships existing between the Ordovician, the volcanosedimentary Ollo de Sapo, the presumed C.X.G. the Porto Serie (Martínez García 1973), several kinds of paragneisses and the deep augengneisses of Morcuera and Miranda do Douro do not permit an accurate comparison at present. The Porto Serie is formed by metaschists, marble lenses and calc-silicate rocks, whose age, first proposed as Cambrian, would be much older, whereas the C.X.G. of Miranda do Douro is difficult to compare with the so-called unit from the Douro, Salamanca and Beiras regions.

Finally, the scheme proposed throughout this work is summarized below. On a metamorphosed Boeturian-like or even older basement, with granitic intrusions, involved in a Pan-African-type collisional belt outcropping southward, where the Ossa-Morena Zone is today, a foreland basin would develop in the Central-Iberian Zone. The first turbiditic fill of the basin is coeval with the initial raising of the chain, and was affected later by advanced phases of the above-mentioned raising (Oretanian Phase). This unit (Lower Alcudian) became folded, and raised together with the already deformed rocks, being vigorously eroded. Afterwards, a distensive faulting event sets up the palaeotopography of the new basin. This is filled by turbiditic, shelf, shore and continental deposits forming the successive cycles of Upper Alcudian, the last episodes of which are consecutively more restricted towards NE and initiate a new instability period, leading to a tardy compressive event (Cadomian Phase). A new distensive faulting event sets up a turbiditic trough whose later outer shelf and shore infill is completed in Lower Cambrian times. Later, a posthumous phase resets prior structures, producing mainly vertical faulting and tilting. Afterwards, a new hypercycle starts; this time it is entirely Palaeozoic.

Acknowledgments. This work was funded by the Spanish project No. 455 of CSIC-CAICYT.

References

Alvarez-Nava H, Robles R (1988) Sobre la existencia de una esquistosidad prehercínica en la zona de Fuenteguinaldo-Pastores, SO de Salamanca (Zona Centro-Ibérica). II Congr Geol Esp, Granada Comun 2: 99-102

Bernardo de Sousa M (1981) Complexo xisto-grauváquico do Douro: A Formação de Bateiras-Facies, ocorrência e significado litoestratigráfico. Cuad Geol Ibérica 7: 645-651

Bernardo de Sousa M (1982) Litostratigrafia e Estrutura do Complexo Xisto-Grauváquico-Grupo do Douro (NE Portugal) Thesis, Univ Coimbra, 222 p

Bernardo de Sousa M (1984) Considerações sobre a estratigrafia do Complexo Xisto-Grauváquico (CXG) e sua relação com o Paleozóico inferior. Cuad Geol Ibérica 9: 9-36

Blatrix P, Burg JP (1981) 40Ar/39Ar dates from Sierra Morena (southern Spain): Variscan metamorphism and Cadomian Orogeny. Neues Jahrb Min Monatsh 10: 470-478

Bouyx E (1970) Contribution à l'étude des Formations Anté-Ordoviciennes de la Méséta Méridionale (Ciudad Real et Badajoz). Mem Inst Geol Min Esp 73: 263 p

Calvet F, Salas R (1988) Tipos de plataformas carbonatadas del Precámbrico Terminal de la zona Centro-Ibérica. II Congr Geol Esp, Granada Comun 1: 59-62

Capote R, Gutiérrez Elorza M, Fernández Casals MJ, Moreno F, Navidad M, Peinado M, Vegas R (1977) The Precambrian in the central part of the Iberian Massif. Estud Geol Madrid 33 (4): 343-355

Carrington da Costa J (1950) Noticia sobre uma carta geológica do Buçaco, de Nery Delgado. Comun Serv Geol Port, 28 p

Conde LEN (1971) Existência em Portugal de uma serie superior a "Formação Xistosa da Beira" e inferior ao Ordovícico. Stud Geol 2: 25-26

Corretgé LG, García de Figuerola LC (1971) Rasgos geológicos de la región comprendida entre el sinclinal de Cañaveral y la Sierra de San Pedro (Cáceres). 1st Congr Hisp-Luso-Am Geol Econ Madrid-Lisboa I (1): 183-195

Crespo V, López Vilchez L, Paz C, Rey J (1971) Contribución al estudio de los sedimentos del Valle de Alcudia (Ciudad Real). 1st Congr Hisp-Luso-Am Geol Econ Madrid-Lisboa I (1): 139

Crespo V, Rey J (1971) Contribución al estudio del Valle de Alcudia. Bol Geol Min 82 (6): 512-515

Dallmeyer RD, Pieren AP (1987) Evidences of Late Proterozoic tectonometamorphic events in SW Iberia. Significance and paleogeographic implications. Univ Georgia, Abstr Ser IGCP Proj 233 Nouakchott Meeting, pp 73-76

Díez Balda MA (1986) El complejo esquisto-grauváquico, las series paleozoicas y la estructura hercínica al Sur de Salamanca. Acta Salmanticensia, Sec Ciencias. Salamanca 52: 1-162

Gama Pereira LC (1983) A Formação dos Amarelos. In: Livro-Guía das Excurções geológicas realizadas em Portugal. VIII Reunião Geol Oeste Peninsular, Coimbra-Salamanca

García de Figuerola LC, Martínez García E (1972) El Cámbrico Inferior de la Rinconada (Salamanca, España Central). Stud Geol 3: 33-41

García-Hidalgo JF (1984) Los materiales carbonatados del Alcudiense Superior en los alrededores de Casas de Miravete. (Prov. de Cáceres). Publ Geol 20: 379-389

García-Hidalgo JF (1988) El anticlinal de Ibor: Estratigrafía, sedimentología, tectónica y ciclos sedimentarios. Simposio sobre Cinturones Orogénicos, II Congr Geol Esp, Granada, pp 101-110

Gil Cid MD (1981) Proplina yochelsoni n sp. Primer monoplacophora del Cámbrico Inferior español. Bol Geol Min 92 (3): 196-202

Gil Cid MD (1988) Redlichiidae del Cámbrico Inferior de Los Navalucillos (Toledo). Bol Geol Min 99 (4): 584-591

Gil Cid MD, Jago JB (1989) New data on the Lower Cambrian trilobites of Cortijos de Malagón (Spain). Estud Geol 45: 91-99

Herranz P (1984) El Precámbrico del NEE de "Ossa-Morena": Planteamiento y estado de la cuestión, unidades, bases para su correlación y esquema evolutivo. Cuad Geol Ibérica 9: 119-211

Herranz P, San José MA, Vilas L (1977) Ensayo de Correlación del Precámbrico entre los Montes de Toledo y el Valle del Matachel. Estud Geol 33 (4): 327-342

Julivert M, Fontboté JM, Ribeiro A, Conde LE (1972) Mapa Tectónico de la Península Ibérica y Baleares, E. 1: 1000000, Memoria Explicativa (1974) Inst Geol Min Esp, 113 p

Liñán E, Palacios T (1987) Asociaciones de pistas fósiles y microorganismos de pared orgánica del Proterozoico, en las facies esquisto-grauváquicas del norte de Cáceres. Consecuencias regionales. Bol R Soc Esp Hist Nat (Geol) 82 (1-4): 211-232

Llopis N, San José MA de, Herranz P (1970) Nota sobre una discordancia posiblemente precámbrica al SE de la provincia de Badajoz, y sobre la edad de las series paleozoicas circundantes. Bol Geol Min 81 (6): 586-592

Lotze F (1945) Zur Gliederung des Varisciden der Iberischen Meseta. Geotekt Forsch 6: 78-92

Lotze F (1956) Über sardischen Bewegungen in Spanien und ihre Beziehungen zur assyntischen Faltung. Geotekt Symp zu Ehren Hans Stille, pp 129-139

Martín Escorza C (1976) Las Capas de Transición, Cámbrico inferior y otras series preordovícicas (¿Cámbrico superior?) en los Montes de Toledo Surorientales: sus implicaciones geotectónicas. Estud Geol 32: 591-613

Martín Escorza C (1977) Nuevos datos sobre el Ordovícico inferior; el límite Cámbrico-Ordovícico y las fases sárdicas en los Montes de Toledo: consecuencias geotectónicas. Estud Geol 33: 57-80

Martínez García E (1973) Deformación y metamorfismo en la zona de Sanabria. Stud Geol 5: 7-106

Martínez García E, Nicolau J (1973) Los terrenos infraordovícicos de la antiforma de Martinamor (Salamanca). Bol Geol Min 84 (6): 407-418

Mitrofanov FP, Timofeiev M (1977) Premier étude des microfaunes du Précambrien de la Péninsule Ibérique. Inst Geol Geocron Precambrian. Inéd Leningrad, 2 p

Nozal F, García Casquero JL, Picart J (1988) Discordancia Intraprecámbrica y series sedimentarias en el sector suroriental de los Montes de Toledo. Bol Geol Min 99 (4): 473-489

Ortega E, González Lodeiro P (1986) La discordancia intraalcudiense en el dominio meridional de la zona centroibérica. Breviora Geol Astúrica 27: 27-32

Ovtracht A, Tamain G (1970) Tectonique en Sierra Morena (España). CR Acad Sci Paris (Ser D) 270: 2634-2636

Pardo MV, Robles R (1988) La discordancia basal del Grupo Valdelacasa en el anticlinal de Valdelacasa (Sector central de los Montes de Toledo). II Congr Geol Esp, Granada, Comun 2: 165-168

Peláez JR, García-Hidalgo JF, Herranz P, Pieren AP, Vilas L, San José MA de (1989) Upper Proterozoic in central Spain. Abstr 28th Int Geol Congr, Washington 2: 590-591

Perejón A (1972) Primer descubrimiento y descripción de Arqueociátidos en la Provincia de Salamanca. Stud Geol 4: 143-149

Perejón A (1984) Bioestratigrafía de los Arqueociatos en España. Cuad Geol Ibérica 9: 213-265

Picart J (1988) Facies e interpretación de los yacimientos fosfatados del Cámbrico Inferior de Fontanarejo, Zona Centro Ibérica (Ciudad Real). II Congr Geol Esp, Granada. Comun 1: 157-160

Pieren Pidal AP, Pineda Velasco A, Herranz Araujo P (1987) Discordancia intra-Alcudiense en el anticlinal de Agudo (Ciudad Real-Badajoz). Geogaceta 2: 26-29

Pieren Pidal AP, Herranz Araujo P (1988) Presencia y caracterización del "Alcudiense Superior" (Vendiense) en el entronque de La Serena y el Valle de Alcudia (Este de la Provincia de Badajoz). Simposio sobre Cinturones Orogénicos, II Congr Geol Esp, Granada, pp 91-100

Rebelo JA, Romano M (1986) A contribution to the lithostratigraphy and palaeontology of the Lower Palaeozoic rocks of the Moncorvo region, northeast Portugal. Comun Serv Geol Port 72 (1/2): 45-47

Redlin K (1955) Stratigraphie und Tektonik der Estremadura im Bereich des Valle de Alcudia. (Spanien). Thesis, Univ Münster, 192 p

Robles R, Alvarez-Nava H (1988) Los materiales Precámbrico-Cámbricos del Domo de Las Hurdes: Existencia de tres series sedimentarias separadas por discordancias, SO de Salamanca (Zona Centro-Ibérica). II Congr Geol Esp, Granada. Comun 1: 185-188

Rodríguez Alonso MD (1979) El complejo esquisto-grauváquico y los materiales ordovícicos al SE de Civdad Rodrigo (Salamanca-España). Stud Geol 14: 7-34

Rodríguez Alonso MD (1984) El complejo esquisto-grauwáckico en el área de Las Hurdes y Sierra de Gata, provincias de Salamanca y Cáceres, España. Cuad Geol Ibérica 9: 37-80

Rodríguez Alonso MD (1985) El Complejo Esquisto-Grauwáckico y el Paleozoico en el Centro-Oeste español. Acta Salmanticensia, Sec Ciencias. Salamanca 51: 1-174

Rölz P (1975) Beiträge zum Aufbau des jungpräkambrischen und altpaläozoischen Grundgebirges in den Provinzen Salamanca und Cáceres (Sierra de Tamames, Sierra de Francia und östliche Sierra de Gata. Spanien. Münstersches Forsch Geol Paläont 36: 1-68

San José MA (1980) Estudio estratigráfico de los materiales preordovícicos del Anticlinal de Navalpino (Provincias de Badajoz y Ciudad Real). Mem final beca Fundación "Juan March". Inéd Madrid, 164 p

San José MA (1983) El complejo sedimentario pelítico-grauváquico. In: Comba JA (ed) Libro Jubilar J. M. Ríos. Inst Geol Min Esp Madrid 1: 91-99

San José MA (1984) Los materiales anteordovícicos del anticlinal de Navalpino. Cuad Geol Ibérica 9: 81-117

Santamaría J (1988) Facies y asociaciones de facies de las fosforitas de Robledo del Mazo (Precámbrico-Cámbrico de los Montes de Toledo). II Congr Geol Esp, Granada. Comun 1: 205-208

Santamaría J, Picart J, Solé FJ (1987a) Association of facies of the Pre-Cambrian phosphorites of Horcajo de Los Montes (Ciudad Real). Spain. Abstr. VIII European Regional Meeting. IAS, Tunis

Santamaría J, Picart J, Solé FJ (1987b) Some sedimentological aspects of the Precambrian phosphorites of Horcajo de Los Montes (Ciudad Real). Spain. Abstr. VIII European Regional Meeting. IAS, Tunis

Szuy K (1961) Das Kambrium Spaniens. Teil (II): Trilobiten. Akad Wiss Lit Ab Math Naturwiss Kl 7-8: 217-408 (499-690)

Tamain G (1970) Guide sommaire de l'excursion de la Commission Internationale de la Carte Tectonique de L'Europe (Péninsule Ibérique, Espagne), et Complément. Impaca-Bibas édit, 2 fasc

Teixeira C (1954) Os Conglomerados do Complexo xisto-grauváquico ante-silúrico. Sua importancia geológica e paleogeográfica. Comun Serv Geol Port 35: 33-49

Teixeira C (1955) Notas sobre geologia de Portugal. O Complexo Xisto Grauváquico Ante-Ordoviciano. (Ed Autor). Emprenta Lit Fluminense. Lisboa, 50 p

Teixeira C (1964) La faune de Lingulellinae des formations anté-ordoviciennes de Marão. Bol Soc Geol Port 15: 117-122

Teixeira C (1979) Le Précambrien portugais. C R som séan Soc Géol Fr 1979(5-6): 228-230

Vegas R (1978) Sedimentation and tectonism in the Iberian Massif prior to the Hercynian deformation (Late Precambrian to Silurian times). In: Geologia de la parte Norte del Macizo Ibérico. Vol Homenaje I. Parga Pondal, Ed do Castro. La Coruña, pp 271-286

Vilas L, Herranz P, San José MA de, Peláez JR (1981) El Precámbrico y sus relaciones con el Paleozoico inferior en el sector centro meridional del Macizo Ibérico. R Acad Cienc Exact Fís Nat Tomo especial Curso de Conferencias sobre el PICG (Mayo de 1979) Madrid I: 76-93

Vilas L, San José MA, García-Hidalgo JF, Peláez JR, Herranz P, Pieren AP (1986a) Modèle évolutif du bassin Précambrien terminal-Cambrien inférieur dans la région de Toledo-Ciudad Real-Extremadura (España). 11 ème Réunion des Sciences de la Terre (RST), Clermont-FD, Soc Géol Fr édit Paris, pp 183

Vilas L, García-Hidalgo JF, San José MA, Pieren AP, Peláez JR, Perejón A, Herranz P (1986b) Stratigraphy and basin evolution in the Upper pre-Cambrian-Lower Cambrian of the SW Spanish meseta. Abstr. Int Conf Iberian Terranes and their Regional Correlation. Oviedo IGCP 233: 46

Vilas L, García-Hidalgo JF, San José MA, Pieren AP, Peláez JR, Perejón A, Herranz P (1987) Episodios sedimentarios en el Alcudiense Superior (Proterozoico) y su tránsito al Cámbrico en la zona centro meridional del Macizo Ibérico. Geogaceta 2: 43-45

2.1.3 Post-Cambrian Palaeozoic Stratigraphy

J. C. Gutierrez Marco[1], M. A. de San Jose[2] and A. P. Pieren[2]

2.1.3.1 *Introduction*

In the Central-Iberian Zone, the most extensive outcrops of Palaeozoic rocks younger than the great basal Ordovician transgression in the whole Hesperian Massif occur. These materials appear generally located either in the core of stretched and narrow synclines, or preserved in tectonic basins related to late or posthumous movements in the Hercynian. Such structures are transversely separated by generally wide antiforms where pre-Ordovician substrata (pre-Cambrian and/or Cambrian) are outcropping. The Palaeozoic post-Cambrian units overlie the basement by means of a first-grade stratigraphic disconformity (refer to the previous chapter or later).

If we consider all the different sequences analyzed, the post-Cambrian Palaeozoic materials of the Central-Iberian Zone show a conspicuous facies homogeneity and temporal development. The sequences can be grouped by means of stratigraphic discontinuities which divide them into different sedimentary megasequences with outstanding palaeogeographic significance, analyzed throughout this subchapter.

However, it should be emphasized that the outcrops of Palaeozoic rocks with Central-Iberian characteristics overstep conventional boundaries established for this zone. Thus, facies and even palaeontological content of successions appearing in the Ollo de Sapo fringe are very similar to those recognized in the Caurel-Truchas Domain of the West-Asturian-Leonese Zone (see Chap. 3), while Palaeozoic rocks of the northern Ossa-Morena Zone (immediately south of Los Pedroches Batholith) present many aspects compatible with those from the neighbouring Central-Iberian "s. str." zone.

The following description has been divided according to the main sedimentary episodes or megasequences forming a post-Cambrian Hypercycle (Fig. 1), in which all processes and materials related to the tectonosedimentary Caledono-Hercynian evolution of the Central-Iberian Zone are recorded. The limits between these episodes are sedimentary discontinuities (gaps or unconformities), whose chronostratigraphic time equivalence has not always been precise, having a larger or shorter diachrony, and only sometimes coincident with those of Systems, Series and Stages. As has been said above (see previous chapter), these episodes can be grouped into two large units, the first one is the Lower Ordovician-Lower Carboniferous, which includes three megasequences of preorogenic and early synorogenic materials, and the second is the Middle and Upper Carboniferous, late and postorogenic. A sedimentological analysis of the Ordovician-Silurian sequence from the southern Central-Iberian Zone permitted the division of the sequence into eight major tectonosedimentary units (Portero and Dabrio 1988).

2.1.3.2 *Ordovician*

Within the Central-Iberian Zone, materials belonging to this period (or linked to it) form a sedimentary megasequence which can be further divided into three successive units:

1. Lower Ordovician. Having a markedly transgressive character, it consists of several sandy units, among which is included the well-known Armorican Quartzite;
2. Middle Ordovician (Tristani Beds). Predominantly muddy, with some sandy intercalations in the middle and upper parts;
3. Upper Ordovician. Discontinuous and more diverse lithologically, with sands alternating with

[1] Departamento de Paleontología, Universidad Complutense de Madrid, Madrid, Spain
[2] Departamento de Estratigrafía, Universidad Complutense de Madrid, Madrid, Spain

R. D. Dallmeyer and E. Martínez García (Eds.)
Pre-Mesozoic Geology of Iberia

shales and culminating in a calcareous level, sometimes absent due to a generalized erosion close to the Ordovician-Silurian boundary.

2.1.3.2.1 The Pre-Ordovician Unconformity and Related Formations

The base of the Ordovician *(s. l.)* in the Central-Iberian Zone is a first-order unconformity everywhere, erroneously equated to the Middle-Upper Ordovician Sardic Unconformity, as stated by Hammann et al. (1982) and Pillola and Gutiérrez Marco (1988), which separates Ordovician strata from their Proterozoic and/or Lower Cambrian substratum, described in the previous subchapter. This pre-Ordovician unconformity, comprised of several successive unconformities (Iberian and Toledanian Phases of Lotze 1956; Martín Escorza 1976, etc.), represents the boundary between two tectonosedimentary hypercycles, the Assyntic-Pan-African and the Caledono-Hercynian, at the end of which a total or partial regrouping of the sialic blocks occurred. However, between the lithological megaunits (Lower Ordovician *s. l.* and its pre-Ordovician basement, Cambrian and/or Proterozoic), some other unconformable units may appear: diverse clastic nearshore or fluviotidal complexes, with an important, mostly acid volcanic component, as described by Martín Escorza (1976) in eastern Montes de Toledo, and in the Os Amarelos Formation, in the Portuguese Beira Litoral Region (Gama Pereira 1983); the equivalent of both may be the Ollo de Sapo Formation. The upper limit of this volcanogenic-terrigenous formation has been assigned to the base of the Ordovician by some authors (Martínez García 1973; Ferragne 1972; Iglesias and Ribeiro 1981; etc.). Because of the lack of palaeontological data, the age of these units is not known precisely, although as they appear unconformably on top of the youngest beds of Lower Cambrian, they can be estimated as at least Upper Cambrian or even basal Ordovician. In other cases, below the dated Lower Ordovician sandy unit, some other units, also unconformable with regard to the substratum, appear locally (Carrascalejo Beds and equivalents). These include conglomerates, sandstones and shales, representing a large facies diversity, changing from alluvial paraconglomerates to intertidal facies (sometimes with layers crowded by *Skolithos*) that could represent the first deposits related to the lower Tremadoc global transgression. The outcrops of these rocks are discontinuous, remaining preserved only in depressions or basins that may have tilted before the most extensive transgression culminated during Arenig, but generally they lie paraconformably or slightly unconformably below the Arenig deposits. We must also consider that in several cases these units are formed by very immature substratum-provided breccio-conglomerates, siltstones and graywackes, in response to local differences in the source area, and therefore they have been incorrectly considered as being a part of the Teixeira's (1954, 1955) Complexo Xisto-Grauváquico ante-Ordoviciano. Consequently, the pre-Ordovician schist and graywacke complex was regarded to be stratigraphically continuous with the Ordovician (McDougall et al. 1987), thereby neglecting the magnitude of the hiatus included in the paraconformity that separates Upper Riphean, Vendian or Lower Cambrian from Ordovician rocks, however geometrically conformable that it may locally seem. In this sense the poorly preserved incomplete remains of trilobites found recently near the top of the Desejosa Fm. in the Moncorvo region (Rebelo and Romano 1986) should be considered most likely either as Lower Cambrian or as Upper Cambrian-Early Ordovician, due to the discontinuity existing between this formation and the Armorican Quartzite. This discontinuity is not considered by these authors in Moncorvo. The Quinta da Ventosa Fm. should then be included in the Lower Ordovician (see also discussion in de San José et al., this Vol.).

2.1.3.2.2 The Lower Ordovician Succession

In most cases, overlying the basal Ordovician main unconformity is a basal unit which includes conglomerates, sandstones and some siltstones, generally weathered to red-purplish or purple colors. This unit is known by several names (Basal Red Series, Sarnelha Fm., Intermediate Beds, Constante Fm., Hito Fm., etc.); towards the NW its equivalents are mainly silty (Puebla Fm., Montes Shales). Its thickness may exceed 350 m, but noticeable variations may occur along a single structure, and the unit can even disappear, thus showing the existence of a palaeotopography prior to the Ordovician transgression. This unit includes fluvial to intertidal, restricted and subtidal facies, and some authors consider its lower part as having continental red bed affinities (Vegas et al. 1977). The recurrence of conglomerates at higher levels, but still in the lower part of the unit, could be related to small unconformities that may indicate reactivation periods within the main transgressive episode. The Lower Ordovician ichnofossils, algal structures and inarticulate brachiopods found (Romano 1974; Moreno et al.

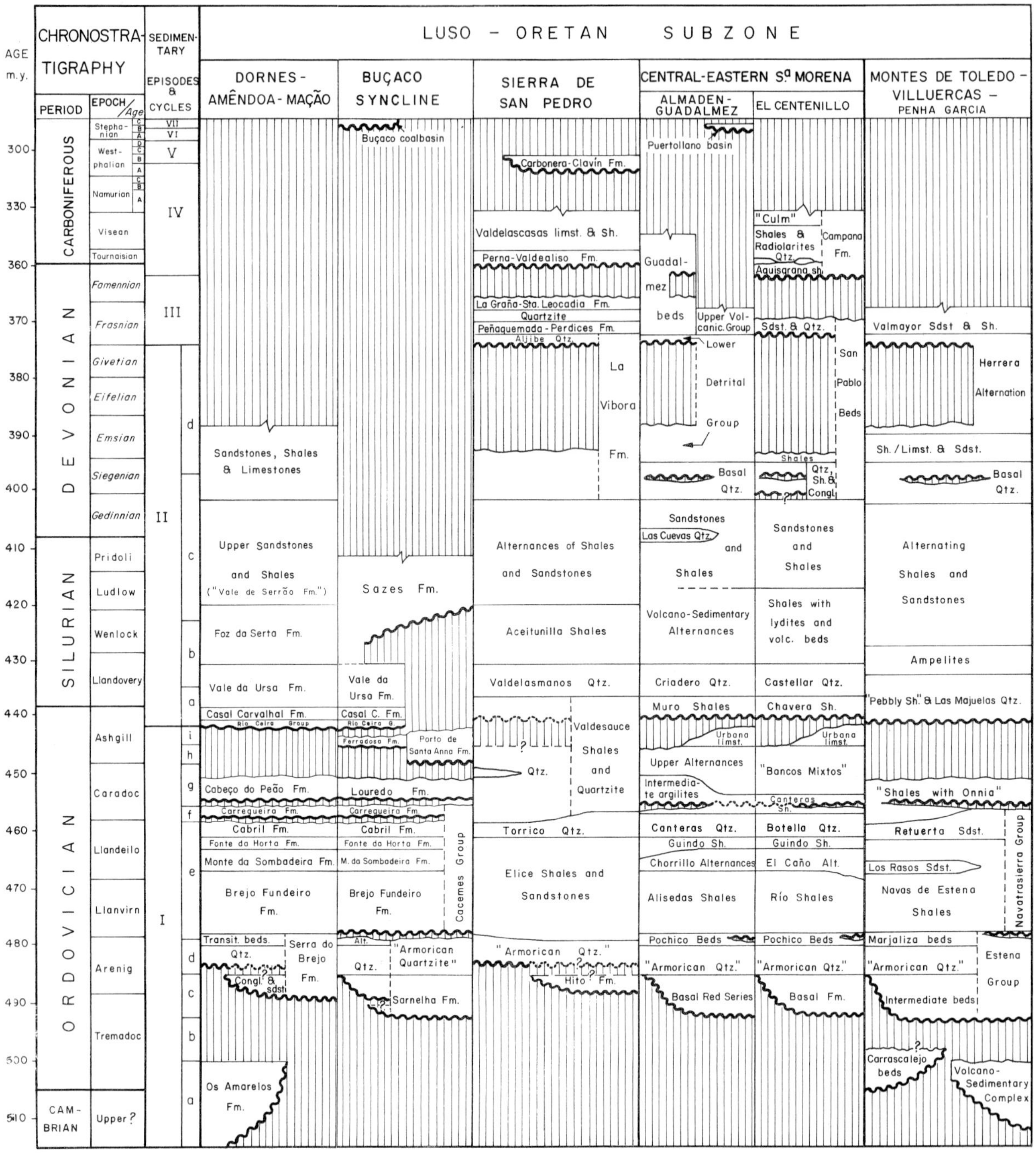

Fig. 1. (pp. 162 and 163) Correlation Chart of post-Cambrian lithostratigraphic units from the Palaeozoic of the Central-Iberian Zone. Units are described from SW represented on the *left*, to NE on the *right*

1976) do not permit precise chronostratigraphical dating; however, several authors assign the unit to the Tremadoc, based on its lithostratigraphical position (below Arenig quartzites). However, if we compare these arguments with the general scheme of the Lower Ordovician global transgressions and regressions (Fortey 1984), all of the units located in continuity or paraconformity below the Armorican Quartzite could have been deposited since the Hunneberg interval (Tremadoc/Arenig), marking the initiation of a generalized transgression, which reached a maximum during Middle Arenig.

Sedimentation during the Lower Ordovician proceeded in the whole Central-Iberian zone with

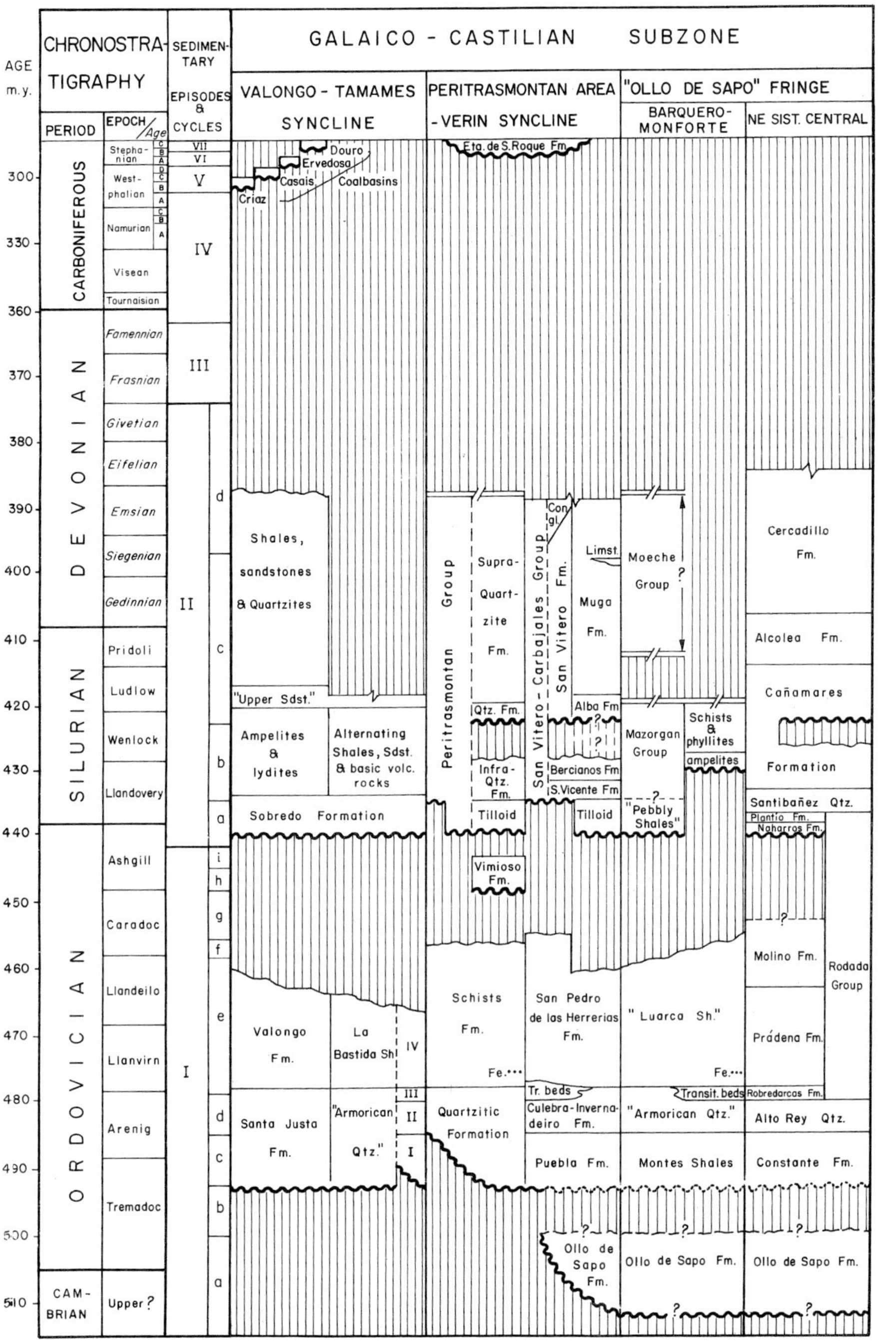

the ubiquitous Armorican Quartzite facies, which now constitutes the higher reliefs in this zone. This unit either received barely local names in some areas (Alto Rey Fm., Culebra-Invernadeiro Fm.), was included in a larger ranked unit (Estena Group) or was considered as the most conspicuous member of some formations (Serra do Brejo Fm., Santa Justa Fm.). The thickness of the Armorican Quartzite facies within the Central-Iberian Zone varies considerably. The thickness generally oscillates between 150 and 300 m, but on rare occasions it can be as low as 5–10 m and can keep this range for almost 100 km (Sierra de S. Pedro) or increase very quickly for palaeogeographic reasons. Sedimentological evidence suggests a nearshore environment, consisting of thick sandy bars interpreted as supratidal, barrier islands and high energy subtidal deposits influenced by currents and some storm-generated reworking. Regarding its age, the ichnofossils (*Cruziana* and *Skolithos* ichnofacies) permit relegating the Armorican Quartzite to the Arenig *s. l.* (Delgado 1886, 1887; Moreno et al. 1976; Cooper and Romano 1982; Pickerill et al. 1984).

The Armorican Quartzite passes gradually into an overlying unit (Pochico Beds, Marjaliza Beds or Transitional Beds) consisting of alternating quartzites and shales, which mark the transition to the Tristani Beds. The thickness of this unit oscillates between 50 and 150 m, although it may occasionally be absent. The sedimentation in this unit is framed within a regressive setting, which very likely already started in the last levels of the Armorican Quartzite; this implies few sedimentary interruptions. The hiatus related to these interruptions, represented by ferruginous microconglomeratic layers with inarticulate brachiopods lumachelles, phosphatic grains and detrital concentrations of radioactive heavy minerals (Bouyx 1970; Tamain 1972; Paris 1981) is difficult to estimate. These layers might be coeval with the genesis of the "Ferruginous Beds" located in the hanging wall of the Quartzitic Formation from the Peritrasmontan area. Facies quickly evolve to inter- and subtidal, heavily bioturbated, storm-dominated environments. In addition to the above mentioned lingulids, the unit contains numerous ichnofossils (*Daedalus, Rouaultia* and *Cruziana*), with scarce trilobites close to the top (Rábano 1984). Available data suggest that the Arenig-Llanvirn boundary should be located within the uppermost part of the unit.

2.1.3.2.3 Middle Ordovician

Lower Ordovician sandy units pass to a very fossiliferous dominantly shaly group, informally known as Tristani Beds (Gutiérrez Marco et al. 1984, p. 299). The series commences gradually with a thick lutitic formation (between 150 and 1000 m), which reaches its major thickness in the north and includes some oolitic ironstones in its lower third (Peritrasmontan and Zamora areas), while southwards it represents a larger proportion of sandy intercalations. These form, at the top of the Tristani Beds, a second mainly quartzitic unit (Torrico, Canteras and Botella Quartzites, Retuerta Sandstones) up to 200 m in thickness. The group was generated under open shelf conditions (offshore shales), with a very low topographic gradient (possibly lower than 0.1°; see Brenchley et al. 1986), representing a very uniform sedimentation regime over large areas (at least in the whole Luso-Oretan Subzone). This is reflected too in a marked similarity of corresponding biofacies. Data furnished by these and by the lithofacies indicate a predominantly transgressive regime during Llanvirn times. At the start of Llandeilo, this tendency is reverted, and the new epoch is generally regressive. This mainly shaly sequence includes interbedded sandy units with hummocky cross stratification (Monte da Sombadeira Fm., Chorrillo and El Caño Alternations, Los Rasos Sandstones). These units, interbedded in the decantation lutitic sequence, are assigned to episodes of storm removal, which may transport sands along a muddy shelf 100 km away from the shoreline (Brenchley et al. 1986).

Considered as a whole, the Tristani Beds are extremely fossiliferous. In the Central-Iberian Zone more than 250 taxa have been determined comprising the following groups: trilobites, graptolites, brachiopods (articulate and inarticulate), molluscs (bivalves, gastropods, cephalopods, rostroconchs, monoplacophorans and hyolithids), ostracods, echinoderms (diploporids, crinoids, ophiuroids, homalozoans) conularids, machaeridians, bryozoans, chitinozoans, acritarchs, scolecodonts and ichnofossils, which have been described in several works summarized by Romano (1982); Hammann (1983); Gutiérrez Marco et al. (1984); Gutiérrez Marco (1986) and Babin and Gutiérrez Marco (in litt.). This abundance of fossils permits the establishment of a precise biozonation, from basal Lower Llanvirn up to Upper Dobrotivian (Llandeilo), and to corroborate that the base of this unit is never older than Lower Llanvirn (Gutiérrez Marco 1986; Romano et al. 1986).

2.1.3.2.4 Upper Ordovician

Upper Ordovician rocks are generally absent in the Galaico-Castilian Subzone, due to pre-Silurian erosion, but the existence of materials which might be attributed to the Lower Caradoc in continuity with the previous sequence cannot be discarded in any area, even with the lack of palaeontological data. On the contrary, in the Luso-Oretan Subzone the Tristani Beds (Navatrasierra Group, Cacemes Group and equivalents) are succeeded by a more lithologically varied sequence (Young 1988). This generally commences with a shaly unit between 15 and 200 m thick (Shales with Onnia, Canteras Shales, Intermediate Argillites, Carregueira Fm.). This shaly unit may include quartzitic sets (Louredo Fm.) in whose base or lower third exists a phosphate-bearing microconglomeratic horizon, representing at least one stratigraphic interruption within the Caradoc. This discontinuity probably correlates with the true Sardic unconformity, a hiatus which has been determined as Harnagian-Soudleyan (p. p.) according to palaeontological data (Paris 1981; Hammann 1983; Fig. 1 in Gutiérrez Marco and Rábano 1987). At the top of this shale unit, the stratigraphic log is interrupted at several points (see Young 1988 and the Correlation Chart), but in east-central Sierra Morena it continues with alternating sands and shales, analogous to the Bancos Mixtos (Mixed Banks) of Tamain (1967, 1972). The upper part of this unit passes to the Urbana Limestone (equivalent to the Ferradosa Fm. of the southern part of the Buçaco syncline), which is the first post-Lower Cambrian carbonate episode registered in the Central-Iberian zone; it is composed of up to 40 m of pure bioclastic limestones. The Bancos Mixtos and Urbana Limestone are very fossiliferous, including frequent lumachelles which have furnished several brachiopods, trilobites, bryozoans, echinoderms, cnidarians, molluscs, conodonts and some graptolites (Fuganti and Serpagli 1968; Hammann 1976, 1983; Hafenrichter 1979, 1980; Gutiérrez Marco and Rábano 1987; Sarmiento in litt.) whose age is between Upper Actonian and Rawtheyan (Upper Caradoc-Middle Ashgill).

As has already been indicated, Upper Ordovician rocks show great variation along the whole Central-Iberian Zone, more in the NW-SE direction than transverse to the structures. Thus, in the Buçaco syncline and in the Peritrasmontan area (in the Luso-Oretan and Galaico-Castilian subzones respectively), the top of the Ordovician sequence is formed by a volcanoclastic unit including important limestone levels (Porto do Santa Anna Fm., Vimioso Fm.), which makes this unit comparable with the Rosan Fm. of the French Armorican Massif. In Buçaco, the base of this unit is marked by a thin layer of red ash (10 cm), as well as an oolitic ironstone bed (35 cm), which seal a stratigraphic discontinuity which is succeeded by fossiliferous levels (Thadeu 1947; Mitchell 1974; Paris 1979, 1981) indicative of a Lower Ashgill age (see discussion in Gutiérrez Marco and Rábano 1987, pp. 651–652). Also, in the subsurface of the El Centenillo Area (Eastern Sierra Morena), the Upper Ordovician "normal" series (see Correlation Chart) include an unusually thick Urbana Limestone (80 m), with important intercalations of acid vulcanites (Pineda 1987). All these characteristics record the existence of a shifting palaeogeography during the Upper Ordovician in the Central-Iberian Zone, related to increasing tectonic instability. This produced a basin fragmentation (controlled by the pre-Ordovician basement: Pieren 1986), and therefore repeated stratigraphic discontinuities, accompanied by important volcanic and volcanosedimentary intercalations. All this corresponds to a tensional-thermal tectonometamorphic setting (Dallmeyer and Pieren 1987).

2.1.3.3 Ordovician-Silurian Boundary

The top of the dated Ordovician sequence is marked by an important stratigraphic discontinuity, related (by several authors) to glacioeustatic erosive events near the Ordovician-Silurian boundary. The erosion was increased by tectonic movements in local areas (Hafenrichter 1980; Iglesias and Robardet 1980; Robardet et al. 1980; Robardet 1981; Pieren 1986; Robardet and Doré 1988). As was suggested before, these phenomena may be responsible for the gross thickness changes in the Urbana Limestone, and for the absence of this and some older units in large areas of the Central-Iberian Zone (see Correlation Chart). However, in spite of its importance, this discontinuity is sometimes barely perceivable in the field, and on rare occasions it is marked by the presence of a ferruginous level with reworked ashgillian fossils (*Heliocrinites* band; Chauvel et al. 1969; Hafenrichter 1980).

The sedimentary record above the Upper Ordovician discontinuity usually comprises a diamictic formation of massive graywackes and shales, with fragments and pebbles of various size and lithologies (Casal-Carvahal Fm., Sobredo Fm., Naharros Fm., Tilloid and Pebbly Shales of the Peritrasmontan and Montes de Toledo Areas, (see Correlation Chart).

This unit has been compared with similar series of clast-bearing formations (diamicts), recorded in southwestern Europe and North Africa (see references in Robardet and Doré 1988) that have in general been interpreted as glaciomarine sediments (ice-drop paratillites). In this regard, the pebbles would be carried by icebergs derived from a Gondwana (Central Saharian) "inlandsis", developed during Late Ordovician glaciation. However, an alternative interpretation would instead consider these pebbly shales as mixtites (gravity-flow sediments) of local origin, produced by a tectonic instability in the sea floor. This reflects a diastrophic episode in the Ordovician-Silurian boundary, which would be quickly flushed by low energy shore facies. Both uncontradictory interpretations are possible because late-tectonic, uprising-triggered glaciations are widespread along the geological record. However, the presence of outsized blocks within lutites in some places (Fortuin 1984; IGME 1986) and dropstone structures (Guadarranque area) argue in favor of a glaciomarine origin. Regarding its age, the graywackes with fragments have not provided datable fossils in any Central-Iberian locality. However, a correlation over a larger area suggests that this unit would essentially correspond to the Hirnantian (Latest Ordovician) age. In this sense a sparse *Hirnantia*-like fauna (without *Hirnantia* itself) has been recently recorded in the Ribeira do Braçal Fm. (Rio Ceira Group) of Buçaco, above the Rawtheyan Porto do Santa Anna Fm. and beneath the glaciomarine Casal Carvalhal Fm. (Young 1988).

Escorting the stratigraphic gaps existing in the Ordovician-Silurian boundary, the formation of ferruginous crusts sealed by the Las Majuelas Quartzite can be observed, as well as a lateral disappearance of units, because of the existence of palaeofaults. These limit the units of graywackes with fragments, formed by a partial dismantling of the shelf. The faults reflect vertical tectonics in a distensive environment (Pieren 1986). Additional evidence includes volcanic emissions (Almadén volcanosedimentary interbeds and chimneys are the best example), peralkaline volcanism and very low-grade metamorphism which particularly affected the entire pre-Cambrian substratum. The absolute Ar40/Ar39 cooling age dating oscillates between 420 and 423 Ma (Dallmeyer and Pieren 1987, recalculated in 1989).

2.1.3.4 Silurian-Lower Devonian

Generally sealing the Middle Caradoc-Llandovery gap, the Silurian record starts in most southern Central-Iberian synclines with a quartzite, commonly known as the Basal Silurian Quartzite. It overlies the Muro Shales or Chavera Shales, Criadero and Castellar Qtzs., or the Infraquartzite Fm. on the Peritrasmontan area, Tilloids, and the Santibáñez Qtz. in the Ollo de Sapo fringe (see Correlation Chart).

The upper member of these quartzites recently provided Upper Aeronian graptolites in the El Centenillo area of Jaén (Gutiérrez Marco and Pineda 1988), and possibly Rhuddanian graptolites at Dornes (central Portugal: Young 1988).

On top of the previous units (Silurian quartzites or Hirnantian clast-bearing formations), separated by a rough contact (in most cases mechanized), is a layer 5 to 25 m thick of graptolite-rich black shales (sapropelites). The black shales are uniform and ubiquitous along the whole Central-Iberian Zone. They pass vertically to micaceous dark siltstones alternating with sandstones and quartzites in centi- to decimetric layers; just to the NE of the Central Ranges, they lead into a thick sandy unit (Alcolea Fm., 750–800 m), even though in Almadén there are some quartzites in a similar position (Las Cuevas Quartzites). In the last area, the presence of important volcanosedimentary interbeds becomes the distinctive feature (Saupé 1971, 1973), which in the El Centenillo and Valongo areas are accompanied by scarce lydite layers in the lowest part of the unit (Tamain 1972; Teixeira and Gonçalves 1980). We would like to emphasize this analogy between areas that are presently in diametrically opposed positions within the Central-Iberian Zone, which will require an adequate palaeogeographic interpretation.

Regarding the age of these units, the oldest faunas obtained in the base of the ampelites are almost invariably indicative of a Middle Llandovery (Aeronian) to Early Upper Llandovery age (between *-gregarius* and *-turriculatus* zones), followed by another association reaching the Llandovery-Wenlock boundary (Haberfelner 1931; Romariz 1962; Saupé 1971; Tamain 1972; Robardet et al. 1980; Gutiérrez Marco and Pineda 1988; Rodríguez Núñez et al. in litt.). The upper sandstone and shale-alternating beds include, in their lower half, several horizons with Wenlock and Lower Ludlow graptolites. These are represented in the Sazes Fm. of the Buçaco Syncline (Paris 1981) whose base also contains true Upper Llandovery graptolites (Romariz 1962). The remaining alternating shales and sandstones have not yet provided classifiable fossils, but their top is marked by an ubiquitous quartzitic level which has, very close to its base, bentonic faunas of Upper Lochkovian-Pragian age. This implies that within the underlying previous interbeds, Ludlow, Pridolian and Lochkovian (Gedinnian) materials should be represented, even though there is no adequate evidence to pinpoint the Silurian-Devonian boundary. However, to the north of the Central System, these units have a much larger palaeontologic record, with Ludlow and Pridolian fossils in the Upper part of the Cañamares Fm. (which is somewhat equivalent to the Upper Sandstones and Shales). The unit is succeeded by diverse horizons with Pridolian and Lower Gedinnian bentonic faunas in the overlying Alcolea Fm. The Silurian-Devonian boundary would then be located within the interbeds overlying the sapropelites (Carls 1977; see synthesis in Fernández Casals and Gutiérrez Marco 1985).

In the NE Central System Ranges, the Devonian sequence continues into the Cercadillo Fm., a shaly unit with important intercalations of limestones and dolostones which reach a thickness of 625–700 m, and whose age is between Lower Gedinnian and Lower Eifelian. The age has been determined by several levels containing brachiopods, conodonts, trilobites, dacrioconarids and ammonoids (data synthesized in Fernández Casals and Gutiérrez Marco, 1985). In the southern Central-Iberian Zone (Luso-Oretan Subzone) the Silurian-Devonian boundary interbeds are overlain by a quartzitic level (improperly called Devonian Base Quartzite, Basal Quartzite, or Siegenian Quartzite) which encloses scarce Gedinnian-Siegenian fossils and intercalated ferruginous crust levels. This suggests the existence of sedimentary interruptions within this unit (Pardo and García Alcalde 1984). The repetitive conglomerate levels which occur in the lower and middle parts of the San Pablo Beds in eastern Sierra Morena (El Centenillo) could correspond to any of such discontinuities. Overlying this there appears a widespread heterogeneous sequence, in which thin beds of sandstones and quartzites (Rivas Quartzite) occur within predominantly silts and reddish shales. The silts and shales are mixed with lenticular intercalations of bioclastic limestones (lumachelles) towards the middle and upper parts of the unit. The entire sequence is very fossiliferous, and it has furnished numerous brachiopods, corals, trilobites, tentaculitids and conodonts from Upper Siegenian to Emsian age (Puschmann 1970; Teixeira and Gonçalves 1980; Pardo and García Alcalde 1984). Above these levels, the sequence continues with similar lithologic characteristics, but in some places an extremely different sedimentary environment; the palaeontological content is also clearly Frasnian, revealing the presence of the Meso-Devonian Discontinuity described by Puschmann (1967, 1970) within these kinds of apparently continuous sequences (La Víbora Fm., Lower Detrital Group, San Pablo Beds, Herrera Interbeds).

As can be seen in the Correlation Chart, the Siluro-Devonian sequences in the Peritrasmontan Area-Verín Syncline and in the Barquero-Monforte Belt differ markedly from the areas described above. Their main characteristics are: very thick sedimentation (even several thousand metres), a great variety of lithological facies (conglomerates and microconglomerates, arkosic sandstones, quartzites, shales, sapropelites, limestones and lydites), important intercalations of acid vulcanites (breccias, tuffs, and rhyolitic-trachytic lavas) and frequent lateral and vertical tectonic contacts. These characteristics confer a sedimentary distinctiveness to this area, which is further confirmed by the existence of important interruptions, as testified by the presence of conglomerates with pebbles of metamorphic rocks, acid vulcanites, limestones, lydites and Ordovician quartzites and other rocks (Martínez García 1972, 1973; Aldaya et al. 1973, 1976). The palaeontological content of the formations included in the San Vitero-Carbajales, Peritransmontan and Mazorgán Groups includes Llandovery and Wenlock graptolites (Romariz 1969; Martínez García 1973; Jiménez Fuentes and Quiroga 1981), Devonian dacrioconarids (Truyóls Massoni and Quiroga 1982), Siluro-Devonian conodonts (Aldaya et al. 1976; Quiroga 1982) and some plant remains (Martínez García 1973; Teixeira and Pais 1973). A possible equivalence between these units and the Moeche Group, although limited by faults and located in a different geodynamic environment (refer to Ribeiro et al., Part IV, Chap. 3.1), is based on the identification of typical lithologies from those units (for example limestones with crinoids, bryozoans, fragments of tabulate corals and some foraminiferans) included as olistolith-like clasts in the Moeche Group.

The tectonosedimentary uniqueness of this area, located in the NW outermost part of the Galaico-Castilian Subzone, can be related to the existence of a system of continental rift-like subsident furrows in the northwestern corner of the Hesperian Massif (Ribeiro 1974; Iglesias and Robardet 1980). This particular area is located next to or surrounding the main outcrops of the allochthonous units (Galaico-Trasmontan Subzone). The area itself is semiallochthonous or parautochthonous. The subsident furrows should have brief continuity towards the SE, and should be likely bonded to the presence of a subcrustal hot spot during Lower Palaeozoic in this area.

2.1.3.5 *Upper Devonian-Lower Carboniferous*

In the Luso-Oretan Subzone, the Devonian sequence continues with fossil-rich Frasnian terrigenous units (Müller 1929; Puschmann 1970; Pardo and García Alcalde 1984), lying disconformably or paraconformably on Lower Devonian rocks (Meso-Devonian Discontinuity). The sequence includes some carbonated lenses, and in the Almadén Syncline it also includes important intercalations of volcanic breccias, tuffs and basic lavas mixed with volcanoclastic sandstones. At Almadén the thickness reaches even 550 m (Upper Volcanic Group), while towards the south and east, the total thickness of beds related to the Frasnian does not exceed 100 m. Further west, the Frasnian sequence in Sierra de San Pedro comprises several terrigenous units which together are 700 m thick (Soldevila, written comm., 1987). In their upper part also, important volcanic sills appear, from diabases to rhyolites. The correlation of all these Frasnian rocks has been improved with the development of a very precise local system of biozones (Pardo and García Alcalde 1984).

In most Palaeozoic synclines of the Central-Iberian Zone, the youngest sediments preserved are Frasnian in age. However, near the boundary of the Ossa-Morena Zone, in Sierra de San Pedro and Guadalmez Synclines and the El Centenillo area, younger terrigenous-carbonated sequences are recognized; generally their thickness surpasses 800 m, and their base is generally not older than the Famennian-Tournaisian boundary. The existence within these sequences of Famennian rocks could not be proven palaeontologically in most cases. Nevertheless, in eastern Sierra Morena (El Centenillo), the base of the Campana Fm. (Aquisgrana Schists) has locally provided some ostracods of Upper Famennian-Strunian age (Charpentier et al. 1977). The base of this unit represents either a paraconformity (Sierra de San Pedro), or a disconformity sealing a hiatus which can even include part of the Silurian (El Centenillo). However, in Guadalmez the hiatus only involves, when larger, the Lower and Middle Tournaisian, because Upper Tournaisian overlies a Famennian condensed sequence. The latter is itself lying on terminal Frasnian-Lower Famennian nodular slates. Remaining Guadalmez Beds contain nodular limestones and shales, well-bedded, fetid and massive limestones, and in the hanging wall a Culm-type shaly-sandy complex with a few detritic limestone lenses. Its age should be (according to the determined flora) between Upper Tournaisian and Middle Viséan (García Alcalde et al. 1984; Pardo and García Alcalde 1984). In El Centenillo, Upper Famennian-Viséan precocious synorogenic rocks are represented in the so-called Campana Fm., which includes sandy schists (Aquisgrana Schists), discontinuous quartzites (Campana Quartzites), schists with siliceous layers and finally a thick flysch-facies complex. The flysch complex consists of Culm-type schists and graywackes, with intraformational conglomeratic lenses. It has provided a flora undoubtedly Viséan in age, but it could even reach the Lower Namurian (Péran and Tamain 1967; Poupon and Tamain 1969; Charpentier et al. 1977). The Devono-Dinantian sequence is represented also in the Sierra de San Pedro by a volcanosedimentary unit (Perna-Valdealiso Fm.) consisting of tuffs and andesitic breccias, sandstones and limestones, with rare lydite layers, whose palaeontological content (brachiopods, trilobites and coral) indicates a Lower Carboniferous age (Tournaisian: Soldevila, written comm., 1987). This unit is overlain by fetid limestone banks topped by shales (Valdelascasas Limestones and Shales) whose age is supposed to be Upper Viséan-Namurian, and which are similar to the shales which appear in the core of the Cáceres syncline (Aldea-Moret Limestones and overlying shales).

2.1.3.6 Middle and Upper Carboniferous

Finally, the youngest Hercynian Palaeozoic rocks present in the Central-Iberian Zone correspond to diverse Middle and Upper Carboniferous intramontane limnic basins. These were preserved from erosion either as protected basins along late shear bands or as fini-tectonic semi-grabens. The rocks bear witness to the main diastrophic episodes of Hercynian Orogeny, which sometimes can be documented along a single line of basins (as in the Portuguese Criaz-Casais-Ervedosa and Douro basins). Fanglomeratic breccias and red to grey fluviotorrential conglomerates and sandstones are generally found. These pass into cyclothematic deposits (productive sometimes) or bituminous shales. Nevertheless, the lithologies and thicknesses of the Carboniferous sequences registered in each and every one of the basins are highly variable. For their description, see Lemos de Sousa and Wagner (1983); Domingos et al. (1983); Wagner (1985); and Seara (1986).

Principal palaeontologic data concerning the fossil floras have been summarized by Wagner and Lemos de Sousa (1983) and Wagner (1983). It can be deduced from these data that some of these basins, like Buçaco, continued their evolution during Lower Permian (Autunian), being relayed by the clearly post-Hercynian red Saxonian Facies, unconformable below the Middle-Upper Triassic extensive fluviatile sediments. More complete sequences have also been studied in the Central System Ranges (Hernando 1977; Sopeña 1979, etc.), which comprise Stephano-Autunian productive limnic grey and red fluviotorrential facies deposited in a fini-Hercynian transpressive regime with synchronous 287 ± 12 Ma calc-alkaline volcanism. The latter are unconformably overlain by Middle Permian (Saxonian) and fini-Permian/Early Triassic red alluvial facies whose distensive setting points out the beginning of the Alpine Mesozoic evolution.

Acknowledgments. We would like to thank Dr. J. Soldevila (Barcelona) for the unpublished information on the Sierra de San Pedro Palaeozoic sequence, and Dr. J. J. Kowalik (Newmont de España) for his comments regarding the English version of the manuscript.

References

Aldaya F, Arribas A, González Lodeiro P, Iglesias M, Martínez Catalán JR, Martínez García E (1973) Presencia de una nueva fase de deformación probablemente pre-herciniana en el noroeste y centro de la Península Ibérica (Galicia Central, Zamora y Salamanca). Stud Geol 6: 29-48

Aldaya F, Carls P, Martínez García E, Quiroga JL (1976) Nouvelles précisions sur l'âge de la série de San Vitero (Zamora, Nord Ouest de L'Espagne). C R Acad Sci Paris (Ser D) 283: 881-883

Babin C, Gutiérrez Marco JC (in litt) Middle Ordovician (Llanvirn-Dobrotiva) bivalve molluscs from Spain. Their phyletic and palaeogeographic significance. Palaeontology

Bouyx E (1970) Contribution à l'étude des formations anté Ordoviciennes de la Méséta Méridionale (Ciudad Real et Badajoz). Mem Inst Geol Min Esp 73: 1-263

Brenchley PJ, Romano M, Gutiérrez Marco JC (1986) Proximal and distal hummocky cross-stratified facies on a wide Ordovician shelf in Iberia. In: Knight RJ, McLean JR (eds) Shelf sands and sandstones. Can Soc Petrol Geol Mem Calgary 11: 241-255

Carls P (1977) The Silurian-Devonian boundary in northeastern and central Spain. In: The Silurian-Devonian boundary. IUGS, Ser A 5: 143-158

Charpentier JL, Lethiers F, Tamain G (1977) Les "Schistes Aquisgrana" à Ostracodes du Dévonien supérieur-terminal en Sierra Morena orientale (Espagne). Ann Soc Géol Nord T. XCVI-XCVII (1976-1977)

Chauvel J, Drot J, Pillet J, Tamain G (1969) Précisions sur l'Ordovicien moyen et supérieur de la "série-type" du Centenillo (Sierra Morena Orientale, Espagne). Bull Soc Géol Fr (7ème Série) 11 (5): 613-626

Cooper AH, Romano M (1982) The lower Ordovician stratigraphy of the Dornes-Figueiró dos Vinhos area, central Portugal, with description of *Merostomichnites* ichnosp. and *Rosselia socialis;* two previously unrecorded trace fossils. Comun Serv Geol Port 68 (1): 73-82

Dallmeyer RD, Pieren AP (1987) Evidences of Late Proterozoic tectonometamorphic events in SW Iberia. Significance and palaeogeographic implications. Univ Georgia, Abstr Ser. IGCP Proj 233 Nouakchott Meeting, pp 73-76

Delgado JFN (1886) Étude sur les Bilobites et autres fossiles des Quartzites de la base du Système Silurien du Portugal. Mém Sec Trab Geol Port: 113 p

Delgado JFN (1887) Étude sur les Bilobites et d'autres fossiles des quartzites de la base du système Silurien du Portugal-(Supplément). Mém Sec Trab Geol Port: 76 p

Domingos LCG, Freire JLS, Gomes da Silva F, Gonçalves F, Pereira E, Ribeiro A (1983) The structure of the intramontane Upper Carboniferous basins in Portugal. Mém Serv Geol Port 29: 187-205

Fernández Casals MJ, Gutiérrez Marco JC (1985) Aspectos estratigráficos de la Cadena Hercínica en el Sistema Central. Rev R Acad Cienc Ex Fís Nat Madrid 79 (4): 487-509

Ferragne A (1972) Le Précambrien et le Paléozoïque de la province d'Orense (Nord-Ouest de l'Espagne). Stratigraphie-tectonique-métamorphisme. Thesis, Univ Bordeaux, 249 p

Fortey RA (1984) Global earlier Ordovician transgressions and regressions and their biological implications: In: Bruton DL (ed) Aspects of the Ordovician System. Palaeontological Contributions, Univ Oslo 295: 37-50

Fortuin AR (1984) Late Ordovician glaciomarine deposits (Orea Shale) in the Sierra de Albarracín, Spain. Palaeogeogr Palaeoclimatol Palaeoecol 48: 245-261

Fuganti A, Serpagli E (1968) Geological remarks on Urbana Limestone and evidence for its Upper Ordovician age by means of conodonts, eastern Sierra Morena, south Spain. Boll Soc Geol Ital 87: 511-521

Gama Pereira LC (1983) A Formação dos Amarelos. In: Livro-Guía das Excurções geológicas realizadas em Portugal. VIII Reun Geol Oeste Peninsular. Coimbra- Salamanca

García Alcalde JL, Arbizu MA, Pardo Alonso MV, García López S (1984) El límite Devónico-Carbonífero en el área de Guadalmez-Santa Eufemia (Provs de Ciudad Real y Córdoba, Sierra Morena, España) I Congr Geol Esp Segovia I: 421-430

Gutiérrez Marco JC (1986) Graptolitos del Ordovícico español. Thesis, Univ Complutense, Madrid 3 vol 701 p (Inéd)

Gutiérrez Marco JC, Rábano I (1987) Trilobites y Graptolitos de las lumaquelas terminales de los "Bancos Mixtos" (Ordovícico superior de la zona Centroibérica meridional): Elementos nuevos o poco conocidos. Bol Geol Min 98 (5): 647-669

Gutiérrez Marco JC, Pineda Velasco A (1988) Datos bioestratigráficos sobre los materiales silúricos del subsuelo de El Centenillo (Jaén) Comun. II Congr Geol Esp, Granada 1: 91-94

Gutiérrez Marco JC, Rábano I, Prieto M, Martín J (1984) Estudio bioestratigráfico del Llanvirn y Llandeilo (Dobrotiviense) en la parte meridional de la zona Centroibérica (España). Cuad Geol Ibérica 9: 287-319

Haberfelner E (1931) Eine Revision der Graptolithen der Sierra Morena (Spanien). Abh Senckenb Naturforsch Ges 43 (2): 19-66

Hafenrichter M (1979) Paläontologisch-Ökologische und lithofazielle Untersuchungen des "Ashgill-Kalkes" (Jungordovizium) in Spanien. Arb Paläont Inst Würzburg 3: 139 p

Hafenrichter M (1980) The lower and upper boundary of the Ordovician System of some selected regions (Celtiberia, eastern Sierra Morena) in Spain. Part II: The Ordovician-Silurian boundary in Spain. Neues Jahrb Geol Paläont Abh 160 (1): 138-148

Hammann NW (1976) Trilobiten aus dem oberen Caradoc der östlichen Sierra Morena (Spanien). Senckenb Lethaea 57 (1): 35-85

Hammann NW (1983) Calymenacea (trilobita) aus dem Ordovizium von Spanien: ihre Biostratigraphie, Ökologie und Systematik. Abh Senckenb Naturforsch Ges 542: 1-117

Hammann W, Robardet M, Romano M, Gil Cid MD, Gutiérrez Marco JC, Herranz P, Marcos A, Martín J, Peláez JR, Pérez Estaún A, Prieto M, Rábano I, San José MA de, Truyóls J, Vegas R, Vilas L, Villas E, Villena J (1982) The Ordovician System in southwestern Europe (France, Spain and Portugal). IUGS Publ 11: 1-47

Hernando Costa S (1977) Pérmico y Triásico de la región Ayllón-Atienza (Provincias de Segovia, Soria y Guadalajara). Semin Estratigr Ser Monogr 2: 1-408

Iglesias M, Ribeiro A (1981) Position stratigraphique de la Formation "Ollo de Sapo" dans la région de Zamora (Espagne)-Miranda do Douro (Portugal). Comun Serv Geol Port 67: 2, 141-146

Iglesias M, Robardet M (1980) El Silúrico de Galicia media (Central): su importancia en la paleogeografía varisca. Cuad Lab Xeol Laxe 1: 99-115

IGME (ed) (1986) Exploración y caracterizacion de pizarras ornamentales en el Sistema Central (provincias de Madrid, Guadalajara y Segovia) 1ª Fase. Mem Inst Geol Min Esp, 110 p

Jiménez Fuentes E, Quiroga JL (1981) Yacimientos de graptolitos del Silúrico de la provincia de Zamora. Bol Soc Geol Port 22: 9-13

Lemos de Sousa MJ, Wagner RH (1983) General description of the terrestrial Carboniferous basins in Portugal and history of investigations. Mem Serv Geol Port 29: 117-126

Lotze F (1956) Über sardischen Bewegungen in Spanien und ihre Beziehungen zur assyntischen Faltung. Geotekt Symp Hans Stille: 129-139

Martín Escorza C (1976) Las "Capas de Transición", Cámbrico inferior y otras Series preordovícias (¿Cámbrico Superior?) en los Montes de Toledo surorientales: sus implicaciones geotectónicas. Estud Geol 32: 591-613

Martínez García E (1972) El Silúrico de San Vitero (Zamora). Comparación con series vecinas e importancia orogénica. Acta Geol Hisp 7: 104-108

Martínez García E (1973)Deformación y metamorfismo en la zona de Sanabria. Stud Geol 5: 7-106

McDougall N, Brenchley PJ, Rebelo JA, Romano M (1987) Fans and fan deltas precursors to the Armorican Quartzite (Ordovician) in western Iberia. Geol Mag 124 (4): 347-359

Mitchell WI (1974) An outline of the stratigraphy and palaeontology of the Ordovician rocks of central Portugal. Geol Mag 111 (5): 385-396

Moreno F, Vegas R, Marcos A (1976) Sobre la edad de las series ordovícicas y cámbricas relacionadas con la discordancia "Sárdica" en el anticlinal de Valdelacasa (Montes de Toledo, España). Breviora Geol Astúrica 20 (1): 8-16

Müller W (1929) Die Fauna der Frasnes-Stufe bei Almadén (Sierra Morena, Spanien) Abh Senckenb Naturforsch Ges 41 (5): 243-282

Pardo MV, García Alcalde JL (1984) Bioestratigrafía del Devónico de la región de Almadén (Ciudad Real, España). Trab Geol 14: 79-120

Paris F (1979) Les Chitinozoaires de la Formation de Louredo, Ordovicien Supérieur du Synclinal de Buçaco (Portugal). Palaeontogr Abt A 164: 24-51

Paris F (1981) Les Chitinozoaires dans le Paléozoïque du sud-ouest de L'Europe (Cadre géologique-Étude systématique-Biostratigraphie). Mém Soc Géol Min Bretagne 26: 412 p

Peran M, Tamain G (1967) La formation "Campana" dans le Nord de la Province de Jaén (Espagne). CR Acad Paris (Ser A) 265: 844-847

Pickerill RK, Romano M, Meléndez B (1984) Arenig trace fossils from the Salamanca area, western Spain. Geol J 19: 249-269

Pieren AP (1986) Mobilité du bassin sédimentaire au Silurien inférieur dans le synclinal de Herrera del Duque. (Badajoz, España). 11ème Réunion des Sciences de la Terre, Clermont-Ferrand. Comun Soc Géol Fr édit Paris 1: 148

Pillola GL, Gutiérrez Marco JC (1988) Graptolites du Tremadoc du sud-ouest de la Sardaigne (Italie). Géobios 21 (5): 553-565

Pineda Velasco A (1987) La Caliza Urbana (Ordovícico superior) y sus tramos volcanoclásticos en el subsuelo del norte de El Centenillo (Jaén). Bol Geol Min 98 (6): 780-793

Portero JM, Dabrio C (1988) Evolución tectosedimentaria del Ordovícico y Silúrico de los Montes de Toledo Meridionales y Campo de Calatrava. II Congr Geol Esp, Granada 1: 161-164

Poupon G, Tamain G (1969) Les "Quartzites Campana" et le Néo-Dévonien Est Marianique (Espagne). CR Acad Sci Paris 268: 475-478

Puschmann H (1967) Zum Problem der Schichtlücken im Devon der Sierra Morena (Spanien). Geol Rundschau 56: 528-542

Puschmann H (1970) Das Paläozoïkum der Nördlichen "Sierra Morena" am Beispiel der Mulde von Herrera del Duque (Spanien). Geol Dtsch 19 (3): 309-329

Quiroga JL (1982) Estudio Geológico del Paleozoico del W de Zamora. Trab Geol 12: 205-226

Rábano I (1984) Trilobites ordovícicos del Macizo Hespérico español: Una visión bioestratigráfica. Cuad Geol Ibérica 9:267-287

Rebelo JA, Romano M (1986) A contribution to the lithostratigraphy and palaeontology of the Lower Palaeozoic rocks of the Moncorvo region, northeast Portugal. Comun Serv Geol Port 72 (1/2): 45-47

Ribeiro A (1974) Contribution à l'étude tectonique de Trás-os-Montes Occidental. Mem Serv Geol Port 24 (Nova Serie) 168 p

Robardet M (1981) Late Ordovician tillites in the Iberian Peninsula. In: Hamberg MJ, Harland WB (eds) Earth's pre-Pleistocene glacial record. Cambridge Univ Press, pp 585-589

Robardet M, Doré F (1988) The late Ordovician diamictic formations from southwestern Europe: North-Gondwana glaciomarine deposits. Palaeogeogr Palaeoclimatol Palaeoecol 66: 19-31

Robardet M, Vegas R, Paris F (1980) El techo del Ordovícico en el centro de la Península Ibérica. Stvd Geol Salm 16: 103-121

Rodriguez Núñez VM, Gutiérrez Marco JC, Sarmiento GN (in litt) Rasgos bioestratigráficos de la sucesión silúrica del Sinclinal del Guadarranque (Provincias de Cáceres, Badajoz y Ciudad Real). COL-PA 42: 24 p

Romano M (1974) The palaeoenvironment and ichnology of the Lower Ordovician rocks at Apúlia, north Portugal. Bol Mus Lab Mineral Geol Fac Ciênc, Lisboa 14 (1): 63-76

Romano M (1982) The Ordovician biostratigraphy of Portugal - a review with new data and reappraisal. Geol J 17: 89-110

Romano M, Brenchley PJ, McDougall D (1986) New information concerning the age of the beds immediately overlying the Armorican Quartzite in central Portugal. Géobios 19 (4): 421-433

Romariz C (1962) Graptolitos do Silúrico portugués. Rev Fac Ciênc, Lisboa (Ser C) 10 (2): 115-312

Romariz C (1969) Graptolitos silúricos do Noroeste Peninsular. Comun Serv Geol Port 53: 107-156

Sarmiento GN (in litt) Conodontos de la Zona Ordovicicus (Ashgill) en la Caliza Urbana, Corral de Calatrava (Ciudad Real). Geogaceta 7

Saupé F (1971) La série Ordovicienne et Silurienne d'Almadén (province de Ciudad Real, Espagne). Point des connaissances actuelles. In: Colloque Ordovicien-Silurien. Brest Mém BRGM 73: 355-365

Saupé F (1973) La géologie du gisement de mercure d'Almadén. Sci Terre Mém 29:342 p

Seara Valero JR (1986) Datación relativa y tectónica de la

Formación de la Ermita de San Roque (Galicia central, NW del Macizo Hespérico). Cuad Lab Xeol Laxe 10: 263-281

Sopeña A (1979) Estratigrafía del Pérmico y Triásico del noroeste de la provincia de Guadalajara. Semin Estratigr Ser Monogr 5: 329 p

Tamain G (1967) El Centenillo, zone de référence pour l'étude de l'Ordovicien de la Sierra Morena Orientale (Espagne). CR Acad Sci Paris (Ser D) 267: 389-392

Tamain G (1971) L'Ordovicien est-marianique (Espagne). Sa place dans la province méditerranéenne. In: Colloque Ordovicien-Silurien. Brest Mém BRGM 73: 403-416

Tamain G (1972) Recherches géologiques et minières en Sierra Morena orientale (Espagne). Thesis, Univ Paris Sud (Centre d'Orsay) 3: 870 p

Teixeira C (1954) Os Conglomerados do Complexo xisto-grauváquico ante-silúrico. Sua importancia geológica e paleogeográfica. Comun Serv Geol Port 35: 33-49

Teixeira C (1955) Notas sobre geología de Portugal. O Complexo Xisto Grauváquico Ante-Ordoviciano. Emprenta Lit Fluminense. Ed Autor, Lisboa, 50 p

Teixeira C, Gonçalves F (1980) Introdução á Geología de Portugal. Inst Nac Invest Cient, Lisboa, 475 p

Teixeira C, Pais J (1973) Sobre a presença de Devónico na região de Bragança (Guadramil e Mofreita) e de Alcañices (Zamora). Bol Soc Geol Port 18: 199-202

Thadeu D (1947) Trilobites do Silúrico de Louredo (Buçaco) Bol Soc Geol Port 6 (3): 217-236

Truyóls Massoni M, Quiroga JL (1982) Tentaculites Dacrioconáridos en el sinforme de Alcañices (Prov Zamora). Cuad Lab Xeol Laxe 2: 171-175

Vegas R, Roiz JM, Moreno F (1977) Significado del Complejo Esquisto-grauváquico en relación con otras series "Pre-Arenig" de España Central. Stud Geol 12: 207-215

Wagner RH (1983) The palaeogeographical and age relationships of the Portuguese Carboniferous floras with those of other parts of the Western Iberian Peninsula. Mem Serv Geol Port 29: 153-177

Wagner RH (1985) Upper Stephanian stratigraphy and palaeontology of the Puertollano Basin, Ciudad Real, Spain. An Fac Ciênc Univ Porto 64 (Spec Vol): 171-231

Wagner RH, Lemos de Sousa MJ (1983) The Carboniferous megafloras of Portugal. A revision of identifications and discussion of stratigraphic ages. Mem Serv Geol Port 29: 127-152

Young TP (1988) The lithostratigraphy of the Upper Ordovician of central Portugal. J Geol Soc London 145: 377-392

2.2 Structure

M.A. DIEZ BALDA[1], R. VEGAS[2]
and F. GONZALEZ LODEIRO[3]

2.2.1 Introduction

The Central-Iberian Zone (CIZ) is considered to be the area contained in the Central-Iberian Zone defined by Julivert et al. (1972), with the exception of the recently defined Galicia-Tras-Os-Montes Zone (Farias et al. 1987; Arenas et al. 1988), which constitutes its allochthonous part. The boundaries of the CIZ have been somewhat modified and include the whole of the Ollo de Sapo Anticlinorium and the area contained between the granitic lineation of Los Pedroches-Cáceres and the Badajoz-Cordoba Shear Zone (Fig. 1).

The southern boundary of the CIZ was situated by Julivert et al. (1972) in the Los Pedroches batholith (and in its continuation towards the NW), perhaps owing to the spectacular linearity of this granitic outcrop, the situation of the original geotectonic divisions established by Lotze (1945), based on the differences in the sedimentation of the Lower Paleozoic and the distribution of areas of intense granitization or high grade metamorphism. However, from both the structural and stratigraphic points of view, the southern limit of the CIZ should be situated in the Badajoz-Cordoba Shear Zone.

Regarding the northern boundary, Julivert et al. (1972) proposed the axial trace of the Ollo de Sapo Anticlinorium. However, no changes occur on both sides of this structure, either in the stratigraphic or in the structural characteristics, whereas important stratigraphic changes do occur towards the E and NE of the Anticlinorium. Consequently, we include the whole of the Ollo de Sapo Anticlinorium in the CIZ and situate its northern boundary on the Vivero Fault and the Sil-Truchas Syncline, as proposed by Martínez Catalán (1985); see West Asturian-Leonese Zone, this Vol.).

Two different tectonic domains can be differentiated in the CIZ, taking into account the megascopic structures of the first Hercynian phase of deformation (Fig. 1):

1. The Domain of Recumbent Folds;
2. The Domain of Vertical Folds.

The outline of the boundary between both domains is imprecise, owing to the large number of granitic intrusions that do not permit the determination of the geometry of some of the first Hercynian-phase folds.

To a certain extent, this subdivision (although the boundaries do not coincide) is reminiscent of the classic division made by Lotze (op. cit.) with the more metamorphic Castillian-Galician Zone to the N and the less metamorphic eastern Portuguese-Alcudian zone to the S. These two zones established by Lotze were later encompassed within the CIZ (Julivert et al., op. cit.), although different tectonic styles have suggested the appropriateness of making an internal subdivision, despite the identity of the Paleozoic and Precambrian series. Since in the present work a structural criterion for the subdivision of the CIZ is used, it is unnecessary to consider the higher grade of metamorphism of the northern sectors (Castillian-Galician Zone of Lotze) nor the epimetamorphic nature of the southern sector (the eastern Portuguese-Alcudian Zone of Lotze) as differentiation criteria. In fact, terrains with a high metamorphic grade do outcrop in the Domain of Vertical Folds in the core of antiforms (Martinamor, Cespedosa, Piedrahita, Toledo), whereas in the Domain of Recumbent Folds, there are extensive zones of low-grade metamorphism (the Alcañices and Majaelrayo synforms, for instance; see Fig. 1).

[1] Departamento de Geología, Universidad de Salamanca, 37008 Salamanca, Spain
[2] Departamento de Geomorfología y Geotectónica Universidad, Complutense de Madrid, 28040 Madrid, Spain
[3] Departamento de Geodinámica Universidad de Granada, 18071 Granada, Spain

R. D. Dallmeyer and E. Martínez García (Eds.)
Pre-Mesozoic Geology of Iberia

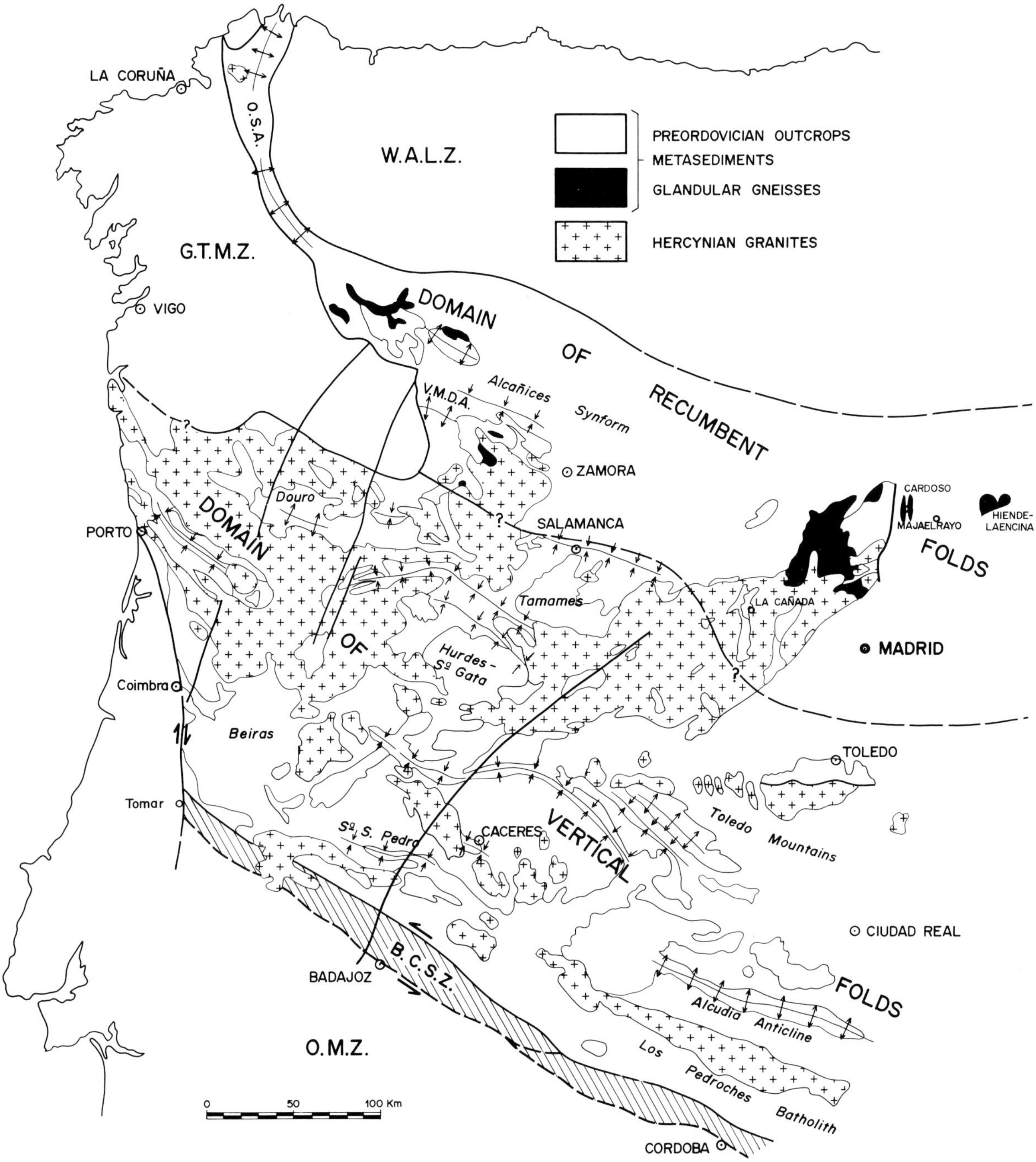

Fig. 1. Structural domains of the Central-Iberian Zone: Domain of Recumbent Folds to the north and Domain of Vertical Folds to the south. *G.T.M.Z.* Galician-Tras-Os-Montes Zone; *W.A.L.Z.* West Asturian-Leonese Zone; *O.S.A.* Ollo de Sapo Anticlinorium; *V.M.D.A.* Villadepera-Miranda Do Douro Antiform; *B.C.S.Z.* Badajoz-Córdoba Shear Zone; *O.M.Z.* Ossa Morena Zone. Only the Hercynian basement has been represented

2.2.2 The Pre-Hercynian Deformation

The pre-Hercynian deformation is deduced from the appearance of two angular unconformities visible at outcrop and map scales. The oldest unconformity separates two stratigraphic units of Upper Proterozoic age; the most recent one is situated between the Lower Ordovician and Lower Cambrian or Upper Proterozoic sediments. Although both pre-Hercynian unconformities are observed in the

Domain of Vertical Folds, in the Domain of Recumbent Folds only that between the Ordovician and its substrate has been identified.

2.2.2.1 The Precambrian Deformation

The Pre-Ordovician occupies almost two thirds of the extension of the Domain of Vertical Folds and has been extremely difficult to date and correlate owing to its relative monotony and the absence of macrofossils. It has received the following names: Ante-Ordovician-Xisto-Grauváquico Complex, in Portugal (Carrington da Cósta 1950; Teixeira 1955) and also in its Spanish prolongation in Salamanca and Cáceres "Complejo Esquisto-Grauváquico" (Schist-Greywacke Complex; García de Figuerola 1970; Rodríguez Alonso 1979; Díez Balda 1980); the Valdelacasa series and Transition Layers (in the sense of transition to the Cambrian) in the Montes de Toledo (Lotze 1961) and the Alcudian in the S part of the CIZ (Tamain 1972). Moreover, the Precambrian series that outcrop to the S of the Pedroches Batholith are known as the Serie Negra (Black Series) (Vegas 1968) and according to Parga and Vegas (1971) and Vegas (1980) are equivalent to the Valdelacasa Series and to the Lower Alcudian. There are two macrounits in the Alcudian separated by a unconformity (Bouyx 1970): the Lower Alcudian, which is a monotonous flysch-type sequence of pelites and greywackes, with a thickness of 6000 to 7000 m and an Upper Rifeiense age, and the Upper Alcudian, which is more varied lithologically, with intercalated conglomerates and limestones, a thickness of 4000 to 5000 m and an Upper Vendian age (Ortega and González Lodeiro 1986).

In the N part of the Domain of Vertical Folds (Salamanca and Cáceres) and to the W, in the Douro region (Portugal), microfossils have been found in the Schist-Greywacke Complex (algae, acritarchs) and it is possible to establish correlations with the Montes de Toledo Units (Rodríguez Alonso 1979, 1985; Díez Balda 1980, 1986; Bernardo de Sousa 1981). Until now, the intra-Alcudian unconformity has not been found in this N part of the domain and there is a gradual transition from the upper part of the complex to the fossiliferous Lower Cambrian. We therefore feel that the Schist-Greywacke Complex of Salamanca, that of the Douro region in Portugal and that of the NW of the province of Cáceres must correspond to the Upper Alcudian and more recent formations, while that of Central Portugal (Las Beiras) and the rest of the province of Cáceres could correspond to the Lower Alcudian.

The Precambrian deformation is revealed by the clearly visible unconformity that separates the Upper Alcudian rocks from those of the Lower Alcudian. This unconformity was described in the Rio Esteras Anticline by Bouyx (1970). A similar unconformity was reported by Crespo and Rey (1971) in the Alcudia Anticline, although the lack of knowledge of the age of the upper rocks (with limestones and conglomerates) led Parga and Vegas (1971) to attribute a younger age (Sardic) for this unconformity. Recently, Ortega and González Lodeiro (1986), Ortega et al. (1988), have described this unconformity in the Tirteafuera (Abenojar), Alcudia and Esteras Anticlines (Fig. 5), pointing to its regional extension and its location between the two sedimentary groups of the Alcudian. This Precambrian deformation would have produced a weak folding unaccompanied by schistosity or regional metamorphism and it must be Lower or Middle Vendian in age (Ortega and González Lodeiro, op. cit.).

2.2.2.2 The Sardic Deformation

This deformation was proposed by Lotze (1956) to account for the occurrence of several series under the Arenigian Quartzites. This author assumed the existence of two unconformities in the area that constitutes the Domain of Vertical Folds: an upper one (Iberian unconformity) between the Armorican Quartzite and the basal red beds of the Lowermost Ordovician (Upper Cambrian, according to Lotze, op. cit.) and another lower one (Toledo unconformity) separating the red beds (or Intermediate Series) from the Lower Cambrian rocks. The lower unconformity has been described as a clearly visible unconformity between the Precambrian and Intermediate Series in the Alcudia Anticline by Richter (1967). Later, Moreno et al. (1976) described it in the Navas de Estena Syncline (Fig. 2), suggesting a Middle-Upper Tremadoc age for the intermediate layers and a Lower Cambrian age for the underlying ones. In a recent work, Roiz and Vegas (1980) have described the regional extension of this unconformity, which also seals an important calc-alkaline volcanism in the Fuente el Fresno and Cortijos de Malagón region (Fig. 6; Muñoz et al. 1985).

Regarding the Upper Sardic unconformity, this has been described in the Cañaveral Syncline (Fig. 2; Martín Herrero and Bascones 1982), where an intermediate series outcrops, and in many other places where the intermediate series of the Tremadoc are missing. However, even in these cases, the basal conglomerate of the Armorican Quartzite may

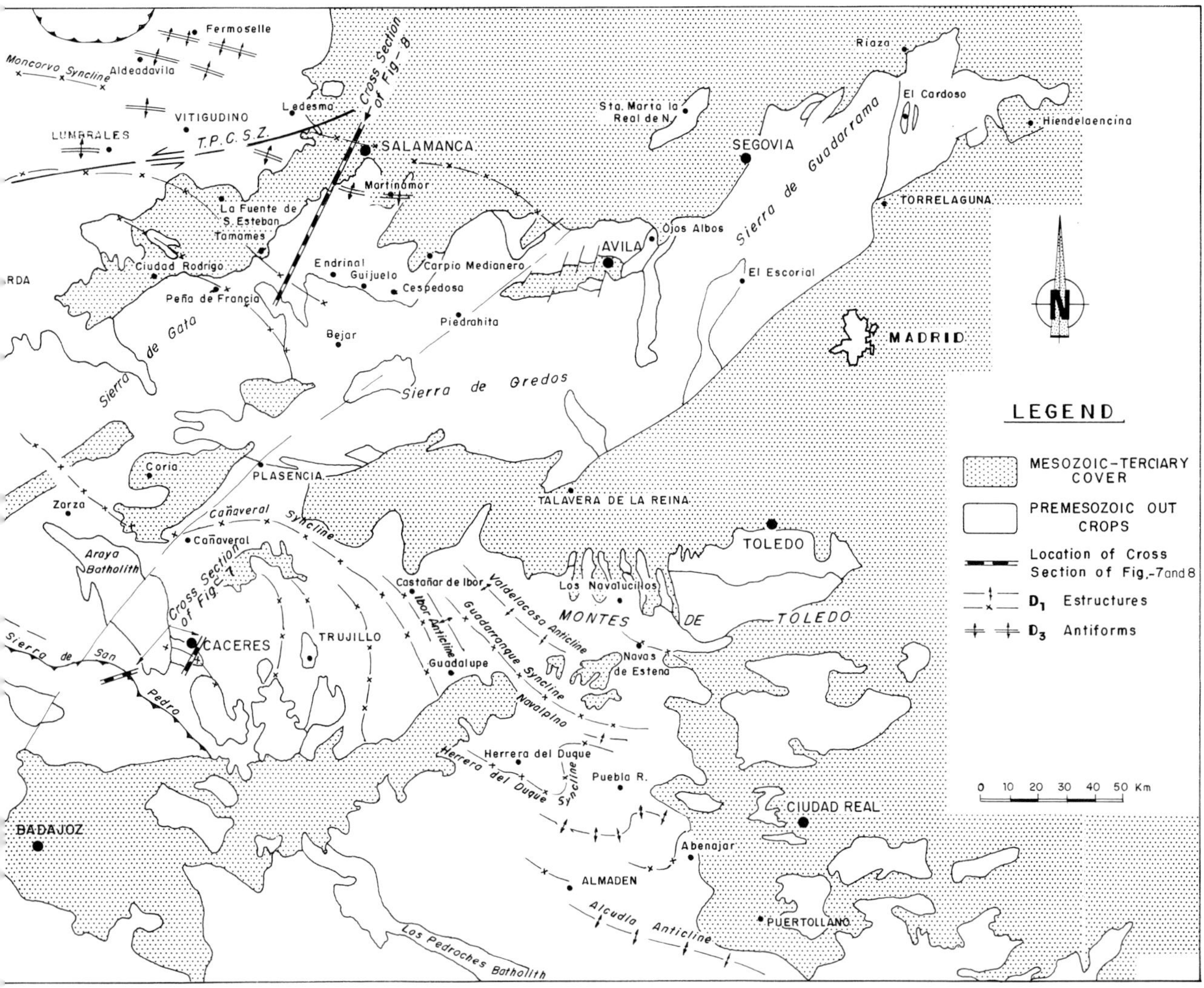

Fig. 2. Simplified geological map for the location of main structures, geographic names of interest cited in the text and location of cross-sections of Figs. 7, 8. *T.P.C.S.Z.* Traguntia-Penalva do Castelo Shear Zone

to a certain extent be a representative of these Series, such as in the Tamames Syncline, for instance. In these cases, variation in thickness and even the occasional disappearance of the conglomeratic series could be related to the presence of the upper unconformity. In any case where the conglomeratic series does exist, their bottom is considered to represent the lower unconformity.

According to such considerations, one can admit that the unconformity at the base of the Tremadoc is the most important one and that it is related to the Sardic movements, which generated folding structures. Concerning the geometry of the Sardic folds, Oen Ing Soen (1970) has described the interference between these folds and those of the Hercynian D_1 in the Las Beiras region, ascribing a N-S to NE-SW direction to the Sardic folds. For the Tras-Os-Montes region, Ribeiro (1974) suggested a box-fold type geometry with large subhorizontal limbs and short subvertical limbs with E-W to NE-SW hinges. Roda (1986) describes the Sardic folds to the south of the Araya batholith (Fig. 2) as being of the angular type, with an interlimb angle of 70° - 120°, a wavelength of about 100 m and an amplitude of 30 m; the Sardic deformation did not generate schistosity and it does not have an associated metamorphism.

In the Domain of Recumbent Folds, the Sardic unconformity has been identified in the Hiendelaencina region (González Lodeiro 1980) in Ojos Albos (Capote 1973) and in Santa Maria la Real de Nieva (Alvarez 1982; Figs. 2, 3). Accordingly, the Sardic unconformity characterizes the whole Central-Iberian Zone, as defined in the present work.

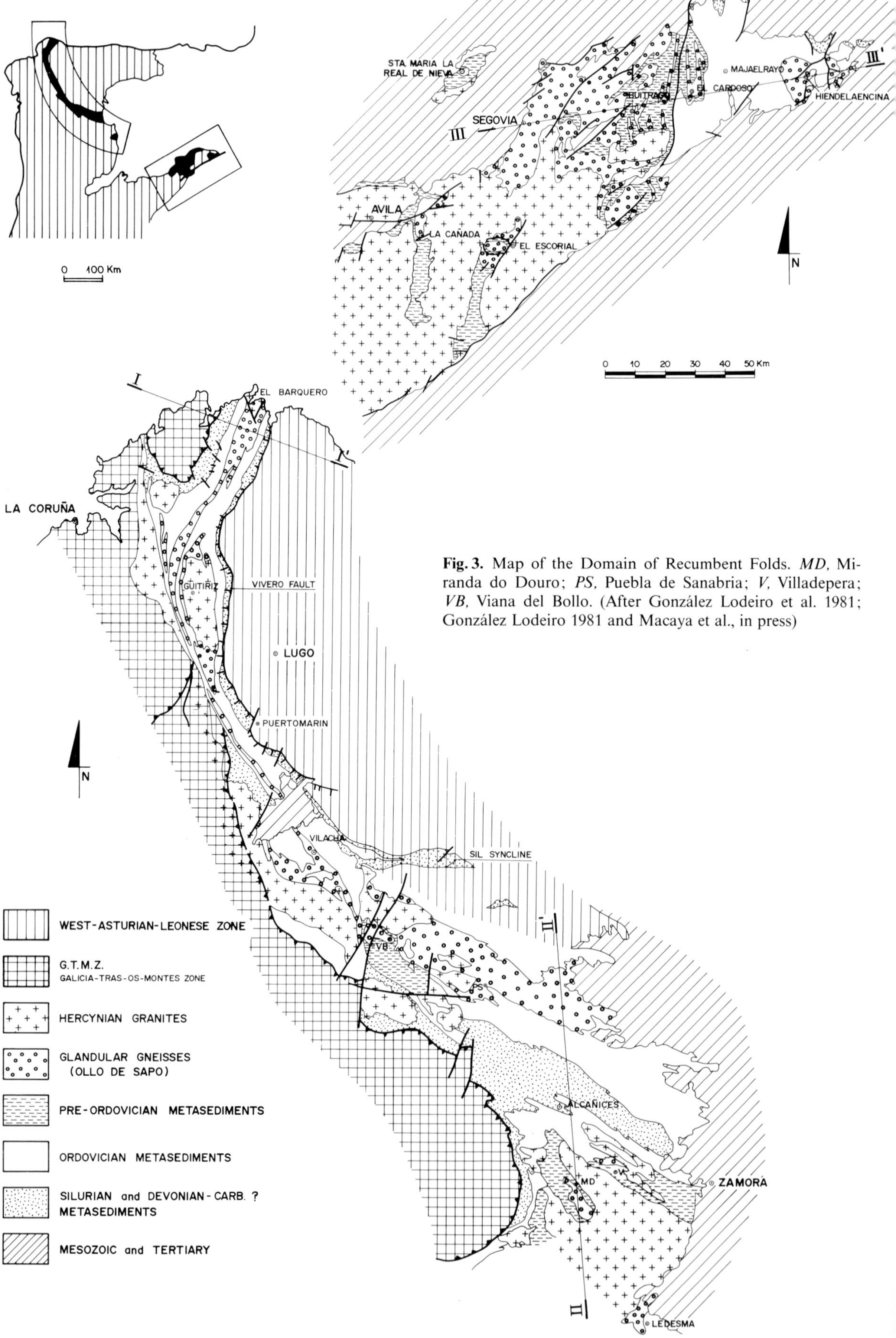

Fig. 3. Map of the Domain of Recumbent Folds. *MD*, Miranda do Douro; *PS*, Puebla de Sanabria; *V*, Villadepera; *VB*, Viana del Bollo. (After González Lodeiro et al. 1981; González Lodeiro 1981 and Macaya et al., in press)

2.2.3 The Domain of Recumbent Folds

The structures comprising this domain outcrop in two sectors: one in the NW of the peninsula and the other in the eastern part of the Spanish Central System (Sierra de Guadarrama and eastern edge of the Sierra de Gredos; Figs. 2, 3). In the former, there are three main structures which, from NE to SW, are as follows: the Ollo de Sapo Anticlinorium, the Alcañices Synform and the Villadepera-Miranda do Douro Antiform (Fig. 1). In the second, from E to W, three large structures can also be recognized: the Hiendelaencina Anticlinorium, the Majaelrayo Synform and a large complex antiformal structure that extends from El Cardoso to the La Cañada outcrops (Figs. 1, 3). The structures of both sectors are, to a large extent, correlatable; in this sense, the Hiendelaencina Anticlinorium and the Majaelrayo Synform could be correlated with the Ollo de Sapo Anticlinorium and the Alcañices Synform respectively.

The most important stratigraphic features of this domain are the existence of glandular gneisses, situated in the lower part of the sequence; infra-Ordovician detritic formations with intercalated dolomites and limestones over which the Lower Ordovician rests unconformably, and mainly detritic Paleozoic formations with a few intercalations of carbonate layers and of acid volcanic rocks in the Upper Ordovician, Silurian and Devonian.

Of all the formations, those that have posed the greatest problems regarding their origin, age and meaning are the glandular gneisses.

In the Ollo de Sapo Anticlinorium, including its prolongation in the Central System (Hiendelaencina Anticlinorium), two types of glandular gneiss formations can be seen: the Ollo de Sapo Formation, so-named by Parga Pondal et al. (1964) and whose name (toad eye) stems from the bluish quartzes observed in it, and another that outcrops in the Viana del Bollo and Hiendelaencina regions and that has received several names: In the Viana del Bollo region: Porto gneisses and San Sebastian gneisses (Martínez García 1973), and, more recently, the Viana del Bollo orthogneisses (Iglesias et al. 1981; Lancelot et al. 1985). In the Hiendelaencina region it was formerly called the Antoñita gneiss (Schäfer 1969).

The Ollo de Sapo Formation is composed of glandular gneisses with small, lenticular, bluish quartz grains and feldspar megacrysts that, in the upper part, intercalate with microglandular gneisses and quartzitic sedimentary rocks (Parga Pondal et al. 1964; González Lodeiro 1980).

Both the macroglandular and microglandular gneisses are of igneous origin. A plutonic or subvolcanic origin has been assigned to the macroglandular gneisses (de Waard 1950; Bischoff et al. 1973; González Lodeiro 1980), while the microglandular gneiss must correspond to subaerial lavas and tuffs deposited in a platform environment together with sandstones and shales (Parga Pondal et al., op. cit.; Riemer 1966; Schäfer 1969; Navidad 1978; González Lodeiro 1980). The intrusion of the macroglandular gneisses has been dated by Bischoff et al. (1986) at 580 Ma. This age was obtained by the U/Pb method in zircons.

The Viana del Bollo and Antoñita gneisses are mesoglandular biotite-bearing gneisses cut by deformed aplite and pegmatite veins; they arise from pre-Hercynian granites (Iglesias et al. 1981; González Lodeiro 1980). The aplites and pegmatites represent their related dykes (González Lodeiro, op. cit.; Navidad and Peinado 1981). The Viana del Bollo gneiss has been dated with the U/Pb method in zircons by Lancelot et al. (1985). These authors propose an age of ± 465 Ma for their emplacement, which shows that the gneiss is intrusive in part of the Ollo de Sapo Formation, as has been suggested by other authors (González Lodeiro 1980; Iglesias and Ribeiro 1981).

In the Villadepera-Miranda do Douro Antiform and in the large antiform to the W of the Majaelrayo Synform (El Cardoso-La Cañada Antiform), several types of gneiss outcrop: those of Villadepera and El Cardoso can be correlated with the microglandular facies of the Ollo de Sapo Formation. Others, such as those of Miranda do Douro (Ribeiro 1974) and Morcuera (Fernández Casals 1974), derive from granite massifs that intruded into the infra-Ordovician metasediments and that are equivalent to the Ollo de Sapo macroglandular facies. According to Lancelot et al. (op. cit.), the Miranda do Douro gneisses were emplaced 618 Ma ago.

In agreement with the foregoing, the radiometric ages point to the existence of two magmatic events. The oldest, calk-alkaline in nature, would have given rise to the emplacement of late kinematic granites related to the Cadomian orogeny and is represented by the Ollo de Sapo type gneisses. The second, alkaline in nature, occurring around 470 Ma ago, would have been related to a pre-Ordovician distensive phase (González Lodeiro 1980; Lancelot et al., op. cit.).

2.2.3.1 *The Structure of the Ollo De Sapo Anticlinorium*

This structure stretches from El Barquero to the Hiendelaencina region and is mainly the result of the superimposition of the D_1 and D_3 Hercynian deformation phases (Figs. 3, 4). In some cases it is also possible to observe bands of heterogeneous ductile deformation and thrust faults attributed to the second Hercynian phase. The D_1 folds have an average N-S to NW-SE direction and E vergence. The D_3 folds have the same direction and an axial surface which varies from subvertical to inclined towards the E. The interference between D_1 and D_3 is of type 3 of Ramsay (1967). The detailed structure of this anticlinorium varies from some sections to others (Figs. 3, 4); in some cases the Ollo de Sapo occupies the core of D_1 anticlines and in others the core of D_3 antiforms (Iglesias 1981; González Lodeiro 1981; Barrera et al, in press; Naval and Iglesias 1981; Pablo Macia et al. 1981).

This anticlinorium between El Barquero (Lugo) and Sanabria (Zamora) is composed of a series of folds which, from N to S, are: El Barquero, Guitiriz, Portomarin, Vilachá, Queija and Sanabria (Figs. 3, 4). The Ollo de Sapo gneisses outcrop in the core of all these structures with the exception of the NE part of the Sanabria Antiform, where infra-Ordovician metasediments, probably of Lower Cambrian or Upper Precambrian age (Viana Series), and the Viana del Bollo orthogneisses (Iglesias 1981) appear beneath the Ollo de Sapo Formation.

In the Sanabria Antiform the structure corresponds to a D_1 fold with NE vergence, affected by a D_2 shear zone in its deepest part that would have

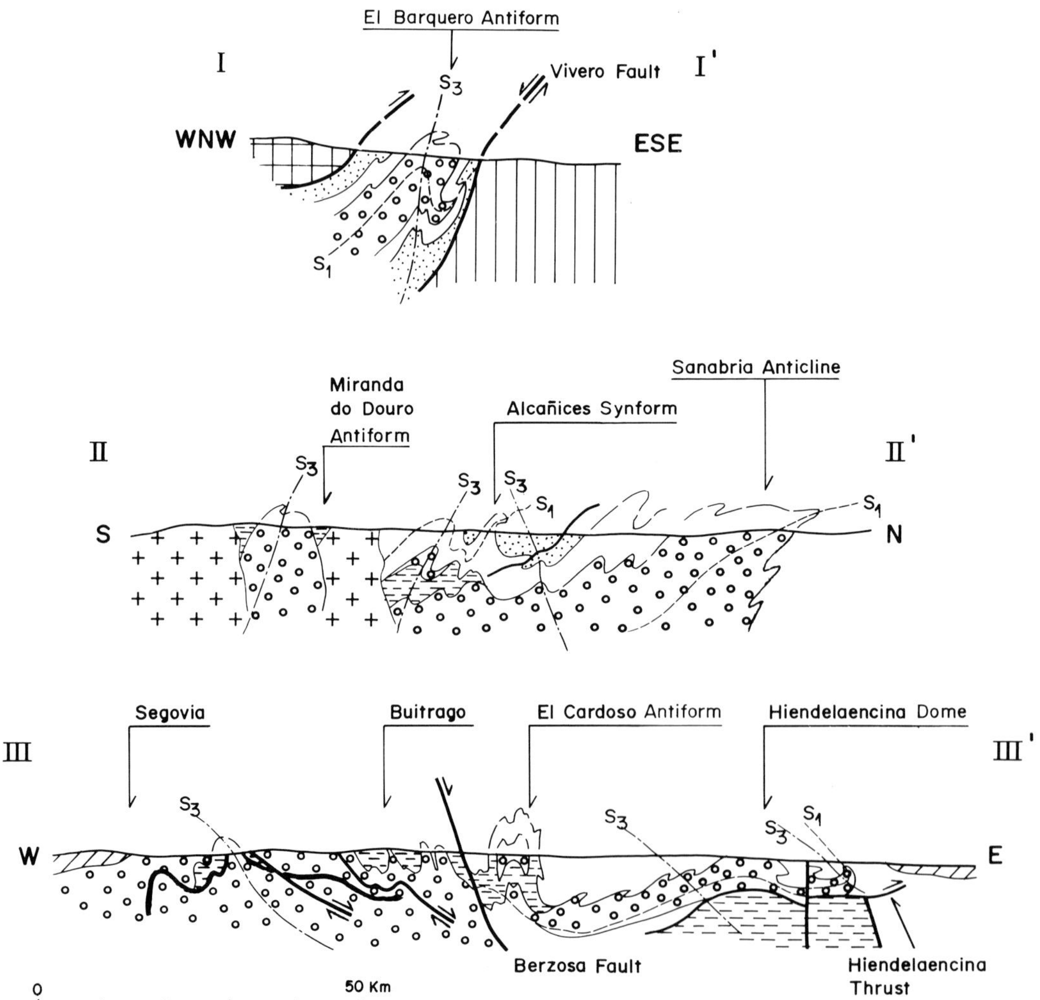

Fig. 4. Cross-sections in the Domain of Recumbent Folds. For location and legend see Fig. 3

deformed the Viana del Bollo orthogneisses and the encompassing metasediments, the whole D_1 plus D_2 structures having been refolded by D_3.

On the E part of the Sierra de Guadarrama (Fig. 2), the Hiendelaencina Anticlinorium, as in previous cases, is a D_1 structure intensely affected by later deformations. The D_1 structure is a recumbent anticlinorium with a N160° E direction and a vergence towards the E; it was affected on its reverse limb by a D_2 shear zone, and was refolded by D_3 and by another very late phase of N130° E direction and a subvertical axial plane, D_4, which gave rise to the creation of a dome (González Lodeiro 1980). In this case the D_2 deformation produced a 1 km thick ductile shear zone associated with a thrust fault (González Lodeiro 1981).

Owing to D_1, minor folds and an S_1 foliation, which varies from a slaty cleavage in the zones of very low or low grade metamorphism to a schistosity in those of intermediate grade, were formed.

As mentioned above, D_2 produced shear bands and a thrust fault in the Hiendelaencina zone. The minor D_2 structures outcrop in this structure and also in the Puebla de Sanabria Antiform (Fig. 3), where strongly asymmetric and isoclinal sheath folds associated with planolinear fabrics developed together with S-C tectonites and milonitic rocks, which point to a sense of movement towards the E.

The minor D_3 structures are distributed heterogeneously and are best developed in the hinge zones of the larger folds. S_3 is a crenulation cleavage distributed in bands.

2.2.3.2 The Structure of the Alcañices and Majaelrayo Synforms

These two synforms can be correlated with each other, since both are adjacent to the Ollo de Sapo Anticlinorium (Figs. 1, 3).

The Alcañices Synform is a D_3 structure that was developed on the normal limb of the Puebla de Sanabria D_1 Anticline which limits it to the NE; rocks ranging in age from Ordovician to Lower Carboniferous outcrop in its core (Quiroga 1981; Vacas and Martínez Catalán 1987; Figs. 3, 4). The normal limb of the D_1 structure is some 15 km in width, and was affected by D_2 structures such as ductile shear zones and thrust. The minor D_1 folds have N120° E direction and their axial surface dips variably as a consequence of the D_3 folding. They show an associated S_1 cleavage. The D_2 structures are only seen in narrow bands associated with thrust faults, with a translation sense towards the E. The D_2 deformation in these bands produced asymmetric folds, a crenulation cleavage and mylonitic rocks. D_3 generated folds associated with a crenulation cleavage that dips towards the NE. They are homoaxial with D_1 folds but opposite in vergence (Vacas and Martínez Catalán 1987). Later, sinistral wrenchfaults appeared parallel to the previous structures.

The Majaelrayo Synform is a D_3 structure with N-S direction and a subvertical axial surface developed on the normal limb of the Hiendelaencina D_1 Anticline. Its core is occupied by Upper Silurian deposits. The D_1 and D_3 structures are similar to those described for the Alcañices Synform, but in this case, no D_2 structures have been described up until now.

2.2.3.3 The Structure of the Villadepera-Miranda Do Douro and The El Cardoso-La Cañada Antiforms

The Villadepera-Miranda do Douro Antiform was described by Iglesias and Ribeiro (1981). According to these authors, it is the result of the superimposition of D_1 and D_3 folds; in this case the reverse limb of the D_1 folds is not visible and only a large refolded normal D_1 limb can be seen. The D_1 folds are recumbent and isoclinal, with a N110° E direction and E-vergence and are accompanied by a foliation parallel to their axial surfaces. The D_3 folds are homoaxial with the former but with an axial surface subvertical or dipping some 70°-80° towards the SW. S_3 is a crenulation cleavage. A subvertical foliation developed during the D_3 phase in the granitic rocks, together with ductile shear zones with a N150° E to N155° E direction; these are subvertical and show a horizontal movement in the dextral sense (Iglesias and Choukroune 1980).

The antiform situated between El Cardoso and the La Cañada massif (Fig. 3) also corresponds to a D_3 structure. Its geometry remains to be elucidated, owing to the large amounts of granitic rocks that intruded into it and because of the Late Hercynian NE-SW wrench faults that were reactivated as faults with vertical movements, which considerably hinder observation. However, the succession of deformation phases is the same as that described in the Hiendelaencina area.

D_1 macrostructures are only seen in the Ordovician rocks of the eastern limb of the large antiform and in the outcrops of Santa Maria la Real de Nieva and Ojos Albos (Fig. 2). They are folds with a N150° E direction, an original vergence to the NE and, in low-grade zones, are associated with an S_1

slaty cleavage in the slates or a rough cleavage in the sandstones. The S_1 is a schistosity in intermediate- and high-grade zones, where it is normally obliterated by the later foliations. D_2 is characterized by the development of discontinuous shear bands that affected the metasediments and orthogneisses and was accompanied by the development of mylonitic foliations (Macaya 1983; Villaseca 1983). The D_2 ductile deformation affects extensive zones but not the upper structural levels (zones of Santa Maria la Real de Nieva and Ojos Albos, for instance). In the metasediments, the D_2 ductile deformation gave rise to strongly asymmetric folds, sometimes of sheath type, and a stretching lineation with a N 130° E direction. In the orthogneisses, which are abundant in the western part of the megastructure, the D_2 ductile deformation gave rise to narrow mylonitic bands. The mechanical behaviour of metasediments and orthogneisses has recently been described by Macaya et al. (in press). According to these authors, the orthogneisses are so abundant that they can be considered as a crystalline basement. The main mechanical differences between this basement and its metasedimentary cover arose during the D_2 phase, in which narrow, mylonitic shear zones developed in the basement, whereas a wide ductile shear zone was created in the lower part of the schistose sequence.

D_3 gave rise to backfolds to which a crenulation cleavage is associated in the metasediments and low grade orthogneisses and a schistosity in the high-grade orthogneisses. The most important D_3 structures are the El Cardoso Antiform, until now interpreted as a D_1 structure (Fernández Casals 1974; González Lodeiro 1980; González Casado 1987), and those that fold the mylonitic bands of the central and western part of the structure (Macaya 1983; Macaya et al., in press); they are folds with N-S to N 130° E direction and a vergence to the west.

Later, the deformation became more brittle and gave rise to normal faults with dextral components (Macaya et al., in press) such as the Berzosa Fault, with N-S direction and a dip towards the east. Previously, this fault had been interpreted as a D_2 thrust (Capote et al. 1977; González Lodeiro 1980; Arenas et al. 1982; González Casado 1987), but the fact that it cross-cut the D_3 folds of El Cardoso indicates a post D_3 origin for this structure (Macaya et al., in press).

Other late structures are gentle folds with a large curvature radius and N-S and E-W direction that led to the formation of interference patterns such as open domes and basins.

2.2.4 The Domain of Vertical Folds

This domain occupies most of the CIZ and stratigraphically is characterized by the large amounts of Precambrian rocks, the presence in these materials of an unconformity that separates two large units (Bouyx 1970; Ortega and González Lodeiro 1986), the absence of an Upper and Middle Cambrian and the transgressive nature of the Lower Ordovician. Deposits of Lower Cambrian age have been dated with trilobite and archaeociate faunas in Endrinal and Tamames (Salamanca), in Los Navalucillos, Urda and Los Cortijos de Malagón (Montes de Toledo) (Figs. 2, 6; Perejón 1972; García de Figuerola and Martínez García 1972; Perejón et al. 1976; Zamarreño et al. 1976; Brasier et al. 1979).

In structural terms, of interest are the narrow synclines well marked by the Armorican Quartzite and the broad antiforms occupied by the pre-Ordovician rocks. Upon examining a geological map of this domain, at first sight the synclines are seen to be highly regular and continuous in their NW part and much more digitate and irregular in their SE part. The Tamames Syncline, for example, in the NW part of the Domain is more than 250 km long while its width is hardly 5 km. In the SE part of the Domain, folding interferences are very frequent and domes and basins and more complicated cartographic shapes can be observed (Julivert et al. 1983).

The pre-Hercynian deformations did not produce any appreciable internal deformation or metamorphism, and it was the Hercynian deformation that was responsible for the strain and metamorphism observed.

The overall structure can be accounted for in terms of the superimposition of three consecutive deformation phases plus some later activity due to subvertical shear zones or to faults. The regional metamorphism is of low or very low grade in most of the Domain, and only in the core of some third-phase antiforms are there intermediate- or high-grade metamorphic rocks, deformed granitic rocks and orthogneisses. Upon studying the metamorphic rocks in these cores, two metamorphic gradients can be seen: the first of intermediate P and the second of lower P and higher temperature (Martínez 1974; Gil Ibarguchi and Martínez 1982).

2.2.4.1 The First Deformation Phase (D_1)

The first Hercynian deformation stage produced subvertical folds with a large wavelength (several km) and an associated foliation, S_1, parallel to their

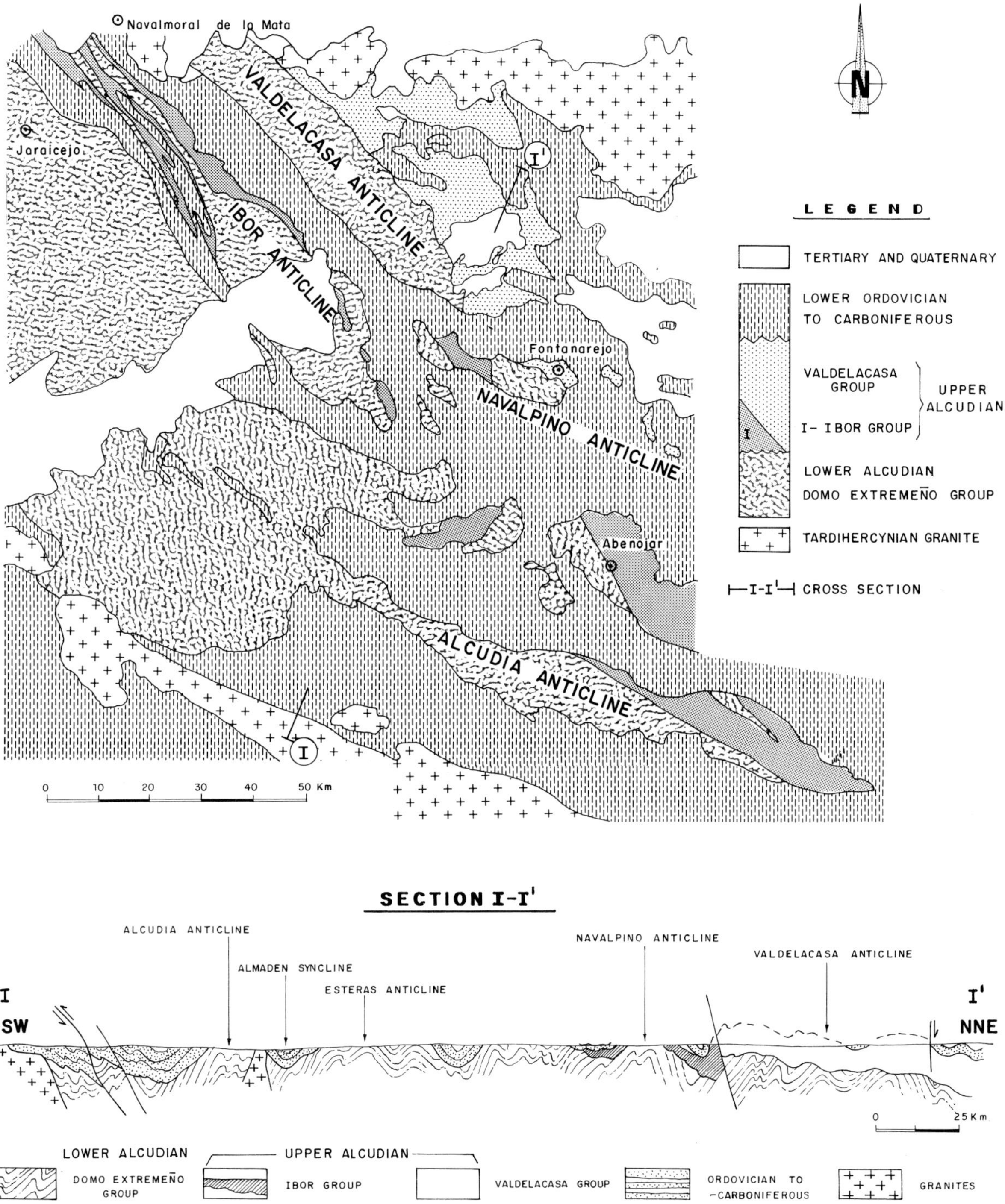

Fig. 5. Geological sketch map and cross-section through the SE part of the Domain of Vertical Folds to show the pre-Hercynian unconformities (After Ortega et al. 1988, modified)

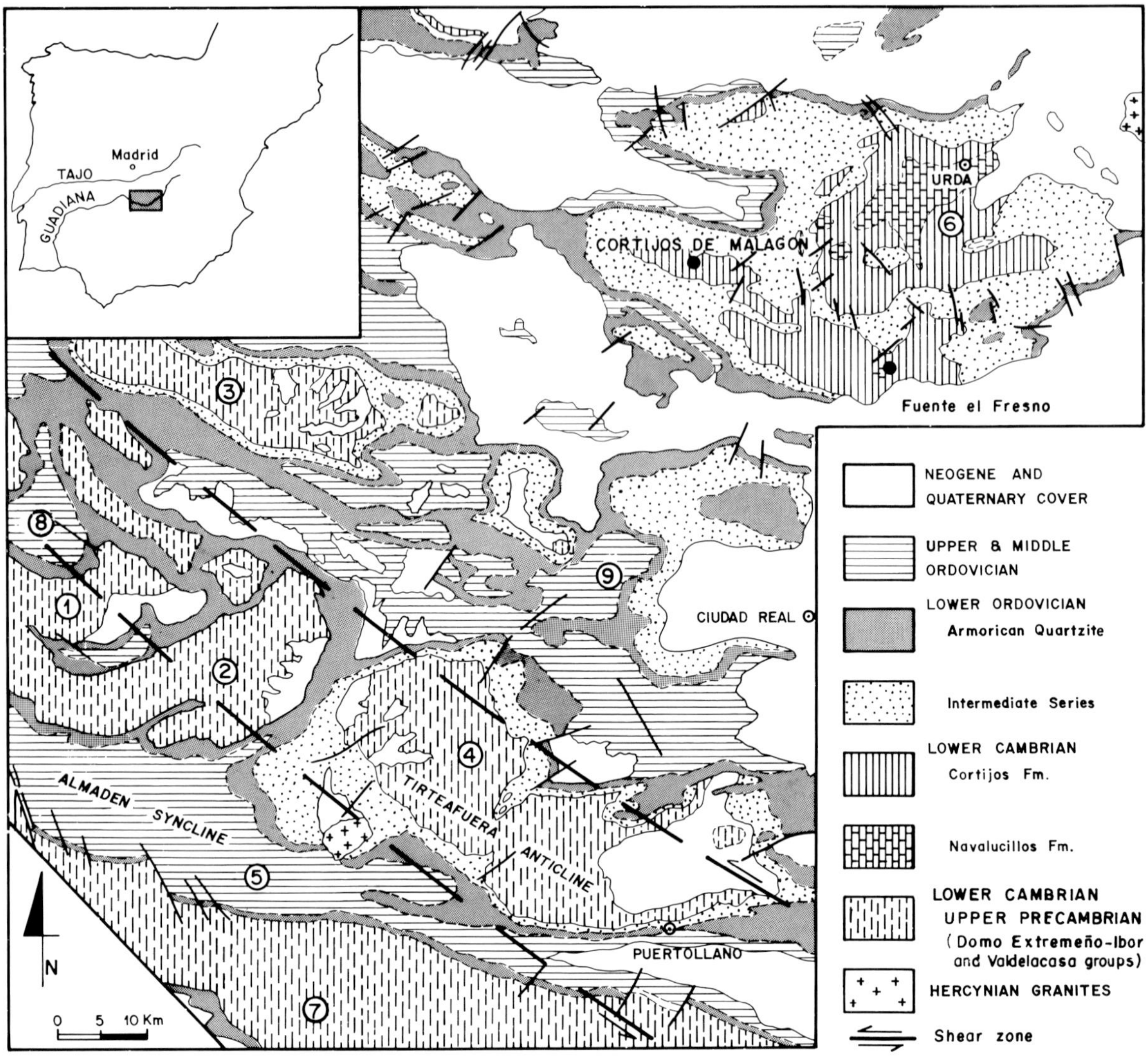

Fig. 6. Structural sketch map of the SE part of the Domain of Vertical Folds. *1* Agudo Anticline; *2* Esteras Anticline; *3* Navalpino Anticline; *4* Tirteafuera Dome; *5* Almadén Syncline; *6* Urda Dome; *7* Alcudia Anticline; *8* Herrera del Duque Syncline; *9* Piedrabuena Basin. (After Julivert et al. 1983; location of vertical shear zone after Ortega, in press)

axial plane that is not always penetrative. This phase gave rise to the most important folds of this domain such as the Synclines of Salamanca, Tamames, Peña de Francia, Cañaveral, Cáceres, Sierra de San Pedro, Guadarranque and Almadén, and the Anticlines of Valdelacasa, Ibor, Urda, Esteras, Trujillo, Navalpino, Tirteafuera and Alcudia (Díez Balda 1981; Macaya 1981; Tena 1980; Martín Herrero and Bascones 1982; Bascones et al. 1980; Vegas and Roiz 1979; Roiz and Vegas 1980; Julivert et al. 1983, Figs. 2, 5, 6, 7, 8).

The direction of the large first-phase folds is variable, with a predominance of NW-SE directions; however, the first foliation and the axial traces may adopt anomalous directions (N-S to NE-SW) over extensive sectors such as to the NW and S of Ciudad Rodrigo, Fig. 2, (Corretgé and López Plaza 1976; Rodríguez Alonso 1985), or to the E of Tamames (Díez Balda 1986), to the E of Cáceres (Castro 1986) or in the region N of Almadén (Roiz 1979; Ortega, in press), due to deviations of the original directions produced by later shearings. In other areas of this sector it is not easy to establish the direction of the first fold axes with any degree of precision, because they are found as almost perfect domes or basins (Julivert et al. 1983; Fig. 6).

The inclination of the axes of the D_1 folds varies considerably; vertical axes are common in the pre-Ordovician, owing to the existence of pre-Hercynian folds, while in the Ordovician and younger rocks

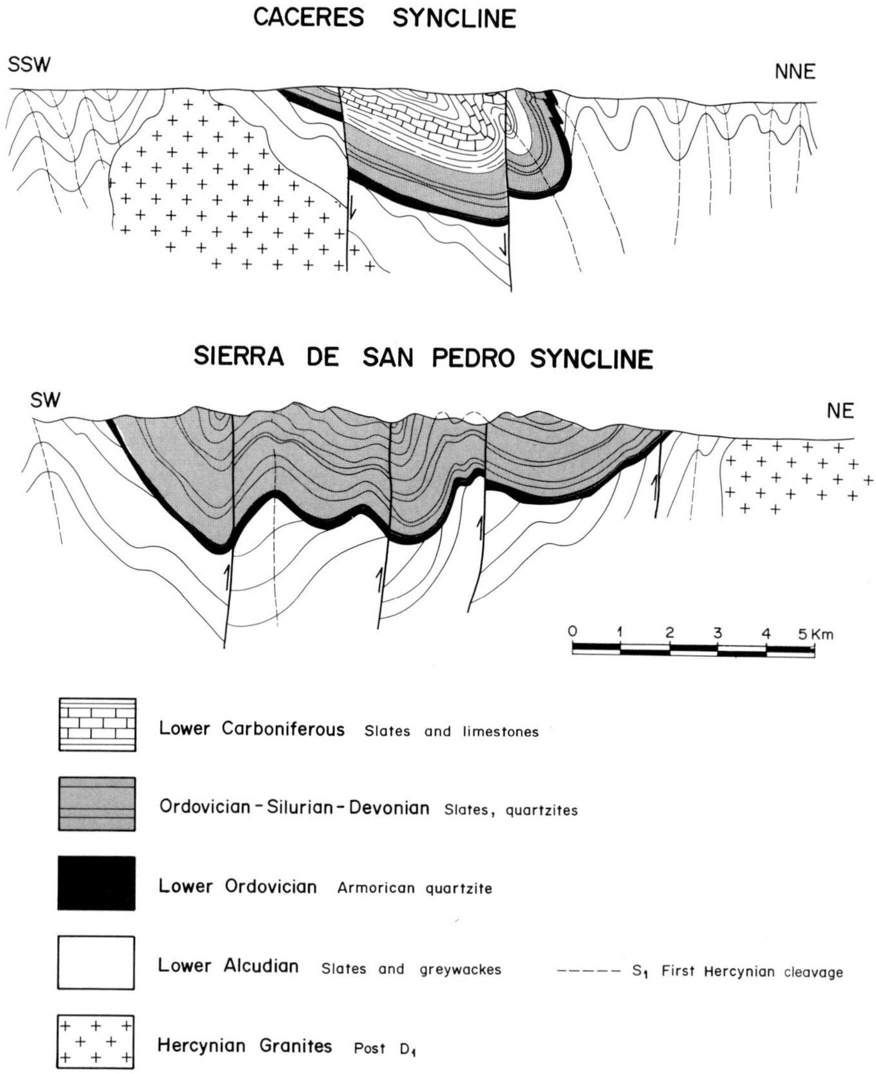

Fig. 7. Cross-sections to the south of the Domain of Vertical Folds after Tena (1980) and Bascones et al. (1980) respectively. For location see Fig. 2

the fold axes are frequently subhorizontal, with the exception of the easternmost zone of the domain (in the Esteras, Tirteafuera and Agudo Anticlines and in the Almadén Syncline for example; Fig. 6).

The S_1 foliation is the most evident planar structure in most of the outcrops of this domain; its characteristics vary considerably, according to the metamorphic grade and type of rock. In the low-grade quartzites, it may be a non-penetrative spaced cleavage or a rough cleavage. In the slates it is commonly a slaty cleavage and ranges to a coarse-grained schistosity in the schists.

Normally, S_1 is almost vertical and approximately parallel to the axial plane of the D_1 folds; however, in some cases the schistosity observed in the outcrop cross-cuts both limbs of the folds, as, for example, in the Almadén Syncline. This kind of arrangement is related to the fact that the schistosity appeared somewhat later than the fold itself, in relation with a sinistral WNW-ESE shearing (Aller et al. 1986).

On the basis of the measurements of internal deformation performed by Díez Balda (1986) in an area of low grade metamorphism to the S of Salamanca, it may be concluded that during the D_1 episode this area underwent a shortening in the NNE-SSW direction, which was reflected in the genesis of NW-SE folds and in the development of a preferentially $S > L$ planar fabric associated with a finite strain ellipsoid with a k value of 0.3 in the

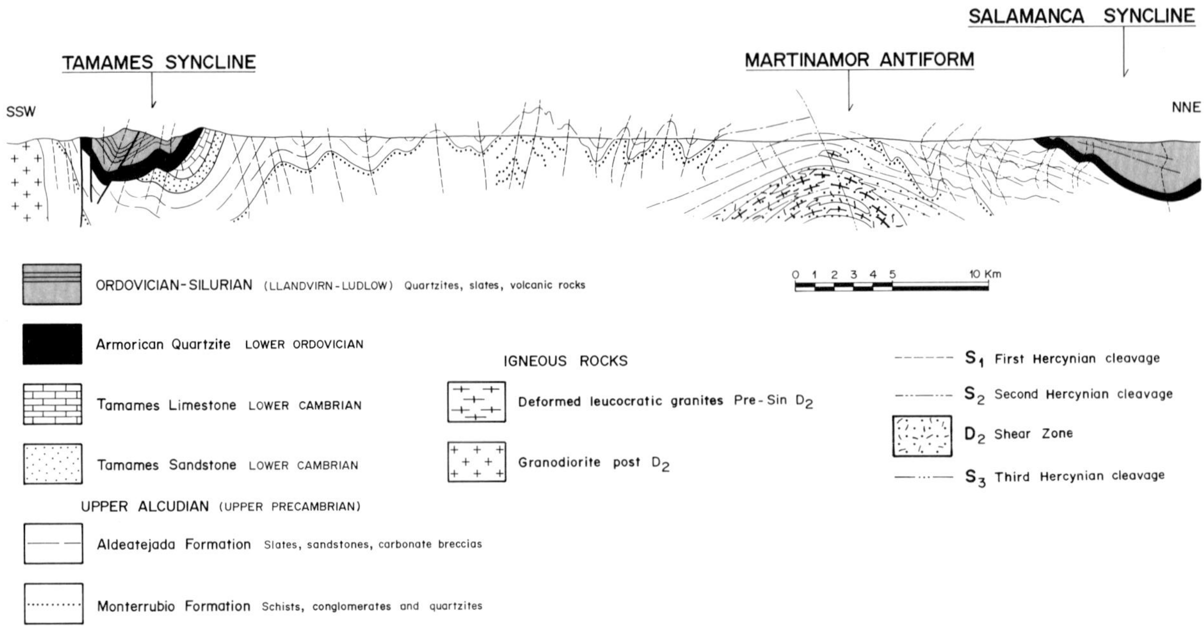

Fig. 8. Cross-section to the north of the Domain of Vertical Folds after Díez Balda (1986). For location see Fig. 2

slates and a planolinear fabric with a predominance of L > S fabric associated with a finite strain ellipsoid with a k value of 1.6 in the conglomerates. The stretching direction is parallel to the axes of the D_1 folds. To the S of Salamanca it has a N 100° E to N 130° E direction, and a plunge of 10° NW. Probably there was a regional gradient, and the D_1 deformation would have been rather heterogeneous. However, at present this cannot be corroborated, owing to the lack of more strain measurements.

2.2.4.2 The Second Deformation Phase (D_2)

In some areas of the domain it is possible to recognize a second deformation phase that originated new structures which were superimposed over those of D_1, introducing into the rocks an internal deformation and fabric that overprinted those already existing.

D_2 is characterized by the generation of wide subhorizontal shear zones that outcrop exclusively in the northern part of this domain, in the antiformal cores of the third-phase folds such as, for example, in the Martinamor Antiform (Figs. 2, 6). In these areas, which show a medium to high metamorphic grade, D_2 led to strongly asymmetric minor folds and sheath folds that affected the S_1 cleavage and developed a new subhorizontal foliation: S_2. D_2 also affected the pre-Hercynian or Early Hercynian granitic rocks, producing a clearly visible subhorizontal foliation and a stretching lineation in them.

The geometry and tightness of the D_2 folds varies from the boundaries to the interior of the shear zones, where very tight sheath folds can be seen.

According to the study of the folded lineations in D_2 folds and also from the application of Hansen's (1971) method to a zone strongly deformed by simple shear to the SE of Salamanca, a flow direction of N 124° E to N 136° E and a dip of 1° to 29° NW are deduced. The flow plane of this simple shear would have been subhorizontal (Díez Balda, op. cit.).

The D_2 deformation was superimposed over the one already existing in the rocks (D_1). Díez Balda (1983) investigated the strain path followed by samples of deformed conglomerates, from areas out of the D_2 shear zone to areas that were highly deformed by D_2, and compared it with theoretical strain paths calculated for certain conditions of superimposition between D_1 and D_2 in pure or simple shear regimes. The k value obtained for the D_2 ellipsoid varies from 0.7 to 1, with an X/Z value of 10. The value of shear would have been at least 3.5 for the most intensely deformed samples. Upon being superimposed over the first, the D_2 ellipsoid gave rise to a marked stretching of the rocks and the development of purely linear fabrics, since the stretching directions of D_1 and D_2 coincided; in the most intensely deformed zones, a subhorizontal flattening plane appeared (Díez Balda 1983).

The S_2 cleavage in the shear zones is frequently the only schistosity in the outcrop visible to the naked eye. In thin sections it can be observed that it ranges from a very strongly developed crenulation cleavage with tectonic banding to a coarse-grained schistosity. In these zones S_2 appears microfolded, due to the effect of later deformations.

In the core of the Martinamor and Carpio-Medianero antiforms (SE of Salamanca) and in other D_3 antiformal folds to the NW of Salamanca (W and S of Fermoselle; Fig. 2), the Hercynian granitic rocks and perhaps even older granitic rocks appear strongly deformed with well-developed fabrics. Reference is made here to the tonalitic gneiss of Fermoselle, the adamellitic San Pelayo gneiss in the Martinamor Antiform and the deformed granitic rocks of Martinamor and Cespedosa (Fig. 2). The foliation of these gneisses appears folded by the third phase and is parallel to the S_2 of the metamorphic host rock. The fabric of these rocks is sometimes preferentially planar and often planolinear, with a well-developed N 100° E to N 130° E stretching lineation; we interpret these rocks as having been affected by the D_2 episode (Díez Balda 1986).

In some outcrops (Cespedosa, Martinamor; Fig. 2), it is possible to observe S-C structures, with subhorizontal intersection between S and C surfaces, which suggest a shear sense of movement towards the E for the upper parts with respect to the lower ones. However, until now no associated thrust fault surface has been detected.

In epizonal areas, very loose angular folds with a rectilinear hinge and subhorizontal axial plane are seen affecting the S_1 cleavage. In these areas the internal deformation due to D_2 is negligible. The weak subhorizontal crenulation cleavage observed in these areas is called S_2, because it is the second cleavage that appears in the rocks. We believe that in these areas D_2 has an extensional character, and that it is related to extension produced above a shear zone in a convergent regime (see for example Platt 1986).

2.2.4.3 *The Third Deformation Phase (D_3)*

D_3 produced folds with a very large wavelength (up to a few tens of kilometres) and a small amplitude, which affect the S_2 schistosity. The folds have a N 100° E to N 120° E direction and a subvertical axial plane, and are accompanied by minor folds and an associated S_3 crenulation cleavage. These folds bend the metamorphic isograds and the foliated granitic rocks. In the core of the antiformal folds of this phase, such as the Martinamor Antiform (Fig. 8), the Antiform to the S of Cespedosa, the Fermoselle Antiform and the Vitigudino Antiform (Salamanca; Fig. 2), there are outcrops of medium- to high-grade metamorphic rocks, migmatites, orthogneisses and numerous granitic rocks, mainly two-mica granites, aplites and pegmatites, more or less deformed by D_2.

S_3 is a weak crenulation cleavage with a N 100° E direction and a subvertical dip. This cleavage can be easily observed in areas where S_2 is well developed; that is, in intermediate- or high-grade micaschists and in granites deformed by D_2. In these cases S_3 is parallel to chevron-type microfolds. In the low-grade areas, however, because S_2 does not appear or is not an anisotropy surface susceptible to folding, and also because of the coincidence in orientation between S_1 and the flattening plane of D_3 in extensive sectors, S_3 is not developed. Only in areas where S_1 is deviated from its general NW-SE regional orientation is it possible to observe a crenulation of the first cleavage. This is, for example, the case to the E of Tamames, where S_1 shows a N 30° E direction and is crenulated by a subvertical N 110° E cleavage that we interpret as being S_3 (see Díez Balda et al. in press). The deviation of S_1 from its regional trend is related to the existence of subvertical, ductile shear zones with wrench components.

It seems that the dominant system of these shear bands have a N 70° E to N 130° E direction and a sinistral sense of displacement; these rotated the D_1 folds and S_1 cleavage and sometimes strongly deformed the local rocks. A well-known example is the Traguntia-Juzbado-Peñalva do Castelo shear zone (Fig. 2) that displaces the Porto-Satao-Tamames Syncline (García de Figuerola and Parga 1968). According to some authors, this shear zone would have been in operation since the first phase (Iglesias and Ribeiro 1981; Regencio Macedo and Portugal Ferreira 1981). Nevertheless, the fact that it cross-cuts the D_1 Syncline points to a post D_1 age. Furthermore, minor sinistral, subvertical shear zones with a direction of N 70° E exist in the Martinamor Antiform affecting the S_2 foliation.

In the Montes de Toledo the traces of the D_1 folds were also frequently folded by the activity of subvertical shear zones; for example, the Herrera del Duque Syncline and the Esteras Anticline (Roiz 1979; Ortega, in press) were affected by a sinistral subvertical shear zone (Fig. 6), which accounts for the interference pattern of these D_1 folds, which include the development of domes and basins. Subvertical, sinistral shear zones are common throughout this domain of the CIZ, and especially in its SW boundary they complicate the structure, as, for example, in the zone of Castelo Branco (Portugal), W of the Araya Batholith (Fig. 2; see Pereira et al 1986). On the other hand, dextral subvertical shears were also active throughout the whole domain; this would have been responsible for the folding of S_1, observed, for example, in Central Extremadura (Castro 1986). The Central-Extremadura dextral shear zone would vary in direction between N 130° E and N 170° E (Martínez Catalán and Díez Balda 1987). An WNW-ESE dextral shear zone has recently been identified to the N of the Tamames Syncline (Díez Balda et al., in press).

The detailed geometry and relative chronology of these subvertical shear zones is the object of our present work. We think that during the D_3 phase of deformation the CIZ has been shortened in the NNE-SSW direction (coherent with the distribution of S_3 crenulation). In our opinion this shortening was partially accommodated by the development of subvertical ductile shear zones.

Another very important dextral shear zone is that of Porto-Tomar, associated with the fault of the same name (Ribeiro et al. 1980).

Besides the shear zones and faults with wrench movements, late faults and shear zones of a normal type developed in the late stages of the orogen. The Toledo Fault, which is an E-W normal fault structure, is the most important of them (Hernández Enrile 1981).

2.2.5 Conclusions

The structural evolution of the CIZ includes a first phase, D_1, which produced the first continuous and penetrative deformation in both the Precambrian and Paleozoic sequences. Two domains can be differentiated according to the rotational (Domain of Recumbent Folds) or essentially non-rotational (Domain of Vertical Folds) character of this deformation phase.

A second phase, D_2, developed, probably as a continuation of D_1, but restricted to subhorizontal ductile bands, wider in the metasediments and narrower in the essentially orthogneissic rocks which occupy the lowermost part of the sequence. This phase shows a pronounced rotational character in the two domains defined. Subhorizontal shear zones and thrust faults have been identified only in the Domain of Recumbent Folds and subhorizontal shear zones in the northern part of the Domain of Vertical Folds, but both structures are lacking in the rest of this latter Domain, probably owing to the present level of erosion. The ductile shear zones outcrop in the core of D_3 antiforms and developed only in medium- to high-grade metamorphic zones. There, a change in metamorphic gradient, from an intermediate- to a low-pressure type of metamorphism, was produced at the beginning of the second phase.

In epizonal areas, the strain due to D_2 is negligible and the weak subhorizontal S_2 may be related to extensional movements.

The latter ductile deformations include the development of open to tight folds, homoaxial with those of the first phase, subvertical or dipping towards the E and NE, which were formed during the third phase of deformation, D_3. To a large extent contemporary with these folds, several subvertical shear zones developed with a sinistral and dextral wrench movement. Finally, later shear zones and faults developed, either with wrench movements or as normal accidents, probably related to the late or post-orogenic crustal thinning stage.

References

Aller J, Bastida F, Ortega E, Pérez Estaún A (1986) Aportación al conocimiento estructural del Sinclinal de Almadén. Bol Geol Min. Inst Geol Min Esp 97: 608–621

Alvarez F (1982) Contribución al estudio tectónico de la región de Santa Maria la Real de Nieva (provincia de Segovia). Temas Geologicos y Mineros. Inst Geol Min Esp Madr 6: 87–108

Arenas R, González Lodeiro F, Peinado M (1982) La zona de cízalla de Berzosa-Riaza en el sector septentrional. Influencia sobre la configuración de las zonas metamórficas. Cuad Lab Xeol Laxe 3: 123–161

Arenas R, Farias P, Gallastegui G et al. (1988) Características geológicas y significado de los dominios que componen la Zona de Galicia-Tras-Os-Montes. Simposio sobre Cinturones Orogénicos. II Congr Geol Esp, Granada, pp 75–84

Barrera JL, Farias P, González Lodeiro F et al. (in press) Mapa Geológico Nacional E. 1:200000 17–27, Orense-Verin. Edic Inst Geol Min Esp Madr

Bascones L, Martín Herrero D, Corretgé LG (1980) Mapa Geológico Nacional E. 1:50000 703, Arroyo de la Luz. Edic Inst Geol Min Esp

Bernardo de Sousa M (1981) Complexo xisto-grauvaquico do Douro: a Formacao de Bateiras: Facies ocorrencia e significado litoestratigráfico. Cuad Geol Ibérica 7: 645–651

Bischoff L, Schäfer G, Schmidt K, Walter R (1973) Zur Geologie der mittleren Sierra de Guadarrama (Zentralspanien). Münster Forsch Geol Paläontol 28: 1–27

Bischoff L, Wildberg H, Baumann A (1986) Uranium/lead ages of zircons from gneisses of the Sistema Central, central Spain. International Conference on Iberian terranes and their regional correlation. Program with abstracts 39. University of Oviedo

Bouyx E (1970) Contribution a l'etude des Formations Anteordoviciennes de la Meseta Meridionale (Ciudad Real et Badajoz). Mem Inst Geol Min Esp 73: 1–263

Brasier MD, Perejón A, San Jose MA (1979) Discovery of an

important fossiliferous Precambrian-Cambrian sequence in Spain. Estud Geol 35: 379-383

Capote R (1973) Estudio geoestructural de los afloramientos metamórficos del norte de la provincia de Avila. Bol Geol Min Inst Geol Min Esp 84: 426-437

Capote R, Fernández Casals MJ, González Lodeiro F, Iglesias M (1977) El limite entre las zonas asturoccidentalleonesa y galaicocastellana en el Sistema Central. Bol Geol Min Inst Geol Min Esp 88: 517-520

Carrington da Costa J (1950) Noticia sobre uma carta geologica do Buçaco, de Nery Delgado. Spec Publ Edic Serv Geol Port, pp 1-27

Castro A (1986) Structural pattern and ascent model in the Central Extremadura batholith, Hercynian belt, Spain. J Struct Geol 8: 633-645

Corretgé LG, López Plaza M (1976) Geología del área granitica y metamórfica al W de Ciudad Rodrigo (Salamanca). I. El Compleyo esquisto-grauváquico: estratigrafía, metamorfismo y deformación. Studia Geol 11: 121-149

Crespo V, Rey J (1971) Contribución al estudio del Valle de Alcudia. Bol Geol Min Inst Geol Min Esp 82: 512-515

Díez Balda MA (1980) La Sucesión del Complejo esquisto-grauváquico al Sur de Salamanca. Estud Geol 36: 131-138

Díez Balda MA (1981) La estructura herciniana entre Salamanca y Sequeros (Zona Centro Ibérica). La superposición de fases y su influencia en la fábrica de las rocas. Cuad Geol Ibérica 7: 519-534

Díez Balda MA (1983) Caracteristicas del elipsoide de deformación finita ligado a la segunda fase hercínica en áreas meso y catazonales del Sur de la provincia de Salamanca. Stud Geol Salmant 18: 65-80

Díez Balda MA (1986) El Complejo Esquisto-Grauváquico, las series paleozoicas y la estructura hercínica al Sur de Salamanca. Edic Univ Salamanca, pp 1-162

Díez Balda MA, García Casquero JL, Monteserín V, Nozal FB, Pardo M, Robles R (in press) Nuevos datos acerca de la estratigrafía y estructura hercínica al Sur de Salamanca. Rev Soc Geol Esp Madr 3: 117-125

Farias P, Gallastegui G, González Lodeiro F et al. (1987) Aportaciones al conocimiento de la litoestratigrafía y estructura de Galicia Central. Anais da Facultade de Ciéncias, Universidade de Porto, Memorias 1: 411-431

Fernández Casals MJ (1974) Significado geotectónico de la formación Gneises de la Morcuera. Stud Geol 7: 87-106

García de Figuerola LC (1970) La existencia de pliegues en el Complejo esquisto-grauváquico de la provincia de Salamanca. Acta Geol Hisp 5: 105-108

García de Figuerola LC, Martínez García E (1972) El Cámbrico inferior de la Rinconada (Salamanca, España Central). Estud Geol 3: 33-41

García de Figuerola LC, Parga JR (1968) Sobre los ortogneises de Tragunitia-Juzbado (Salamanca) y su significado tectónico. Acta Geol Hisp 3: 69-72

Gil Ibarguchi JI, Martínez FJ (1982) Petrology of garnet-Cordierite-Sillimatine gneisses from the El Tormes thermal dome, Iberian hercynian fold belt (NW Spain). Contrib Min Petrol 80: 14-24

González Casado JM (1987) Estudio geológico de la zona de cizalla de Berzosa-Honrubia (Sistema Central Español). Thesis, Univ Madrid

González Lodeiro F (1980) Estudio geológico estructural de la terminación oriental de la Sierra del Guadarrama (Sistema Central Español). Thesis, Univ Salamanca

González Lodeiro F (1981) La estructura del anticlinorio del "Ollo de Sapo" en la región de Hiendelaencina (extremo oriental del Sistema Central Español). Cuad Geol Ibérica 7: 535-545

González Lodeiro F, Hernández Urroz J, Klein E, Martínez Catalán JR, Pablo Maciá JG de (1981) Mapa Geológico Nacional E. 1:200000 8, Lugo. Spec Publ, Edic Inst Geol Min Esp Madr

Hansen E (1971) Strain facies (Minerals and rocks 2). Springer, Berlin Heidelberg New York

Hernández Enrile JL (1981) Evolución microestructural de rocas cuarzofeldespáticas como resultado del aumento de la deformación en la milonita de Toledo. Cuad Geol Ibérica 7: 547-566

Iglesias M (1981) Mapa Geológico Nacional E. 1:50000 9-11, Viana del Bollo. Spec Publ, Edic Inst Geol Min Esp Madr

Iglesias M, Choukroune P (1980) Shear zones in the Iberian arc. J Struct Geol 2: 63-68

Iglesias M, Ribeiro A (1981) Position stratigraphique de la formation Ollo de Sapo dans la region de Zamora (Espagne) - Miranda do Douro (Portugal). Comun Serv Geol Port 67: 141-146

Iglesias M, González Lodeiro F, Díez Balda MA, Martínez Catalán JR (1981) Some examples of ortogneisses in Central Galicia. Abstract. Leidse Geol Meded 52: 1, 117

Julivert M, Fontboté JM, Ribeiro A, Nabais Conde LE (1972) Mapa Tectónico de la Península Ibérica y Baleares E. 1:1000000. Inst Geol Min Esp Madr

Julivert M, Vegas R, Roiz JM, Martínez Rius A (1983) La estructura de la parte SE de la Zona Centroibérica, con metamorfismo de bajo grado. Libro Jubilar JM Rios, Geología de España, vol 1. Geol Min Esp, pp 477-490

Lancelot JR, Allegret A, Iglesias Ponce de León M (1985) Outline of Upper Precambrian and Lower Paleozoic evolution of the Iberian Peninsula according to U-Pb dating of zircons. Earth Planet Sci Lett 74: 325-337

Lotze F (1945) Zur Gliederung der Varisziden der Iberischen Meseta. Geotekt Forsch 6: 78-92 (Traducido an: Publicaciones Extranjeras sobre Geol Esp 5: 149-166)

Lotze F (1956) Über sardische Bewegungen in Spanien und ihre Beziehungen zur assyntischen Faltung. Geotect Symp Hans Stille, pp 129-139

Lotze F (1961) Das Kambrium Spaniens Teil I. Stratigraphie. Traducido por J Gomez de Llarena. Mem Inst Geol Min Esp 75: 1-256

Macaya J (1981) Estudio Geológico estructural de la Sierra de Francia (provincia de Salamanca y Cáceres). Cuad Geol Ibérica 7: 567-576

Macaya J (1983) Bandes miloníticas plegadas en los materiales metamórficos del Sur de Segovia. Stud Geol Salmant 18: 93-106

Macaya J, González Lodeiro F, Martínez Catalán JR, Alvarez F (in press) Continuous deformation, ductile thrusting and backfolding in the basement of the hercynian orogen and their relationships with structures in the metasedimentary cover in the Sierra de Guadarrama (Spanish Central System). Tectonophysics

Martín Herrero D, Bascones L (1982) Estudio geológico de los materiales anteordovicicos y paleozoicos del Sinclinal de Cañaveral (Cáceres) Hojas 622 y 623. Inst Geol Min Esp

Martínez FJ (1974) Petrografía estructura y geoquímica de los diferentes tipos de granitoides del NW de Salamanca (Cordillera Herciniana, España). Trab Geol 7: 61-141

Martínez Catalán JR (1985) Estratigrafía y estructura del domo de Lugo (Sector Oeste de la zona Asturoccidental-leonesa). Corpus Geol Gallaeciae 2° Serie 2

Martínez Catalán JR, Díez Balda MA (1987) Structural pattern and ascent model in the Central Extremadura batholith, Hercynian belt, Spain: Discussion. J Struct Geol 9: 381-382

Martínez García E (1973) Deformación y metamorfismo en la zona de Sanabria. Stud Geol 5: 7-106

Moreno F, Vegas R, Marcos A (1976) La edad de las series ordovicicas y Cámbricas relacionadas con la discordancia "sárdica" en el Anticlinal de Valdelacasa (Montes de Toledo, España). Breviora Geol Astúrica 20: 8-16

Muñoz M, Vegas R, Roiz JM (1985) El vulcanismo cámbrico del centro de España. Programas e Resumos 9 Reuniao de Geologia Oeste Peninsular. Porto

Naval A, Iglesias M (1981) Mapa Geológico Nacional E. 1:50000. 46, Oza de los Rios. Spec Publ, Edic Inst Geol Min Esp Madr

Navidad M (1978) Las series glandulares "Ollo de Sapo" en los sectores nord-occidental y centro-oriental des Macizo Ibérico. Estud Geol 34: 511-528

Navidad M, Peinado M (1981) Ortogneises y metasedimentos de la formación infrabasal al Ollo de Sapo (Macizo de Hiendelaencina, Guadarrama oriental). Cuad Geol Ibérica 7: 183-199

Oen Ing Soen (1970) Granite intrusion, folding and metamorphism in central northern Portugal. Bol Geol Min Inst Geol Min Esp 81: 271-298

Ortega E (in press) Geology and metallogeny of the Almaden area, Centroiberian Zone, Spain. 2nd Remote Sensing European Workshop. Publicaciones de la Comunidad Económica Europea

Ortega E, González Lodeiro F (1986) La discordancia intra-Alcudiense en el dominio meridional de la Zona Centroibérica. Breviora Geol Astúrica 27: 27-32

Ortega E, Hernández Urroz J, González Lodeiro F (1988) Distribución paleogeográfica y control estructural de los materiales anteordovicicos en la parte suroriental del autóctono de la Zona Centro Ibérica. Simposio sobre Cinturones Orogénicos. II Congr Geol Esp, pp 85-89

Pablo Maciá JG de, Martínez Catalán JR, Iglesias Ponce de León M, Hilgen JD (1981) Mapa Geológico Nacional E. 1:50000 71, Sobrado de los Monjes. Spec Publ, Edic Inst Geol Min Esp Madr

Parga JR, Vegas R (1971) Problems and discussion on Precambrian series of the Hesperic Massif (Westerniberian Peninsula). Geol Rundschau 61: 44-69

Parga Pondal I, Matte P, Capdevila R (1964) Introduction à la géologie de l'"Ollo de Sapo", Formation porphyroide antesilurienne du Nord Ouest de l'Espagne. Notas y Comunicaciones Inst Geol Min Esp 76: 119-153

Pereira AJSC, Pereira LCG, Macedo CAR (1986) Os plutonitos da zebreira (Castelo Branco): idade e enquadramento estructural. Mem Not Publ Mus Lab Geol Univ Coimbra 101: 21-31

Perejón A (1972) Primer descubrimimiento y descripción de Arqueociatos en la provincia de Salamanca. Stud Geol 4: 143-149

Perejón A, Moreno F, Vegas R (1976) Datación de las calizas del Cámbrico inferior de los Navalucillos (Montes de Toledo) faunas de Arqueociatos. Breviora Geol Astúrica 20: 33-49

Platt JP (1986) Dynamics of orogenic wedges and the uplift of high-pressure metamorphic rocks. Geol Soc Am Bull 97: 1037-1053

Quiroga JL (1981) Estudio geológico del Paleozoico del W de Zamora. Thesis. Univ of Oviedo

Ramsay JG (1967) Folding and fracturing of rocks. McGraw Hill, New York

Regencio Macedo CA, Portugal Ferreira M (1981) O sinclinal paleozoico de Esposende (Porto) - Sequeros (Salamanca): O troco entre Maroía e o Satao (Portugal Central). Cuad Geol Ibérica Madr 7: 633-644

Ribeiro A (1974) Contribution a l'etude tectonique de Tras Os Montes Oriental. Mem Serv Geol Port Nova Serie. Lisboa, pp 1-168

Ribeiro A, Pereira E, Severo L (1980) Análise da deformacao da zona de cisalhamento Porto-Tomar na transversal de Oliveira de Azeméis. Comun Serv Geol Port 66: 3-9

Richter P (1967) Stratigraphie und Tectonik in der Sierra de San Andrés. (ostiche Sierra Morena/Spanien). Münster Forsch Geol Paläontol 3: 1-144

Riemer W (1966) Datos para el conocimiento de la estratigrafía de Galicia. Notas Comun Inst Geol Min Esp 81: 7-20

Roda J (1986) Nuevos datos sobre la fase de deformación sárdica. Geometría de los pliegues prehercinicos del Rio Salor (Cáceres). Geogaceta, Soc Geol Esp 1: 13-15

Rodríguez Alonso MD (1979) El Complejo esquisto-grauváquico y los materiales ordovicicos y anterodovicicos de la zona SE de Cludad Rodrigo (Salamanca España). Stud Geol 14: 7-34

Rodríguez Alonso MD (1985) El complejo esquisto-grauwackico y el paleozoico en el centro oeste español. Edic Univ Salamanca

Roiz JM (1979) La estructura y la sedimentación herciniana, en especial el Precámbrico superior; en la región de Ciudad Real-Puertollano. Thesis, Univ Complutense de Madrid

Roiz JM, Vegas R (1980) Formaciones ordovicicas y anteordovicicas del Anticlinal de Tirteafuera (Sur de la provincia de Ciudad Real). Stud Geol Salmant 14: 27-36

Schäfer G (1969) Geologie und Petrografie im östlichen Kastilischen Hauptscheidegebirge (Sierra de Guadarrama, Spanien). Münster Forsch Geol Paläont H 10, 207 pp

Tamain G (1972) Recherches geologiques et minieres en Sierra Morena Orientale (Espagne). Thesis, Paris Sud (Centre d'Orsay)

Teixeira C (1955) Notas sobre Geologia de Portugal: O Complexo xisto-grauváquico ante-ordoviciano. Empresa Literaria Fluminense Lda, Lisboa, pp 1-50

Tena M (1980) Mapa Geológico Nacional E. 1:50000, 704/11-28, Cáceres. Inst Geol Min Esp Madr

Vacas JM, Martínez Catalán JR (1987) El sinforme de Alcañices en la transversal de Manzanal del Barco. Stud Geol Salmant 24: 151-175

Vegas R (1968) Sobre la existencia de Precámbrico en la Baja Extremadura. Estud Geol 24: 85-89

Vegas R (1980) Nuevos elementos de correlación del Precámbrico Superior del Centro y Suroeste de España y sus equivalentes en Portugal. Bol Soc Geol Port 22: 135-139

Vegas R, Roiz JM (1979) La continuación hacia el E de las estructural hercínicas de Las Villuercas, Guadalupe y Almadén (Zona Luso-Oriental-Alcudiana). Techniterrae 28: 6-10

Villaseca C (1983) Bandes miloníticas en las áreas suroccidentales de Segovia (Guadarrama Central). Stud Geol Salmant 18: 81-92

Waard D De (1950) Palingenetic structures in augen gneiss of the Sierra de Guadarrama, Spain. Bull Comun Geol Finland 23: 51-66

Zamarreño I, Vegas R, Moreno F (1976) El nivel carbonatado de los Navalucillos y su posición en la sucesión cámbrica de los Montes de Toledo occidentales (Centro de España). Breviora Geol Astúrica 20: 56-64

2.3 Granites as a Paradigm of Genetic Processes of Granitic Rocks: I-Types vs S-Types

J. M. Ugidos[1]

2.3.1 Introduction

The Central-Iberian Massif (CIM) forms part of the so-called Central-Iberian Zone (Lotze 1945; later modified by Julivert et al. 1974), in which the granites and high-grade metamorphic rocks occupy an extension of approximately 10000 km^2. The present study also deals with areas of the western part of the CIM where data of interest have been found (Fig. 1).

The range of granite (s. l.) types is broad and has variable and complex spatial and temporal relationships, although the largest volumes correspond to granodiorites and monzogranites (Streckeisen 1976) emplaced late in the Hercynian orogen. Other petrographic types such as leucogranites, quartzdiorites, diorites, gabbros and ultramafic rocks are also represented but are of less quantitative importance.

In general it may be said that, with respect to the metamorphic events, there are two large groups of granitic rocks: Early Hercynian rocks that were formed subsequent to an intermediate-pressure metamorphism and later Hercynian rocks, associated with a typical low-pressure metamorphism. Actually, such a division is not quite as simple, but it is useful to the extent that it establishes two points of temporal reference, since the metamorphism-granite relationships are not easy to pinpoint in all cases and are still problematic. Possible convergence processes involved in the evolution of the intrusive bodies also complicate attempts to systematize in petrographic, chemical and petrogenetic terms, and the scarcity or absence of isotopic data is an important limitation in making an interpretation.

[1] Departamento de Geología, Facultad de Ciencias, Universidad de Salamanca, 37008 Salamanca, Spain

Despite this, the excellent quality of many outcrops and the considerable body of evidence culled regarding many areas of the CIM have led to the accumulation of specially relevant data from which it is possible to offer a general hypothesis for the set of igneous and metamorphic events. This hypothesis may be applied to other zones, be they Hercynian or not, with similar granite typologies. The aim of the present work is thus double: to offer the most significant data regarding the granitic rocks in the CIM and to propose a model that may help to explain aspects of certain types of granitic rocks characterized by the sometimes abundant presence of Al-minerals (cordierite, andalusite, etc.), such as is the case with a large part of the CIM Hercynian rocks.

These mineralogical characteristics that in principle suggest an S origin for the granites (Chappel and White, 1974) are analyzed with respect to other types of data (granites-metamorphic host rocks interactions, petrographic criteria applied to the aluminous minerals present in both groups of rocks, types of enclaves, some isolated isotopic data, general considerations, etc.), and an alternative proposal is made. From the author's point of view, the granites and associated rocks of the CIM represent a paradigm in the petrogenesis of granitic rocks, and the present study is based on such a perspective, highlighting three main points:

1. The evolution of Hercynian metamorphism and its relationships with granite intrusions;
2. The significance of Al-minerals in the granites;
3. Some chemical data.

2.3.2 Premetamorphic Rocks in the CIM

These are mainly represented by Upper Precambrian-Lower Cambrian pelites and graywackes, among which there are levels and lenses of conglomerates, limestones, rocks with volcanoclastic participation and quartzites (Capote and Fernández Casals 1977; San José 1983; Rodríguez Alonso 1984, 1985). The

R. D. Dallmayer and E. Martínez García (Eds.)
Pre-Mesozoic Geology of Iberia

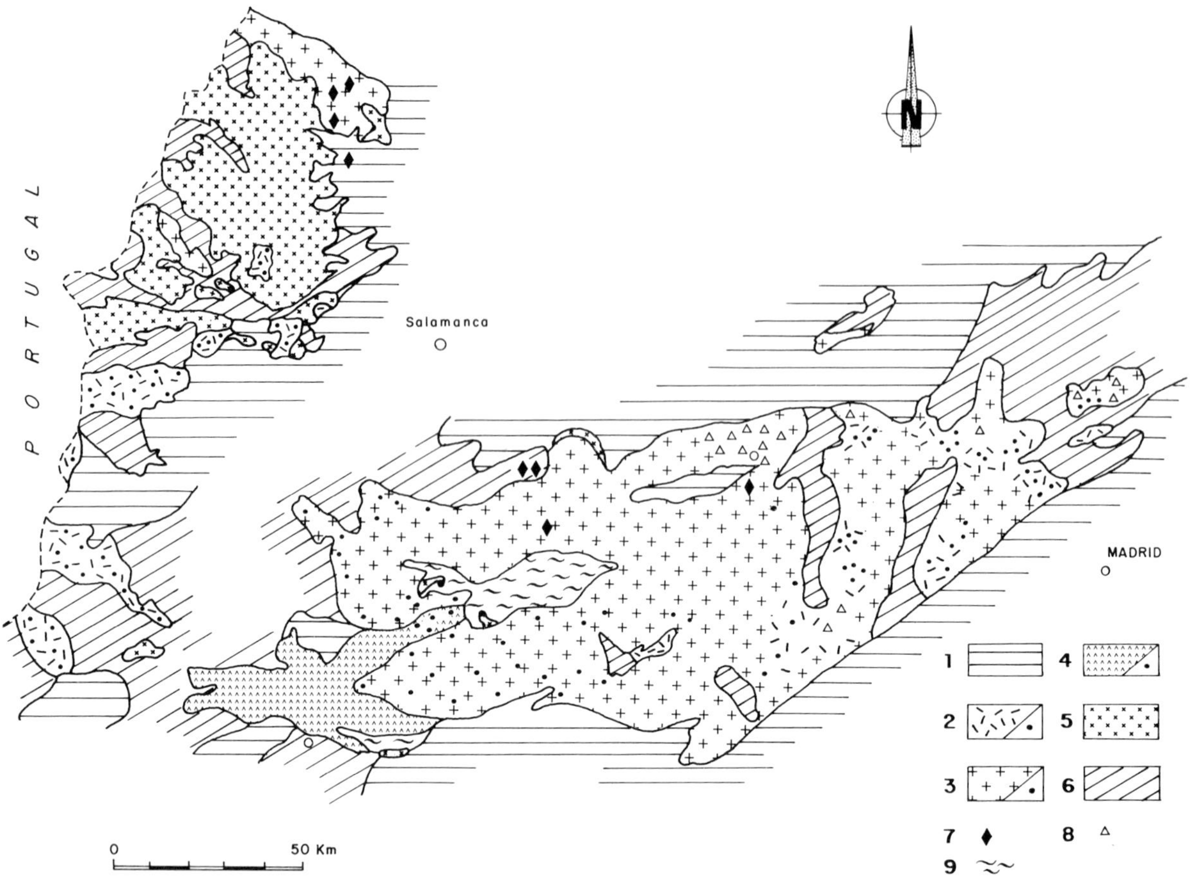

Fig. 1. Geological scheme of the granites in the CIM and central-western areas of Spain. Based on Aparicio et al. (1975), López Ruiz et al. (1975) and the Department of Petrology of the University of Salamanca (1983). *1* Post-Palaeozoic sediments; *2* biotite-muscovite granites/ ± cordierite *(dots)*; *3* biotite granites/ ± cordierite *(dots)*; *4* biotite ± muscovite ± sillimanite ± andalusite granites/ ± cordierite *(dots)*; *5* leucogranites and two-mica granites; *6* metamorphic rocks; *7* mafic and ultramafic rocks; *8* amphibole-bearing biotite granites; *9* cordierite migmatites

presence of some orthogneisses (500–470 m. y.; in Serrano Pinto et al. 1988) has been reported in these rocks (Navidad and Peinado 1981; Fuster et al. 1981; Peinado and Alvaro 1981) and also of important levels of volcanosedimentary origin (see synthesis: Capote 1983 a) characterized by the frequent presence of megacrysts of alkaline feldspar encompassed with a schistose mesostasis of quartz, feldspars and micas. Their chemical compositions vary widely (Navidad 1976), as do the proportions of felsic and femic minerals.

All these rocks were affected during Hercynian times by two or three main deformation phases, the first corresponding to the development of the regional schistosity that was affected by the two other phases to a variable extent, depending on the areas and levels considered (see Díez Balda 1986; Capote 1983 b, and references therein). In the text, reference is made to early stages (post-phase I) and late stages (post-phase II or III, depending on the authors) in order to avoid confusion for the reader.

2.3.3 Hercynian Metamorphism

The metamorphic mineral parageneses began their development in association with the regional schistosity, reaching conditions corresponding to intermediate pressure (Grt, St, Ky, Sill + Mus, Sill + Kfeld) during early stages (Garcia Cacho 1973; Fuster et al. 1974; López Ruiz et al. 1975; Navidad and Peinado 1981). These parageneses and the distribution of isograds are only appreciable in the eastern sectors of the CIM (Sierra de Guadarrama), in which important masses of late granitic intrusions are absent, whereas in other areas this metamorphic episode can only be deduced from relict crystals of staurolite, garnet and kyanite, more or less destabilized, found isolated in rocks with later parageneses (Babin 1974, 1975; Martínez 1974a; López Ruiz et al. 1975; García de Figuerola and Franco 1975; García de Figuerola et al. 1983; Kilmurray 1978; Gil Ibarguchi and Martínez 1982; Martínez and Gil Ibarguchi 1983).

During later stages, the metamorphic conditions correspond to a low-pressure metamorphism (Cord + K feld ± And ± Sill; Ugidos 1974a; Carnicero 1980, 1980–1981; and the authors above) associated with basic intrusions (Franco 1980) and large volumes of granite intrusions (granodiorites and monzogranites, Ugidos, op. cit.; Ugidos and Bea 1978), around which there are migmatites, migmatitic hornfelses and hornfelses of pyroxene hornfels facies; the outcrops of these rocks are close to each other and are common in the western sectors of the CIM (Ugidos 1987a, b). In general terms, the first metamorphism can be said to be situated above the intersection point determined by the following reaction:

$$\text{Mus} + \text{Ab} + \text{Q} \longrightarrow \text{Al}_2\text{SiO}_5 + \text{K feld} + \text{H}_2\text{O},$$

with the curve of the granitic melt, while the second metamorphism is situated below the reactions (Fig. 2):

$$\text{Biot} + \text{Mus} + \text{Ab} + \text{Q} \longrightarrow \text{Cord} + \text{K feld} + \text{H}_2\text{O},$$
$$\text{Mus} + \text{Ab} + \text{Q} \longrightarrow \text{Al}_2\text{SiO}_5 + \text{K feld} + \text{H}_2\text{O},$$
$$\text{Biot} + \text{Sill} + \text{Q} \longrightarrow \text{Cord} + \text{K feld} + \text{H}_2\text{O/melt},$$

in the case of hornfelses and cordierite-bearing migmatites (Ugidos, op. cit.). A noteworthy modification in the pressure conditions of the early to late metamorphic stages may therefore be deduced in zones to the W of the CIM, also reported by Gil Ibarguchi and Martínez (op. cit.) from the equilibrium evidence:

$$\text{Biot} + \text{Sill} + \text{Grt} + \text{Q} \longrightarrow \text{Cord} + \text{K feld} + \text{H}_2\text{O}.$$

These data allow one to consider that the Hercynian metamorphism, as a whole, is plurifacial and multiphasic, as established by Fuster et al. (1974). Between the metamorphic stages cited, intermediate low-pressure conditions have been described (Martínez and Gil Ibaraguchi 1983), suggesting a continuous evolution of the Hercynian metamorphism (considered as a whole in the Iberian Peninsula) from intermediate-pressure conditions to low-pressure conditions.

Synthesizing the more relevant aspects with respect to the objectives of the present work, the outstandings features are:

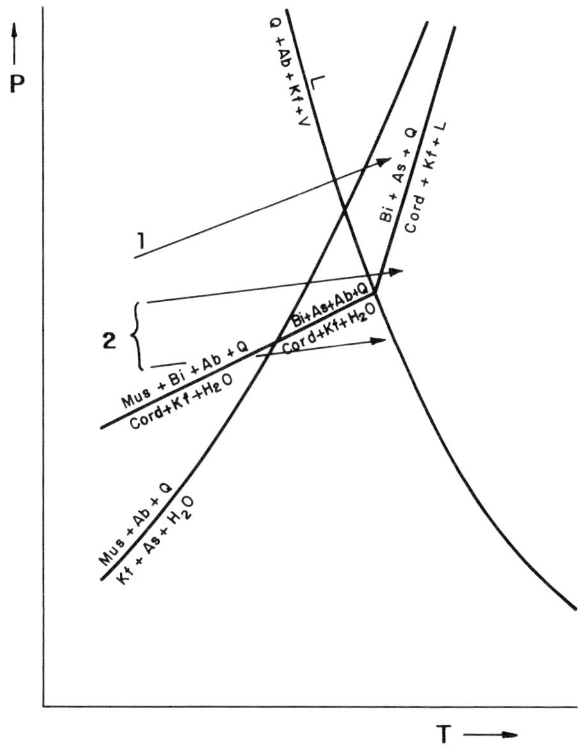

Fig. 2. Schematic representation of mineral equilibria with respect to the metamorphism of the CIM. Reactions according to Hoffer (1976, 1978) and Thompson (1982). *1* Thermal gradient during the initial stages of metamorphism; *2* range of conditions for the low-pressure stages (cordierite migmatites, hornfelses of pyroxene-hornfels facies and migmatitic hornfelses). See also Martínez (1974a), Gil Ibarguchi and Martínez (1982) and Ugidos (1987a)

1. The small extensions of migmatites and associated rocks produced in the initial metamorphic-anatectic stages compared with the extension shown by migmatites and other rocks within the zones of low-pressure mineral associations. The small extension of all these rocks compared with that of the late granites;
2. The spatio-temporal coincidence between low-pressure metamorphism and most of the late granite intrusions (Ugidos and Bea op. cit.). In general it may be stated that the high-grade metamorphic rocks with low-pressure parageneses are always associated with granite intrusions, although these later are also emplaced in levels of low-grade metamorphism where they develop hornfelses with typical poikiloblastic textures;
3. The continuous evolution of the initial mineral parageneses towards a cordierite and andalusite-bearing paragenesis (the former according to the above-mentioned reactions and the latter partly

originated by inversion of fibrolite in late metamorphic rocks; Ugidos 1987 a);
4. The evolution of cordierite crystals to prismatic forms of varying sizes in migmatites, migmatitic hornfelses and hornfelses of pyroxene-hornfels facies (granoblastic textures). These rocks contain abundant idiomorphic cordierite crystals, without inclusions of sillimanite (specially in the case of hornfelses where the cordierite may reach percentages close to 50%) or subidiomorphic-idiomorphic crystals with and without sillimanite inclusions (in migmatites whose mean cordierite contents are often in the 10–20 range). This textural evolution has been partly attributed to possible sintering phenomena favoured by the high concentration of ferromagnesian and aluminous phases in the schistosity planes of rocks rich in pelitic component (Ugidos 1985, 1987 a);
5. The relative abundance among the cordierite migmatites of nebulitic types with a granodioritic composition rich in cordierite. These rocks are often found immediately surrounding intrusive granodiorites and monzogranites that exhibit cordierite prisms and frequent migmatitic enclaves.

Table 1. Chemical data relating to leucogranites and two-mica granites[a]

	1	2	3	4	5
SiO_2	73.48	73.40	72.00	71.73	73.85
TiO_2	0.11	0.18	0.36	0.32	0.19
Al_2O_3	15.02	14.51	14.97	15.22	14.34
Fe_2O_3	0.40	–	–	–	–
FeO	0.81	*1.40[b]	*1.85	*1.55	*1.28
MgO	0.26	0.45	0.50	0.46	0.27
CaO	0.48	0.62	1.15	1.03	0.76
Na_2O	3.35	3.00	2.71	2.93	3.19
K_2O	4.83	4.82	5.84	5.74	5.21
P_2O_5	0.29	0.19	0.11	0.13	0.12
Rb	214	252	298	338	343
Sr	41	65	110	78	56
Ba	759	269	478	343	218
K/Rb	193	181	175	143	128

[a] 1 Leucogranites (n = 15; Bea and Maldonado 1980–81).
2 Two-mica granites (n = 46; Carnicero 1980; Bea et al. 1988).
3 Autochthonous two-mica granites (n = 9; Martínez 1974 b; Bea 1975; Bea et al. 1988).
4 Subautochthonous two-mica granites (n = 18; Martínez op. cit.; Bea, op. cit.; Bea et al. op. cit.).
5 Allochthonous two-mica granites (n = 30; Martínez, op. cit.; Bea, op. cit.; Bea et al. op. cit.).
[b] * FeO as total Fe.

2.3.4 Granitic Rocks

2.3.4.1 *Leucogranites and Two-Mica Granites*

Leucogranites (aplitic-medium grained; biotite <5%) and two-mica granites (biotite >5%, muscovite ≥ biotite) are the most significant group of early granites. They are autochthonous-subautochthonous and in most cases are closely related to volcanosedimentary (quartz feldspar-rich) premetamorphic levels and/or orthogneisses (Martínez 1974 a, b; Carnicero 1980, 1980–81; García de Figuerola and Franco 1975; García de Figuerola et al. 1983; Bea and Maldonado 1980–81; Peinado and Alvaro 1981; González del Tanago and Bellido 1981; Fuster and Villaseca 1979; Villaseca 1983). They are peraluminous and have a mineralogical composition of quartz, albite (oligoclase), alkali feldspar, muscovite, biotite ± sillimanite ± andalusite ± cordierite ± garnet. In these granites, microgranular enclaves are absent or very rare and they may exhibit small enclaves of biotite + sillimanite and/or nodules of cordierite. Aplitic and pegmatitic rocks are frequently associated with these granites. Chemical data are given in Table 1.

In many cases, these granites and their migmatitic context are found in zones in which the metamorphism has evolved towards low-pressure parageneses, which is consistent with the presence of andalusite and cordierite in them.

The origin of these granites has been attributed to the melting of quartz feldspar-rich levels and in any case their mesocrustal origin is widely accepted (see authors cited).

Granites to the W of the CIM show ages of between 300–326 m. y. (the granites of Lumbrales and Barruecopardo; in Serrano Pinto et al., op. cit.).

2.3.4.2 *Biotite Granites ± Muscovite*

These represent the most abundant group of granitic rocks in the CIM, and their emplacement was late compared with the above-mentioned leucogranites and two-mica granites. The group includes monzogranites and granodioritic types that are characterized by variable amounts of Al-minerals. Apparently, they could be situated along the same line of processes (mesocrustal anatexis) that generated the leucogranites and two-mica granites, owing to their Al-mineralogy. However, several data suggest a more complex interpretation, as will be detailed below. The following groups may be established:

1. Biotite monzogranites ± muscovite ± sillimanite ± andalusite. Porphyritic or not, peraluminous, the muscovite predominantly associated with sillimanite and andalusite, the predominant plagioclase is oligoclase with occasional nuclei of andesine. They show small enclaves of biotite + sillimanite, biotite + andalusite, macro-meso enclaves of cordierite migmatites, macro-meso enclaves of hornfelses of pyroxene hornfels facies (Cord + K feld ± And ± Sill) and a relative frequency of mafic enclaves (gabbros and tonalites). In one case, the gabbro enclaves show spinel and cummingtonite (Ugidos et al. 1985).

2. Cordierite-bearing biotite (± muscovite) monzogranites/granodiorites. In general porphyritic, they are weakly peraluminous, with occasional sillimanite needles. Muscovite is predominantly associated with the alteration of other minerals. The plagioclases (andesine-basic oligoclase) are characterized by their oscillatory zoning and form frequent synneuses. The cordierite (prismatic, very rare without alteration) content varies considerably (< 4% in general; higher in some points close to enclaves of cordieritite, or contacts with cordierite migmatites; sometimes predominates in marginal zones; its size varies from some millimeters to 2 cm; sizes between 1-2 cm are very characteristic. There are frequent tonalitic or basic enclaves (Ugidos 1974 b; Aparicio et al. 1975; Corretgé and López Plaza 1977; Barrera et al. 1981; Aparicio et al. 1983; Brandebourger et al. 1983; Bea and Moreno Ventas 1985), and several cases have been described in which the microgranular biotitic tonalitic enclaves show cordierite prisms sizing up to 1 cm (Ugidos et al. 1985). Other enclaves whose abundance depends on the proximity to the host rocks are cordierite migmatites, hornfelses of pyroxene-hornfels facies (Ugidos 1987 b) and surmicaceous enclaves (Aparicio et al. 1975), with cordierite, andalusite and sillimanite. The latter seem to be closely related to the areas in which the granites exhibit these Al-minerals (Aparicio et al., op. cit. p. 67-68) as isolated crystals.

This group of granites exhibits interaction with the migmatitic host rock and cordierite-rich anatectic rocks (Ugidos 1974 b) and gradual transitions to biotite granites (without Al-minerals) and to amphibole-bearing biotite granites (Brandebourger et al., op. cit., Aparicio et al., op. cit.).

3. Biotite-muscovite (± cordierite ± andalusite ± sillimanite ± garnet) monzogranites. They are distinguished from the above-mentioned rocks in their greater abundance of muscovite, their more leucocratic aspect and a larger grain size; they are rarely porphyritic and the plagioclases have lower contents of anorthite (oligoclase-albite); marked peraluminous character. Cordierite shows the same characteristics as in the previous group; andalusite is less abundant (< 0.5%) and is often idiomorphic-subidiomorphic.

Basic and tonalitic enclaves are scarce or absent (García de Figuerola and Franco, op. cit.; Corretgé and López Plaza 1977; Franco 1980; Carnicero 1980; García de Figuerola and Franco (in Vargas et al. 1982); Aparicio et al. 1983; Brandebourger et al., op. cit.; Brandebourger 1984). Transitions to cordierite-bearing biotite granites can be observed.

Absolute datings of these granites are not abundant. The most frequent ages are between 290-280 m. y. (in Serrano Pinto et al., op. cit.). It should be noted that the intitial Sr isotopic ratio of this group of peraluminous granites often shows intermediate values. This is the case of the Villar de Ciervo granite (0.7077, García Garzón and Locutura 1981), the granites from Villacastin-Espinar (0.7065), Morazarzal (0.7083), La Cabrera (0.7073 and 0.7094) and La Pedriza (0.7072) (Vialette et al. 1981; Ibarrola et al. 1986).

Data for similar types of peraluminous granites from Portuguese areas are also known (Serrano Pinto 1983); the Alva granite (0.7061), Lamas granite (0.7078), Calde granite (0.7093); the Lamelas amphibole-bearing biotite granodiorite (0.7063).

Chemical data are presented in Table 2; REE patterns in Figs. 3, 4.

2.3.4.3 Biotite and Amphibole-Bearing Biotite Granodiorites/Monzogranites

These are of variable porphyritic character (predominant in the biotite facies); they are metaluminous to weakly peraluminous and exhibit abundant basic and tonalitic enclaves. Plagioclases are zoned and synneusis textures are frequent. Allanite and sphene are characteristic accessory minerals (Franco 1980; García de Figuerola and Franco, op. cit.; Aparicio et al., op. cit. Fernández et al. 1982; Hernández et al. 1982). These granites show transition to cordierite-bearing biotite facies. In one case the presence of a macroenclave of cordieritite (60-80% of cordierite sizing up to 3.5-4 cm) has been reported in biotite granites and in its immediate neighbourhood a cordierite-muscovite facies has developed (Ugidos 1988 a). Chemical data can be found in Table 2; REE patterns in Figs. 3, 4.

Table 2. Data for typical Late Hercynian granites from Central Iberian areas[a]

	1	2	3	4	5
SiO_2	65.29	73.14	70.01	67.02	66.05
TiO_2	0.78	0.20	0.41	0.68	0.62
Al_2O_3	15.56	14.00	14.85	15.37	15.57
Fe_2O_3	6.09	1.67	3.13	4.27	4.13
MgO	2.80	0.38	0.85	1.27	1.50
CaO	0.96	0.97	2.15	2.56	3.06
Na_2O	2.41	3.42	3.28	3.42	3.49
K_2O	3.44	4.57	4.26	4.25	3.83
P_2O_5	0.20	0.11	0.12	0.27	0.23
Rb	148	261	193	184	157
Sr	172	81	130	155	171
Ba	562	271	483	531	592
La	41.69	21.10	28.85	38.15	30.63
Ce	80.12	44.17	64.51	79.00	65.88
Nd	32.28	20.37	30.98	35.25	27.75
Sm	7.22	4.80	6.53	7.49	6.33
Eu	1.52	0.50	0.91	1.12	1.18
Gd	5.18	3.96	5.56	5.79	4.75
Dy	4.48	4.35	4.84	5.03	4.37
Er	2.35	2.51	2.67	2.47	2.25
Yb	2.62	2.34	2.35	2.68	2.49
Lu	0.38	0.34	0.36	0.37	0.34

[a] 1 Cordierite nebulites (n = 10; Ugidos, unpubl.).
2 Biotite-muscovite (+/− cordierite) granites (n = 9; Brandebourger 1984).
3 Cordierite bearing (+/− muscovite) biotite granites (n = 13; Brandebourger, op. cit.)
4 Biotite granites (n = 14; Ugidos, unpubl.).
5 Amphibole-bearing biotite granites (n = 13; Ugidos, unpubl.).

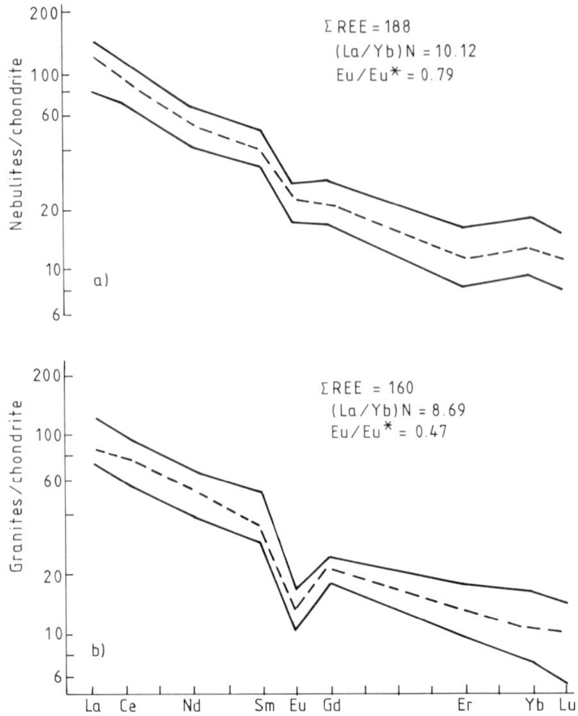

Fig. 3. Range and average of REE abundances normalized to chondrite values (Haskin et al. 1968). **a** Cordierite nebulites. **b** Cordierite-bearing biotite granites

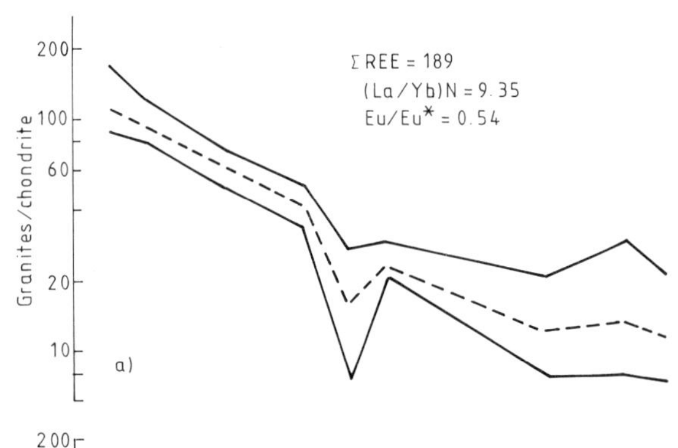

Fig. 4. Range and average of REE abundances normalized to chondrite values. **a** Biotite granites. **b** Amphibole-bearing biotite granites

2.3.4.4 Quartzdiorites and Tonalites

A large number of small bodies of these rocks is known throughout the CIM, some of which predate the large masses of granite rocks while others intruded contemporaneously (Capdevila et al. 1973; Brandebourger et al. op. cit.; Martínez 1974a, b; Aparicio et al. 1975; López Plaza and Corretgé 1979; López Plaza 1982; Fuster and Rubio 1980). In this situation they are commonly found as enclaves. Essential minerals in quartzdiorites and tonalites are quartz, plagioclases, biotite and amphiboles (not always). Occasionally pyroxene is found as an accessory mineral.

In areas to the south of those considered here, intrusive quartzdiorites have produced a thermal aureole in which hornfelses of the pyroxene-hornfels facies have developed, in which cordierite crystals are frequent. This mineral is also found in the quartzdiorites-forming polycrystalline aggregates (sometimes associated with quartz), in which the crystals of cordierite have the same shape and size as in the hornfelses of the pyroxene hornfels facies (in both cases, without inclusions of other minerals (Castro 1984). The xenogenic origin of the cordierite therefore seems to be evident.

This also suggests that the cordierite-bearing tonalitic enclaves are fragments of early quartzdiorites-tonalites incorporated into later intrusive bodies.

2.3.5 Mafic and Ultramafic Rocks

This group is represented by diorites, gabbros, olivine pyroxenites, pyroxene-olivine hornblendites and associated rocks in western areas of the CIM (Fig. 1) where there is a large number of small outcrops. These rocks are found both in metasedimentary rocks and in the form of macro-meso enclaves within granodioritic/monzogranitic rocks (Franco 1980, 1980–81; Franco and García de Figuerola 1986; García de los Rios 1981; Franco and Sánchez 1988). Some of them predate the granodiorite an mozogranite intrusions while others intruded simultaneously (Capdevila et al., op. cit. and above authors).

2.3.6 Volcanic Rocks

To the NE of the CIM amphibole- and pyroxene-bearing amphibole, andesites have been described together with intermediate petrographic types of Permian age (287 ± 12 m. y., Hernando et al. 1980). They are characterized by the presence of abundant microgranular enclaves (diorites and quartzdiorites) and a wide distribution of metamorphic enclaves with sizes mainly around 1 cm but generally smaller than 4 cm (Hernán et al. 1981).

In mineralogical terms, these enclaves are composed of: biotite, plagioclase and garnet as the most frequent minerals. Sillimanite, cordierite and occasionally corundum, spinel, kyanite, orthopyroxene and K feldspar are present. Garnet is also found as an isolated mineral within the andesite mesostasis and its optical and textural characteristics are identical to those garnets of the metamorphic enclaves (Hernán et al., op. cit.).

Chemical data show the peraluminous nature of the andesites (mean value of normative corundum: 0.81, Ancochea et al. 1980) and similar compositions for garnets in the metamorphic enclaves and in the andesite mesostasis (Aparicio and García Cacho 1984).

The authors cited deduce an xenogenic origin for the garnet and suggest the assimilation of metamorphic rocks and a possible relationship between the andesites and the calc-alkaline plutonism of the CIM.

2.3.7 Significance of Al-Minerals in the Granites

As has been pointed out in previous sections, andalusite, sillimanite and in particular prismatic cordierite are common in the late granites, and interpretation of the significance of these minerals is a key point in genetic studies of granitic rocks.

The first problem is whether these minerals are of primary magmatic origin or whether they are residual. Comparison of the petrographic characteristics among the minerals in metamorphic and granitic rocks point to the following aspects (Ugidos 1987 a, b):

1. There are no textural differences among the minerals of either group of rocks. Likewise, the types of optically appreciable inclusions are the same;
2. Hornfelses of pyroxene-hornfels facies show high percentages of prismatic cordierite with no or scanty inclusions of biotite (Plates 1, 2). In the same rocks there are cordierite prisms included in plagioclases, quartz and alkali feldspar. If these textural relationships were appreciated in an igneous rock, one would probably propose a magmatic origin for the cordierite. Since these textures are also found in metamorphic rocks, petrographic arguments cannot be used as sure criteria in favour of a magmatic origin;
3. Migmatitic hornfelses and nebulitic migmatites also show cordierite prisms that in many cases are free of sillimanite inclusions (Plate 3);
4. Andalusite is developed in high-grade metamorphic rocks by inversion of fibrolite, which is sometimes preserved associated or within andalusite. This kind of relationship can also be observed in the granites. In both groups of rocks the andalusite crystals are often idiomorphic-subidiomorphic and inclusion-free. Moreover, in andalusite-bearing granites metamorphic enclaves with this mineral (idioblastic and without inclusions) can also be found (associated or not to cordierite);
5. The presence of cordierit enclaves (60–80% of prismatic cordierite sizing up to 3.5–4 cm; always pinitized; Plates 3, 4) with varying contents in andalusite prisms (sizes up to 5–7 mm; Ugidos 1988 a). The outcrop of cordieritite occupies some m^2 in surface area. It has been drilled up to 10.5 m, without cutting the lower contact with the granite (Ugidos 1988 a). Biotite granites present a

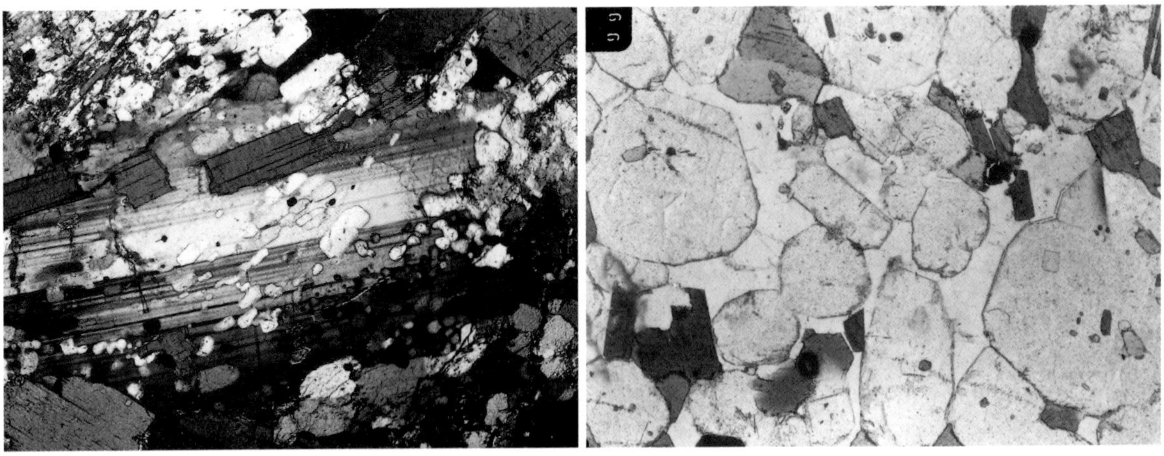

1 2

Plate 1. *1* Cordierite prisms. Hornfels of pyroxene hornfels facies; × 50. *2* Cordierite crystals. Hornfels of pyroxene hornfels facies; × 50

3 4

Plate 2. *3* Cordierite prisms. Hornfels of pyroxene hornfels facies; × 50. *4* Cordierite crystals included in a plagioclase. Hornfels of pyroxene hornfels facies; × 90

cordierite-bearing mineralogy only in the neighbourhood of a cordieritite enclave (Plate 4).

All these data, apart from petrological considerations (Ugidos, op. cit.) suggest that these minerals do not have a generalized primary magmatic origin in the granites. The aspects summarized above to a large extent reduce the validity of petrographic criteria (idiomorphism of cordierite and andalusite; absence of inclusions) commonly used in support of the direct magmatic origin of these minerals when they are found in igneous rocks.

A second problem consists in establishing in which cases these minerals are residual in the anatectic process that generated the granite containing them, and in which cases they were incorporated as xenocrysts by assimilation from the host rocks (restites and/or hornfelses and migmatitic rocks induced in the thermal aureole). The greatest difficulty in solving this problem lies in the fact that in the granites of the CIM unaltered crystals of cordierite are rare. In any case, both situations are possible, as shown by the data regarding autochthonous granites, intrusive granites in migmatitic levels, intrusive granites with cordieritite enclaves and andesites rich in metamorphic enclaves.

2.3.8 Chemical Data

Table 1 represents typical compositions of early granites. On the whole they are characterized by their high content of SiO_2, K_2O and Rb, and by low content of CaO, FeO, MgO and Sr.

5 

 6

Plate 3. *5* Cordierite in a cordierite migmatite; ×32. *6* Fragment of cordieritite included in biotite granite; scale in cm

7 

8

Plate 4. *7* Aspect of the cordierite facies (biotite granite) close to cordieritite outcrop. *8* Drill cores of cordieritite included in biotite granite. *Numbers* indicate depth in m

Table 2 represents compositions of more typical late granites (REE patterns in Figs. 3, 4). Special emphasis is given on chemical data of the late granites because they are relevant to the interpretation proposed. Data on migmatites (cordierite nebulites) are also included, since these rocks are sometimes spatially related to cordierite-bearing biotite granites.

A hypothesis worth considering is that the cordierite nebulites are the source rocks of cordierite-bearing biotite granites. The most notable aspects (Ugidos 1988 b) are synthesized as follows:

1. The chemical data of the granites normalized with respect to the average nebulite composition show that (Fig. 5) the granites:

a) have similar patterns;

b) are much richer in CaO (more than twice as much) than the nebulites;

c) have a lower content of Sr, Eu and Eu/Eu* and a similar content of HREE and LREE.

As it is difficult to accept an anatectic process that enriches the melts selectively in CaO at the same time as it weakens them in Sr, Eu (and similar REE values are maintained), it seems clear that compositions close to that of the average nebulite cannot be a source of cordierite-bearing biotite granites.

Moreover, cordierite nebulites and biotite granites show a similar REE pattern when normalized with respect to chondrite values (Fig. 5). Hence the possible mixing (assimilation) of biotite granites and cordierite nebulites does not significantly modify the REE pattern of the resulting granites (cor-

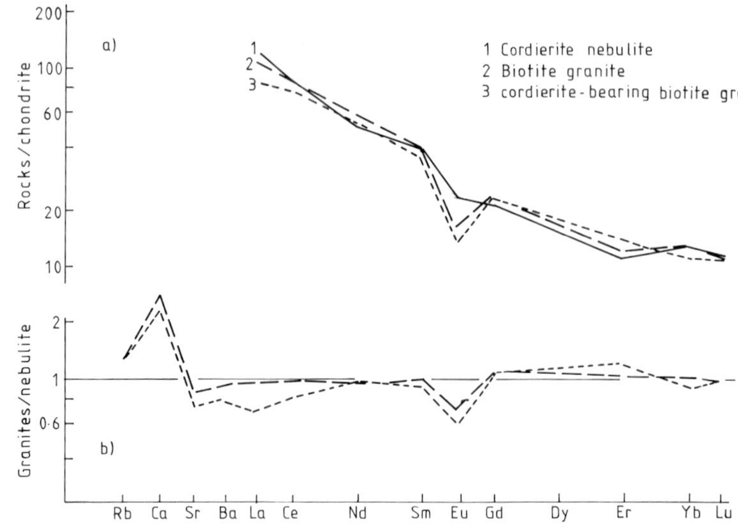

Fig. 5 a. Average value of REE abundances of cordierite nebulites, biotite granites and cordierite-bearing biotite granite normalized to chondrite. **b** Average value of REE abundances of biotite granites and cordierite-bearing biotite granites normalized to average nebulite

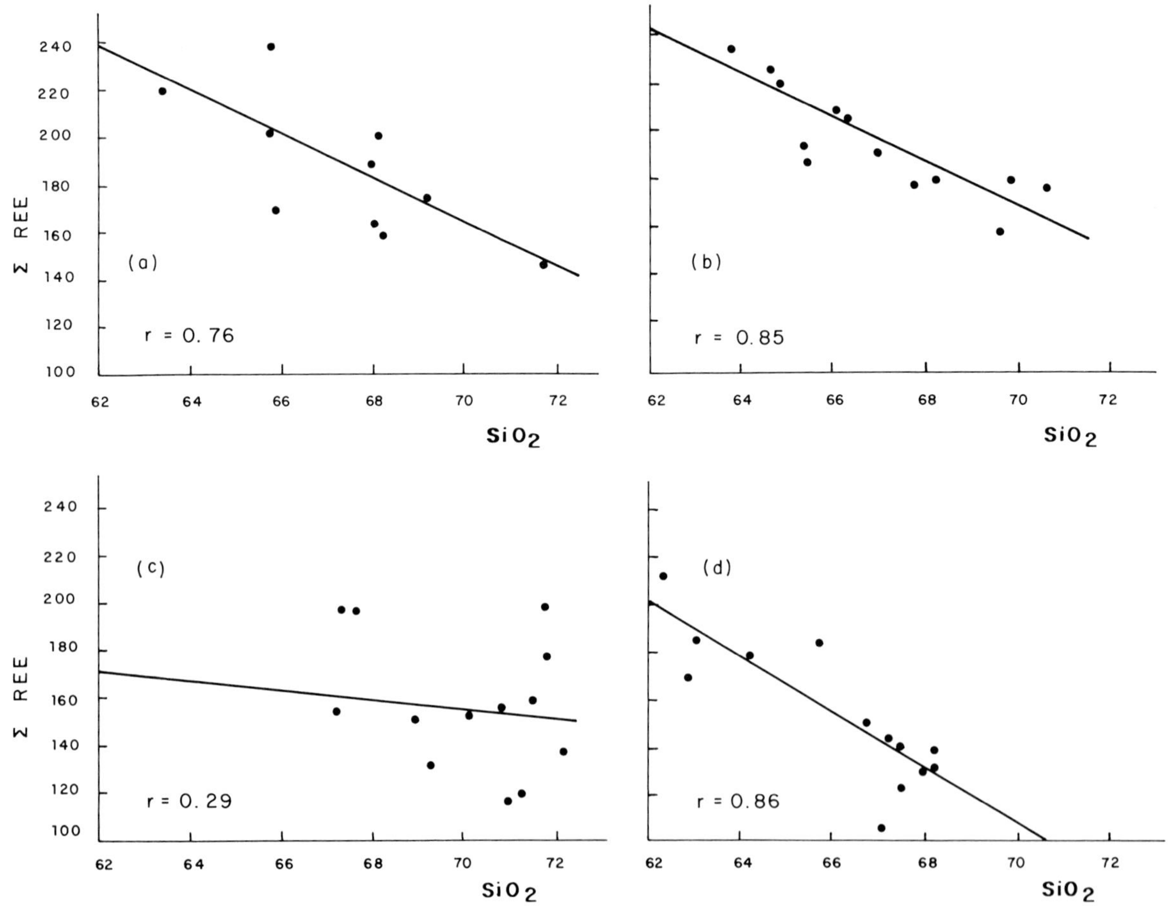

Fig. 6. $\sum$REE-SiO_2 diagrams. **a** Cordierite nebulites; **b** biotite granites; **c** cordierite-bearing biotite granites; **d** amphibole-bearing biotite granites

dierite nebulites show a $\sum$REE-SiO_2 correlation similar to that shown by the granites in Fig. 6).

2. Simple mixing equations have been applied for the content range in Fe_2O_3 and MgO in biotite granites, cordierite nebulites and cordierite-bearing biotite granites (the Fe_2O_3 and MgO content is very different in nebulites and biotite granites; Table 2) in order to determine whether the assimilation is compatible with these chemical data. The results show that more than 40% of the Fe_2O_3 and MgO values in cordierite-bearing biotite granites are compatible with simple mixing processes (biotite granite: 97.5 to 75%, nebulite: 2.5 to 25%). It seems logical to accept that other Fe_2O_3 and MgO values may result from pre- or post-assimilation differentiation processes (although it should be taken into account that other rocks, cordieritites, may be assimilated).

The $\sum$REE-SiO_2 diagrams show a clear trend for an increasing SiO_2 in biotite granites but there is no trend for cordierite-bearing biotite granites (Fig. 6). It is suggested that:

(a) Cordierite is not produced in a magmatic differentiation process of biotite granites and
(b) Cordierite-bearing biotite granites show no correlation because the composition of each particular sample depends (if really produced by assimilation) on: the degree of differentiation of the biotite granites in pre-assimilation stages; possible post-assimilation differentiation; percentage of assimilated component (and its composition). Considering these circumstances, two trends are to be expected in a $\sum$REE-SiO_2 diagram: 1. a trend with negative slope (post-assimilation differentiation) and 2. a trend with positive slope (addition of nebulite to granitic compositions, which is poorer in REE than nebulites, resulting from pre-assimilation differentiation).

The joint representation of chemical data corresponding to two possible populations of samples (trend 1 and trend 2) hinders the appreciation of a specific trend (Fig. 6) for cordierite-bearing biotite granites.

It is concluded that the chemical data are compatible with assimilation processes (biotite granites + cordierite nebulites).

Biotite granites and amphibole-bearing biotite granites show parallel REE patterns (Fig. 4) but different differentiation trends (Fig. 6). Probably both types of granite originated from magmas generated (different percentages of melting) in the lower crust (protholiths of intermediate composition).

It is not the aim of this article to discuss the origin of biotite granites but some considerations are made in this sense. The REE patterns suggest a deep origin, since they are very close to post-Archean upper crust REE patterns (see Taylor 1979; Taylor and McLennan 1985). If the upper crust had its origin in the meteorization, etc. of igneous rocks originating from the melting of intermediate compositions (Taylor, op. cit.; Taylor and McLennan, op. cit.) it seems logical to accept that the CIM granites originated from similar compositions (underplated previously to the Hercynian collision?; island arc andesite-type compositions?). In this case, the extraction of important volumes of granite melts (as is the case of CIM) must have left behind a residue composed principally of minerals with low REE distribution coefficients (pyroxenes, plagioclase and probably amphiboles; absence of garnet). However the seismic data (Banda et al. 1981, 1983) exclude mafic compositions in the lower crust of the CIM. This characteristic of the crust is general for European Hercynian areas and other zones (Caledonian areas of Great Britain and eastern areas of North America; see Meissner 1986) in which the crust has (as in the CIM) a thickness of 32–35 km and granite rocks are abundant.

The apparent relationship between the thickness of the crust, the abundance of granites and absence of mafic rocks in the lower crusts in the cases mentioned cannot be casual. Possibly the mafic residue is not detected seismically due to the development (via eclogitization of the residue) of a new crust-mantle discontinuity on shallower crustal levels (32–35 km) in post-orogenic times (for evolution and reduction of the thickness of the crust see Meissner, op. cit.).

2.3.9 Discussion and Genetic Processes: I-Types vs S-Types

Previous data and considerations underlie the discussion: the evolution of Hercynian metamorphism, field data, the significance of Al-minerals in granites, chemical data, assimilation processes and to a certain extent the intermediate values of the initial Sr isotopic data in peraluminous granites.

The thermal models for the continental crust in orogenic domains exclude the possibility that low-pressure metamorphisms can be originated by thermal conduction phenomena or as a result of radiogenic heat (see for example Brown and Hennessy 1978; Hildreth 1981; England and Thompson 1984; Thompson and England 1984). This implies that the cause of such metamorphism must necessarily lie in the heat supplied by intrusions of magmas originating in the lower crust and/or mantle (ibid).

Seismic data do not allow a deduction of the presence of significant masses of basic rocks for the CIM (Payo and Ruiz de la Parte 1977; Banda et al., op. cit.), and field evidence shows that the volume of basic rocks in the CIM is very small compared with the large extensions of granitic rocks. Accordingly, it is possible that granitic magmas have played a great part in the thermal supply necessary to develop the low-pressure metamorphism, although the initial development and ascent of the isotherms is caused by the thermal relaxation subsequent to crustal thickening and/or the thermal supply from the mantle to the lower crust (see for example, Hildreth, op. cit.; Thompson and England, op. cit.).

The temperatures typical of granodioritic-tonalitic magmas (800–900 °C, Brown and Hennessy, op. cit.) are insufficient to induce in their surroundings high-grade parageneses at relatively shallow levels corresponding to low-pressure metamorphism, and in general terms it is unlikely that they can be sufficient to cause the digestion of the host rocks in their emplacement levels such that significant modifications will occur in the compositional characteristics of the intrusive magmas. There is evidence that in many areas the igneous rocks of the types cited do not induce high-grade metamorphic parageneses in their host rocks at shallow levels of emplacement (although other factors such as the degree of crystallization and volume are also important). However, if the host rock has been preheated the situation is completely different. In this sense, numerous examples are known in which the country rocks present medium high-grade parageneses and the thermal contribution of the intrusions caused parageneses of higher temperature and lower pressure, finally leading to migmatization. Such is the case of some Caledonian granites in Great Britain (Ashworth and Chinner 1978; Ashworth and Tyler 1983; Pattison and Harte 1985). These examples show that a high intrusion temperature, such as those that correspond to basaltic magmas are not necessary for the host rocks to generate high-temperature parageneses and partial melting although preheating of these rocks may be necessary prior to the granite intrusions.

In the case of the Iberian Massif and in particular in the CIM, metamorphism evolved steadily from intermediate- to low-pressure conditions, as has been pointed out, which implies on one hand a rapid uplift and on the other the necessary thermal contribution (see for example Chamberlain and England 1985) which can be spatially related to the late granite intrusions. Under these circumstances, intrusions reach relatively shallow levels (lower pressure) at times during which the country rocks still conserve temperatures similar to the metamorphic climax of the first syntectonic stages. Accordingly, host rocks receive a heat supplement in the surroundings of the intrusion (intrusion heat + crystallization heat) that permits the increase in temperature and/or the maintenance of the initial high temperatures during an episode of rapid decrease in pressure. Country rocks in upper amphibolite facies (corresponding to an intermediate-pressure metamorphism) above the dehydration reactions involving biotite (Fig. 7) cross these reactions as a result of the decrease in pressure, favouring the continuation of the initial anatectic phenomena (increasing the volume of the melt), with the development of cordierite and/or garnet more or less synchronous with the emplacement of late granite intrusions. It is evident that the assimilation of host rocks would be easier if the ascent of the intrusive masses (and/or their differentiation products) continued and the problem of contamination were mainly mechanical and not thermal. Furthermore, and depending on the intensity of the process, one would expect the schistosity defined by the biotite + sillimanite to disappear, and thus none or very few of these textural relicts would remain in the resulting granite.

The initial syntectonic metamorphism produced anatectic melts and associated granitic rocks (leucogranites and two-mica granites), which also involves the presence of restitic rocks in subjacent levels. These restites must be composed of biotite + sillimanite, cordierite and/or garnet or combinations of these minerals (apart from possible smaller amounts of felsic minerals), according to whether PTX conditions permit the development of dehydration reactions involving the first mineral pair. In any case, the biotite + sillimanite pair tends to produce the other minerals with a decrease in pressure if high temperatures are maintained. It is unlikely that this reaction (in the restites) would cause a significant melt fraction but it would produce, at least, an interstitial fluid enriched in elements retained by the biotite (with low K/Rb ratios, for example). If this type of rock is incorporated into the intrusive granites, again the assimilation problem would be mechanical (disgregation of the rock is favoured by the presence of an interstitial fluid) and not thermal, and the result would be the acquisition of an S character in both mineralogical and chemical terms.

It should be taken into account that a low assimilation percentage of the restitic rock has important consequences in the mineralogy and content of normative corundum and in the possibility of producing secondary minerals (muscovite, andalusite; see Abbot and Clarke 1979).

The final result is that the granites of deep origin (hybrid or not) and/or their differentiation products may acquire typical S characteristics if their ascent takes place subsequent to the regional metamorphism, as proposed. The possibility of mixing with induced anatectic rocks (derived from metasedimentary, orthogneissic or volcanoclastic protholiths) and/or restitic compositions is quite feasible, and the evidence that low-pressure metamorphic rocks (hornfelses of pyroxene hornfels-facies, migmatites and restites) are associated with the granites in metamorphic contexts such as those mentioned makes the interpretation proposed very plausible. Accordingly, with the aspects considered in the first part of this work, the following sequence of igneous-metamorphic events could be proposed:

1. An initial intermediate-pressure metamorphism (post-crustal thickening thermal relaxation). Anatexis and production of granites (leucogranites and two-mica granites) in mesocrustal levels. Production of granitic magmas in deep parts of the crust.
2. Small intrusions of mafic-ultramafic rocks, quartzdiorites and tonalites. Possible lower crust-mantle interaction (magma hybridization).
3. Subsequent intrusions of large volumes of granodiorites and monzogranites from deep origin (incorporation of enclaves of preceding igneous rocks) associated with a rapid uplift. Thermal contribution to more and more shallow levels. Telescoping of initial regional isograds: low-pressure metamorphism in the neighbourhood of the intrusive masses. Dehydration-melting reactions as a result of a decrease in pressure.
4. Assimilation and acquisition of an S appearance by these intrusions and/or their differentiation products.

In simplified form: the assimilation-mixture depends mainly on the temporal relationship between metamorphism and the granite intrusion from deep levels. Accordingly, the following situations are possible (Ugidos and Bea, op. cit.; Ugidos 1988 c):

a) Ascent of the granites (from deep origin) is faster than the ascent of regional isotherms. Important assimilation phenomena are unlikely, since the host rocks are not preheated.

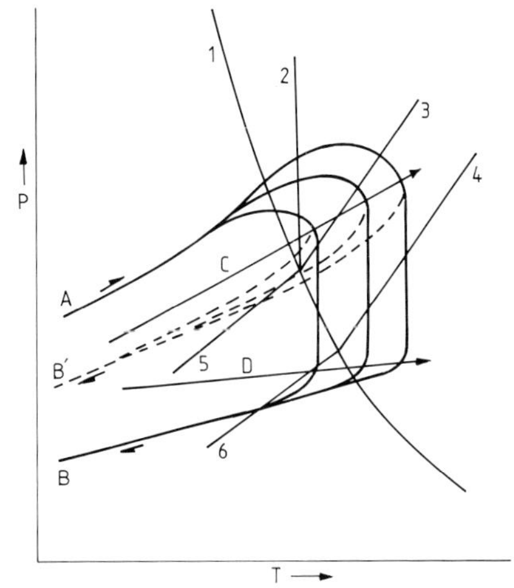

Fig. 7. *AB* P-T-t path for metamorphic rocks from initial stages (intermediate pressure; *C* "metamorphic gradient"). *AB* P-T-t path for the case of a metamorphism of simple evolution. Schemes based on Thompson and England (1984). In the case of AB the thermal conditions are maintained (or increased) as a result of the proximity of the granite intrusions from deep levels while pressure decreases owing to a rapid uplift rate. In this way, reactions *4, 5* and *6* occurred (together with the reaction Mus+Bi+Ab+Q=Cord+Kf+H_2O in Fig. 2) with the production of cordierite in restites, migmatites and hornfelses of pyroxene-hornfels facies.
1 Ab+Kf+Q+H_2O=L;
2 Mus+Ab+Kf+Q+H_2O=Al_2SiO_5+L;
3 Mus+Ab+Q=Kf+Al_2SiO_5+L;
4 Bi+Al_2SiO_5+Ab+Q=Kf+Cord+L;
5 Mus+Ab+Q=Kf+Al_2SiO_5+H_2O;
6 Bi+Al_2SiO_5+Ab+Q=Kf+Cord+H_2O. Reactions after Thompson and Tracy (1979) and Thompson (1982)

b) The regional isotherms preceded the granite intrusions and these are emplaced during periods in which relatively high temperatures are maintained in the country rocks. The assimilation and mixture with anatectic products are favoured, especially if the decrease in pressure (uplift) leads to the production of dehydration melting reactions at times close to the intrusive events. Possible assimilation of restitic phases left behind by granitic melts that have migrated previously;

c) The emplacement of the granites takes place after cooling of the metamorphic rocks. As in case a), if it does take place, assimilation is of lesser importance.

The consequences of situations similar to b) permit a wide possible range of granitic rocks, depending on the type of protoliths affected (and assimilated) during the evolution of the metamorphism and the degree of differentiation of the granites participating in the processes of assimilation. The true genesis of the resulting granites may be very difficult to establish, since the possibilities of assimilation include both anatectic metasedimentary rocks of pelitic composition and previous ortho-derived rocks (or with ortho-derived participation).

These genetic possibilities lead to a situation in which granites rich in Al-minerals, emplaced in shallow non-metamorphic levels such as is the case with part of the granites of the CIM, may be the result of the convergence of different processes. Current data do not permit one to discriminate rigorously between the different possibilities, but the few isotopic data regarding Sr, together with petrographic information and chemical and field data, suggest, in the opinion of the author, that the assimilation hypothesis (which does not preclude other possibilities of lesser importance) is the most suitable to generate most of the late Al-mineral-bearing granites in the CIM. A possible, maxim would be: "too much cordierite in too many granite types in excessive outcropping extensions".

Data from other orogenic areas are significant with respect to the interpretation proposed. In this sense, in Argentina, important volumes of restitic rocks have been described (cordieritites: 70–90% cordierite) associated with biotite granites that also bear cordierite (Gordillo 1979). Large batholiths, as the South Mountain Batholith (Nova Scotia) of S appearance (cordierite, muscovite, etc.) exhibit chemical and isotopic characteristics that are incompatible with a simple origin from metasedimentary rocks, such that a more complex genesis has been proposed involving the participation of deep materials (lower crust and possible participation of the mantle) together with metasedimentary material from shallower levels (Clarke and Halliday 1980; Maillet and Clarke 1985). The data from these authors also show that the cordierite in the granites does not exhibit petrographic or chemical differences compared with the cordierite from metamorphic rocks, such that the interpretation of a xenogenic origin for this mineral is preferred by the authors. Although the possibility (in part) of a primary origin cannot be ruled out: "if both types do occur, we have not yet found any criteria by which to distinguish between them" (Maillet and Clarke, op. cit. p. 702). The initial Sr isotopic ratio of the granites of the South Mountain Batholith show intermediate values, as in the CIM (0.7076–0.7094; Clarke and Halliday, op. cit.).

In a generalization of the proposed model it is evidently necessary to take into account the main variables considered in this paper. In this sense, if the temporal relationship between the initial regional metamorphism and intrusions is different (a greater lapse of time between both events), the country rocks are in conditions of lower temperature and the heat supplied by the late granites would not be sufficient to induce anatexis. Therefore, if assimilation takes place it is restricted and local. Accordingly, in the case of very late granites, regarding the regional metamorphism it is unlikely that circumstances would be favourable for the assimilation of host rocks, since the country rocks would have cooled to some extent at the time intrusions occurred.

Therefore, and in accordance with the analysis put forward above, the author suggests the possibility that the large batholithic bodies really have their initial origin in deep zones (lower crust-mantle) and that the structural characteristics of the crust and its evolution (thickening by shortening or by underplating) together with other variables (start and rate of the uplift and temporal relationships between regional metamorphism and intrusive episodes) condition the possibility of interactions with the host rocks in such a way that the final appearance of the granites is very different from one orogenic domain to another.

If one accepts (as has been done in this work) the thermal regime models of the kind proposed by England and Thompson (1984) and Thompson and England (1984), with the connotations inherent to them, the internal logic of such proposals together with data such as those exemplified by the CIM do make the model proposed above feasible; to a large extent it is compatible with the models suggested by Halliday (1983) and Hildreth (1981), among other authors.

One therefore envisages the possibility that the differences between Andean, Caledonian and Hercynian granites (Pitcher 1982) would have been partly conditioned by the temporal relationship between metamorphism and the intrusions. In the Andean case, the ascent of magmas took place relatively quickly through favourable structures, and the crustal thickening would have been mainly due to underplating (Pitcher 1982). This situation is radically different from that of contexts in which crustal thickening took place by shortening, and the sequence of events (post-thickening thermal relaxation, evolution of metamorphism-intrusive episodes, etc.) and the gross structure and composition of the crust are unlikely to be the same.

In the case of the Caledonian and Hercynian granites, the difference probably lies in that the latter were emplaced during periods temporally proximal to the regional metamorphic event and the former at latter times. However, cases are known of Caledonian granites in Great Britain, very late with respect to the regional metamorphism, which have produced hornfelses of pyroxene-hornfels facies and migmatites in their aureoles (Ashworth and Chinner 1978; Tyler and Ashworth 1983; Pattison and Harte 1985; Frost and O'Nions 1985), and also cases in which genetic processes that involve interaction between I-type magmas and metasediments have been proposed (Brown et al. 1979; Stephens and Halliday 1979; Stephens et al. 1985). If the temporal metamorphism-intrusive episode dephasing had been less pronounced (also with respect to the uplift) it is possible that a great number of granites with S appearance would have developed, as in the case of the CIM.

Data from different areas indicate the isotopic compositional continuity between I granites and cordierite-bearing granites, which supports the ideas expressed above. Cases of this (to quote some examples) are some of the SE Australian granites (Gray 1984) and the Loch Doon granites (Halliday et al. 1980; Tindle and Pearce 1981).

The geological context is thus important (see Pitcher 1982), but the cause of the generation of granites of apparently S origin does not lie in the sequence: metamorphism-anatexis-S-type granite batholiths (in fact many authors throw doubt on the possibility of the generation of granites at batholithic dimensions of migmatitic-anatectic origin; see for example, Brown and Hennessy, op. cit.; Pitcher 1983), but rather in the sequence: metamorphism-restricted anatexis (leucogranites and related rocks)-uplift, associated in time with intrusions of deep origin (maintenance of high temperatures during the decrease in pressure)-assimilation/mixture of host rocks (acquisition of S characteristics) at shallower levels.

Acknowledgements. The author would like to thank N. S. D. Skinner for supervising the translation of this paper. This work benefited from the comments of an anonymous reviewer.

References

Abbot RN, Clarke DB (1979) Hypothetical liquidus relationships in the subsystem Al_2O_3-FeO-MgO projected from quartz, alkali feldspar and plagioclase for $a(H_2O) \leq 1$. Can Mineral 17: 549-560

Ancochea E, Perni A, Hernán F (1980) Caracterización geoquímica del vulcanismo del área de Atienza (provincia de Guadalajara. España). Estud Geol 36: 327-337

Aparicio A, García Cacho L (1984) Quimismo de los principales componentes minerales de las rocas volcánicas paleozoicas del área de Atienza (provincia de Guadalajara). Bol Geol Min Esp 95: 80-89

Aparicio A, Barrera JL, Carballo JM, Peinado M, Tinao JM (1975) Los materiales graníticos hercínicos del Sistema Central español. Mem Inst Geol Min Esp 88: 145 pp

Aparicio A, Bellido F, Brandle JL, García Cacho L, Santos V (1983) Caracterización de los granitoides hercínicos del sector centrooriental del Sistema Central español. Estud Geol 39: 271-306

Ashworth JR, Chinner GA (1978) Coexisting garnet and cordierite in migmatites from the Scottish Caledonides. Contrib Mineral Petrol 65: 379-394

Ashworth JR, Tyler IM (1983) The distribution of metamorphic temperatures around the Strontian granodiorite. Geol Mag 120: 281-290

Babin RB (1974) Materiales metamórficos y plutónicos presentes en la región de Piedrahita-Barco de Avila-Béjar. Stud Geol Salmant 7: 41-61

Babin RB (1975) Estudio estructural de los afloramientos metamórficos de la región de Piedrahita-Barco de Avila-Béjar (Sierra de Gredos). Thesis, Univ Complutense de Madrid, Spain

Banda E, Suriñac E, Aparicio A, Sierra J, Ruiz de la Parte E (1981) Crust and upper mantle structure of the Central Iberian Meseta (Spain). Geophys J R Astron Soc 67: 779-789

Banda E, Udias A, Mueller St, Mezcua J, Boloix M, Gallart J, Aparicio A (1983) Crustal structure beneath Spain from deep seismic sounding experiments. Phys Earth Planet Inter 31: 277-280

Barrera JL, Bellido F, Peinado M (1981) Variaciones mineralógicas y químicas de enclaves microgranulares de los granitoides tardihercínicos del Sistema Central español. Cuad Geol Ibérica 7: 235-253

Bea F (1975) Caracterizacion geoquimica y esquema petrogenético de los granitoides hercínicos del área Salamanca-Cáceres. Thesis, Univ Salamanca

Bea F, Maldonado C (1980-81) El granitoide del área Puente del Congosto-Barco de Avila-Candelario (España). Su geología, geoquímica y petrogénesis. Bol Soc Geol Port 22: 77-87

Bea F, Moreno Ventas I (1985) Estudio petrológico de los granitoides del área centro-norte de la Sierra de Gredos (Batolito de Avila; Sistema Central español). Stud Geol Salmant 20: 137-174

Bea F, Sánchez JG, Serrano Pinto M (1988) Una compilación geoquímica (elementos mayores) de los granitoides del Macizo Hespérico. In: Bea F, Carnicero A, Gonzalo JC, López Plaza M, Rodrígnez Alonso MD (eds) Geología de los granitoides y rocas asociadas del Macizo Hespérico (libro homenaje a García de Figuerola). Editorial Rueda, Madrid, pp 87-193

Brandebourger E (1984) Les granitoides hercyniens tardifs de la Sierra de Guadarrama (Systeme Central, Espagne). Petrographie et Géochimie. Thése, Centre de Recherches Pétrographiques et Géochimiques. Vandoeuvre-les-Nancy

Brandebourger F, Casquet C, Debon F, Fuster JM, Ibarrola E, Isnard P, Le Fort P, Navidad M, Peinado M, Stussi JM, Villaseca C (1983) Nota previa sobre la petrología y geoquímica de los granitoides de la Sierra de Guadarrama (España). Stud Geol Salmant 18: 251-264

Brown GC, Hennessy J (1978) The initiation and thermal diversity of granite magmatism. Philos Trans R Soc Lond A 288: 631-643

Brown GC, Cassidy J, Tindle AG, Hughes DJ (1979) The Loch Doon granite: an example of granite petrogenesis in the British Caledonides. J Geol Soc Lond 136: 745-753

Capdevila R, Corretgé LG, Floor P (1973) Les granitoïdes varisques de la Meseta Ibérique. Bull Soc Géol Fr (7) 15: 209-228

Capote R (1983 a) Formaciones porfiroides. In: Geología de España (libro jubilar JM Rios). Inst Geol Min Esp 1: 84-91

Capote R (1983 b) La estructura de la extensión SE de la zona centroibérica en el Sistema Central. In: Geología de España (libro jubilar JM Rios). Inst Geol Min Esp 1: 467-476

Capote R, Fernández Casals MJ (1977) Las series anteordovícicas del Sistema Central. Bol Geol Min Esp 86: 581-596

Carnicero MA (1980) Estudio petrológico del metamorfismo y los granitoides entre Cipérez y Aldea del Obispo (Oeste de la provincia de Salamanca). PhD Thesis, Univ Salamanca

Carnicero MA (1980-81) Granitos anatexíticos de la zona de Lumbrales (Salamanca). Bol Soc Geol Port 22: 107-110

Castro A (1984) Los granitos y la estructura hercínica en Extremadura Central. PhD Thesis, Univ Salamanca

Chamberlain CP, England PC (1985) The Acadian thermal history of the Merrimack syclinorium in New Hampshire. J Geol 93: 593-602

Chappel BW, White AJR (1974) Two contrasting granite types. Pac Geol 8: 173-174

Clarke DB, Halliday AN (1980) Strontium isotope geology of the South Mountain Batholith, Nova Scotia. Geochim Cosmochim Acta 44: 1045-1058

Corretgé LG, López Plaza M (1977) Geología del área granítica y metamórfica al oeste de Ciudad Rodrigo (Salamanca). II: Las rocas graníticas. Stud Geol Salmant 2: 47-73

Departamento de Petrología (1983) Mapa 1:200000 de la síntesis geológica del basamento. Zona del Centro-Oeste español. Universidad de Salamanca

Díez Balda MA (1986) El complejo esquisto-grauváquico. Las series paleozoicas y la estructura hercínica al sur de Salamanca. Ediciones Universidad de Salamanca, Acta Salmant 52: 162 pp

England PC, Thompson AB (1984) Pressure-temperature-time paths of regional metamorphism. I. Heat transfer during the evolution of regions of thickened crust. J Petrol 25: 894-928

Fernández J, Olivé A, Carreras F, Hernández A, Aguilar MJ, Capote R (1982) Mapa Geológico de España. E. 1:50000. Hoja nº 505 (15-20) Mirueña de los Infantes. Inst Geol Min Esp

Franco MP (1980) Estudio petrológico de las formaciones plutónicas y metamórficas de la depresión del Corneja-Amblés (Sierra de Avila). PhD Thesis, Univ Salamanca

Franco MP (1980-81) Las rocas básicas y ultrabásicas localizadas al N de la depresión del Corneja (Avila-España). Bol Soc Geol Port 22: 101-104

Franco MP, García de Figuerola LC (1986) Las rocas básicas y ultrabásicas en el extremo occidental de la Sierra de Avila (Provincias de Avila y Salamanca). Stud Geol Salmant 23: 193-218

Franco MP, Sánchez T (1988) Características petrológicas del área de El Mirón (N del Valle del Corneja. Provincia de Avila). In: Bea F, Carnicero A, Gonzalo JC, López Plaza M, Rodríguez Alonso MD (eds) Geología de los granitoides y rocas asociadas del Macizo Hespérico (libro homenaje a García de Figuerola). Editorial Rueda, Madrid, pp 293-313

Frost CD, O'Nions RK (1985) Caledonian magma genesis and crustal recycling. J Petrol 26: 515-544

Fuster JM, Rubio JI (1980) El afloramiento granodiorítico-tonalítico de Ventosilla (Guadarrama Central). Bol Geol Min Esp 91: 494-502

Fuster JM, Villaseca C (1979) Migmatización en las series metamórficas de la vertiente norte de la Sierra de Guadarrama. Estud Geol 32: 285-290

Fuster JM, Navidad M, Villaseca C (1981) Relaciones entre ortoneises y series volcanosedimentarias en el macizo de El Caloco (Guadarrama Central). Cuad Geol Ibérica 7: 161-171

Fuster JM, Aparicio A, Casquet C, García Cacho L, Mora A, Peinado M (1974) Interacciones entre los metamorfismos plurifaciales y polifásicos del Sistema Central español. Geol Min Esp 85: 83-88

García Cacho L (1973) Evolución temporal del metamorfismo y procesos de blastesis sucesiva en el sector oriental del Sistema Central español. PhD Thesis, Univ Complutense de Madrid

García de Figuerola LC, Franco MP (1975) Las formaciones infraordovícicas y el borde de las granodioritas al este de Guijuelo (Salamanca). Estud Geol 21: 487-500

García de Figuerola LC, Franco MP, Castro A (1983) Características petrológicas del complejo laminar pegmatoide (Serie del Alamo) de las provincias de Salamanca y Avila. Stud Geol Salmant 19: 33-77

García Garzón J, Locutura J (1981) Datación por el método Rb-Sr de los granitos de Lumbrales-Sobradillo y Villar de Ciervos-Puerto Seguro. Bol Geol Min Esp 92: 68-72

García de los Rios, JI (1981) Estudio petrológico-estructural de la región granítica situada al noreste de Almeida (provincia de Zamora). Thesis, Univ Salamanca

Gil Ibarguchi JI, Martínez FJ (1982) Petrology of garnet-cordierite-sillimanite gneisses from the El Tormes thermal dome, Iberian Hercynian foldbelt (NW Spain). Contrib Mineral Petrol 80: 14-24

González del Tanago J, Bellido F (1981) Estudio de los granitos de dos micas del cortejo pegmatítico asociado en las inmediaciones del Cerro de S Pedro (Madrid). Cuad Geol Ibérica 7: 295-308

Gordillo CE (1979) Observaciones sobre la petrología de las rocas cordieríticas de la Sierra de Córdoba. Bol Acad Nac Cien Argentina 53: 3-44

Gray CM (1984) An isotopic mixing model for the origin of granitic rocks in southeastern Australia. Earth Planet Sci Lett 70: 47-60

Halliday AN (1983) Crustal melting and the genesis of isotopically and chemically zoned plutons in the Southern Uplands of Scotland. In: Atherton MP, Gribble CD (eds) Migmatites, melting and metamorphism. Shiva Geology Series. Cheshire, United Kingdom, pp 54-61

Halliday AN, Stephens WE, Harmon RS (1980) Rb-Sr and O isotopic relationships in 3 zoned Caledonian granitic plutons, Southern Uplands, Scotland: evidence for varied sources and hybridization of magmas. J Geol Soc Lond 137: 329-349

Haskin LA, Frey FA, Wildeman TR (1968) Relative and absolute terrestrial abundances of rare earths. In: Ahrens LH (ed) Origin and distribution of the elements. Pergamon Press, New York, pp 889-912

Hernán F, Perni A, Ancochea E (1981) El vulcanismo del área de Atienza. Estudio petrológico, Estud Geol 37: 13-25

Hernández A, Carreras F, Fernández J, Olivé A, Aguilar JM, Capote R (1982) Mapa Geológico de España, E: 1:50000. Hoja nº 506 (16-20). Cardeñosa. Inst Geol Min Esp

Hernando S, Schott JJ, Thuizart R, Montigny R (1980) Age des andesites et des sediments interstratiphiés de la region d'Atienza (Espagne): étude stratigraphique, geochronologique et paléomagnetique. Sci Geol Bull 32: 119-128

Hildreth W (1981) Gradients in silicic magma chambers: implications for lithospheric magmatism. J Geophys Res 86: B 11: 10153-10192
Hoffer E (1976) The reaction sillimanite + biotite + quartz = cordierite + K-feldspar + H_2O and partial melting in the system K_2O-FeO-MgO-Al_2O_3-SiO_2-H_2O. Contrib Mineral Petrol 55: 127-130
Hoffer E (1978) Melting reactions in aluminous metapelites: stability limits of biotite + sillimanite + quartz in the presence of albite. Neues Jahrb Mineral Monatsh 9: 396-407
Ibarrola E, Villaseca C, Vialette Y, Fuster JM, Navidad M, Peinado M, Casquet C (1986) Dating of Hercynian granites in the Sierra de Guadarrama (Spanish Central System) (Abstr). In: Int Conf Iberian Terranes and their regional correlation, 31, Univ Oviedo
Julivert M, Fontboté JM, Ribeiro A, Conde LE (1974) Memoria explicativa del Mapa Tectónico de la Península Ibérica y Baleares. E 1:1000000. Inst Geol Min Esp
Kilmurray J (1978) Distena relicta en la vertiente septentrional de la Sierra de Guadarrama. Estud Geol 34: 359-361
López-Plaza M (1982) Contribución al conocimiento de la dinámica de los cuerpos graníticos en la penillanura salmantino-zamorana. PhD Thesis, Univ Salamanca
López Plaza M, Corretgé LG (1979) Cizallas en relación con movimientos ascensionales en el domo del Tormes (Salamanca, España). Publ Mus Lab Min Geol Fac Ciênc Port 91, 4ª serie: 111-118
López Ruiz J, Aparicio A, García Cacho L (1975) El metamorfismo de la Sierra de Guadarrama. Sistema Central español. Mem Inst Geol Min Esp 86: 127 pp
Lotze F (1945) Zur Gliederung der Varisziden der Iberischen Meseta. Traducción española por JM Rios: Observaciones respecto a la división de las Varíscides de la Meseta Ibérica. Publicaciones extranjeras sobre geología de España 5: 149-166 (1950)
Maillet LA, Clarke DB (1985) Cordierite in the peraluminous granites of the Meguma Zone, Nova Scotia, Canada. Mineral Mag 49: 695-702
Martínez FJ (1974 a) Estudio del área metamórfica del NW de Salamanca (Cordillera herciniana, España). Trab Geol 7: 3-59
Martínez FJ (1974 b) Petrografía, estructura y geoquímica de los diferentes tipos de granitoides del NW de Salamanca (Cordillera herciniana, España). Trab Geol 7: 61-141
Martínez FJ, Gil Ibarguchi JI (1983) El metamorfismo en el Macizo Ibérico. In: Geología de España (libro jubilar JM Rios). Inst Geol Min Esp 1: 555-569
Meissner R (1986) The continental crust. A geophysical approach. Int Geophys Ser Academic Press, London, 34, 426 pp
Navidad M (1976) Significado petrológico y geoquímico de las series glandulares en los sectores noroccidental y central del Macizo Ibérico. PhD Thesis, Univ Complutense de Madrid
Navidad M, Peinado M (1981) Ortogneises y metasedimentos de la formación infrabasal al Ollo de Sapo (macizo de Hiendelaencina, Guadarrama oriental). Cuad Geol Ibérica 7: 183-199
Pattison D, Harte B (1985) A petrogenetic grid for pelites in the Ballaculish and other Scottish thermal aureoles. J Geol Soc Lond 142: 7-28
Payo G, Ruiz de la Parte E (1977) El primer perfil sísmico de la Meseta. Bol Geol Min Esp 88: 543-551
Peinado M, Alvaro M (1981) Magmatismo pre e intrahercínico en el sector de El Escorial (Sistema Central español). Cuad Geol Ibérica 7: 201-215
Pitcher WS (1982) Granite type and tectonic environment. In: Hsü KJ (ed) Mountain building processes. Academic Press, London, pp 19-40
Pitcher WS (1983) Granite: typology, geological environment and melting relationships. In: Atherton MP, Gribble CD (eds) Migmatites, melting and metamorphism. Shiva Geology Series, Cheshire, United Kingdom, pp 277-285
Rodríguez Alonso MD (1984) El complejo esquisto-grauváquico en el área de las Hurdes y Sierra de Gata, provincias de Salamanca y Cáceres, España. Cuad Geol Ibérica 9: 37-80
Rodríguez Alonso MD (1985) El complejo esquisto-grauváquico y el Paleozoico en el Centro-Oeste español. Acta Salmanticensia 51, Ediciones Univ Salamanca, 174 pp
San José MA de (1983) El complejo sedimentario pelítico-grauváquico. In: Geología de España (libro jubilar JM Rios). Inst Geol Min Esp 1: 91-99
Serrano Pinto M (1983) Geochronology of Protuguese granitoids: a contribution. Stud Geol Salmant 18: 277-306
Serrano Pinto M, Casquet C, Ibarrola E, Corretgé LG, Portugal Ferreira (1988) Síntese geocronológica dos granitóides do Maciço Hespérico. In: Bea F, Carnicero A, Gonzalo JC, López Plaza M, Rodríguez Alonso MD (eds) Geología de los granitoides y rocas asociadas del Macizo Hespérico (libro homenaje a García de Figuerola), Editorial Rueda, Madrid, pp 68-86
Stephens WE, Halliday AN (1979) Compositional variation in the Galloway plutons. In: Atherton MP, Tarney J (eds) Origin of granite batholiths. Geochemical evidence. Shiva Publishing Limited, Kent, United Kingdom, pp 9-17
Stephens WE, Whitley JE, Thirlwall MF, Halliday AN (1985) The Criffel zoned pluton: correlated behaviour of rare earth element abundances with isotopic systems. Contrib Mineral Petrol 89: 226-238
Streckeisen A (1976) To each plutonic rock its proper name. Earth Sci Rev 12: 1-33
Taylor SR (1979) Chemical composition and evolution of the continental crust: the rare earth element evidence. In: Mc Elhinny MW (ed) The earth: its origin, structure and evolution. Academic Press, London, pp 353-376
Taylor SR, McLennan SM (1985) The continental crust: its composition and evolution. Blackwell Scientific Publications, Oxford, 312 pp
Thompson AB (1982) Dehydration-melting of pelitic rocks and the generation of H_2O undersaturated granitic liquids. Am J Sci 282: 1567-1595
Thompson AB, England PC (1984) Pressure-temperature-time paths of regional metamorphism II. Their inference and interpretation using mineral assemblages in metamorphic rocks. J Petrol 25: 929-955
Thompson AB, Tracy RJ (1979) Model systems for anatexis of pelitic rocks. II. Facies series melting and reactions in the system CaO-$KAl0_2$-SiO_2-H_2O. Contrib Mineral Petrol 70: 429-438
Tindle AG, Pearce JA (1981) Petrogenetic modelling of in situ fractional crystallization in the zoned Loch Doon pluton, Scotland. Contrib Mineral Petrol 78: 196-207
Tyler IM, Ashworth JR (1983) The metamorphic environment of the Foyers granitic complex. Scott J Geol 13: 271-285
Ugidos JM (1974 a) Características del metamorfismo en el área Béjar-Plasencia (Salamanca-Cáceres). Bol Geol Min Esp 85: 73-81
Ugidos JM (1974 b) Los granitos biotíticos ± cordierita de Béjar y áreas adyacentes. Bol Geol Min Esp 85: 214-222
Ugidos JM (1985) Textural types of cordierite in migmatites and granites from W of the Iberian Central Massif (Abstr). Rend Soc Ital Mineral Petrol 40: 426

Ugidos JM (1987 a) Características del metamorfismo de baja presión en zonas centro-occidentales españolas y su significado en relación con minerales alumínicos en los granitos. IX Reunião de Geologia do Oeste Peninsular. Porto, Portugal. Actas e Comunica ções 1: 149-170

Ugidos JM (1987 b) Cordierita, andalucita y sillimanita en las rocas graníticas. Fases magmáticas o residuales? IX Reunião de Geologia do Oeste Peninsular. Porto, Portugal. Actas e Comunica ções 1: 383-398

Ugidos JM (1988 a) New aspects and considerations on the assimilation of cordierite-bearing rocks. Rev Soc Geol Esp 1: 129-133

Ugidos JM (1988 b) Estudio geoquímico de elementos de las tierras raras en granitos y nebulitas de áreas occidentales del Macizo Central español: consecuencias petrológicas. Thesis, Univ Salamanca

Ugidos JM (1988 c) Asimilación en los granitos hercínicos: aspectos básicos e implicaciones. In: Bea F, Carnicero A, Gonzalo JC, López Plaza M, Rodríguez Alonso MD (eds) Geología de los granitoides y rocas asociadas en el Macizo Hespérico (libro homenaje a García de Figuerola). Editorial Rueda Madrid, pp 315-320

Ugidos JM, Bea F (1978) Ensayo sobre la génesis de las rocas graníticas del Macizo Hespérico. Stud Geol Salmant 14: 35-77

Ugidos JM, Rodríguez Alonso MD, Martin Herrero D, Bascones Alvira L (1985) Mapa geológico de España E 1: 50000. Hoja nº 575 (12-23). Hervás. Inst Geol Min Esp

Vargas I, Peón A, García de Figuerola LC, Franco MP (1982) Mapa geológico de España E 1:50000. Hoja nº 504 (14-20) Alba de Tormes. Inst Geol Min Esp

Vialette Y, Bellido F, Fuster JM, Ibarrola E (1981) Données geochronologiques sur les granites de la Cabrera. Cuad Geol Ibérica 7: 327-335

Villaseca SM (1983) Rocas estictolíticas en la Sierra de Guadarrama (España). Stud Geol Salmant 18: 265-176

2.4 Distribution, Characteristics and Evolution of Metamorphism

F. J. MARTINEZ[1], L. G. CORRETGE[2] and O. SUAREZ[3]

The Central-Iberian Zone (CIZ) is the part of the Chain in which granitic plutonism and metamorphism are more developed and widespread. In this CIZ, which also includes the so-called Galicia-Tras Os Montes subzone, the distribution of isograds in bands or belts is closely related to granite plutonism in the same way it was described in the chapter about the West Asturian-Leonese Zone (WALZ), with the western part of which this CIZ shares many features. However, the individuality of these bands is less obvious than in WALZ, due to the existence of more connections and coalescence between the different belts, which is caused by a larger occurrence of syn- and post-metamorphic plutonism.

A number of these bands or belts can be defined (see Fig. 1) in CIZ, the most conspicuous of which are as follows: Puentes de Garcia Rodriguez-Sanabria, Santa Comba-Santiago-Bande, Finisterre-La Guardia, Vila Real-Moncorvo-Vitigudino (continuing in the Guadarrama area of the Spanish Central System) and Oporto-Viseu. All these belts show cross-cutting relations between isograds, which proves the overlapping of metamorphic events. The latter of these episodes seems clearly related with the intrusion of palingenetic granites (two-micas, S-type granites). These rocks intruded during a long process, in which the older Hercynian isograds, which show prograde assemblages with garnet, staurolite and sometimes kyanite, are cut across by new ones. These later isograds correspond to minerals like andalusite, cordierite and sillimanite, which are closely related to granite intrusion. This process of continuous intrusion and cross-cutting of former isograds was called plutonometamorphism by Oen (1970, 1974), and appears particularly evident in bands 7 and 8 of Fig. 1 (northern Portugal).

As a consequence of this isograd overlapping process the present metamorphic picture, in most of the CIZ belts, represents what can be called the final metamorphic state. This final state would result from a long evolving process where a Barrovian episode, similar to the first metamorphism in the WALZ (see Suárez et al., this Vol.), is the oldest event. This event with garnet, staurolite and sometimes kyanite assemblages in pelites took place some time after P_1 (S_1 schistosity), and continued up to P_2 (S_2 is a penetrative crenulation cleavage that locally obliterates S_1).

The metamorphism continues evolving later in a process connected with I and S plutonism that is centering more and more the metamorphic zoning around the elongated granite belts. The peak of granite intrusion in the belts occurs roughly between 330–315 m.y. (Pinto 1983; Priem and Den Tex 1984) in the Namurian and Westphalian, and is related to P_3, a dome-stage refolding of earlier S_1 and S_2 cleavages.

Mineral growth in the above-mentioned metamorphic belts happened at different times, depending on the type of metamorphism. For instance, relic kyanite from the earlier Hercynian episode has been found inside andalusite porphyroblasts (Atherton et al. 1974). The mineral assemblages with garnet, staurolite or kyanite, which correspond to the oldest Hercynian Barrovian events (Oen 1970, 1974), developed between the P_1 and P_2 deformation phases. The assemblages with andalusite, cordierite and sillimanite developed during the subsequent thermal plutonometamorphics events related with the syn P_2 to post P_3 granite intrusions.

This temporal thermal increase in the evolution of metamorphism may take place with or without a pressure decrease. For example biotite-staurolite-andalusite assemblages (Godinho 1974; Martínez

[1] Departamento de Geología, Facultad de Ciencias, Universidad Autonoma de Barcelona, 08193 Bellaterra, Barcelona, Spain
[2] Departamento de Geología, Universidad de Oviedo, 33080 Oviedo, Spain

R. D. Dallmeyer and E. Martínez García (Eds.)
Pre-Mesozoic Geology of Iberia

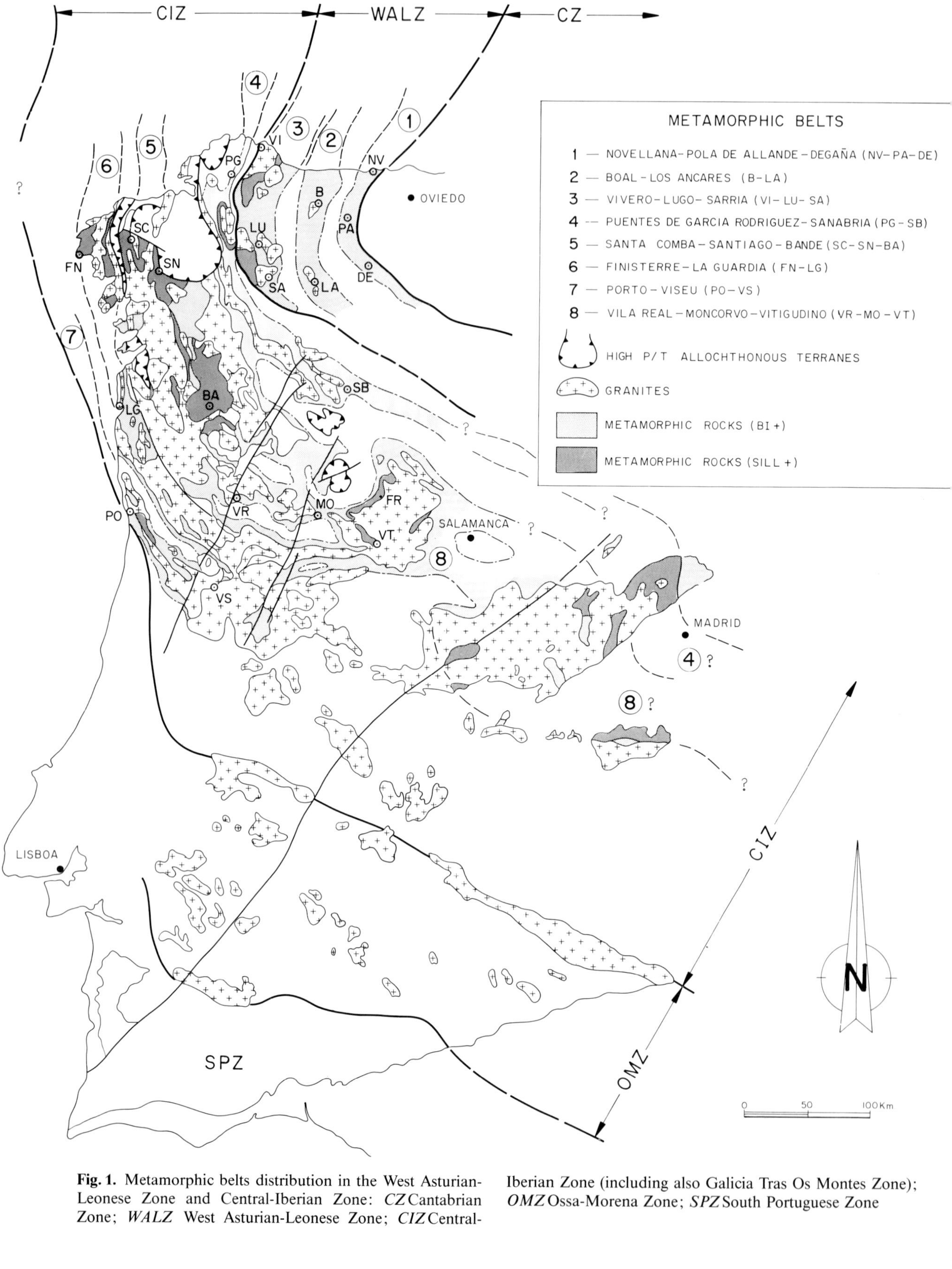

Fig. 1. Metamorphic belts distribution in the West Asturian-Leonese Zone and Central-Iberian Zone: *CZ* Cantabrian Zone; *WALZ* West Asturian-Leonese Zone; *CIZ* Central-Iberian Zone (including also Galicia Tras Os Montes Zone); *OMZ* Ossa-Morena Zone; *SPZ* South Portuguese Zone

1974; López Ruiz et al. 1975; Minningh 1975; Gil Ibarguchi 1982; Marquínez and Klein 1982) are locally found that may be produced by the continuous reaction:

St + Musc + Q — And/Sill + Bi + H_2O (Martínez and Rolet 1988).

In general, the P, T, t paths in the areas where biotite-andalusite-sillimanite ± (cordierite) grow over previous garnet-staurolite-kyanite assemblages produce a final metamorphic gradient represented by a concave curve towards the temperature axis (Martínez and Rolet 1988).

On the other hand, the occurrence of large migmatitic areas in the centre of some of the belts, especially in the area of Finisterre-La Guardia, Celanova-Bande, Sanabria-Fermoselle-Vitigudino, the Spanish Central System and Toledo zone, is a well-known fact. Following the logic of the description in the previous section, these migmatites could be considered in a traditional way, as produced by the latest higher-temperature metamorphic episode which was closely connected with the S and I type granite intrusion climax during P_3.

However, in some migmatitic areas, f. i. Fermoselle-Miranda (FR in Fig. 1) (Gil Ibarguchi and Martínez 1982; Martínez et al. 1988), restitic pelitic granulitic gneisses occur that show decompression garnet resorption into cordierite plus plagioclase coronas when intruded by S-type granitic veins. This suggests that migmatite development preceded widespread granite injection. Pre-garnet resorption P (and T?) estimates from garnet cores (Fig. 2) including rutile give values of more than 7 kb and 750 °C for the migmatitic episode, whereas syn-post P_3 intrusion seems to represent a well-defined step in a likely P-T-t path, recording conditions of 5 to 6.6 kb. This path, and others resembling it for similar areas, show a syn doming stage decompression and, to a lesser extent, a temperature decrease (Fig. 2).

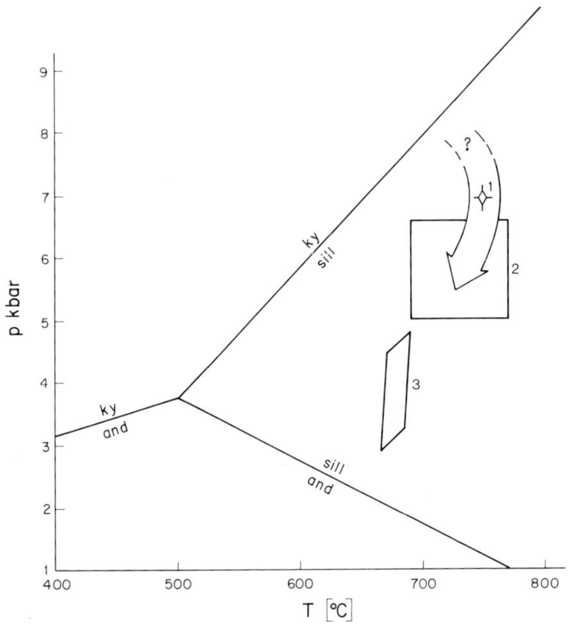

Fig. 2. Scheme of the P, T, t evolution in the core of the Fermoselle-Miranda belt (Fig. 1). *Small upper rhomb (1)* represents most likely P conditions for garnet-core growth in migmatitic restites (7 kb ~750 °C). *Large square (2)* represents garnet resorption conditions (690–770° and 5–6.6 kb) in the same rocks, related to Namuro-Westphalian granitic vein injection. *Arrow* denotes highest-grade area trajectories in the Hercynian belts (data from Fermoselle-Miranda belt; Martínez et al. 1988). *Large rhomb (3)* represents conditions (~660–680 °C and ~3.0–4.5 kb) for the granite intrusion related event with replacement of staurolite for $Al_2Si_2O_5$-biotite in the Fermoselle-Miranda belt (garnet-biotite thermometry plus garnet-sillimanite-plagioclase-quartz barometry) Al_2SiO_5 polymorphs stability after Holdaway (1971)

Temporal Evolution of Metamorphism

Some problems arise now: for instance, whether the above-described evolution is a continuous process or not, and what would the age for every episode be and upon what kind of proofs would this age be based?

As for the restitic granulitic gneisses in migmatite areas, they can be considered either as belonging to a pre-Hercynian (see later discussion about ages) crystalline basement or as representing only an early stage in the Hercynian metamorphic evolution. An argument against a pre-Hercynian age is that their location always coincides with the highest-grade cores of the Hercynian belts, since such granulitic gneisses have never been found unconformably located and never show a sharp metamorphic contrast with respect to other younger materials from outside the belts.

However, it should be mentioned here that very similar migmatitic gneisses in the Hercynian basement of the Pyrenees have been considered as representing either a Precambrian basement in the Castillon massif (Zwart 1979) or low-pressure Hercynian (Stephanian) granulites by Pin and Vielzeuf (1983) in the Agly massif.

The development of the older Barrovian assemblages with garnet, staurolite or kyanite mentioned before can give information about the temporal evolution of metamorphism in the low- to middle-grade parts of the belts. This Barrovian event in

CIZ can be reasonably compared with the development of the same minerals in the Mondoñedo fold nappe in WALZ (see Suárez et al., this Vol.), and therefore a post-Mid-Devonian age for the beginning of metamorphism is assumed. This age would also be supported by some local (and still scarce) evidence like the Rb-Sr whole-rock Devonian age (387 ± 16 Ma, Marquínez 1984) of a low-grade, S_1-affected orthogneiss in the Santa Comba-Santiago-Bande belt (Fig. 1, *5*).

The formation of the previously referred to garnet cores, which record conditions around 750 °C and 7 kb in the Fermoselle-Miranda area (FR in Fig. 1; Martínez et al. 1988) might well correspond to the former Barrovian event. This is the older recrystallization unraveled so far in the authochthonous terrains of CIZ. If the correlation with the Mondoñedo nappe for these Barrovian events were correct, they would be post-Mid-Devonian all over CIZ, and therefore a pre-Hercynian age for the Barrovian metamorphism would be precluded in the CIZ autochthonous terrains.

The superposition of the low-pressure event related with granite plutonism over the earlier Barrovian one is evident in a number of places like Porto-Viseu. The age of this low-pressure event should be about 330–315 Ma (Namurian-Westphalian), that is, the age of the granites (Pinto 1983; Priem and Den Tex 1984) to which cordierite, andalusite and sillimanite are closely related. In consequence, the whole process of isoclinal folding and nappe emplacement ($P_1 + P_2$) related with the Barrovian episode, as well as the subsequent nappe refolding related a thermal event connected with granite plutonism, would have occurred within a time period between the Lower Namurian and the Westphalian. The lower boundary of this time period is deduced from the fact that in the Vivero-Lugo-Sarria belt (Fig. 1 of Suárez et al., see this Vol.), some Barrovian episode related granites affected by S_2 have yielded Rb-Sr whole-rock Namuro-Westphalian ages (Bastida et al. 1986). Further radiometric ages, as well as a more careful study of granite-S_2 cleavage relationships, are still needed, but an eventual low Carboniferous age corroboration for the Barrovian episode would suppose that the whole metamorphic evolution might be considered as a process completed in around 20 Ma.

It should be pointed out that in the Hercynian Spanish Central System northeast of Madrid (Fig. 1, *4*?), the same formal facts have been indicated: a first Barrovian episode with almandine-staurolite-kyanite and sillimanite roughly syn S_1 and a second episode with andalusite and sillimanite as early as the shearing phase causing the S_2 schistosity (Capote et al. 1982; Casquet and Navidad 1985). The essentially post-metamorphic climax third (P_3) folding phase originates a retrovergence with an attitude change of the former S_1 and S_2 foliations.

In conclusion, the relationship between geotectonic and metamorphic evolution in CIZ could be explained as follows: Once Lower-Middle Carboniferous syn-collisional deformation ($S_1 + S_2$ cleavages) has taken place, further shortening is resolved refolding the earlier $P_1 + P_2$ structures. This younger intracontinental deformation should have happened at approximately the same time (Namurian-Westphalian) all over the inner parts of the Massif and probably a little later in the eastern tip of the West Asturian-Leonese Zone (formation of the Narcea Antiform). It should then be at this Namurian-Westphalian time when the dome-stage P_3 refolding and the linked granite intrusion took place, producing the last thermal supply which, added to the former Barrovian zoning, defines the present-day final metamorphic picture now observed.

References

Atherton MP, Atkin BP, Naggar MH (1974) Kyanite in the Hercynian metamorphic rocks of the Porto-Viseu belt, North Portugal. Geol Mijnbouw 533: 189–192

Bastida F, Martínez Catalán JR, Pulgar JA (1986) Structural, metamorphic and magmatic history of the Mondoñedo nappe (Hercynian belt, NW Spain). J Struct Geol 8: 415–430

Capote R, Casquet C, Fernández Casals MJ (1982) Los grandes complejos estructurales del Sistema Central: modelo de evolución termometamórfica. Rev R Acad Cienc Exactas Fis Nat Madr LXXVI: 313–331

Casquet C, Navidad M (1985) El metamorfismo en el Sistema Central Español. Comparación entre el Sector Central y el sector oriental en base al zonado del granate. Rev R Acad Sienc Exactas Fis Nat Madr LXXIXI: 523–548

Gil Ibarguchi JI (1982) Metamorfismo y plutonismo en la región de Muxia-Finisterre (NO de España). Thesis, Univ Autón Barcelona, 255 p. Public en: Corpus Geol Gallaeciae (Lab Geol Lage), 2ª Serie, I. La Coruña

Gil Ibarguchi JI, Martínez FJ (1982) Petrology of garnet-cordierite-sillimanite gneisses from the El Tormes Thermal dome, Iberian Hercynian foldbelt (NW Spain). Contrib Mineral Petrol 80: 14–24

Godinho MM (1974) Sobre o plutonometamorfismo da regiao de Guardao (Caramulo-Portugal). Mem Not Publ Mus Lab Min Geol Univ Coimbra 78: 37–77

Holdaway MJ (1971) Stability of andalusite and the aluminium silicate phase diagram. Am J Sci 271: 97–131

López Ruiz J, Aparicio A, García Cacho L (1975) El metamorfismo de la Sierra de Guadarrama, Sistema Central español. Mem Inst Geol Min Esp 85: 1–127

Marquínez JL (1984) La geología del area esquistosa de Galicia Central (Cordillera Herciniana, NW de España). Mem Inst Geol Min Esp 100: 1–231

Marquínez JL, Klein E (1982) Evolución temporal y distribución del metamorfismo en el area esquistosa de Galicia Central (NO de España). Trab Geol 12: 227-242

Martínez FJ, Rolet J (1988) Late Paleozoic metamorphism in the northwestern Iberian peninsula, Brittany and related areas in southwestern Europe. In: Harris AL, Fettes DJ (eds) The Caledonian-Appalachian Orogen. Geol Soc Publ 38: 611-620

Martínez FJ, Julivert M, Sebastian A, Arbolcya ML, Gil Ibarguchi I (1988) Structural and thermal evolution of high-grade areas in the northwestern parts of the Iberian Massif. Am J Sci 288: 969-996

Minnigh LD (1975) Tectonic and petrographic aspects of an area SW of the Lalin unit (Prov Orense and Pontevedra, NW Spain). In: Julivert M et al. (eds) Geologia de la parte Norte del Macizo Ibérico, Tomo Homenaje. I. Parga Pondal (Laboratorio Geologico de Lage). La Coruña, pp 205-207

Oen IS (1970) Granite intrusion, folding and metamorphism in central Portugal. Bol Geol Min 81: 271-298

Oen IS (1974) A note on lower pressure and higher pressure metamorphism belts in North Portugal. Geol Mijnbouw 53: 193-194

Pin C, Vielzeuf D (1983) Granulites and related rocks in Variscan median Europe: a dualistic interpretation. Tectonophysics 93: 47-74

Pinto MS (1983) Carboniferous granitoids of Portugal: some geochemical and geochronological aspects. Papers on the Carboniferous of the Iberian Peninsula. An Fac Cienc Porto Suppl 64: 15-33

Priem HNA, Tex E den (1984) Tracing crustal evolution in the NW Iberian Peninsula through the Rb-Sr and U-Pb systematics of Paleozoic granitoids: a review. Phys Earth Planet Interiors 35: 121-130

Zwart HJ (1979) The geology of the central Pyrenees. Leidse Geol Meded 50: 1-74

2.5 Metallogeny

P. Gumiel[1] and A. Arribas[2]

2.5.1 Introduction

The Central-Iberian Zone, which is the inner and most stable part of the Hercynian Massif, is an important metallogenic province. The famous Almadén mercury deposit, the strata bound antimony deposit of San Antonio (Badajoz), the old Pb-District of Linares-La Carolina (Jaen), and some important Sn, W, U, Cu, P, Li, Pb, Zn, Sb and Au mineralizations related to the collisional granitoids are all located in this zone.

The first metallogenic event in the Central-Iberian Zone is represented by the syngenetic phosphate mineralization, concentrated in oncolites, which occur in a shelf environment in the Upper Precambrian, and is related to the Cadomian synorogenic sequences (the Fontanarejo deposit, Fig. 1, *1*).

The ore deposits have been classified according to the different tectonic cycles: the pre-Hercynian cycle, in which the Hg and Sb mineralizations are related to preorogenic volcanism, and the Hercynian cycle in which the remaining metals occur as veins and lodes in extension fractures related to collisional magmatism (Fig. 1).

There is a special group of stratabound sheelite deposits of skarnoid type which occur in the Schist-Greywacke Complex of Upper Precambrian age; here, the mineralization is related to the regional metamorphism (Fig. 1; Table 1).

The main tectonic and metallogenic features which characterize the most representative ore deposits of the Central-Iberian Zone are described according to their geological setting and to their relationship with the main tecto magmatic cycles.

[1] Dir. Recursos Minerales, I.T.G.E., Rios Rosas 23, Madrid, Spain

[2] D. G. XII, 200 Rue de la Loi, Brussels, Belgium

R. D. Dallmeyer and E. Martínez García (Eds.)
Pre-Mesozoic Geology of Iberia

2.5.2 Exhalative Deposits Related to Preorogenic Volcanism

In the Central-Iberian Zone the volcanism is especially important from the metallogenic viewpoint, because the most important antimony and mercury deposits in the Iberian Peninsula are related to it.

Antimony Deposits

Antimony mineralization related to preorogenic volcanism occurred in the Early Palaeozoic, especially in Silurian, Devonian and Lower Carboniferous times (Gumiel 1982; Gumiel and Arribas 1987):

During the Lower Ordovician (Arenigian) an important volcanic activity, both felsic and mafic, took place over the whole area. The stibnite deposits of Sierra Morena (La Nazarena Mine, Fig. 1, *2*, and some occurrences in the Alcudian Valley) are related to this period of bimodal volcanism.

At the Ordovician-Silurian boundary, preorogenic Late Ordovician volcanism is considered responsible for some antimony deposits in the Upper Ordovician of the Alcudian Valley. The stibnite deposits located in the Lower Silurian (Criadero Quartzite of Llandoverian age) near Almadén (Fig. 1, *4*), probably formed as exhalative deposits on the seafloor, associated with Ordovician-Silurian volcanism. This volcanic activity is very important near Almadén, where mercury mineralization is related to it.

In the Devonian, this volcanism is especially important to the antimony mineralization, because the largest antimony deposit in the Iberian Peninsula, San Antonio (Fig. 1, *3*), is related to it. This deposit was studied by Gumiel (1982) and Arribas and Gumiel (1984). The mineralization of San Antonio occurs in a calcareous belt made up of black limestones, intraformational breccias and calcareous shales, as well as some siliceous layers of Devonian age.

The main ore minerals present at the San Antonio deposit are stibnite, scheelite, berthierite and pyrite; arsenopyrite and antimony are the main accesories; stibiconite, cervantite and goethite are the

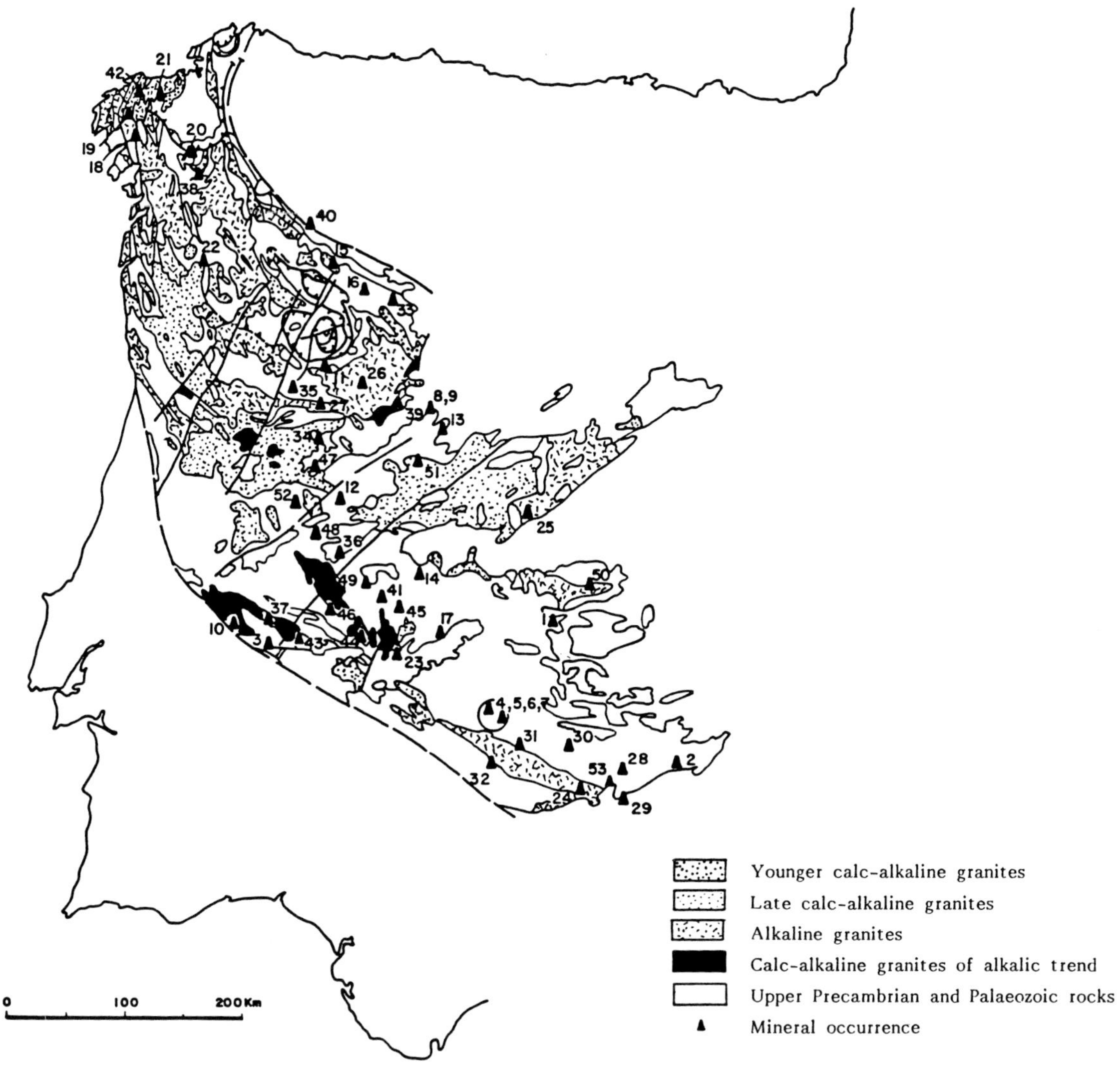

Fig. 1. Location of the main deposits and occurrences in the Spanish Central-Iberian Zone. *1* P, Fontanarejo (Ciudad Real); *2* Sb, La Nazarena, Almuradiel (Ciudad Real); *3* Sb-W-(Hg), San Antonio (Badajoz); *4* Hg, Almaden (Ciudad Real); *5* Hg, El Entredicho, Almaden (Ciudad Real); *6* Hg, Vieja Concepción (Ciudad Real); *7* Hg, Las Cuevas (Ciudad Real); *8* W, Alegria Mine, Morille (Salamanca); *9* W, Explotada Mine (Salamanca); *10* Au, La Codosera (Badajoz); *11* W, Barruecopardo (Salamanca); *12* W, La Lapa (Cáceres); *13* W, Santa Genoveva (Salamanca); *14* Sn-W, Teba (Cáceres); *15* Sn-W, Calabor (Zamora); *16* Sn-W, Arcillera (Zamora); *17* Sn-W-P, Logrosán (Cáceres); *18* Sn-W, San Finx (La Coruña); *19* Sn-W, Santa Comba (La Coruña); *20* Sn-W, Fontao; *21* Sn-W, Monteneme (La Coruña); *22* Sn-W, Lovios (Orense); *23* Sn-W, La Parrilla (Badajoz); *24* U, La Virgen (Jaen); *25* U, Escalona (Toledo); *26* U, Peralonso (Salamanca); *27* U, Sobradillo (Salamanca); *28* Pb-Zn, Linares (Jaen); *29* Pb-Zn, La Carolina (Jaen); *30* Pb-Zn, Diógenes (Ciudad Real); *31* Bi, Conquista (Cordoba); *32* Bi, Pozoblanco (Cordoba); *33* Sb, Losacio (Zamora); *34* Sn-Li, Castillejo (Salamanca); *35* Sn-Li, La Fregeneda (Salamanca); *36* Sn-Li, Las Navas (Cáceres); *37* Sn-Li, Tres Arroyos (Badajoz); *38* Sn-Li, Lalin (Pontevedra); *39* Sn, Golpejas (Salamanca); *40* Sn, Penouta (Orense); *41* Sn, El Trasquilón (Cáceres); *42* Au, Corcoesto (La Coruña); *43* U, Alburquerque (Badajoz); *44* U, Albalá (Cáceres); *45* U, Trujillo (Cáceres); *46* U, Cabeza de Araya (Cáceres); *47* U, Fe Mine, Ciudad Rodrigo (Salamanca); *48* P, Zarza La Mayor (Cáceres); *49* P, Aldea Moret (Cáceres); *50* Pb, Ag, Guajaraz (Toledo); *51* Pb, Ag, Bejar (Salamanca); *52* Pb, Ag, Cilleros (Cáceres); *53* Sb, San Agustín (Jaen)

Table 1. Metallogeny of the Spanish Central-Iberian Zone

5 *Mineralizations related to post-orogenic volcanism*
Antimony-lead. San Agustin (53)

Hercynian

4 *Ore deposits related to collisional magmatism*

4.1 Alkaline granites (two-mica leucogranites)
Tungsten: Barruecopardo (11), La Lapa (12), Santa Genoveva (13).
Tungsten-Tin: Teba (14), Calabor (15), Arcillera (16), Logrosán (17), San Finx (18), Santa Comba (19), Fontao (20), Monteneme (21), Lovios (22), La Parrila (23),
Uranium: La Virgen (24), Esacalona (25), Peralonso (26), Sobradillo (27).
Lead-Zinc-Silver: Linares (28), La Carolina (29), Diógenes (30).
Bismuth: Conquista (31), Pozoblanco (32).
Antimony: Losacio (33).
Tin-Lithium pegmatites: Castillejo (34), La Fregeneda (35), Las Navas (36), Tres Arroyos (37), Lalin (38).

4.2 Apogranites
Tin: Golpejas (39), Penouta (40), El Trasquilón (41).
Gold: Corcoesto (42).

4.3 Calc-alkaline granites of alkalic trend
Uranium: Alburquerque (43), Albalá (44), Trujillo (45), Cabeza de Araya (46). Fe mine (47) in the exocontact.
Phosphates: Alburquerque (43), Albalá (44), Zarza La Mayor (48), Logrosán (17), Aldea Moret (49).

4.4 Late-Hercynian calc-alkaline granites
Lead-Silver: Guajaraz (50), Bejar (51), Cilleros (52).

3 *Mineralization related to regional metamorphism*
Exhalative deposits associated with clastic sediments: Scheelite-bearing calc-silicate assemblages (skarnoid type): Alegría mine (8), Explotada mine (9) in Morille (Salamanca).
Gold remobilizations related to Hercynian shear zones: La Codosera area (10) in Badajoz.

Pre-Hercynian

2 *Passive margin. Exhalative deposits related to pre-orogenic volcanism*
Antimony: Antimony stratabound deposits: La Nazarena mine (2), San Antonio mine (3).
Mercury: Mercury stratabound deposits: Almadén type (4), El Entredicho (5), Vieja Concepción (6), Las Cuevas (7).

Cadomian orogeny
Phosphate mineralizations in the Upper Precambrian. Fontanarejo deposit (1).

most important secondary minerals, whereas very scarce quartz and calcite are the sole constituents of the gangue. This mineral assemblage belongs to the Sb-W-Hg association as defined by Maucher (1965). The mineralization occurred in an outer continental shelf sedimentary environment, at the shelf break or the inner continental slope. This geological setting became the site for intraformational breccias resulting from the tectonic activity to which such a zone would be prone, given the active fault system that created the Devonian shelf itself. Along these faults, submarine hydrothermal processes related to the preorogenic intra-Devonian mafic volcanism gave place to Sb-W (Hg) mineralization.

Mercury Deposits

The Almadén district (Fig. 1, *4*) is the most famous and largest mercury producer in the world. Almadén has produced more than 8 million Hg flakes. The relationship between the mercury mineralization and the volcanic activity has been studied by Saupé (1973) and Hernández (1984). The volcanism is mostly basaltic, consisting of lavas, which occasionally contain ultramafic xenoliths and pyroclastic rocks, related to explosive vents that produced chimneys and craters, and dykes and sills, which are more abundant near the volcanic centers.

The Almadén mineralization is monometallic, made up of cinnabar, native Hg and pyrite. Quartz, dolomite and baryte are the constituents of the gangue. The Hg mineralization occurs as stratabound disseminations in the Lower Silurian Quartzite (Criadero Quartzite) in Almadén (Fig. 1, *4*), El Entredicho (Fig. 1, *5*) and Vieja Concepción, (Fig. 1, *6*) and filling fissures and veins in volcanic tuffs and lavas in Las Cuevas (Fig. 1, *7*). The first type is related to the early phases of the explosive volcanism, so that the Hg was trapped in the Criadero Quartzite before diagenesis. In this type of deposit, the palaeochannels have played an important role in the distribution and concentration of the mineralization. The second type has been considered as hydrothermal in a fissural phase rich in silica and sulphides.

Support for a volcanic origin of the Hg mineralization is also given by the S34/S32 ratio of the cinnabar, which suggests a magmatic origin (Hernández 1984).

2.5.3 Mineralization Related to Regional Metamorphism

Scheelite Deposits in Calc-Silicate Rocks

There are several scheelite occurrences in the Central-Iberian Zone, mainly in the Salamanca area, which are related to the regional metamorphism. The mineral association is, in fact, a pre-Variscan mineral assemblage which has been metamorphosed and reworked during the Hercynian cycle (Arribas 1978, 1980). These scheelite-bearing calc-silicate rocks of the skarnoid type are found in strongly folded, medium to highly metamorphosed rocks of the Schist-Greywacke Complex of Upper Precambrian age, and they are never found in direct contact with intrusive rocks.

Two types of deposits can be recognized: strata-bound and vein-like. The strata bound ore bodies consist of plagioclase, quartz, hornblende, diopside, tremolite-actinolite, clinozoisite, titanite, malyaite, grossularite, idocrase, epidote, biotite, muscovite, calcite and occasionally piedmontite. The scheelite is disseminated among these minerals, together with apatite, rutile, albite, pyrite, arsenopyrite, chalcopyrite, pyrrotite, ilmenite, magnetite and fluorite.

Boudins of these calc-silicate rocks, ranging in thickness and length from several centimeters to some meters, are frequently found in all mineralized areas. A metasomatic diffusion of Ca, Mg, Al and Si, which took place at the contact of the marly and pelitic layers, is responsible for the development of an amphibolite rim at the margin of the boudins, which consist mainly of plagioclase, zoisite and garnet.

The primary origin of these scheelite strata-bound mineralizations (Arribas 1980) is probably synsedimentary, with tungsten being concentrated in specific horizons during the regional metamorphism. Later on, in the Hercynian orogeny, a retrograde metamorphism gave place to a mineral assemblage of a lower metamorphic grade, and to the remobilization and recrystallization of the scheelite in quartz veins filling extension fractures originated during the late tectonic phases, e. g. the Alegria mine (Fig. 1, *8*) in the Morille and Explotada mines (Fig. 1, *9*) in Tornadizos (Salamanca).

Gold Deposits

An interesting type of gold mineralization occurs in the La Codosera area (Fig. 1, *10*) in Extremadura. Here, the gold-bearing quartz veins are hosted by Devonian quartzites and slates, and are developed in NE-SW extension faults which cross-cut the Hercynian structures. These faults, developed at high angle to the local structures, show pinnate veins, dilational jogs and slickensides indicative of late oblique strike-slip movements. The ultimate origin of the gold mineralization is still doubtful, although it was probably contained in the host rocks and remobilized during the regional metamorphism and concentrated in reactivated structures (faults).

2.5.4 Ore Deposits Related to Collisional Magmatism

The Central-Iberian Zone is characterized by an important collisional magmatic activity which gave place to the intrusion of different types of granitoids. Thus, a large and diverse group of ore deposits is genetically related to granites intruded during the Hercynian orogeny. Some deposits were originated from ore-forming fluids produced during the emplacement of the granitoids, whereas others formed indirectly as a result of the remobilization of preexisting sedimentary and sedimentary-exhalative deposits.

The most significant magmatic deposits are grouped according to their association with different types of granites (Arribas et al. 1988).

2.5.4.1 Alkaline Granites (Two-Mica Leucogranites)

In general, most of the tungsten and tin deposits are related to syn- and late-kinematic granites. They occur as veins, lodes and stockworks which traverse the granite and the adjoining schists, or as greisen-bordered veins in the extremely altered apices of the granitic bodies.

Tungsten. These deposits appear as intragranitic scheelite veins (Barruecopardo, La Lapa and Santa Genoveva, Fig. 1, *11, 12, 13* respectively) in porphyritic two-mica leucogranites of alkalic trend, corresponding to highly evolved granite (with albitization, K-alteration and greisenization processes) of probably mesocrustal origin which contains numerous aplitic, pegmatitic and micrograntic bands, subparallel to the magmatic foliation and to the roof of the intrusives.

The paragenesis is made up by scheelite, wolframite, pyrite and abundant arsenopyrite. Occasionally, there is also some chalcopyrite, emplektite and molybdenite, as well as traces of gold. The gangue minerals consist mainly of quartz and muscovite.

Barruecopardo may be characterized as tungsten porphyry containing low-grade, high-tonnage tungsten distributed in numerous veins, which requires the following conditions for its formation (Arribas 1979): (1) Pre-existing tungsten in the sedimentary or metamorphic rocks that eventually are to be granitized or intruded by granites and (2) presence of a cover of unfractured rocks to prevent the escape of the mineralized solutions during the consolidation of the fertile granites.

There are other deposits belonging to the Barruecopardo type, among them the tungsten-bearing granites of La Lapa (Fig. 1, *12*) in Cáceres province, in which the mineralization appears in sheeted veins trending NE (Gumiel 1979, 1981), and Santa Genoveva (Fig. 1, *13*) in Salamanca province.

Tungsten-Tin. Such deposits of the Central-Iberian Zone show the following features in common:

1. The mineralization appears generally in NE-trending extension veins and as stockworks surrounding outcropping or near surface granites. Thickness of the veins ranges from 1 cm to 1 m;
2. Tin is more abundant in some deposits (Teba *14*, Calabor *15*, Arcillera *16*, and Logrosán *17* in Fig. 1), but usually the veins have both minerals, wolframite and cassiterite, it being possible to recognize mineral zoning in some deposits (e. g. San Finx, Fig. 1, *18*);
3. The mineral assemblage in most of the ore deposits is made up of high-temperature minerals (cassiterite, wolframite, and scheelite), with an intermediate temperature sulphide stage deposited in two phases: (1) molybdenite, arsenopyrite, stannite, bismuth and bismuthinite and (2), the final stage, with pyrrotite, chalcopyrite, sphalerite and pyrite. Quartz is formed continuously during the whole deposition sequence. This metallogenic sequence has been observed in most of the deposits in this zone (Fig. 1): San Finx *18* (Mangas 1987), Santa Comba *19* and Fontao *20* (Nesen 1981), Monteneme *21* (Gouanvic 1979), Lovios *22* (Cottard 1979), La Parrilla *23* (Gumiel 1981; Gumiel and Pineda 1981), Teba *14* (Gumiel 1981), Arcillera *16* (Arribas 1979; Gumiel 1984);
4. The main hydrothermal alterations affecting the host-rocks close to the veins are characteristic; greisenization in granitic rocks and tourmalinization in metamorphic rocks. Silicification is also observed.

Uranium. Important uranium deposits, mostly veins in the granites and surrounding metasediments and some interesting pegmatites, which contain beryllium, niobium, tantalum, lithium and rare earth minerals, occur in the Central-Iberian Zone. In the first case, there is an important structural control of the mineralization, because most of the veins are filling NE-trending extensional fractures.

Three mineral associations have been established by Arribas (1979): (1) Quartz-uranium-copper sulphides (La Virgen, Fig. 1, *24*). This association characterizes the northern border of the Pedroches batholith where the uranium minerals accompany the copper sulphides in veins within the granites or at the exocontact, (2) quartz-pitchblende-polymetallic sulphides (Escalona, Fig. 1, *25*). This association is quite similar to the quartz-pitchblende-Fe sulphide which characterizes the uranium deposite associated with the calc-alkaline granites, but differs in the amounts of sphalerite and galena, which is higher in the quartz-pitchblende-polymetallic sulphide association and (3) a quartz-coffinite-fluorite association which characterizes the Peralonso deposit (Fig. 1, *26*). Here, the most important uranium mineral is coffinite, which occurs in a strongly brecciated granodiorite intruded along a regional tectonic structure (Arribas 1975).

In the Sobradillo deposit (Fig. 1, *27*) the uranium is related to an episyenite (Arnaiz et al. 1984) originated from a leucogranite by hydrothermal alteration. The episyenite is made up of microcline, sericitized plagioclase, chlorite and muscovite, these two minerals were produced by hydrothermal alteration of the biotite. The coffinite, which is the most important uranium mineral, appears in open-space filling structures. The origin of the mineralization is related to hydrothermal alteration (episyenitization) of the granite.

Lead-Zinc-Silver. The Pb-Zn, Pb-Zn-Cu and Pb-Ag ores occur as sulphide vein-type deposits. They lie both within the granite and at some distance from the igneous rocks, and they are nearly always controlled by extensional faults. The most important deposits in this zone are the Linares *28* and La Carolina *29* mining districts, which for a long time were the most important lead producers in Europe. Most of the orebodies are located on the eastern and southern margin of the Pedroches batholith. The mineral association consists of quartz, carbonates, barite, silver-bearing galena, sphalerite and minor amounts of pyrite, marcasite and chalcopyrite (Arribas 1979). The Los Guindos vein in La Carolina (Fig. 1, *29*) is one of the largest in the district. It is 10 km long and, in some places, the ore shoots of silver-rich galena appear in dilation zones, more than 2 m across. Galenas gave ages ranging between 190 and 250 m. y. (Arribas 1971). This suggests that the ore-forming solutions were much younger than the Hercynian granites, and therefore belong to a Late-Variscan stage. Diógenes mine *30*, in Solana del Pino (Ciudad Real), is the only deposit belonging to the carbonate-quartz-stibnite-galena-silver association. This orebody is located in the Alcudian Valley, in Precambrian formations which crop out in the southern flank of the Alcudian anticlinorium. The metallogenic sequence indicates that the crystallization of pyrite, ullmanite and chalcopyrite was followed by an enrichment of the ore solutions in Pb and Pb-Sb, which gave place first to the deposition of galena as the major mineral, then to the formation of lead sulphoantimonides. Finally, during a later stage of brecciation, Sb-Ag sulphosalts and stibnite were deposited filling out voids and fissures (Gumiel and Arribas 1987).

Bismuth. This is found in NE-trending quartz-carbonate veins which cross-cut the contact between the Pedroches granodiorite and the surrounding Carboniferous schists. The small deposits are located on both the northern (Conquista, Fig. 1, *31*) and southern (Pozoblanco, Fig. 1, *32*) borders of the Pedroches batholith. The mineral assemblage consist of quartz, carbonates, gersdorffite, cobaltite, emplektite, tetrahedrite, pyrite and arsenopyrite, as well as traces of galena, gold and pitchblende.

Antimony. The totally weathered Losacio antimony deposit (Fig. 1, *33*) is spatially related to the albitized leucogranite of Losacio. The host rocks are an alternating sequence of quartzites and slates of Ordovician age. The mineralization occurs in step E-W trending veins, and the mineral association is made up of antimony oxides, namely stibiconite, cervantite and senarmontite (Gumiel 1982).

Tin-Lithium Pegmatites. These Sn-Li pegmatites are always associated with other cassiterite deposits of the Central-Iberian Zone. Some pegmatites are bordered by stocksheiders, as in Castillejo *34* (Salamanca province), and the mineral assemblage is made up of K-feldspar, quartz, muscovite, ambligonite, topaz and lepidolite. In La Fregeneda (Fig. 1, *35;* Salamanca) a large pegmatite dyke with Li-muscovite and lepidolite is associated with a stockwork of quartz-cassiterite veins cutting the metamorphic Precambrian rocks.

The Las Navas deposit (Fig. 1, *36*), in Cáceres province, consists of a swarm of N-trending pegmoaplitic dykes cross-cutting the Schist-Greywacke Complex of Upper Precambrian age. The tin mineralization appears as cassiterite in open space filling structures. The lithium minerals are lepidolite and spodumene. Similar pegmatitic bodies occur at Tres Arroyos mine (Fig. 1, *37*) in Badajoz province, where NW-trending pegmatitic dykes are related to albitic facies of the Alburquerque batholith.

In Lalin (Fig. 1, *38*), Pontevedra province, the pegmatitic dykes extend in a zone more than 5 km long, made up of garnet and staurolite mica schists of Ordovician or Silurian age. They consist of quartz, microcline, albite, muscovite and beryl, with large crystals of spodumene up to 20 cm long. The dykes are crossed by veinlets of quartz containing cassiterite, molybdenite, pyrite and chalcopyrite.

2.5.4.2 Apogranites

Tin. These types of Sn deposits are related to the apogranites originated during the final stages of the magmatic evolution of some muscovite leucogranites. These granitic bodies occur as dykes, stocks and sheets connected with cupolas. The albitite-greisen mineralization, mainly cassiterite, is disseminated in the albite granites, which are usually strongly greisenized and silicified. These microgranites have been worked for their tin and tantalum content. The most important deposits (Fig. 1) are Golpejas *39* (Arribas et al. 1981; Mangas and Arribas 1984), Penouta *40* (Mangas 1987) and El Trasquilón *41* (Gumiel 1981; González Aguado and Gumiel 1984), in Salamanca, Orense and Cáceres provinces respectively. In all these deposits the mineral assemblage is made up of quartz, albite, K-feldspar, muscovite, cassiterite, tantalite, tapiolite, ilmenite, rutile, apatite and fluorite.

Gold. The Corcoesto gold deposit (Fig. 1, *42*) in La Coruña province, is located in the Malpica Shear Zone. The host rock is a two-mica leucogranite with a strong C-S fabric, which shows extensive muscovitization and tourmalinization next to the veins. The ore is made up of arsenopyrite, pyrite, chalcopyrite and scheelite. Pyrrothite and gold appear as inclusions in the arsenopyrite (Arribas and Iglesias 1981). The mineralization occurs as disseminations in the granite host, and the gold-bearing quartz veins occupy extensional fractures developed during the late stages of ductile-brittle deformation.

2.5.4.3 Calc-Alkaline Granites of Alkalic Trend

The intrusion of these important types of granites, the so-called younger granites, took place in the Central-Iberian Zone after the third Hercynian tectonic phase. Some important uranium and phosphate occurrences are related to these granites.

Uranium. The uranium deposits are vein-like and occupy NE-trending extensional fractures, and are located both in the granites and the peribatholitic schists. Among the first (Fig. 1), the Alburquerque *43*, Albalá *44*, Trujillo *45* and Cabeza de Araya *46* occurrences, all located in Extremadura region, are the most important. The mineral association of quartz-pitchblende-Fe sulphides, which characterizes the intragranitic deposits, consists almost exclusively of pitchblende, coffinite, pyrite, marcasite and melnikovite in a gangue of quartz and jasper, usually hematitic, with traces of calcite (Arribas 1975, 1978). The secondary U minerals are very abundant in all the orebodies. Among the peribatholitic occurrences, the most important is the Fe mine (Fig. 1, *47*) in Salamanca province, which is also the biggest uranium deposit in Spain. The mineralization, made up of coffinite and pitchblende, fills breccias and fractures in the Schist-Greywacke Complex of Upper Precambrian age, but the origin has not been convincingly explained yet. Since 1959, the mineralization has been successively explained by the concentration of U in fractured and brecciated zones of the schists by one of several processes: (1) a *magmatic* process in which U is transported in hydrothermal fluids related to the evolution and emplacement of the Hercynian granites; (2) a *supergene* process whereby U is released from the Hercynian granites during the weathering and erosion processes that produced the Pliocene peneplain; (3) a *segregation* process that leached U from plutonic rocks during Late- and/or post-Hercynian tectonic movements (faulting) and (4) a *thermal diffusion or hydrothermal flow* process which concentrated U from fertile metasediments. More recently, Arribas (1985), based upon the high geochemical U content (30 to 200 ppm) of the carbonaceous shales of the Schist-Greywacke Complex; the nature of the alteration processes, chloritization and hematitization of the host rocks; the radiometric age of the pitchblende, 37 to 57 m.y; the low temperature mineral association; the peculiar gravitational textures of the ore minerals; the fluid inclusions data from the carbonates; and the shallow tectonic activity which gave place to the hydraulic fracturing and the fault breccias which control the ore deposition, has attributed the origin of the mineralization to seismic pumping of the U contained in the carbonaceous host rocks. According to this latter model, Lower-Middle Tertiary strain in the Hercynian basement developed around shear zones resulting from the Alpine tectonic activity could have given rise to episodic remobilizations of the U by hydrothermal fluids.

Phosphates. This peculiar and widespread type of deposit occurs in granites of the southern Central-Iberian Zone (Extremadura) where they are called fosforites. The NE-trending veins are up to 1 km long, 1 to 3 m wide and up to 100 m deep. They consist almost exclusively of quartz and apatite (mainly the radial-fibrous variety dahllite) (Arribas 1978; Aizpurua et al. 1982).

The veins are hosted both in the endocontact of some calc-alkaline leucogranites (Alburquerque *43*, Albalá *44*, Zarza La Mayor *48* in Fig. 1) which are themselves frequently tin- and uranium-bearing, or in the surrounding metamorphic Precambrian (Logrosán, Fig. 1, *17*) and palaeozoic rocks (Aldea Moret, Fig. 1, *49*).

2.5.4.4 Late-Hercynian Calc-Alkaline Granites

The lead-silver occurrences related to these granites are of less economic significance than those related to the alkaline granites. The deposits occur as veins and filling fractures in the granites or in the surrounding metamorphic rocks. In the Guajaraz deposit (Fig. 1, *50*), Toledo province, the mineralization which occurs in late, E-W trending extensional veins is made up of sphalerite, pyrite and silver-bearing galena (Arribas 1971). Other deposits belonging to this type are Béjar *51* and Cilleros *52* (Fig. 1).

2.5.5 Mineralizations Related to Postorogenic Volcanism

The postorogenic felsic volcanism is of lesser importance than the preorogenic in the whole Hercynian Massif. The best example is the San Agustín mine (Fig. 1, *53*) near La Carolina (Jaén), where antimony and also lead ores are spatially and probably genetically related to rhyodacitic subvolcanic dykes. These dykes which intrude rocks of Lower Carboniferous age (Culm) represent a postorogenic magmatic event.

References

Aizpurua J, Gumiel P, Pineda A (1982) Introducción al estudio de los yacimientos de fosfatos del Macizo Iberico Meridional. Bol Geol Min Esp XCIII: 390–414

Arnaiz J, Arribas A, García L (1984) Caracteres geológicos, prospección y evaluación de las episienitas uraniferas de Sobradillo (Salamanca). VIII Con Int Min Metal Barcelona 1: 311–326

Arribas A (1971) Distribución geoquímica de los elementos en trazas de los yacimientos españoles del tipo BGPC. Programas de investigacion, Funcacion Juan March, 704 p

Arribas A (1975) Caracteres geológicos de los yacimientos españoles de uranio. Su importancia economica e interés en el desarrollo energético del pais. Stvd Geol 9: 7–63

Arribas A (1978) Mineral paragenesis in the Variscan metallogeny of Spain. Stvd Geol 14: 223-260
Arribas A (1979) Les gisements de tungstene de Barruecopardo. Chron Rech Min 450: 42-49
Arribas A (1980) Los yacimientos de tungsteno de la zona de Morille (prov. de Salamanca). Bol Geol Min 91-2: 3-19
Arribas A (1985) Origen, transporte y deposición del uranio en los yacimientos en pizarras de la provincia de Salamanca. Estud Geol 41: 3-16
Arribas A, Gumiel P (1984) First occurrence of a stratabound Sb-W-Hg deposit in the Spanish Hercynian Massif. In: Wauschkuhn A, Kluth C, Zimmerman RA (eds) Syngenesis and epigenesis in the formation of mineral deposits. Springer, Berlin Heidelberg New York, pp 468-481
Arribas A, Iglesias M (1981) Observaciones sobre el control estructural de las mineralizaciones auriferas de la zona de Corcoesto (La Coruña). III Reun Geol Min Lab Xeoloxico Laxe Res do No. Peninsolar. Lab Xeoloxico Laxe Res (abstr)
Arribas A, Conzalo F, Iglesias M (1981) Génesis de una mineralización asociada a una cupula granítica. El yacimiento de estaño de Golpejas (Salamanca). Cuad Lab Xeol Laxe 3: 563-592
Arribas A, Arribas Rosado A, Gumiel P, Martin Izard A, Reguilon R (1988) Caracteres metalogénicos de los yacimientos minerales asociados a los granitoides del Macizo Hesperico. Homenaje a LC Garcia de Figuerola, pp 233-263
Cottard F (1979) Petrologie structurale et metallogenie du complexe granitique de Lovios Geres. Le modele de mise en place de la mine de Las Sombras (Sn-W-Mo-B) (Sud Galice-Espagne). Thesis, University of Nancy I, 217 p
González Aguado MT, Gumiel P (1984) El yacimiento de estaño del Trasquilon (Caceres): una mineralizacion asociada a una cupula granitica. Bol Geol Min 95-4: 354-373
Gouanvic Y (1979) Approche structurale d'un lineament de Verin (Galice Espagne). DEA Geochim Pet Metall Univ Nancy I, 41 p
Gumiel P (1979) Estudio del yacimiento intragranitico de scheelita de La Lapa (Hernan Perez) Caceres. Bol Geol Min 90-1: 21-35
Gumiel P (1981) Essay sur la classification typologique des principaux gisements de Sn-W d'Extremadure (Espagne). Chron Rech Miniere 463: 5-26
Gumiel P (1982) Metalogenia de los yacimientos de antimonio de la Peninsula Iberica. Thesis, University of Salamanca, 324 p
Gumiel P (1984) Tipologia de los yacimientos de estaño y wolframio del Macizo Iberico. I Congr Esp Geol Segovia, April 1984, 5: 183-216
Gumiel P, Arribas A (1987) Antimony deposits in the Iberian Peninsula. Econ Geol 82: 1453-1463
Gumiel P, Pineda A (1981) Estudio del yacimiento de scheelita de La Parrilla (Cáceres-Badajoz). Tecniterrae 39: 16-38
Hernández A (1984) Estructura y génesis de los yacimientos de mercurio de la zona de Almadén. Thesis, University of Salamanca, T-C-3701, 64 p
Mangas J (1987) Estudio de las inclusiones fluidas en los yacimientos españoles de estaño asociados a granitos hercinicos. Thesis, University of Salamanca, 644 p
Mangas J, Arribas A (1984) Evolución y caracteristicas de las fases fluidas asociadas a los filones de cuarzo del yacimiento estannifero de Golpejas (Salamanca). I Congr Geol Segovia, April 1984, 2: 551-564
Maucher A (1965) La formation Sb-W-Hg et ses rapports avec le magmatisme et la geotectonique. Freiberger Forsch H Dtsch 186: 173-187. Tras BRGM n° 4945
Nesen G (1981) Le modele exogranite-endogranite a stockscheider et la metallogenese Sn-W. Thesis, University of Nancy, 327 p
Saupe F (1973) La geologie du gisement de mercure d'Almaden. Sci Terre ENSG Nancy 29: 343

3 Allochthonous Sequences

3.1 Structure in the Northwest of the Iberian Peninsula

A. Ribeiro,[1,2] E. Pereira[2] and R. Dias[1]

3.1.1 Introduction

The NW segment of the Iberian Peninsula is characterized by the emplacement of a pile of thrust sheets on top of the Autochthon of the Central-Iberian Zone. A traverse along the coastal section in Galicia and Asturias shows a consistent vergence to the E in the different tectonic units, implying that the nappes come from W. But a traverse along the present southern limit of the nappe system, located along the right bank of the Douro River, shows a more complex structure (Fig. 1). We will describe below the structure of the autochthon and allochthon, the mechanism of emplacement and probable root zone of each nappe, and we will refer briefly to the tectonic evolution after nappe buildup.

The northern traverse, mainly situated in Spain, has been described recently in detail (Arenas et al. 1986), so we will emphasize the southern traverse, mainly situated in Portugal, and we will present an alternative tectonic interpretation of the Allochthonous Structures of NW Iberia.

[1] Departamento de Geologia, Faculdade de Ciências, Universidade de Lisboa, Lisboa, Portugal
[2] Serviços Geológicos de Portugal

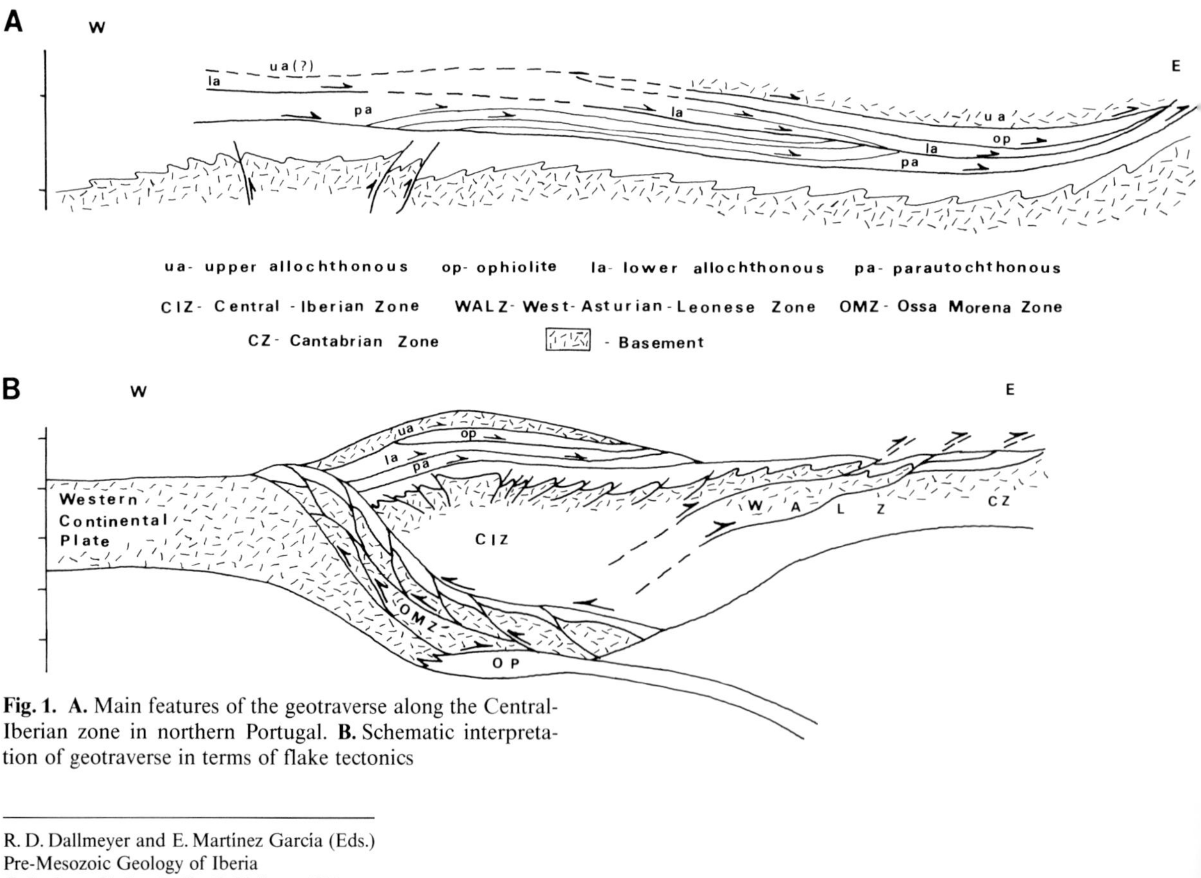

Fig. 1. A. Main features of the geotraverse along the Central-Iberian zone in northern Portugal. **B.** Schematic interpretation of geotraverse in terms of flake tectonics

R. D. Dallmeyer and E. Martínez García (Eds.)
Pre-Mesozoic Geology of Iberia

3.1.2 Structural Description

3.1.2.1 Autochthon (Central-Iberian Zone)

The youngest sediments in the Autochthon of the Central-Iberian Zone are, in its western segment, Lower Devonian platform sequences preserved in the Douro-Beira Carboniferous trough, an important ductile fault zone to which we will refer later on; flyschoid deposition started probably at the same time further to the NE (Pereira 1987), announcing the beginning of the Variscan Orogeny in the more internal zones. In the eastern part of the Central-Iberian Zone condensed platform facies of Lower and Middle Devonian age are followed by Upper Devonian or Lower Carboniferous proximal flysch and wild flysch sequences already containing pebbles and exotic klippe derived from the allochthonous units (Ribeiro 1974).

The structure of the Central-Iberian autochthon is mainly due to the first Variscan deformation event, D_1 (Figs. 2, 5); the sedimentary sequence previously described is affected by D_1, which is therefore of post-Lower Devonian age and approximate-

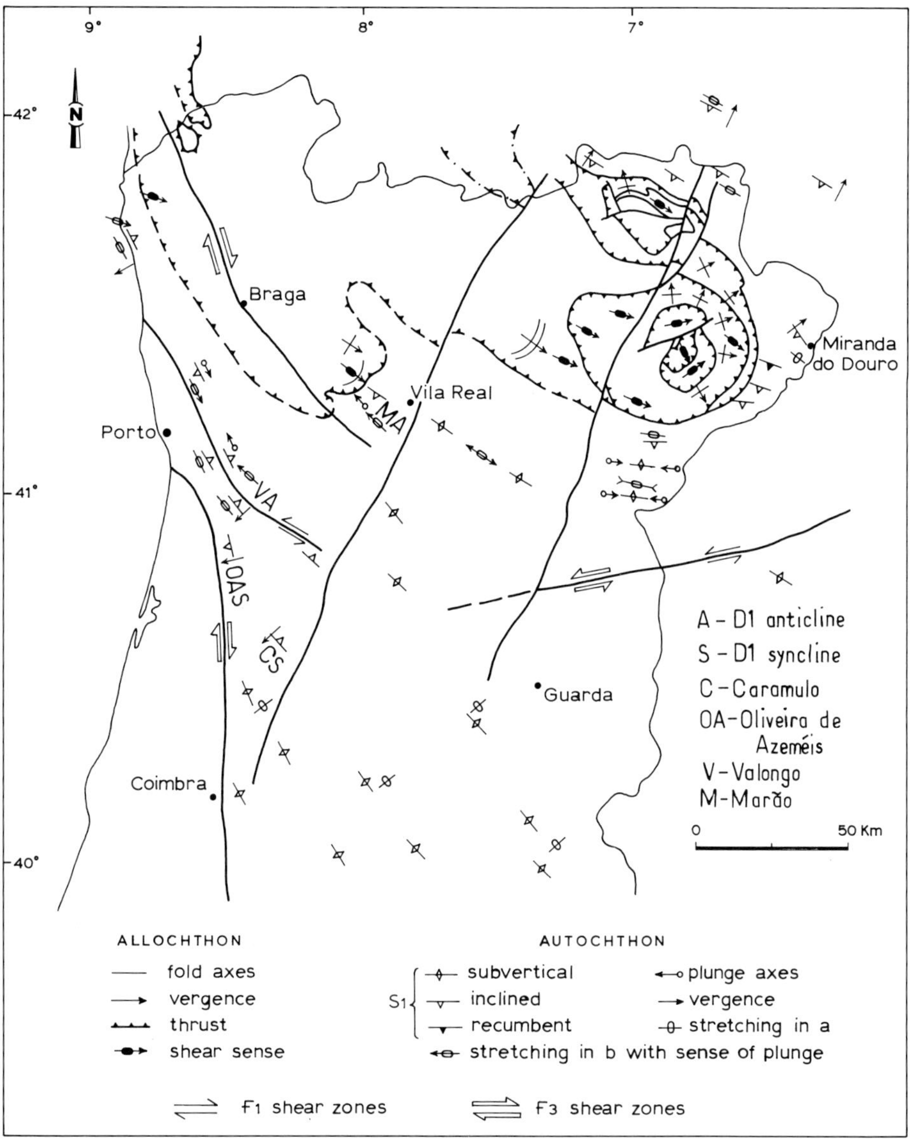

Fig. 2. Major D_1 structures in autochthon and allochthon. The original geometry of D_1 structures in Autochthon and Allochthon is shown, referring to S_1 axial plane cleavage, plunge of axes, stretching lineation X_1 and vergence of D_1 folds

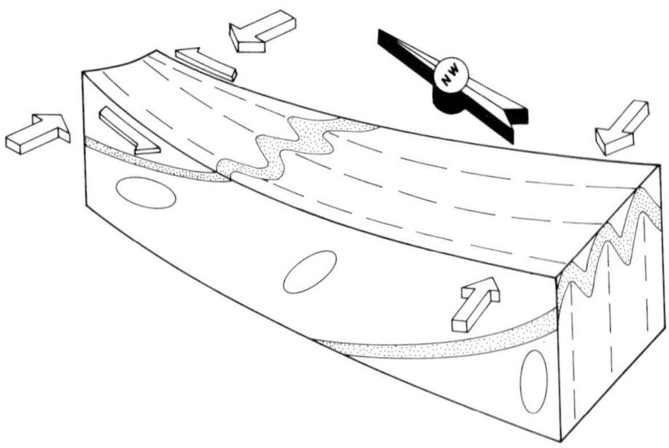

Fig. 3. Variation from left lateral transpression regime to pure compression regime in the slate area between Porto and Guarda

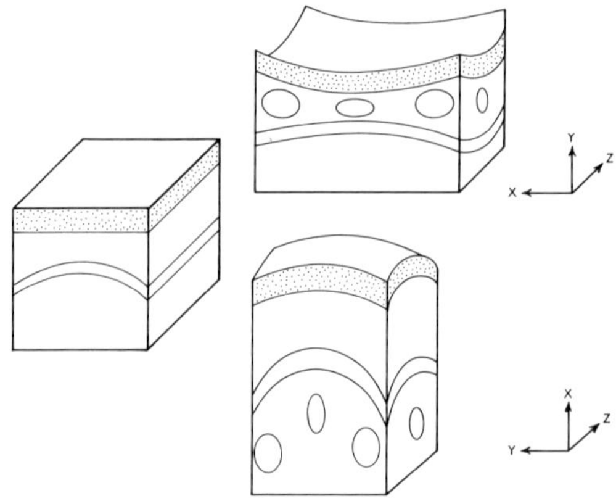

Fig. 4. Attenuation *(above)* and amplification *(below)* of sardic folds by Variscan D_1 structures in the case of stretching in b and in a respectively

ly synchronous with the Upper Devonian or Lower Carboniferous flysch in the Bragança-Alcanices area.

The D_1 event is characterized by the presence of a zone with steep slaty cleavage, axial planar to folds that turn from E-W inland, to N-S direction near the coast, delineating the typical arc pattern of the Variscan Fold Belt in the NW Iberian Peninsula. A fan-like geometry is clearly depicted between the Valongo anticline and the Serra do Marão anticlinorium. On the NE side of this steep zone, D_1 folds verge to the NE, with progressively less-inclined axial planes to the NE. On the SW side of the same steep zone, D_1 folds verge to the SW and become recumbent with kilometric inverted limbs in the Oliveira de Azemeis and Serra do Caramulo areas.

Along the steep belt between Valongo and Marão the D_1 fold axes plunge gently, and these and the stretching lineation, X_1, are subparallel (stretching in the *b* geometric axis). Subvertical shear zones are present along the whole belt; the regional S_1 cleavage rotates from approximately 20° consistently more to the N towards parallelism with the shear zone plane as one approaches those shear zones. This D_1 geometry and deformation pattern can be explained by a transpression regime (Sanderson and Marchini 1984), in which a horizontal shortening perpendicular to structures is accompanied by sinistral wrench. The resultant strain plots in the flattening field, but not very far from the plane-strain line (Dias and Ribeiro, in press).

There is a general tendency for increased strain intensity along the arc as one moves from SE to NW, approaching the sector where the curvature of the arc also increases. The S_1 cleavage converges toward higher strained domains and due to stretching in *b* a depression of fold axes results (Ribeiro 1974). This explains the general tendency of D_1 fold axes to plunge to NW and the preservation of allochthonous units in the same direction.

If one leaves the NW segment of Iberia and approaches the Central segment, the curvature of the arc decreases and the D_1 folds become upright everywhere, with a steep axial plane cleavage oriented consistently in the NW-SE direction; the stretching lineation, X_1, on the subvertical S_1 axial plane cleavage becomes down-dip, as in many slate belts; this indicates a change of the deformation regime from transpression to pure compression.

The transition between both regimes can be observed only in restricted areas because of intervening granitoid intrusion after D_1; the X_1 stretching lineation changes progressively from subhorizontal to subvertical as one proceeds from NW to SE in the slate area between Porto and Guarda (Fig. 3). This change from stretching in *b* to stretching in *a* is clearly shown by the interference between precleavage Sardic folds and D_1 syncleavage folds in the slate and greywacke complex; in fact the Sardic folds are attenuated by stretching in *b* and amplified by stretching in *a* (Fig. 4).

3.1.2.2 Allochthon

The NW Iberia Allochthon consists of a pile of four main thrust sheets (Table 1 and Fig. 5), which are, from bottom to top (Iglesias et al. 1983):

1. Parautochthonous thrust complex (PTC);
2. Lower allochthonous thrust complex (LATC);
3. Ophiolite thrust complex (OTC);
4. Upper allochthonous thrust complex (UATC).

Table 1. Correlation between the main units/complexes and formations in the NW of the Iberian Peninsula

	WESTERN GALICIA-MINHO	WESTERN TRÁS-OS-MONTES-MIDDLE GALICIA	EASTERN TRÁS-OS-MONTES: MORAIS	EASTERN TRÁS-OS-MONTES: BRAGANÇA	EASTERN GALICIA-CABO ORTEGAL
UPPER ALLOCHTHONOUS	LA PIOZA UNIT (?)	ORDENES SCHISTS MELLID ORTHOGNEISS SOBRADO DUPLEX	LAGOA SCHISTS LAGOA ORTHOGNEISS AZIBO T. VINHAS-CAMINHO VELHO DUPLEX	BRAGANÇA COMPLEX	CARIÑO SCHISTS CABO ORTEGAL COMPLEX
OPHIOLITE		LAS CRUCES COMPLEX	MORAIS UNIT LIMÃOS T. IZEDA-REMONDES UNIT	SOEIRA UNIT MOFREITA-AVELEDA TECTONIC MÉLANGE	ESPASANTE COMPLEX
LOWER ALLOCHTHONOUS	NOYA UNIT	LALIN UNIT VILA NUNE UNIT	MACEDO DE CAVALEIROS SIMPLEX CENTRO-TRANSMONTAN COMPLEX		
PAR-AUTOCHTHONOUS	PHYLLITES (Silurian) VILA VERDE T.	VILA NUNE T. MOUQUIM UNIT MOUQUIM T. CANADELO UNIT CANADELO T.	PERITRANSMONTAN COMPLEX MAIN TRÁS-OS-MONTES T.		
AUTOCHTHONOUS	Lower Dev. Sil. Ordov.	Lower Dev. (?) Sil. Ordov.	Sil. Ordov.	Upper Dev. Lower Dev. Sil. Ordov.	Sil. Ordov. Ollo de Sapo
	SLATE AND GRAYWACKE "COMPLEX"				
	FOZ DO DOURO GNEISS "COMPLEX"		MIRANDA DO DOURO GNEISS COMPLEX		

The general succession of tectonostratigraphic units is masked by out-of-sequence late thrusting only in very restricted areas, as in the western margin of the Ordenes Massif and the eastern margin of the Cabo Ortegal Massif (Arenas et al. 1986), in the northwestern margin of the Morais Massif and along the antiform axis that subdivides the Bragança Massif (Ribeiro 1983).

The tectonic style of the Allochthon is dominated by the thrust regime related to nappe emplacement. So, contrary to what happens in the Autochthon where strain increases downwards, in the Allochthon there is a discontinuous increase of strain upwards, with zones of higher strain gradients delineating the main thrust planes. Those isolate sheets with a thickness of 1 to 2.5 km and length over 100 km, measured parallel to transport direction (Fig. 5).

In the Allochthon one can find two Variscan deformation events in Paleozoic rocks. D_1 produces recumbent folds with axial plane cleavage, S_1. The fold hinges are frequently curved and sheath folds can occur at various scales; the mesoscopic sheath folds are indirectly proved by bending of intersection lineations between bedding and S_1 cleavage L_1 ($S_0\hat{}S_1$) and by a constant stretching lineation, X_1, along which the differential displacement takes place.

D_2 also produces recumbent folds with axial plane crenulation cleavage or schistosity, S_2; in the highly deformed zones, usually along thrust planes, S_2 can retranspose S_1, becoming the regional cleavage, and the D_2 fold axes become increasingly curved.

The study of deformation-recrystallization relationships during Variscan events shows that the peak of metamorphism was reached during D_2 (Ribeiro 1974) with slightly prograde conditions deduced from parageneses coeval with D_1 and D_2 respectively (Munhá et al. 1984; Ribeiro 1976; Farias et al., in press).

The nappes and thrusts are deformed by later Variscan deformation events (see Sect. 3.3, below).

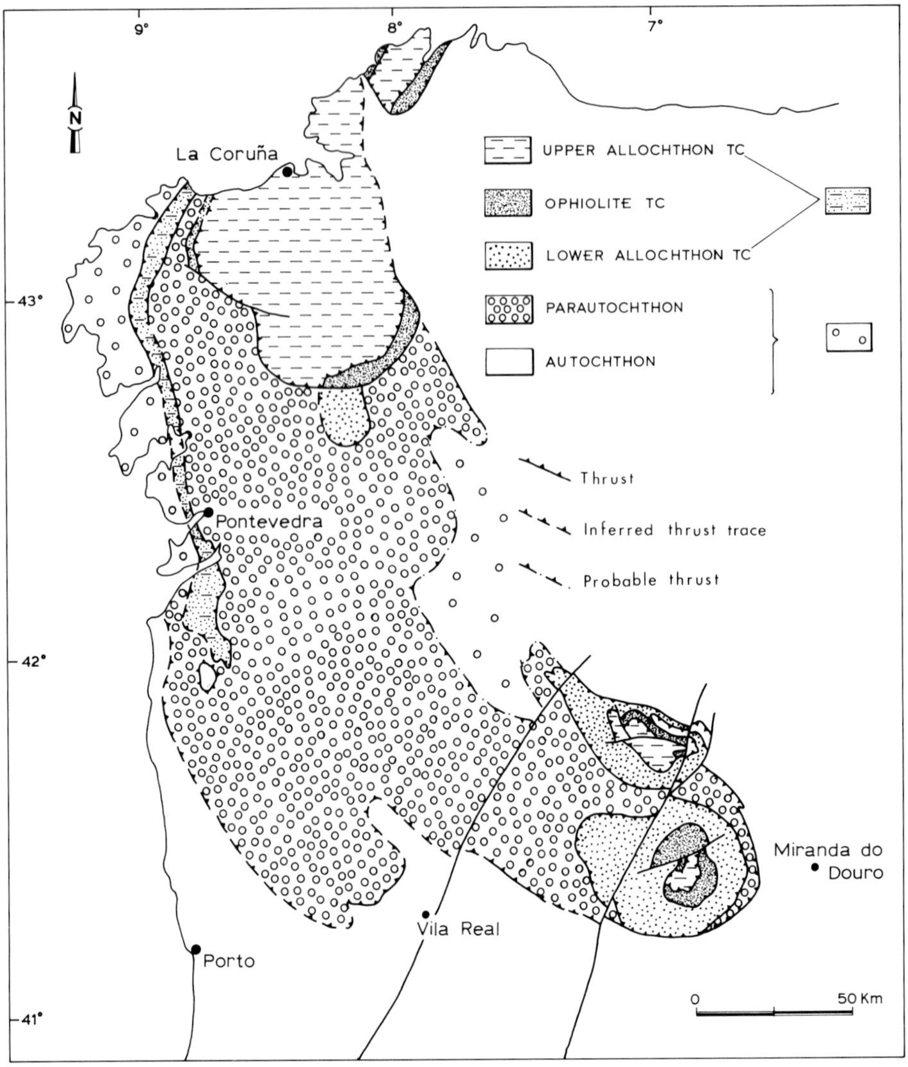

Fig. 5. Distribution of allochthonous units in the NW of the Central-Iberian Zone

3.1.2.2.1 Parautochthonous Thrust Complex

The Parautochthonous Thrust Complex (PTC) includes, for the major part, metasediments (phyllites, graywackes, quartzites and cherts), with close paleogeographic affinities with the Autochthon; its age is Silurian with possibly Lower Devonian flysch towards the top.

In Eastern Trás-os-Montes (Ribeiro 1974) the Allochthon rests on top of the Autochthon through an intervening thrust, the Main Trás-os-Montes Thrust (MTMT). Where it is exposed the MTMT is clean-cut, with a thin zone (1–10 cm) of cataclasis and fault gouge. There is a clear discontinuity in tectonic style between the Autochthon, with steep S_1 cleavage and stretching in *b* and the Parautochthon, with flat-lying S_1 and S_2 cleavages and stretching rotated towards the transported direction; this discontinuity is of course obscured where post-nappe deformation events are more intense. In some domains S_2, which becomes more penetrative along the thrust plane, is steeper in the Autochthon, suggesting that the PTC was emplaced during D_2, but movements along MTMT persisted until very late. In the Autochthon the D_1 sinistral shear zones can be reactivated as high-angle thrusts during D_2, as in the sector north of Moncorvo-Carviçais antiform (Dias 1986), a segment of the larger Vila Real-Moncorvo antiform (Fig. 12).

The PTC ist cut by second-order thrust planes in a variable number of thrust slices (one to three). The internal deformation in each slice is shown by

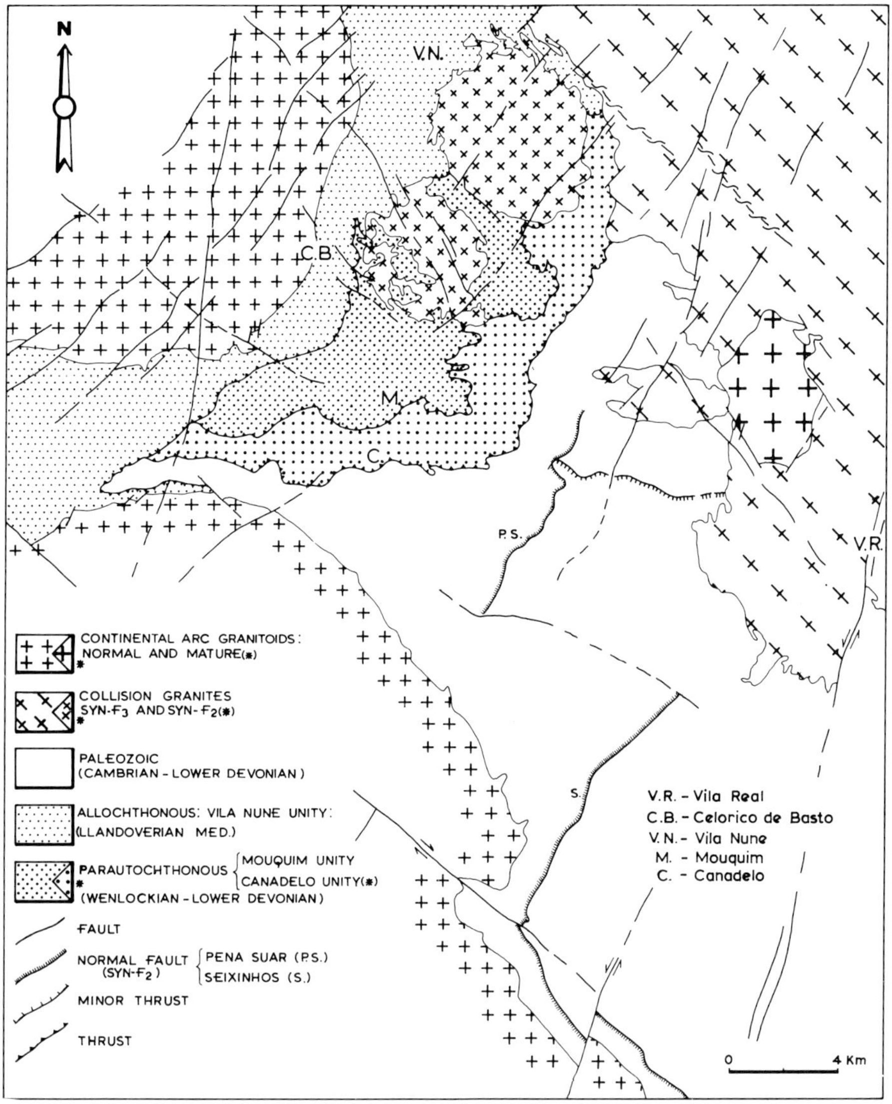

Fig. 6. Structure of the Serra do Marão region: thrust system in the NW area and extensional system in the SE area

the presence of recumbent folds with inverted limbs up to 8 km in Serra de Santa Comba, where a macroscopic sheath fold could be mapped with noses pointing to SE, which is therefore the transport direction during D_1 in this sector.

In Middle Galicia (Farias et al., in press) the PTC has been described; it corresponds to the lower part of the so-called Schistose Domain of Galicia - Trás-os-Montes. From bottom to top the sequence includes: augengneisses derived from granites, phyllites and quartzphyllites with felsic metavolcanics, the Santabaia Group, of unknown age and finally, a sequence of black phyllites with thin lenses of black cherts, the Nogueira Group, of Silurian age.

In Western Trás-os-Montes, NW sector of Serra do Marão (Pereira 1987), the style variation from Autochthon to Allochthon can be put in evidence. Three thrust units are superposed; the two lower ones, Canadelo and Mouquim, belong clearly to PTC, and the uppermost, Vila Nune, has a paleogeographic character transitional to the Lower Allochthonous Thrust Complex but is similar to PTC from the metamorphic regime point of view, and should therefore be transitional between both thrust complexes.

As the basal thrust plane is approached, the S_1 cleavage in the Autochthon changes gradually from a steep NW-SE attitude to subhorizontal; the fold axes become increasingly curved upwards but the

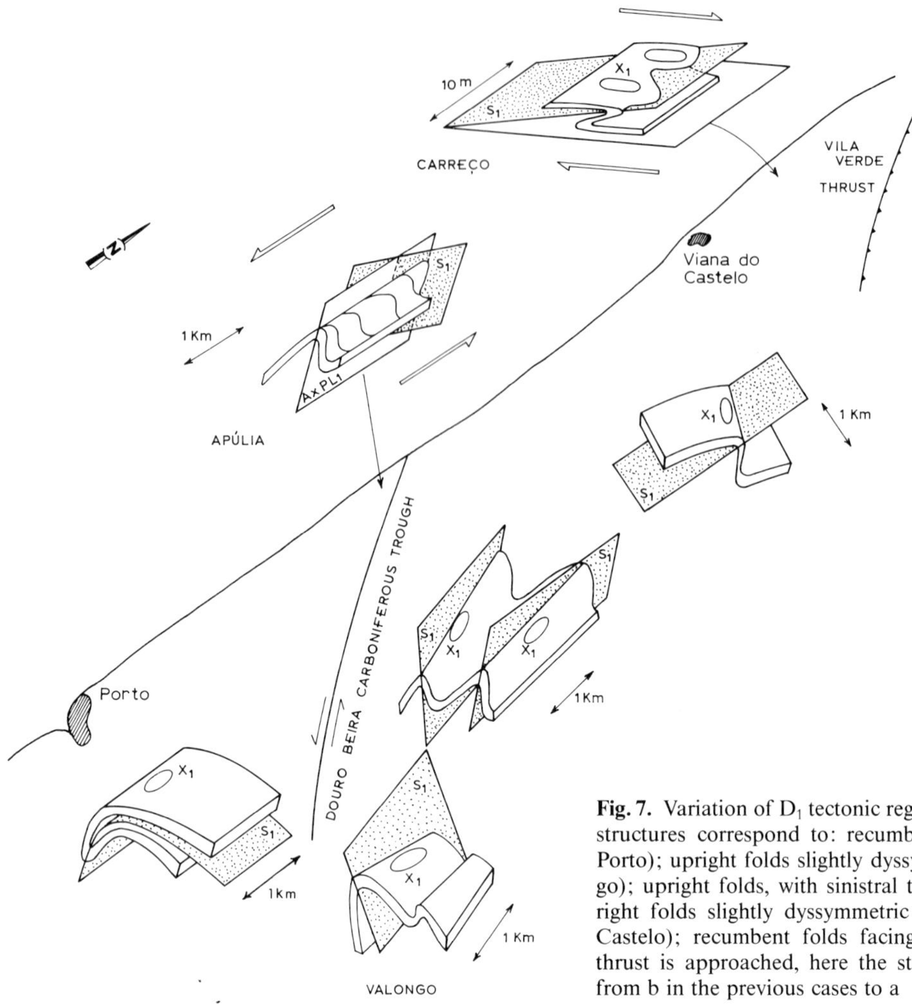

Fig. 7. Variation of D_1 tectonic regime north of Porto. The D_1 structures correspond to: recumbent folds facing W (E of Porto); upright folds slightly dyssymmetric to the W (Valongo); upright folds, with sinistral transpression (Apúlia); upright folds slightly dyssymmetric to the E (S of Viana do Castelo); recumbent folds facing E when the Vila Verde thrust is approached, here the stretching lineation changes from b in the previous cases to a

stretching lineation, X_1, remains approximately constant and oriented NW-SE. The presence of the thrusts ist also marked by the progressive development of D_2 structures with curved hinges and a stretching lineation oriented approximately N-S.

In the Serra do Marão thrust system the transport direction can be deduced from the geometry of imbrication (Butler 1982); the trailing branch line resulting from intersection of the Vila Nune and Mouquim thrusts is oriented NE-SW, and the trailing branch line resulting from the intersection of Vila Nune and Canadelo is oriented NNE-SSW (Pereira 1987). So, the transport direction varies downwards from NW-SE to WNW-ESE, making a small angle with D_1 folds in the Autochthon (Fig. 6).

In the NW sector of the Minho province, the presence of PTC has been recently proved (Ribeiro and Pereira 1986), where this complex rests on top of Autochthon by means of the Vila Verde thrust. The transition between the deformation regime in the Autochthon and the Allochthon can be mapped in the coastal zone between Porto and the Minho river (Fig. 7). From Valongo to the NW the axial plane of D_1 folds changes its dip in a direction subparallel to the structures; at Valongo the axial plane dips steeply, 60° to the NE, then becomes subvertical and, W of Famalicão, it dips steeply (70°) to the SW. The stretching lineation, X_1, remains subhorizontal at least in some domains as in Apúlia (35 km N of Porto), the S_1 cleavage transects D_1 folds and is slightly shifted northwards with respect to the axial plane of folds. This can be explained by a sinistral component of strike-slip along the Douro-Beira Carboniferous trough during the D_1 deformation event. As one moves further NW and Viana do Castelo is approached, the S_1 cleavage becomes less inclined towards the WSW, if we remove the effects of later deformation events; the stretching lineation, X_1, remains NW-SE and subhorizontal. Further N the S_1 cleavage becomes rapidly flat-lying and axial planar with respect to recumbent mesoscopic sheath folds with noses pointing to the which is the direction of a prominent stretching lineation, X_1. The reason for this is the vicinity of the Vila Verde thrust plane dipping moderately to the E, which cuts slightly obliquely the structures in the Autochthon; where it is exposed the thrust plane is a crush zone with abundant D_2 shear zones showing sense of shear to the east.

All the above observations can be explained by a change of transpression regime in the Autochthon to a thrust regime in the Allochthon.

The Parautochthon above the Vila Verde thrust plane is mainly composed of Silurian metasediments; the Autochthon is composed of slate and graywacke complex and Ordovician metasediments. The thrust plane coincides with a highly ductile horizon composed of carbonaceous metapelites that served as a decollement horizon. Towards the N, on the left bank of the Minho river, just E of Caminha, the Arenig quartzites are engaged in the Allochthon, suggesting that in this direction the thrust plane cuts down the section.

The present geometry of the thrust zones we described in Eastern Trás-os-Montes, Middle Galicia NW of Serra do Marão and NW of Minho shows that there is a major, more or less continuous, thrust plane at the base of Allochthon, folded by D_3 structures and cut by postnappe Variscan granites. Therefore, the Serra do Marão thrust system and the Vila Verde thrust are lateral equivalents of MTMT.

3.1.2.2.2 Lower Allochthonous Thrust Complex

The Lower Allochthonous Thrust Complex (LATC) is put on top of the PTC by means of a major thrust plane. The stratigraphic sequence and preorogenic magmatism are very distinct from those of the PTC and Autochthon, proving that we are dealing with a far-travelled nappe; in fact the Lower Paleozoic rocks of this unit contain a bimodal magmatic assemblage, including peralkaline rhyolites and granites, which are absent from lower thrust units. The metamorphic regime is also peculiar because there are some relicts of high-pressure metamorphism (Ribeiro 1976; Munhá et al. 1984; Schermerhorn and Kotsch 1984) later overprinted by Barrovian greenschists, in the E, to amphibolite facies metamorphism, in the W (Ferreira 1964; Arenas et al. 1986); this contrasts with the predominantly lower-pressure metamorphism of the PTC.

This thrust complex covers large areas in Eastern Trás-os-Montes, corresponding to the so-called Centro-Transmontan complex, that outcrops around the Bragança and Morais massifs; it also extends around the Cabo Ortegal and Ordenes massifs and forms at least the lower part of the Malpica-Vigo Band.

The general sequence in the Centro-Transmontan Thrust Complex (LATC) includes, from bottom to top:

1. Saldanha augengneisses, derived from a megacryst rhyolitic tuff;
2. Phyllito-quartzitic complex;
3. Quartzites, frequently sericite-rich;
4. Volcano-siliceous complex, including basic metavolcanics, rhyolitic tuffs associated with phyllites containing a member of hematitic phyllites and some limestones lenses;
5. Basic metavolcanics and quartz-phyllites.

The hematitic phyllites contain lenses of cherts with Monograptus, implying a Silurian age (Ribeiro 1974); the quartzites are very distinct from the Armorican orthoquartzites (Lower Ordovician) but quite similar to Caradoc and Silurian impure quartzites of the Central-Iberian and Ossa-Morena Zones; so (1) and (2) above could be from Lower Silurian to pre-Caradoc in age. Concerning the uppermost formation the age could be Upper Silurian or Lower Devonian.

The hematite phyllites correspond to an oxidizing environment that contrasts sharply with the reducing environment of the black shales of similar age rocks in the Parautochthon and Autochthon.

The Centro-Transmontan Thrust Complex comprises two main units and a hanging wall Horse or Simplex, the Macedo de Cavaleiros Simplex (Ribeiro 1986, 1987); in this the peralkaline rhyolites occur at the base of the volcanic-siliceous complex; they are associated with alkaline-transitional metabasalts. The metabasic volcanics of the uppermost formation are oceanic tholeiite basalts. This magmatic evolution indicates that a Lower Silurian intracontinental rift system evolved into an oceanic ridge in Upper Silurian or Lower Devonian times. So the Macedo de Cavaleiros area constitutes a transitional domain between the Centro-Transmontan realm, where the magmatism is of an essentially intracontinental type, and the oceanic domain represented by the Morais ophiolite thrusted on top of the Centro-Transmontan Duplex.

In Western Galicia the LATC also form at least the lower part of the Malpica-Tuy and Santiago Complexes (Arenas et al. 1986).

The Malpica-Tuy Band includes orthogneisses of various types, with ages ranging from 470 to 430 m. y; there are biotite-rich aluminous gneisses derived from granites; felsic gneisses of volcanic to subvolcanic origin; and alkaline to peralkaline orthogneisses (Floor 1966). These gneisses are intercalated in albitic schists and paragneisses and micaschists with thin bands of carbonaceous schists, carbonate lenses and calc-silicate rocks, deriving from flysch-like sequences.

Lenses and boudins of mafic rocks occur in all the rocks except in alkaline to peralkaline orthogneisses; these are in eclogite facies generally retrograded to amphibolite facies (Gil and Ortega 1985).

There is some internal imbrication on the Malpica-Tuy band (Arenas et al. 1986), and the possibility exists that its upper part could be correlated with the UATC (see below) and not with the LATC (Table 1).

In Middle Galicia the LATC can be correlated with the Lalin-Forcarei Group that rests on the Parаño Group (Farias et al., in press). Both Groups should be thrusted over the Parautochthonous. The Paranõ Group consists mainly of metagraywackes and phyllites, as the Phyllito - quartzitic complex of Eastern Trás-os-Montes and the Lalin - Forcarei Group consists of bimodal volcanics including some with peralkaline tendency, like the volcano-siliceous complex of Eastern Trás-os-Montes, with quartzites between both groups (Farias et al., in press).

The Early Variscan deformation phases, D_1 and D_2, related to nappe emplacement, are better preserved in a large tectonic basin surrounding the Morais Massif, the Morais Basin (Ribeiro 1974). The D_1 fold hinges are curved at all scales and tend to become parallel to the X_1 stretching lineation by the usual mechanism of sheath fold generation. The shear sense is towards the E and SE. But around the Morais Massif the X_1 stretching lineation, parallel to fold hinges, becomes itself curved at the macroscopic scale, contouring the SE bord of the Morais Massif. This suggests that the thrusting of the ophiolite on top of the Lower Allochthonous unit took place prior to full development of D_1 internal deformation in the ophiolite complex, and must have been an early generation structure during the development of the entire thrust belt. The D_2 deformation event shows anomalies of geometry and kinematics in some regions; in fact D_2 folds verge down-dip and curve around the Morais Massif. This suggests that D_2 folding and the generation of the basin are related. The thrust planes were active during D_2, because near them the S_1 axial plane cleavage is gradually retransposed by the S_2 axial plane crenulation cleavage.

In the Malpica-Tuy Band the same deformation events can be seen. D_1 is a subhorizontal schistosity, mostly obliterated by D_2, also subhorizontal, vergent to the east, with sheath fold geometry. Fold axes of both phases are statistically subparallel, gently plunging to SSW (Arenas et al. 1986).

3.1.2.2.3 Ophiolite Thrust Complex

The Ophiolite Thrust Complex (OTC) is present in the massifs of Cabo Ortegal, Ordenes, Bragança and Morais; the original sequence is best displayed in the last massif where the following sequence can be seen, from top to bottom: amphibolites; sheeted dyke complex; flaser gabbro; ultramafics.

The amphibolites are highly deformed and the original textures are almost destroyed; they can derive from the upper part of a sheeted dyke complex or from metabasalts. The sheeted dyke complex is transposed by S_1 schistosity, but its origin is still recognizable by the presence of multiple dykes with a fine grain, more mafic compositions cutting coarse-grained and more leucocratic host rock grading into the lower flaser gabbro. So the lower part of the sheeted dyke complex and the underlying gabbros are well characterized; some pods of isotropic underformed gabbro are present, even in the contact with underlying ultramafics, where layering is still recognizable. Petrological and geochemical studies by various authors confirm the ophiolitic character of the complex (for a review see Arenas et al. 1986).

In the Morais Massif the ophiolite is repeated by the Limãos thrust in the Monte de Morais - Talhinhas upper ophiolite unit and the Remondes - Izeda lower ophiolite unit; the latter rests on top of the Centro-Transmontan Thrust Complex by means of the Basal Morais Thrust (Fig. 8). The metamorphic grade attains amphibolite facies, with retrogradation into greenschists facies in a wide zone, approximately 15 m thick, along the Basal Thrust.

In the Bragança Massif the original sequence is obscured by repeated thrusting and by a stronger retrogradation (greenschists facies) of the entire Soeira Unit. This is due to the fact that the ophiolite thrust complex thins out towards the front of the Allochthon, which should be located in the Bragança domain along this geotraverse.

The OTC also occurs in Middle and Eastern

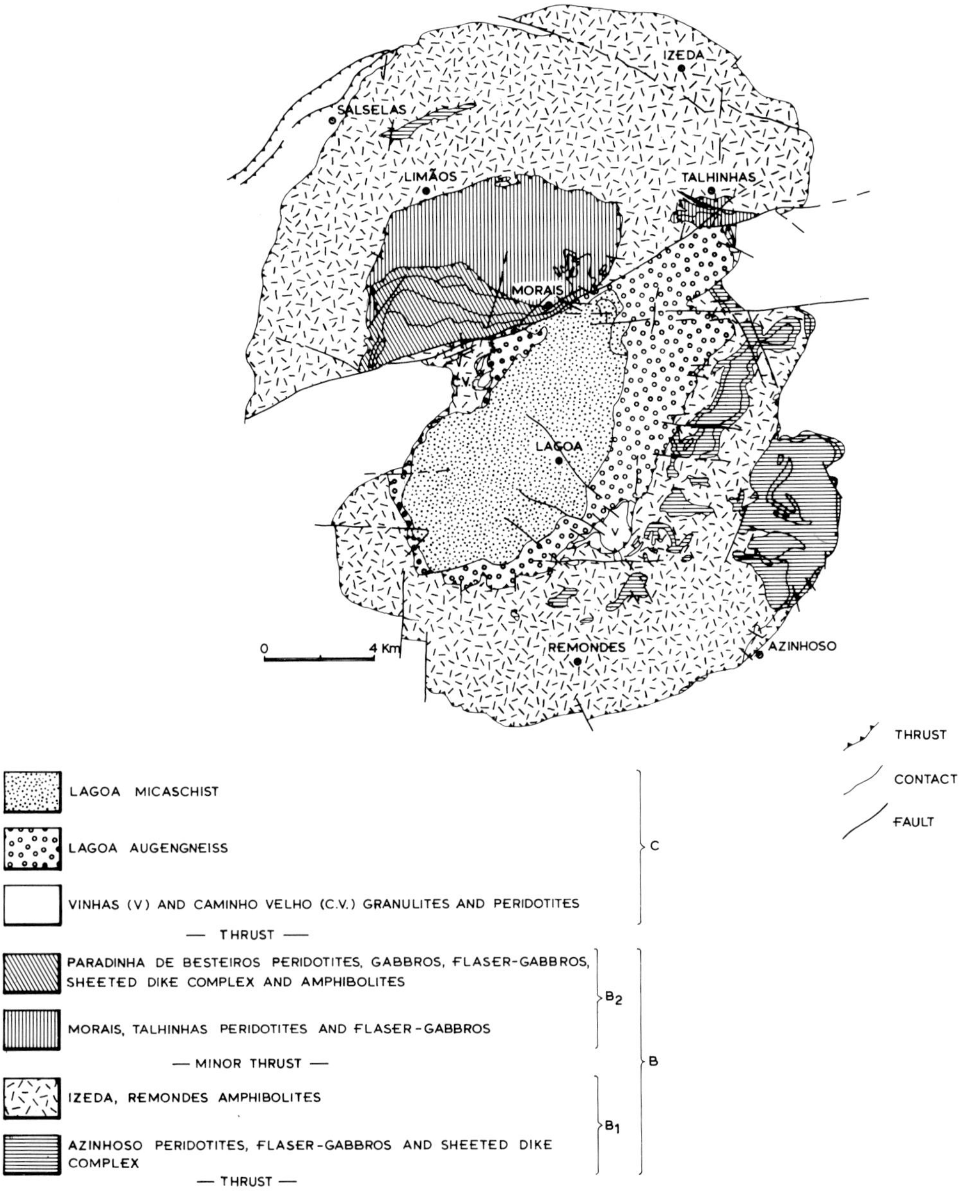

Fig. 8. Morais Massif: upper allochthonous thrust complex *(C)* and ophiolite thrust complex *(B)*

Galicia, around the Ordenes and Cabo Ortegal Complexes respectively (Arenas et al. 1986).

In the Ordenes Complex the OTC occur mainly in the Las Cruces Complex, of greenschist grade, composed of metabasic volcanic rocks, felsic metavolcanic rocks and plagiogranites with some ultramafics and metasediments.

In the Cabo Ortegal Complex the OTC occur in the imbricated Espasante Complex, consisting of tectonic mixing of metadiabases, metavolcanics, metagabbros, metaplagiogranites and serpentinites, all in the greenschist facies metamorphic grade (Arenas et al. 1986) and in the Purrido – Peña Escrita Amphibolite Unit; these units are separated by a thin thrust sheet of the UATC (see Sect. 2.2.4) emplaced by out-of-sequence thrusting.

The age of the ophiolite complex can not be proved by direct evidence, due to the absence of any fossils or isotopic dates. But the tectonostratigraphic couple constituted by the OTC and Macedo de Cavaleiros Simplex, with transport directions almost parallel, suggest that both complexes come from the same paleographic realm.

In that case the formations inside de Macedo de

Cavaleiros Simplex correspond to a rifting continental margin and the OTC to the adjacent ocean. So the Upper Silurian and/or Lower Devonian age of the tholeiites of the Macedo de Cavaleiros Formation points to a slightly younger age for the ophiolite.

The internal deformation of the ophiolite thrust complex is more clearly displayed in the Morais Massif. D_1 is marked by a flat schistosity S_1 containing an E–W ±30° stretching lineation; shear-sense criteria are best preserved in the flaser gabbros showing a general sense of thrusting to the E. Along the Limãos thrust the foliation of the Monte de Morais serpentinites rotates from NW–SE in the massive inner part of the ultramafic body to E–W in the intensely foliated domain along the thrust plane. This also implies a sense of shear to the E, subparallel to the prominent stretching lineation visible in the same domain. D_2 produces polyclinal and conjugate folds with subhorizontal axial planes and subhorizontal axes. The vergence is mainly towards the center of the almost circular Morais Basin, due to tilting of S_1 schistosity. The Basin is asymmetrically subdivided by the Morais Fault. D_3 corresponds to a slight ondulation without penetrative axial plane cleavage; D_4 produced open kinks and chevrons with NE–SW subvertical axial planes, which contribute to the slight elongation of the Basin in the same direction.

3.1.2.2.4 Upper Allochthonous Thrust Complex

On top of the OTC, and separated from it by a major thrust complex, lies the Upper Allochthonous Thrust Complex (UATC), well represented in the Cabo Ortegal, Ordenes, Morais and Bragança Massifs (and along the upper part of the Malpica-Tuy band?).

In the Morais Massif the sequence is, from top to bottom (Fig. 8):

3. Lagoa micaschists;
2. Lagoa augengneisses;
   -- Thrust --
1. Mafic granulites and metaperidotites of Vinhas and Caminho Velho.
   -- Thrust --

The Lagoa micaschists are a flyschoid sequence containing microfossils of Upper Cambrian to Lower Ordovician age (Arenas et al. 1986). Lithologically they resemble the slate and greywacke complex. The Lagoa augengneisses show relict textures characteristic of a megacryst granite. The Lagoa micaschists are affected by a main D_1 flat schistosity, S_1; a curved intersection lineation with bedding L_1 ($S_0{}^1S_1$) and a constantly oriented NNW–SSE stretching lineation are visible in the schistosity plane. The Lagoa augengneisses show the same L-S fabric as the micaschists: a flat S_1 foliation and a prominent stretching lineation with a very constant attitude in the NNW–SSE direction. A consistent sense of shear to the SSE is present, proved by S/C planes and curved tails around asymmetric augen.

Both the Lagoa micaschists and gneisses are cut by mafic dykes and sills that are transformed in amphibolites by the Variscan deformation and regional metamorphism. In the Lagoa micaschists the dykes share with their host rock the same S_1 subhorizontal fabric, showing they are of the – D_1 age. One mafic dyke cuts the micaschists – augengneisses contact, proving that there is no tectonic break between both rock types during D_1. In the Lagoa augengneisses more complex field relationships can be observed. In a localized domain clearly exposed along the Sabor Valley, 500 m N of the Remondes Bridge, the gneisses show an anomalous L-S fabric; in the flat foliation, an E–W stretching lineation can be seen with S/C planes defining a sense of shear to the W. The dykes are very heterogeneously deformed and boudinaged, allowing the preservation of relict ophitic textures in the less deformed domains. The dyke walls cut clearly the foliation of the gneisses. We conclude that in this particular sector the L-S fabric of the gneisses predates the intrusion of the mafic dykes and that we are located in a lensoid domain where the Variscan deformation was anomalously low in the gneisses. The previous L-S fabric of the gneisses was slightly reoriented with some internal deformation of the dykes, expressed in incipient foliation. We conclude that a pre-D_1 fabric is locally preserved in this sector. Elsewhere, the Lagoa augengneisses contain boudinaged lenses of amphibolites derived from mafic dykes whose walls made angles of less than 10° with the foliation of the gneisses; the original intrusive relationship of dykes or gneisses have been obliterated by more intense D_1 Variscan deformation (Fig. 9).

The above means that a pervasive foliation existed in the Lagoa augengneisses prior to the deposition of the Lagoa micaschists, which never show a pre-Variscan schistosity. The Lagoa augengneisses and the Lagoa micaschists show then a basement – cover relationship. Given the age of the micaschists and the fact that a significant time interval is needed to deposit sediments on top of foliated granites, the conclusion that the Lagoa augen-

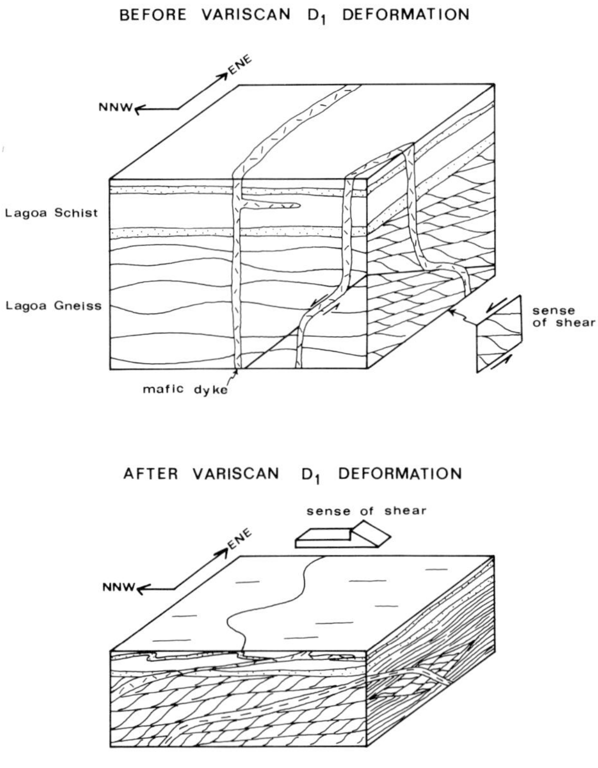

Fig. 9. Geometric relationships between mafic dykes and the Lagoa gneisses and schists. Before Variscan deformation *(above)* the mafic dykes intrude Lagoa schists and cut already foliated Lagoa gneiss with C-S fabric, indicating sense of shear to the W. After Variscan D_1 deformation *(below)*, the mafic dykes share a common S_1 with Lagoa schists and are affected by a C-S D_1 fabric with sense of shear to SSE in the Lagoa gneisses; the pre-Variscan C-S fabric is preserved only locally

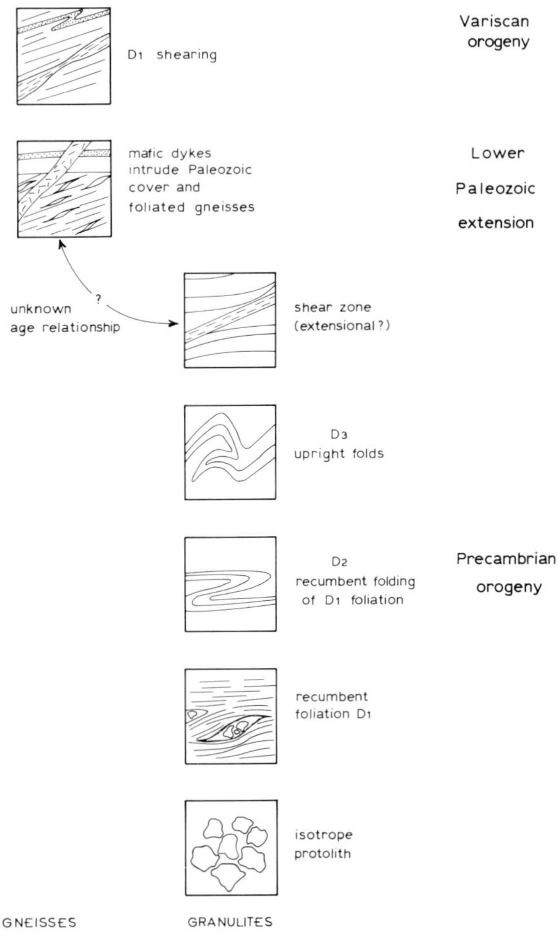

Fig. 10. Structures in the granulites and gneisses of the Morais Massif: Precambrian orogeny, Lower Paleozoic extension and Variscan orogeny

gneisses are part of a Precambrian basement is inescapable.

Further evidence of pre-Variscan tectonometamorphic events can be found in the mafic granulites of Vinhas and Caminho Velho. Along the contact with the Lagoa augengneisses one finds a 1 m wide zone of mylonites derived from the gneisses, showing that there is a minor thrust plane, because the mylonite zone on top of the OTC is much thicker (± 15 m). This is confirmed by the fact that the Vinhas and Caminho Velho duplexes are aligned parallel to the NNW-SSE stretching lineation in the Lagoa augengneisses and micaschists. Therefore, they are footwall duplexes of the Lagoa Nappe, tectonically attached to it. Away from the thrust planes the more massive mafic granulites show complex structures due to polyphase deformation; those include an earlier flat-lying foliation leaving porphyroclasts and tectonic inclusions of a former isotropic protolith; this foliation is folded by recumbent and later on by upright folds; shear zones, with sense of movement to the W, formed afterwards. The whole set is cut by dilation veins (with hornblende) with long axes parallel to the stretching lineation, X_1, of the Lagoa augengneisses and micaschists. The structures described previously then are of pre-D_1 age; they transformed the original granulites into retrograded granulites with development of blastomylonites in amphibolite facies (Fig. 10). The conditions during the D_1 Variscan event correspond to upper amphibolite facies (Munha et al. 1984).

In the Bragança Massif, rocks similar to those in the Caminho Velho duplex cover a much larger area; they are tectonically imbricated with paragneisses that contain lenses of kyanite eclogites (Ferreira, 1964; Anthonioz, 1972; Ribeiro 1974); detailed structural and petrological work on the Bragança Massif is currently in progress. (Munha and Marques 1987).

In Galicia the Massifs of Ordenes and Cabo Ortegal show lithologies and field relationships that are similar to the Massifs of Morais and Bragança respectively.

In the Ordenes Massif (Kuijper and Arps 1983; Martinez et al. 1984; Arenas et al. 1986) the sequence is, from top to bottom:

3. Ordenes micaschists;
2. Mellid-Sobrado orthogneisses;
   -- Thrust --
1. Mafic granulites and metaperidotites of Mellid-Sobrado.
   -- Thrust --

The Ordenes micaschists, tentatively correlated with the Lagoa micaschists, are deformed by three deformation events: D_1 produces a slaty cleavage to schistosity, axial planar to large recumbent folds with N-S axis and E vergence (Matte and Capdevilla 1978); D_2 is a subhorizontal crenulation and D_3 a subvertical crenulation cleavage associated with N to NNE folds.

The orthogneisses from Mellid-Sobrado intrude the micaschists and are from tonalitic to granodioritic composition; they carry the D_1 Fabric. Radiometric U-Pb analyses indicate a Lower Ordovician protholith age (Kuipjer and Arps 1983).

Metabasic rocks of various types cut both rock types. A Barrovian-type metamorphism ranging from kyanite to chlorite grade affects this unit, with a peak during D_1.

In Cabo Ortegal (Vogel et al. 1983; Marcos et al., 1984) the UATC is composed of high-pressure eclogites and granulites; metaperidotites and felsic gneisses with metabasic inclusions in eclogite facies (banded gneisses), high-pressure granulite facies (Chimparra gneisses) and amphibolite facies (Cariño gneisses). The structure is extremely complex, with earlier thrusts refolded by recumbent N-S folds cut by later thrusts.

The catazonal rocks have suffered the following tectonometamorphic evolution (Arenas et al. 1986): D_1 is the high-pressure eclogite to granulite facies metamorphism of Lower Ordovician to Lower Silurian age; D_2 is blastomylonitization of amphibolite facies grade of Upper Silurian to Lower Devonian age; D_3 produces large recumbent folds correlated with the first Variscan schistosity of Upper Devonian to Lower Carboniferous age; D_4 produces subhorizontal shear zones in greenschists facies conditions, and is correlated with the second Variscan schistosity of Tournaisian to Viséan age; and D_5 produces a steep crenulation in greenschists facies correlated with the third Variscan episode, also a steep crenulation cleavage. All the Variscan events are approximately homoaxial with N-S axes.

The UATC has been interpreted in two very different ways.

In one interpretation (synthesized in Arenas et al. 1986) the earliest metamorphic assemblages are considered to be of Lower Paleozoic age, based on available isotopic ages. The UATC is then subdivided by a major thrust plane in the so-called Upper Catazonal Unit, at a lower structural level, and an uppermost unit of metabasites, metasediments and orthogneisses, at an upper structural level.

Based on geochemical characteristics both units are interpreted as having a geotectonic setting of marginal or back-arc basin type. They are a single terrane or two separate terranes, assembled during Lower Paleozoic and transported during Upper Paleozoic to their present position, in an essentially continuous tectonometamorphic cycle.

We favour another interpretation. The Lower Paleozoic ages for the earliest metamorphic assemblages are dispersed, suggesting resetting, and they also are in conflict with field evidence that points to a Precambrian age for the polymetamorphic rocks older than the Lagoa micaschists and their lateral equivalents, the Ordenes micaschists in the Ordenes massif. We suggest that the Precambrian basement rocks have been reset during Lower Paleozoic by low-pressure, high-temperature, extensional regional metamorphism, probably related to the opening of the ocean from which the Lower Paleozoic ophiolite was obducted during the Variscan orogeny. Possibly some of the structures that caused the retrogradation of the polymetamorphic rocks by blastomylonitization developed during this extensional episode. The pre-D_1 shear zones cutting earlier recumbent and upright folds of the retrograded mafic granulites are likely candidates, as are some of the pre-D_1 shear zones cutting the already foliated Lagoa augengneisses prior to the emplacement of the mafic dykes.

In this interpretation two orogenic cycles have affected the UATC rocks; one Precambrian and one Variscan (390-280 m. y.), separated by a period of crustal thinning and oceanization during Lower Paleozoic (500-400 m. y.). The bimodal magmatic assemblage of the LATC would be essentially of the same age.

We conclude that the Upper Allochthonous thrust complex corresponds to a basement nappe of Precambrian age, thrusted on top of the ophiolite. The Lagoa and Ordenes micaschists corresponds to the Lower Paleozoic sedimentary cover of this basement; the Lagoa and Mellid orthogneisses correspond to upper granitic crust; the mafic granulites to the lower and the metaperidotites, sometimes with garnet (Anthonioz 1972; Kuijper and Arps 1983) to infracontinental upper mantle. Studies on basement nappes would be as important to recognize the nature of continental crust and upper mantle as studies on ophiolites were important to recognize the nature of oceanic crust and upper mantle.

3.1.3 Tectonic Interpretation

3.1.3.1 Nappe Development

Having described the geometry of the Autochthon and Allochthon, we can now approach the problem of interpretation of the genesis of the nappe structures, with regard to age, succession, sense of movement and possible root zone for each thrust complex.

The age of thrusting can be constrained only in terms of a large time window and taking into account its probable diachronism. The youngest sediments inside the Allochthon are the Upper Silurian and/or Lower Devonian basic metavolcanics and quartz-phyllites on top of the LATC. On the other hand, fragments of already cleaved Allochthonous lithotypes are present in an Upper Devonian Flysch deposited in the front of the Thrust Complex in the Bragança-Alcanices area (Ribeiro 1974). The beginning of tectonic instability is generally marked by the onset of flysch deposition coming from more Internal Zones, but the age of this flysch in the Autochthon, PTC and LATC is poorly constrained (Lower to Upper Devonian?).

The sequence of thrusting is probably complex. The thrust zones become increasingly ductile as one moves upwards in the thrust pile, suggesting a piggy-back sequence of events, with transport of nappes to S (UATC over OTC) followed by transport of nappes to W (OTC over LATC), in very general terms. But the transport direction for each thrust unit swings markedly and does not follow a continuous path of rotation in the same sense. This suggests that at least some thrusting should be out-of-sequence. And to reconstruct the transport direction in the first stages of thrusting we have to replace the nappes that moved later back to their root zones. In a curved thrust belt, as in the Ibero-Armorican Arc, the simultaneous thrust and strike-slip components should be compatible (Ramsay and Huber 1987, p. 545). The time relationships of the wrench regime in the Autochthon and the thrust regime in the Allochthon suggest that during the latest stages of the D_1 Variscan event the transport direction was initially oriented towards SE, with the macroscopic sheath folds in the PTC pointing their noses in that direction and the ductile sinistral strike-slip faults acting as lateral ramps inside the thrust system (Fig. 5).

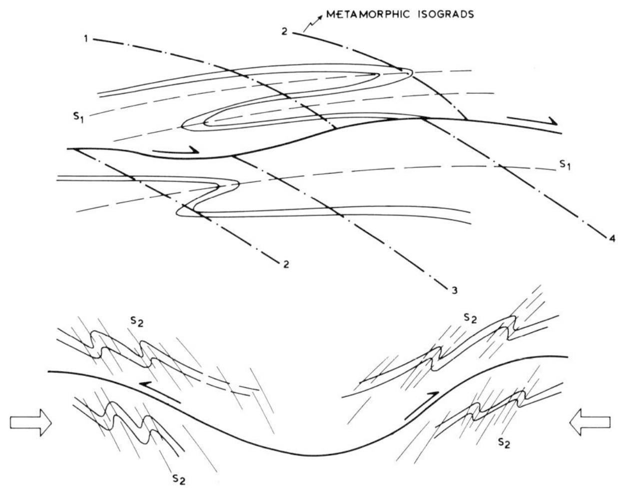

Fig. 11. Backthrusting in the trailing edge of a thrust system induced by differences in ductility caused by different metamorphic paths in the hanging wall and footwall

Later on, mostly during D_2, the thrust regime becomes predominant in the NW Iberian sector of the Variscan arc. Resheaving along thrust planes does occur; in fact, in some of the klippe the sense of movement deduced from D_2 structures can be opposed to the general sense towards E or SE depicted by D_1 structures; this can be explained by the fact that the hanging wall is more metamorphic and less ductile than the footwall, and continued compression produces backthrusting in the trailing edge of the thrust system (Fig. 11).

We can now address the problem of the root zone of each thrust complex. The PTC shows a lithostratigraphic sequence similar to that of the Autochthon, as referred to above; it comprises mainly Silurian (and Lower Devonian?) sequences. This can be explained by the presence of a decollement horizon along the black shales of the basal Silurian. We conclude that the PTC has Central-Iberian paleogeographic affinities and that the amount of displacement could be variable from a few tens of kilometers in the lowermost units to something in the order of 100 km on the uppermost units.

The LATC shows close affinities with the Ossa-Morena Zone in terms of sedimentary and magmatic assemblages. In fact, no peralkaline rhyolites and

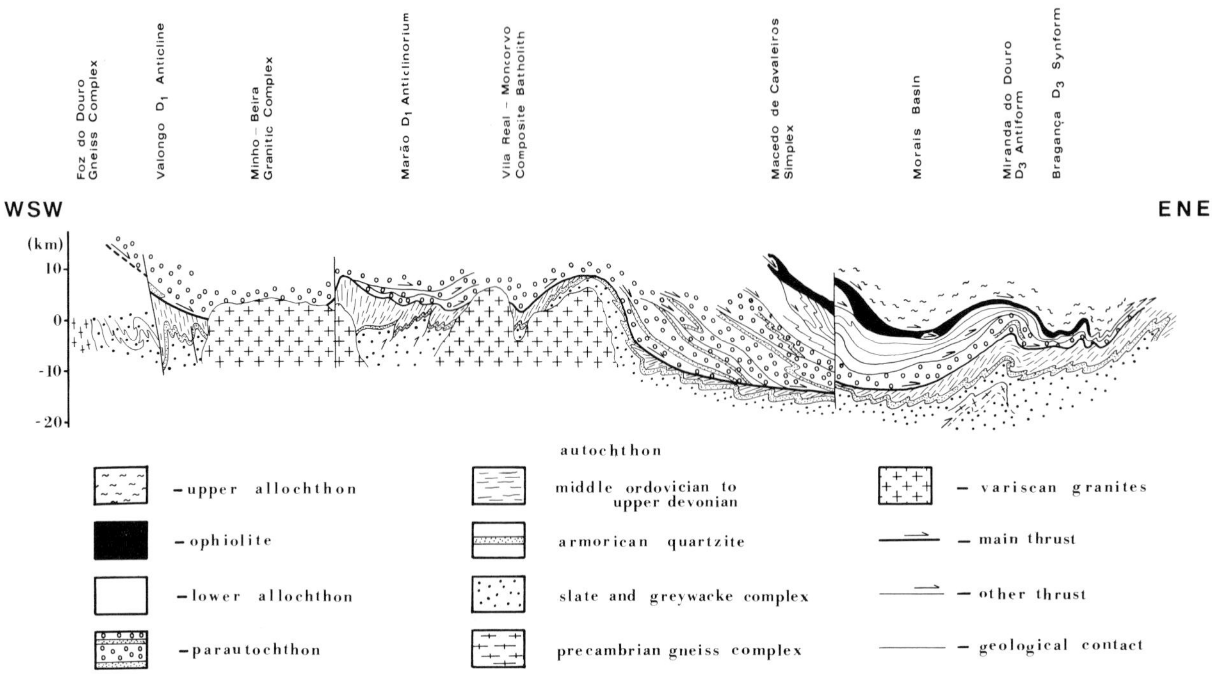

Fig. 12. Geotraverse Miranda do Douro (ENE) – Porto (WSW)

granites are known in the Central-Iberian Autochthon or the PTC but this rock type is present in the Ossa-Morena Zone, both as intrusions or extrusions of Upper Cambrian to Lower Silurian age. The general stratigraphic sequence in the LATC also has affinities with the post-Lower Cambrian stratigraphic sequence in the Ossa-Morena Zone. The latter is thrusted over the Central-Iberian Zone along the Portalegre-Tomar Thrust Zone (Julivert et al. 1974; Burg et al. 1981), with sense of movement to the NW. The northward extension of this thrust, displaced 100 km to the N along the Porto-Tomar dextral shear zone, should be the root zone of the LATC, and it should lie offshore, to the west of Porto. The minimum amount of displacement in the LATC is then 200 km (Fig. 12).

The OTC was obducted from an oceanic domain situated to the W of the northward extension of the Ossa-Morena Zone. This domain should be in some way connected with Beja-Acebuches Ophiolite (Munha et al. 1986) between the Ossa-Morena and South-Portuguese Zones.

The UATC must come from a continent situated to the W or NW of the ocean from which the OTC was obducted. To clarify the relationships between this continent and the surrounding oceans we have to integrate the geology of the NW Iberian in the larger context of the Variscan Fold Belt in Europe.

Having identified the root zone for each thrust complex, the pattern of variation of the amplitude of displacement emerges clearly: the amount of intracontinental overthrusting increases around the Ibero-Armorican Arc as one follows the D_1 structures in a dextral sense. In fact, the PTC disappears to the SW; the amount of displacement of the LATC decreases in the same direction from more than 200 km in the Porto-Miranda do Douro traverse to less than 20 km along the Portalegre thrust zone, in the boundary between the Ossa-Morena and Central-Iberian Zones. This suggests dextral transpression developed because of an oblique angle of collision between the continental players (Ribeiro et al. 1987). This picture is confirmed by sense of younging movement in a clockwise sense in the Cantabrian Zone (Perez-Estaun et al. 1987). The global dextral shear between forelands of the Variscan Fold Belt is compatible with opposed, more localized, senses of shear in the Iberian, sinistral, and Armorican, dextral, branches formed around the Cantabrian indenter.

3.1.3.2 Implications for Terrane Analysis

The tectonic interpretation we developed above has some implications in terms of Terrane Analysis of the NW Iberian Variscides, as follows. The boundary between the Central-Iberian Allochthon and Ossa-Morena Zones is the Tomar-Badajoz-Cordoba shear, a major fracture zone with components of sinistral strike-slip and NE thrusting of the Ossa-Morena Zones over the Central-Iberian Zone (Burg et al. 1981).

This is probably an intracontinental transpressive zone (Lefort and Ribeiro 1980) rather than a cryptic suture (Ribeiro et al. 1983), because no trace of a former oceanic domain could be found between both zones; the apparently different paleomagnetic signature between both domains has to be further investigated. The paleogeographic and magmatic relationships suggest that the Ossa-Morena Zone is simply the attenuated continental crust between an oceanic domain healed in the Beja-Acebuches ophiolite, in the SW, and normal continental crust, in the NE, the Central-Iberian Autochthon. In this interpretation the two allochthonous terranes would be the Ophiolite Thrust Complex and the Upper Allochthonous Thrust Complex (Ribeiro et al. 1987), sitting on top of the Iberian Terrane that includes the Cantabrian, West Asturian-Leonese, Central-Iberian and Ossa-Morena Zones. The ophiolite thrust complex of Cabo Ortegal, Santiago de Compostela, Morais and Bragança is probably connected with the Beja-Acebuches ophiolite and Pulo do Lobo Terrane, deriving all these units from a single oceanic domain between the Iberian and the South Portuguese Terranes.

3.1.3.3 Tectonic Evolution After Nappe Emplacement

After nappe emplacement the tectonic evolution of the study area changes from a thrust-wrench regime to a predominantly wrench regime during a D_3 deformation event.

At the same time, large volumes of syntectonic and late-syntectonic granitoids are emplaced both in the Autochthon and Allochthon.

But it could be demonstrated recently that a period of normal faulting occurred between D_2 and D_3 in the Serra do Marão area (Pereira 1987). In fact, a system of normal faults running (NNE-SSW) and with variable dips (60°-20°) to the W cuts the structures of the Autochthon. In the north, the Pena Suar fault runs between Pardelhas and the Freitas topographic beacon; it is undulated by metric D_3 folds, and quartz fibres along the accident are oriented E-W. To the south the Seixinhos fault extends from Paradela through Teixeiró to Arões; the trace of the fault plane changes from NNE in the N to NW in the S, due to later D_3 folding with an anticline hinge near Várzea (Teixeiró) with axial plane N40W, 70° SW and axis plunging 50° to SW. The D_3 fold is slightly offset by the Late Variscan dextral movement along the Amarante-Mesão Frio strike-slip fault. The two faults are possibly connected by a lateral ramp oriented approximately in the E-W direction (Fig. 6).

The extension direction in the Pena Suar System is approximately opposite to the thrust direction of transport in the Parautochthon of Serra do Marão. In fact, due to movement in the thrust system, the thickening related to emplacement of successive horses of parautochthon can be estimated to be at least 5 km. The maximum compressive stress can then change from the subhorizontal orientation and become subvertical, causing extension tectonics (Burchfiel and Royden 1985).

The D_3 folds have subvertical axial planes, curving gently around the Ibero-Armorican Arc, and subhorizontal fold axes. The strain intensity is variable and the regional flattening is heterogeneously distributed along conjugate shear zones, making obtuse angles with the maximum compressive stress (Iglesias and Ribeiro 1981). The D_3 folds form en échelon above these shear zones, and the stretching lineation is subparallel to fold axes, showing the predominant wrench regime. The amount of movement in the shear zones is in the order of km or tens of km but becomes negligible in some D_3 structures, where C planes oblique to the main trend indicate the non-coaxial character of the strain path regime.

The age of the D_3 event is well constrained in the 300 ± 10-million-year range by the emplacement of syn to late D_3 granitoids. The D_3 deformation is attributed to plastic indentation by the rigid Cantabrian Block. The wrench regime W of this Block clearly postdates nappe emplacement in the same area, because D_3 folds can be traced, through thrust planes, from the uppermost thrust units to the Autochthon; but it is coeval with the thrust regime in the Cantabrian Block itself, which is Upper Carboniferous in age.

Acknowledgements. This work is dedicated to the memory of Isidro Parga Pondal, who initiated the combined work of all geologists interested in NW Iberia.

We benefitted from discussions with M. Iglesias and A. Perez-Estaun and R. D. Dallmeyer. The Director of Geological Survey of Portugal, Dr. Delfim de Carvalho, and the Director of the Serviço de Cartografia of the Survey, Prof. Miguel Ramalho, gave full support to work in the field and in the Lab. A. Ribeiro and R. Dias were supported by Modelo Sismotectonofísico de Portugal, a Research Project from Direcção Geral de Qualidade do Ambiente. Fernando Barriga corrected the English manuscript; Mr. José Oliveira drafted the figures and Miss Maria de Jesus Sousa and Miss Carmen Diego typed two versions of the manuscript with great care.

References

Anthonioz PM (1972) Les complexes polymétamorphiques précambriens de Morais et Bragança (NE du Portugal): étude pétrographique et structurale. Mem Serv Geol Port 20: 192

Arenas R, Gil Ibarguchi J, González Lodeiro F, Klein E, Martínez Catalán JR, Ortega Girones E, Pablo Maciá JG, Peinado M (1986) Tectonostratigraphic units in the complexes with mafic and related rocks of the NW of the Iberian Massif. Hercynica II, 2: 87–110

Burchfiel BC, Royden LH (1985) North-south extension within the convergent Himalayan region. Geology (Boulder) 13: 679–682

Burg JP, Iglesias M, Laurent P, Matte P, Ribeiro A (1981) Variscan intracontinental deformation. The Coimbra-Cordoba shear zone (SW Iberian Peninsula). Tectonophysics 78: 161–177

Butler R (1982) The terminology of structures in thrust belts. J Struct Geol 4 (3): 239–245

Dias R (1986) Estudo de um sector do autoctone de Trás-os-Montes oriental a ENE de Torre de Moncorvo. M. Sci Thesis, Lisbon University, 153 p

Dias R, Ribeiro A (in press) Finite strain analysis in a transpression regime (Variscan autochthon, NE Portugal). Submitted to Tectonophysics and Abstract to Conference on Plate Tectonics and Deformation, Oviedo, 1987

Farias P, Gallastegui G, González Lodeiro F, Marquínez J, Martini Parra LM, Martínez Catalán JR, Pablo Maciá JG de, Rodríguez Fernández LR (in press) Aportaciones al conocimiento de la litoestratigrafia y estructura de Galicia Central. An Fac Cienc Porto

Ferreira MRP (1964) Geologia e Petrologia da região de Rebordelo-Vinhais. Mem Not Mus Lab Min Geol Univ Coimbra 58: 292 pp

Floor P (1966) Petrology of an aegirine-riebeckite gneiss-bearing part of the Hesperian Massif: the Galineiro and surrounding areas Vigo, Spain. Leidse Geol Meded 36: 1–203

Gil Ibarguchi JI, Ortega Girones E (1985) Petrology, structure and geotectonic implications of glaucophane-bearing eclogites and related rocks from the Malpica-Tuy (MT) Unit, Galicia, northwest Spain. Chem Geol 50: 145–162

Iglesias M, Ribeiro A (1981) Zones de cisaillement ductile dans l'arc ibéro-armoricain. Comun Serv Geol Port 67 (1): 85–87

Iglesias M, Ribeiro ML, Ribeiro A (1983) La interpretación aloctonista de la estrutura del Noroeste Peninsular. Geol Esp Inst Geol Min Esp Madr 1: 459–467

Julivert M, Fontboté JM, Ribeiro A, Conde L (1974) Memoria Explicativa. Mapa Tectónico de la Península Ibérica y Baleares, esc. 1:1 000 000. Inst Geol Min Esp, pp 1–113

Kuijper RP, Arps CES (1983) Los complejos de Ordenes, Lalín y Forcarey. Geol Esp 1: 422–430

Lefort JP, Ribeiro A (1980) La Faille de Porto-Badajoz-Cordove a-t-elle controlée l'évolution de l'ocean paléozoique sud-armoricain? Bull Soc Géol Fr (7 Sér.), 22 (3): 455–462

Marcos A, Marquínez J, Pérez Estaún A, Pulgar JA, Bastida F (1984) Nuevas aportaciones al conocimento de la evolución tectonometamórfica del Complejo de Cabo Ortegal (NW de España). Cuad Lab Xeol Laxe 7: 125–137

Martínez Catalán JR, Klein E, Pablo Maciá JG de, González Lodeiro F (1984) El Complejo de Ordenes: Subdivisión y discusión sobre su origen. Cuad Lab Xeol Laxe 7: 139–210

Matte P, Capdevilla R (1978) Tectonique en grands plis couchés et plissements superposés d'âge hercynien dans la série de Ordenes-Betanzos (Galice Occidentale). Cuad Semin Estud Ceram Sargadelos 27: 193–201

Munha J, Marques F (1987) A model for the Lower Continental Crust (Morais and Bragança Massifs, NE Portugal): II Petrology. Terra Cognita 7 (2–3): 161

Munha JM, Ribeiro A, Ribeiro ML (1984) Blueschistes in the Iberian Variscan chain (Trás-os-Montes: NE Portugal). Comun Serv Geol Port 70 (1): 31–53

Munha J, Oliveira JT, Ribeira A, Oliveira V, Quesada C, Kerrich R (1986) Beja-Acebuches ophiolite: characterization and geodynamic significance. Maleo, Bol Inf Soc Geol Port 2 (13): 31

Pereira E (1987) Estudo geológico-estrutural da área de Celorico de Basto e sua interpretação geodinâmica. PhD Thesis, Lisbon Univ, 274 p

Pérez Estaún A, Bastida F, Alonso JL et al. (1987) The Cantabrian Zone: an interpretation for an arcuate foreland thrust belt. Conf Deformation and plate tectonics, Abstr, Oviedo, p 92

Ramsay JG, Huber M (1987) The techniques of modern structural geology, vol 2. Folds and fractures. Academic Press, London, 700 pp

Ribeiro A (1974) Contribution à l'étude tectonique de Trás-os-Montes oriental. Mem Serv Geol Port 24 (Nova Série): 177

Ribeiro A (1983) Los Complejos de Bragança y Morais. Geol Esp 1: 450–458

Ribeiro A, Pereira E (1986) Flake tectonics in the NW Iberia Variscides. Maleo, Bol Inf Soc Geol Port 2 (13): 38

Ribeiro A, Iglesias M, Ribeiro ML, Pereira E (1983) Modéle Géodynamique des Hercynides Ibériques. Comun Serv Geol Port 69 (2): 291–293

Ribeiro A, Quesada C, Dallmeyer D (1987) Tectonostratigraphic terranes and the geodynamic evolution of the Iberian Variscan Fold Belt. Conf Deformation and plate tectonics, Abstr, Oviedo, pp 60–61

Ribeiro ML (1976) Considerações sobre uma ocorrência de crossite em Trás-os-Montes oriental. Mem Not Mus Lab Mineral Geol Univ Coimbra 82: 1–16

Ribeiro ML (1986) Geologia e Petrologia da região a SW de Macedo de Cavaleiros (Trás-os-Montes oriental). PhD Thesis, Lisbon University, 202 pp

Ribeiro ML (1987) Petrogenesis of early Paleozoic peralkaline rhyolites from the Macedo de Cavaleiros region (NE Potugal). Geol Rundschau 76/1: 147–168

Sanderson JD, Marchini WRD (1984) Transpression. J Struct Geol 6 (5): 449–458

Schermerhorn LJG, Kotsch S (1984) First occurrence of lawsonite in Portugal and tectonic implications. Comun Serv Geol Port 70 fasc. 1: 23–29

Vogel DE, Engels JP, Tex E den (1983) El Complejo de Cabo Ortegal. Geol Esp 1: 440–449

3.2 Metamorphic Evolution of the Allochthonous Complexes from the Northwest of the Iberian Peninsula

J. I. Gil Ibarguchi[1] and R. Arenas[2]

3.2.1 Introduction

The complexes of Cabo Ortegal and Ordenes (Galicia, NW Spain), Bragança and Morais (Trás-os-Montes, N Portugal) and the Malpica-Tuy Band (W Galicia), constitute remnants of an important allochthonous structure, thrust upon the materials of the Schistose Domain of Galicia-Trás-os-Montes and the Central-Iberian Zone (Fig. 1). This structure is made up of different units separated by thrusting surfaces. The tectonometamorphic and lithologic characteristics of these units may vary greatly from one another, having been the subject of a recent systematization by Arenas et al. (1986). From bottom to top these authors suggest the following disposition: (1) Basal Unit of orthogneisses, metasediments and locally eclogitized metabasites; (2) Epizonal Metasedimentary Unit; (3) Lower Epizonal Ophiolitic Unit; (4) Lower Catazonal Unit; (5) Upper Epi-Mesozonal Ophiolitic Unit; (6) Upper Catazonal Unit; and (7) Uppermost Unit of metabasites, metasediments and orthogneisses. According to the same authors, some of these units may in fact be the same terrane, though dismembered and duplicated due to amalgamation and emplacement processes. That is most likely the case for the two Catazonal and the Uppermost Units. The Basal and Epizonal Metasedimentary Units may in turn be considered as belonging to the margin of the Gondwana continent, which is represented now by the autochthonous of the thrusting structure.

[1] Departamento de Mineralogía y Petrología, Universidad del País Vasco, P. O. Box 644, 48080 Bilbao, Spain
[2] Departamento de Petrología y Geoquímica, Facultad de Ciencias Geológicas, Universidad Complutense, 28040 Madrid, Spain

One of the most striking features of these remnants of the allochthonous structure is the evidence for a rather different metamorphic evolution from that recorded by the Paleozoic materials forming their autochthonous. Thus, while a middle- to low-pressure/low- to high-temperature metamorphism is usually observed in the autochthonous (Martínez et al., this Vol.), high-pressure granulites and different types of eclogites may be found within the complexes and Malpica-Tuy Band.

The available geochronological data, along with considerations concerning the crystal growth-deformation relationships, permit the distinction of three main episodes of regional metamorphism and two of thermal or contact metamorphism within the different units making up the complexes and band. The episodes of contact metamorphism are related to the emplacement of pre-Hercynian and Hercynian intrusives. Pre-Hercynian intrusives are mainly biotitic granitoids and gabbros emplaced around 460 to 480 Ma ago (Rb/Sr WR isochrons, García Garzón et al. 1981; Kuijper et al. 1982), which induced hornblende- to pyroxene-hornfels facies conditions in the host rocks. This contact metamorphism took place before the final emplacement of the band and complexes and before the second regional metamorphism as well, although its relationship to the first regional episode has not been accurately established yet. Hercynian intrusives were mainly granites to granodiorites and their emplacement took place through a wide time span (280 to 340 Ma, cf. Serrano Pinto and Gil Ibarguchi 1987) inducing mainly hornblende- to albite and epidote-hornfels recrystallizations in the host rocks.

As for the regional metamorphism, the first episode is represented by a relatively high-pressure metamorphism of Lower Ordovician age that locally reached eclogite - or high-pressure granulite-facies conditions (older, mainly Precambrian, ages have been suggested too for this high-grade metamorphic event, see below). Secondly, and more or less synchronous with the amphibolitic retrogradation of some of the former parageneses, a prograde metamorphism of Silurian to Lower Devonian age

R. D. Dallmeyer and E. Martínez García (Eds.)
Pre-Mesozoic Geology of Iberia

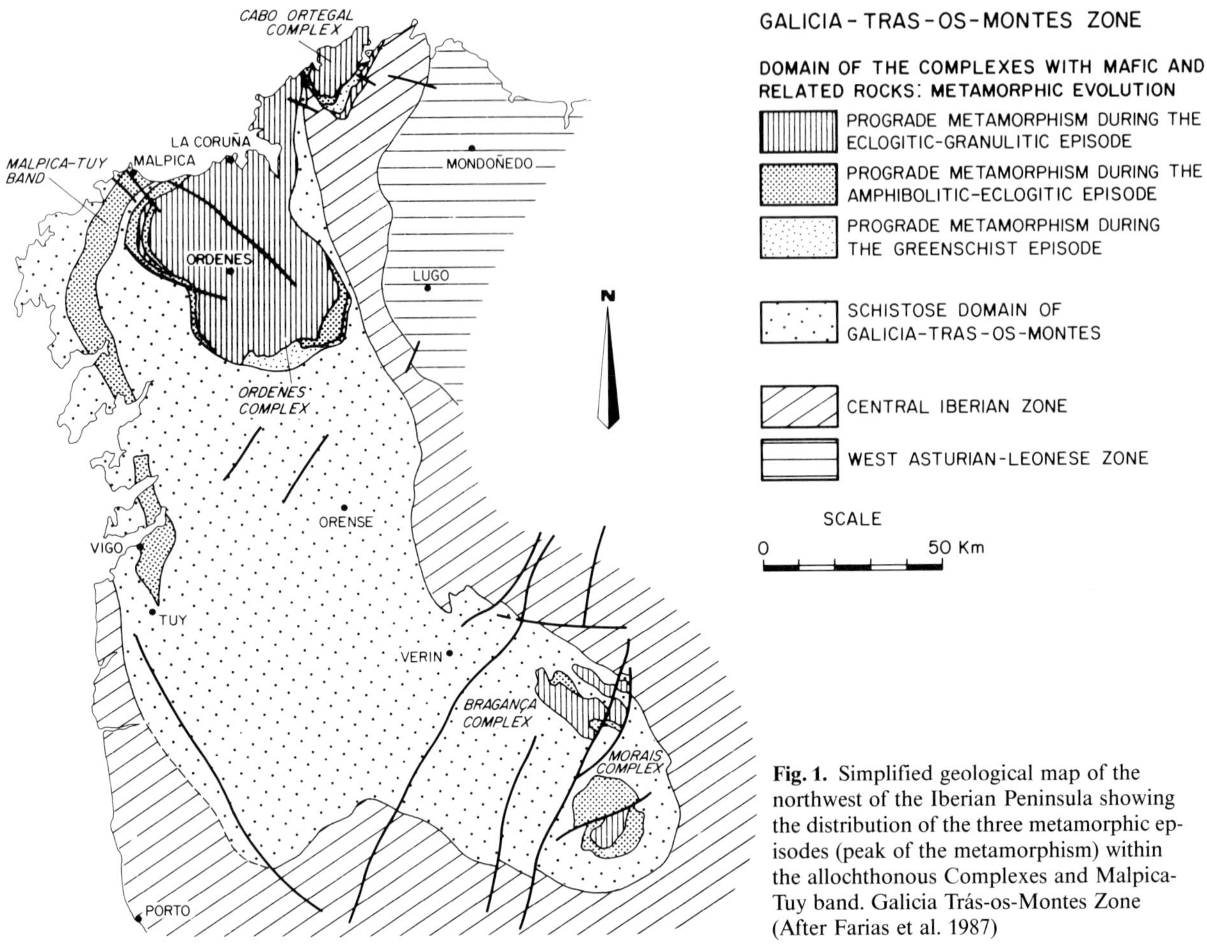

Fig. 1. Simplified geological map of the northwest of the Iberian Peninsula showing the distribution of the three metamorphic episodes (peak of the metamorphism) within the allochthonous Complexes and Malpica-Tuy band. Galicia Trás-os-Montes Zone (After Farias et al. 1987)

took place in several units, giving rise to the development of a great number of amphibolite-facies rocks and, locally, to low- to intermediate-temperature eclogites (C-type). Finally, and related to the final emplacement of the whole allochthonous structure into its present position during Hercynian times, a low-grade episode of Upper Devonian to Lower Carboniferous age variably retrograded the previously metamorphosed materials to greenschists, or amphibolite (only occasionally), facies conditions. Some of the allochthonous units, such as the Lower Ophiolitic and the Epizonal Metasedimentary Units, now appearing in tectonic mélange-like structures and intermingled with more highly-metamorphosed rocks, are likely to have been affected only by this last metamorphism. Figure 1 represents the distribution of the three episodes of regional metamorphism.

In the following paragraphs a brief description of the main characteristis of the regional metamorphic episodes will be given. It follows a discussion of the P-T-t metamorphic paths and their geodynamic meaning, along with some comments about the implications for the metamorphic evolution of the autochthonous, as related to the final emplacement of the allochthonous structure.

3.2.2 The Eclogitic-Granulitic Episode

The materials affected by this metamorphism are rather abundant in the Cabo Ortegal, Ordenes and Bragança Complexes, while they are very scarce in Morais and totally absent in the Malpica-Tuy Band (Fig. 1). They comprise the two Catazonal Units and likely the Uppermost Unit of metabasites, metasediments and orthogneisses from the classification of Arenas et al. (1986). On the basis of U/Pb determinations for zircons and monazites from gneisses of the Ordenes Complex and zircons from eclogites, granulites and gneisses from the Cabo Ortegal Complex (Kuijper et al. 1982, Bernard Griffiths et al. 1985, Peucat et al. 1990), the age of this metamorphism may be established as Lower Ordovician (475–490 Ma). On the basis of field correlations and structural and petrological interpretations, a Precambrian age has been proposed as well for this metamorphic event (e. g. Ries and Shackleton 1971; Iglesias et al. 1983; Ribeiro et al. 1987).

The most important lithologies showing evidence of this metamorphism include eclogites, basic granulites, metaperidotites, orthogneisses, metasediments, amphibolites and metagabbros to metadiorites. Prograde recrystallization previous to the metamorphic climax, i. e. what may be called the pre-eclogitic/granulitc stage, is represented by minerals included in garnets of eclogites and of some metasediments. Included minerals comprise quartz, epidote-clinozoisite, plagioclase, kyanite, biotite, muscovite, rutile and opaques; the innermost zone of some of the metapelitic garnets, which reaches molecular contents of spessartine of up to 31.4 mol %, should be a remnant of that metamorphic stage also.

The peak of the first regional metamorphism was reached during the first deformation phase. In the highest-grade rocks, such as eclogites and granulites, only a preferred orientation of clinopyroxenes remains from that deformation (Engels 1972).

Metabasites show a wide variety of mineral parageneses and textures, depending upon the intensity of the metamorphism they underwent. Among the most characteristic, there are eclogites and garnet-clinopyroxene granulites (pirygarnites and plagiopirygarnites). Mutual relationships between these two types of rocks suggest that they might have been formed at the same time, although starting from different protoliths. Thus, eclogites tend to form smaller and more homogeneous outcrops, of undoubted MORB compositions in some cases (Bernard Griffiths et al. 1985), while granulites show a great diversity of composition (mafic to felsic) and form much greater outcrops. The mineral associations found in eclogites include garnet, omphacite, quartz, zoisite, kyanite and rutile, while granulites show the same mineralogy but with plagioclase (locally scapolite too) and sodic augite instead of omphacite. Analyses of garnets and clinopyroxenes from the eclogites and granulites have been published by Vogel (1967); Anthonioz (1972); Hubregtse (1973); Kuijper (1979); Arenas and Peinade (1984) and Gil Ibarguchi et al. (1988, in press), among others. On the basis of the last authors' results, P-T conditions for garnet-clinopyroxene equilibria have been estimated at ca. 790 °C, > 17 kbar for the eclogitic climax and slightly higher T and slightly lower P (13.5 kbar) for granulites.

Other metabasites include layered amphibolites and metagabbros to metadiorites. The amphibolites are often intercalated within the schists, which suggests they may derive from volcanic rocks or from tabular intrusions. They are composed of amphibole, plagioclase and quartz, with variable amounts of biotite, epidote, garnet, rutile and opaques. The metagabbros to metadiorites may pass to flaser gabbros and amphibolites or constitute barely deformed isolated bodies of considerable size in some instances (e. g. the Monte Castelo Gabbro of ca. 15 km in diameter). The latter often show subophitic textures and are composed of variable amounts of ortho- and clinopyroxene, olivine, plagioclase and brown hornblende. Garnet coronas are found in some places between plagioclase and pyroxene, ilmenite or amphibole, which suggests these rocks could locally undergo granulite-facies conditions.

The metasediments vary greatly too: from slates or low-grade schists to migmatitic paragneisses. Metamorphic minerals include staurolite, kyanite, phengite, rutile and zoisite, which, together with the presence of intercalations of rutile-bearing amphibolites, eclogites and granulites, indicate that they underwent a high-pressure regime under variable conditions of temperature.

The ortho-derived acid rocks range in composition from tonalitic to granitic. They may represent big intrusions, such as the Corredoiras orthogneiss in the Ordenes Complex (35 × 15 km), often with conspicuous porphyritic textures, or minor intercalations of plagiogranites within some granulites and amphibolites. Their mineral associations are rather simple, including variable amounts of quartz, plagioclase, K-feldspar, biotite, muscovite, amphibole, garnet, rutile and ilmenite.

The metaperidotites are mainly spinel-bearing harzburgites. The presence of secondary pargasitic hornblende in most of these rocks may be an indicator of their being previously rich in clinopyroxene, in which case they should be considered instead as spinel-bearing lherzolites. Other lithologies often found are dunites and pyroxenites. A strong plastic deformation with development of folds and a schistosity parallel to that of the nearby granulites or eclogites may be observed in some cases. Besides that, there often exist garnet coronas around orientated spinel in the pyroxenites, which suggests that these rocks underwent a process of orogenic deformation and recrystallization coeval with that of the rocks described above.

3.2.3 The Amphibolitic-Eclogitic Episode

The highest-grade rocks formed during the previous eclogitic/granulitic episode show evidence of a second episode of regional metamorphism, contemporary with the second deformation phase, which represents a retrogradation under amphibolite-facies conditions. Obviously, the lower-grade rocks are rarely affected by the new episode, except for recrystallizations associated with the development of crenulation cleavages and some subhorizontal mylonitic bands. The age of the amphiboles formed by retromorphoses of granulitic and eclogitic parageneses clusters around 390–400 Ma (K/Ar, Van Calsteren et al. 1979; $^{39}Ar/^{40}Ar$, Peucat et al. 1990), which suggests Silurian age for this amphibolitic episode.

More or less at the same time, the Upper Epi-Mesozonal Ophiolitic Unit outcropping in the Cabo Ortegal, Ordenes, Bragança and Morais Complexes (Fig. 1) was affected by prograde metamorphism, reaching amphibolite-facies conditions in most cases. The age of this metamorphism is likely to be Silurian too, since (prograde) amphiboles have been dated at ca. 390 Ma (Dallmeyer and Gil Ibarguchi, 1990; Peucat et al. 1990).

Other parts of the main allochthonous structure, such as the Basal Unit (Malpica-Tuy Band), with features of a continental margin, were affected by prograde metamorphism that produced amphibolites, C-type eclogites and high-pressure/low- to intermediate-temperature parageneses in acid rocks and metapelites. The age of this metamorphism is not precisely known yet; however, it should not be very different from that of the above-mentioned metamorphism of the ophiolitic materials, since secondary micas from the eclogitic rocks have been dated at 360–370 Ma (K/Ar and Rb/Sr, Van Calsteren et al. 1979, and $^{39}Ar/^{40}Ar$, Dallmeyer, written comm. 1987), while orthogneisses yielded Rb/Sr whole-rock isochrons of ca. 470 Ma (García Garzón et al. 1981; Kuijper et al. 1982).

This series of mainly amphibolitic tectonometamorphic processes has been tentatively correlated by Arenas et al. (1986), with the first Hercynian deformation phase in the autochthonous. It would have been responsible for the stacking of the different units within a single structure before its final emplacement upon the materials of the Schistose Domain of Galicia-Trás-os-Montes and the Central-Iberian Zone during the second Hercynian deformation phase.

Regarding the metamorphic evolution, the retrograde effects on the rocks affected by the previous eclogitic/granulitic episode, and the prograde recrystallization in the Upper Ophiolitic and Basal Units, will be considered separately.

In the first case, the eclogites firstly developed clinopyroxene (diopsidic)-plagioclase symplectites after primary omphacite (actually, this episode seems to be static and could be related to passive uplifting prior to the amphibolitic regional recrystallization); later on, typical amphibolitic associations were formed with hornblende, either symplectitic or granoblastic, or even as big secondary porphyroblasts more or less orientated. Other changes include the development of hornblende keliphitic rims around garnet, and the appearance of clinozoisite/epidote, sphene and, in some instances, phengite and phlogopitic mica. Basic granulites developed abundant secondary hornblende, epidote and sphene and, as in the case of eclogites, the final evolution led in many cases to the development of garnet-bearing amphibolites which may be seen to pass gradually to less-retromorphosed eclogites or granulites. A most conspicuous feature of this metamorphic stage, rather implying drastic increase of the P_{fl}/P_t ratio, is the formation of leucocratic veins of trondhjemitic to granodioritic composition, with the aspect of injected mobilizates cross-cutting the foliation of the rocks.

Other old high-grade rocks, such as the migmatitic metasediments, show less-marked evidence of this episode. The main changes are related to resorbtion of garnet, replaced by biotite + muscovite aggregates, indicating retrograde evolution from high-pressure conditions (Vielzeuf 1984), and the replacement of phengitic micas by muscovites, and of kyanite, zoisite and rutile by clinozoisite, epidote and sphene respectively. As for the metaperidotites, their amphiboles could effectively be formed after primary clinopyroxenes during this stage, as it is further suggested by their Rb/Sr age (380 Ma, Van Calsteren et al. 1979). However, the textures are not conclusive, since quite often amphibole and phlogopite of up to 5 cm longer appear, irregularly distributed and forming veins or pods or dispersed through the mass of the rocks, thus suggesting an origin in the mantle or lower crust by metasomatic impregnation of the ultramafic protolith.

The prograde recrystallizations during this episode are best observed in the Upper Ophiolitic and Basal Units. In the first unit, low- to high-grade amphibolitic parageneses including hornblende, plagioclase, epidote/clinozoisite, garnet and rutile were commonly formed after the ophiolitic metabasic rocks (gabbros, dolerites and basalts). Nevertheless, some orthopyroxene, clinopyroxene and spinel in reaction rims between olivine and plagioclase have been found in some gabbros and ultramafic rocks (Warnaars 1967), which may indicate that, locally, granulite-facies conditions were attained. The ophiolitic ultramafics, mainly harzburgites, websterites and spinel/plagioclase-bearing wehrlites with rather well preserved primary mineralogies and mantelic textures, quite often developed secondary amphiboles, not very different from those of the high-grade metaperidotites, during this metamorphic episode.

In the Basal Unit, the most abundant rocks are the orthogneisses (biotite-rich to riebeckite- and aegirine-bearing peralkalines), which, although often recrystallized without important mineralogical changes, in some cases show garnet coronas between plagioclase and biotite, along with phengitic micas, zoisite and rutile developed during this metamorphism. The metasediments were transformed into micaschists and non-migmatitic paragneisses, with variable proportions of garnet, phengite, biotite, chloritoid, kyanite, rutile and zoisite as the most characteristic minerals. Zoned plagioclase porphyroblasts along well-defined levels are a conspicuous feature of many of these rocks, suggesting a volcanosedimentary ascent of part of this unit. The metabasites were transformed either into amphibolites with hornblende and plagioclase and variable amounts of garnet, biotite, rutile, opaques, sphene and epidote, or into C-type eclogites composed of garnet and omphacite, with variable proportions of quartz, rutile, zoisite and kyanite. Eclogitic acid rocks (eclogitic schists, with more than 64 wt % of SiO_2) have been found at one point (La Pioza outcrop) carrying quartz, garnet, omphacite, kyanite and zoisite. Barroisitic amphibole or glaucophane in eclogites and phengite and paragonite in these rocks and in the eclogitic schists are typical post-eclogitic minerals, suggesting a relatively high-pressure/low- to intermediate-temperature regime during the decompression processes (Fig. 3). Chemical analyses of garnets and clinopyroxenes of eclogitic rocks from the Basal Unit have been published by Van der Wegen (1978) and Gil Ibarguchi and Ortega Gironés (1985). On the basis of the last authors' results, P-T conditions calculated for the peak of this episode in that unit are around 600 °C, > 14 kbar.

3.2.4 The Greenschist Episode

All the materials of the complexes and Malpica-Tuy Band underwent a final episode of recrystallization under medium-pressure greenschist-facies conditions. This episode presents different characteristics, depending upon the kind of protoliths affected, i. e. whether or not they had a previous metamorphic history. Thus, it is a retrograde episode for the lithologies previously affected by the eclogitic/granulitic and/or amphibolitic/eclogitic episodes, whereas it is prograde for those other lithologies unaffected by those metamorphisms, namely the Epizonal Metasedimentary and the Lower Epizonal Ophiolitic Units. The absence of isotopic and/or precise stratigraphic data preclude an accurate statement of the age of the greenschist episode. Nevertheless, the structures associated to the recrystallization in the allochthonous pile, that is, subhorizontal shear zones and zones of maximum deformation of uprighted folds, are recognized in the relative autochthonous too, where their Hercynian age (Upper Devonian - Lower Carboniferous) is well established (Bastida et al. 1986).

The development of the episode under greenschist-facies conditions is closely related to the final emplacement of the units that compose the main allochthonous structure. The basal thrusting surfaces of those units are zones of intense mylonitic deformation and important fluid activity, which favoured widespread recrystallization. This process is particularly evident when the mylonitized and recrystallized materials were previously affected by the granulitic/eclogitic episode. In this case, the high-grade rocks have often been transformed into mylonitic to ultramylonitic greenschists, with rare relics of previous parageneses. Good examples of this kind of lithology, on the other hand quite common elsewhere in the allochthonous pile, have been described in the Cabo Ortegal (Arenas 1985); Ordenes (Díaz García 1986) and Trás-os-Montes (Anthonioz 1972) Complexes.

An inversion of the metamorphic sequence took place coeval with the emplacement of the ensemble of allochthonous units into its present position. This fact led to the disposition of the highest-grade lithologies in the upper structural levels (Arenas et al. 1986), thus helping its retrogradation. As a consequence, even outside of the highly deformed areas, granulitic/eclogitic rocks recorded the retrograde metamorphism, giving rise to the development of secondary parageneses.

Extreme retrogradation of metabasites (eclogites, mafic granulites, amphibolites and metagabbros or metadiorites) produced fine-grained greenschists with chlorite, actinolite, epidote/clinozoisite, quartz, albite, sphene, opaques ± sericite ± biotite ± stilpnomelane.

The same association may be found in the metabasites (metabasalts, metadiabases and metagabbros) of the Lower Epizonal Ophiolitic Unit, but in this case as a result of a prograde recrystallization. It may be noted that some materials of that unit were originally affected by ocean floor metamorphism which led to the local development of amphibolitic parageneses; in one case, even prograde coronitic garnet around opaques has been found as well in the metabasites of that unit (basaltic pillow brecciae near Espasante, Arenas 1985).

Various types of retrograde associations were developed in the previously metamorphosed ultrabasic rocks. The most common are made up of a combination of serpentine, chlorite, talc, tremolite, carbonates and magnetite together with pre-existent phases (opx, cpx, oliv, etc.). The ultrabasic rocks from the Lower Epizonal Ophiolitic Unit, which were affected only by the low-grade metamorphism, are generally completely transformed into serpentinites; this is the case, for instance of the big outcrops of serpentinites found SE of the Cabo Ortegal Complex (Moeche) as dismembered parts of that unit. Monomineralic aggregates of tremolite, talc or chlorite and calcic skarns are not uncommon within the serpentinitic bodies; in one case, rodingitic parageneses with hydrogrossular, zoisite/clinozoisite, muscovite, chlorite, carbonates and sphene have also been observed (Arenas 1985).

The retromorphosis of the different types of orthogneisses only changed slightly the primitive mineralogies. Gil Ibarguchi and Ortega Gironés (1985) and Díaz García (1986) have described retrograde greenschist-facies parageneses with quartz, albite, chlorite, sphene and epidote in orthogneisses from the Malpica-Tuy Band and from the Lower Catazonal Unit respectively. Analogous rocks in the Cabo Ortegal Complex present the same association with K-feldspar and stilpnomelane. The prograde parageneses formed during this episode are best observed in plagiogranites from the Lower Epizonal Ophiolitic Unit; they include plagioclase, quartz, sericite, opaques ± K-feldspar ± chlorite stilpnomelane ± biotite ± actinolite ± epidote/clinozoisite ± sphene.

The metasediments from the Epizonal Metasedimentary Unit are metapelites (slates and phyllites), metasandstones, metaconglomerates and marbles. The greenschist episode produced prograde parageneses with variable amounts of quartz, chlorite, white mica, albite and stilpnomelane. In the already metamorphosed metasediments from the other units, the retrograde parageneses are not very different from those just described, although their development is usually more restricted and is accompanied by destabilization of previous phases such as garnet, biotite, kyanite, staurolite and plagioclase.

P-T conditions for the greenschist episode are rather constant through all sectors. On the basis of chemical analyses of characteristic phases and calculated stabilities of mineral associations, Arenas (1985) suggested this episode corresponded to a medium-pressure gradient with an average T of 375–425 °C and P of 2.5–3.5 kb (Fig. 1).

3.2.5 Main Characteristics of the Pressure-Temperature-Time Paths and Their Geodynamic Implications

The pressure-temperature-time (P-T-t) paths of the lithologies affected by the first and second metamorphic episodes show rather similar trends, being characterized by drastic decompression and moderate cooling trajectories (Figs. 2, 3). This evolution took place after the reaching of the metamorphic peak, which in most cases erased previous parageneses, thus preventing the reconstruction of the prograde segments of the path. The trajectories for the materials affected only by the greenschist episode are opposite to the previous ones, since they indicate burying and prograde metamorphism during exhumation and retrogradation of the higher-grade units. The trajectories are related to the formation of the allochthonous pile and to the position that the different thrust units occupy within it. Moreover, the trajectories reflect the metamorphic history of each unit and the previous episode related to the metamorphic climax, as well as that coeval with the final emplacement.

The oldest eclogitic associations (475–490 Ma or older, see above) were equilibrated at pressures of 17 kbar at least, which corresponds to depths of ca. 60 km. These great depths, along with the lack of evidence for a zone of thickened Ordovician Crust and the low T/P gradients for the eclogitic metamorphism (around 15 °C/km), suggest that the high-P metamorphism took place in a subductive context (Ernst 1973; Franz et al. 1986). The strong decompressions recorded by the P-T-t trajectories of the eclogites are interpreted as a result of uprising and progressive emplacement of the subducted lithologies. This process ended during Upper Devonian to Lower Carboniferous, that is, during the main period of the Hercynian collision. As in the case of other mountain belts, e. g. the Alpine Chain of Western Europe or the Betic Chain in SE Spain,

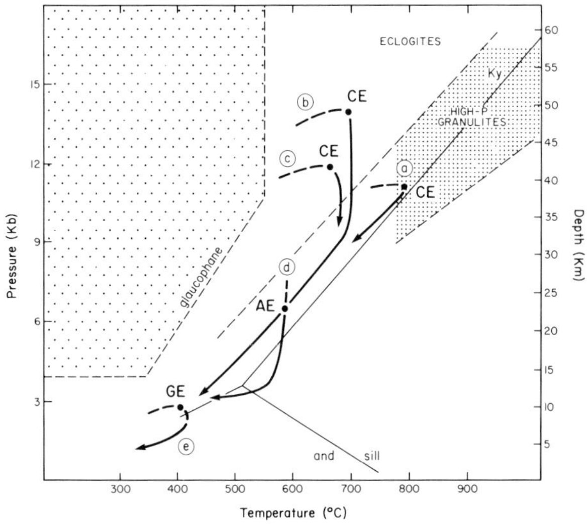

Fig. 2. P-T-t paths for the different units of the Cabo Ortegal Complex: **a** and **b**, mafic granulites and eclogites, respectively, of the Upper Catazonal Unit; **c**, eclogites of the Lower Catazonal Unit; **d,** amphibolites of the Upper Epi-Mesozonal Ophiolitic Unit; **e,** lithologies of the Lower Epizonal Ophiolitic Unit and Epizonal Metasedimentary Unit. *CE* eclogitic/granulitic episode (475–490 Ma). *AE* amphibolitic/eclogitic episode (ca. 400 Ma). *GE* greenschists episode (350–325 Ma). Mineral reactions after Green and Ringwood (1967); Holdaway (1971) and Maresch (1977)

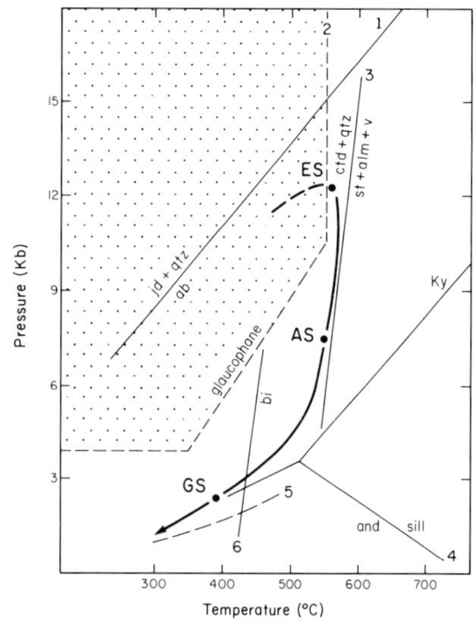

Fig. 3. P-T-t paths for eclogites formed in the Basal Unit (Malpica-Tuy Band) during the amphibolitic/eclogitic episode (Silurian). *ES* Eclogitic stage; *AS* amphibolitic stage; *GS* greenschist stage. Mineral reactions: **1** Holland (1980); **2** glaucophane 'out', Maresch (1977); **3** Rao and Johannes (1979); **4** Al_2SiO_5 polymorphs, Holdaway (1971); **5** phengite $Si_{3.2}$, Velde (1967); **6** stilp + pheng = biot + ctd + qtz + H_2O, Nitsch (in Winkler 1976)

the ages recorded for the different metamorphic stages indicate that exhumation of high-pressure lithologies started before the final collision (Platt 1986).

An anologous interpretation is suggested for the eclogitic rocks related to the Silurian (ca. 400 Ma), mainly amphibolitic stage. The lithologies eclogitized during this episode formed part of the continental margin of the Gondwana Continent (Arenas et al. 1986), being subducted in more recent times. However, the new subduction must only affect a part of the preserved materials, since during this episode most lithologies were metamorphosed under amphibolite-facies conditions with a medium T/P regime.

The segments of the P-T-t trajectories related to the greenschist episode (Fig. 4), were originated during the final emplacement of the allochthonous units. In the lithologies affected by the first and second metamorphic episodes, these segments follow the tendency towards decompression and cooling started previously. In those materials not affected by previous recrystallizations (Epizonal Metasedimentary and Lower Epizonal Ophiolitic Units), the paths reflect the compression and heating that took place when those materials were overridden by other allochthonous units.

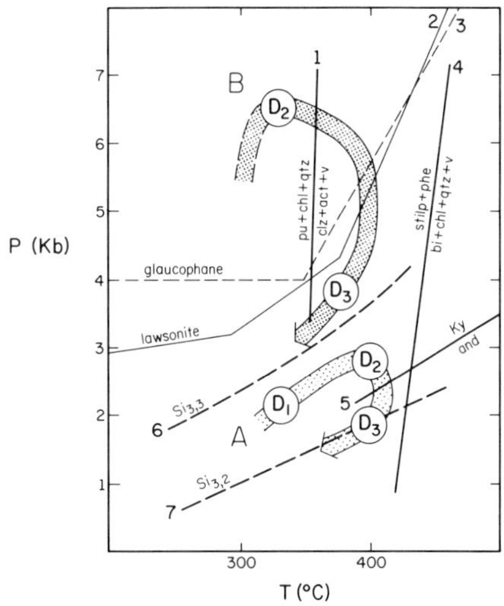

Fig. 4. P-T-time paths for the metamorphism developed under the Complexes of Cabo Ortegal (**A** metapelites) and Bragança-Morais (**B** blueschists) (cf. text). D_1, D_2, and D_3: Hercynian deformation phases. Mineral reactions: *1* Nitsch (1971); *2* lawsonite 'out', Newton and Kennedy (1963); Liou (1971); *3* glaucophane 'out', Maresch (1977); *4* Nitsch (in Winkler 1976); *5* Al_2SiO_5 polymorphs, Holdaway (1971); *6* and *7* phengites $Si_{3.3}$, respectively, Velde (1967)

One of the main problems that arises from this interpretation for the tectonometamorphic evolution of the complexes and Malpica-Tuy Band is that of its possible relationship to one or several orogenic cycles. The P-T-t trajectories show a simple geometry, compatible with a single-cycle evolution (Selverstone 1985; Arenas, in press). There are no complex trajectories, with recurrences or long periods of time without recrystallization, like those that characterize regions affected by a polycyclic evolution (Austrheim and Griffin 1985; Austrheim 1987). However, an important difficulty remains, and that is the long duration of the orogenic cycle: from Lower Ordovician to Lower Carboniferous (it may be noted that, in the regional context of the northwestern part of the Iberian Massif, there is no clear evidence for a major orogenic event during Ordovician times).

If the metamorphic evolution were produced by a single orogenic cycle, say the Hercynian Orogenic Cycle, its development should be analogous to that proposed for other orogenic belts. It must start coinciding with compressive tectonics, with development of subduction in the continental margin or in the island arc system (after Arenas et al. 1986, part of the allochthonous units should be originated in relation to an arc system). The subduction episode gave rise to an eo-Hercynian high-pressure metamorphism (Pin and Vielzeuf 1983; Pin and Peucat 1986), the age of which extends for quite a long time span (up to Silurian). The continuation of the compressive tectonics led to the progressive uprising of the lithologies affected by the first two metamorphic episodes and to their final emplacement onto the continental margin (in part as obducted flakes) during the final continental collision that closed the Cycle in the Lower Carboniferous.

3.2.6 The Relationship with the Metamorphism in the Relative Autochthonous

The sequence of deformations that produced during Devonian to Carboniferous times the emplacement of a number of relatively thick allochthonous units in the northwest of the Iberian Massif, was responsible for the formation of a zone of strongly thickened crust in that area. A widespread Hercynian regional metamorphism is one of the main features of this area (Martínez et al., this Vol.). The metamorphic evolution of the relative autochthonous at the base of the Complexes depended in great part on the original thickness of the allochthonous megastructure. Some information about that original thickness may be gathered from the study of those autochthonous materials. The relationship between the metamorphism in the allochthonous and autochthonous obeys some rather simple principles: whereas most lithologies from the first ensemble underwent decompression and cooling as they were uprised and approached their present position, the autochthonous materials were compressed and heated as the different allochthonous sheets overrode them.

In the materials of the Central-Iberian Zone placed at the base of the Cabo Ortegal Complex, the climax of the Hercynian metamorphism was reached after the emplacement of the allochthonous complex. The sequence of peak metamorphic zones recognized at different levels is characteristic of a medium-pressure metamorphic regime, including chlorite, biotite, almandine, staurolite and sillimanite zones, with the local presence of kyanite-bearing parageneses (Capdevila 1969; Bastida et al. 1984). P-T equilibrium conditions at the chlorite zone, immediately under the Cabo Ortegal Complex, (path A, Fig. 4) may be estimated at ca. 400 °C and 2.5–3 kb (Arenas 1985). These figures suggest that the original thickness of the allochthonous structure, of which the Cabo Ortegal Complex was forming part, was no more than 8 to 10 km. An analogous conclusion may be obtained for the Ordenes Complex, as deduced from the medium-pressure metamorphic regime recognized under that Complex in the Schistose Domain of Galicia-Trás-os-Montes (Marquínez 1984).

The characteristics of the metamorphism in the materials situated at the base of the Bragança and Morais Complexes (North Portugal) are slightly different from those described under the Galician Complexes. Munhá et al. (1984) have described glaucophane-bearing schists from Silurian to Lower Devonian materials of the Centro-Transmontana Sequence (a part of the Schistose Domain of Galicia-Trás-os-Montes) outcropping under the Portuguese allochthonous complexes. According to these authors, the parageneses of the blueschists (including albite, crossite-glaucophane, epidote, phengite, chlorite, magnetite, hematites, quartz and carbonates) were equilibrated during the second phase of Hercynian deformation, being produced by the overload induced by the allochthonous structure of Bragança and Morais. Schermerhorn and Kotsch (1984) have described metadacitic tuffs in an analogous position, containing lawsonite, quartz, albite, phengite, calcite and leucoxene; the accompanying metabasites carry magnesio-riebeckite which substitutes for actinolite. These parageneses are preceded by a metamorphic episode under greenschist-facies conditions and were developed at the base of the thrusting units during Lower Carboniferous. The above-described associations suggest that the empilement of allochthonous sheets in Trás-os-Montes led to the development of a relatively HP/LT metamorphism under the complexes (path B, Fig. 4). The P-T conditions proposed for this metamorphism are 400 °C and 6 kb (Munhá et al. 1984), which suggest a thickness for the original allochthonous structure of ca. 22 km. This thickness decreased towards the north to reach the 8–10 km proposed for the Galician Complexes.

References

Anthonioz PM (1972) Les complexes polymétamorphiques Precambriens de Morais et Bragança (NE du Portugal): étude pétrographique et structurale. Mem Serv Geol Port 20: 1–112

Arenas R (in press) Opposite P, T, t paths of Hercynian metamorphism between the upper units of the Cabo Ortegal Complex and their substratum (NW of the Iberian Massif). Tectonophysics

Arenas R (1985) Evolución petrológica y geoquímica de la unidad alóctona inferior del complejo metamórfico básico-ultrabásico de Cabo Ortegal (Unidad de Moeche) y del Silúrico paraautóctono, Cadena Hercínica Ibérica (NW de España). Thesis, Univ Complutense Madr 543 pp

Arenas R, Peinado M (1984) Petrología de las granulitas de alta presión y de las eclogitas de tipo B del Macizo de Cabo Ortegal, Galicia, NW de España. I Congr Esp Geol 2: 27–48

Arenas R, Gil Ibarguchi JI, González Lodeiro F et al. (1986) Tectonostratigraphic units in the complexes with mafic and related rocks of the NW of the Iberian Massif. Hercynica II, 2: 87–110

Austrheim H (1987) Eclogitization of lower crustal granulites by fluid migration through shear zones. Earth Planet Sci Lett 81: 221–232

Austrheim H, Griffin WL (1985) Shear deformation and eclogite formation within granulite-facies anorthosites of the Bergen Arcs, western Norway. Chem Geol 50: 267–281

Bastida F, Marcos A, Marquínez J, Martínez Catalán JR, Pérez Estaún A, Pulgar JA (1984) Mapa geológico de España, 1:200 000, Hoja nº 1 (La Coruña). IGME Madrid

Bastida F, Martínez Catalán JR, Pulgar A (1986) Structural, metamorphic and magmatic history of the Mondoñedo Nappe (Hercynian belt, NW Spain). J Struct Geol 8, 3/4: 415–430

Bernard Griffiths J, Peucat JJ, Cornichet J, Iglesias Ponce de Leon M, Gil Ibarguchi JI (1985) U-Pb, Nd isotope and REE geochemistry in eclogites from the Cabo Ortegal Complex, Galicia, Spain: an example of REE inmobility conserving MORB-like patterns during high-grade metamorphism. Chem Geol (Isotope Geosci Sect) 52: 217–225

Calsteren PWC van, Boelrijk NAIM, Hebeda EH, Priem HNA, Tex E den, Verdurmen EATH, Verschure RH (1979) Isotopic dating of older elements (including the Cabo Oretal mafic-ultramafic complex) in the Hercynian Orogen of NW Spain: manifestations of a presumed Early Paleozoic mantle-plume. Chem Geol 24: 35–56

Capdevila R (1969) Le métamorphisme regional progressif et les granites dans le segment herycinien de Galice Nord-oriental (NW de l'Espagne). Thesis, Univ Montpellier, 430 pp

Dallmeyer RD, Gil Ibarguchi JI (1990) Age of the amphibolitic metamorphism in the ophiolitic unit of Morais (North Portugal). Implications for the early stages of the Hercynian Orogenesis in the Iberian Massif. I Ged Soc London 147: 5

Díaz García F (1986) La Unidad de Agualada. Borde W del Complejo de Ordenes (NW de España). Trab Geol 16: 3–14

Engels JP (1972) The catazonal poly-metamorphic rocks of Cabo Ortegal (NW Spain), a structural and petrofabric study. Lcidse Geol Meded 48: 83–133

Ernst WG (1973) Blueschist metamorphism and P-T regimes in active subduction zones. Tectonophysics 17: 255–272

Farias P, Gallastegui G, González Lodeiro F et al. (1987) Aportaciones al conocimiento de la litoestratigrafia y estructura de Galicia Central. Mem Fac Cienc Univ Porto 1: 411–431

Franz G, Thomas S, Smith DC (1986) High-pressure phengite decomposition in the Weissenstein eclogite, Münchberger Gneiss Massif, Germany. Contrib Mineral Petrol 92: 71–85

García Garzón J, Pablo Maciá JG de, Llamas Borrajo JF (1981) Edades absolutas obtenidas mediante el método Rb-Sr de dos cuerpos de ortogneises en Galicia Occidental. Bol Geol Min Madr 92: 463–466

Gil Ibarguchi JI, Ortega Girones E (1985) Petrology, structure and geotectonic implications of glaucophane-bearing eclogites and related rocks from the Malpica-Tuy Unit, Galicia, northwest Spain. Chem Geol 50: 145–162

Gil Ibarguchi JI, Abalos B, Campillo A et al. (1988) Asociaciones con granate-clinopiroxeno en la 'unidad catazonal superior' del complejo del Cabo Ortegal (NW España). Cuad Lab Xeol Laxe 12: 165–182

Gil Ibarguchi JI, Mendia M, Girardean J, Pencat JJ (in press) Petrology of eclogites and clinopyroxene – garnet metabasites from the Cabo Ortegal Complex (northwestern Spain). Lithos (submitted)

Green DH, Ringwood AE (1967) An experimental investigation of the gabbro to eclogite transformation and its petrological applications. Geochim Cosmochim Acta 31: 767–833

Holdaway MJ (1971) Stability of andalusite and the aluminum silicate phase diagram. Am J Sci 271: 97–131

Holland TJB (1980) The reaction albite = jadeite + quartz determined experimentally in the range 600–1200 °C. Am Min 65: 129–134

Hubregtse JJMW (1973) Distribution of elements in some basic granulite-facies rocks. Konink Ned Akad, Amsterdam 27 (1): 68

Iglesias M, Ribeiro ML, Ribeiro A (1983) La interpretación aloctonista de la estructura del Noroeste Peninsular. Geol Esp I (IGME Ed. Madr) 1: 459–467

Kuijper RP (1979) U-Pb systematics and the petrogenetic evolution of infracrustal rocks in the Paleozoic basement of western Galicia (NW Spain). Thesis Univ Leiden, 101 pp

Kuijper RP, Priem HNA, Tex E den (1982) Late Archaean-Early Proterozoic source ages of zircons in rocks from the Paleozoic Orogen of Western Galicia, NW Spain. Precambrian Res 19: 1–29

Liou JG (1971) P-T stability fields of laumontite, wairakite, lawsonite and related minerals in the system Ca-$Al_2Si_2O_8$-SiO_2-H_2O. J Petrol 7: 221–245

Maresch WV (1977) Experimental studies on glaucophane: an analysis of present knowledge. Tectonophysics 43: 109–125

Marquínez JL (1984) La Geología del área esquistosa de Galicia Central (Cordillera Herciniana, NW de España). Mem Inst Geol Min Esp Madr 100: 231

Munha J, Ribeiro A, Ribeiro ML (1984) Blueschists in the Iberian Variscan Chain (Trás-os-Montes: NE Portugal). Comun Serv Geol Port 70: 31–53

Newton RC, Kennedy GC (1963) Some equilibrium reactions in the join $CaAl_2Si_2O_8$-H_2O. J Geophys Res 68: 2967–2983

Nitsch KH (1971) Stabilititatsbezichungen von Prehnit und Pumpellyit-haltigen Paragenesen. Contrib Min Petrol 30: 240–260

Peucat JJ, Bernard Griffiths J, Dallmeyer RD, Gil Ibarguchi JI, Iglesias M, Menot P (1990) Geochemical and geochronological cross section of the deep Hercynian crust: the Cabo Ortegal high-pressure nappe (NW Spain). Tectonophysics 177: 263–292

Pin C, Peucat JJ (1986) Ages des épisodes de métamorphisme paléozoïques dans le Massif Central et le Massif Armoricain. Bull Soc Géol Fr 8, II, 3: 461–469

Pin C, Vielzeuf D (1983) Granulites and related rocks in Variscan median Europe: a dualistic interpretation. Tectonophysics 93: 47–74

Platt JP (1986) Dynamics of orogenic wedges and the uplift of high-pressure metamorphic rocks. Geol Soc Am Bull 97: 1037–1053

Rao BB, Johannes W (1979) Further data on the stability of staurolite + quartz and related assemblages. Neues Jahrb Min Monatsh 437–447

Ribeiro A, Pereira E, Marques F, Meirelles C (1987) A model for the lower continental crust (Morais and Bragança massifs, NE Portugal): I. Tectonics. Terra Cognita 7: 161

Ries A, Shackleton RM (1971) Catazonal complexes of northwest Spain and north Portugal, remnants of a Hercynian thrust plate. Nat Phys Sci 234: 65–68

Selverstone J (1985) Petrologic constraints on imbrication, metamorphism and uplift in the SW Tauern Window, eastern Alps. Tectonics 7: 687–704

Serrano Pinto MC, Gil Ibarguchi JI (1987) Revisión de datos geocronológicos e isotópicos de granitoides hercínicos de la región Galaico-Castellana. Mem Fac Cienc Univ Porto 1: 171–186

Velde B (1967) Si^{4+} content of natural phengites. Contrib Mineral Petrol 14: 250–258

Vielzeuf D (1984) Relations de phases dans le facies granulite et implications geodynamiques. L'exemple des granulites de Pyrenées vol 1. Thesis, University of Clermont-Ferrand, 288 pp

Vogel DE (1967) Petrology of an eclogite and pyrigarnite bearing polymetamorphic rock complex at Cabo Ortegal, NW Spain. Leidse Geol Meded 40: 121–213

Warnaars FW (1967) Petrography of a peridotite-amphibolite and gabbro-bearing polyorogenic terrain NW of Santiago de Compostela (Spain). Thesis, Univ Leiden, 208 pp

Wegen G van der (1978) Garnet-bearing metabasites from the Blastomylonitic Graben, Western Galicia, Spain. Scr Geol Leiden 45: 1–95

Winkler HGF (1976) Petrogenesis of metamorphic rocks (fourth edition). Springer, Berlin Heidelberg New York, 254 pp

Part V Ossa-Morena Zone

1 Introduction

C. QUESADA[1]

Since its first definition by Lotze (1945) in his subdivision of the Iberian Massif (Fig. 1), mainly on the basis of its outstanding stratigraphical and structural differences with respect to the neighbouring zones, the Ossa-Morena Zone has been one of the most controversial units within the Iberian Hercynian Foldbelt. The discussion, currently centred on the geotectonic significance of its singularities as compared with the rest of the Iberian Massif, was preceded, and this is still partly going on, by intensive discussions on its boundaries, major units, stratigraphy and correlations. But undoubtedly the most passionate argument has been dealing, in the last two decades, with the existence or not of a Precambrian orogenic cycle affecting this zone. This possibility, suggested for the first time by MacPherson (1879) and Gonzalo y Tarín (1878), was, however, forgotten or even disputed by the work of Lotze's team (1939–50) and lately by French and Spanish researchers (Fabriés 1963; Vegas 1968;

[1] I. T. G. E., Rios Rosas, 28003 Madrid, Spain

a

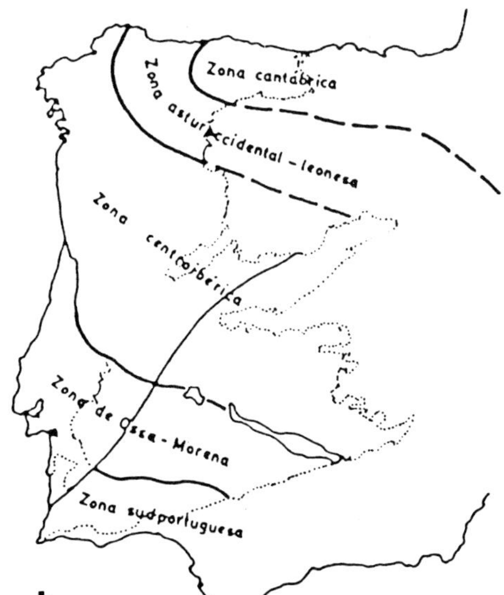

b

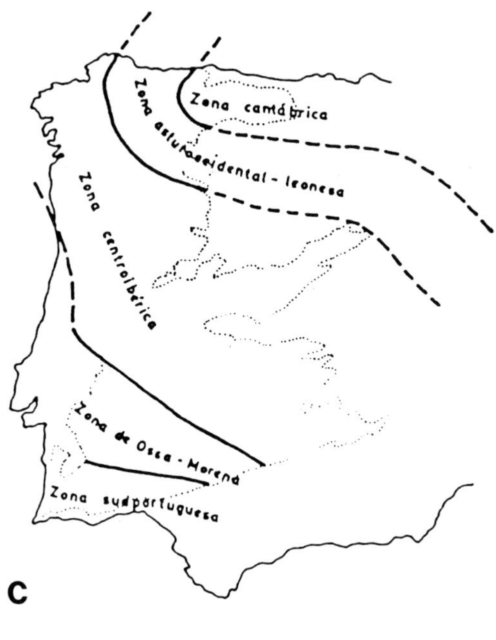

c

Fig. 1. Zonal subdivision of the Iberian Massif. **a** After Lotze 1970, modified from Lotze 1945; **b** after Julivert et al. 1974; **c** after Robardet 1976

R. D. Dallmeyer and E. Martínez García (Eds.)
Pre-Mesozoic Geology of Iberia

Bard 1969; Laurent and Bladier 1967; etc.). The problem, raised again by Delgado (1971), has been the main subject of numerous recent works (Garrote 1976; Pérez Lorente 1979; Chacón 1979; Eguiluz and Quesada 1980; Pascual 1981; Cueto et al. 1983; etc.), the result of which is the unequivocal recognition of the existence of a polyphased orogenic cycle in Late Precambrian times, although it is only partly understood, mainly due to extensive reactivation and dispersion during the Paleozoic evolution. The existence of such a Precambrian basement in the Ossa-Morena Zone constitutes precisely one of its most characteristic singularities.

On the other hand, the aforementioned contributions, together with the 1:50000 scale mapping of most part of the zone, undertaken in the last decade by the Spanish Geological Survey, has resulted in a substantial knowledge enhancement on the Paleozoic evolution as evidenced by numerous papers dealing with the subdivision of the Ossa-Morena into various tectonosedimentary units: the so-called domains in the regional literature. The domainal recognition and characterization have represented a fundamental hit, which has allowed the analysis of individual units to be undertaken, thus obtaining an approximate idea of their sedimentary and structural evolution. Important contributions under these perspectives are those by Bard (1969); Chacón et al. (1974); Robardet (1976); Delgado et al. (1977); Quesada (1983); Florido and Quesada (1984); etc.

In the following description, however, the exhaustive analysis of the main characteristics of every domain is not included. A general interpretative overview has been preferred, provided the synthetic review nature of this book, although the details of relevant data justifying the proposed interpretations are obviously given. In every following chapter dealing with different geological subjects of the Ossa-Morena Zone, the description is arranged according to an evolutionary pattern which follows the major evolutionary events as outlined in Fig. 2. Two major cycles are thus distinguished: a Precambrian one and a Paleozoic one. The Precambrian cycle is subdivided into a preorogenic stage and a

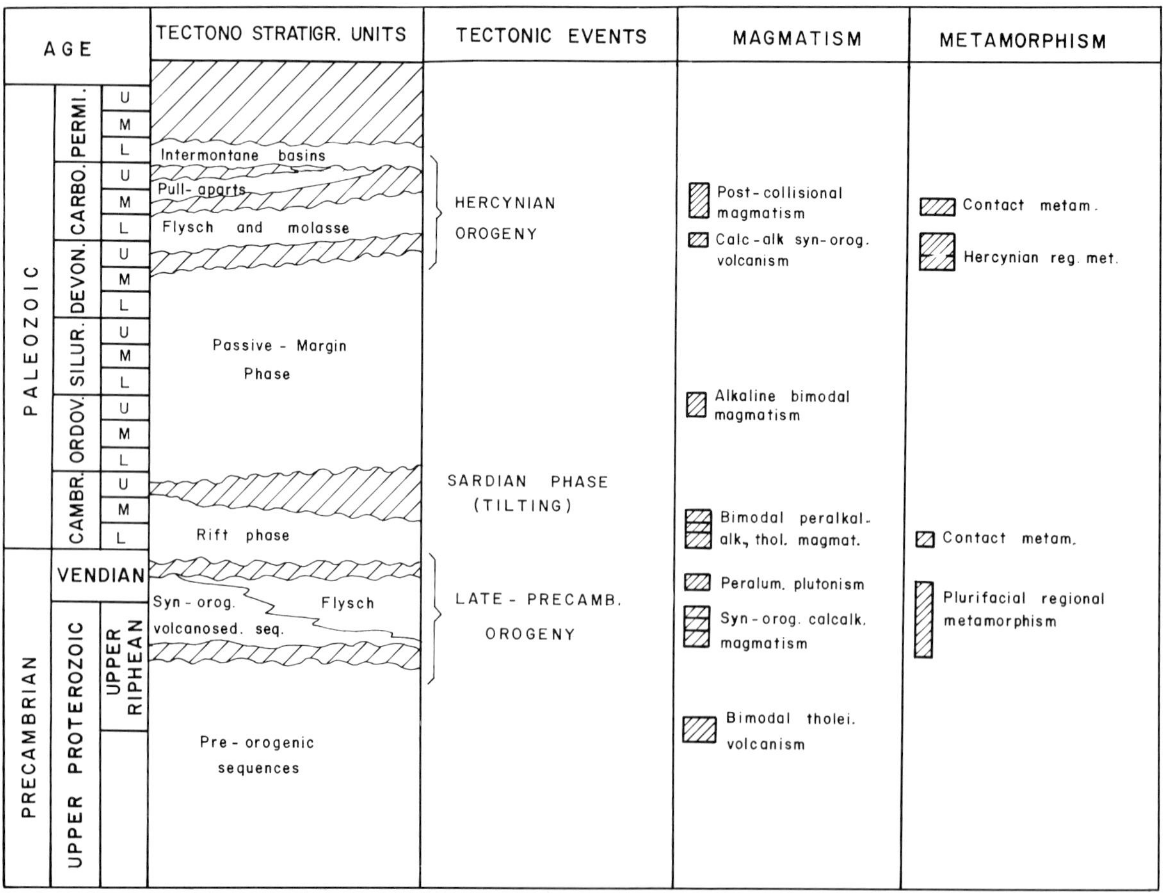

Fig. 2. Tectonostratigraphic chart of the Ossa-Morena Zone

synorogenic *sensulato* stage, whereas the Paleozoic cycle, much more accessible to study, may be more accurately subdivided into: (1) initial continental rifting stage (2) passive margin stage and (3) synorogenic stage, in turn suitable for further subdivision.

With respect to the geological boundaries of the Ossa-Morena Zone and their significance, it is worthwhile to remember that these have been a subject of intensive debate in the recent past, and, though moderately, it is still going on (Robardet 1976; Delgado et al. 1977; Chacón and Pascual 1979; Florido and Quesada 1983; etc.). Without entering the discussion here, as it will be at least in part dealt with further on, and in order to coordinate this chapter with those referring to the neighbouring zones, the following boundaries are arbitrarily considered: (1) northeastern boundary: Portalegre Thrust (in the western half) and Pedroches Batholith (in the eastern half); (2) southeastern boundary: Guadalquivir valley Tertiary cover; (3) southern boundary: Ficalho-Almonaster Thrust; and (4) western boundary: Porto-Tomar wrench fault system (see below for location).

References

Bard JP (1969) Le métamorphisme régional progressif des Sierras d'Aracena en Andalousie occidentale. Sa place dans le segment hercynien sud-ibérique. Thesis, Univ Montpellier

Chacón J (1979) Estudio geológico del sector central del anticlinorio Portalegre-Badajoz-Córdoba (Macizo Hespérico Meridional). Thesis, Univ Granada

Chacón J, Pascual E (1979) El anticlinorio Portalegre-Badajoz-Córdoba divisoria entre las zonas Centro Ibérica y Ossa-Morena (Sector SW del Macizo Ibérico). Cuad Geol Univ Granada 8: 21-35

Chacón J, Delgado M, Garrote A (1974) Sobre la existencia de dos diferentes dominios de metamorfismo regional en la banda Elvas-Badajoz-Córdoba (Macizo Hespérico Meridional). Bol Geol Min 85 (6): 713-717

Cueto LA, Eguiluz L, Llamas FL, Quesada C (1983) Lagranodiorita de Pallares, un intrusivo precámbrico en la alineación Olivenza-Monesterio (Zona de Ossa-Morena). Comun Serv Geol Port 69 (2): 219-226

Delgado M (1971) Esquema Geológico de la Hoja número 878 de - Azuaga (Badajoz). Bol Geol Min 82: 277-286

Delgado M, Liñan E, Pascual E, Pérez F (1977) Criterios para la diferenciación de dominios en Sierra Morena Central. Stud Geol 12: 75-90

Eguiluz L, Quesada C (1980) La sucessión precámbrica de la transversal de Monesterio (Badajoz). Nota preliminar. Temas Geol Min 4: 123-141

Fabriés J (1963) Les formations cristallines et métamorphiques du NE de la province de Seville (Espagne). Essai sur le métamorphisme des roches éruptives basiques. Thesis, Univ Nancy

Florido P, Quesada C (1984) Estado actual de concimientos sobre el macizo de Aracena. Cuad Lab Xeol Laxe 8: 257-277

Garrote A (1976) Asociaciones minerales del núcleo metamórfico de Sierra Albarrana (Prov. de Córdoba). Sierra Morena Central. Mem Not Univ Coimbra 82: 17-39

Gonzalo y Tarin J. (1878) Reseña físico-geológica de la provincia de Badajoz. Bol Com Mapa Geol Esp 6: 389-412

Julivert M, Fontboté JM, Ribeiro A, Conde L (1974) Mapa Tectónico de la Península Ibérica y Baleares. Memoria Explicativa. IGME, pp 1-101

Laurent P, Bladier Y (1976) L'évolution tectonométamorphique de l'axe Badajoz-Córdoba: un exemple de réactivation dynamique d'un socle ancien. Comun Serv Geol Port 60: 269-271

Lotze F (1945) Zur Gliederung der Varisciden der Iberischen Meseta. Geotekt Forsch 4 (6): 78-92

Lotze F (1970) El Cámbrico en España. Mem IGME 75: 1-256

MacPherson J (1879) Estudio geológico y petrográfico del N de la provincia de Sevilla. Bol Com Mapa Geol Esp 10: 97-269

Pascual E (1981) Investigaciones geológicas en el sector Córdoba-Villaviciosa de Córdoba (Sector Central de Sierra Morena). Thesis, Univ Granada, pp 1-521

Pérez Lorente F (1979) Geología de la zona Ossa-Morena al - Norte de Córdoba (Pozoblanco-Belmez-Villaviciosa de Córdoba). PhD Thesis, Univ Granada 281: 1-340

Quesada C (1983) El Carbonífero de Sierra Morena. In: C. Martínez (ed): Carbonífero y Pérmico de España. IGME, pp 243-278

Robardet M (1976) L'originalité du segment hercynien sud-ibérique au Paléozoïque inférieur: Ordovicien, Silurien et Dévonien dans le nord de la province de Séville (Espagne). CR Acad Sci Paris D 283: 999-1002

Vegas R (1968) Sobre la existencia de Precámbrico en la Baja Extremadura. Estud Geol 24: 85-89

2 Stratigraphy

2.1 Precambrian

C. Quesada[1], O. Apalategui[2], L. Eguiluz[3], E. Liñan[4] and T. Palacios[5]

2.1.1 Introduction

The existence of Precambrian rocks in the Ossa-Morena Zone was shown for the first time by MacPherson's (1878, 1879, 1883) works for the Commission of the Geological Map of Spain. Since then, this has been a subject of intensive debate, in particular the attribution of certain outcrops or sequences to the Precambrian and, above all, the possible existence of tectonic events of such age (Mallada 1985; Lotze 1929, 1956; Alvarado 1935; Fricke 1941; Hernández Pacheco 1953; Fabriès 1957, 1963; Alía 1963; Carvalhosa 1965; Bard 1969; etc.). New works undertaken in the late 1960s by a young generation of researchers (Gutiérrez Elorza and Hernández Enrile 1965; Vegas 1968, 1970; Hernández Enrile 1971; Gutiérrez Elorza et al. 1971; Delgado 1971) outstandingly enhanced the current knowledge on the Ossa-Morena's Precambrian and stimulated the interest of researchers from various universities (Garrote 1976; Liñán 1978; Pérez Lorente 1979; Chacón 1979; Eguiluz and Quesada 1980; Pascual 1981; Herranz 1984; Eguiluz 1987; etc.). This, together with the elaboration of the 1:50000 scale National Geological Map of the region in the last decade, has contributed to form the present state of the art which is synthesized below.

The profusion of recent research obviously has not solved every problem under discussion, as new problems have emerged, implying deeper knowledge levels. However, some problems still persist concerning the attribution to the Precambrian of several generally metamorphic azoic sequences, among which those constituting the Sierra Albarrana domain (see Apalategui et al., this Vol; Chacon et al. 1974; Delgado et al. 1977) are most significant, due to their very large areal extension. Although they have been included in the Precambrian by most authors since the early paper by Delgado (1971), their ambiguous relationships with Cambrian rocks north of Ahillones (Apalategui and Higueras 1983) as well as their problematical correlation with well-known Precambrian sequences advise leaving the question open to further research. Nevertheless, their description will be included here as being their Precambrian attributions most accepted, and their correlation with Paleozoic sequences are equally problematical.

The Precambrian rocks within the Ossa-Morena Zone (Fig. 1) constitute various successions that may be grouped together and subdivided, according to their relationships with the tectonic events that the region underwent in Late Precambrian times: (1) preorogenic successions and (2) synorogenic *(sensu lato)* successions (Quesada et al. 1987). Both groups consist in turn of a number of sequences (Fig. 2) whose mutual relationships are not always known in detail.

2.1.2 Preorogenic Successions

Three different preorogenic successions are known within the Precambrian of the Ossa-Morena Zone (Fig. 2), with a highly variable areal representation and always separated by tectonic contacts from one another. This fact, together with the almost absolute lack of biostratigraphic data, is the reason why their mutual primary relations, if they have ever existed, are unknown. The successions are as follows:

1. The most largely represented is the so-called Valencia de las Torres – Cerro Muriano Supergroup (Quesada et al. 1987), which consists, in turn, of two distinct groups; from bottom to top:

[1] I. T. G. E., Rios Rosas 23, 28003 Madrid, Spain
[2] Ingemisa, Avda. Virgen de las Augustias 16, 14006 Córdoba, Spain
[3] Departament of Geology, Universidad Pais Vasco, Bilbao, Spain
[4] Universidad Zaragoza, Spain
[5] Departament of Geology, Universidad Extremadura, Badajoz, Spain

R. D. Dallmeyer and E. Martínez García (Eds.)
Pre-Mesozoic Geology of Iberia

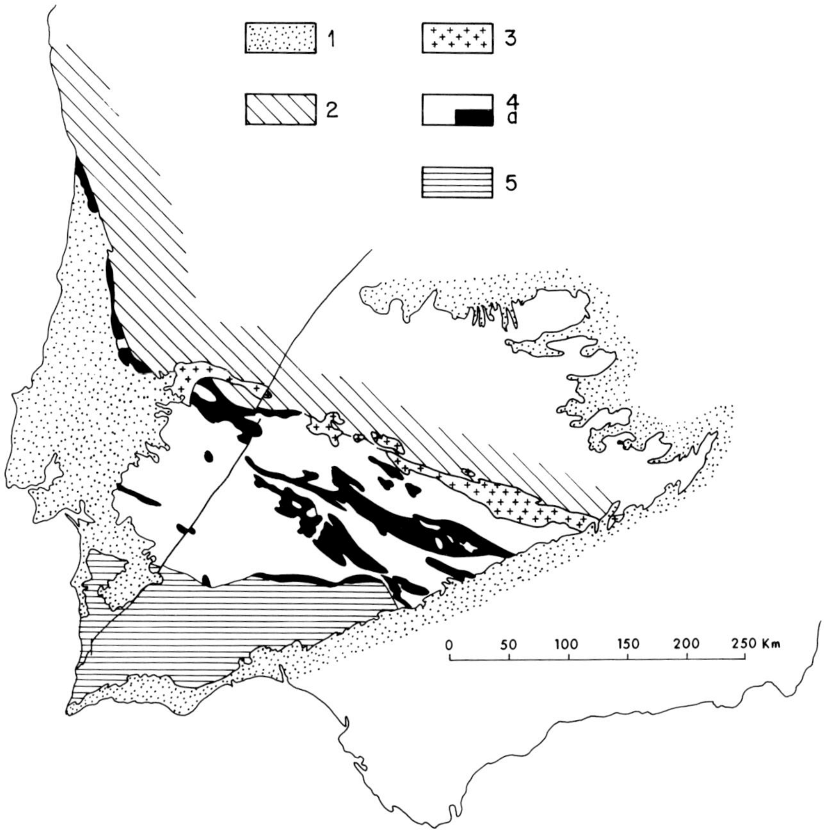

Fig. 1. Distribution of Precambrian occurrences in the Ossa-Morena Zone. *1.* Recent cover; *2.* Central-Iberian Zone; *3.* Pedroches Batholith; *4.* Ossa-Morena Zone; *4a.* Precambrian occurrences; *5.* South Portuguese Zone; *A* Olivenza-Monesterio antiform

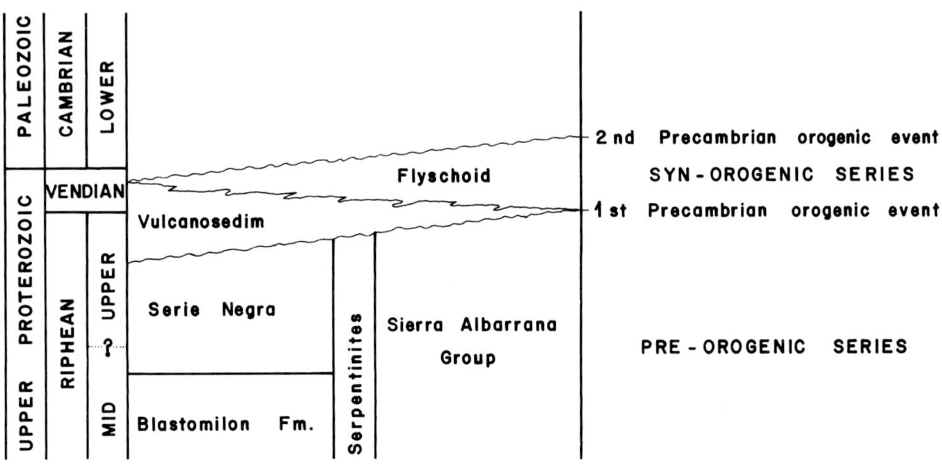

Fig. 2. Stratigraphic chart of the Precambrian successions in the Ossa-Morena Zone

Blastomylonitic Formation (Delgado 1971). This is also known by other local names, and mainly consists of probably volcano-derived (Arriola et al. 1983) gneissic rocks (Azuaga Gneiss; Delgado, op. cit.) and amphibolites (Las Mesas Member; Delgado, op. cit.). Minori lithologies include biotite-bearing paragneisses (mainly concentrated towards the top of the unit), black cherts and marbles, locally. Partly retrograted eclogite and granulite boudins are also scattered throughout the occurrences. The rocks in this group always appear under medium- and high-grade metamorphic conditions and are extremely deformed. For this reason, any attempt to establish their original characteristics is unavoidably speculative. The volcano-derived rock chemistry, mainly the amphibolites that have been recently characterized as continental tholeiites (García Casquero, in press), together with their bimodal nature (basalt/rhyolite), suggest an origin related to a continental crustal extension (rifting) event. This sequence occurs exclusively in the northernmost units of the Ossa-Morena zone.

Serie Negra Group (Carvalhosa 1965). This overlies the former, constituting the uppermost sequences in the Valencia de las Torres - Cerro Muriano Supergroup. The nature of the contact between both sequences is not clear and has been interpreted either as an unconformity separating the Serie Negra from an older basement (Pascual 1981) or as a gradual passage (Apalategui and Higueras 1983). According to this latter interpretation, which is preferred here, the Blastomylonitic formation would represent an initial developing stage of the basin where the Serie Negra would be subsequently laid down (Quesada et al. 1987).

The Serie Negra group is widely represented throughout the Ossa-Morena Zone, but it is best known in the central area of the Olivenza - Monesterio antiform (A in Fig. 1), where Eguiluz (1987) has distinguished two sequences: the lower one (Montemolín succession), 1000-2000 m thick, consists of graphite-rich pelites and quartzwackes, with abundant amphibolites towards the top as major constituents. Besides those, there are frequent black chert lenticular interbeds and, locally, marbles. The upper sequence (Tentudía succession), 2-3 cm thick, overlies the former, apparently with a conformable contact, and the main distinction with respect to it resides in the scarcity of amphibolites and the volcano-derived, markedly greywacke nature of the sandy horizons. A restricted shallow marine sedimentary environment has been considered by Eguiluz (op. cit.) as most likely for the Serie Negra, mainly on the basis of black chert characteristics (graphite-bearing cherts with possible cyanobacteria; Gonçalves and Palacios 1984) and the sparse algae-bearing carbonate lenses.

Biostratigraphic data are as yet very scarce, thus making it impossible to establish the age of both groups precisely. Acritarches have been reported from various Serie Negra horizons, which allow an imprecise attribution to the Middle-Upper Riphéan (Chacón 1979; Chacón et al. 1984) to be made. Thus, the lower group (blastomylonitic formation) should be older than the Middle Riphéan. Nevertheless, Palacios (pers. comm., 1987) considers, on the basis of recent taxonomic revisions and his own palynological results on these rocks, that at least a part of the material cited but not figured by the aforementioned authors could well be not acritarches but cyanobacteria, whereas another part consists undoubtedly of inorganic structures, frequently considered as Acritarches in the current literature.

2. A second preorogenic unit, interpretable with difficulty and with very limited areal representation, consists of scarce serpentinite bodies occurring in the northern limb of the Olivenza-Monesterio antiform (A in Fig. 1) near Calzadilla de los Barros. The Precambrian age of these rocks is out of the question since they constitute olistoliths and clasts within synorogenic volcanoclastic Precambrian sequences. Other serpentinite occurrences included in the Badajoz-Córdoba shear zone (see Apalategui et al., this Vol.) could be related to these, but their tectonic relations to other sequences inable their direct correlation.

In the Calzadilla area there are three main large outcrops (Fig. 3) and others, on the decimetrical to metrical scale, enclosed by andesitic volcaniclastic material belonging to the Precambrian synorogenic successions (Fig. 2). There is no doubt about the clastic nature of the small outcrops, but the westernmost occurrence within Fig. 3, which is the largest as well, shows mylonitic contacts with the enclosing rocks which has led Eguiluz (1987) to consider it as a (probably) ophiolitic thrust sheet. The alpine-like character of these rocks is strongly suggested by the absence of contact metamorphism around them, the existence of pod and lense-like chromite deposits, and their mainly serpentinitic composition, although a certain degree of zonation with clorite and tremolite-rocks exists near the outcrop border, that according to the alteration textures may derive from different ultrabasic rock types (Aguayo 1985).

The correlation between these serpentinites and other Precambrian successions in the Ossa-Morena Zone or elsewhere is not possible with the available data. Nevertheless, they must be borne in mind in any attempt to reconstruct the Precambrian history of this region which, if their ophiolitic nature is confirmed, should have taken place under plate tectonic regimes similar to those active in recent orogens.

3. The third preorogenic Precambrian ensemble in the Ossa-Morena Zone is the so-called Sierra Al-

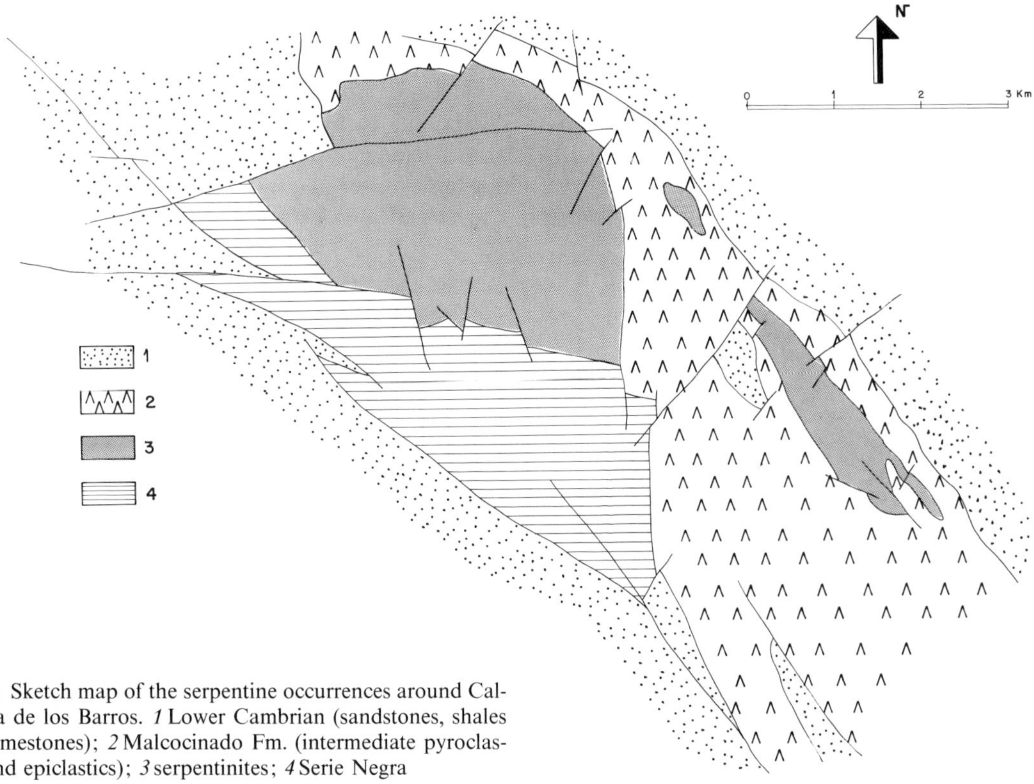

Fig. 3. Sketch map of the serpentine occurrences around Calzadilla de los Barros. *1* Lower Cambrian (sandstones, shales and limestones); *2* Malcocinado Fm. (intermediate pyroclastics and epiclastics); *3* serpentinites; *4* Serie Negra

barrana group (Apalategui et al. 1985; Quesada et al. 1987) which mainly occurs within the Sierra Albarrana structural domain (see Apalategui et al., this Vol.). This group is most controversial due to the lack of reliable biostratigraphic data, the current ignorance of its detailed stratigraphy and its complex relationships with other sequences, both Precambrian and Paleozoic. Moreover, even its Precambrian age is disputed (Apalategui et al. 1985) on the basis of a dubious boundary between one of its integral sequences (Azuaga Formation; Delgado 1971) and Lower Cambrian rocks north of Ahillones. On the other hand, its correlation with dated Paleozoic successions is not straightforward either. The currently admitted attribution to the Precambrian is mainly founded on the indirect evidence given by alkaline and peralkaline Lower Paleozoic plutons intruding into these sequences posteriority to a synmetamorphic tectonic event; namely the Ribera del Fresno pluton (423 +/− 38 Ma; Rb-Sr whole-rock isochron; García Casquero et al. 1985) and the Almendralejo pluton (470 +/− 26 Ma; Rb-Sr whole rock isochron; García Casquero op. cit.). Provided that no Caledonian event has ever been recognized in the region, the pre-intrusion fabric should have been developed in Precambrian times.

Although very poorly known or discussed in detail, the stratigraphic sequence of this group consists of: (1) basal micaschists (El Hoyo Micaschists; Delgado 1971; Pérez Lorente 1979), (2) intermediate feldspathic quartzite sequence (Albarrana Formation; Garrote et al. in press) with abundant scolithus and other fossil burrows, and (3) upper complex high-alumina micaschist sequence (Albariza Formation and others, Garrote et al., in press; Atalaya Formation, Chacón 1979). Apalategui et al. (1985) proposed that the former sequence was reversed and correlated the intermediate quartzite formation with the Lower Ordovician Armorican Quartzite.

2.1.3 Synorogenic Successions

The onset of compressional tectonic activity in this region in the Upper Proterozoic (Upper Riphéan) produced dramatic changes in the scenario which led to the development of other successions intimately controlled by such a tectonic environment. As it was mentioned above, two types of successions may be recognized: (1) volcanosedimentary complexes and (2) flysch-like complexes (Fig. 2). These share the following common characteristics:

1. Both have a markedly synorogenic character: calc-alkaline andesitic volcanism and flysch respectively.
2. They overlie unconformably the preorogenic Precambrian successions, partly previously deformed.
3. Both are affected by the late tectonothermal events of the Late-Precambrian orogeny.

The relationship between both complexes is mainly a facies change, appearing in the same vertical section in many localities.

2.1.3.1 Volcanosedimentary Complex

This occurs widely in the Ossa-Morena Zone, where it consists mainly of pyroclastic lava, epiclastic and sedimentary rocks. The volcanic constituents range in composition from basalt to rhyolite, but andesite and dacite intermediate rocks largely predominate. It is worthwhile mentioning that volcanic rocks *sensu stricto* (lavas and pyroclastics) are indeed minor, probably as a result of their low preservation potential in an active environment like this. On the other hand, the volcanosedimentary rocks range in grainsize from gravel to lutite, but sand and lutite-sized rocks are by far predominant. From a sedimentological point of view, debris flow and turbidite beds alternate with minor volcanic horizons (pyroclastic and lava) as well as with pelictic (partly tuffitic) and carbonate beds. The latter are locally thick (in excess of 300 m), namely at Fuente del Maestre, Fuente del Arco, etc. In the easternmost part of the Ossa-Morena, conglomerate lenses are frequent, including clasts with a granitic, carbonatic, Blastomylonitic formation and Serie Negra group source (Liñán 1978). Although the possibility that the volcanism was (at least in part) subaerial cannot be discarded, the preserved deposits correspond to marine sediments, as it may be deduced from their sedimentological characteristics. Nevertheless, the fact that stromatolitic facies of carbonates are by far major strongly suggests shallow marine environments.

Dioritic to granitic plutonic rocks in spatial and genetic association with these sequences are widely scattered throughout the region, showing evidence for a very shallow crustal emplacement. From a geochemical point of view, both plutonic and volcanic rocks are markedly calc-alkaline in nature (Apalategui et al. 1986). This fact, together with the general tectonic scenario, would support an interpretation as subduction-related synorogenic magmatism, although the lack of enough reliable data does not enable us to decide whether it was related to an island arc or to an Andean-type active margin.

The time span covered by the rocks within this complex is collectively admitted to range from the Upper Riphéan to the Upper Vendian, although the available biostratigraphic evidence, which is summarized below, is very restricted. Firstly, Liñan and Schmitt (1980) recorded ribulariacean cyanobacteria and the stromatolite cf. *Tungussia cumata* Bertrand Sarfati 1972, that they considered as Riphéan in age, from carbonate intercalations of unknown stratigraphic context, However, these data cannot be regarded as conclusive for the definitive inclusion of these rocks within the Riphéan. On the other hand, organic-wall microfossils have been reported from the uppermost part of the complex (San Jerónimo Formation, see below; Liñán and Palacios 1983; Liñán et al. 1984), which include numerous specimens of *Bavlinella faveolata* (Shepeleva 1962) Vidal 1976 and other planktonic croococcales cyanobacteria (Palacios, in press). This type of fossiliferous association is characteristic, on the global scale, of the Varangerian glacial stage and specially of the Valdainian post-glacial stage (Middle Vendian – Upper Vendian). The latter coincides with a global transgression during which an important increase in nutrient supply to the oceans took place, resulting in their eutrophication and the subsequent thriving of planktonic cyanobacteria. The San Jerónimo Formation lies disconformable under the Torreárboles Formation, which is considered to represent the lowermost Early Cambrian (Fedonkin et al. 1983; Liñán et al. 1984).

In order to help potential readers of the abundant literature dealing with this complex, it is worthwhile mentioning that the rocks within it have been given many different local names, among which the most widely used are as follows: Malcocinado Formation (Fricke 1941; andesites); Jabugo porphyroids (Bard 1969; rhyolites, dacites, andesites and basalts); Bodonal porphyroids (Hernández Enrile 1971; rhyolites and dacites); San Jerónimo Formation (Liñán 1978; Liñán et al. 1984; andesites, sandstones, lutites and conglomerates); etc.

2.1.3.2 Flyschoid Complex

This denomination is applied to generally turbiditic sequences that occur in the Ossa-Morena Zone, having been referred to in the current literature under various local names, e. g. Azuaga Formation (Delgado 1971), Sierra Velita Formation (Chacón 1979), etc. In most cases it consists of very thick (several kilometers) successions, almost exclusively terrigenous in nature, in which unimportant carbonate and generally reworked volcanic intercalations are found locally. From a lithological point of view they are fairly monotonous pelite - greywacke alternations mostly displaying turbidite characteristics, whereas sedimentologically, deep sea fan facies as well as slope and shelf facies may be recognized (Gabaldón, pers. comm. 1987). However, the lack of reliable stratigraphical and sedimentological data, hinders any attempt of basin paleogeographic reconstruction as well as the analysis of its evolution with time.

The only published reference concerning the fossil content of this complex corresponds to non-figured Acritarches, attributed to the Riphéan (Mitrofanov and Timofeev, in Chacón 1979), and to endogenous trace fossils: *Planolites* sp. and *Cochlichnus* sp. (Liñán, pers. comm. 1987) whose existence in Precambrian strata would be indicative of the Upper Vendian (Crimes 1987). Anyhow, a Riphean age should be disregarded for any intensively bioturbated beds.

These flysch-like sequences have been recently correlated with the so-called Schist and Greywacke Complex which largely occurs in the adjacent Central-Iberian zone (Quesada et al. 1987). Both complexes would represent different parts of the infilling of a foreland basin, formed onto the Iberian continental margin in latest Proterozoic times as a consequence of the tectonic events (Late Precambrian, Pan-African or Cadomian, orogeny; see Apalategui et al., this Vol.) happening simultaneously.

References

Aguayo J (1985) Rocas ultramáficas en el sector de Calzadilla de los Barros (Badajoz). Thesis, Univ Bilbao, 110 pp

Alia M (1963) Rasgos estructurales de la baja Extremadura. Bol R Soc Esp Hist Nat (Geol) 20: 247-262

Alvarado A (1935) Terreno Arcaico. Explicación del nuevo mapa geológico de España. Mem IGME I: 235-280

Apalategui O, Borrero JD, Higueras P (1985) División en grupos de rocas en Ossa-Morena oriental. Temas Geol Min 7: 73-80

Apalategui O, Garrote A, Roldan JF, Sánchez R (1986) Mapa Geológico de España escala 1:50.000. Hoja N° 878: Azuaga. IGME

Apalategui O, Higueras P (1983) Mapa Geológico de España escala 1:50.000. Hoja N° 855: Usagre. IGME

Arriola A, Chacón J, Eraso A, Eguiluz L, Garrote A, Sánchez R, Vargas I (1983) Mapa Geológico de España escala 1:50.000. Hoja N° 829: Villafranca de los Barros. IGME

Bard JP (1969) Le métamorphisme régional progressif des Sierras d'Aracena en Andalousie occidentale (Espagne). Sa place dans le segment hercynien sud-ibérique. Thesis, Univ Montpellier

Bertrand Sarfati J (1972) Stromatolites columnaires du Précambrien supérieur. Sahara Nord-occidental. CNRS, Serie Géol 14: 1-245

Carvalhosa A (1965) Contribuição para o conhecimiento geológico da região entre Portel y Ficalho (Alentejo). Mem Serv Geol Port II: 1-130

Chacón J (1979) Estudio geológico del sector central del anticlinorio Portoalegre-Badajoz-Córdoba (Macizo Ibérico Meridional). Thesis, Univ Granada

Chacón J, Delgado M, Garrote A (1974) Sobre la existencia de dos diferentes dominios de metamorfismo regional en la banda Elvas-Badajoz-Córdoba (Macizo Hespérico Meridional). Bol Geol Min 85 (6): 713-717

Chacón J, Fernández J, Mitrofanov F, Timofeev BV (1984) Primeras dataciones microfitopaleontológicas en el sector Valverde de Burguillos-Jerez de los Caballeros (Anticlinorio de Olivenza-Monesterio). Cuad Lab Xeol Laxe 8: 211-220

Crimes TP (1987) Trace fossils and correlation of late Precambrian and early Cambrian strata. Geol Mag 124 (2): 121-133

Delgado M (1971) Esquema Geológico de la Hoja número 878 de Azuaga (Badajoz). Bol Geol Min 82: 277-286

Delgado M, Liñán E, Pascual E, Pérez F (1977) Criterios para la diferenciación de dominios en Sierra Morena Central. Stud Geol 12: 75-90

Eguiluz L (1987) Petrogénesis de rocas ígneas y metamórficas en el antiforme Burguillos-Monesterio (Macizo Ibérico Meridional). Thesis, Univ Bilbao

Eguiluz L, Quesada C (1980) La sucesión precámbrica de la transversal de Monesterio (Badajoz). Nota preliminar. Temas Geol Min 4: 123-141

Fabriès J (1957) El problema del "estrato-cristalino" en el NE de la provincia de Sevilla (España). Not Comun IGME 48: 195-203

Fabriès J (1963) Les formations cristallines et métamorphiques du NE de la province de Seville (Espagne). Essai sur le métamorphisme des roches éruptives basiques. Thesis, Univ Nancy

Fedonkin M, Liñan E, Perejón A (1983) Icnofósiles de las rocas precámbrico-cámbricas de la Sierra de Córdoba. Bol R Soc Esp Hist Nat (Geol) 81 (1-2): 125-138

Fricke W (1941) Die Geologie des Grenzgebietes zwischen nordöstlicher Sierra Morena und Extremadura. Thesis, Univ Berlin

García Casquero JL (in press) Continental trondhjemites in displaced terranes of the Oporto-Portalegre-Badajoz-Córdoba belt (Hesperian Massif). Geologische Rundschau

García Casquero JL, Boelrijk NAIM, Chacón J, Priem HNA (1985) Rb-Sr evidence for the presence of Ordovician granites in the deformed basement of the Badajoz-Córdoba Belt, SW Spain. Geol Rundschau 74 (2): 379-384

Garrote A (1976) Asociaciones minerales del núcleo metamórfico de Sierra Albarrana (Prov. de Córdoba). Sierra Morena Central. Mem Not Univ Coimbra 82: 17-39

Garrote A, Delgado M, Contreras MC (in press) Mapa Geológico de España, escala 1:50.000 Hoja N° 900: La Cardenchosa. IGME

Gonçalves F, Palacios T (1984) Novos elementos paleontológicos e estratigráficos sobre o Proterozoico Português da Zona de Ossa-Morena. Mem Acad Ciênc Lisboa 25: 225-235

Gutiérrez Elorza M, Hernández Enrile JL (1965) Notas geológicas de la región septentrional de la provincia de Huelva. Bol R Soc Esp Hist Nat (Geol), 63: 289-297

Gutiérrez Elorza M, Hernández Enrile JL, Vegas R (1971)

Los grandes rasgos geológicos del Sur de la provincia de Badajoz y Norte de la de Huelva. Bol Geol Min 82 (3-4): 269-273

Hernández Enrile, JL (1971) Las rocas porfiroides del límite Cámbrico-Precámbrico en el flanco meridional del anticlinal Olivenza-Monesterio (Badajoz). Bol Geol Min 82 (3-4): 143-154

Hernández Pacheco, F (1953) Edad de las formaciones con facies estratocristalina en la provincia de Badajoz. Not Comun IGME 31: 3-33

Herranz P (1984) El Precámbrico y su cobertera paleozoica en la región centro-oriental de la provincia de Badajoz. Thesis, Univ Complutense Madrid

Liñán E (1978) Bioestratigrafía de la Sierra de Córdoba. Thesis, Univ Granada 191 p

Liñán E, Palacios T (1983) Aportaciones micropaleontológicas para el conocimiento del limite Precámbrico - Cámbrico en la Sierra de Córdoba, España. Comun Serv Geol Portugal 69 (2): 227-234

Liñán E, Schmitt M (1980) Microfósiles de las calizas precámbricas de Córdoba (España). Temas Geol Min 4: 171-194

Liñán E, Palacios T, Perejón A (1984) Precambrian - Cambrian boundary and correlation from southwestern and central parts of Spain. Geol Mag 121 (3): 221-228

Lotze F (1929) Stratigraphie und Tektonik des Keltiberischen Grundgebirges (Spanien). Abh Ges Göttingen Math 14 (2): 1-320

Lotze F (1956) Das Präkambrium Spaniens. Neues Jahrb Geol Paläontol Monatsh 8: 373-380

MacPherson J (1878) Sobre la existencia de la fauna primordial en la provincia de Sevilla. An Soc Esp Hist Nat 7: 281-284

MacPherson J (1879) Estudio Geológico y Petrográfico del N de la provincia de Sevilla. Bol Com Mapa Geol Esp 10: 97-269

MacPherson J (1883) Sucesión estratigráfica de los terrenos arcaicos de España. An R Soc Esp Hist Nat 12: 3-40

Mallada L (1895) Explicación del Mapa Geológico de España. Rocas hipogénicas y sistema estrato-cristalino. Mem Com Mapa Geol Esp 1: 107-420

Palacios T (1989) Microfósiles de pared orgánica del Proterozoico Superior (Región Central de la Península Ibérica). Mem Mus Paleontol Univ Zaragoza 3

Pascual E (1981) Investigaciones geológicas en el sector Córdoba-Villaviciosa de Córdoba (Sector Central de Sierra Morena). Thesis, Univ Granada, pp 1-521

Pérez Lorente F (1979) Geología de la zona Ossa-Morena al Norte de Córdoba (Pozoblanco-Belmez-Villaviciosa de Córdoba). PhD Thesis, Univ Granada, 281: 1-340

Quesada C, Larrea FJ, Florido P et al, (1987) Mapa Geológico-Minero de Extremadura. Dirección General Industria, Energía y Minas. Junta de Extremadura

Shepeleva ED (1962) Plant? fossils of unknown taxonomic position from the deposits of the Bavlinskaya series in the Volga-Urals oil province. Alcad Nauka SSSR Poklady (English translation in Earth Science Translations, 1964, 142: 1-170)

Vegas R (1968) Sobre la existencia de Precámbrico en la Baja Extremadura. Estud Geol 24: 85-89

Vegas R (1970) Formaciones precámbricas de la Sierra Morena Occidental. Relación con las series anteordovícicas de Almadén, Don Benito y Cáceres. Estud Geol 26: 225-231

Vidal G (1976) Late Precambrian microfossils from the Visingsö beds in southern Sweden. Fossil Strata 9: 57 pp

2.2 Rift Phase (Cambrian)

E. LIÑAN[1] and C. QUESADA[2]

2.2.1 Introduction

As pointed out in the introductory paragraph (Quesada, this Vol.) the Paleozoic evolution of the Ossa-Morena Zone may be considered as consisting of three major successive phases: (1) rift phase (Cambrian), (2) passive margin phase (Ordovician-Devonian) and (3) synorogenic phase (Mid-Devonian-Lower Permian). From a stratigraphic point of view every one of them is well characterized by the corresponding sedimentary sequences, quite different from one another, and the aim of this section is to focus on the relevant stratigraphic features developed during the initial rifting which affected this zone for most of the Cambrian Period. The stratigraphy of the two other phases will be dealt with in detail in the next two chapters, by Robardet and Gutiérrez Marco (this Vol., Part V, Chap. 2.3) and Quesada et al. (this Vol., Part V, Chap. 2.4) respectively.

In general terms the most salient characteristics of the sedimentary record of this extensional phase are the existence of remarkable facies and thickness variations, and the existence of an important bimodal volcanic activity, both providing strong evidence of the tectonic instability which characterized this phase. It is worthwhile mentioning that the domainal subdivision of the Ossa-Morena (see Quesada this Vol., Part V, Chap. 1 and Apalategui et al., this Vol.) was mainly based on the outstanding stratigraphic differences of the Cambrian sedimentary sequences of individual tectonic units. This fact is better understood by considering that many of the structures developed during the Cambrian rifting phase (presumably listric normal faults) were subsequently reactivated (as thrust or wrench faults) during the Upper Paleozoic Hercynian orogeny.

With respect to the simultaneous magmatism (volcanism and plutonism), which is dealt with in detail elsewhere in this section (Sánchez Carretero et al., this Vol.), we include here only a reference to its most relevant characteristics with paleogeographical significance. In this sense, the most significant aspect of the Cambrian igneous rocks, the volcanic part of which locally constitutes thick piles interbedded within the sedimentary sequences, is provided by their compositional evolution both in time and space (southward) from bimodal (basalt/rhyolite) alkaline rocks, with peralkaline end members, to tholeiite basalt suites (Quesada 1987). This evolutionary trend is consistent with the rifting axis being located somewhere south of the Ossa-Morena Zone. The existence of an oceanic terrane south of it, the Pulo do Lobo Unit (see Oliveira, this Vol., Part VI, Chap. 2) which includes the Beja-Acebuches ophiolite (Munhá et al. 1986), provides strong support for this interpretation.

2.2.2 Cambrian Stratigraphy and Facies

As stated above, the Cambrian sedimentary sequences of the various structural units within the Ossa-Morena Zona differ markedly in both thickness and facial development. Thicknesses vary between a minimum of several hundred meters and a maximum of several thousand meters, which is indicative of the existence of distinct belts with remarkable differential subsidence, most likely related to downthrow and tilting along active growth faults. This particular arrangement of the Cambrian rocks within fault-bounded troughs had not been noticed until very recent times (Liñán and Perejón 1981; Liñán and Mergl 1982; Liñán 1984b). Its recognition imposes the need of a detailed reevaluation of the lithostratigraphic and biostratigraphic characteristics of the Ossa-Morena Cambrian sequences, as well as a thorough nomenclatural revision, currently based on the classical (static) correlation with the two only sequences (Córdoba and

[1] Departament of Paleontology, Universidad Zaragoza, Spain
[2] I. T. G. E., Rios Rosas 23, 28003 Madrid, Spain

R. D. Dallmeyer and E. Martínez García (Eds.)
Pre-Mesozoic Geology of Iberia

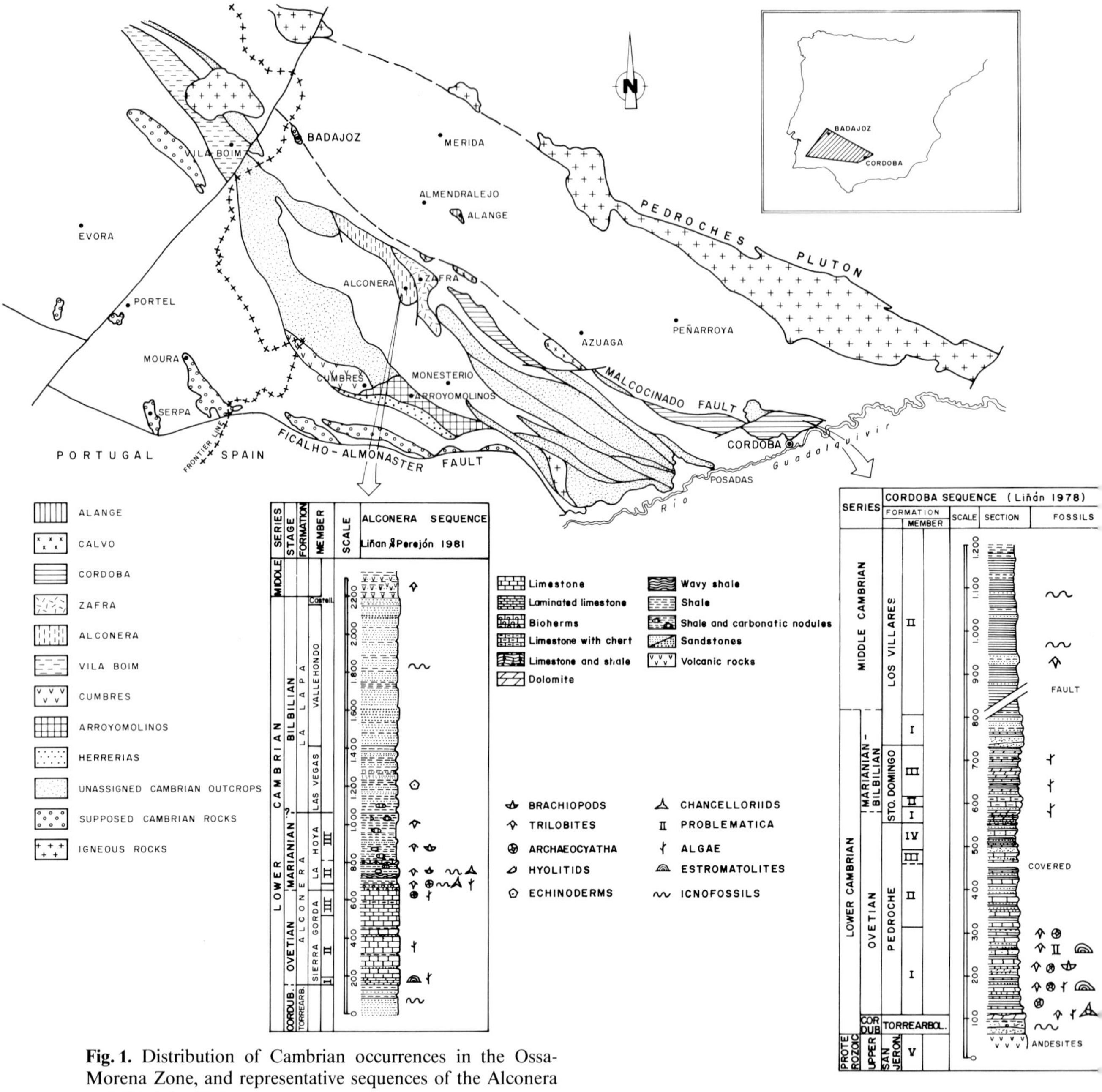

Fig. 1. Distribution of Cambrian occurrences in the Ossa-Morena Zone, and representative sequences of the Alconera and Córdoba Units

Alconera troughs, Fig. 1) in which their component stratigraphic units have been formally defined. Although this reevaluation is just beginning to be undertaken, a tentative sketch chart of several Cambrian sequences, representative of various structural units, is shown in Fig. 2 for correlation purposes. A paleogeographical reconstruction of the Cambrian basin is, however, far from being achieved, due to the subsequent structural complication added during the Hercynian orogeny.

In general terms, every individual sequence consists of three major lithosomes: (1) lower terrigenous, (2) middle carbonate and (3) upper terrigenous.

1. The lower terrigenous lithosome unconformably overlies the Precambrian basement and is formally named Torreárboles Formation (Liñán 1978), although other informal names also appear in the literature (e. g. Usagre Arkoses, Chacón 1979). It

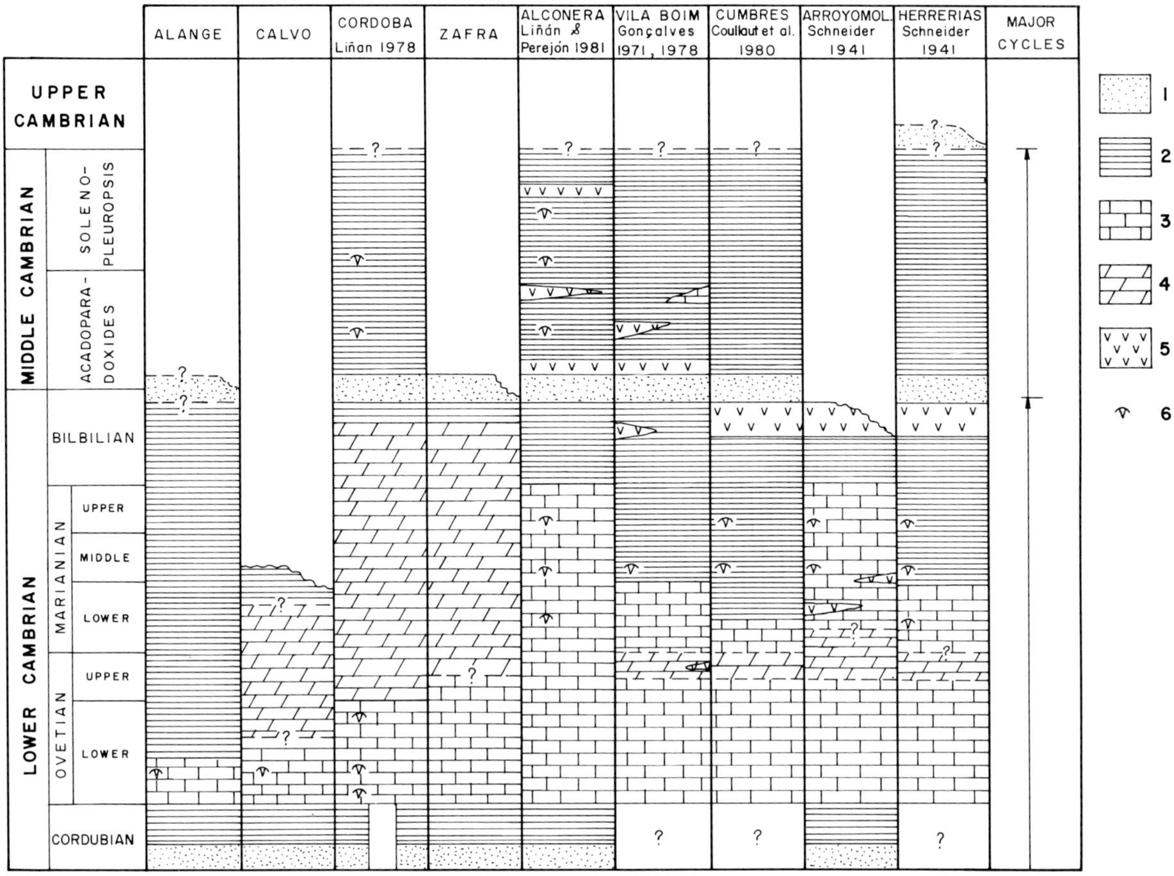

Fig. 2. Correlation chart of various Cambrian sequences of the Ossa-Morena Zone. *1* Sandstones; *2* mudstones; *3* limestones; *4* dolostones; *5* volcanic rocks; *6* fossiliferous horizons

comprises sandstones and conglomerates (at the bottom) and bioturbated red shales with interbedded sandstones (at the top) that were deposited under shallow subtidal and intertidal conditions (Brasier 1980; Fedonkin et al. 1983; Liñán and Fernández Carrasco 1984). Thickness varies from zero to 350 m, depending on a preexistent topography. Deposition of this formation coincided with the onset of the Cambrian transgression, constituting the base of the first major stratigraphic cycle ("grand cycle", in the sense of Aitken 1966). Trace fossils and soft body metazoa are present within the Torreárboles Formation.

2. The intermediate, carbonate lithosome generally overlies conformably the lower terrigenous lithosome, but locally it rests directly on the Precambrian basement (namely at Sierra de las Ermitas locality; Córdoba trough), because of the above mentioned peculiar paleotopography. This lithological unit, in which several formations have been formally defined (Liñán 1988; Liñán and Perejón 1981), consists of mixed siliciclastic/carbonate sediments: stromatolitic limestones, oncolitic limestones, archaeocyatid bioherms and biostromes, chert, algal limestomes, dolomites, nodular limestones, and interbedded mudstones and fine-grained sandstones. Synsedimentary slumped beds and volcanic horizons are locally frequent in this carbonate sequence. Thickness ranges between 50 m (Alange trough) and 1000 m (Alconera trough).

Sedimentological studies have been undertaken only at the Córdoba and Alconera localities. In Córdoba, they suggest that the lower part of the carbonate lithosome with polymeroid trilobites, brachiopods, archaeocyatid bioherms, calcareous algae, stromatolites and problematica (Liñán et al. 1981) was deposited in a subtidal environment (Zamarreño 1977, 1978; Moreno-Eiris 1987a, b, c) whereas the upper part, mainly dolomitic, was laid down in an intertidal to subtidal environment (García Hernández and Liñán 1983). Subtidal environments have been also considered for the deposition of the Alconera carbonate sequence (Moreno Eiris

1987a, b, c). The analysis of trilobite assemblages (polymeroids/miomeroids) and facies suggest that deposits on an unstable slope and/or deep basin must not be put aside in the other distinguished troughs.

3. The overlying upper terrigenous lithosome includes shales, sandstones and volcanic rocks. Thickness is largely variable from one sequence to another, because of an important erosional event after the Middle Cambrian. It ranges between 400 m at Alange and 1200 m at Alconera. The sporadic existence of an intermediate horizon of sandstones and conglomerates, showing evidence for deposition in a foreshore environment, has been interpreted as pointing out the maximum reached by a regressive Lower Cambrian pulse (Liñán 1984b). Shallow marine shelf deposition was restored with the beginning of a Middle Cambrian transgression which was accompanied by widespread tectonic instability, as demonstrated by abrupt facies changes and important volcanic activity in the southern Ossa-Morena Zone, resulting in the interfingering of trilobite facies with alkaline volcanic rocks.

2.2.3 Cambrian Fauna and Biostratigraphy

The existence of Cambrian rocks in Sierra Morena has been known since the fourth quarter of the last century, when the first archaeocyatans were discovered at El Pedroso, near Sevilla (MacPherson 1878; Roemer 1878). The large surface covered by Cambrian rocks in this region has been subsequently precised through active paleontological research undertaken in Portugal (Delgado 1904; Teixeira 1953; Gonçalves 1971, 1978) and Spain (Hernández Pacheco 1917; Hernández Sampelayo 1935; Richter and Richter 1927, 1940, 1941; Simon 1939; Meléndez 1941a, b, 1943; Rosso de Luna and Hernández Pacheco 1955a, b; Henningsmoen 1958; Lotze 1961; Sdzuy 1961, 1962; Debrenne and Lotze 1963; Bard 1964, 1965; Laus 1968), and more recently by the works of Gil Cid, Perejón, Liñán, Moreno-Eiris and co-workers, mainly on archaeocyathans and trilobites, during the 1970s and 1980s. The biostratigraphic framework of the Ossa-Morena Cambrian has, however, not been definitely established, mostly due to the structural complexity of this region.

Lotze (1961) proposed 32 trilobite levels or bands to subdivide the Iberian Peninsula Cambrian System, on the basis of his own stratigraphic data and Sdzuy's (1961) paleontological results. Six of them (bands 7 to 12) characterized the Ossa-Morena Lower Cambrian. These were:

12 *Saukianda-guillermoi* Band *(S. andalusiae, Perrector perrectus, Alanisia guillermoi)*
11 *alanisana-Saukianda* Band *(Saukianda andalusiae, Ellipsostrenua alanisana)*
10 *cf. speciosus-morenica* Band *(Serrodiscus* cf. *speciosus, Aldonaia morenica)*
9 *serratus* Band *(Eodiscus (Serrod.) serratus, Strenvaeva* cf. *vigilans*
8 *souzai-schneideri* Band *(Delgadella souzai, "Protolenus" scheneideri, Hicksia elvensis)*
7 *cornuta-sevillana* Band *(Andalusiana cornuta, Termierella sevillana)*

Most of these trilobite taxa were first established by R. and E. Richter (1940, 1941) being subsequently revised by Sdzuy (1962), who enlarged the number of trilobite species known. The existence of Middle Cambrian rocks was in turn first pointed out by Bard (1964) after discovery of trilobites near Zafra (studied by Hupé).

Sdzuy (1971a) proposed the division of the Iberian Peninsula Lower Cambrian Series into four stages: (1) not named, (2) Ovetian, (3) Marianian and (4) Bilbilian, and correlated them with chrono/biostratigraphic units in Newfoundland, Comley, Anti-Atlas and Siberia. Trilobites were the main foundation of this division, in which the Marianian Stage was defined with data supplied by the Ossa-Morena Lower Cambrian (Mariani was the Roman name of the Sierra Morena mountains). Stage and substage divisions of the Middle Cambrian Series were also published by Sdzuy (1971b), who distinguished the Acadoparadoxides, Solenopleuropsidae and Solenopleuropsidae-free stage, which he correlated with similar units in France, Sweden and Newfoundland.

Liñán and Perejón (1981) recorded the species *Saukianda* underneath *Termierella, Hicksia* and *Serrodiscus* occurrences in the Alconera section (Fig. 1). This anomalous record, bound to the structural complexity and facies variation of this region, made them advise the cautious usage of Lotze's bands for correlation, and recommended the usage of Sdzuy's biostratigraphic scale for this purpose. Liñán (1984b) subdivided, on the basis of trilobite data, the Marianian Stage into three biostratigraphic units, and called Cordubian (from the Roman name of Córdoba) the formerly unnamed lowermost Lower Cambrian Stage of Sdzuy (1971a), which he characterized by trilobite trace fossils (Fig. 3).

Finally, Perejón (1984) proposed eleven archaeocyathan zones to characterize the Iberian Lower Cambrian Series and subdivided the Ovetian Stage

SERIES	STAGE	SUBSTAGE Sdzuy 1971	TRILOBITA (Liñán, 1984) Agnostida	Polimera	ARCHAEOCYATHA (Perejón, 1984) Zone	Stage
MIDDLE CAMBRIAN	SOLENOPLEUROPSIDAE	Solenopleur.				
		Pardailhania	Condylopyge	Parabailiella, Solenopleurina, Paradoxides, s.l. Ctenocephalus, Pardailhania, Sao		
		Badulesia		Badulesia, Agraulos Eccaparadoxides, Ellipsocephalus, Parabailiella		
	ACADOPARADOXIDES	Acadolenus		Acadolenus, Ellipsocephalus, Paradoxides, s.l.		
		Conocoryphe ovata	Peronopsella	Jincella, Bailiella Skreiaspis Paradoxides, Ellipsocephalus		
LOWER CAMBRIAN	BILBILIAN				XI	ELANKIAN
	MARIANIAN	UPPER	Hebediscus, Serrodiscus, Chelediscus	Termierella, Strenuaeva Triangulaspis	X	BOTOMIAN
		MIDDLE	Delgadella	Protaldonaia, Hicksia, Atops? Strenuaeva, Rinconia Saukianda* Eops?, Triangulaspi Andalusiana* Callavia? Longiandia?	X	BOTOMIAN
		LOWER	Delgadella	Saukianda, Gigantopygus, Alanisia, Perrector, Eops,	IX	BOTOMIAN
					VIII	BOTOMIAN
	OVETIAN	UPPER			VII	BOTOMIAN
					VI, V, IV	ATDABANIAN
		LOWER		Lemdadella, Serrania Bigotina	III, II, I	ATDABANIAN
	CORDUBIAN		Rusophycus, Monomorphichnus, Phycodes, pedum			TOMMOTIAN

Fig. 3. Ossa-Morena biostratigraphy founded of trilobites and archaeocyathids; * unknown biostratigraphic position

into two parts on the basis of previous and new data on trilobites, stromatolites and archaeocyathans. This author included his zones within the Sdzuy's Lower Cambrian Stage subdivision and made a correlation with the Russian zonation. Six of these zones (I-III and VIII-X) are supported by results obtained in Sierra Morena (Fig. 3).

In summary, the biostratigraphy of the Ossa-Morena Cambrian rocks is well established at the stage or substage level. At zone level it is only partially known with archaeocyathans (Lower Ovetian and Lower-Middle Marianian), because of an irregular record in the sequence. With respect to the trilobite fauna, the presence of different species within different sections, the difficulty of finding continuous sequences not disturbed by tectonic phenomena and the poor preservation of the specimens (highly deformed) can be invoked as well, to explain the limited knowledge of its biostratigraphic units.

Some paleontological characteristics of the Cambrian Epochs could be synthesized as follows: during Cordubian times the whole region constituted a terrigenous marine platform where soft-body metazoa communities were increased with the progressive transgression advance. Twenty-five ichnotaxa from two selected sections have been recorded (Fedonkin et al. 1983; Liñán 1984a) and three ichnofacies may be differentiated within the overall succession. The first one is represented by *Skolithos, Cochlichnus* and *Planolites*. The second assemblage is defined by Arthropoda trace fossils outstanding the first record of *Phycodes pedum, Bergaueria* and *Teichichnus*. The third one is characterized by trilobite-like trace fossils. *Rusophycus* gr. *radwanski, Monomorphichnus bilinearis* and *Treptichnus bifurcus* are the best biostratigraphic ichnotaxa of this assemblage. According to Crimes (1987) his zone II (Lower Tommotian) and zone III (Upper Tommotian-Lower Atdabanian) could be represented.

During Ovetian times a carbonate platform was widespread in the region, together with algae and archaeocyathid bioherms which persisted until Marianian times (Debrenne 1958; Debrenne and Lotze 1963; Debrenne and Rozanov 1983; Perejón 1971, 1973, 1975a-c, 1976a-b, 1977, 1984; Perejón et al. 1981; Zamarreño and Debrenne 1977, Moreno-Eiris 1987b, d), trilobites (Liñán and Sdzuy 1978; Liñán 1978), stromatolites (Schmitt 1982), brachiopods (Liñán and Mergl 1982), soft-body metazoa (Cabanas and Meléndez 1966) and problematica (Liñán 1978). No trilobite and archaeocyathan record is known during Upper Ovetian times.

In the Marianian Stage, the carbonate platform extent was slowly decreasing and became restricted to the central part of the Ossa-Morena Zone (Fig. 2). The algae-archaeocyathid communities were also progressively reduced and disappeared by the end of the Middle Marianian. However, brachiopod and trilobite communities developed in the southern and central parts during the whole Marianian Stage (Richter and Richter 1940, 1941; Henningsmoen 1958; Sdzuy 1961, 1962; Gil Cid 1971, 1973, 1978; Gil Cid and Melou 1988; Liñán and Mergl 1982), enhanced by prevailing open ocean conditions, as is suggested by the record of miomeroid trilobites, in accord with Palmer's (1973) and Robison's (1976) interpretations for the Great Basin of North America.

Terrigenous platform sedimentation became fully widespread during the Bilbilian Stage in which siliciclastic strata and interbedded volcanic rocks were deposited. Scarce trace fossils are the only known paleontological data. The open terrigenous platform conditions continued in the Middle Cambrian Epoch. It is characterized by a predominant trilobite community that has only been recorded in Alconera (Bard 1964; Gil Cid 1981, 1982a, b, 1984) and Córdoba (Liñán 1978; Gil Cid 1985), because of strong subsequent erosion. The trilobite taxa found allow the proposal of the existence of a communication with the Middle Cambrian marine platform of the Cantabrian and West Asturian-Leonese Zones during the *Conocoryphe ovata* and *Badulesia* times.

2.2.4 Concluding Remarks

Recent advances in the knowledge of structural, sedimentological and stratigraphical aspects of the Ossa-Morena Zone have shown the need for much more detailed studies and a thorough reevaluation of the available data, in order to clear up the paleogeographical evolution of this area during the Cambrian rift phase. Nevertheless, the available information, as synthesized above, allows a provisional approach to its stratigraphic evolution, the most salient features of which may be summarized, as a kind of conclusion, as follows:

1. The existence of a basal unconformity (angular and erosional) over the local basement (see Quesada et al., this Vol. Part V, Chap. 2.1)
2. The existence of two "grand cycles" (Aitken 1966), interpreted as two transgressive-regressive episodes, coinciding aproximately with the Lower and Middle Cambrian Epochs. Salient features of both cycles are summarized in the following points.
3. Initial transgression that led to the widespread establishment of shallow marine, shelf sedimentary conditions throughout the area in earliest Cambrian times. The relative timing of carbonate platform sedimentation development across the Ossa-Morena Zone, as inferred by the trilobite fauna, strongly suggests that this Early Cambrian transgression proceeded from south to north.
4. Mixed clastic/carbonate shelf deposition in the lower part of the Lower Cambrian (Ovetian and Marianian stages), during which the distribution of facies and their relative thicknesses were probably controlled by the differential input of clas-

tics, with only a mild influence of synsedimentary extensional tectonics.

5. Onset of widespread tectonic instability in Late Lower Cambrian times (Bilbilian Stage), as evidenced by appearance of volcanic activity, abrupt vertical and lateral facies variations, tilting, local unconformities and, finally, tectonically controlled local transgression-regression pulses variably affecting the different parts of the compartmented basin.
6. The existence of a generalized regression in uppermost Lower Cambrian times followed by a generalized transgression in Middle Cambrian times is most likely related to regional scale (rather than local) tectonic processes and/or eustatic processes. In this sense, an evolution to milder climatic conditions during the Middle Cambrian, whether related to regional tectonics or not, could probably account for the latter event.
7. Final regression leading to the integral exposure to subaerial conditions of the Ossa-Morena Zone during the Late Cambrian (no Upper Cambrian rocks are known in the entire region). Overlying rocks correspond to Early Ordovician shelf deposits, which characterize the initiation of the next, passive margin phase of the tectonostratigraphic evolution of this zone (see Robardet and Gutierrez Marco, this Vol.)

Acknowledgements. O. Apalategui, A. Perejón and F. Gonçalves have largely contributed, through long constructive discussions, to a better understanding of the problems presented here. The efficient collaboration by R. Fernández and P. Asensio (typing) and J. Vallejo (drawing) is deeply appreciated.

References

Aitken JD (1966) Middle Cambrian to Middle Ordovician cyclic sedimentation, southern Rocky Mountains of Alberta. Can Petrol Geol Bull 14: 405-441

Bard JP (1964) Observaciones sobre la estratigrafía del Paleozoico de la región de Zafra (Prov. de Badajoz, España). Not Comun IGME 76: 175-180

Bard JP (1965) Introduction à la géologie de la chaîne hercinienne dans la Sierra Morena occidentale (Espagne). Rev Géogr Phys Géol Dynam 7 (4): 323-337

Brasier MD (1980). The Lower Cambrian transgression and glauconite-phosphate facies in western Europe. J Geol Soc (Lond) 137: 695-703

Cabanas R, Meléndez B (1966) Notas estratigráficas de la provincia de Córdoba, con una nota sobre un nuevo fósil del Cambriano. Not Comun IGME 90: 77-84

Chacón J (1979) Estudio geológico del sector central del Anticlinorio Portoalegre-Badajoz-Córdoba (Macizo Ibérico Meridional). PhD Thesis, Univ Granada, pp 1-728

Coullaut JL, Fernández Carrasco J, Gil Cid MD, Perejón A (1984). Flanco S del anticlinorio Olivenza-Monesterio, núcleo metamórfico de Valuengo y unidad de Alconera. Cuad Lab Xeol Laxe 8: 97-119

Crimes TP (1987) Trace fossils and correlation of late Precambrian and early Cambrian strata. Geol Mag 124 (2): 97-119

Debrenne F (1958) Sur un répresentant de la famille Syringoonematidae de la région d'Alanís. CR Somm Soc Géol Fr, pp 58-59

Debrenne F, Lotze F (1963) Die Archaeocyatha des spanischen Kambriums. Akad Wiss Lit, Abh Math-Naturwiss Kl, Wiesbaden, 2: 109-143

Debrenne F, Rozanov AY (1983) Paleogeographic and stratigraphic distribution of Regular Archaeocyatha (Lower Cambrian fossils). Geobios 16 (6): 727-736

Delgado JFN (1904) Faune cambrienne du Haut Alentejo (Portugal). Comm Comis Serv Geol Port 5: 307-374

Fedonkin M, Liñan, E, Perejón A (1983) Icnofósiles de las rocas precámbrico-cámbricas de la Sierra de Córdoba. Bol R Soc Esp Hist Nat (Geol) 81 (1-2): 125-138

García Hernández M, Liñán E (1983) Estromatolitos y facies asociadas en la Formación Santo Domingo (Cámbrico inferior de la Sierra de Córdoba). Libro Jubilar JM Ríos, IGME 3: 125-132

Gil Cid MD (1971) Nuevo yacimiento de trilobites en el Cámbrico inferior de Huelva (Sierra Morena). Estud Geol 27: 293-296

Gil Cid MD (1973) Nota preliminar sobre el contenido faunístico y edad del Cámbrico de Zafra y Alconera (Badajoz). Bol Geol Min 84 (1): 26-31

Gil Cid MD (1978) El género Hicksia y su interés bioestratigráfico en el Cámbrico inferior de España. Estud Geol 34: 29-32

Gil Cid MD (1981) Los Conocoryphidae del Cámbrico medio de Zafra (Badajoz). COL-PA 36: 13-39

Gil Cid MD (1982a) Los Solenopleuropsidae del Cámbrico medio de Zafra (Badajoz). Bol Geol Min 93 (1): 19-25

Gil Cid MD (1982b) Hallazgo de *Paradoxides (Eccaparadoxides) brachyrhanchis* Linnarsson 1883, en el Cámbrico medio de Zafra (Badajoz). Bol Geol Min 93 (6): 470-474

Gil Cid MD (1984) *Jincela? sulcata* SDZUY 1967. Trilobites solenopleurido del Cámbrico medio de Zafra (Badajoz). Mem Not 97: 1-10

Gil Cid MD (1985) Trilobites Agnóstidos del Cámbrico medio de la Sierra de Córdoba. Bol Geol Min 96 (4): 367-371

Gil Cid MD, Melou, M (1988) Brachiopodes du Cambrien moyen de Zafra (Province de Badajoz, Espagne). Géol Méditer 12-13 (3-4): 197-205

Gonçalves F (1971) Subsidios para o conhecimento geológico do Nordeste alentejano. Mem Serv Geol Port (Nova serie) 18: 1-62

Gonçalves F (1978) Estado actual do conhecimento geológico do nord-este alentejano. IV Curso Extensão Univ Cienc Geol Fac, Cienc Lisboa, pp 1-23

Henningsmoen G (1958). Los trilobites de las capas de Saukianda. Cámbrico inferior en Andalucía. Estud Geol 14 (35/36): 251-272

Hernández Pacheco E (1917) La fauna primordial de la Sierra de Córdoba. Asoc Esp Prog Cienc Congr Sevilla 2: 76-84

Hernández Sampelayo P (1935) Explicación del nuevo mapa geológico de España a escala 1:1.000.000. El Sistema Cambriano. Mem IGME 1: 291-528

Laus L (1968) Stratigraphie und Tektonik im südlichen Teil der Provinz Badajoz (Spanien). Thesis, Univ Münster, pp 1-131

Liñán E (1978) Bioestratigrafia de la Sierra de Córdoba. Ph D Thesis, Univ Granada pp 1-191

Liñán E (1984a) Los icnofósiles de la Formación Torreárboles (Precámbrico?-Cámbrico inferior) en los alrededores de Fuente de Cantos, Badajoz. Cuad Lab Xeol Laxe 8: 283-314

Liñán E (1984b). Introduccional problema de la Paleogeografia del Cámbrico en Ossa Morena. Cuad Lab Xeol Laxe 8: 283-314

Liñán E; Fernández Carrasco J (1984) La Formación Torreárboles y la Paleogeografia del limite Precàmbrico-Cámbrico en Ossa-Morena (flanco norte de la alineación Olivenza-Monesterio. Cuad Lab Xeol Laxe 8: 315-328

Liñán E, Mergl M (1982) Lower Cambrian brachiopods of Sierra Morena, SW Spain. Bol R Soc Esp Hist Nat (Geol) 8 (1-4): 207-220

Liñán E, Moreno-Eiris E, Perejón A, Schmitt M (1981) Fossils from the basal level of the Pedroche Formation, Lower Cambrian (Sierra Morena, Córdoba, Spain). Bol R Soc Esp Hist Nat (Geol) 79: 277-286

Liñán E, Perejón A (1981) El Cámbrico inferior de la "Unidad de Alconera", Badajoz (SW de España). Bol R Soc Esp Hist Nat (Geol) 79: 125-148

Liñán E, Sdzuy K (1978) A trilobite from the Lower Cambrian of Córdoba (Spain), and its stratigraphical significance. Senckenb Lethaea 59 (4-5): 387-399

Lotze F (1961) Das Kambrium Spaniens. Teil I: Stratigraphie. Akad. Wiss Lit Abh Naturwiss Kl 6: 285-498

MacPherson J (1878) Sobre la existencia de la fauna primordial en la provincia de Sevilla. An Soc Esp Hist Nat 7: 281-284

Meléndez B (1941a) El yacimiento de Arqueociátidos de Alconera (Badajoz). Bol R Soc Esp Hist Nat 390: 231-239

Meléndez B (1941b) Los terrenos cámbricos de los alrededores de Zafra (Badajoz). An Cienc Nat 2: 1-7

Meléndez B (1943) Los terrenos cámbricos de la Península Hispánica. Trab Inst. Cienc Nat "José Acosta" Ser Geol 1: 1-179

Moreno Eiris E (1987a) Los montículos arrecifales de Algas y Arqueociatos del Cámbrico inferior de Sierra Morena. I: Estratigrafía y facies. Bol Geol Min 98 (3): 295-317

Moreno Eiris E (1987b) Los montículos arrecifales de Algas y Arqueociatos del Cámbrico inferior de Sierra Morena. II: Las algas calcáreas. Bol Geol Min 98 (4): 449-459

Moreno Eiris, E (1987c) Los montículos arrecifales de Algas y Arqueociatos del Cámbrico inferior de Sierra Morena. III: Microfacies y diagénesis. Bol Geol Min 98 (5): 591-621

Moreno Eiris E (1987 d) Los montículos arrecifales de Algas y Arqueocitatos del Cámbrico inferior de Sierra Morena. IV: Bioestratigrafía y Sistemática de Arqueociatos. Bol Geol Min 98 (6): 729-779

Munha J, Oliveira JT, Ribeiro A, Oliveira V, Quesada C, Kerrich R (1986) Beja-Acebuches Ophiolite: characterization and geodynamic significance. Maleo 2 (13): 31

Palmer AR (1973) Cambrian trilobites. In: A. Hallam (ed): Atlas of paleobiogeography. Elsevier, Amsterdam pp 3-11

Perejon A (1971). *Pachecocyathus,* nuevo género de Archaeocyathidos del Cámbrico español. Estud Geol 26 (1): 81-83

Perejón A (1973) Contribución al conocimiento de los Arqueociátidos de los yacimientos de Alconera (Badajoz). Estud Geol 29 (2): 179-206

Perejón A (1975a). Arqueociatos de los subórdenes Monocyathina y Dokidocyathina. Bol R Soc Esp Hist Nat (Geol) 73 (1-4): 125-145

Perejón A (1975b) Arqueociatos Regulares del Cámbrico inferior de Sierra Morena (SW de España). Bol R Soc Esp Hist Nat (Geol) 73 (1-4): 147-193

Perejón A (1975c) Nuevas formas de Arqueociatos del Cámbrico inferior de Sierra Morena (I). Tecniterrae 8: 8-29

Perejón A (1976a) Nuevas formas de Arqueociatos del Cámbrico inferior de Sierra Morena (II). Tecniterrae 9: 7-24

Perejón A (1976b) Nuevos datos sobre los Arqueociatos de Sierra Morena. Estud Geol 32 (1): 5-33

Perejón A (1977) Arqueociatos con túmulos en el Cámbrico inferior de Córdoba. (Sierra Morena Oriental). Estud Geol 33 (6): 545-555

Perejón A (1984) Bioestratigrafía de los Arqueociatos en España. Cuad Geol Ibérica 9: 213-265

Perejón A, Moreno Eiris E, Herranz, P (1981) Datación con Arqueociatos del Cámbrico Inferior al norte de Llerena, Badajoz (SW de España). Estud Geol 37 (1-2): 89-96

Quesada C (1987) Lower Paleozoic rifting and subsequent miogeocline development in SW Iberia. Int Conf on West-African Orogens, Nouakchott (Mauritania). Extended abstract.

Richter R, Richter E (1927) Eine Crustacee (*Isoxis carbonelli* n. sp.) in den *Archaeocyathus*-Bildungen der Sierra Morena. Senckenb Lethaea 9 (5): 188-195

Richter R, Richter E (1940) Die *Saukianda*-Stufe von Andalusien, eine fremde Fauna im europäischen Ober-Kambrium. Abh Senckb Naturforsch Ges 450: 1-88

Richter R, Richter E (1941) Die Fauna des Unter-Kambriums von Cala im Andalusien. Abh Senckb Naturforsch Ges 455: 1-90

Robison RA (1976) Middle Cambrian trilobite biostratigraphy of the Great Basin. Geol Stud 23 (2): 93-109

Roemer F (1878) Über *Archaeocyathus marianus* n. sp. Z Dtsch Geol Ges 30: 369-370

Rosso de Luna I, Hernández Pacheco F (1955a) Mapa Geológico de España. Escala 1:50.000 (1a. serie). Explicación de la Hoja num. 854: Zafra (Badajoz). IGME, pp 1-142

Rosso de Luna I, Hernández Pacheco F (1955b) Mapa Geológico de España. Escala 1:50.000 (1a. serie). Explicación de la hoja num. 853: Burguillos del Cerro. IGME, pp 1-108

Schmitt M (1982) Columnar stromatolites from the lower Cambrian Formación Pedroche, Sierra Morena, S-Spain. Bol R Soc Esp Hist Nat (Geol) 80 (1-4): 5-23

Sdzuy K (1961) Das Kambrium Spaniens. Teil II: Trilobiten. Akad Wiss Lit Abh Naturwiss Kl 7-8: 499-690

Sdzuy K (1962) Trilobiten aus Unter-Kambrium der Sierra Morena (S-Spanien). Senckenb Lethaea 43 (3): 181-229

Sdzuy K (1971a) Acerca de la correlación del Cámbrico Inferior en la Peninsula Ibérica. I. Congr Hispano-Luso-Americano Geol Econ 2, 1: 753-768

Sdzuy K (1971b) La subdivisión bioestratigráfica y la correlación del Cámbrico Medio. I Congr Hispano-Luso-Americano Geol Econ 2, 1: 769-782

Simon W (1939) Archeocyathacea: I. Kritische Sichtung der Superfamilie. II. Die Fauna im Kambrium der Sierra Morena (Spanien). Abh Senckb Naturforsch Ges 448: 1-87

Teixeira C (1953) Notas sobre geología de Portugal. O Sistema Cámbrico. Ed. Author, Lecture booklet, Lisboa, pp 1-22

Zamarreño I (1977) Early Cambrian algal carbonates in southern Spain. In: Flügel (ed) Fossil algae. Springer, Berlin Heidelberg New York, pp 360-365

Zamarreño I (1978) Tipos y distribución de facies en el nivel carbonatado del Cámbrico de España, Cuad Sem Est Ceram Sargadelos 27: 287-311

Zamarreño I, Debrenne F (1977) Sédimentologie et biologie des constructions organogènes du Cambrien inférieur du sud de l'Espagne. Mem BRGM 89: 49-61

2.3 Passive Margin Phase (Ordovician-Silurian-Devonian)

M. Robardet[1] and J. C. Gutierrez Marco[2]

2.3.1 Introduction

Stratigraphical studies in the Ossa-Morena Zone (OMZ) are particularly difficult because of the general scarcity of fossiliferous levels. The age of the Paleozoic formations has frequently been based upon lithological similarities or deduced from their location in the structural units (Carbonnel 1926). Such stratigraphical attributions are far from unquestionable considering the relative monotony of the terrigenous facies and the local structural complexity. However, stratigraphical studies based upon paleontological data have been completed in a few areas of the OMZ where parts of the Paleozoic succession have been established (see references below). These reference areas (Barrancos, Venta del Ciervo, Valle and Cerron del Hornillo synclines) are separated by large regions where stratigraphical attributions are rather uncertain and sometimes erroneous (see Bard 1966; Wiechert 1987 unpublished work; Mette 1987 unpublished work).

2.3.2 Ordovician

In general, the Ordovician succession and faunas are poorly known (see Julivert et al. 1983, p. 238) and geological maps frequently do not distinguish Ordovician from Silurian sequences.

In the northern part of the OMZ (between the Los Pedroches Batholith and the Badajoz-Cordoba Shear Zone) the Ordovician succession is strongly similar to that of the Central-Iberian Zone (CIZ). Here Ordovician rocks overlie transgressively Cambrian or Late Precambrian rocks. The succession includes the Armorican Sandstone Fm. (Kelch 1957 unpublished work; Mayorga Fm. auct) and overlying terrigenous formations (dark shales, siltstones and sandstones) with *Neseuretus tristani* and other trilobites and graptolites of Llanvirn to Caradoc age (Maas 1958 unpublished work; Perdigao 1965; Gutiérrez-Marco et al. 1988). Upper portions of the succession may include Ashgill levels (Pérez Lorente 1979; Herranz 1984) but the whole sequence is not known precisely.

In the southern part of the OMZ (south of the Badajoz-Cordoba Shear Zone), the reference areas are those of Barrancos (Delgado 1908), Venta del Ciervo (Schneider 1939 unpublished work, 1951) and the Valle syncline (Simon 1951; Fig. 1). Recent studies have provided additionnal stratigraphical information and new biostratigraphical data (Perdigao 1967; Gutiérrez Elorza 1970; Robardet 1976; Hafenrichter 1979, 1980; Paris 1981; Perdigao et al. 1982; Gutiérrez Marco 1982, 1986 unpublished work; Gutiérrez Marco et al. 1984; Mette 1987 unpublished work; Erdtmann et al. 1987; Hammann and Rabano 1988). However, detailed paleontological studies are rather rare and have only focused on graptolites (Perdigao 1967; Gutiérrez Marco 1986 unpublished work; Erdtmann et al. 1987), trilobites (Hammann and Rabano 1988) and microfossils (Paris 1981; Mette 1987 unpublished work).

From a general point of view, the Ordovician succession still remains poorly known. Several authors have described Ordovician sequences in different localities (Venta del Ciervo: Schneider 1939 unpublished work, 1951; Villanueva del Fresno Kalthoff 1963 unpublished work; Encinasola: Assmann 1959 unpublished work), but most stratigraphical attributions remain uncertain and some of them have been shown to be erroneous. Correlations between the different regions (Fig. 4) are

[1] Laboratoire de Paléontologie et de Stratigraphie, Université de Rennes I, 35042 Rennes Cedex, France. U. R. A. 1364 "Paléozoique marin gondwanien" associée au CNRS

[2] U. E. I. de Paleontologia, Instituto de Geologia Economica (C. S. I. C. - U. C. M.), Facultad de Ciencias Geologicas, 28040 Madrid, Spain

R. D. Dallmeyer and E. Martínez García (Eds.)
Pre-Mesozoic Geology of Iberia

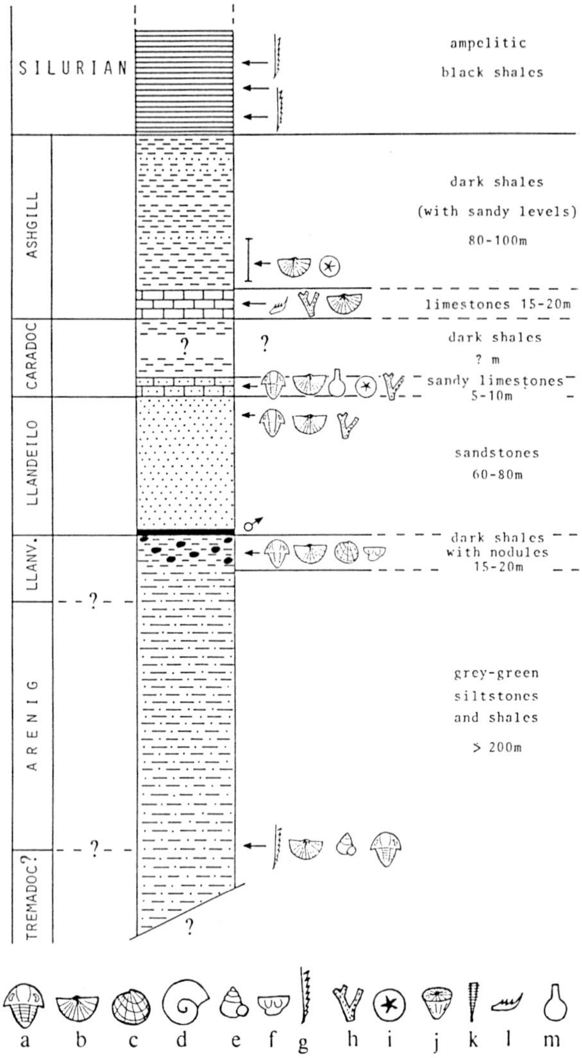

Fig. 1. The Ordovician succession in the Valle syncline (After Gutiérez Marco et al. 1984). *a* Trilobites; *b* brachiopods; *c* bivalves; *d* cephalopods; *e* gastropods; *f* ostracods; *g* graptolites; *h* bryozoans; *i* echinoderms; *j* corals; *k* tentaculitids; *l* conodonts; *m* chitinozoans

therefore only tentative. In spite of the rather poor knowledge of the Ordovician rocks in the OMZ, it must be emphasized that regions south of the Badajoz-Córdoba Shear Zone strongly differ in paleontological affinities from those situated immediately north (northern OMZ and CIZ). In these regions the Ordovician succession and constituent faunas are almost homogeneous and have close affinities to those of the Armorican Massif (see sedimentary and faunal domains).

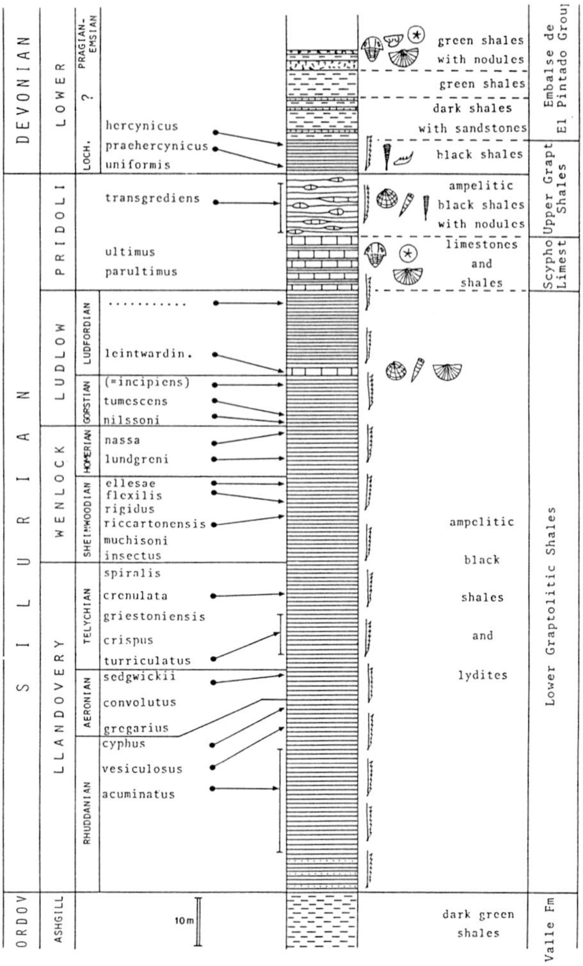

Fig. 2. The Silurian succession in the Valle and Cerrón del Hornillo synclines (After Jaeger and Robardet 1979). *Symbols* as in Fig. 1

2.3.3 Silurian

Silurian rocks have been known for a long time in the OMZ and numerous workers have cited Silurian black shales and graptolite faunas from many localities (see references in Truyóls and Julivert 1983, p. 263–264). The best understanding of the Silurian rocks has resulted from detailed studies (Jaeger and Robardet 1979) in the Valle and Cerron del Hornillo synclines where these rocks had previously been recognized (Hernandez-Sampelayo 1934, 1942; Simon 1951; Fabries 1963). The lithology and the faunas are almost identical in the two synclines: the Silurian and earliest Devonian sequence (Fig. 2) consists of ca. 150 m of black argillaceous shales and siliceous slates with subordinate sandy levels in the lowermost part and calcareous beds in the upper part of the succession. The occurrence of 21 graptolite zones indicates that the sequence ranges from the true base of the Llandovery (*acuminatus* Zone) up to the top of the Pridoli (*transgrediens* Zone) and extends into the earliest Devonian (Lochkovian: *hercynicus* Zone). The only important lithological change takes place in the Pridoli with the *Scyphocrinites* limestone (10–15 m of limestone-shale alternations) which divides the black shale succession into two parts. The continuity of sedimentation from the base of the Llandovery up to the Lochkovian is demonstrated by the very complete graptolite succession.

In the other parts of the OMZ, the Silurian sequence is not as precisely known, but the graptolite faunas listed in numerous publications indicate that the Silurian fossiliferous sequence starts off with the Lower Llandovery and extends up to the Ludlow or the Pridoli. Most probably Pridolian and Lochkovian strata are represented to the northwest (Barrancos area) by chloritoid banded slates (cf. Perdigao et al. 1982). Despite the absence of the *Scyphocrinites* limestone which has never been observed in the western areas, the Silurian sequence of the Valle syncline is most probably representative of the whole OMZ.

On several points this type of Silurian succession differs from what is known in the other regions of the Iberian Peninsula. In most regions of central and northern Iberia the lowermost parts of the succession consist of sandy or quartzitic formations (Valent or Llandoverium Quartzite of the German geologists: Kalthoff 1963 unpublished work; Hafenrichter 1980). In these regions the earliest graptolitic black shales are of Middle Llandovery age, and most probably indicate a stratigraphical gap at the Ordovician-Silurian boundary.

In upper parts of the succession in the central and northern regions, a thick arenaceous sedimentation developed in the Ludlow or Pridoli (Furada, San Pedro, Luesma, Alcolea Fms). In the OMZ euxinic conditions persisted during the whole Silurian Period, as in Catalonia and Pyrenees (Julivert et al. 1985).

2.3.4 Devonian

Devonian rocks are known in northern parts of the OMZ from the wide synclinorium located south of the Los Pedroches batholith and extending from Portalegre to the north of Córdoba (see references in Julivert et al. 1983, p. 288–289 and Peinado et al. 1985 unpublished work). In the southern areas they occur near Barrancos (Delgado 1908; Pruvost 1914; Perdigao et al. 1982), Venta del Ciervo (Schneider 1939 unpublished work, 1951) and in the Cerron del Hornillo and Valle synclines in the southeast where they have recently been recognized (Robardet 1976) and partly described (Racheboeuf and Robardet 1986; Racheboeuf 1985).

In the northern parts of the OMZ the lithology and the faunas have much in common with those of the CIZ. The succession begins with Gedinnian sandstones overlain by shales and siltstones with calcareous levels (Pragian, Siegenian). The whole sequence seems to be of Lower Devonian age (up to Emsian). Middle Devonian rocks are apparently lacking everywhere even in the Portalegre area where faunas of supposed Middle Devonian age (Perdigao 1967) have proved to be Upper Emsian (Racheboeuf and Robardet 1986).

In southern areas the most complete succession (Fig. 3) is known from the Cerron del Hornillo and above all the Valle synclines. The earliest Devonian rocks (Lochkovian) are graptolitic black shales which are conformably overlain by green to brown shales and siltstones with benthic faunas (brachiopods, trilobites, ostracods, etc.) of Pragian and Emsian age, which also occur in the Venta del Ciervo and Barrancos areas (Racheboeuf and Robardet 1986). As it has previously been indicated, Middle Devonian rocks are generally unknown in the OMZ (Puschmann 1967, 1970). Montemor-o-Novo (Portugal) is the only locality where Middle Devonian faunas (conodonts) have been recorded (Boogaard 1972) in limestones which had previously been regarded as Upper Devonian or Lower Carboniferous (Pruvost 1914).

In several regions, flysch-like or molassic-type Upper Devonian-Lower Carboniferous rocks correspond to the earliest synorogenic phases of the evolution of the OMZ (see below). However, the Devonian succession in the Valle syncline includes Famennian (black shales and limestones) rocks which conformably overlie a Lower Devonian sequence (Racheboeuf and Robardet 1986). Paleontological and biostratigraphical studies have focused on brachiopods of the Lower Devonian and several faunizones have been established and are useful for regional correlation (*Ctenochonetes robardeti; Plectodonta (Dalejodiscus) minor minor; Lissatrypa villosa; "Reticulariopsis"* sp.; *Andalucinetes hastatus:* Racheboeuf 1985; Racheboeuf and Robardet 1986). Some information has also been reported for ostracods and trilobites (in Mette 1987 unpublished work). These Lower Devonian faunas of the southern part of the OMZ (Fig. 3) are of dominant Hercynian type and differ strongly from those of Rhen-

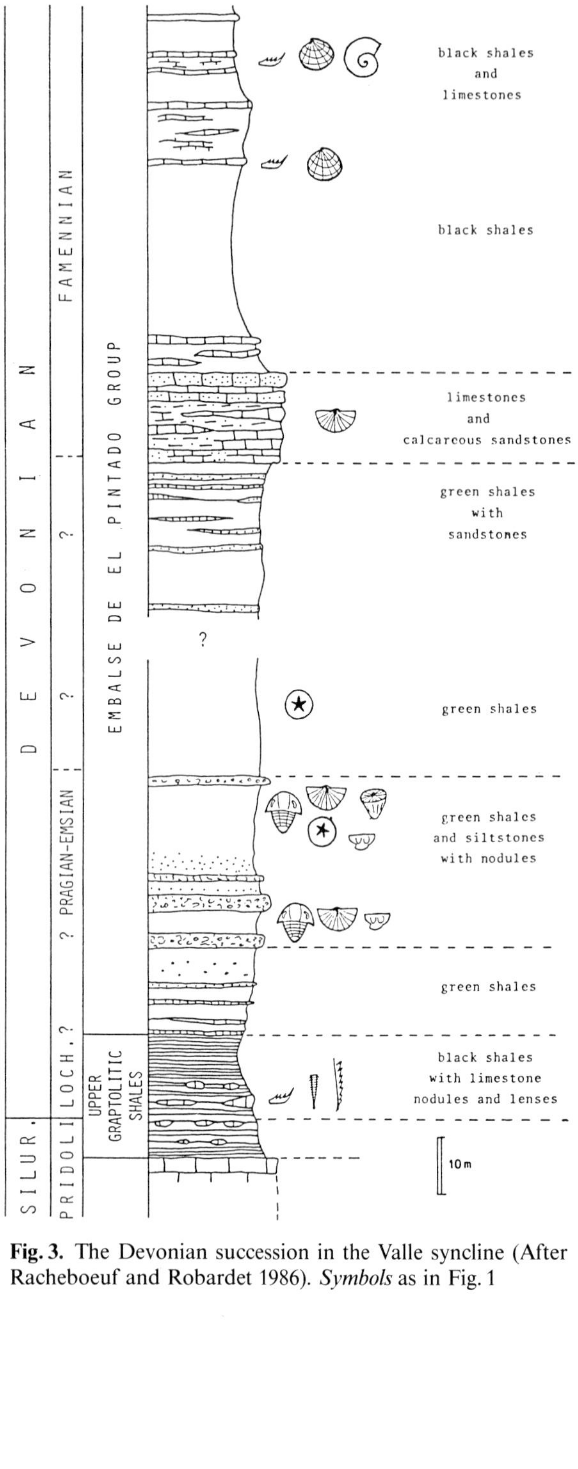

Fig. 3. The Devonian succession in the Valle syncline (After Racheboeuf and Robardet 1986). *Symbols* as in Fig. 1

ish type, which occur in the northern OMZ and in the Central-Iberian Zone (Julivert et al. 1983; Pardo Alonso and García Alcalde 1984).

2.3.5 Conclusion

The entire Ordovician-Silurian-Devonian succession in the OMZ (Fig. 4) consists of shallow-water sediments which were likely deposited under shelf conditions. The rather condensed euxinic Silurian sequence may be due to a smaller influx of terrigenous particles and perhaps to climatic and paleogeographical factors. Some levels in the Lower Devonian (green shales and siltstones with nodules) might correspond to slope deposits.

Several authors have asserted either that "Caledonian" events occurred at the Silurian-Devonian boundary (Gutiérrez-Elorza 1970) or that the Hercynian tectonometamorphic episodes began as early as the end of the Ordovician (Apalategui 1980). These interpretations mainly resulted from erroneous stratigraphical attributions and must be abandoned. Notwithstanding incomplete knowledge of the stratigraphical record in the whole OMZ, it can be reasonably assumed that the shelf passive margin sedimentation developed almost continuously from the earliest Ordovician up to the end of the Lower Devonian. Absence of Middle Devonian rocks in the central and southern parts of the Iberian Peninsula may partly result from erosional processes related to the earliest episodes of the Hercynian orogeny which preceded synorogenic deposits of Upper Devonian-Lower Carboniferous (Oliveira et al. 1986). However, the absence of Middle Devonian rocks over such a large area, and especially in localities (Valle syncline: Racheboeuf and Robardet 1986), where a Lower Devonian-Upper Devonian succession occurs, is apparently indicative of a stratigraphical gap.

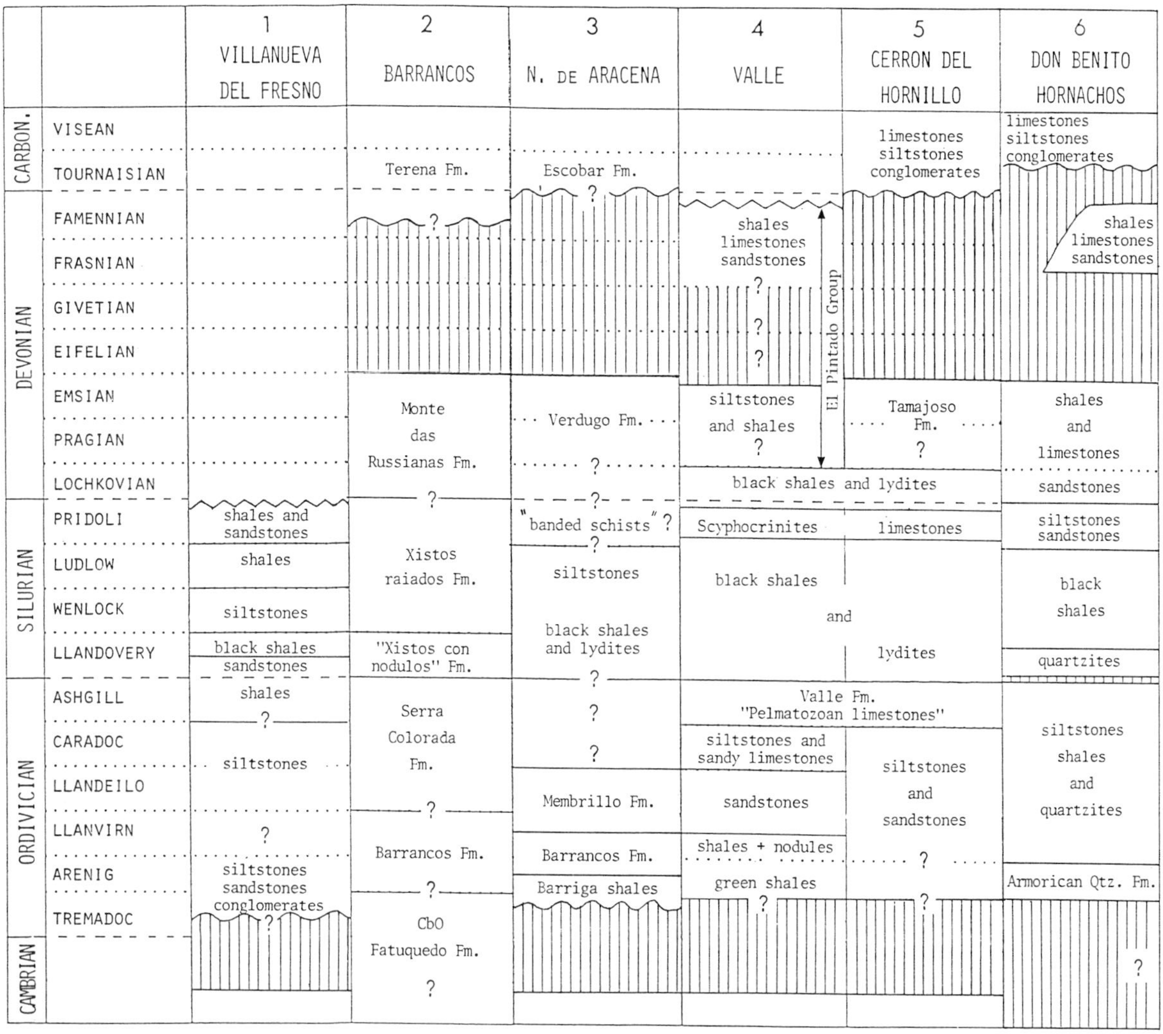

Fig. 4. Tentative correlation chart of the Ordovician-Silurian-Devonian successions in the Ossa-Morena Zone (references cited in the text)

References

Apalategui O (1980) Consideraciones estratigráficas y tectónicas en Sierra Morena occidental. Temas geologicos y mineros. Inst Geol Min Esp, pp 23-41

Bard JP (1966) Quelques précisions sur la lithologie du "Silurien" de la région de Aracena (Huelva, Espagne). Not Comun Inst Geol Min Esp 83: 93-98

Boogaard M van den (1972) Conodont faunas from Portugal and southwestern Spain, part 1. A Middle Devonian fauna from near Montemor o Novo. Scr Geol 13: 1-11

Carbonnel A (1926) Depositos considerados como cámbricos en el Sur de España que deben pasar al Culm y al Devoniano. Contribución al estudio de las series paleozoicas de la Sierra Morena. 24th Int Geol Congr Madr (Ed 1927): 751-767

Delgado JFN (1908) Systeme silurique du Portugal. Comun Serv Geol Port, pp 1-245

Erdtmann BD, Maletz J,Gutiérrez Marco JC (1987) The new Early Ordovician (Hunneberg Stage) graptolite genus *Paradelograptus* (Kinnegraptidae), its phylogeny and biostratigraphy. Paläontol Z 61 (1-2): 109-131

Fabries J (1963) Les formations cristallines et métamorphiques du Nord-Est de la province de Séville (Espagne). Sci Terre 4: 1-270

Gutiérrez Elorza (1970) Estudio geológico-estructural de la region de Aracena-Cumbres Mayores (Provincia de Huelva y Badajoz). Junta Energ Nucl, pp 1-170

Gutiérrez Marco JC (1982) Descubrimiento de nuevos niveles con Graptolitos ordovicicos en la unidad "Pizarras con Didymograptus" - Schneider 1939 - (Prov. Huelva, S. W. de Espana). Comun Serv Geol Port 68 (2): 241-246

Gutiérrez Marco JC, Apalategui O, Rabano I (1988) Fósiles ordovicicos del N. O. de Adamuz (Cordoba): aportación paleobiogeográfica a la concepcion del límite entre las Zonas Centroibérica y Ossa Morena durante el Paleozoico. Geogaceta 3: 22-24

Gutiérrez Marco JC, Rabano I, Robardet M (1984) Estudio

bioestratigráfico del Ordovícico en el Sinclinal del Valle (Provincia de Sevilla, S. O. de Espana). Mem Not Coimbra 97: 12-37

Hafenrichter M (1979) Paläontologisch-okologische und lithofazielle Untersuchungen des "Ashgill-Kalkes" (Jungordovizium) in Spanien. Arb Paläontol Inst Würzburg 3: 1-139

Hafenrichter M (1980) The lower and upper boundary of the Ordovician system of some selected regions (Celtiberia, eastern Sierra Morena) in Spain. Part II: The Ordovician-Silurian boundary in Spain. Neues Jahrb Geol Palaeontol Abh 160, 1: 138-148

Hammann W, Rabano I (1988) Morphologie und Lebensweise der Gattung *Selenopeltis* (Hawle and Corda, 1847) (Trilobita), und ihr Vorkommen im Ordovizium von Spanien. Senckenb Lethaea 68, 1-4: 91-137

Herranz P (1984) Las discontinuidades estratigraficas principales en el sector central del N. E. de "Ossa-Morena": rango y significado tectosedimentario. Mem Not Coimbra 97: 51-80

Hernández Sampelayo P (1934) El Sistema Cambriano. Graf reun Edit Madr, pp 1-240

Hernández Sampelayo P (1942) El Sistema Siluriano. Mem Inst Geol Min Esp II, 1: 1-592

Jaeger H, Robardet M (1979) Le Silurien et le Dévonien basal dans le Nord de la Province de Séville (Espagne). Géobios 12 (5): 687-714

Julivert P, Truyols J, Verges J (1983) El Devónico en el Macizo Iberico. In Geologia de Espana, Libro jubilar JM Rios. Inst Geol Min Esp 1: 265-311

Julivert M, Duran H, Rickards RB, Chapman AJ (1985) Siluro-Devonian graptolite stratigraphy of the Catalonian Coastal Ranges. Acta Geol Hisp 20 (3-4): 199-207

Oliveira JT, García Alcalde JL, Linan E, Truyóls J (1986) The Famennian of the Iberian Peninsula. Ann Soc Géol Belg 109: 159-174

Pardo Alonso M, García Alcalde JL (1984) Bioestratigrafía del Devónico de la región de Almaden (Ciudad Real, Espana). Trab Geol 14: 79-120

Paris F (1981) Les Chitinozoaires dans le Palézoïque du Sud-Quest de l'Europe (Cadre géologique - Etude systématique - Biostratigraphie). Mém Soc Géol Min Bretagne 26: 1-412

Perdigao JC (1965) Sobre a descoberta de Didymograptus na regiao de Portalegre. Comun Serv Geol Port 48: 157-160

Perdigao JC (1967) Descoberta de Mesodevonico em Portugal (Portalegre). Comun Serv Geol Port 52: 27-48

Perdigao JC, Oliveira JT, Ribeiro A (1982) Noticia explicativa da folha 448 Barrancos. Comun Serv Geol Port, pp 1-52

Pérez Lorente F (1979) Geología de la Zona Ossa-Morena al Norte de Córdoba (Pozoblanco-Belmez Villaviciosa de Cordoba). Thesis, University of Granada 281: 1-340

Pruvost P (1914) Observations sur les terrains dévoniens et carbonifères du Portugal et sur leur faune. Comun Serv Geol Port 10: 1-21

Puschmann H (1967) Zum Problem der Schichtlucken im Devon der Sierra Morena (Spanien). Geol Rundschau 56: 528-542

Puschmann H (1970) Eine Palaeogeographie des Devons auf der Iberischen Halbinsel. Z Dtsch Geol Ges 120: 107-118

Racheboeuf P (1985) *Andalucinetes hastatus* n. g., n. sp., nouveau Chonetacea (Brachiopoda) dévonien du sud-ouest de l'Espagne. Géobios 18, 5: 659-664

Racheboeuf P, Robardet M (1986) Le Pridoli et le Dévonien inférieur de la Zone d'Ossa Morena (Sud-Ouest de la Péninsule Ibérique). Etude des Brachiopodes. Geol Paleontol 20: 11-37

Robardet M (1976) L'originalité du segment hercynien sud-ibérique au Paléozoïque inférieur: Ordovicien, Silurien et Dévonien dans le Nord de la Province de Séville, Espagne. CR Acad Sci Paris D 283: 999-1002

Schneider H (1951) Das Palaeozoikum im Westteil der Sierra Morena (Spanien). Z Dtsch Geol Ges 103: 134-135

Simon W (1951) Untersuchungen im Palaeozoikum von Sevilla (Sierra Morena, Spanien). Abh Senckenb Naturforsch Ges 485: 31-62

Truyóls J, Julivert M (1983) El Silúrico en el Macizo Ibérico. In Geología de Espana, Libro jubilar JM Rios, Inst Geol Min Esp 1: 246-265

2.4 Synorogenic Phase (Upper Devonian - Carboniferous - Lower Permian)

C. QUESADA,[1] M. ROBARDET[2] and V. GABALDON[1]

2.4.1 Introduction

The Ossa-Morena Paleozoic synorogenic deposits (Fig. 1) may be divided into three, progressively younger, main groups, witnessing the regional progressive evolution from a more or less stable passive continental margin during most of the Lower Paleozoic, to its final uplift and emergence as a response to the convergence tectonic processes that affected the Iberian continental margin throughout the Upper Paleozoic. Such groups may be named, according to their relationships with the tectonic processes, as follows:

1. Foreland basin infilling flyschs and molasses;
2. Terrestrial fill of synorogenic intermontane basins;
3. Terrestrial fill of late-orogenic intermontane basins.

[1] I. T. G. E., Rios Rosas 23, 28003 Madrid, Spain
[2] Institut de Géologie, Université de Rennes I, Campus de Beaulieu, 35042 Rennes Cedex, France

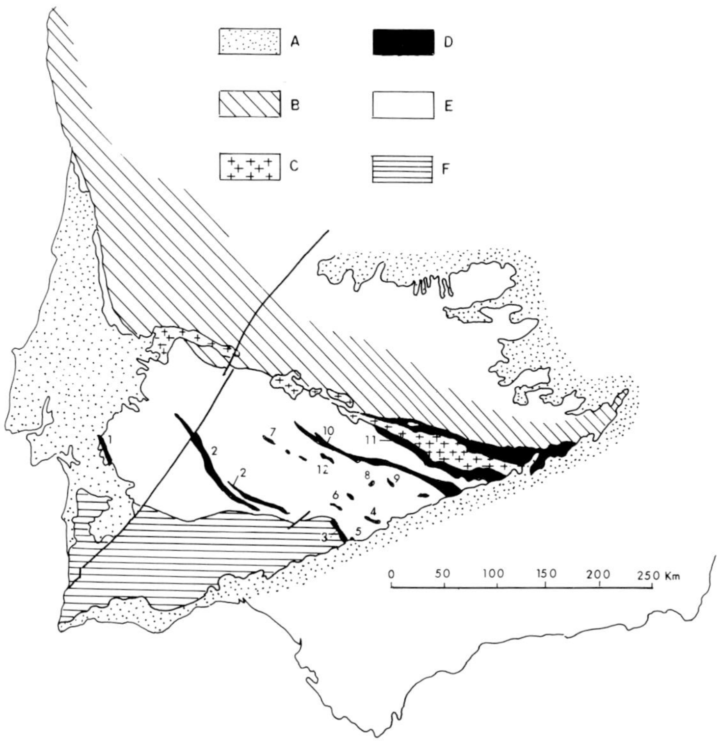

Fig. 1. Distribution of synorogenic deposits in the Ossa-Morena Zone (Upper Devonian-Carboniferous-Lower Permian). *A* Recent Lower; *B* Central-Iberian Zone; *C* Pedroches Batholith; *D* Synorogenic deposits; *E* Ossa-Morena Zone; *F* South Portuguese Zone. *1* Cabrela Fm. and Santa Susana Basin; *2* Terena Flysch; *3* Viar Basin; *4* Cerrón del Hornillo Syncline; *5* Villanueva del Río Basin; *6* Guadalcanal Basin; *7* Los Santos de Maimona Coalfield; *8* Valdeinfierno Basin; *9* Benajarafe Coalfield; *10* Guadiato Belt and Peñarroya-Belmez Basin; *11* Pedroches Belt; *12* Berlanga occurrence

R. D. Dallmeyer and E. Martínez García (Eds.)
Pre-Mesozoic Geology of Iberia

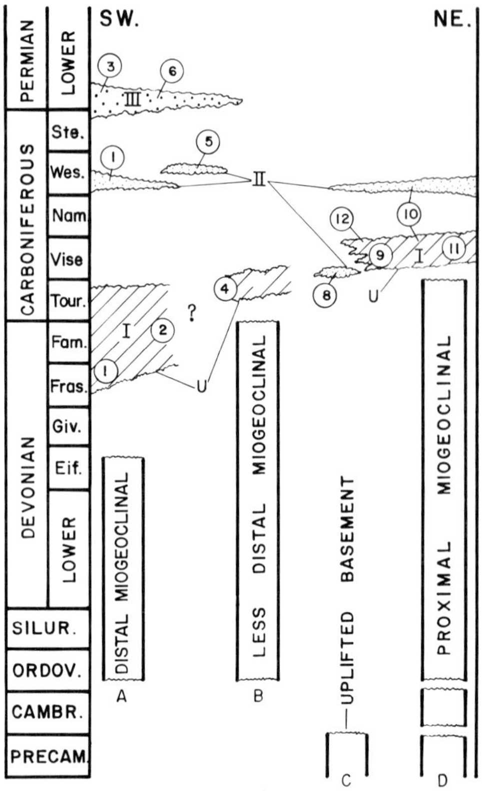

Fig. 2. Time distribution and relationship of Late Paleozoic synorogenic deposits in the Ossa-Morena Zone, with indication of the youngest underlying strata in various units. **I** Foreland basin marine flysch and terrestrial molasse; **II** terrestrial fill of synorogenic intermontane pull-aparts; **III** late-orogenic intermontane basins; transition miogeocline foreland basin unconformity. *A* Beja Massif; *B* Olivenza-Monesterio antiform northern limb; *C* Badajoz-Córdoba Shear Zone; *D* Pedroches Carboniferous Basin. *Arabic numbers* have the same meaning as in Fig. 1

The time distribution of all three groups in different Ossa-Morena areas is sketched on Fig. 2, in which the respective youngest underlying strata are indicated as well.

2.4.2 Upper Devonian - Lower Carboniferous Flyschs (Locally Molasses)

This is the best represented group despite the fact that in central Ossa-Morena areas it constitutes small-sized scattered occurrences, in which the preserved sequences are most commonly fairly thin, which usually impedes correlation. This group consists of generally marine successions, the characteristics and age of which vary from one unit to another. These include the so-called Terena flysch, Cabrela Formation (Beja Massif) and the mainly terrigenous fill of the following Carboniferous basins: Cerrón del Hornillo, Los Santos de Maimona, Bienvenida, Casas de Reina, Berlanga, Matachel, Benajarafe, Guadiato, Guadalbarbo, Pedroches, etc. (see Fig. 1 for location). The available biostratigraphic information is as yet very limited and, together with the existing physical disconnection among occurrences, in many cases prevents their detailed correlation. In spite of these facts, it is possible to recognize that the flyschoid successions within the group are progressively younger from southwest to northeast, which has been interpreted as indicating that the simultaneous deformation migrated in the same direction (Apalategui and Quesada 1987).

Common features shared by every sequence in this group (Quesada 1983) are summarized as follows:

1. Existence of a basal unconformity, the importance of which also decreases in a northeasterly direction (in the northernmost occurrences it is actually a disconformity).
2. Presence of conglomerates including previously deformed and metamorphosed clasts, derived from the underlying Lower Paleozoic and/or Precambrian successions.
3. Evolution, in the most completely preserved sequences, from variably deep flyschoid marine facies to shallow marine and/or terrestrial molasse facies.
4. Simultaneous existence of volcanic activity, located along certain alignments inside the basins.

Given the synthetic character of this review, exhaustive descriptions of each and every outcrop cannot be given. The stress is put on bringing out significant features of three representative selected areas, which may help the reader to obtain an overall perspective. These areas are: (1) the southern units (Beja-Aracena Massif neighbourhood); (2) the central area (Cerrón del Hornillo Syncline) and (3) the northern area (Pedroches Basin).

2.4.2.1 *Southern Units*

Synorogenic sequences in this area correspond to the Cabrela Formation (Ribeiro 1983), occurring in the western end of the Beja Massif, and the Terena Flysch (Perdigão et al. 1982), which occurs in the Terena Syncline, north of the same Massif (see Fig. 1 for location). It is likely that both once belonged to the same basin, but their accurate correlation is as yet hindered by geographical disconnec-

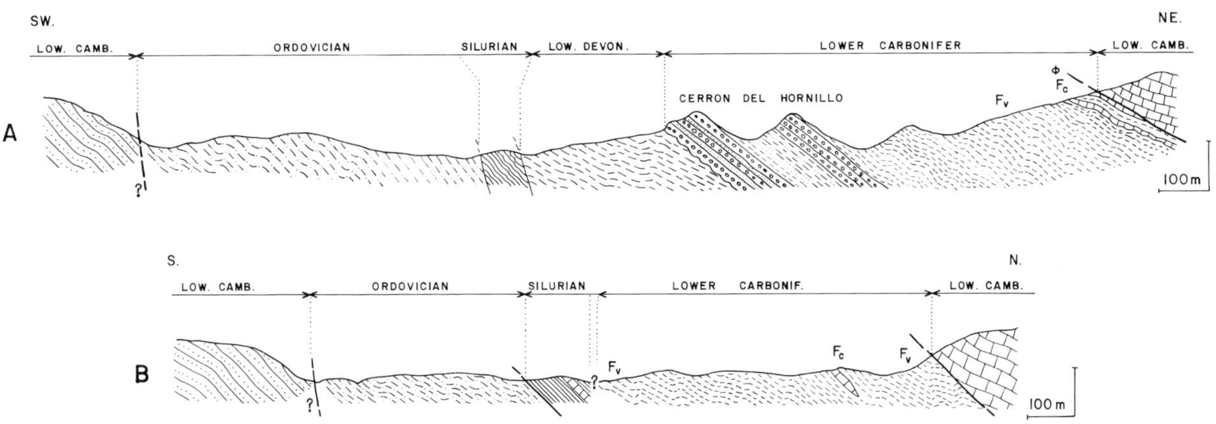

Fig. 3. Sketch sections across the Cerrón del Hornillo syncline with indication of fossil localities (Robardet et al. 1986). *A* Western Cerrón del Hornillo structure; *B* eastern section

tion and the lack of biostratigraphic data from the latter. The presence of rich conodont assemblages of Frasnian age (van den Boogaard, in Ribeiro 1983) in carbonate lenses near the base of the Cabrela Formation, which consists of mudstone-siltstone-greywacke alternations with minor carbonate, conglomerate and keratophyre interbeds, gives a special relevance to this formation in order to constrain the timing of the first Hercynian deformation phase in this area, as in a quarry near the Cabrela railway station, this formation overlies, with an angular unconformity, recumbently folded and cleaved Eiffelian limestones (van den Boogaard 1972).

The Terena Flysch in its turn predominantly consists of turbidites (at least 1000 m thick) and minor volcanic and conglomeratic intercalations, generally deposited below wave-base level. In the Santa Olalla de Cala area (eastern end of the occurrence) the turbidites show an upward gradual passage to shelf deposits which culminate with conodont-bearing carbonates of Tournaisian - Lower Viséan age and deposited in high energetic near-shore marine environments (van den Boogaard and Vázquez 1981). These data may be interpreted as indicative of migration of these deposits also along the strike, from west to east, which is not difficult to understand if the oblique nature of the simultaneous convergent orogenic processes is taken into account (see Apalategui et al., this Vol.).

2.4.2.2 Cerrón del Hornillo Syncline

The small Carboniferous outcrops which constitute the core of the synclinal structure were unknown until very recent times. In fact they were described for the first time by Robardet et al. (1986), who presented as well abundant biostratigraphic information from them. Apart from recording the existence of these marine deposits in an area where they are very scarce, the importance of this contribution lies in the Mid-Tournaisian age assigned to these rocks on the basis of their fossil content (conodonts and plant remains). Although the fossiliferous horizons are located some 100 m above the bottom of the sequence, the remarkable facies continuity between them and the basal strata makes unlikely a large time separation, particularly in such a mobile synorogenic environment. This allows one reasonably to consider that the base of this sequence may be older than that of the southern units, from which it is separated by a large zone lacking equivalent successions, where significant shortening presumably occurred during the Hercynian orogeny. On the other hand, the marine Carboniferous facies in this occurrence overlie unconformably Lower Devonian (Emsian) rocks (Fig. 3). However, in a relatively close locality (Valle Syncline), a Lower Paleozoic succession, reaching the Fammenian, is exposed (Racheboeuf and Robardet 1986), thus strongly suggesting that the base of the synorogenic deposits may be heterochronous.

2.4.2.3 Northern Area

This is the part within the Ossa-Morena zone where more and larger occurrences of synorogenic flyschoid successions exist. All of them constitute the so-called northern Ossa-Morena Lower Carboniferous basin (Gabaldón et al. 1985) in which various sedimentary environments have been identified (Fig. 4). The most salient feature of this basin is that

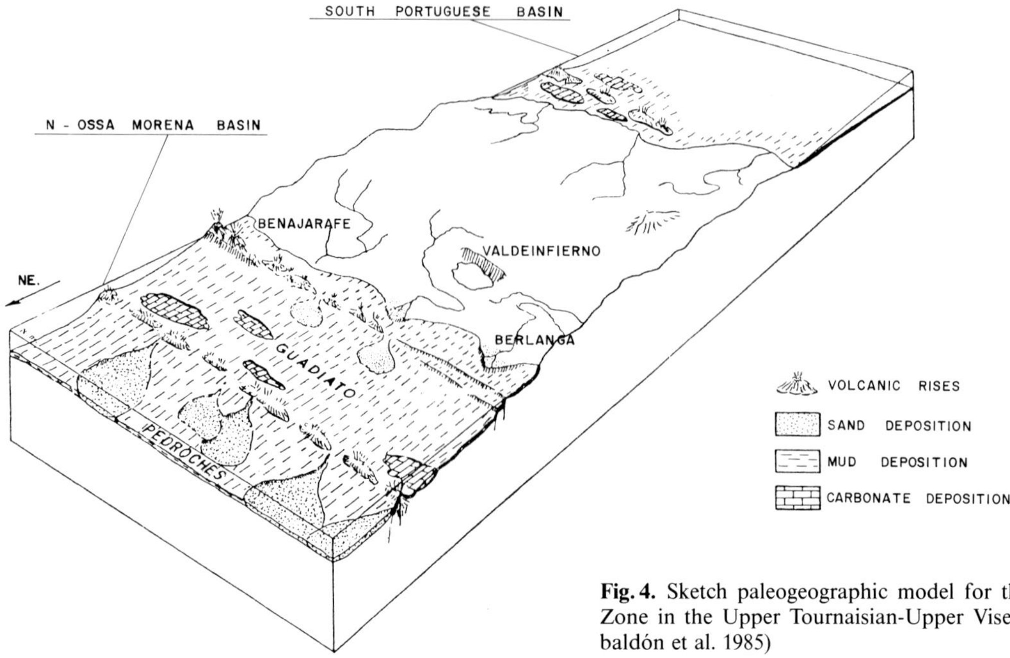

Fig. 4. Sketch paleogeographic model for the Ossa-Morena Zone in the Upper Tournaisian-Upper Visean interval (Gabaldón et al. 1985)

its constituent sequences extend to the north beyond the boundary with the adjacent Central-Iberian zone, thus providing a link between both zones at that time. The oldest known rocks within this basin are as young as the Upper Tournaisian, which is in agreement with the admitted heterochronism of these deposits, successively younging in a northerly direction.

Basin fill took place according to the following general pattern: an initial transgressive process, with its maximum in the Upper Viséan, was succeeded by a regression which led to the final basin emergence in the basal Namurian, characterized by terrestrial deposits immediately preceding the deformation phase that affected them penetratively (Quesada et al. 1987). This general evolutionary trend is typical of most foreland basins. Sedimentary facies, eminently terrigenous, are diagnostic of storm-dominated shelf deposition, although different highly varied paleoenvironments may be recognized (Fig. 4). Storm-dominated shelf conditions were persistent all along the basin evolution, with the only exception being ephemeral periods of low sediment input, during which several attempts of carbonate shelf development were soon aborted by renewed sediment supply into the basin (Gabaldon et al. 1983; Quesada and Garrote 1983).

The existence of several volcanic alignments (Fig. 4) approximately parallel or slightly oblique to the basin boundaries and at least partly emerged (Garrote and Sánchez Carretero 1983; Gabaldón et al. 1985) probably constitutes the most outstanding feature of this basin, in which the variability of sedimentary subenvironments was directly related to such linear volcanic highs. The basin southern boundary (trending approximately WNW-ESE) can be fairly precisely defined by linking the near-shore and deltaic facies which constitute the Benajarafe, Berlanga and Matachel occurrences, in the Upper Tournaisian - Upper Viséan interval (see Fig. 1 for location). A migration of this southern boundary in a northerly direction is evidenced by the basal Namurian deltaic facies which overlie Viséan open-shelf deposits in the Guadiato occurrence. The presence of an emerged area south of this marine basin is neatly demonstrated (Garrote and Broutin 1979) for the Upper Tournaisian period, by the terrestrial fluviolacustrine Valdeinfierno Basin, which will be dealt with in some detail further on. Nevertheless, the absence of younger deposits in the intervening area between this and the Terena Flysch occurrences enables a precise evaluation of the emerged land-mass magnitude.

2.4.3 Synorogenic Intermontane Basins

As a direct consequence of the convergent tectonic processes affecting the Ossa-Morena zone throughout the Upper Paleozoic, this underwent progressive uplift which led to its total subaerial exposure by Namurian times. Intermontane basins were de-

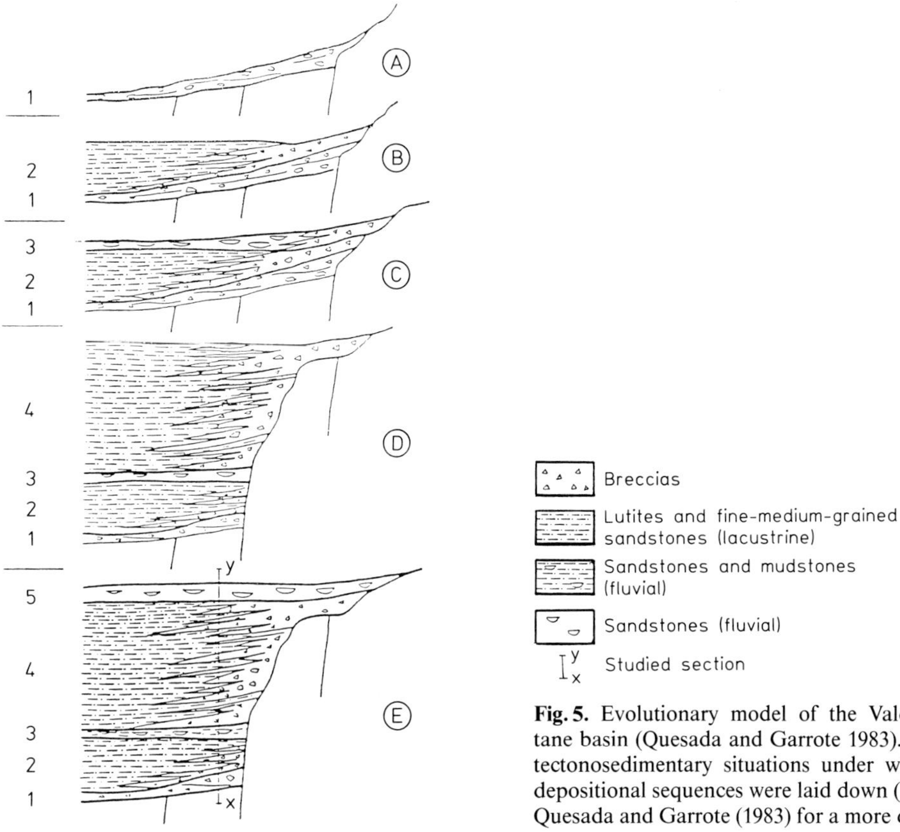

Fig. 5. Evolutionary model of the Valdeinfierno intermontane basin (Quesada and Garrote 1983). **A-E** Five successive tectonosedimentary situations under which five basin-wide depositional sequences were laid down (*1-5* respectively); see Quesada and Garrote (1983) for a more detailed explanation

veloped simultaneously under a generally transtensional regime (pull-apart type basins). Their life was usually somewhat ephemeral, being subsequently deformed under transpressional conditions as less intense the younger the basins were (Gabaldón and Quesada 1986). Four different basins of this type are known in the Ossa-Morena Zone, all of them located in the vicinity of major fracture zones and with quite different ages. The oldest is the aforementioned Valdeinfierno Basin, Upper Tournaisian in age (Wagner 1978; Quesada and Garrote 1983), which is located in the emerged intervening block separating the South-Portuguese and Northern Ossa-Morena Basins, fairly close to the Badajoz-Córdoba shear zone (see Apalategui et al., this Vol.). Sedimentation within this basin took place under fluviolacustrine conditions and had a marked structural control (Gabaldón et al. 1983; Roldán 1983). An evolutionary sketch of the sedimentation within the Valdeinfierno Basin is shown in Fig. 5. The other representatives of this basin type (Fig. 1): Villanueva del Río Basin (Westphalian A; Wagner et al. 1983); Peñarroya-Belmez Coalfield (Westphalian B; Lemos de Sousa and Wagner 1983; Wagner 1983); and Santa Susana Basin (Westphalian D-Stephanian A; Conçalves 1983), also placed near important fracture zones, underwent a comparable evolution, the best known being the Peñarroya-Belmez Basin, because of the intensive coal mining centred on it since the turn of the century. For this long and narrow graben, Gabaldón and Quesada (1986) proposed an evolutionary model (Fig. 6) which with slight variations may be considered as representative of this type of basin.

2.4.4 Late-Orogenic Intermontane Basins

A new set of intermontane basins was developed, under a markedly extensional regime, accompanying the Hercynian orogeny final stages during the Upper Stephanian-Lower Autunian time span. Tectonics, clearly waning by that time, hardly had even a mild influence on their evolution (Gabaldón and Quesada 1986). The only preserved remains correspond to the Viar basin and the small occurrences around Gualdalcanal (Fig. 1). Deposits in these basins, fluviolacustrine as well, fossilize an abrupt paleorelief (Broutin 1981; Quesada and Garrote 1983) and exhibit two remarkable characteristics,

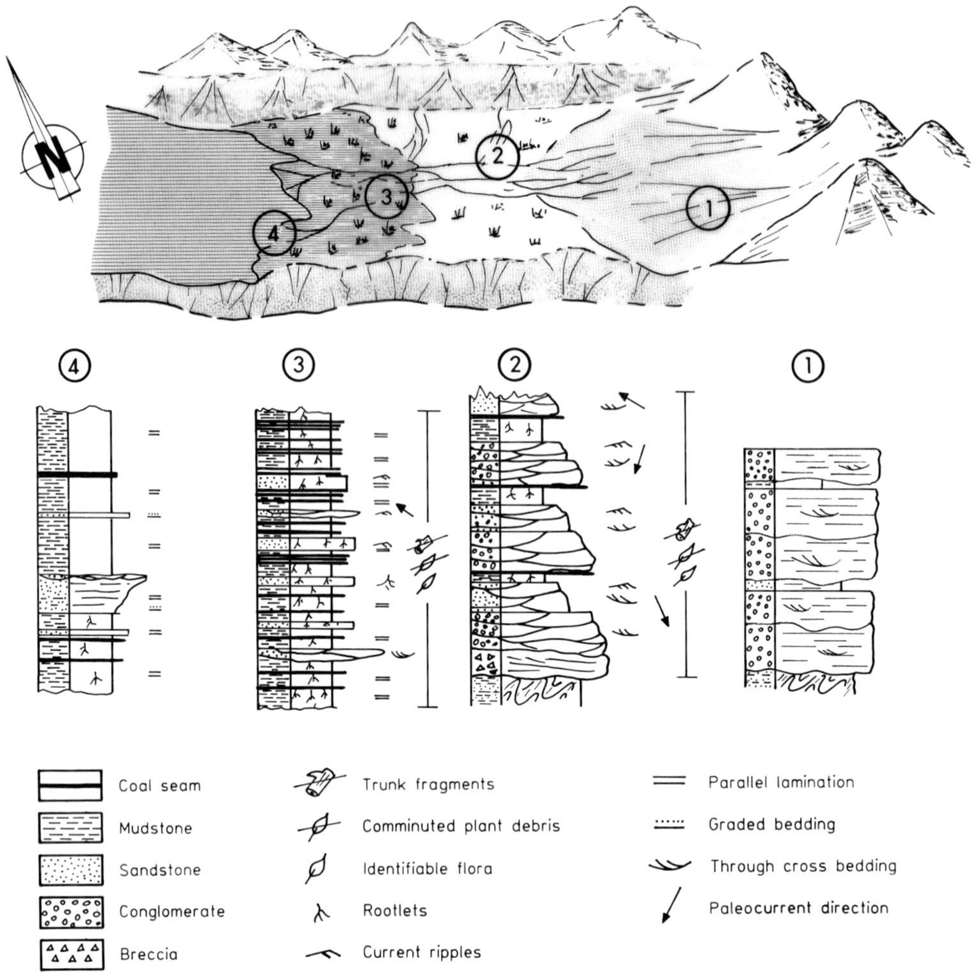

Fig. 6. Interpretative paleogeographic model of the Peñarroya-Belmez Basin and representative successions of different zones. Without scale. *1* Alluvial fans domain; *2* fluviatile domain; *3* fluvio-palustrine domain; *4* fluvio-lacustrine domain (slightly modified from Gabaldón and Quesada 1986)

important to establish the paleogeography of the Ossa-Morena Zone:

1. Presence of fluviatile red beds in the Viar Basin, of Lower Autunian age (Broutin 1981; Simancas 1985), forerunners of what was widespread later (in Upper Permian and Triassic times) throughout the Iberian Peninsula
2. Existence of a floral mélange of Euroamerican and Gondwanan species, which represents the oldest known locality witnessing the arrival of African flora in Europe. This fact is largely recorded throughout the continent from Triassic times (Broutin 1981, 1982)

Apart from their value as paleolatitude indicators, the second characteristic, which has never been recorded in time-equivalent basins elsewhere (northwards) in the Iberian Peninsula, has two important paleogeographic connotations:

1. Existence of a terrestrial connection between the Ossa-Morena zone and Africa at that time
2. Existence of a paleophytogeographical barrier, north of the Ossa-Morena zone, preventing the northerly migration of African flora. This barrier, according to the available geological evidence, could be most likely an important mountain belt (Quesada and Garrote 1983).

Acknowledgements. Substantial comments by F. Simancas are kindly acknowledged. Professional assistance given by R. Fernández Toledo (typing) and J. Vallejo (drawing) is also very much appreciated.

References

Apalategui O, Quesada C (1987) Transversal geológica Zona Ossa-Morena. Excursion Guidebook, pp 1-90

Boogaard M van den (1972) Conodont faunas from Portugal and Southwestern Spain. Part 1. A Middle Devonian fauna near Montemor-o-Novo. Scr Geol 13: 1-11

Boogaard M van den, Vazquez F (1981) Conodont faunas from Portugal and southwestern Spain. Part 5. Lower Carboniferous conodonts at Santa Olalla de Cala (Spain). Scr Geol 61: 1-8

Broutin J (1982) Importance paléobiogéographique de la découverte dune flore continentale mixte dans le Permien inférieur du Sud-Ouest de la Péninsule Ibérique. C R Acad Sci Paris II, 295: 419-422

Broutin J (1981) Etude paléobotanique et palynologique du passage Carbonifère-Permien dans les bassins continentaux du Sud-Est de la zone d'Ossa-Morena (environs de Guadalcanal, Espagne du Sud). Implications paléogéographiques et stratigraphiques, 2 Vol. Thesis, Univ Pierre et Marie Curie-Paris VI

Gabaldón V, Quesada C (1986) Exemples de bassins houillers limniques du sud-ouest de la Péninsule Ibérique: évolution sédimentaire et contrôle structural. Mém Soc Géol Fr NS 149: 27-36

Gabaldón V, Garrote A, Quesada C (1983) Las cuencas de Valdeinfierno y Benajarafe (Tournaisiense-Viseense). Caracterización sedimentológica e implicaciones regionales. Dominio de Sierra Albarrana (Zona de Ossa-Morena). Comun Serv Geol Port 69 (2): 209-218

Gabaldón V, Garrote A, Quesada C (1985) El Carboníferro inferior del Norte de la zona de Ossa-Morena (SW de España). CR 10th Int Carboniferous Congr Madr 3: 173-186

Garrote A, Broutin J (1979) Le bassin Tournaisien de Benajarafe (province de Cordoue, Espagne). Géologie et prémieres données paléobotaniques et palynologiques. CR 104 Congr Nat Soc Savantes Bordeaux 1: 175-184

Garrote A, Sánchez Carretero R (1983) Materiales volcanoclásticos en el Carbonífero inferior, al S-SW de Villaviciosa de Córdoba (Zona de Ossa-Morena). Comun Serv Geol Port 69 (2): 249-257

Gonçalves F (1983) Santa Susana basin. Comun Serv Geol Port 69 (2): 201-208

Lemos de Sousa MJ, Wagner RH (1983) General description of the terrestrial Carboniferous basins in Portugal and history of investigations. In: Lemos de Sousa MJ, Oliveira JT (eds) The Carboniferous of Portugal. Mem Serv Geol Port, pp 117-126

Perdigão J, Oliveira JT, Ribeiro A (1982) Noticia explicativa da Folha 44-B (Barrancos). Serv Geol Port, pp 1-52

Quesada C (1983) El Carbonífero de Sierra Morena. In: Martínez C (coord): Carbonífero y Pérmico de España. IGME, pp 243-278

Quesada C, Garrote A (1983) Carboniferous geology of the Sierra Morena. Guidebook of field trip D. 10th Int Carboniferous Congr Madr, pp 1-104

Quesada C, Larrea FJ, Floridc P et al. (1987) Mapa Geológico-Minero de Extremadura. Dirección General de Industria, Energía y Minas. Junta de Extremadura

Racheboeuf PR, Robardet M (1986) Le Pridoli et le Dévonien inférieur de la zone d'Ossa-Morena (sud-ouest de la Péninsule Ibérique). Etude des Brachiopodes. Geol Paleontol 20: 11-37

Ribeiro A (1983) Relações entre formações do Devónico superior e o Maciço de Evora na região de Cabrela (Vendas Novas). Comun Serv Geol Port 69 (2): 267-269

Robardet M, Weyant M, Laveine JP, Racheboeuf PR (1986) Le Carbonifère inférieur du synclinal du Cerrón del Hornillo (province de Seville, Espagne). Rev Paléobiol 5 (1): 71-90

Roldan JF (1983) La Cuenca Tournaisiense de Valdeinfierno. Interpretación tectónica y paleoambiental (Córdoba, España). In: Lemos de Sousa MJ (ed) Contributions to the Carboniferous geology and paleontology of the Iberian Peninsula. Univ Porto, pp 41-50

Simancas JF (1985) Estudio estratigráfico de la cuenca del Viar. Temas Geol Min 7: 7-17

Wagner RH (1978) The Valdeinfierno sequence (Prov. Córdoba): its tectonic, sedimentary and floral significance. Ann Soc Géol Nord 98: 59-66

Wagner FH (1983) *Neuropteris guadiatensis,* a new species from the Westphalian B of the Peñarroya-Belmez Coalfield in the Province of Córdoba, SW Spain. In: Lemos de Sousa MJ (ed) Contributions to the Carboniferous geology and palaeontology of the Iberian Peninsula. Univ Porto, pp 93-100

Wagner RH, Coquel R, Broutin J (1983) Mississipian floras of the Sierra Morena, SW Spain: a progress report. In: Lemos de Sousa MJ (ed) Contributions to the Carboniferous geology and palaeontology of the Iberian Peninsula. Univ Porto, pp 101-126

3 Structure

O. Apalategui,[1] L. Eguiluz[2] and C. Quesada[3]

3.1 Introduction

The Ossa-Morena Zone, as compared from a structural point of view to the adjacent Central-Iberian and South Portuguese zones, exhibits a much more complex image, in which a tight fracture network results in a wedge-shaped compartmentation of the region at all scales. The frequently complicated internal structure of these wedges is quite varied in style and history, creating an additional difficulty to determining the correlation between structural units. This complex picture is the result of superposition of various structural elements which were developed as a consequence of the different tectonic events that the region underwent throughout its history, from the Upper Proterozoic to the Upper Paleozoic, when it finally became stabilized.

As stated elsewhere in this book (Quesada et al., Part V, Chap. 2.1) the existence of a Precambrian orogenic event in this area has been a matter of debate since the first reference by MacPherson (1879), indicating its existence. However, the most extended opinion in this century, until a fundamental pro paper by Delgado (1971), was that only the Late Paleozoic Hercynian orogeny was to be held responsible for the deformation of the Iberian Massif (Mallada 1985; Lotze 1929, 1956; Hernández Pacheco 1953; Fabriès 1957, 1963; Alía 1963; Vegas 1968, 1970, etc.), although some discrepant voices could also be heard (Alvarado 1935; Fricke 1941). In the last decade the amount of new evidence proving the existence of a Late-Precambrian polyphase orogenic event has increased substantially (Quesada 1975; Garrote 1976; Pérez Lorente 1979; Chacón 1979; Eguiluz and Quesada 1980; Pascual 1981; Herranz 1984, etc.), the discussion now being centred on the degree to which the Hercynian events have overprinted the Precambrian structures of individual areas or units.

The present compartmentation of the region reflects in most cases the structure developed during the last stages of the Hercynian orogeny. On the other hand, the internal structure of individual blocks was formed generally in older, Paleozoic and/or Precambrian events. The structural correlation between adjacent blocks is, however, an ardous task because of the uncertainties generated, among others, by the absence of Paleozoic deposits in many basement blocks, the poor expressiveness of the structure itself and the lack of radiometric data of syntectonic metamorphic minerals.

The outstanding differences between blocks, not only concerning the structure but also stratigraphic and petrological aspects, gave birth to the concept of structural domain (Chacón et al. 1974; Delgado et al. 1977) that was applied in profusion between 1974 and 1984. Figure 1 shows a synthetic domainal subdivision of the Ossa-Morena Zone compiled from the many proposals launched in that period. The dissection of the zone into domains has allowed their exhaustive analysis as separate entities to be undertaken, resulting in a substantial improvement of the general understanding of their stratigraphy, petrology, structure, etc. With regard to the structure, such an analysis has allowed a Hercynian important thrust structure to be assessed. This thrust structure, obscured frequently by the domain bounding Late-Hercynian wrench faults, separates variably allochthonous units that have been recognized in many cases within various domains. This fact minimizes the tectonostratigraphic significance originally given to the domains (Apalategui and Pérez Lorente 1983; Apalategui et al. 1985; Apalategui and Quesada 1987). As an example, a sketch structural map of the central part of the Ossa-Morena Zone is shown in Fig. 2, where the thrust structure, which involves both the Precambrian basement and its Paleozoic cover, can be seen.

In the following description the Precambrian and Hercynian structures will be dealt with separately, and, finally, a model for the tectonic evolution of the southwestern Iberian Massif will be outlined as a kind of conclusion.

[1] INGEMISA, Avda. Virgen de las Augustias 16, 14006 Córdoba, Spain
[2] Dpt. Geology, Universidad Pais Vasco, Bilbao, Spain
[3] I. T. G. E., Rios Rosas 23, 28003 Madrid, Spain

R. D. Dallmeyer and E. Martínez García (Eds.)
Pre-Mesozoic Geology of Iberia

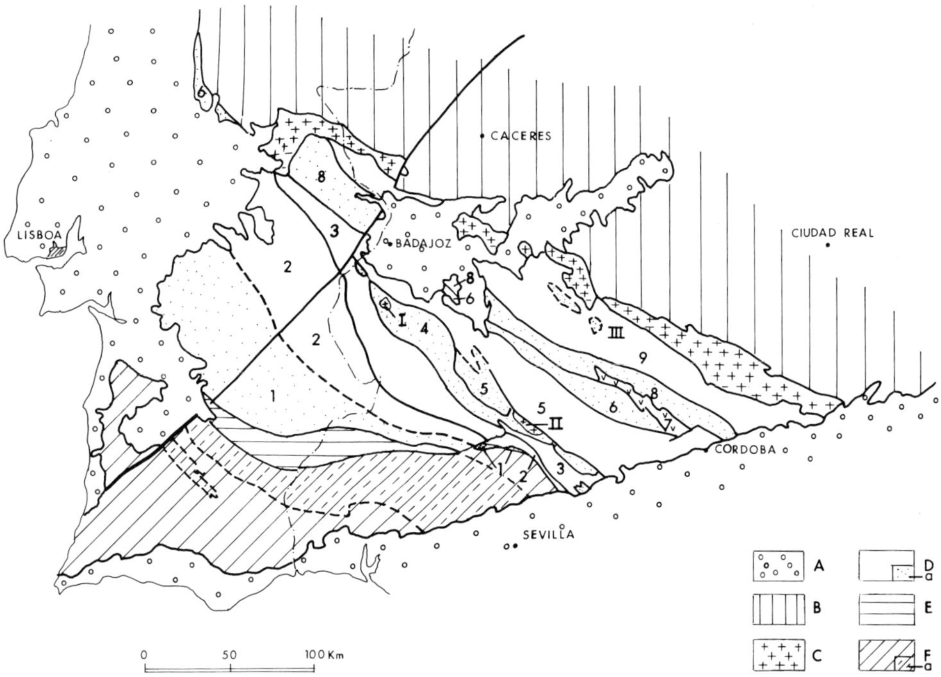

Fig. 1. Domain subdivision of the Ossa-Morena Zone (Modified after Quesada 1983). *A* Post-Paleozoic cover; *B* Central-Iberian Zone; *C* Pedroches Batholith; *D* Ossa-Morena Zone (*a* metamorphic belts): *1* Beja-Aracena Domain; *2* Barrancos-Hinojales Domain; *3* Elvas-Cumbres Mayores Domain; *4* Arroyomolinos Domain; *5* Zafra-Monesterio Domain; *6* Sierra Albarrana Domain; *7* Villaviciosa-La Coronada igneous belt; *8* Valencia de las Torres-Cerro Muriano Domain (cf. Badajoz-Córdoba Shear Zone); *9* Obejo-Valsequillo-Puebla de la Reina Domain; *E* Pulo do Lobo unit (including Beja-Acebuches ophiolite); *F* South Portuguese Zone (*a* Pyrite Belt). **I** Barcarrota pluton; **II** Pallares pluton; **III** Peraleda del Saucejo antiform. Area shown in Fig. 2

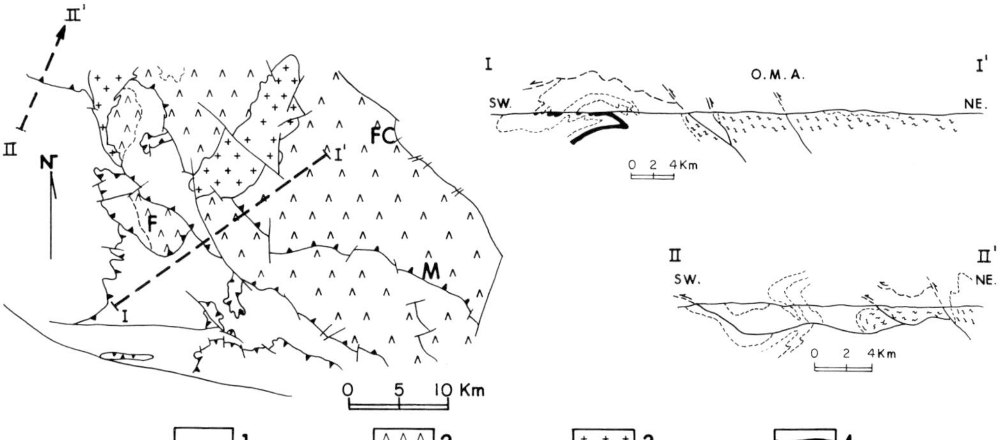

Fig. 2. Imbricated Hercynian structure in the central part of the Ossa-Morena Zone (Modified after Vauchez 1976 and Apalategui et al. 1983 a). See Fig. 1 for regional location. *1* Paleozoic (Cambrian through Silurian); *2* Precambrian basement; *3* plutons; *4* Middle Cambrian metabasites; *F* Fregenal; *FC* Fuente de Cantos; *M* Monesterio; *OMA* Olivenza-Monesterio antiform

3.2 Precambrian Structure

It becomes clear from the preceding introductory discussion that it is not an obvious thing to distinguish, within basement occurrences, Precambrian from Hercynian structures. Moreover, in many places it is impossible as yet to decide whether the meso- or microscopic fabric was developed in Precambrian or Hercynian times. It is only in areas with a mildly deformed Paleozoic cover and/or with radiometrically dated older plutons where direct pieces of evidence exist which allow demonstration of the fact that the fabric displayed by the basement rocks was formed prior to such referential elements. Two areas are known best from this point of view (Fig. 1): the Arroyomolinos unit (approximately) equivalent to the former Arroyomolinos Domain), and the basement occurrences which constitute the base of the imbricated thrust sheet sequence characterizing the Obejo-Valsequillo-Puebla de la Reina Domain.

In the Arroyomolinos unit, where the Hercynian deformation is limited to a gentle refolding of the Precambrian structure (including the metamorphic isograds), the Precambrian age of the penetrative fabric is demonstrated by three main lines of evidence. Firstly, the complex interference pattern of folding shown in Fig. 3, as drawn by discontinuous black chert horizons, is unconformably overlain by Late Proterozoic volcanosedimentary rocks (Fernández et al. 1983). Secondly, the contact metamorphic aureole which overgrows the penetrative structure of the Arroyomolinos unit around the Pallares pluton (572+/−74 Ma, Rb-Sr whole-rock isochron, Cueto et al. 1983) and the Barcarrota pluton (505+/−10 Ma, hornblende K-Ar cooling age, Galindo et al. 1986, Fig. 1) and, thirdly, the existence in basal Cambrian and Late Precambrian synorogenic conglomerates of clasts with a previously deformed and metamorphosed Serie Negra provenance (Eguiluz et al. 1984; Fig. 4). Nevertheless, the Arroyomolinos unit is a notable exception to the generally poor chartographic expression of the Precambrian structures, mainly due to the lack of continuous marker beds and to their apparent parallelism to the Hercynian ones. According to the available data (Quesada 1975; Garrote et al. 1983; Fernández et al. 1983; Eguiluz and Ramón Lluch 1983; Eguiluz 1987), the Precambrian structuration of this unit is the result of a first deformation phase which developed E-W-trending isoclinal folds and associated very penetrative axial plane schistosity/slaty cleavage. A second folding phase, N 60–80 E in trend, responsible for the refolding that may be observed in Fig. 3, could also be Precambrian in age, postdating the synorogenic successions in this case, although the lack of both Paleozoic cover in this unit and radiometric dating makes this a debatable interpretation. Both generations of structures are refolded by a third one, N 120 E in trend and SW-verging, undoubtedly Hercynian in age as demonstrated by its relationship with the Monesterio thrust (Fig. 3) which constitutes the northern boundary of this unit, and its geometrical congruency with the Hercynian structures in neighbouring

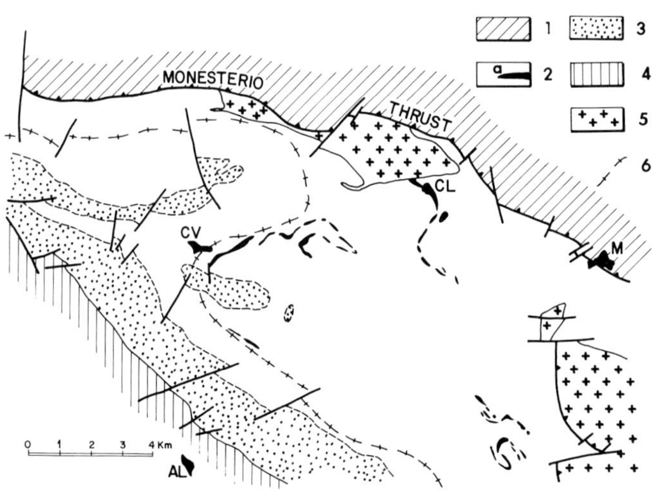

Fig. 3. Sketch map of the central part of the Arroyomolinos unit (Modified after Fernández et al. 1983). *1* Monesterio unit; Arroyomolinos unit: *2* Serie Negra (*a* black cherts and marbles); *3* Late Precambrian synorogenic volcanoclastics; *4* Lower Cambrian; *5* plutons; *6* biotite in isograd; *AL* Arroyomolinos de León; *CL* Calera de León; *CV* Cabeza la Vaca; *M* Monesterio

Fig. 4. Interference folding pattern shown by a black chert clast included in lower Cambrian conglomerates

units. The interference of the three folding phases accounts for the present geometry of the internal structure of this Arroyomolinos unit, the analysis of which, by applying the homogeneous zones method (Turner and Weiss 1963), has been undertaken by Eguiluz and Ramón Lluch (1983). They drew and compared a structure contour map after ideal marker horizons (Fig. 5 A) with a computer model map obtained by superposing three folding phases with the aforementioned characteristics (Fig. 5 B). The outstanding geometrical similarity between both the natural and the theoretical models strongly supports the conclusions drawn after the field observations.

The other area in the Ossa-Morena Zone where the Precambrian deformation is easily identifiable is the so-called Obejo-Valsequillo-Puebla de la Reina Domain (Fig. 1), which occupies the interviewing area between the Pedroches Batholith and the Badajoz-Córdoba shear zone. The macroscopic structure of this domain, which will be described in detail further on, consists of a NE-verging thrust sheet forward sequence, the stratigraphic succession of which includes a variably thick Paleozoic cover and, only locally, a Precambrian basement. Although no detailed structural studies are as yet available on this domain and occurrence is very poor, the existence of an angular unconformity between the Paleozoic succession (Lower Ordovician, locally Lower Cambrian) and different Precambrian units is indubitable, with a conspicuous contrast in metamorphic grade in many localities (Santos and Casas 1980; Apalategui and Quesada 1987; Apalategui et al. in press). Only locally, however, the existence of an older unconformity between different Precambrian successions has been tested. This is the case in the Urra area, near the Spanish-Portuguese border, where Gonçalves (1971) and Santos and Casas (1980) have described an unconformity between the synorogenic successions and the underlying Serie Negra: the presence of deformed Serie Negra clasts in conglomerates within the syn-orogenic sequence together with chartographic relations are envisaged by those authors as diagnostic of the unconformable relationships, although the lack of detailed structural studies in this area prevents the Precambrian structure from being characterized. Eastwards in this domain a Serie Negra correlative medium-grade metamorphic sequence occurs in the core of the Peraleda antiform, underlying non-metamorphic Ordovician rocks. These include conglomerates within which abundant clasts of the underlying metamorphic rocks are conspicuous and display a cremulated schistosity in

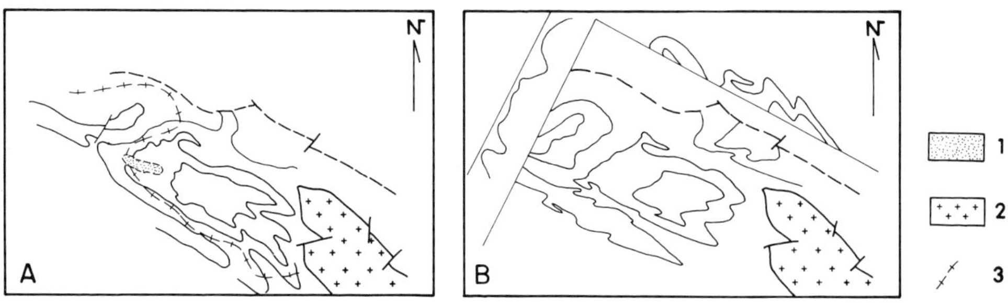

Fig. 5. Comparison of **A** natural and **B** theoretical structural contour maps of the Arroyomolinos unit (central part) slightly modified after Eguiluz and Ramón Lluch (1983). *1* Precambrian synorogenic volcanic rocks; *2* Castillo pluton; *3* biotite- in isograd

the appropriate lithologies. The internal structure of the basement within this Peraleda antiform consists of a first generation of synmetamorphic E-W trending isoclinal folds and axial plane schistosity, which is very penetratively deformed by a second N-S trending set of folds and crenulation cleavages. Both generations of structures are overlain unconformably by Ordovician rocks, thus being Precambrian in age. Nevertheless, the absence of synorogenic Precambrian successions in this domain makes it impossible to elucidate whether the crenulation is older or younger than them. Another interesting feature of this domain is the existence of numerous scattered Late-Precambrian plutons (unconformably overlain by Lower Paleozoic rocks) which intruded into the Serie Negra after its penetrative deformation and metamorphism. They are interpreted to be genetically correlative to the synorogenic calc-alkaline volcanics and are variably (generally little) deformed in turn (Quesada et al. 1987; Apalategui et al. in press). Finally, in this domain the existence of Precambrian metamorphic and deformational events has been radiometrically demonstrated: 550+/−10 Ma, a 40 Ar-39 Ar cooling age for Serie Negra muscovites within the Peraleda antiform (Blatrix and Burg 1981).

Apart from the two areas described above where the evidence for Precambrian deformational events is clearer than elsewhere, there are also others where such events can be inferred by means of indirect criteria. This is the case in the areas where deformed Precambrian successions are intruded by Lower Paleozoic granitoids (see Quesada et al., Part V, Chap. 2.1), or the case of the serpentinite occurrences near Calzadilla de los Barros, interpreted as olistholiths within the synorogenic Precambrian successions, derived from a dismembered ophiolite sheet (Aguayo 1985; Quesada et al. 1987) and hardly understandable if not regarded in direct connection with the orogenic process.

3.3 Hercynian Structure

The Hercynian deformation, responsible for the large-scale structural grain as well as for most of the penetrative structures within the Ossa-Morena Zone, shows great variations in style and intensity across the strike, being, however, relatively homogeneous along the strike. For this reason, the following description has been undertaken by dividing the zone into more or less homogeneous bands, although with remarkable differences from one another. From south to north these are (Fig. 6):

1. Southern crystalline massifs (Evora-Beja-Aracena-Almadén de la Plata);
2. Ossa-Morena south-central belt;
3. Ossa-Morena north-central belt;
4. Sierra Albarrana domain;
5. Badajoz-Córdoba Shear Zone;
6. Obejo-Valsequillo-Puebla de la Reina domain.

In general terms the Hercynian structures which are diagrammatically shown in Fig. 7 are progressively younger from southwest to northeast and present a southwesterly vergence, with the only exception being the Obejo-Valsequillo-Puebla de la Reina domain, where they verge towards the northeast. Evidence for the development of the Hercynian structures under a transpressional regime exists at all scales within the Ossa-Morena Zone.

3.3.1 Southern Crystalline Massifs

This southernmost band of the Ossa-Morena Zone is composed of a series of juxtaposed metamorphic (see Quesada and Munha, this Vol.) complexes which are thrust to the south over the Beja-Acebuches ophiolite and Pulo do Lobo unit (see Oliveira, Part VI, Chap. 2; Silva et al., this Vol). The internal structure of these metamorphic massifs, the stratigraphic sequence of which includes Precambrian and Lower Paleozoic rocks, shows the effects of two superposed deformation phases (Bard 1969; Apalategui et al. 1983 a, 1984), but locally also a third, intermediate phase of deformation is recognized. The first phase generated S/SW-verging isoclinal recumbent folds and associated axial plane slaty cleavage/schistosity. The second phase was responsible for the local development of N 20/50 E - trending, SE verging, asymmetric recumbent folds and associated axial planar crenulation cleavage. Finally, the third phase gave rise to N 100/150 E - trending folds, with an open cylindrical geometry and a subvertical to slightly SW-verging axial plane.

The metamorphic massifs which characterize this band appear nowadays dispersed by important ductile and brittle shear zones (both reverse and normal) with an important sinistral component of displacement (Crespo 1987; Crespo and Orozco 1988), and postdating the penetrative deformation of the individual complexes. This is undoubtedly older than the Middle Devonian, as demonstrated by the fauna described by Conde and Andrade (1974) from limestones unconformably overlying

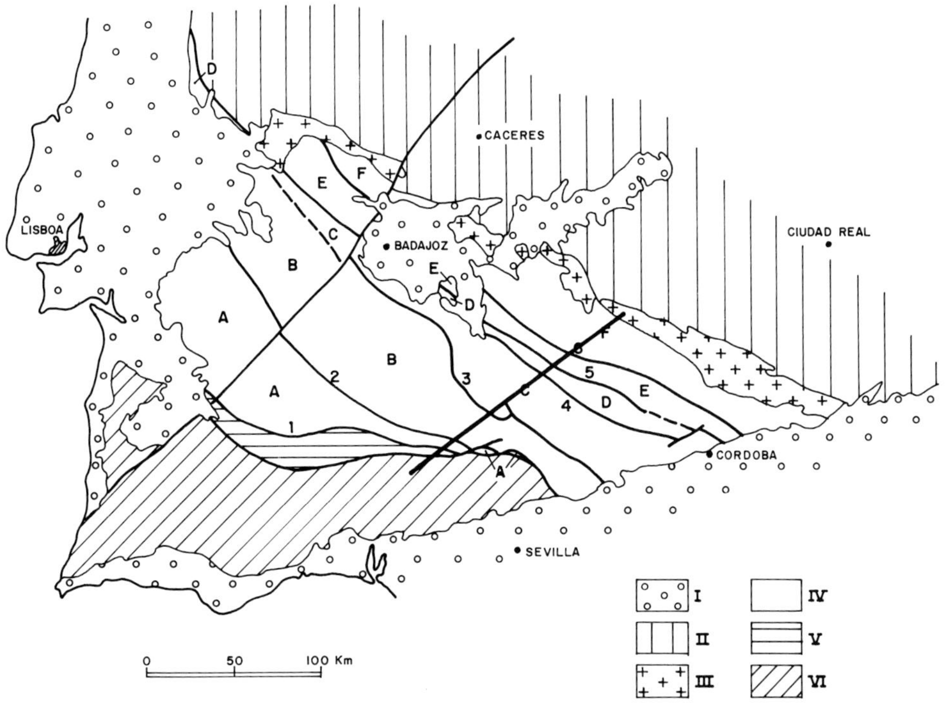

Fig. 6. Major belts in the Ossa-Morena Zone with different structural style and/or deformation age. **I** Post-Paleozoic cover; **II** Central-Iberian Zone; **III** Pedroches Batholith; **IV** Ossa-Morena Zone: *A* southern crystalline complexes; *B* south-central belt; *C* north central belt; *D* Sierra Albarrana Domain; *E* Badajoz-Córdoba Shear Zone; *F* Obejo-Valsequillo-Puebla de la Reina Domain; **V** Pulo do Lobo unit (including Beja-Acebuches ophiolite); **VI** South Portuguese Zone; *1* Ficalho-Almonaster thrust; *2* Beja-Valdelarco Fault; *3* Monesterio thrust; *4* Malcocinado fault; *5* Azuaga Fault; *6* Hornachos-Villaharta Fault. *Thick line:* location of the cross-section shown in Fig. 7

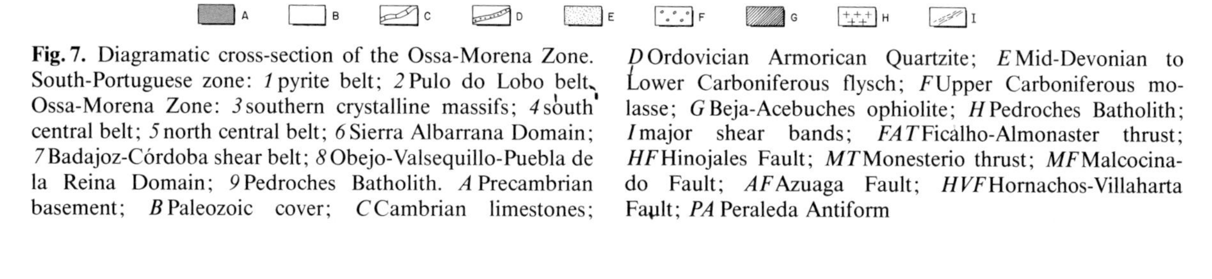

Fig. 7. Diagramatic cross-section of the Ossa-Morena Zone. South-Portuguese zone: *1* pyrite belt; *2* Pulo do Lobo belt. Ossa-Morena Zone: *3* southern crystalline massifs; *4* south central belt; *5* north central belt; *6* Sierra Albarrana Domain; *7* Badajoz-Córdoba shear belt; *8* Obejo-Valsequillo-Puebla de la Reina Domain; *9* Pedroches Batholith. *A* Precambrian basement; *B* Paleozoic cover; *C* Cambrian limestones; *D* Ordovician Armorican Quartzite; *E* Mid-Devonian to Lower Carboniferous flysch; *F* Upper Carboniferous molasse; *G* Beja-Acebuches ophiolite; *H* Pedroches Batholith; *I* major shear bands; *FAT* Ficalho-Almonaster thrust; *HF* Hinojales Fault; *MT* Monesterio thrust; *MF* Malcocinado Fault; *AF* Azuaga Fault; *HVF* Hornachos-Villaharta Fault; *PA* Peraleda Antiform

the Beja gabbros, which intrude into the Beja metamorphic complex posterior to its penetrative deformation. The thrusting of the metamorphic massifs over the southerly units must be interpreted as a relatively late event, going along with the aforementioned shear zones, in as much as it affects rocks as young as uppermost Devonian.

3.3.2 Ossa-Morena South-Central Belt

This term is applied to the area comprised between the previously described crystalline massifs and the Monesterio thrust and associated strike-slip faults (Fig. 6). The most salient feature of this belt, in terms of structure, is the superposition of several flat-lying thrust sheets and associated kilometrical scale recumbent folds. Locally, structurally much simpler parautochthonous units occur as tectonic windows underneath the nappe pile. Figure 2 depicts representative sections across this belt, the internal structure of which is thus characterized by a first generation of recumbent folds with an incipient to non-existent axial plane slaty/pressure-solution cleavage. The overturned limbs of these folds evolved, as deformation, and progressed to flat-lying thrusts with a southwesterly sense of displacement. This first folding phase took place in the interval Middle Devonian (age of the youngest affected successions, Quesada et al. 1987) - Upper Devonian (pre-Upper Tournaisian; presumed age of the oldest non-affected sequences, van den Boogaard and Vázquez 1981). Subsequently, but prior to the lower Westphalian age of the oldest overlying unconformable deposits (Quesada 1983), a second deformation phase gave rise to the refolding with a NW-SE trend and a slight southwesterly vergence of the previous structures. This second phase also produced a locally very penetrative slaty/crenulation cleavage which transects sinistrally the co-genetic folds, thus providing strong evidence for the transpressional nature of the deformation.

3.3.3 Ossa-Morena North-Central Belt

This belt is bounded by the Monesterio thrust and the Malcocinado Fault (Fig. 6) and appears extremely split by many late strike-slip faults. Its internal structure is also two-phased and quite similar to that of the former belt, from which it differs mainly in style and age. The most characteristic structures developed during the first deformation phase are decollement nappes, recumbent folds are rarely found. The area presents alternating bands with flat-lying structures and incipient cleavage, and bands with southwesterly verging structures and very penetrative cleavage. This fact is interpreted as representing respectively ramps and flats on the basal detachment.

Rocks younger than the Middle Cambrian only occur in very few localities, where it is possible to recognize that the youngest successions affected by the first deformation phase are Famennian in age (Racheboeuf and Robardet 1986). On the other hand, the oldest overlying unconformable strata are Mid-Tournaisian (Robardet et al. 1986), proving thus that the Hercynian deformation arrived at this belt later than at the previously described ones.

Structures developed during the second deformation phase are very similar to those in southernmost units and their age is as yet unknown, due to the lack of overstepping deposits and radiometrically dated plutons.

3.3.4 Sierra Albarrana Domain

As discussed elsewhere in this book (Quesada et al., Part V, Chap. 2.1) this is probably the least known unit in the whole Ossa-Morena Zone; the structure not being an exception to this general situation. It is bounded by the Malcocinado Fault in the southwest and the Azuaga Fault in the northeast, two left-lateral strike-slip faults which at least in part postdate the internal structure of this domain. This includes late N 170 E - trending, westerly verging macroscopic structures of indubitable Hercynian age, since they fold the metamorphic isograds, radiometrically demonstrated to be Hercynian (Dallmeyer and Quesada 1989). There are also older structures, only recognizable at the meso- and microscopic scales, whose relationship to the metamorphic recrystallization have been studied in detail by Garrote (1976). The only time constraint to the age of the structure in this domain is provided by the fact that the entire penetrative fabric was achieved by the Upper Tournaisian age of the uncleaved terrestrial deposits of the Valdeinferno coalfield which overlie unconformably the metamorphic succession (Garrote 1976; Garrote et al. in press).

A significant fact, disclosed by the available geochronological data (Dallmeyer and Quesada 1989) is the existence of an important uplift event of this unit and the neighbouring Badajoz-Córdoba shear zone, in the interval Upper Devonian - Lower Carboniferous, responsible for the occurrence at the surface in the Upper Tournaisian (age of the overlying unconformable deposits) of metamorphic rocks yielding 40 Ar-39 Ar cooling ages for hornblendes in the range 365-370 Ma (closing temperature of Ar within hornblendes is nearly 500 °C, which approximately corresponds to ca. 15 km depth, if average geothermal gradients are considered).

3.3.5 Badajoz-Córdoba Shear Zone

Bounded by the Azuaga Fault in the southwest and the Portalegre thrust (NW of the Guadiana Tertiary basin) and Hornachos-Villaharta Fault (SE of that basin) in the northeast, this is a very complex belt, where Precambrian rocks with a long and compli-

Fig. 8. Large-scale subhorizontal S-C fabric developed in the eastern part of the Badajoz-Córdoba Shear Zone

cated tectonothermal history occur, together with Lower Paleozoic (García Casquero et al. 1985) igneous intrusives. As stated above, this is an uplifted block, the most salient feature of which is the widespread presence of mylonitic fabrics of all kinds, showing a great variation along strike. Medium grade (amphibolite facies) flat-lying mylonitic structures predominate in the southeastern half of the belt with a prominent along-strike stretching lineation (oblique or orthogonal stretching lineations are only locally found). This subhorizontal mylonitic structure which also affects the Lower Paleozoic plutons, developed S-C-C' structures at all scales (Fig. 8) and is affected in turn by late spaced shear zones with the same orientation but with a subvertical dip, lower grade (in the ductile-brittle transition) and a neat sinistral sense of shearing (Chacón 1974, 1979). Geometrically associated to the subhorizontal mylonitic fabric there exists a superposition of (thrust?) sheets which results in the imbrication of the various units constituting this belt. These include locally scattered remnants of a serpentite sheet (Apalategui et al. 1983 b).

In the western half of the belt, on the other hand, the shear zone structure is exclusively subvertical, also with a sinistral sense of shearing and developed under slightly cooler metamorphic conditions (greenschist-amphibolite facies transition).

The interpretation of this intracontinental shear zone, frequently regarded as a suture (Burg et al. 1981; Matte 1983, 1986; etc.), still remains a subject of debate. The fundamental question is posed around the significance of the subhorizontal fabric predominant in the eastern half and its relationship with the subvertical fabric widespread throughout the belt. The visible geometrical relations are compatible with an intracontinental subvertical sinistral shear zone progressively passing in depth (with a listric-like geometry) to a subhorizontal mid-crustal detachment (Ribeiro, pers. comm. 1987) where the fabric developed would be flat-lying as well (Fig. 9). If a gentle dip of the basal decollement to the northwest is admitted, then the general sinistral displacement could account for the progressive uplift of the unit as well as for the late development of subvertical shear zones in shallower conditions. Nevertheless, this interpretation, although integrating the most part of the geometrical features of the structure, must be treated with caution, due to the lack of geochronological evidence proving the cogenetic character of both subhorizontal and subvertical structures: their separation in time as well as their formation as a response to unrelated tectonic processes cannot be disregarded yet.

With respect to the amount of displacement involved, it is probably worthwhile to emphasize the complexity of the deformational history of the Badajoz-Córdoba shear zone involving both sinistral and (late) dextral shearing, which, together with the lack of macroscopic markers, renders highly diffi-

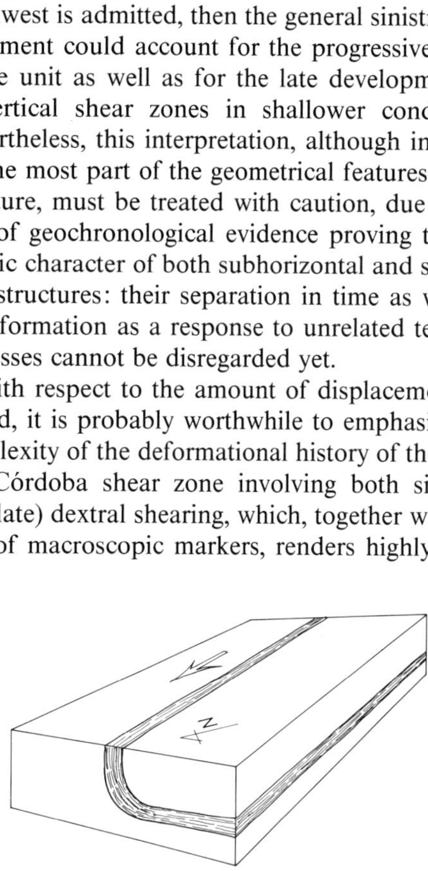

Fig. 9. Interpretative model for the Badajoz-Córdoba Shear Zone if rooted in a mid-crustal, flat-lying detachment

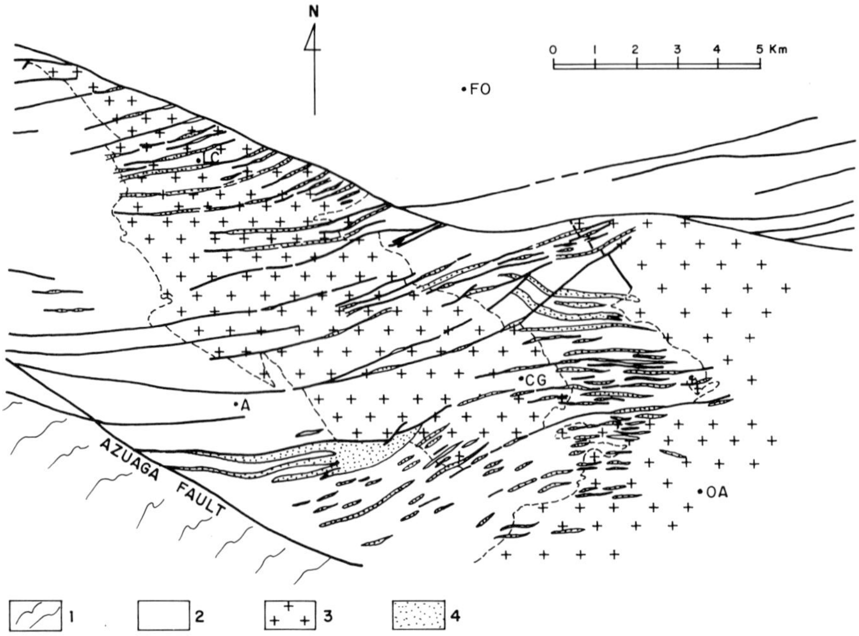

Fig. 10. *En échelon* fault and dyke pattern in the NW extremity of the Villaviciosa-La Coronada igneous belt, displaying sigmoidal geometries consistent with sinistral strike-slip (Modified after Apalategui et al. 1986). *1* Sierra Albarrana Domain; *2* Precambrian rocks of the Badajoz Córdoba Shear Zone; *3* igneous rocks of the Villaviciosa-La Coronada Belt; *4* dykes; *A* Argallón; *CG* Cañada del Gamo; *FO* Fuente Obejuna; *LC* La Coronada; *OA* Ojuelos Altos

cult any kind of calculation. On the basis of finite strain measurements on deformed objects and angular relations between S and C planes, Burg et al. (1981) estimated a minimum sinistral displacement of ca. 72 km. This figure, however, must be considered much smaller than the real one if the remarkable differences (both structural and stratigraphic) between the bounding units and the proven evidence of late deformation (thrusting and dextral shearing), producing substantial changes in the shear zone geometrical characteristics (width, angular relations, etc.), are taken into account.

Disregarding the possibility that the subhorizontal fabric was an older unrelated element, the available radiometric data (Dallmeyer and Quesada 1989) demonstrate that the Badajoz-Córdoba Band was active as a sinistral strike-slip shear zone at least since the Upper Devonian to the Lower Carboniferous (330 +/− 5 Ma 40 Ar-39 Ar muscovite cooling ages from subvertical mylonites; Dallmeyer and Quesada op. cit.). During this time span the generalized uplift of the belt took place. The extension into later times of the sinistral strike-slip activity, already under brittle conditions, is neatly characterized by the development of *en échelon* fractures, taking advantage of which the intrusion of the Carboniferous Villaviciosa-La Coronada igneous rocks was achieved (Fig. 10).

3.3.6 Obejo-Valsequillo-Puebla de la Reina Domain

This unit is frequently included within the Central-Iberian zone (Robardet 1976; Chacón and Pascual 1977; Quesada et al. 1987) and shows remarkable differences from the above - described belts. Apart from stratigraphic variations, the main structural difference resides in the northeasterly vergence of its structures as opposed to the generalized southwesterly vergence of the structures throughout the Ossa-Morena Zone. From the Portalegre-Hornachos-Villaharta Fault Zone, which thrusts the Badajoz-Córdoba shear zone over this domain, the structure consists of a fold and thrust forward sequence cross-cut by the Pedroches Batholith, to the north of which an autochthonous sequence typical of a foreland environment appears. The age of this structure is constrained between the basal Namurian age of the youngest affected deposits, and the Westphalian B age of the oldest overlying unconformable strata, thus being younger in this domain than in the southern belts. An Upper Carboniferous second phase gave rise to the refolding of this unit under a neatly dextral transpressional regime.

No correlation has been established as yet between the structure in this domain and that of the southern units. The link between both zones is provided by the Lower Carboniferous marine deposits of the northern Ossa-Morena basin (Gabaldón et al. 1985) which oversteps the boundary, postdating the first deformation phase in the Ossa-Morena units, but being affected by the first deformation in this Obejo-Valsequillo-Puebla de la Reina domain.

3.4 Discussion: Tectonic Evolution of the Ossa-Morena Zone

The data presented in the preceding sections together with all the available stratigraphical, petrological, etc. information (see respective sections) outline a very complex history for the Ossa-Morena Zone.

Starting in Precambrian times, this includes at least two orogenic cycles (Late Precambrian and Hercynian) before cratonization following the latter. A detailed reconstruction is, however, far from being reached, because of the scarcity of data, mainly geochemical and radiometrical. This is particularly true for the Precambrian part of the history, otherwise complicated by variably intensive Hercynian reactivation. Although highly speculative, the main lines of the tectonic evolution of the Ossa-Morena Zone during the Precambrian may be summarized as follows: Starting with the pre-orogenic evolution, the known pre-orogenic sequences (Quesada et al., Part V, Chap. 2.1) seem to assess an initial continental rifting proccess leading firstly to the formation of continental tholeiites (Blastomylonitic Formation; Mata and Munha 1986; García Casquero, in press), then evolving secondly to MORB - like tholeiites (Serie Negra amphibolites; Eguiluz 1987). This rifting proccess would have led eventually to oceanic crust formation, now represented by sparse serpentinite occurrences. The rifting event was presumably succeeded by a passive margin stage, during which a miogeoclinal wedge was built up onto the continental margin (Serie Negra deposition) and, finally, the previously formed oceanic basin would start to close in Upper Riphéan times. According to the available structural and petrologic data (the regional metamorphic evolution in particular; Quesada and Munhá, this Vol.) it seems likely that the Late-Precambrian orogeny took place in several phases, involving firstly subduction and subsequently collision. The synorogenic calc-alkaline volcanism would have been developed during the subduction stage and probably constituted an island arc or an Andean-type active margin (Quesada et al. 1987). On the other hand, continental margin loading by thrusting following collision presumably gave rise to the formation of a foreland basin where the syn-orogenic flysch-like deposits would have been laid down (Quesada et al., op. cit.). If the present-day spatial arrangement of synorogenic successions is taken into account (volcanosedimentary in the SW; flyschoid in the NE), then the Benioff Zone should have dipped in a southwesterly direction. Finally, decompression related to isostatic readjustments, once the orogenic activity had been completed, is to be held responsible for magma generation and ascent and the subsequent development of the thermal domes scattered throughout the region.

Despite the aforementioned paucity of data, the Paleozoic tectonic evolution of the Ossa-Morena Zone can be established, on the other hand, with much more precision than the Precambrian one. An initial rifting event is recognized in this zone which, starting in the Cambrian, led to the formation of an oceanic basin to the present SW of the Iberian craton. The area under consideration underwent a passive margin type of evolution, from the Lower Ordovician and covering most of the Lower Paleozoic, representing a distal area of the Iberian continental margin (Ribeiro et al., Part VIII). Neither the moment when sea-floor spreading ceased nor when the basin first started to close is known as yet. The oldest evidence for Paleozoic orogenic activity in the Ossa-Morena Zone corresponds to the Beja-Acebuches ophiolite obduction, achieved by the Middle Devonian (Apalategui and Quesada 1987). From this moment and until the Upper Carboniferous, the region underwent deformation under what seems to be an oblique collisional regime, which affected progressively areas located more and more to the northeast. A precollisional subduction process presumably gave rise to an island arc, the characterization of which is still poorly evidenced (probably the Southern Crystalline Massifs, the Beja Massif in particular, with a profusion of igneous rocks, could represent root zones of such an arc, uplifted as a consequence of subsequent post-collisional intracontinental deformation). The oblique nature of the subduction/collision processes is strongly indicated by a generalized transection of folds by the associated cleavage as well as by the development of pull-apart type basins in which the Upper Paleozoic synorogenic sequences were deposited. The South Portuguese basin, dealt with in detail elsewhere in this book (Silva et al.), and which probably lies near the suture, is the most important of these basins, both for its large size and tectonic significance.

Acknowledgements. Comments and suggestions made by J. B. Silva and an anonymous reviewer concerning a first version has greatly improved this contribution. R. Fernández Toledo, who carefully typed the various versions of the manuscript and J. Vallejo, who drew all the art work, are most gratefully acknowledged.

References

Aguayo J (1985) Rocas ultramáficas en el sector de Calzadilla de los Barros (Badajoz). Thesis, Univ Bilbao, 110 pp

Alia M (1963) Rasgos estructurales de la baja Extremadura. Bol R Soc Esp Hist Nat (Geol) 20: 247-262

Alvarado A (1935) Terreno Arcaico. Explicación del nuevo mapa geológico de España. Mem IGME 1: 235-280

Apalategui O, Pérez Lorente F (1983) Nuevos datos en el borde meridional de la zona Centro-ibérica. El dominio Obejo-Valsequillo-Puebla de la Reina. Stud Geol Salmant 18: 193-200

Apalategui O, Quesada C (1987) Transversal geológica Zona Ossa-Morena. Excursion Guidebook, pp 1-90

Apalategui O, Barranco E, Contreras F, Delgado M, Roldan FJ, Quesada C (1983 a). Mapa Geológico de España, escala 1:50000. Hoja nº 916: Aroche. IGME

Apalategui O, Higueras P, Quesada C, Cueto LA (1983 b) Mapa Geológico de España escala 1:50000. Hoja nº 855: Usagre. IGME

Apalategui O, Barranco E, Contreras F, Roldan FJ, Quesada C (1984) Mapa Geológico de España, escala 1:50000. Hoja nº 917: Aracena. IGME

Apalategui O, Borrero JD, Higueras P (1985) División en grupos de rocas en Ossa-Morena oriental. Temas Geol Min 7: 73-80

Apalategui O, Garrote A, Roldan FJ, Sánchez R, Cueto LA, Quesada C (1986) Mapa geológico de España, escala 1: 50000. Hoja nº 879: Peñarroya-Pueblonuevo. IGME

Apalategui O, Contreras F, Jorquera A, Villalobos M, Eguiluz L, Quesada C, Cueto LA (1989) Mapa Geológico de España, escala 1:50000. Hoja nº 804: Oliva de Mérida. IGME

Bard JP (1969) Le métamorphisme régional progressif des Sierras d'Aracena en Andalousie occidentale (Espagne). Sa place dans le segment hercynien sud-ibérique. Thesis, University of Montpellier

Blatrix P, Burg JP (1981) ^{40}Ar-^{39}Ar dates from Sierra Morena (Southern Spain): Variscan metamorphism and Cadomian orogeny. Neues Jahrb Mineral Monatsh 1981 (10): 470-478

Boogard M van den, Vázquez F (1981) Conodont faunas from Portugal and southwestern Spain. Part 5: Lower Carboniferous conodonts at Santa Olalla de Cala (Spain). Scr Geol 61: 1-8

Burg JP, Iglesias M, Laurent P, Matte P, Ribeiro A (1981) Variscan intracontinental deformation: the Coimbra-Córdoba Shear zone (SW Iberian Peninsula). Tectonophysics 78: 161-177

Chacón J (1974) Metamorfismo dinámico en la banda Badajoz-Córdoba: los gneises milonitas de Higuera de Llerena (provincia de Badajoz). Bol Geol Min 85 (6): 700-712

Chacón J (1979) Estudio geológico del sector central del anticlinorio Portalegre-Badajoz-Córdoba (Macizo Ibérico Meridional). Thesis, Univ Granada, 1-728

Chacón J, Pascual E (1977) El anticlinorio Portalegre-Badajoz-Córdoba divisoria entre las zonas Centro-Ibérica y Ossa-Morena (Sector SW del Macizo Ibérico). Cuad Geol 8: 21-35

Chacón J, Delgado M, Garrote A (1974) Sobre la existencia de los diferentes dominios de metamorfismo regional en la banda Elvas-Badajoz-Córdoba (Macizo Hespérico Meridional). Bol Geol Min 85 (6): 713-717

Conde LN, Andrade AAS (1974) Sur la faune méso et ou neodevonienne des Calcaires des Monte das Cortes, Odivelas (Massif de Beja). Mem Not 78: 141-146

Crespo A (1987) El Macizo de Aracena (Macizo Ibérico Meridional): propuesta de división sobre la base de nuevos datos estructurales y petrográficos. Bol Geol Min 98 (4): 507-515

Crespo A, Orozco M (1988) The Southern Iberian Shear Zone: a major boundary in the Hercynian Folded Belt. Tectonophysics 148: 221-227

Cueto LA, Eguiluz L, Llamas FL, Quesada C (1983) La granodiorita de Pallarés, un intrusivo precámbrico en la alineación Olivenza-Monesterio (zona de Ossa-Morena). Comun Serv Geol Port 69 (2): 219-226

Dallmeyer RD, Quesada C (1989) Geochronological constraints to the structural development of the Badajoz-Cordoba shear belt (southwest Iberia). Terra Abstr 1 (1): 366

Delgado M (1971) Esquema geológico de la hoja número 878 de Azuaga (Badajoz). Bol Geol Min 82: 277-286

Delgado M, Liñan E, Pascual E, Pérez F (1977) Criterios para la diferenciación de dominios en Sierra Morena Central. Stud Geol 12: 75-90

Eguiluz L (1987) Petrogénesis de rocas ígneas y metamórficas en el antiforme Burguillos-Monesterio. Macizo Ibérico Meridional. Thesis, Univ Bilbao, pp 1-694

Eguiluz L, Quesada C (1980) La sucesión precámbrica de la transversal de Monesterio (Badajoz). Nota preliminar. Temas Geol Min 4: 123-141

Eguiluz L, Ramon Lluch R (1983) La estructura del sector central del dominio de Arroyomolinos. Anticlinorio Olivenza-Monesterio, Ossa-Morena. Stud Geol Salmant 18: 171-192

Eguiluz L, Fernández J, Garrote A, Liñan E, Quesada C (1984) Sucesiones estratigráficas del anticlinorio de Olivenza-Monesterio en la transversal Montemolín-Arroyomolinos. Cuad Lab Xeol Laxe 8: 117-123

Fabriès J (1957) El problema del "estrato cristalino" en el NE de la provincia de Sevilla (España). Not Comun IGME 48: 195-203

Fabriès J (1963) Les formations cristallines et métamorphiques du NE de la province de Seville (Espagne). Essai sur le métamorphisme des roches éruptives basiques. Thesis, Univ Nancy

Fernández J, Coullaut JL, Eguiluz L, Garrote A, Quesada C, Cueto LA (1983) Mapa Geológico de España, escala 1: 50000. Hoja nº 897: Monesterio. IGME

Fricke W (1941) Die Geologie des Grenz-gebietes zwischen nordöstlicher Sierra Morena und Extremadura. Thesis, Univ Berlin

Gabaldón V, Garrote A, Quesada C (1985) El Carbonífero inferior del Norte de la zona de Ossa-Morena (SW de España). CR 10th Int Carboniferous Congr Madr 3: 173-186

Galindo C, Casquet C, Portugal Ferreira M, Macedo CAR (1986) O complexo Plutónico de Táliga-Barcarrota-un complexo intrusivo com idades caledónica e hercínica. Maleo 2 (13): 22

García Casquero JL (in press) Continental trondhjemites in displaced terranes of the Oporto-Portalegre-Badajoz-Córdoba Belt (Hesperian Massif). Geol Rundschau

García Casquero JL, Boelrijk NAIM, Chacón J, Priem HNA

(1985) Rb-Sr evidence for the presence of Ordovician granites in the deformed basement of the Badajoz-Córdoba Belt, SW Spain. Geol Rundschau 74 (2): 379-384

Garrote A (1976) Asociaciones minerales del núcleo metamórfico de Sierra Albarrana (Prov. de Córdoba). Sierra Morena Central. Mem Not Univ Coimbra 82: 17-39

Garrote A, Arriola A, Eguiluz L, Sánchez R, Quesada C, Cueto LA (1983) Mapa Geológico de España, escala 1:50000 Hoja número 876, Fuente de Cantos. IGME

Garrote A, Delgado M, Contreras MC (in press) Mapa Geológico de España escala 1:50000. Hoja número 900: La Cardenchosa. IGME

Gonçalves F (1971) Subsidios para o conhecimiento geológico do Nordeste alentejano. Mem Serv Geol Port Nova Serie 18: 1-62

Hernández Pacheco F (1953) Edad de las formaciones con facies estratocristalinas en la provincia de Badajoz. Not Comun IGME 31: 3-33

Herranz P (1984) El Precámbrico y su cobertera paleozoica en la región centro-oriental de la provincia de Badajoz. Thesis, Univ Complutense Madr, 1220 pp

Lotze F (1929) Stratigraphie und Tektonik des Keltiberischen grundgebierges (Spanien). Abh Ges Göttingen Math 14 (2): 1-320

Lotze F (1956) Das Präkambrium Spaniens. Neues Jahrb Geol Paläontol Monatsh 8: 373-380

MacPherson J (1879) Estudio geológico y petrográfico del N de la provincia de Sevilla. Bol Com Mapa Geol Esp 10: 97-269

Mallada L (1895) Explicación del Mapa Geológico de España. Rocas hipogénicas y sistema estrato-cristalino. Mem Com Mapa Geol Esp 1: 107-420

Mata J, Munhá J (1986) Geodynamic significance of high-grade metamorphic rocks from Degolados-Campo Maior (Tomar-Badajoz-Córdoba shear zone). Maleo 2 (13): 28

Matte P (1983) Two Geotraverses across the Ibero-Armorican Variscan Arc of western Europe. In: Rast N, Delany M (eds) Profiles of orogenic belts. N AGU Geodynam Ser 10: 53-81

Matte P (1986) Tectonics and plate tectonics model for the Variscan Belt of Europe. Tectonophysics 126: 329-374

Pascual E (1981) Investigaciones geológicas en el sector Córdoba-Villaviciosa de Córdoba (Sector Central de Sierra Morena). Thesis, Univ Granada, pp 1-521

Pérez Lorente F (1979) Geología de la zona Ossa-Morena al Norte de Córdoba (Pozoblanco-Belmez-Villaviciosa de Córdoba). PhD Thesis, Univ Granada 281: 1-340

Quesada C (1975) Geología de un sector de la parte central del anticlinorio Olivenza-Monesterio. Alrededores de Monesterio (Badajoz). Thesis, Univ Granada, 128 pp

Quesada C (1983) El Carbonífero de Sierra Morena. In: Martínez C (coord) Carbonífero y Pérmico de España. IGME, pp 243-278

Quesada C, Larrea FJ, Florido P et al. (1987) Mapa Geológico Minero de Extremadura. Dirección General Industria, Energía y Minas. Junta de Extremadura

Racheboeuf PR, Robardet M (1986) Le Pridoli et le Dévonien inférieur de la Zone d'Ossa-Morena (Sud-Ouest de la Peninsule Ibérique). Etude des Brachiopodes. Geol Palaeontol 20: 11-37

Robardet M (1976) L'originalité du segment hercynien sud-ibérique au Paléozoique inférieur: Ordovicien, Silurien et Dévonien dans le nord de la province de Séville (Espagne). CR Acad Sci Paris D 283: 999-1002

Robardet M, Weyant M, Laveine JP, Racheboeuf P (1986) Le Carbonifère inférieur du Synclinal du Cerrón del Hornillo (province de Seville, Espagne). Rev Paléobiol 5 (1): 71-90

Santos JA, Casas J (1980) Caracterización de unas series precámbricas al N de Badajoz. Temas Geol Min 4: 213-233

Turner FJ, Weiss LE (1963) Structural analysis of metamorphic tectonites. McGraw-Hill, New York, 545 pp

Vauchez A (1976) Les structures hercyniennes dans la région de Fregenal-Oliva de la Frontera (Badajoz, Espagne). Un exemple de tectoniques tangéantielles superposées. Comun Serv Geol Port 60: 261-265

Vegas R (1968) Sobre la existencia de Precámbrico en la Baja Extremadura. Estud Geol 24:85-89

Vegas R (1970) Formaciones precàmbricas de la Sierra Morena Occidental. Relación con las series anteordovícicas de Almadén, Don Benito y Cáceres. Estud Geol 26: 225-231

4 Igneous Rocks

R. Sanchez Carretero,[1] L. Eguiluz,[2] E. Pascual[3] and M. Carracedo[1]

4.1 Introduction

Peculiarities of the Ossa-Morena Zone (OMZ) are not only tectonic, stratigraphic and metamorphic; magmatism is also different from that of the neighbouring zones in the Iberian Massif. Banded structuration NW–SE is also marked by distribution of igneous bodies forming magmatic alignments (Bard 1971; Bard and Fabries 1970; Aparicio et al. 1977, etc.; Fig. 1).

The main differential, general features of OMZ magmatism are small size of plutonic bodies (in contrast with huge Central-Iberian Zone, CIZ, granitic batholiths), common appearance of basic (gabbroic) rocks, frequent time-space association of basic and acid rocks (bimodal magmatism) and important volcanic and subvolcanic events.

[1] Departamento de Mineralogía y Petrología, Universidad del País Vasco, Apartado 644, 48080-Bilbao, Spain
[2] Departamento de Estratigrafía, Geodinámica y Paleontología, Universidad del País Vasco, Apartado 644, 48080-Bilbao, Spain
[3] Sección de Geología de La Rábida, Universidad de Sevilla, Palos de la Frontera, 21819-Huelva, Spain

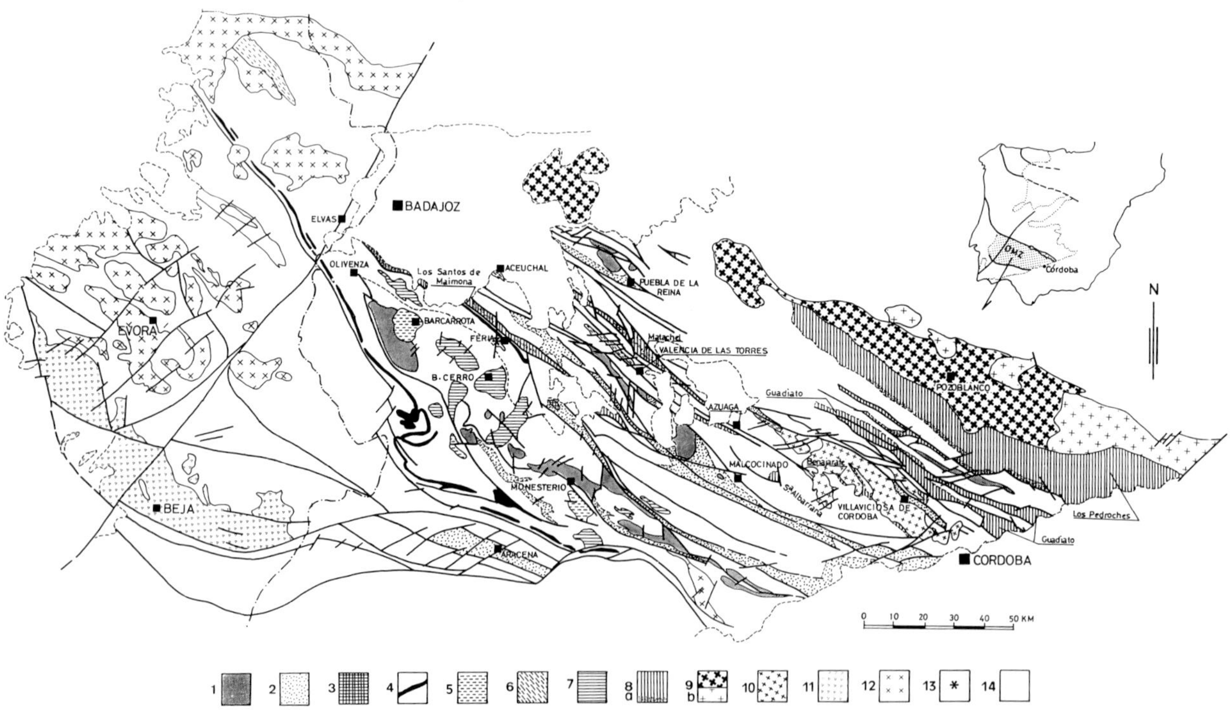

Fig. 1. Geological map of the main outcrops of igneous rocks from the Ossa-Morena Zone (Iberian Massif). Late-Precambrian (Upper Riphean-Vendian): *1* Calc-alkaline granitic rocks; *2* volcanosedimentary synorogenic series (Malcocinado Fm., Bodonal-Cala porphyries, etc.); *3* serpentinized ultramafic rocks. Lower to Middle Cambrian: *4* basic to intermediate alkaline volcanic rocks; *5* alkalic granitoids. Lower to Middle Ordovician: *6* gneissic alkaline granites. Early Hercynian: *7* calc-alkaline granitoids (gabbros, diorites, and granites). Hercynian (Carboniferous): *8* Lower Middle Carboniferous basins (lutites and sandstones); *8 a* acid and basic volcanic rocks; *9* Los Pedroches granodiorite; *9 a* Los Pedroches granite; *10* Villaviciosa de Córdoba-La Coronada Magmatic Alignment (acid-intermediate volcanic rocks and basic, intermediate and acid intrusive rocks); *11* Beja Complex (gabbros, diorites and granites); *12* granitic rocks s.l. Post-Hercynian (Permian): *13* Feria-Sierra Vieja Massif (alkaline granites). Precambrian-Paleozoic: *14* undifferentiated metamorphic and sedimentary series

R. D. Dallmeyer and E. Martínez García (Eds.)
Pre-Mesozoic Geology of Iberia

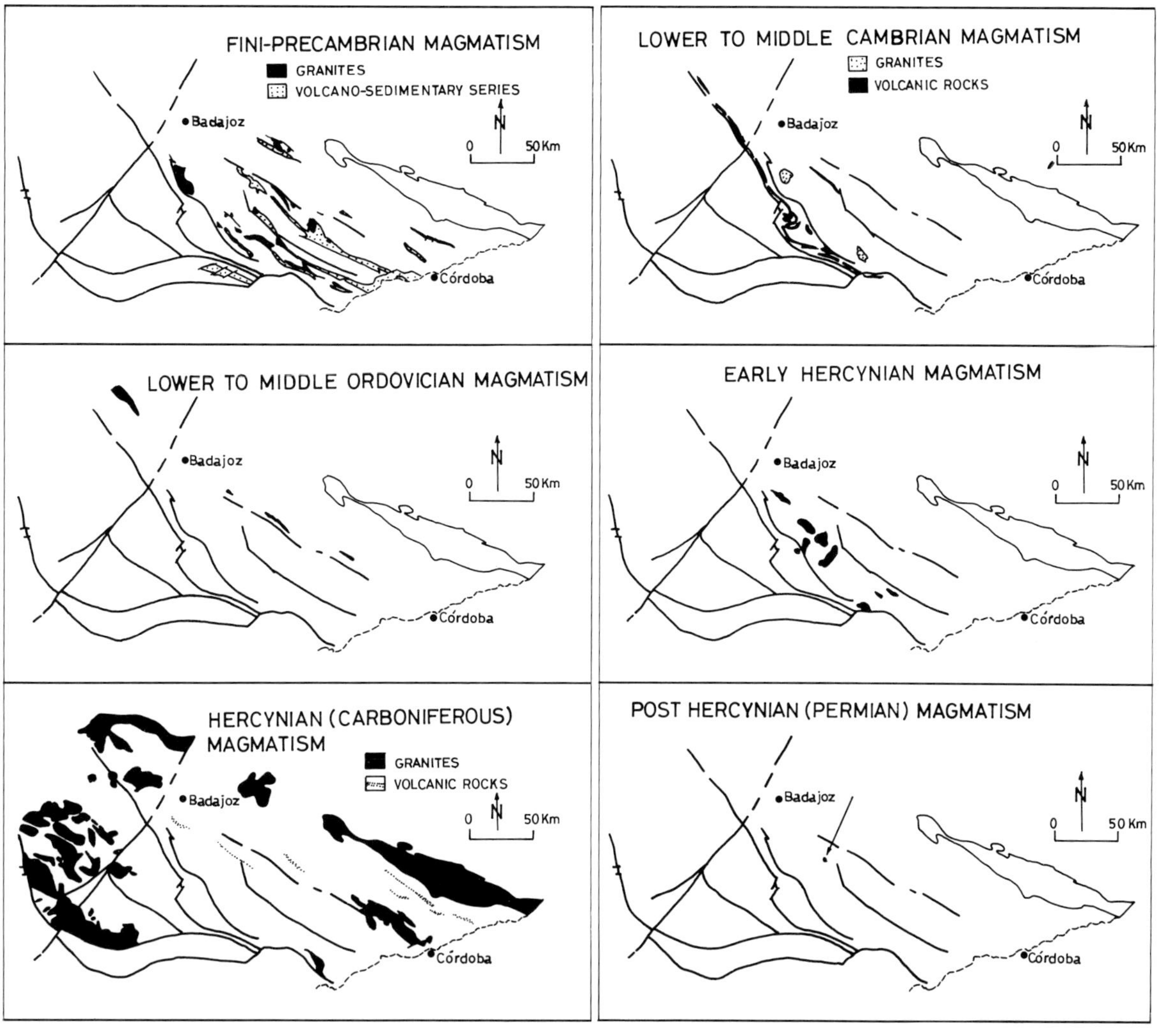

Fig. 2. Simplified geological maps of the OMZ showing the outcrops of igneous rocks related to the main magmatic events mentioned in the text

The variety of petrographic types with ages ranging from Middle-Upper Proterozoic to Permian, scarce and/or unreliable radiometric data and insufficient geochemical knowledge of many of these igneous rocks make systematic studies and synthesis difficult. There are, however, some of such works previous to ours at Iberian Massif or OMZ scale (Bard and Fabries 1970; Bard 1971; Capdevila et al. 1973). In many instances, they consider together igneous rocks of different ages and significance.

The main purpose of the present work is to sketch a classification of OMZ igneous rocks based on age or tectonomagmatic cycles, stressing specially volcanic events, which in general are poorly considered in previous syntheses. Precambrian magmatism is thus related to the Fini-Precambrian orogeny, either pre- or synorogenic. Similarly, Paleozoic igneous rocks prior to the Hercynian cycle are grouped as pre-Hercynian Paleozoic magmatism. Hercynian magmatism s. s. and post-orogenic, peralkaline magmatism are the other groups in the classification proposed (Fig. 2).

4.2 Precambrian Magmatism

We include in this group a wide variety of igneous rocks of Precambrian age, more or less affected by Precambrian tectonometamorphic events.

Following the present book scheme, Precambrian materials in the OMZ are classified into two groups concerning Late-Precambrian orogeny

(Upper Riphean-Vendian), namely preorogenic and synorogenic groups. Igneous rocks are distributed among these as follows:

Preorogenic Magmatism: Bimodal magmatism of the Valencia de las Torres-Cerro Muriano Terrane (Middle to Upper Proterozoic) and

Synorogenic Magmatism: anatectic granites related to metamorphic cores, Upper Riphean to Vendian volcanics and subvolcanics and serpentinized ultramafic rocks of uncertain age.

4.2.1 Bimodal Magmatism of the Valencia De Las Torres-Cerro Muriano Terrane (Middle to Upper Proterozoic)

Although gneisses and amphibolites are widespread generally in the Proterozoic rocks, they occur mostly in the lower part of the series (Delgado Quesada 1971). They consist of mylonite gneisses (paragneisses and some ortho-derived, in some places pegmatoid, masses) with interbedded amphibolites variable in thickness. These amphibolites often show relict mineral assemblages of granulite facies. Metamorphic country rocks are carbonate, calc-silicate rocks and black quartzites. Development of mylonitic fabrics usually make the recognition of initial textural and especially field relationships difficult.

Original series may be interpreted as a volcano-sedimentary sequence related to bimodal volcanism. However, some of the granite bodies appear to be related to anatectic processes during the Precambrian metamorphic climax (e. g. Monesterio core).

Orthogneisses form masses variable in size and shape, proceeding from granites and/or acid vulcanites. They consist of quartz, K-feldspar (sometimes perthitic), acid oligoclase, biotite and minor zircon, opaques, apatite and titanite. Garnet is seldom present. They show mylonitic gneissic textures, often porphyroblastic.

The main evidence of an igneous origin of some of the gneisses is their relative uniformity at outcrop scale, since relationships with "country" rocks are obliterated by deformation and metamorphism. Igneous character is also supported by available

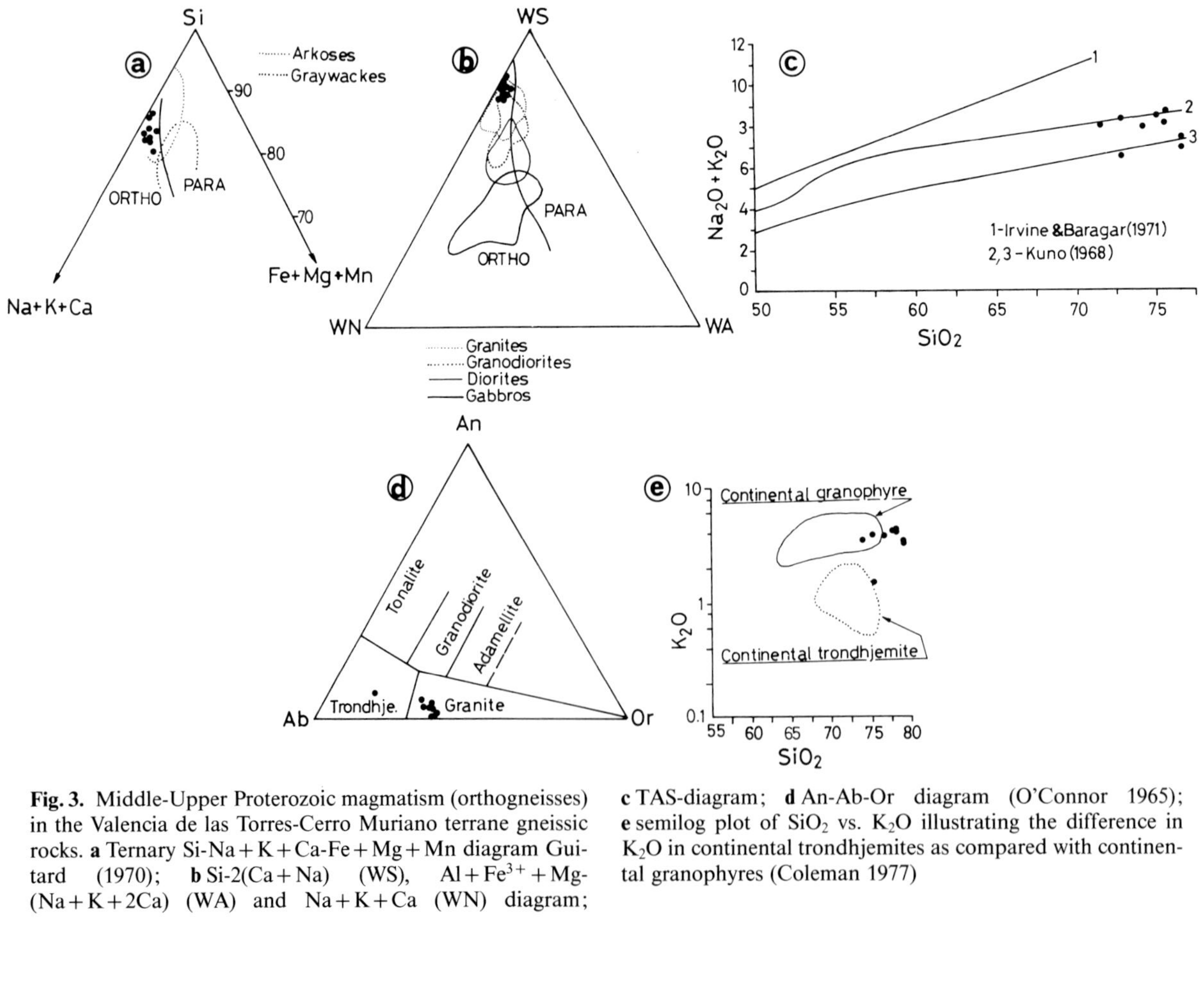

Fig. 3. Middle-Upper Proterozoic magmatism (orthogneisses) in the Valencia de las Torres-Cerro Muriano terrane gneissic rocks. **a** Ternary Si-Na+K+Ca-Fe+Mg+Mn diagram Guitard (1970); **b** Si-2(Ca+Na) (WS), Al+Fe^{3+}+Mg-(Na+K+2Ca) (WA) and Na+K+Ca (WN) diagram; **c** TAS-diagram; **d** An-Ab-Or diagram (O'Connor 1965); **e** semilog plot of SiO_2 vs. K_2O illustrating the difference in K_2O in continental trondhjemites as compared with continental granophyres (Coleman 1977)

chemical data (Fig. 3 a, b). Their plotting in the TAS diagram (Fig. 3 c) evidences a subalkaline character, with granite composition, except for a sample of trondhjemitic composition (Fig. 3 d and Table 1). K contents (100 K/K + Na average ratios ranging 49; Coleman 1977) let them be characterized as continental granophydic rocks (Fig. 3 e). But it should be stressed that chemical characters may have been modified by metamorphism, especially under granulite conditions. García Casquero (1988) points out that some rocks of the Valencia de las Torres Fm. define a trondhjemitic-tholeiitic association due to amphibolite partial melting.

Orthoamphibolites crop out mainly in the Azuaga gneisses and in the Serie Negra and they are particularly abundant in the upper part of the Azuaga gneisses (Las Mesas subunit, Delgado Quesada 1971). They are tabular masses consisting of several types of rock (flaser gabbros, garnet-bearing amphibolites, etc.). They have been metamorphosed in the amphibolite facies, with mylonitic fabric development. Granulite relics are often conspicuous.

These amphibolites are composed of plagioclase and amphibole. Garnet is very often present, whilst pyroxene appears locally. Plagioclase An-content, up to 45–48, suggests a premetamorphic character, at least in part.

In the Serie Negra, amphibolites appear at the top of the lower unit (Montemolín unit, Eguiluz 1988), where they are interbedded with biotite schists and black quartzites. There they consist of plagioclase, amphibole (actinolite to brown hornblende), quartz and biotite. In high-grade metamorphic zones pyroxene is also present; it has been recognized that both volcanic and subvolcanic protoliths have been transformed into amphibolites.

Figure 4 shows a plot of chemical data of amphibolitic rocks from both Azuaga gneisses and Serie Negra. In all cases amphibolites are derived from igneous rocks (Fig. 4 a). The An content of "pre-metamorphic" plagioclase as well as the abundance of sphene also suggest this conclusion.

SiO_2 content varies from 47 to 61%, the lowest contents corresponding to the Montemolín unit (Table 1). Subalkaline character of these rocks is clear (Fig. 4 b), as all samples plot in the high-Al and tholeiitic fields of the diagram of Kuno (1968). A tholeiitic character could have been envisaged from the diagrams of Miyashiro (1975) (Figs. 4 d, e).

The tectonic environment corresponding to this magmatism is not well defined. Eguiluz (1988) suggests that amphibolites from the Serie Negra would represent tholeiitic magmatism of oceanic affinities.

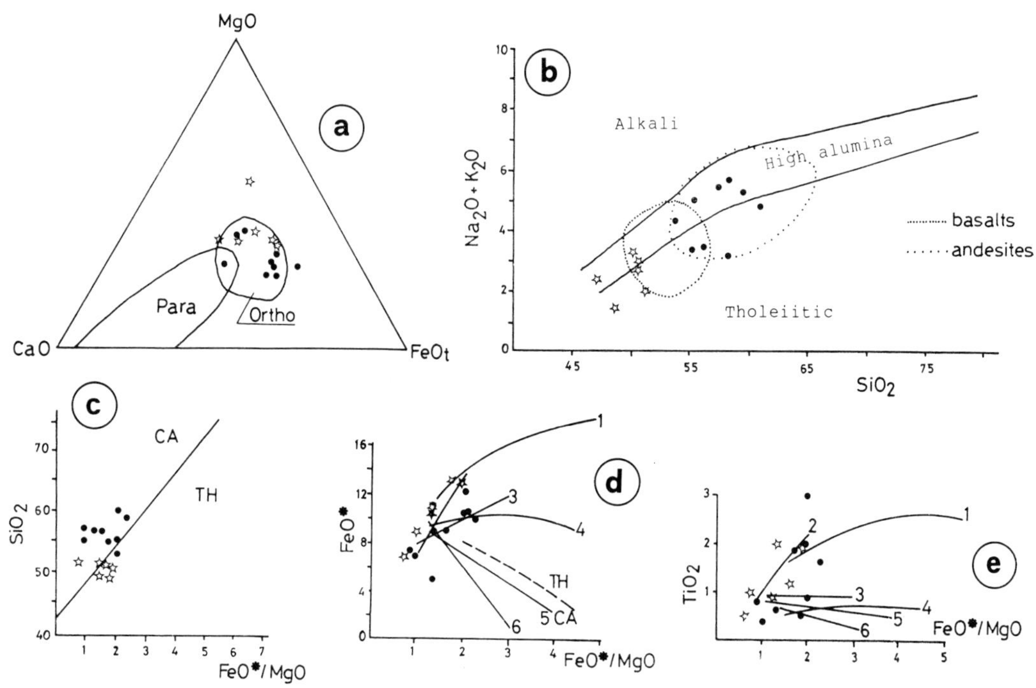

Fig. 4. Middle-Upper Proterozoic magmatism (amphibolites) in the Valencia de las Torres-Cerro Muriano terrane gneissic rocks *(dots)* and Serie Negra *(stars)*. **a** MgO-CaO-FeOt diagram (Walker et al. 1960); **b** TAS diagram (Kuno 1968); **c, d** and **e** SiO_2, FeO* and TiO_2 vs FeO*/MgO diagram (*1* Skaergaard; *2* Abyssal tholeiites; **3** Macauley; *4* Tofua; *5* Asama; *6* Amagi (Miyashiro 1975)

Table 1. Preorogenic (Middle-Upper Proterozoic) magmatism

	Montemolin Amphibolites (n=8)		Azuaga Amphibolites (n=9)		Azuaga Orthogneisses (n=9)	
	Average	Std. Dev.	Average	Std. Dev.	Average	Std. Dev.
SiO_2	49.03	1.44	56.94	2.10	77.02	1.71
Al_2O_3	15.69	0.54	15.38	2.42	13.11	1.41
Fe_2O_3	12.06	1.65	1.97	1.08	0.49	0.26
FeO	n. d.	-	7.39	1.89	0.45	0.21
MgO	7.92	0.90	5.85	1.60	0.11	0.08
CaO	10.84	2.41	5.49	1.28	0.62	0.50
Na_2O	2.19	0.53	3.64	0.68	4.24	0.37
K_2O	0.48	0.47	0.82	0.40	3.54	0.79
TiO_2	1.59	0.89	1.32	0.86	0.06	0.02
MnO	0.02	0.08	0.18	0.03	0.34	0.94
P_2O_5	0.18	0.11	n. d.	-	n. d.	-

Table 2. Anatectic granitic rocks

	Monesterio Granodiorite (n=8)		Leucogranites (n=28)	
	Average	Std. Dev.	Average	Std. Dev.
SiO_2	69.37	1.45	75.27	2.11
Al_2O_3	15.17	0.31	14.08	0.76
Fe_2O_3	4.12	0.55	0.9	0.91
FeO	1.74	0.36	0.43	0.35
CaO	1.95	0.62	0.65	0.38
Na_2O	3.15	0.37	3.13	0.84
K_2O	3.29	0.54	4.90	1.32
TiO_2	0.54	0.19	0.12	0.14
MnO	0.05	0.02	0.02	0.01
P_2O_5	0.21	0.04	0.14	0.08

4.2.2 Anatectic Granites Related to Metamorphic Cores

More or less allochthonous granitic rocks appear in direct relation with migmatitic zones. They are best exposed in the Monesterio metamorphic core, where leucogranites and granodiorites form discrete outcrops. Other masses, such as Pallares or Central del Pintado granodiorites, also are included in this group. They are all heterogeneous bodies including leucogranites and two-mica granites as well as granodiorites.

Leucogranites are composed of different petrographic facies, generally closely related, forming dykes and stratoid or irregular masses of small size intruding the granodiorite. They consist of pegmatites, muscovite granites, two-mica leucogranites, granites with biotite or cordierite nodules, etc. They are composed of quartz, K-feldspar and acid plagioclase. Biotite and muscovite are present in variable amounts, and cordierite, andalucite and sillimanite may also appear. In more evolved rocks garnet, possibly igneous, and tourmaline can also be present.

Granodiorites form either medium-sized bodies such as Pallares granodiorite (572±74 m.y., Rb/Sr, Cueto et al. 1983), or smaller, as in the Monesterio area. They are very often porphyritic rocks with feldspar megacrystals up to several cm in length. They contain many metamorphic enclaves. All granodiorites consist of quartz, zoned plagioclase (andesine to acid oligoclase), K-feldspar and biotite. They may also contain minor sillimanite, andalucite, cordierite and, sometimes, amphibole. Besides accessories such as zircon, apatite and allanite can also be found.

Chemical composition for all anatectic rocks corresponds to granites and granodiorites. SiO_2 contents range from 62 to 72% in granodiorites, exceeding 72% in leucogranites (Table 2). Both rock groups show continuous oxide against silica and Rb/Sr vs Sr (Fig. 5) variation trends (although a compositional gap may be seen in some cases), suggesting a single magma source. The evolution trend is calc-alkaline, both on AFM and FeO-Al_2O_3-MgO diagrams (Fig. 6 a, b). According to Eguiluz (1988), granites have a clear peraluminous character and show a silicic-potassic trend similar to that found in Higher Himalayan belt rocks (Debon and Le Fort 1983).

Plot of chemical data in normative diagrams Q-Ab-Or and Ab-An-Or (Fig. 7 a, b) indicate that these rocks are produced by low-pressure crustal melting. Leucogranite plots, very close to minimum melting point, are well distinguished from granodiorites. This fact can be interpreted in terms of a higher restite content in the latter in view of the whole geological data, which suggest an autochthonous character and do not support high-T melts far from the minimum melting point.

Other metamorphic cores containing similar granitic rocks in the OMZ are those of Evora (Carvalhosa 1983) and Lora del Río (Arriola and Eguiluz 1983), both probably related to Precambrian metamorphism.

4.2.3 Late-Proterozoic Orogenic Vulcanism

During the Cadomian or Pan-African (Upper Riphean to Vendian) orogenic cycle, widespread volcanism is developed, both in OMZ and southern

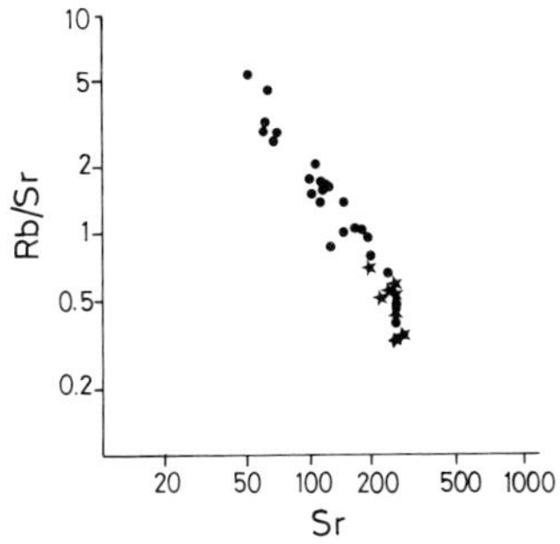

Fig. 5. Rb/Sr vs Sr variation diagram for Monesterio core rocks (*dots* granodiorites; *stars* leucogranites; after Eguiluz 1988)

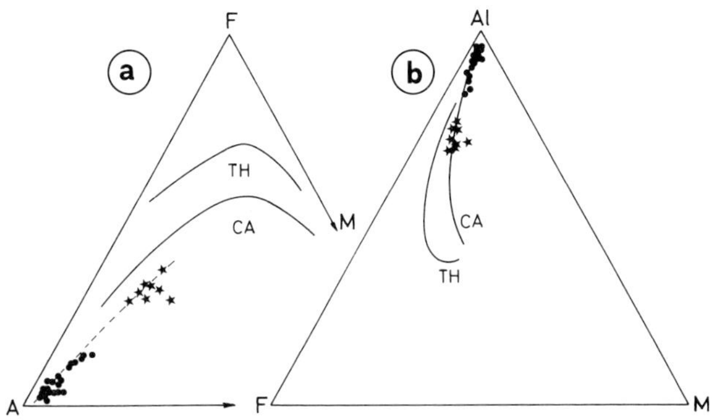

Fig. 6. Monesterio core rocks. **a** AFM diagram; **b** Al_2O_3-Fe_2O_3+FeO-MgO diagram. *CA* calc-alkaline; *TH* tholeiitic. (*dots* granodiorites; *stars* leucogranites; after Eguiluz 1988)

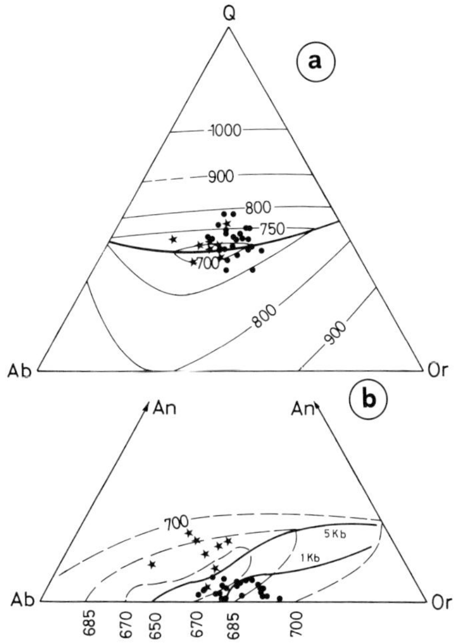

Fig. 7. Monesterio core rocks. **a** Normative Q-Ab-An diagram with isothermal and cotectic lines for 2 Kb; **b** normative AN-Ab-Or diagram with isothermal lines for 5 Kb and cotectic lines for 5 and 1 Kb (*dots* granodiorites; *stars* leucogranites; after Eguiluz 1988)

CIZ, forming thick lava flows and volcanoclastic deposits closely followed by several subvolcanic granite plutons.

Volcanic rocks crop out interbedded with detritic rocks (epiclastites, arkoses, pelites) and carbonate rocks, forming a volcanosedimentary sequence widely spread throughout the OMZ (see Figs. 1, 2). First data about these materials are shown by Fricke (1941). Later works (Delgado Quesada 1971; Hernández Enrile 1971; Bard 1969; Liñán 1978; Florido and Quesada 1984; Apalategui et al. 1984, 1985) afford details on petrography and spatial distribution. First chemical data (Sánchez Carretero et al. 1989 a) show a calc-alkaline character for this magmatism.

Calera and Tablada granites (Figs. 1, 2) belong to this group. They are often hydrothermally altered and do not show a clear contact aureole. Textures range from coarse-grained to porphyroid and cataclastic. Field and textural features are indicative of shallow emplacement.

Geochemistry. Petrographic and geological evidence suggest that these rocks correspond to a calc-alkaline series: scarce basaltic rocks, common andesite texture and mineral composition (although more or less changed through hydrothermal alteration), dacitic and rhyolitic terms common and explosive volcanism are thought to be significant features.

The calc-alkaline character of this magmatism (Fig. 8 a, b, c, d) can be interpreted in terms of an active margin (Andean type; Sánchez Carretero et al. 1989 a), but it must be remembered that there exist only major-element data (Table 3). Other tectonic criteria (Apalategui et al., in litt., a) suggest a Precambrian subduction zone to the north of OMZ (Obejo-Valsequillo-Puebla de la Reina zone), with Benioff plane dipping to the S (?). This subduction would also be responsible for metamorphism and deformation of Precambrian series, including the volcanosedimentary rocks themselves.

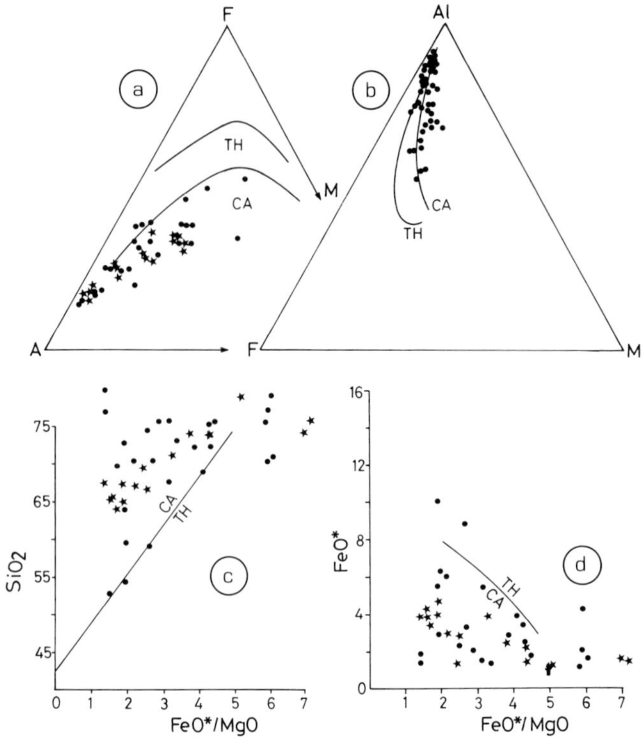

Fig. 8. Upper Riphean-Vendian (synorogenic) magmatism. **a** AFM diagram (*CA* calc-alkaline; *TH* tholeiitic); **b** Al_2O_3-Fe_2O_3+FeO-MgO diagram; **c, d** SiO_2 and FeO* vs FeO*/MgO diagram (*dots* volcanic rocks; *stars* granites; after Sánchez Carretero et al. 1989 a)

Table 3. Synorogenic (Upper Riphean-Vendian) magmatism

	Andesites (n=3)		Dacites (n=4)		Rhyolites (n=19)		Granites (n=16)	
	Aver.	Std. Dev.	Aver.	Std. Dev.	Aver.	Std. Dev.	Aver.	Std. Dev.
SiO_2	57.56	2.29	67.52	2.20	74.55	2.85	70.01	4.35
Al_2O_3	17.56	0.85	14.16	2.61	14.08	1.70	15.33	2.14
Fe_2O_3	3.63	0.34	0.99	0.53	1.56	0.73	0.87	0.44
FeO	5.17	1.82	3.78	1.03	1.2	1.06	2.19	1.01
MgO	3.96	0.93	1.93	0.68	0.93	0.72	1.35	0.91
CaO	2.37	1.19	1.96	1.45	0.54	0.66	2.04	1.15
Na_2O	3.75	0.55	2.92	0.88	2.01	1.88	3.83	0.86
K_2O	1.61	1.43	2.90	0.83	3.52	1.63	2.51	1.22
TiO_2	1.16	0.60	0.42	0.09	0.22	0.12	0.23	0.10
MnO	0.14	0.04	0.11	0.04	0.05	0.05	0.06	0.02

4.2.4 Serpentinized Ultramafic Rocks

They appear as little masses, partly or completely serpentinized, different in origin and age, which can be classified as follows:

1. Related to Azuaga gneisses and amphibolites;
2. Related to large strike-slip faults;
3. Related to volcanoclastic such as Malcocinado Formation.

Rocks from group 1, cropping out, for example, in Villafranca de los Barros Petrographic and field study allow the division of igneous rocks into three groups: lava flows, volcanoclastics and intrusive rocks.

Lava Flows. They have an intermediate (metaandesitic) or acidic (rhyolitic to dacitic) composition. Texture is porphyritic fluidal and varies from microcrystalline to aphanitic. The former consist of plagioclase (oligoclase to andesine) and mafic (hornblende and pyroxene) phenocrystals in a schistose to aphanitic groundmass containing plagioclase, mafic minerals and alteration products such as chlorite, epidote, titanite, calcite, opaques and sericite. Acidic volcanites are made up of quartz, K-feldspar, oligoclase and biotite, wich accessory zircon, titanite and tourmaline and deuteric sericite.

Volcanoclastic Rocks. They include pyroclastics, epiclastites and hybrid rocks, and considered as a whole represent about the 60% volume of these magmatic rocks. Fragment size is in all classes extremely variable, ranging from ash-clay/silt to metric blocks.

Most frequent types are acid meta-tuffs containing mono-mineralic and volcanic fragments. Metacinerites and/or phyllites are variable in colour and consist of fine-grained volcanic material more or less reworked, with textures ranging from granolepidoblastic to lepidoblastic.

Andesitic meta-tuffs are composed of plagioclase fragments, altered mafics, quartz, andesite and others. Matrix is fine-grained and chloritic, with a schistose porphyroclastic texture.

Regional metamorphism affecting all these rocks, in part implicit in the previous description, correspond to very low-grade to low-grade conditions, attaining at most the chlorite zone.

Intrusive Rocks. They are plutonites ranging from granodiorite to granite, closely related to volcanites. Ahillones, La Bomba-Los Parrados, Mosquil, Palomas, Sierra Padrona, (Arriola et al. 1983 a) and Usagre (Apalategui and Higueras 1983; Chacón and Velasco 1981) consist of olivine and pyroxene relics, serpentinite and subordinate talc, actinolitic amphibole, chlorite and magnetite. A tentative genetic relation with tholeiitic amphibolites (see Sect. 2.1) is proposed.

Masses from group 2 have even more controversial relationships, and some have been considered Hercynian (Chacón and Velasco 1981). They consist of serpentinites and pargasitic dunites, and may not be so intensely deformed as the "country" rocks.

Type 3 masses crop out near Calzadilla de los Barros (Badajoz), related to Late-Precambrian vulcanites (Malcocinado Formation; Arriola et al. 1983 b, c). They form three peridotitic bodies with 0.5, 3 and 10 km width. They have been mechanically emplaced subconcordant with the NW-SE structure of country rocks (Figs. 1, 2).

Several rock types have been distinguished, including serpentinites, chloritites, tremolitites, chromitites, talc-carbonate rocks, esteatitites and basic dykes (Aguayo 1985). Although masses are not zoned as a whole, chlorite and tremolite rocks are located near the contacts, being then probably related to alteration during emplacement. Olivine and pyroxene relics are very scarce. Rocks show very low- to low-grade metamorphism, similar to that existing in country rocks.

Available data suggest a tectonic, cold emplacement with coeval and/or late serpentinization. Quesada (1987) has also suggested them to represent an oceanic fragment obducted during the Late Precambrian orogeny.

4.2.5 Pre-Hercynian Paleozoic Magmatism

4.2.5.1 Introduction; Limit Choosing

This section involves a number of different igneous rocks with different composition, time of emplacement and genesis. Despite all these differences, they have been grouped together for two reasons:

1. They are, from the oldest to the youngest, well distinguished both from Precambrian and Hercynian s.s. igneous rocks. Both petrographic and geochemical differences are clear.
2. All rocks grouped together have features suggesting an orogenic character.

Lower and upper limits chosen in this work are surely provisional, especially the latter, and have been tentatively marked to separate events occurring in different space and time. It should be emphasized that a major difficulty in doing this accurately is the lack of reliable geochemical and radiometric data.

4.2.5.2 Lower to Middle Cambrian (Alkaline) Magmatism

We include here volcanic rocks cropping out in southern segments of the OMZ. Available radiometric data indicate that the Barcarrota pluton is to be included in this group. However, such radiometric data are in contradiction with geological data, as discussed below, classification for this pluton is hence doubtful.

Volcanic Rocks

Apart from the early works by MacPherson (1879) and by Schneider (1939), who defined the Umbria-Pipeta volcanic rocks and assigned them a Middle Cambrian age, these rocks have been studied by Fabries (1963); Bard (1969); Dupont (1979); Oliveira (1982); Apalategui et al. in litt. (b); Eguiluz (1988) and Galindo (pers. comm.). Best outcrops are located in the Alconera syncline, southern flank of Monesterio antiform and Aracena Massif (Figs. 1, 2).

Rocks in most outcrops are quartz-keratophyres underlain by limestones and covered by spilites. The common acid rocks are tuffs and other volcanoclastics, although lavas and porphyries are also present. They consist of albitic plagioclase and/or K-feldspar phenocrystals (or phenoclasts) in a quartz-feldspathic altered groundmass. In the San Amaro region some alkaline to peralkaline porphyries occur (Galindo, pers. comm.).

Basic rocks form stratoid bodies, very common in the southern flank of the Monesterio antiform. Pillow lavas and breccias are the most frequent varieties. Transition to amphibolites is known in metamorphic areas. Despite textural variation, mineralogical composition is rather uniform, consisting of altered (albitized) plagioclase, augitic pyroxene, opaques and secondary minerals (chlorite, epidote, quartz, carbonates, sericite, oxides, etc.) widely developed on other minerals and fractures or amygdales.

Alkaline to peralkaline rocks cropping out in the Alconchel-Juromenha region are mainly microsyenites and trachytes (Galindo, pers. comm.; Oliveira 1982; Apalategui et al., in litt., b).

Mata and Munhá (1986) distinguish a first oceanic tholeiitic magmatism during the Lower Cambrian from a younger transitional intraplate alkaline volcanism. This sequence in time would correspond to a rifting process then stopped by migration of the expansion axis.

Plutonic Rocks

Barcarrota Granite. This circular undeformed pluton is included here because of its K/Ar radiometric age of 510 ± 10 m. y. (Galindo et al. 1986). This age, however, is in contradiction to that of Táliga granite pluton, which, although cut by Barcarrota pluton, gives a K/Ar age of 380 ± 10 m. y. This apparent contradiction may be easily explained if the 380 m. y. age is considered as the age of the metamorphism, which has been widely established in the OMZ (Quesada, pers. comm.).

This zoned pluton has a core of basic rocks surrounded by a narrow band of acid plutonic rocks (Carnicero and Castro 1982). The former consist of hypersthene, titaniferous clinopyroxene, kaersutitic amphibole, biotite, opaques, apatite, titanite and alteration minerals. Chemical and mineralogical composition is controlled, at least in part, by crystal cumulation processes, including amphibole cumulates (Ruiz de Almodovar et al. 1984).

Acid plutonites vary in composition from alkali-feldspar granites to syenites (syenodiorites). They are made of a single perthitic feldspar, quartz, biotite, hastingsitic amphibole, ilmenite, magnetite, titanite and zircon. Only syenodiorites also contain oligoclase.

Basic rocks contain normative nepheline, while acid rocks are acmite normative (Bea et al. 1987). Rock association suggests a rifting environment, in which "A" granitoids in contrasted association with basic rocks are usually produced (Pitcher 1982).

Table 4. The Castillo granite (n = 6)

	Averge	Std. Dev.
SiO_2	71.14	0.90
Al_2O_3	14.02	0.20
Fe_2O_3	3.66	0.52
MgO	0.34	0.19
CaO	0.83	0.49
Na_2O	3.44	0.22
K_2O	5.05	0.17
TiO_2	0.27	0.03
MnO	0.04	0.02
P_2O_5	0.04	0.02

The Castillo Granite and the Valera Porphyries. In the Monesterio area, a small granite pluton is found which exhibits mineralogical and chemical characters similar to Lower- to Middle-Cambrian volcanics. It is a medium-grained rock, locally deformed, made of quartz, alkali feldspar, biotite and hastingsitic amphibole. High K and Na contents (Table 4) and a marked Zr (~555 ppm) positive anomaly is shown by the Castillo granite (Eguiluz 1988). The Valera porphyry is made of microgranites and porphyries spatially related to the Arroyomolinos spilites and shows a marked Zr positive anomaly as well.

4.2.5.3 Lower to Middle Ordovician Plutonism in Northern OMZ

A number of alkali-granites are emplaced throughout this time almost exclusively in the northern part of the OMZ, although some, clearly different, appear also in the south at the same time. Apart from other minor areas, main outcrops are those of Las Minillas, Ribera del Fresno, Aceuchal and Almendralejo (Figs. 1, 2). They are all elongated parallel to Hercynian structures NW–SE. Rocks of the same age and characters exist also in Portugal, as in the Portalegre region (Pinto 1984; Carvalhosa 1970).

Rocks consist of two feldspars, biotite and/or alkaline amphibole (riebeckite), quartz and accessories in variable amounts. They are also characterized by a well-developed plano-linear fabric, whose mylonitic textures are considered to be related to Hercynian deformation. (Chacón 1974; García Casquero et al. 1985). These same authors also argue that plutonic rocks have been affected by two high-grade regional metamorphic phases related to two S_1 and S_2 cleavages. Thus, following their suggestions, all regional metamorphism in the northern OMZ should be Hercynian.

Table 5. Burguillos del Cerro complex[a]

	1		2		3		4		5		6	
	Avr.	Std. Dev.	Avr.	Std. Dev.	Avr.	Std. Dev.	Avr.	Std. Dev.	Avr.	Std. Dev.	Avr.	Std. Dev.
SiO_2	46.77	2.59	44.85	0.89	45.33	2.29	45.43	4.49	47.64	2.16	57.52	4.69
Al_2O_3	9.60	2.88	18.31	3.46	16.90	0.84	17.89	1.81	20.12	1.77	16.77	0.96
Fe_2O_3	1.51	0.71	4.81	3.14	5.60	3.58	8.18	5.89	6.80	3.54	4.27	3.12
FeO	7.45	0.80	6.46	0.65	6.42	0.55	6.51	1.24	5.40	0.87	4.56	2.27
MgO	19.95	5.59	8.93	1.63	8.71	0.60	6.19	0.82	3.93	0.78	3.33	1.40
CaO	8.69	1.74	14.23	0.95	13.35	1.31	11.23	1.00	9.01	1.02	6.08	1.61
Na_2O	1.57	0.52	1.56	0.37	1.89	0.42	2.47	0.58	3.45	0.37	3.98	0.43
K_2O	0.64	0.23	0.39	0.13	0.76	0.17	0.97	0.34	1.78	0.58	2.71	0.90
TiO_2	0.55	0.19	0.90	0.26	1.27	0.33	1.77	0.66	1.92	0.50	1.14	0.39
MnO	0.16	0.01	0.15	0.03	0.14	0.02	0.13	0.02	0.09	0.02	0.11	0.03
P_2O_5	0.18	0.06	0.19	0.08	0.29	0.16	0.47	0.38	0.68	0.24	0.28	0.10

[a] *1* Cumulates (n = 4); *2* olivinic gabbros (n = 9); *3* "Ojo de Culebra" gabbros (n = 13); *4* blues gabbros (n = 13); *5* diorites (n = 15); and *6* marginal unity (n = 14). Data source: J. Pons (1982)

There is no mineralogical evidence, however, of a regional metamorphism affecting these plutonic rocks, and a polyphasic deformation is also doubtful (see Chacón 1979). We consider these rocks to be Lower-Palaeozoic intrusive in previously metamorphosed Precambrian rocks.

Rb/Sr available data (García Casquero et al. 1985) range from 423 and 470 m. y., with an unusually high initial Sr_{87}/Sr_{86} ratio, 0.720–0.722. Apart from inherent age uncertainties, this proposed range may agree with other geological data.

Mineralogical evidence, such as frequent appearance of riebeckite, also suggests here an acid-alkaline anorogenic plutonism related to a tensional tectonic environment. It is to be emphasized that traces of a similar pre-Hercynian magmatism are found elsewhere to the NW of the OMZ (e. g. Trasos-Montes and Malpica-Tuy Band, Ribeiro 1987; Gil Ibarguchi and Ortega 1985).

4.2.5.4 Pre-Hercynian to Early Hercynian Plutonism in Southern OMZ Domains

In contrast with relatively numerous radiometric data for Hercynian and Caledonian granites in northern OMZ (Pinto 1984), the same data are scarce for many plutonic rocks from the Monesterio-Olivenza alignment to the south. Among these, basic rocks are abundant (Dupont et al. 1981). Best examples of these plutons are those of Evora, S. Manços, Reguengos, Burguillos del Cerro, Valencia del Ventoso, Brovales, Salvatierra de los Barros and Jerez de los Caballeros (Figs. 1, 2).

Available references dealing with these rocks are of unequal significance. Most complete regional works are those of Carvalhosa (1970, 1983) in Portugal, and Bard and Fabries (1970) and Pons (1982) in the Spanish OMZ. Besides them, Pinto (1984) gives a good critical summary of radiometric data.

Although many of these plutons have been firstly described as granites, Pons (1982) has pointed out the heterogeneous, zoned character of most of them. Despite this heterogeneity, chemical data (Table 5) of basic rocks are markedly similar for several intrusions (Dupont 1979; Pons 1982). Dominant petrographic varieties are biotite-amphibole diorites; olivine gabbros and granitoides appear in a lesser amount. In zoned plutons, cores are in general more basic than shells.

Pons (1982) suggests that the Burguillos del Cerro, Valencia del Ventoso, Brovales and Salvatierra plutons have been formed respectively by fractional crystallization of basaltic, dioritic, granodioritic and granitic magmas. The last three magma types would be originated from the first through fractional crystallization in several steps. Gechemical data indicate that initial basaltic magma would be intraplate tholeiitic (Pons 1982), with a calc-alkaline evolution trend (Fig. 9).

K/Ar radiometric data of these same plutons, ranging from 423 ± 30 m. y. for Burguillos del Cerro to 305 ± 10 m. y. for Brovales (Dupont et al. 1981), make it difficult to consider magmatism to be rigorously Hercynian. This is why we have preferred to group all of them in this more ambiguous way.

It must also be remarked that, despite the detailed geochemical data existing for some plutons (Pons 1982), many uncertainties still remain con-

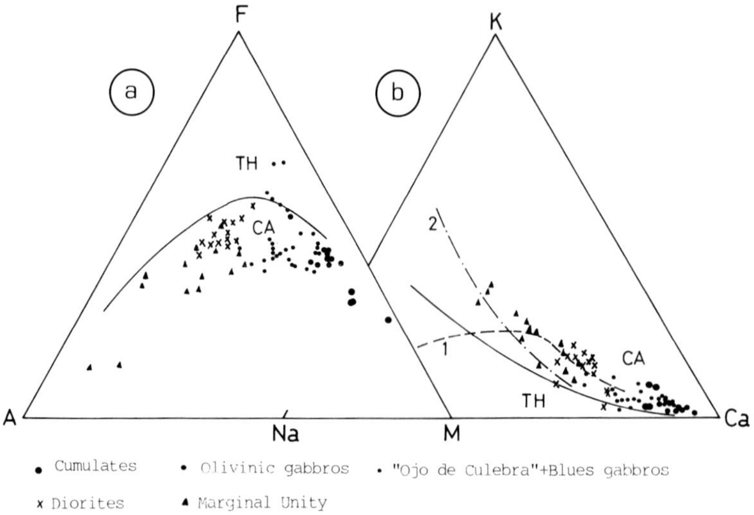

Fig. 9. Burguilos del Cerro Complex. **a** AFM diagram; **b** Na-K-Ca diagram (*1* trondhjemitic trend; *2* calc-alkaline trend; Barker and Arth 1976; data source, Pons 1982)

cerning magma types and tectonic environment. It is possible that in such a wide time span more than one magmatic type would have been successively generated, each one having a particular significance. Present classification has to be considered, therefore, as a provisional approach to a complex, poorly known matter.

4.3 Hercynian Magmatism

4.3.1 Localization and Subdivision

The different igneous materials generated between Late Devonian to Lower Carboniferous and Lower Permian (345–280 m. y.) are clustered here.

The different magmatic episodes may be structured as follows, when related to the Hercynian deformation phases, although some uncertainties relating to its age all over the Ossa-Morena Zone (OMZ) still remain:

1. First episode: Upper Tournaisian to Middle Viséan volcanics (pre- to syn-DI);
2. Second episode: basic to intermediate plutonic activity of Upper Viséan to Namurian age (post-DI);
3. Third episode: Middle Westphalian (syn- to late-DII);
4. Fourth episode: Stephanian-Lower Permian (post-DII).

Either the lavic or the intrusive rocks outcrop distributions are closely related to the position shown by Carboniferous basins, following NW–SE bands or alignments (Bard and Fabries 1970; Bard 1971; Capdevila et al. 1973; Aparicio et al. 1977; Gabaldón et al. 1985; etc.).

The available geochronologic data (Mendes 1968; Penha and Arribas 1974; Prost-Dame 1980; Bellon et al. 1979; Dupont et al. 1981) and the micro-paleontological dating resulting for those materials interbedded with lava flows at different Carboniferous basins (Garrote and Broutin 1979; Pérez Lorente 1979; Armengot and Martínez 1972; Quesada 1985; etc.), as well as the cross-cut relations deduced from cartography (Pascual and Pérez Lorente 1975; Garrote and Sánchez Carretero 1983; Andrade 1983; Simancas 1983; Delgado Quesada et al. 1985), let us establish a stepped magmatic sequence around the previously signaled episodes: Upper Tournaisian-Middle Viséan (345–335 m. y.), Namurian (325–320 m. y.), Middle Westphalian (305 m. y.) and Stephanian-Lower Permian (290–280 m. y.).

From NE to SW the magmatic alignments show the following disposition (Figs. 1, 2):

1. Igneous activity related to the Valle de los Pedroches basin:
   a) Los Pedroches plutonic alignment;
   b) Varas-Guadalbarbo volcanic alignment.
2. Volcanic activity related to the Guadiato River basin.
3. Magmatism related to the Benajarafe-Matachel basin:
   a) Villaviciosa de Córdoba-La Coronada magmatic alignment.
4. Volcanic activity related to the Bienvenida-Los Santos de Maimona-Valverde de Leganés basins.
5. Beja Complex.

The absence of accurate radiometric ages makes it difficult to establish a specific and particular treatment for the different igneous bodies as related to the previously established episodes. Nevertheless, it may be pointed out that volcanic activity precedes plutonism, being the first pre-(locally syn)-DI, whereas plutonism is post-DI.

4.3.2 Pre-Di Volcanic Activity

The different products referred to this first OMZ Hercynian igneous event are widespread at a group of Carboniferous basins which, from NE to SW, are the following:

Los Pedroches Basin
The Varas-Guadalbarbo volcanic alignment occupies the bottom of the Los Pedroches Culm sedimentary sequence, and is composed of both acidic and basic volcanics (Pérez Lorente 1979; Crousilles and Dixsaut 1977; Figs. 1, 2).

Basic rocks are basalts showing spilitic characteristics (albite-oligoclase, pyroxene, chlorite, calcite, epidote, prehnite, opaques, etc.) interbedded with shale-lutitic materials, in the form of pillow-lava flows, as well as with reworked epiclastics. Textures vary from microlitic to subophitic. Major element geochemical study reveals a tholeiitic character for these rocks (Crousilles and Dixsaut 1977).

Laterally, mafic volcanics pass into hematite dust, towards the NW; and upwards, the sedimentary sequence, into acidic volcanics and pyroclastics (tuffs, welded tuffs and cinerites; Pérez Lorente 1979).

The radiometric (K/Ar) age for spilites averages 348 ± 18 and 334 ± 16 m. y. (Crousilles and Dixsaut 1977).

Guadiato Basin
This is a punctual and less developed volcanism. It is composed of both spilitic basalts and rhyolites. Mafic terms, showing pillow structures, are located at the basin's southern band (Alhondiguilla Culm, Middle-Upper Viséan, Pérez Lorente 1979).

Rhyolites, with small thickness, are located at the central band (clastic-carbonate unit, Upper Viséan-Namurian), also building necks or plugs between Peñarroya and Fuenteobejuna.

Benajarafe-Matachel Basin
The igneous rocks filling this basin, mainly located between La Coronada and Villaviciosa de Córdoba, constitute a 70×9 km NW-SE alignment traditionally known as the Villaviciosa de Córdoba-La Coronada Magmatic Alignment (VC-LC) (Pascual and Pérez Lorente 1975; Garrote and Broutin 1979; Garrote and Sánchez Carretero 1983; Gabaldón et al. 1985; Delgado Quesada et al. 1985; Pascual 1985).

The different outcropping igneous products show a wide mineralogic textural and outcrop pattern variability, giving rise to the establishment of a group of units or complexes: Los Ojuelos Complex (Delgado Quesada 1971; Pascual and Pérez Lorente 1975), Alcornocal Complex (Delgado Quesada 1971), Erillas Volcanic Complex; La Campana Volcanic Complex (Garrote and Sánchez Carretero 1983: Delgado Quesada et al. 1985).

The VC-LC Magmatic Alignment building is the result of four stages of activity cycles superposed during a time span and clearly differentiated, both from a petrogenetic or emplacement point of view (Delgado Quesada et al. 1985).

The chronologic order shown by them is:

1. Pre-DI volcanic activity:
   a) La Campana and Erillas Volcanic Complexes.
2. Post-DI plutonic activity:
   a) Mafic to intermediate plutonism (Ojuelos Complex);
   b) Acidic shallow plutonism;
   c) Mafic subvolcanism.

La Campana Volcanic Complex. This represents the primary igneous activity within the Alignment, and is composed of volcanic rocks (andesites and dacites), subvolcanics (quartz-monzodioritic porphyries and granodiorites) and volcanoclastics. Its outcrops are localized along the eastern half of the alignment (Fig. 1). Volcanic rocks lie concordantly over Carboniferous slaty materials, and discordantly on Cambrian. Laterally, they pass into or are interbedded with volcanoclastics (epiclastics), formed because of their own mechanical removal. This fact, together with the presence of sedimentary structures proper of shores and beaches within the epiclastic materials, let us suppose that volcanic buildings were partly merged (Garrote and Sánchez Carretero 1983; Gabaldón et al. 1985).

Andesites show porphyritic and/or microlitic, locally flow, textures with a microcrystalline to vitreous matrix. They are composed of zoned plagioclase phenocrystals (andesine and labrodorite, An_{62} at the cores), hornblende and augite, as well as of K-feldspar (in the matrix); quartz, apatite, sphene and opaques are accessory minerals.

Dacites show reduced outcrops, approx. 4 km^2 in width, and are porphyritic rocks composed of quartz, plagioclase (oligoclase), hornblende and biotite within a fine-grained quartz-feldspathic matrix.

Quartz-dioritic porphyries show diffuse contacts to the previously described rocks. They are made of oligoclase-andesine, either as phenocrystals or within the matrix, chloritized green-brown hornblende phenocrystals, uralitized augite and scarce biotite, plus quartz, K-feldspar, apatite, opaques and zircon.

The Buenagua granodiorite and granodioritic porphyries are inequigranular rocks. They consist of 23% quartz, 43% zoned plagioclase (An_{54-11}), K-feldspar (12%), biotite (7%), hornblende (6%), magnetite (1%), apatite (0.1%), epidote (3%) and chlorite (5%). They form a subvolcanic intrusive with sometimes sharp contacts with porphyries. Its assignment to the present magmatic episode is a matter of controversy (see Pascual and Pérez Lorente 1975; Garrote and Sánchez Carretero 1983; Delgado Quesada et al. 1985; Pascual 1985; Pacual and Pérez Lorente 1987; Sánchez Carretero et al. 1989 b).

Chemical features, indicated in Fig. 10 a, b, c, d, and Table 6 reveal that this is an igneous series made of acid andesites and relatively K-rich dacites/granodiorites (58.73–67.47% SiO_2, 1.82–4.10% K_2O), with calc-alkalic to potassic-calc-alkaline characteristics (Sánchez Carretero et al. 1989 b).

Erillas Volcanic Complex. The Erillas Volcanic Complex crops out at the VC-LC central part. It consists of volcanic rocks (rhyolites and rhyodacites) interbedded with sedimentary materials (shales), volcanoclastic rocks and tuffs. A great deal of dyke swarms intrude the lava flows as well as their host materials.

Lava flows overlap either the shales, the La Campana Complex materials or the Precambrian rocks. The actual thickness of the volcanic complex reaches more than 350 m. They are effusive rocks, with microcrystalline, porphyritic, orbicular, flow, aglomeratic, welded, etc. textures, being quite widespread hematite traces at small beds and/or stock-work-forming veins. These rocks show great evi-

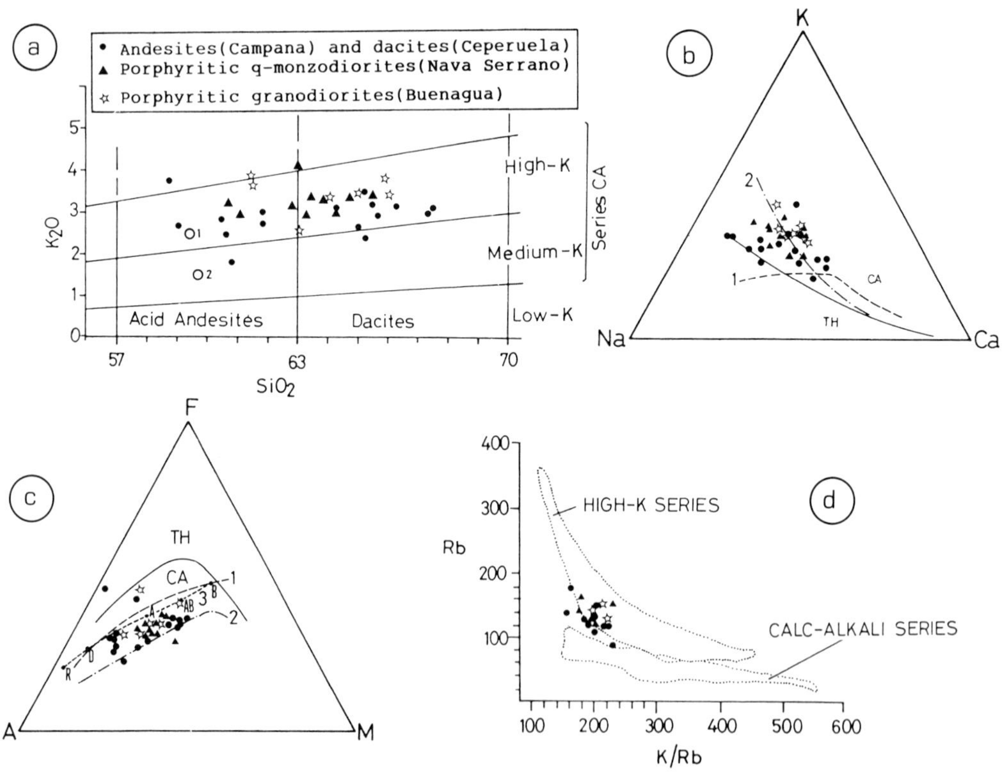

Fig. 10. La Campana volcanic complex (Hercynian orogenic volcanism). **a** SiO_2 vs K_2O classification diagram (*1* average of orogenic acid andesites high-K; *2* average of orogenic acid andesites medium-K, Gill 1981 Table 5.1, p. 99); **b** ternary K-Na-Ca diagram (*1* trondhjemitic trend; *2* calc-alkaline trend; Barker and Arth 1976); **c** AFM diagram; tholeiitic-calc-alkaline fields (after Irvine and Baragar 1971) (*1* trondhjemitic trend; Barker and Arth 1976; *2* Daly trend; *3* cascade trend; *B* Basalt, *AB* Basaltic andesite, *A* Andesite, *D* Dacite and *R* rhyolite); **d** Rb vs K/Rb diagram in relation to the subdivisions of Ewart (1979); (After Sánchez Carretero et al. 1989 b)

Table 6. La Campana volcanic complex

	Andesites (n=7)		Dacites (n=9)		Porph. N. Serrano (n=10)		Granodiorite (n=7)	
	Average	Std. Dev.	Average	Std. Dev.	Average	Std. Dev.	Average	Std. Dev.
SiO_2	60.48	1.13	65.76	1.01	63.27	1.41	63.81	1.79
Al_2O_3	15.12	1.32	15.07	0.78	14.72	1.21	14.98	1.54
Fe_2O_3	5.33	3.11	2.99	2.01	4.12	2.48	4.65	2.83
FeO	4.77	0.20	3.27	0.43	3.57	0.28	1.67	2.18
MgO	4.51	1.10	2.08	0.90	3.85	0.61	3.49	0.60
CaO	3.37	1.25	2.56	0.75	2.69	0.71	2.93	0.61
Na_2O	3.18	0.56	4.40	0.65	3.63	0.49	3.36	0.46
K_2O	2.76	0.54	3.01	0.31	3.27	0.33	3.45	0.41
TiO_2	0.60	0.10	0.44	0.13	0.66	0.07	0.70	0.05
MnO	0.14	0.04	0.08	0.02	0.11	0.02	0.12	0.05

dence of hydrothermal alteration processes (silicification, sericitization). The characteristics shown by these mineral deposits permit the Lahn Dill-type deposits (Baeza Rojano et al. 1978) to be ascribed to them. Moreover, there exist stratiform polymetallic sulphide deposits, potentially workable, to the acid volcanism top and to the bottom of hybrid tuffs (Baeza Rojano et al. 1981).

The age of the interbedded slaty materials, characterized by spores, is Lower Carboniferous (Garrote and Broutin 1979). Later datings (Broutin, Coquel, in Gabaldón et al. 1985) yield ages ranging between Upper Tournaisian and Upper Viséan, that is pre-(syn)-DI.

Volcanism Related to the Bienvenida, Los Santos de Maimona and Valverde de Leganés Basins
Within these small basins, both basic and acid volcanics are interbedded with detritic and carbonate deposits. They overlap unconformably the Córdoba-Alanís Domain of Lower Cambrian age.

They consist of lava flows and subvolcanic intrusives (sills and/or dykes) made of basaltic andesites, dacites, rhyolites and tuffs. In a simplified way, basic rocks are located towards the lower stratigraphic terms, while acid rocks predominate towards the sequence top.

The age of these basins, mainly that of the Los Santos de Maimona basin, would be between Dinantian (Wagner, pers. comm., 1985) and Upper Viséan (Martínez, pers. comm., 1985).

The characteristics just signaled for this volcanism allow it to be related with the volcanism present to the (Pedroches, Benajarafe). To the north-west, near Olivenza, acid volcanic rocks, of unknown age, crop out, being correlated on the basis of their lithology and structural position (possibly a prolongation of the Los Santos de Maimona basin) with a probable Carboniferous age (Eguiluz et al., in litt.).

4.3.3 Post-Di Plutonism and Hipo-Abyssal Rocks

These rocks, as well as volcanics, crop out favoured by NW-SE alignments, being the main outcrops localized to the northeast of the OMZ (Los Pedroches Batholith and VC-LC Magmatic Alignment) and to the SW boundary (Beja Massif).

Los Pedroches Batholith
It crops out along 200 km, following the NW-SE Hercynian trends, between the Guadalquivir river basin to the SE and the Guadiana river to the NW. Its width ranges between 8–25 km (Fig. 1).

Although it has traditionally been regarded as a strictly plutonic massif, the presence of either andesites, epiclastic andesites, rhyolites or acid tuffs at the NW region (Zalamea de la Serena, Eguiguren 1983; Arriola et al. 1985), overlapping the granitoids and clearly affected by contact metamorphism, reveals the existence of a previous volcanic stage during this alignment structuration. The age of this volcanic activity, on the basis of its lithological similarity with rocks located much more southwards (VC-LC Magmatic Alignment, Garrote and Sánchez Carretero 1983), could be Pre-Middle Viséan and, thus, the intrusive rocks making up the batholith should be the same age or younger. These relations are also evident at the central and southern sectors, where granitic rocks metamorphose the previously deformed, during Hercynian DI, Lower Carboniferous.

Much of the outcrop surface (60%) corresponds to a granodioritic rock (Eraso and Garrote 1984; Garrote et al. 1985; Sánchez Pérez 1986) of a 5–10-mm grain size, locally porphyritic, and composed of quartz, plagioclase, K-feldspar, biotite and seldom hornblende, plus apatite, zircon, rutile and scarce opaques. Micro-equigranular enclaves of either tonalitic, quartz-dioritic or dioritic composition are widespread, as well as enclaves of host rocks.

Fine- and coarse-grained porphyritic granites occupy wide areas, mainly to the southeast of the batholith, and consist of quartz, plagioclase (An_{19-23}) K-feldspar, biotite, minor amounts (locally significant) of apatite, zircon and cordierite, plus sphene and opaques. Muscovite, although widespread, always shows a deuteric origin.

There are also leucogranites composed of quartz, porphyritic and micropertitic K-feldspar, albite, muscovite and variable amounts of apatite, tourmaline, cordierite and andalusite. The two latest may attain high proportions. Their outcrops are closely related to porphyritic granites.

At the northwest limit of the batholith, small bodies made up of very altered gabbros, pyroxene relicts-bearing diorites and orthopyroxene-bearing granodiorites crop up (Arriola et al. 1985).

At the central-eastern region of the batholith an important dyke swarm appears. Two main groups are distinguished: one of N160E to N–S strike, and other of N130–140E. The first group, of quartz-monzodioritic composition, is chemically and mineralogically closely related to the granodiorite and never cross-cuts the granites, while the second group, of granitic composition, is chemically, petrographically and spatially related to granites, with which they show intrusive contacts, also intruded are both the granodiorite and the N–S dyke swarm.

The available chemical data show the presence of an incipient compositional gap between the granodiorite, the enclaves and, probably, with the mafic rocks as a first term, and the granites, granitic dykes and leucogranites as a second term (Table 7). The above signaled difference, previously outlined by Prost-Dame (1980) and Pascual (1984), has been specified by the assignment of an alumino-cafemic trend for granodiorites and of an aluminic and/or alumino-cafemic for granites (Sánchez Pérez 1986; Larrea 1987; García Casco et al. 1987 a, b). The latter authors, on the basis of recently published geochemical studies (Pitcher 1982; Batchelor and Bowden 1985), define the alumino-cafemic series as post-collision I-type, and the aluminic as late-orogenic S-type (Fig. 11 A, B).

Los Pedroches Batholith intrudes culm slaty materials at its central-eastern region, and older Palaeozoic materials at its western border. Contact metamorphism reveals an aureole of 1000–1500 maximum extent, varying areally, with albite-epidote, hornblende and pyroxene metamorphic facies materials, the latter in the closest sectors of the batholith-host external rocks boundary.

Pyroxene-facies metamorphism exists, related both to granodiorite as well as to granites, thus implying a high intrusion temperature, necessary to account for the magma rise towards the epizonal levels where the host rocks are located. As a consequence of these data (Sánchez Pérez 1986; García Casco 1986; García Casco and Pascual 1987) temperatures higher than 650 °C at rock pressures no more than 1 kbar should be suggested for the batholith emplacement (García Casco and Pascual 1987).

Radiometric ages, as well as petrographic and field data reveal a post-DI emplacement (Upper Viséan, Pérez Lorente 1979) for granodioritic intrusives, while granites are syn- to post-DII, as is established below (Eraso and Garrote 1984):

1. Granodiorite emplacement: Upper Viséan (approx. 330 m. y.);
2. Porphyritic granite: Middle Westphalian (approx. 305 m. y.).

Table 7. Los Pedroches batholith

	Leucogranites (n = 19)		Monzogranites (n = 22)		Granodiorites (n = 22)		Enclaves (n = 3)	
	Average	Std. Dev.	Average	Std. Dev.	Average	Std. Dev.	Average	Std. Dev.
SiO_2	74.46	0.98	71.67	1.54	69.11	3.023	64.54	0.25
Al_2O_3	14.38	0.61	15.29	0.89	14.97	1.19	16.29	0.41
Fe_2O_3	0.22	0.19	0.38	0.18	2.74	0.80	1.68	0.06
FeO	0.68	0.29	1.59	0.25	1.70	0.60	5.33	0.00
MgO	0.11	0.05	0.36	0.13	1.34	0.64	1.38	0.06
CaO	0.23	0.05	0.82	0.55	2.27	0.77	1.77	0.00
Na_2O	3.72	0.37	3.44	0.35	3.60	0.27	3.34	0.50
K_2O	5.65	0.77	5.76	0.68	3.89	0.83	5.22	0.37
TiO_2	0.13	0.06	0.33	0.08	0.41	0.20	1.04	0.16
MnO	0.02	0.01	0.04	0.02	0.27	1.03	0.09	0.18
P_2O_5	0.14	0.06	0.15	0.04	0.21	0.05	0.35	0.04

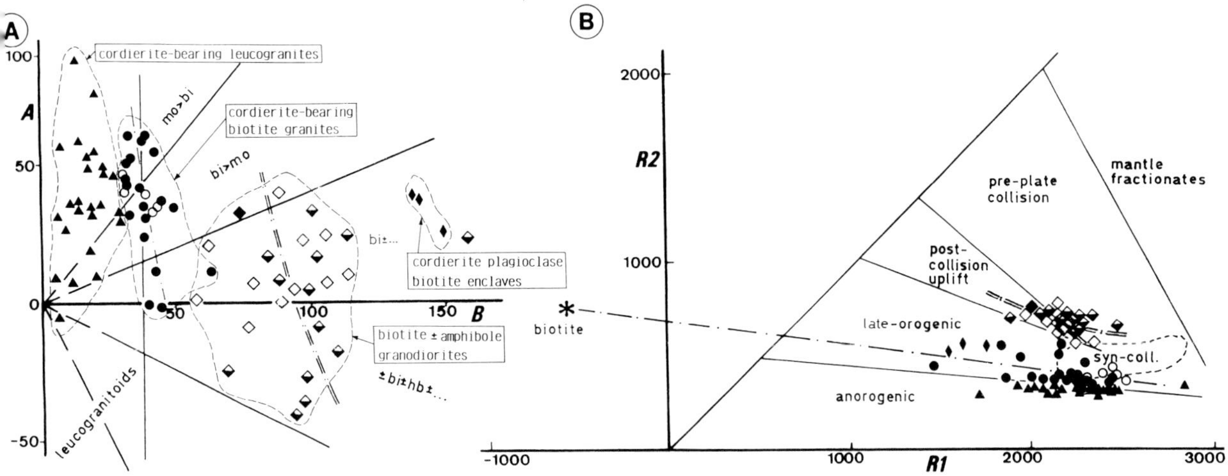

Fig. 11. Multicationic diagrams for plutonic rocks from Los Pedroches batholith. **A** A/B diagram (Debon and Le Fort 1983); **B** R_1/R_2 diagram (De La Roche and Leterrier 1973; Batchelor and Bowden 1985). *Triangles* cordierite-bearing aplitic leucogranites; *circles* cordierite-bearing porphyritic biotite mozogranites; *squares* biotite ± amphibole granodiorites; *rhombs* biotite, plagioclase, cordierite enclaves. Data source: García Casco et al. 1987 a *(filled symbols)*; Prost-Dame 1980 *(open symbols)*; Sánchez Pérez 1986 *(partly filled symbols)*

VC-LC Magmatic Alignment

Los Ojuelos Basic-Intermediate Intrusive Complex. First defined by Delgado Quesada (1971) to include rocks ranging from mafics to granites, this complex was later redefined in order to include only basic rocks (gabbros and diorites) (Pascual and Pérez Lorente 1975; Burgos and Pascual 1976; Pascual 1981). Later studies (Delgado Quesada et al. 1985; Sánchez Carretero et al., 1989 b) reveal the acidic character shown by some of their terms (granodiorites).

These rocks are located within three major massifs and at small masses all over the VC-LC Alignment.

The different petrographic types forming these intrusives are: olivinic gabbros, gabbro-norites, gabbros, porphyritic gabbros, diorites, quartz-diorites, monzodiorites, quartz-monzodiorites, granodiorites and adamellites. Although field and cartographic relations between them are complex, small plagioclase-rich olivinic gabbros, with cumulate characteristics, may be differentiated inside the more common gabbros, showing apparently intrusive contacts. These rocks could be interpreted as being early cumulates carried as megaenclaves by the main gabbroic mass (Pascual 1981).

Gabbronoritic rocks are very scarce and show characteristics similar to those of the common gabbros.

Clinopyroxene-bearing gabbros constitute the bulk of the outcrop area, presenting intergranular to ophitic, medium-grained textures. Some edge zone masses (N of Villaviciosa de Córdoba) show frequent cataclastic textures.

Table 8. Los Ojuelos intrusive complex

	Olivinic Gabbros (n=4)		Porph. Gabbros (n=4)		Gabbros and Diorites (n=15)	
	Average	Std. Dev.	Average	Std. Dev.	Average	Std. Dev.
SiO_2	55.29	2.07	56.06	1.03	55.76	4.51
Al_2O_3	17.47	1.92	19.02	1.18	15.43	1.77
Fe_2O_3	1.86	1.31	0.98	0.24	4.35	4.52
FeO	4.42	0.71	5.19	0.72	6.95	1.26
MgO	7.66	1.31	6.28	0.63	5.26	1.58
CaO	8.86	0.50	6.19	0.67	6.17	2.26
Na_2O	2.34	0.55	2.41	0.35	3.86	0.72
K_2O	0.30	0.12	1.22	0.65	1.22	0.41
TiO_2	0.53	0.12	0.61	0.13	1.54	0.40
MnO	0.11	0.01	0.11	0.02	0.15	0.04

The major element geochemistry of these rocks (Table 8) shows tholeiitic affinities (Fig. 12 a, b), which are inferred from the relatively plane REE spectra as well (Sánchez Carretero et al. 1989 b).

Thus, basic massifs seem to have been built up at two stages. The first of them would correspond to a fractioned crystallization of a basic magma at depth, followed by a shallow emplacement with fast cooling of a dioritic magma. Other hypotheses, implying subsolid conditions, do not seem to be probable, due to the structural relationships shown

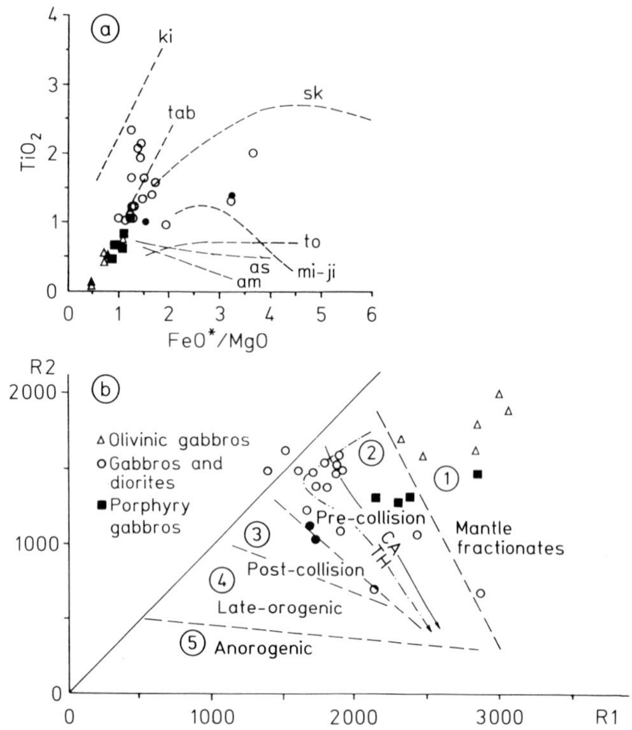

Fig. 12. Los Ojuelos intrusive complex (Hercynian orogenic plutonism). **a** TiO_2 vs FeO*/MgO diagram (*ki* Kilauea, *tab* abyssal tholeiites, *sk* Skaergaard, *mi-ji* Miyake-jima, *to* Tofua, *as* Asama, and *am* Amagi, after Miyashiro 1975). **b** R_1/R_2 diagram (De La Roche and Leterrier 1973; Batchelor and Bowden 1985); (After Sánchez Carretero et al. 1989 b)

by these rocks, being those reduced to the Villaviciosa pluton cataclastic textures.

Acid Shallow Plutonism. This consists of granites whose main characteristics are: shallow emplacement, scarce melanocratic minerals (biotite and locally amphibole), fine-grained porphyritic and granophyric varieties, fluorine and barium deposits, contact metamorphism with occasional pyroxenic hornfelses (Delgado Quesada et al. 1978), micropertitic textures, fluorite and barite of primary origin (Pascual 1985).

These granites cross-cut all the preceding rocks, giving rise, mainly when affecting gabbros, to stopping, silicification, potassification and local epidotization phenomena.

These are low-CaO and relatively alkali-rich granites (Table 9), which gives them an alkaline affinity (Fig. 13 a) close to A-type granites (Loiselle and Wones 1979). The late-kinematic character of the emplacement of these granites is likewise supported by their chemical features (lateorogenic-anorogenic granites, Fig. 13 b). In the same way, alkal-

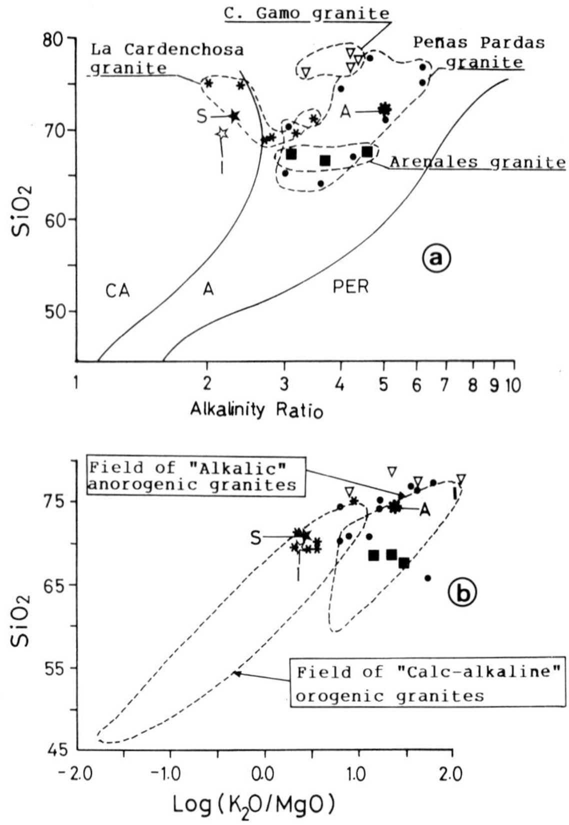

Fig. 13. Acid shallow plutonism (post-anorogenic Hercynian granites). **a** Alkalinity ratio (Al_2O_3 + CaO + total alkalis)/(Al_2O_3 + CaO-total alkalis) vs SiO_2. *CA* Calc-alkaline, *A* alkaline and *PER* peralkaline (boundaries between fields are taken from Wright 1969). Data source: La Cardenchosa granite (Garrote and Sánchez Carretero 1979), Arenales granite (Pascual 1981), Peñas Pardas and Cañada del Gamo granites (Sánchez Carretero et al. 1989 b). *S* Average of S-type granite; *I* average of I-type granite, and *A* average of A-type granite (data source Whalen et al. 1987). **b** Log (K_2O/MgO) vs SiO_2 plot for sampled granites. Fields of calc-alkaline orogenic and alkalic anorogenic granites are taken from Rogers and Greenberg (1981)

Table 9. Acid shallow plutonism (granites, n = 25)

	Average	Std. Dev.
SiO_2	72.59	3.62
Al_2O_3	13.39	1.01
Fe_2O_3	2.22	1.58
FeO	1.69	0.99
MgO	0.73	0.80
CaO	1.05	1.34
Na_2O	3.47	1.06
K_2O	4.37	1.06
TiO_2	0.31	0.24
MnO	0.02	0.01

ic minerals such as hastingsite, have been pointed out (Pascual 1985).

The available radiometric ages (K/Ar) from Los Arenales granite (Fig. 1) which yielded 316± 16 m.y. (conspicuous variety) and 332±17 m.y. (myarolitic variety) (Bellon et al. 1979) locate its intrusion at Upper Viséan-Namurian times, thus post-DI. Such results are in agreement with field relationships, as well as with chemical and mineralogical data (barite and fluorite between accessory minerals) relating to post-kinematic alkaline granites, according to Marmo (1971) or of A-type granites as pointed out by Pitcher (1982). This typification also agrees with the character of the associated basic rocks and the type of mineral deposits (F-veins and Sn-bearing skarns).

Basic Subvolcanism. This is characterized by the intrusion, under subvolcanic conditions, of basic rocks. Its relation with the rocks described above is as follows: they cross-cut the rocks of the Los Ojuelos Complex and of the acid-shallow plutonism. They were emplaced under anorogenic conditions, showing a slight alkaline affinity.

Beja Complex

At the Portuguese area of the OMZ, and immediately to one side of the Ferreira-Ficalho accident limit between the OMZ and the South-Portuguese Zone (SPZ), a wide plutonic massif composed of different basic rocks and minor amounts of acidic porphyries crops out. The igneous products forming the complex have been divided into several units (Silva et al. 1970; Carvalho et al. 1971; Andrade and Gama Pereira 1980; Andrade 1972, 1977, 1979, 1983). According to Andrade (1983), the Beja Complex is made up of two units, from top to bottom: the Cupa Group and the Ferreira Group.

The Ferreira Group is a basic-ultramafic association formed by two subgroups: Monbeja and Odivelas (Andrade 1983). The first is integrated by peridotites, gabbros and basalts, disturbed by important tectonometamorphic changes. The Odivelas subgroup is mainly a basic plutonovolcanic assemblage, also containing subordinate ultramafic and acidic terms. Plutonic rocks are banded gabbros, with cumulate textures, and a great deal of petrographic types (olivinic gabbros, gabbros, anorthosites, norites troctolites, etc.). Gradually but quickly, this unit passes into an upper group of hornblende-bearing gabbros. The latter rocks consist of poikilitic amphibole-bearing dolerites, basaltic lava flows and basic pyroclastics.

Both groups, Odivelas and Mombeja, show tholeiitic character, although with alkalic and calc-alkaline variations (Andrade 1983).

The Cuba Group is formed by two subgroups: Alvito and Baleição. The first is mainly dioritic, although with a markedly petrographic variability (gabbros, diorites, granodiorites, etc.) and texture (equigranular to porphyritic). Chemically, they reveal calc-alkaline affinities. Baleição porphyries are subvolcanic granitic rocks with variable textures (microgranular, granophyric and porphyritic).

The Beja region plutonism has been correlated to that present at the Castilblanco-Campofrío region, from the South Portuguese Zone (SPZ). The proposed analogy is supported by the following criteria (Simancas 1983):

1. Cartographic inspection suggests that both complexes represent two fragments of the same unit, separated by a sinistral Late-Hercynian fault.
2. The global sequence is the same in both complexes: banded gabbros at the bottom, and hornblende-bearing gabbros at the top.
3. Similar petrographic and textural features.
4. Chemical resemblance, with a rather defined tholeiitic evolution.

Nevertheless, there are some problems, especially with the Odivelas Group volcanic products (mainly dykes) (Simancas 1983). If these materials are genetically related to gabbros, Castilblanco-Campofrío correlation would be problematic. Firstly, because the signaled association is a good reason to support the Andrade (1977, 1979, 1983) ophiolitic hypothesis. Secondly, the Middle to Upper Devonian age of this volcanism (Conde and Andrade 1974) disagrees with the Middle Westphalian age of Castilblanco gabbros.

In our opinion, the Ferreira Group lower part, that is, the Mombeja subgroup, constitutes a clearly differentiated unit which, because of its lithology, oceanic chemical character, metamorphism (see following chapter, this Vol.), and deformation, as well as because of its outcrop pattern, is perfectly correlated to the Acebuches amphibolites (Aracena Massif). The latter present a tholeiitic-abyssal character (Bard and Moine 1979), show ophiolitic features and have been correlated with those of Beja (Beja-Acebuches ophiolite, Munha et al. 1986). Thus, Andrade's (1977, 1979, 1983) ophiolitic hypothesis seems to acquire, day by day, a greater entity. The age of these rocks still remains doubtful (Silurian), but it ought to be unraveled from the resting massif, of Upper Devonian to Carboniferous age, which may be correlated to the Castilblanco-Campofrío Massif.

4.4 Post-Hercynian Magmatism

Post-Hercynian magmatism has received a different interpretation, according to the available radiometric data, although its presence is commonly accepted on the basis of geologic data. Here we consider post-Hercynian those rocks of post-Westphalian age, in order to avoid any discussion related to Late Hercynian tectonic phase dating.

According to this prerequisite, only few rocks exist within the OMZ: Feria-Sierra Vieja massif and Alconera dikes, Pavia and Tancos granites and Santa Eulalia granite.

Sierra Vieja Granite is a small stock 1×2 km wide. It is very homogeneous, consisting of alkaline feldspar megacrystals (sometimes quartz megacrystals) within a mainly quartz-feldspathic matrix also containing poikilitic arfvedsonite and aegirine. This is an hypersolvus granite with peralkalic chemical features, following Bonin's (1977) and Dupont and Bonin's (1981) nomenclature.

The Feria complex intrudes Lower Cambrian materials, and contains a number of host xenoliths, sometimes giving rise to continuous breccia bands at the intrusive contact. It consists of two rocks: alkalic syenites and albitites. The alkalic character is supported by two arguments:

1. Amphibole compositions (hastingsite and calcic-barroisite), as well as biotite compositions.
2. Total rock chemical compositions, strongly meta-aluminic and with variable CO_2 contents. Chemical data are also very similar to those of other syenites of ring complexes (Niger, Nigeria) (Dupont and Bonin 1981).

Alconera dyke swarms are considered to be related to the previously described syenitic rocks, to which they show obvious petrographic resemblance.

The rest of the cited outcrops (see Pinto 1984, for critical references) are likely post-Hercynian too, but their age is doubtful. Thus, a late- rather than post-tectonic character can not be disregarded. In any case, their alkalic character has also been signaled (Carvalhosa 1970).

As a result, it seems that the whole post-tectonic magmatism is alkaline, of a tensional setting and similar to that described at rifting zones (Pitcher's 1982 A-type granites, or Loiselle and Wones 1979).

References

Aguayo J (1985) Rocas ultramáficas en el sector de Calzadilla de los Barros (Badajoz). Thesis, Univ del País Vasco, 110 p

Andrade A (1972) Problemes géologiques et métallogéniques de l'Alentejo central (Portugal). Mem Not Univ Coimbra 81: 75–83

Andrade A (1977) The Beja Ophiolitic Complex in southern Portugal: a preliminary description. Mem Not Univ Coimbra 84: 49–57

Andrade A (1979) Aspectos geoquímicos do ofiolitoide de Beja. Comun Serv Geol Port 64: 39–48

Andrade A (1983) Contribution a l'analyse de la nature hercynienne de Beja (Portugal), perspectives metallogeniques. Thesis, Institut National Polytechnique de Lorraine, 137 p

Andrade A, Game Pereira LC (1980) Intrusões básicas no borde de Ossa-Morena: observaçoes. Temas Geol Min 4, pp 5–9

Apalategui O, Higueras P (1983) Mapa y memoria explicativa de la Hoja 1:50 000, N° 855 (Usagre) del Mapa Geológico Nacional (MAGNA). Inst Geol Min Esp (IGME) Madr

Apalategui O, Barranco E, Contreras FJ, Delgado Quesada M, Roldán FJ (1984) Mapa y memoria explicativa de la Hoja 1:50 000, N° 917 (Aracena) del Mapa Geológico Nacional (MAGNA). Inst Geol Min Esp (IGME) Madr

Apalategui O, Borrero J, Delgado Quesada M, Roldán FJ, Garrote A, Eguiluz L (1985) Mapa y memoria explicativa de la Hoja 1:50 000, N° 899 (Guadalcanal) del Mapa Geológico Nacional (MAGNA). Inst Geol Min Esp (IGME) Madr

Apalategui O, Jorquera A, Villalobos M, Eguiluz L Sánchez Carretero R (in litt a) La Zona de Ossa-Morena: Su relación con la Zona Centro Ibérica. Temas Geol Miner

Apalategui O, Jorquera A, Villalobos M, Eguiluz L (in litt b) Mapa y memoria explicativa de la Hoja 1:50 000, N° 801 (Olivenza) del Mapa Geológico Nacional (MAGNA). Inst Geol Min Esp (IGME), Madr

Aparicio A, Barrera JL, Casquet C, Peinado M, Tinao JM (1977) Caracterización geoquímica de plutonismo postmetamórfico de SW del Macizo Hespérico. Stud Geol 12: 9–39

Armengot J, Martínez C (1972) El Carbonífero al Sur del batolito de Los Pedroches. Rev Esp Micropaleontol Número extraordinario, pp 211–213

Arriola A, Eguiluz L (1983) Migmatitas de Lora del Rio (Sevilla): Estructurología y deformación. Comun Serv Geol Port 69 (2): 259–263

Arriola A, Chacón J, Eguiluz L, Eraso A, Garrote A, Sánchez Carretero R, Vargas I (1983 a) Mapa y memoria explicativa de la Hoja 1:550 000, N° 829 (Villafranca de los Barros) del Mapa Geológico Nacional (MAGNA). Inst Geol Min Esp (IGME) Madr

Arriola A, Cueto LA, Fernández Carrasco J, Garrote A (1983 b) Serpentinitas y mineralizaciones de cromo asociadas en el Proterozoico Superior de Ossa-Morena. Cuad Lab Xeol Laxe 8: 137–146

Arriola A, Garrote A, Eguiluz L, Sánchez Carretero R, Fernández Carrasco J (1983 c) Mapa y memoria explicativa de la Hoja 1:50 000, N° 876 (Fuente de Cantos) del Mapa Geológico Nacional (MAGNA). Inst Geol Min Esp (IGME) Madr

Arriola A, Eguiguren E, Garrote A (1985) El batolito de Los Pedroches en la transversal de Esparragosa de la Serena (Badajoz). V Reun GOM Temas Geol Min, pp 19–28

Baeza Rojano LJ, Ruíz García C, Ruiz Montes M (1978) Presencia de formaciones vulcano-sedimentarias y mineralizaciones de hierro asociadas en el eje magmático La Coronada-Villaviciosa de Córdoba. Bol Geol Min 89: 431-437

Baeza Rojano LJ, Ruíz García C, Ruíz Montes M, Sánchez A (1981) Mineralizaciones exhalativo-sedimentaria de sulfuros polimetálicos en la Sierra Morena Cordobesa (España). Bol Geol Min 93-3: 203-216

Bard JP (1969) Le métamorphisme régional progressif des Sierras d'Aracena en Andalousie Occidentale (Espagne). Sa place dans le segment hercynien Sud-Ibérique. Thesis, Univ Montpellier, 398 p

Bard JP (1971) Sur l'alternance des zones métamorphiques et granitiques dans le segment hercynien sud-ibérique; comparaison de la variabilité des caracteres géotectoniques de ces zones les orogènes "orthotectoniques". Bol Geol Min 82-3-4: 324-345

Bard JP, Fabries J (1970) Aperçu pétrographique et structural sur les granitoides de la Sierra Morena occidentale (Espagne). Bol Geol Min 81-2-3: 226-241

Bard JP, Moine B (1979) Acebuches amphibolites in the Aracena Hercynian metamorphic belt (southwest Spain): geochemical variations and basaltic affinities. Lithos 12: 271-282

Barker F, Arth JG (1976) Generation of trondhjemitic-tonalitic liquids and Archean bimodal trondhjemite-basalt suites. Geology (Boulder) 4: 596-600

Batchelor R, Bowden P (1985) Petrogenetic interpretations of granitoid rock series using multicationic parameters. Chem Geol 48: 43-55

Bea F, Sánchez González JG, Serrano Pinto M (1987) Una compilación geoquímica (elementos mayores) para los granitoides del Macizo Hespérico. In: Bea F, Carnicero A, Gonzalo JC, Lopez Plaza M, Rodriguez Alonso MD (eds) Geologia de los granitoides y rocas asociadas del Macizo Hespérico. Libro homenaje a L. C. García Figuerola. Editorial Rueda, Madrid, pp 87-193

Bellon H, Blachere H, Crousilles M (1979) Radiochronologie, évolution tectono-magmatique et implications métallogeniques dans les Cadomo-Variscides du Sud-Est Hesperique. Bull Soc Géol Fr 21: 113-120

Bonin B (1977) Les complexes granitiques subvolcaniques de Corse: caracteristiques, signification et origene. Bull Soc Géol Fr 19: 865-871

Burgos JC, Pascual E (1976) El stock básico del Norte de Villaviciosa de Córdoba (Complejo Los Ojuelos-La Coronada). Sierra Morena (España). Cuad Geol 7: 69-122

Capdevila R, Corretge G, Floor P (1973) Les granitoides Varisques de la Meseta Iberique. Bull Soc Géol Fr 15: 209-228

Carnicero A, Castro A (1982) El complejo básico de Barcarrota: su petrologia y su estructura. Bol Geol Min 93: 165-171

Carvalho D, Goingas J, Oliveira V, Ribeiro A (1971) Observaçoes sobre a geologia do Sur Portugal e consecuencias metalogéticas. Estud Not Trav Serv Fomento Min 20: 153-199

Carvalhosa BA (1970) Roches granitiques de l'Alentejo (Portugal). Bol Geol Min 81-2-3: 211-214

Carvalhosa BA (1983) Esquema geológico do maciço de Evora. Comun Serv Geol Port 69: 201-208

Chacón J (1974) Metamorfismo dinámico en la banda Badajoz-Córdoba: Los gneises milonitas de Higuera de Llerena (prov. de Badajoz). Bol Geol Min 84-5: 700-712

Chacón J (1979) Estudio geológico del sector central del anticlinorio Portalegre-Badaloz-Córdoba (Macizo Ibérico Meridional). Thesis, Univ Granada, 728 p

Chacón J, Velasco F (1981) Rocas ultrabásicas metamorfizadas en el Grupo de Azuaga (Anticlinorio Badajoz-Córdoba). Cuad Geol Ibérica 7: 151-159

Coleman (1977) Ophiolites ancient oceanic lithosphere? Springer, Berlin Heidelberg New York, 229 pp

Conde LEN, Andrade A (1974) Sur la faune méso et/ou néodevonienne des calcaires du Monte das Cortes, Odivelas (Massif de Beja). Mem Not Univ Coimbra 78: 144-146

Crousilles M, Dixsaut C (1977) L'association ophiolitique du Varas-Guadalbarbo (Cordoue, Espagne). Aspects structural et métallogénique. Thesis, Orsay, 289 pp

Cueto LA, Eguiluz L, Llamas FL, Quesada C (1983) La granodiorita de Pallares, un intrusivo Precámbrico en la alineación Olivenza-Monesterio (Zona de Ossa-Morena). Comun Serv Geol Port 69 (2): 219-226

Debon F, Fort P le (1983) A chemical-mineralogical classification of common plutonic rocks and associations. Trans Soc Edinb. Earth Sci 73: 135-149

Delgado Quesada M (1971) Esquema geológico de la hoja N° 878 de Azuaga (Badajoz). Bol Geol Min 82-3-4: 277-286

Delgado Quesada M, Pascual E, Fenoll Hach-Ali P (1978) A geological and metallogenic study of some occurrences of magnetite and sulphides in Sierra Morena (NNW Córdoba, Spain). Estud Geol 34: 461-474

Delgado Quesada M, Garrote A, Sánchez Carretero R (1985) El magmatismo de la alineación La Coronada-Villaviciosa de Córdoba en su mitad oriental (Zona de Ossa-Morena). Comunicaciones de la V Reunión del GOM (1983). Temas Geol Min 7: 41-64

Dupont R (1979) Cadre géologique et metallogènese des gisements de fer du province de Badajoz (Sierra Morena Occidentale, Espagne). Thesis, Institute National Polytechnique de Lorrain, 369 p

Dupont R, Bonin B (1981) Le massif alcalin de Feria-Sierra Vieja (Sierra Morena occidentale, Province de Badajoz, Espagne): un jalon du magmatisme anorogénique de Méditerranée occidentale. Bull Soc Géol Fr 5: 477-485

Dupont R, Linares E, Pons J (1981) Premières datations radiométriques par le méthode potassium-argon des granitoides de la Sierra Morena Occidentale (Province de Badajoz, Espagne): conséquences géologiques et métallogéniques. Bol Geol Min 92-5: 370-374

Eguiguren E (1983) El batolito de Los Pedroches en la transversal de Esparragosa de la Serena (Badajoz). Thesis, Univ del País Vasco, 76 p

Eguíluz L (1988) Petrogénesis de rocas ígneas y metamórficas en el antiforme de Burguillos-Monesterio. Macizo Ibérico Meridional. Thesis, Univ del País Vasco, 694 pp

Eguíluz L, Sánchez Carretero R, Apalategui O (in litt) Las rocas volcánicas de Valverde de Leganés (Anticlinorio Olivenza-Monesterio): Nota preliminar. Comunicación a la VII Reuniónd del Grupo Ossa Morena (GOM), 1985

Eraso A, Garrote A (1984) El batolito de Los Pedroches en la transversal de Cardeña, Córdoba. Cuad Lab Xeol Laxe 8: 235-256

Ewart A (1979) A review of the mineralogy and chemistry of Tertiary-Recent dacitic, latitic, rhyolitic, and related salic volcanic rocks. In: Barker F (ed) Trondhjemites, dacites and related rocks. Elsevier, Amsterdam, pp 13-121

Fabries J (1963) Les formations cristallines et metamorphiques du Nord Est de la province de Seville (Espagne). Essai sur le métamorphisme des roches èruptives basiques. Thesis, Univ Nancy, 267 p

Florido P, Quesada C (1984) Estado actual de conocimiento sobre el Macizo de Aracena. Cuad Lab Xeol Laxe 8: 257-277

Fricke W (1941) Die Geologie des Grenzgebietes zwischen nordöstlicher Sierra Morena und Extremadura. Thesis, Univ Berlin, 91 p

Gabaldón V, Garrote A, Quesada C (1985) Geología del Carbonífero inferior del Norte de la Zona de Ossa-Morena. Introducción a la Excursión de la V Reunión del Grupo de Ossa-Morena (GOM). V Reunión del GOM. Temas Geol Min, pp 101-113

Galindo C, Casquet C, Portugal Ferreira M, Regencio Macedo CA (1986) O complexo plutonico de Taliga-Barcarrota. Un complexo intrusivo con idades Caledonica e Hercinica. Abstract Maleo 2 (13): 22

García Casco A (1986) Petrologia, geoquímica y mineralizaciones de W asociadas del plutón Santa Eufemia (Batolito de Los Pedroches, Córdoba). Thesis, Univ Granada, 303 p

García Casco A, Pascual E (1987) El plutón de Santa Eufemia (batolito de los Pedroches, Córdoba, España): un granitoide epizonal de tipo S. Geogaceta 2: 56-59

García Casco A, Pascual E, Castro A (1987 a) La asociación magmática del batolito de los Pedroches: ensayo de caracterización. Geogaceta 2: 59-61

García Casco A, Pascual E, Fenoll P (1987 b) Petrology of a cordierite-bearing monzogranite and leucogranite pluto of "Los Pedroches" batholith, Hercynian Massif, Spain. Granites and their surroundings. Soc Ital Mineral Petrol, pp 100-101

García Casquero JL (1988) Trondhjemitas continentales en terrenos desplazados de eje Badajoz-Córdoba. II Congr Geol Esp Comun 2: 19-22

García Casquero JL, Boelrijk NAIM, Chacón J, Priem HNA (1985) Rb-Sr evidence for presence of Ordovician granites in the deformed basement of the Badajoz-Cordoba belt, SW Spain. Geol Rundschau 74-2: 379-384

Garrote A, Broutin J (1979) Le bassin Tournaisien de Benajarafe (province de Cordoue, Espagne). Géologie et premières données paléobotaniques et palynologiques. 104 Congr Nat Soc Savantes, Bordeaux I: 175-184

Garrote A, Sánchez Carretero R (1979) Granitos postcinemáticos de tendencia alcalina en Ossa-Morena: el stock de La Cardenchosa (prov. de Córdoba). Acta Geol Hisp. Homeneje L Solé Sabaris 14: 90-96

Garrote A, Sánchez Carretero R (1983) Materiales volcanoclásticos en el Carbonífero inferior, al S-SW de Villaviciosa de Córdoba (Zona de Ossa Morena). Comun Serv Geol Port 69 (2): 249-257

Garrote A, Sánchez Carretero R, Tijero F (1985) El Batolito de los Pedroches en la transversal de Villanueva de Córdoba. V Reunión del GOM. Temas Geol Min, pp 29-39

Gil Ibarguchi JI, Ortega Girones E (1985) Petrology, structure and geotectonic implications of glaucophane-bearing eclogites and related rocks from Malpica-Tuy (MT) Unit, Galicia, northwest Spain. Chem Geol 50: 145-162

Gill J (1981) Orogenic andesites and plate tectonics. Springer, Berlin Heidelberg New York, 390 pp

Guitard G (1970) Le métamorphisme hercynien mésozonal et les gneis oeillés du Massif du Canigou (Pyrénées orientales). Mém Bur Rech Geol Min (BRGM) 63: 1-311

Hernández Enrile J (1971) Las rocas porfiroides del límite Cámbrico-Precámbrico del flanco meridional del anticlinal Olivenza-Monesterio (Badajoz). Bol Geol Min 82: 143-154

Irvine TN, Baragar WRA (1971) A guide to the chemical classification of the common volcanic rocks. Can J Earth Sci 8: 523-548

Kuno H (1968) Differentiation of basalt magmas. In: Hess, Poldervart (eds) Basalts. Wiley Interscience, New York, pp 623-668

Larrea FJ (1987) Estudio geológico del Batolito de Los Pedroches en la transversal de la Virgen de la Cabeza, Andújar (Jaén). Thesis, Univ del País Vasco, 237 p

Liñán E (1978) Bioestratigrafía de la Sierra de Córdoba. Thesis, Univ Granada, 212 p

Loiselle MC, Wones DR (1979) Characteristics of anorogenic granites. Abstr SAAGM, 539 pp

McPherson J (1879) Estudio petrológico y petrográfico del N de la provincia de Sevilla. Bol Com Mapa Geol Esp 6: 97-268

Marmo V (1971) Granite petrology and the granite problem. Elsevier, Amsterdam, 244 pp

Mata J, Munhá J (1986) Geochemistry of Cambrian metavolcanic rocks from the Cordoba-Elvas domain (Ossa Morena Zone). Abstract. Maleo 2 (13): 28

Mendes F (1968) Contribution à l'étude géocronologique par la méthode du strontium de formations cristallines du Porgugal. Bol Mus Lab Min Geol Fac Ciênc Univ Lisboa 11 (1): 1-50

Miyashiro A (1975) Classification, characteristics, and origin of ophiolites. J Geol 83: 249-281

Munhá J, Oliveira JJ, Ribeiro A, Oliveira V, Quesada C, Kerrich R (1986) Beja-Acebuches ophiolite: characterization and geodynamic significance. Abstract. II Congr Geol Port, Maleo 2 (13): 31

O'Connor JT (1965) A classification for quartz-rich igneous rocks based on feldspar ratios. US Geol Surv Res Prof Pap 525: 79-84

Oliveira V (1982) Contribuçao para o conhecimento geologico-minero da regiao de Alandroal-Juromenha (Alto Alentejo). EN T SFM 26: 103-126

Pascual E (1981) Investigaciones geológicas en el sector Córdoba-Villaviciosa de Córdoba. Thesis, Univ Granada, 521 p

Pascual E (1984) El Batolito de Los Pedroches en el sector Hinojosa del Duque-Santa Eufemia. Mem Not Univ Coimbra 97: 39-50

Pascual E (1985) El complejo anular del S de Villaviciosa de Córdoba (Córdoba, España): datos petrológicos y geoquímicos y relaciones con otras rocas ígneas del área. V Reunión del GOM. Temas Geol Min, pp 81-89

Pascual E, Pérez Lorente F (1975) El magmatismo ácido superficial al sur de Villanueva del Rey-Villaviciosa de Córdoba (Sierra Morena, Córdoba). Cuad Geol 6: 15-30

Pascual E, Pérez Lorente F (1987) La alineación o eje magmático de Villaviciosa de Córdoba-La Coronada. In: Bea F, Carnicero A, Gonzalo JC, López Plaza M, Rodríguez Alonso MD (eds) Geología de los granitoides y rocas asociadas del Macizo Hespérico. Libro homenaje a L. C. García Figuerola, pp 365-376

Penha MA, Arribas A (1974) Datación geocronológica de algunos granitos uraníferos españoles. Bol Geol Min 85-3: 271-273

Pérez Lorente F (1979) Geología de la Zona de Ossa Morena al Norte de Córdoba (Pozoblanco-Belmez-Villaviciosa de Córdoba). Thesis, Univ Granada, 340 p

Pinto MS (1984) Granitóides caledónicos e hercínicos na zona de Ossa-Morena (Portugal). Nota sobre aspectos geocronológicos. Mem Not Univ Coimbra 97: 81-94

Pitcher WS (1982) Granite type and tectonic environment. In: Hsu KJ (ed) Mountain building processes. Wiley Interscience, New York, pp 19-40

Pons J (1982) Un modèle d'évolution des complexes plutoiques: gabbros et granïtoides de la Sierra Morena occidentale. Thesis, Univ Paul Sabatier, 451 p

Prost-Dame V (1980) Geologie et gitologie de la région de Castuera (Badajoz-España). Thesis, Univ Paris Sud, Centre d'Orsay, 287 p

Quesada C (1985) El Carbonífero de Sierra Morena. In: Carbonífero y Pérmico de España. X Congr Int Estratigr Geol Carbonífero Madr, pp 245-278

Quesada C (1987) Mapa Geológico-minero de Extremadura E = 1:300 000. Junta de Extremadura, 131 pp

Ribeiro ML (1987) Petrogenesis of Early-Paleozoic peralkaline rhyolites from the Macedo de Cavaleiros region (NE Portugal). Geol Rundschau 76 (1): 147-168

Roche H de la, Leterrier J (1973) Transposition du Tétraédre minéralogique de Yoder et Tilley dans un diagramme chimique de classification des roches basaltiques. CR Acad Sci Ser D 276: 3115-3118

Rogers JJW, Greenberg JK (1981) Trace elements in continental-margin magmatism. Part III. Alkali granites and their relationship to cratonization: summary. Geol Soc Am Bull 92: 6-9

Ruiz de Almodovar G, Galan E, Pascual E (1984) Las rocas plutónicas del sector noroeste del anticlinorio Olivenza-Monesterio y su posible relación con mineralizaciones de tipo Sn-W. Mem Not Univ Coimbra 98: 1-16

Sánchez Carretero R, Carracedo M, Eguíluz L, Garrote A, Apalategui O (1989 a) El magmatismo calcoalcalino del Precámbrico terminal en la Zona de Ossa-Morena (Macizo Ibérico). Rev Soc Geol Esp 2: 7-21

Sánchez Carretero R, Carracedo M, Gil Ibarguchi JI, Ortega Cuesta LA (1989 b) Unidades y datos geoquímicos del magmatismo hercínico de la "Alineación de Villaviciosa de Córdoba-La Coronada" (Ossa-Morena Oriental). Comunicación a la X Reunión sobre la Geología del NW Peninsular. Stud Geol 4: 105-130

Sánchez Pérez JM (1986) Contribución a la petrología, mineralogía y geoquímica del Batolito de Los Pedroches en la transversal de Pozoblanco (Córdoba, España). Thesis, Univ del País Vasco, 176 p

Schneider H (1939) Altpaleäozoikum bei Cala in der Westlichen Sierra Morena (Spanien). Diss Math Naturwiss Rak Univ Berlin, 72 p

Silva LC, Quadrado R, Ribeiro L (1970) Nota previa sobre la existencia de unha extructura zonada e de anortositas no macizo gabro-dioritico de Beja. Bol Mus Lab Min Geol Univ Lisboa 11: 223-232

Simancas JF (1983) Geología de la extremidad oriental de la Zona Sud Portuguesa. Thesis, Univ Granada, 433 p

Walker KR, Joplin GA, Covering JF, Green R (1960) Metamorphic and metasomatic convergence of basin igneous rocks and limemagnesio sediments of the Precambrian of north-western Queensland. J Geol Soc Aust 6: 149-177

Whalen JB, Currie KL Chappel BW (1987) A-type granites: geochemical characteristics, discrimination and petrogenesis. Contrib Mineral Petrol 95: 407-419

Wright JB (1969) A simple alkalinity ratio and implications to the question of non-orogenic granite genesis. Geol Mag 87: 1587-1598

5 Metamorphism

C. QUESADA[1] and J. MUNHA[2]

5.1 Introduction

The regional metamorphic pattern within the Ossa-Morena Zone is characterized by an alternation of low and medium-to high-grade belts which retain more or less homogeneous features along the strike (Fig. 1). As already stated by Quesada et al. (this Vol.) it is now unanimously recognized that this metamorphic framework developed during at least two main tectonothermal cycles: Late Precambrian and Hercynian. Nevertheless, major uncertainties still exist concerning which event is to be held responsible for the metamorphic evolution of specific areas; this is mainly due to our still poor knowledge of the relationships pertaining to recrystallization-deformation and the lack of geochronological constraints on these events. Moreover, the few available absolute ages on these (sometimes) highly complex metamorphic rocks are not amenable to straightforward interpretation. The great majority of the data was obtained by the K-Ar or ^{40}Ar-^{39}Ar methods; thus, we are dealing with cooling ages, which means that the ages determined for the metamorphic rocks represent the timing when those rocks were brought up to shallower crustal levels and/or to P-T conditions below the closing exchange temperature of their respective isotopic systems. Therefore, what the data is telling us is the ages of the main uplift tectonic events and not necessarily the timing of either their metamorphic recrystallization and/or deformation.

The purpose of this chapter is to present a synthetic review of the metamorphic features in some key areas of the Ossa-Morena Zone and to attempt to use the information they provide to constrain the tectonothermal evolution of the SW branch of the Iberian Variscan Foldbelt.

[1] I. T. G. E., Rios Rosas 23, 28003 Madrid, Spain
[2] Department of Geology. University of Lisbon, 1294 Lisbon, Portugal

R. D. Dallmeyer and E. Martínez García (Eds.)
Pre-Mesozoic Geology of Iberia

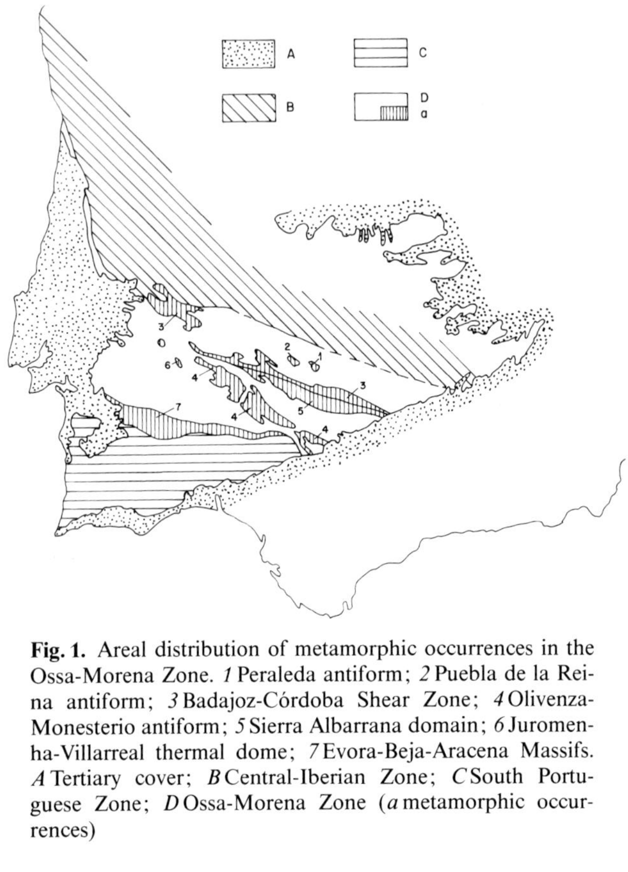

Fig. 1. Areal distribution of metamorphic occurrences in the Ossa-Morena Zone. *1* Peraleda antiform; *2* Puebla de la Reina antiform; *3* Badajoz-Córdoba Shear Zone; *4* Olivenza-Monesterio antiform; *5* Sierra Albarrana domain; *6* Juromenha-Villarreal thermal dome; *7* Evora-Beja-Aracena Massifs. *A* Tertiary cover; *B* Central-Iberian Zone; *C* South Portuguese Zone; *D* Ossa-Morena Zone (*a* metamorphic occurrences)

5.2 Regional Metamorphic Pattern

In this section we have arranged the description of selected key areas following an order of progressive younging of the metamorphic event during which the highest P-T conditions were reached.

5.2.1 Precambrian Metamorphic Complexes

5.2.1.1 Peraleda and Puebla de la Reina Antiformal Structures (Fig. 1, *1*, *2*)

Amphibolite facies pelitic to greywacke and amphibolite rocks belonging to the Serie Negra Group

(Quesada et al., Part V, Chap. 2.1) crop out within the Obejo-Valsequillo-Puebla de la Reina domain (Delgado et al. 1977; see also Fig. 1, Apalategui et al., this Vol.), comprising the cores of several antiformal structures, the most important of which (in terms of areal extent) are those referred to above.

Metamorphic grade ranges from greenschist (chlorite zone) to amphibolite facies (garnet-staurolite zone) and the most characteristic mineral assemblages are as follows (Apalategui et al. 1985; Apalategui et al., in press):

1. Pelitic to greywacke rocks:
   Quartz-chlorite
   Quartz-biotite-chlorite
   Quartz-biotite
   Quartz-muscovite-biotite
   Quartz-muscovite-biotite-garnet
   Quartz-muscovite-biotite-garnet-staurolite
2. Amphibolites:
   Albite-epidote-chlorite
   Plagioclase-hornblende-biotite

The Precambrian age of this metamorphic event is proved by both radiometric dating (Blatrix and Burg 1981: 550 ± 10 Ma, ^{40}Ar-^{39}Ar muscovite cooling age) and field relationships. Staurolite garnet-bearing pelitic pebbles occur in conglomerate horizons at the base of the uncleaved (and anchymetamorphic) Paleozoic succession (basal Ordovician; locally Lower Cambrian). Moreover, in the Precambrian Serie Negra rocks these porphyroblasts are rotational, relative to a first penetrative fabric being flattened by a late (but of course still Precambrian) crenulation cleavage.

5.2.1.2 Badajoz-Córdoba Shear Zone (Fig. 1, *3*)

This linear belt of highly deformed Precambrian metamorphic rocks, as well as Lower Paleozoic (Upper Cambrian to Upper Ordovician; Priem et al. 1970; Allegret and Lancelot, in Burg et al. 1981; García Casquero et al. 1985) intrusives (ranging from undersaturated to oversaturated alkaline to peralkaline felsic rocks), has played a major role in recent geodynamic modelling of the Variscan Foldbelt (Burg et al. 1981; Brun and Burg 1982; Ribeiro et al. 1983; Matte 1983, 1986; etc). This belt was the locus of intensive left-lateral shear movements (reaching up to ca. 100 km displacement; Burg et al. 1981) during Hercynian times, causing mylonitization and retromorphosis (under greenschist to amphibolite facies) of all previous structures and mineral assemblages. These cataclastic events clearly affected the Lower Paleozoic intrusives, which in turn postdate the metamorphic fabric and mineral assemblages of the Precambrian rocks; therefore, providing strong evidence for the existence of Precambrian metamorphism. This is demonstrated by the intrusive relationship of the former in the latter, the pre-nature of the high-grade parageneses with respect to the mylonitic fabric and the consistency of metamorphic grade associated with the mylonitization, resulting in the partial retrogradation of both the preintrusion metamorphic assemblages and the igneous mineralogy of the intrusive rocks.

The Precambrian rocks outcropping in this belt have been described already by Quesada et al. (Part V, Chap. 2.1), corresponding to the Valencia de las Torres-Cerro Muriano Supergroup, including both the Serie Negra Group and the underlying Blastomylonitic Formation. The former includes mainly garnet-biotite (with rare staurolite) micaschists, amphibolites (sometimes garnet-bearing) and minor intercalations of calc-silicate and black cherty rocks, whereas the latter includes mainly ortho and paragneisses (sometimes migmatitic) and amphibolites. The metamorphic evolution of the Blastomylonitic Formation has been the subject of several studies (Muñoz and Vegas 1974; Laurent 1974; Bladier 1974; Chacón 1974; 1979; Pérez Lorente 1978; Pascual 1981, 1985). The following description is a summary of the results of a recent microprobe study (Mata and Munhá 1986) of these rocks from the Degolados-Campo Maior region.

The gneisses carry abundant quartz and plagioclase, with frequent garnet and biotite, and ubiquitous kyanite, although sillimanite is the more common Al-silicate; very rare staurolite relicts also occur, but the gneisses carry no primary muscovite and no K-feldspar (although it is reported in the literature Pascual 1981). The following reactions were probably overstepped as a result of P, T variations during high-grade metamorphism and anatexis:

1. Staurolite + muscovite + quartz → Al_2SiO_5 + biotite + garnet + H_2O.
2. Kyanite → sillimanite.
3. Garnet + sillimanite + quartz → cordierite.

In mafic rocks the characteristic metamorphic assemblage consists of garnet + clinopyroxene + hornblende + plagioclase + quartz + rutile + ilmenite. Clinopyroxene is always present as symplectitic intergrowths with plagioclase (at various stages of replacement by hornblende) and garnet is surrounded by coronas of plagioclase + hornblende.

The textures present are typical of those occurring in many eclogites and it is inferred that an eclogitic garnet-omphacite assemblage once existed, but has since undergone partial retrogression to amphibolite facies according to reactions of the form:

4. Omphacite $\xrightarrow{\text{unmixing}}$ diopside + plag (1).
5. Garnet + diopside + plag (1) + quartz + rutile + H_2O → hornblende + plag (2) + ilmenite.

The eclogitic assemblage is not cofacial with sillimanite-bearing assemblages in surrounding migmatitic gneisses, but the water required for amphibolitization may have been produced by the gneisses along a common segment on the metamorphic path.

Geothermobarometric data (Mata and Munhá 1986) for both felsic and mafic rocks indicate that the P, T-time metamorphic path was characterized by initial heating and decompression, from eclogite facies to 750–800 °C and 9–10 kb, followed by almost isothermal (700–750 °C) decompression to 5 kb; cooling (<600 °C), and further decompression is also suggested by the occurrence of late-stage andalusite-sillimanite-bearing granitic segregations. Presumably, even this last cooling and decompression part of the P, T-t path took place in Precambrian times, as suggested by the preexistence of the andalusite-sillimanite-bearing granitic segregations with respect to the mylonite events.

The Precambrian parageneses underwent severe tectonic reactivation during Hercynian times, giving rise to widespread retrogradation, reaching, in many cases, temperature conditions of up to 500 °C. This is suggested by the reequilibration of radiogenic Ar in several analyzed hornblendes (Dallmeyer and Quesada 1989) showing Hercynian ages.

5.2.1.3 Olivenza-Monesterio Antiform (Fig. 1, *4*)

This structure comprises two units separated by a major thrust (Monesterio Thrust; Fernández et al. 1983). The lower unit is composed of low-grade (greenschist and lower) Serie Negra Group rocks, folded and metamorphosed before the deposition of synorogenic uppermost Precambrian volcaniclastics (Fig. 3 in Apalategui et al., this Vol.). The upper unit, which includes the same stratigraphy, has been metamorphosed to higher grades reflecting the effects of a well-defined thermal dome (Quesada 1975; Eguiluz and Quesada 1980; Eguiluz et al. 1984). Detailed mapping of metamorphic isograds is hampered by the rarity of index minerals; however, a schematic metamorphic zonation of this

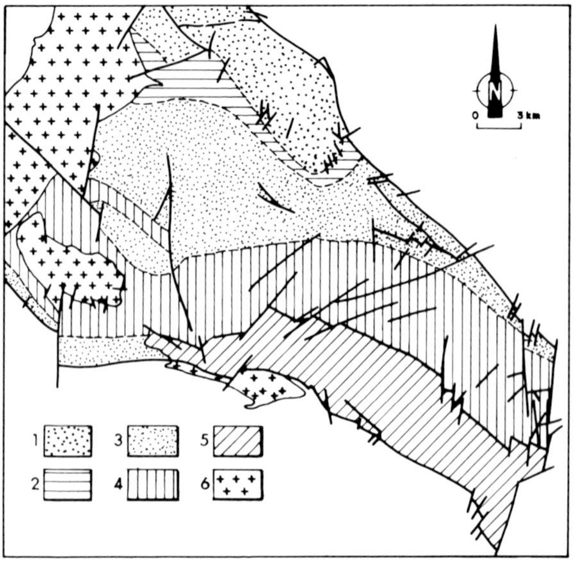

Fig. 2. Metamorphic zonation of the thermal dome constituting the central part of the Olivenza-Monesterio antiform. *1* Very low-grade; *2* lower greenschist; *3* upper greenschist; *4* lower amphibolite; *5* upper amphibolite; *6* igneous rocks (After Arriola et al. 1984)

high-grade unit is shown in Fig. 2. Metamorphism upgrades towards the from greenschist facies (biotite zone) up to high rank amphibolite facies [sillimanite-K-feldspar-cordierite-(rare) garnet-bearing anatectic granodiorites].

The common occurrence of andalusite and cordierite in middle- and high-grade rocks, together with the scarcity of garnet, is strongly indicative of high geothermal gradients.

Hercynian tectonic reactivation (mainly within the vicinity of the Monesterio Thrust) resulted in retrogradation under greenschist facies conditions.

5.2.2 Hercynian Metamorphic Complexes

5.2.2.1 Sierra Albarrana Thermal Dome (Fig. 1, *5*)

Both stratigraphic and tectonic relationships in this unit are still highly debatable (see Quesada et al., Part V, Chap. 2.1). Although there is clear evidence for Precambrian tectonothermal events (Chacón 1979; Garrote et al., in press; Apalategui et al., this Vol.) affecting some of the rocks in the area, available geochronological data (Dallmeyer and Quesada 1989) indicates that the main metamorphic episode responsible for the development of this thermal dome (and other smaller ones in the same unit) took place during Hercynian times. Litholo-

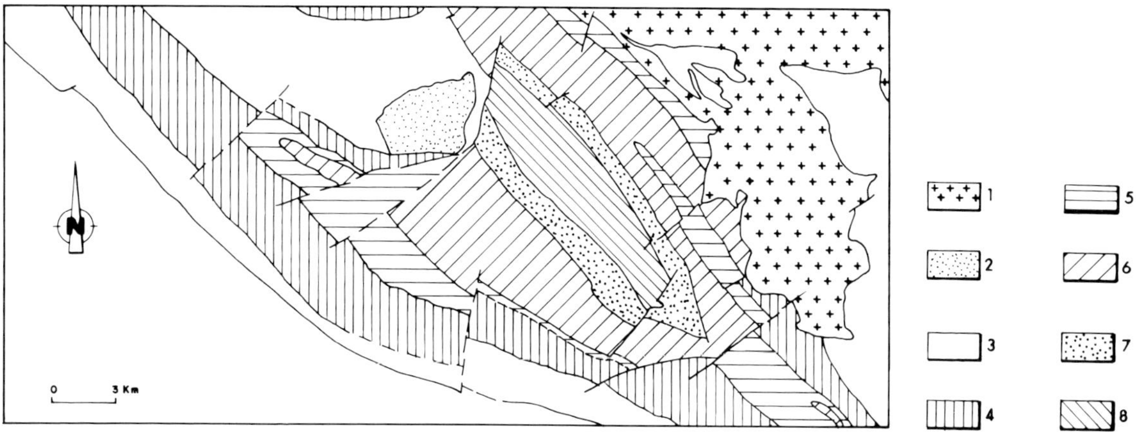

Fig. 3. Metamorphic zones within the Sierra Albarrana thermal dome. *1* Igneous rocks; *2* post-metamorphic Carboniferous occurrences; metamorphic zones; *3* chlorite zone; *4* biotite zone; *5* garnet zone; *6* staurolite-andalusite zone; *7* sillimanite-muscovite zone; *8* sillimanite-K-feldspar zone (Modified after Garrote 1976)

gies in the area include mainly high-Al_2O_3 pelites, quartz-feldspathic sandstones and minor intercalations of amphibolites.

The metamorphic zonation is described in detail by Garrote (1976; see also Fig. 3) from greenschist (chlorite zone), to high rank amphibolite facies (sillimanite-K-feldspar zone). The rare occurrence of kyanite and the late widespread development of andalusite ± cordierite assemblages suggest a progressive increase of geothermal gradients, most likely during an essentially decompressive tectonic event.

5.2.2.2 Juromentha-Villarreal Thermal Dome (Fig. 1, *6*)

Within the Elvas-Cumbres Mayores Domain (Quesada 1983; see also Fig. 1 in Apalategui et al., this Vol.) the great majority of the Paleozoic lithotypes have been metamorphosed under low greenschist facies conditions (mainly chlorite zone). Locally, however, one can observe relatively small thermal domes (mainly concentrated in the NW part of the belt) which have reached amphibolite facies conditions during and postdating the second Hercynian deformation phase. The Juromenha-Villarreal thermal dome is composed of pelitic and psammitic rocks (Cambrian-through-Silurian in age) which upgrade from the regional lower greenschist facies (chlorite + chloritoid), through the biotite zone, to garnet-staurolite-bearing micaschists outcropping at the core of the metamorphic complex.

5.2.2.3 *Evora-Beja-Aracena Massif* (Fig. 1, *7*)

This highly complex unit comprises several medium to high-grade metamorphic complexes which are scattered within a low-grade regional framework. Some of these complexes undoubtedly include lithologies that were metamorphosed during Precambrian times and then intensely reactivated during the Hercynian cycle. Therefore, in the areas where the Hercynian metamorphic effects reached medium- to high-grade conditions it is difficult to clear up the aforementioned chronological relationships.

Perhaps the best-known region of this Massif, and also the one where metamorphism reached its highest grade, is the Sierra de Aracena area (Bard 1969). The Aracena metamorphic complex (Fig. 4) consists of two units separated by a NW-SE trending major normal shear zone (1–2 km wide; Crespo 1987) which juxtaposes granulite facies rocks (towards the SW) to greenschist facies rocks (towards the NE). The lithologies are highly heterogeneous and include some Precambrian inliers (Serie Negra Group) scattered in an uppermost Precambrian to lowermost Paleozoic mostly volcanoclastic succession. The higher grade unit shows homogeneous metamorphic conditions over the whole area reaching the granulite facies. The widespread occurrence of cordierite + orthopyroxene + garnet assemblages suggests geothermal gradients in excess of 70 °C/km (Bard 1969). The very rare occurrence of garnet-clinopyroxene relict assemblages (Crespo, pers. comm., 1985) points towards a complex evolution of this unit, probably involving significant uplifting of deep crustal levels.

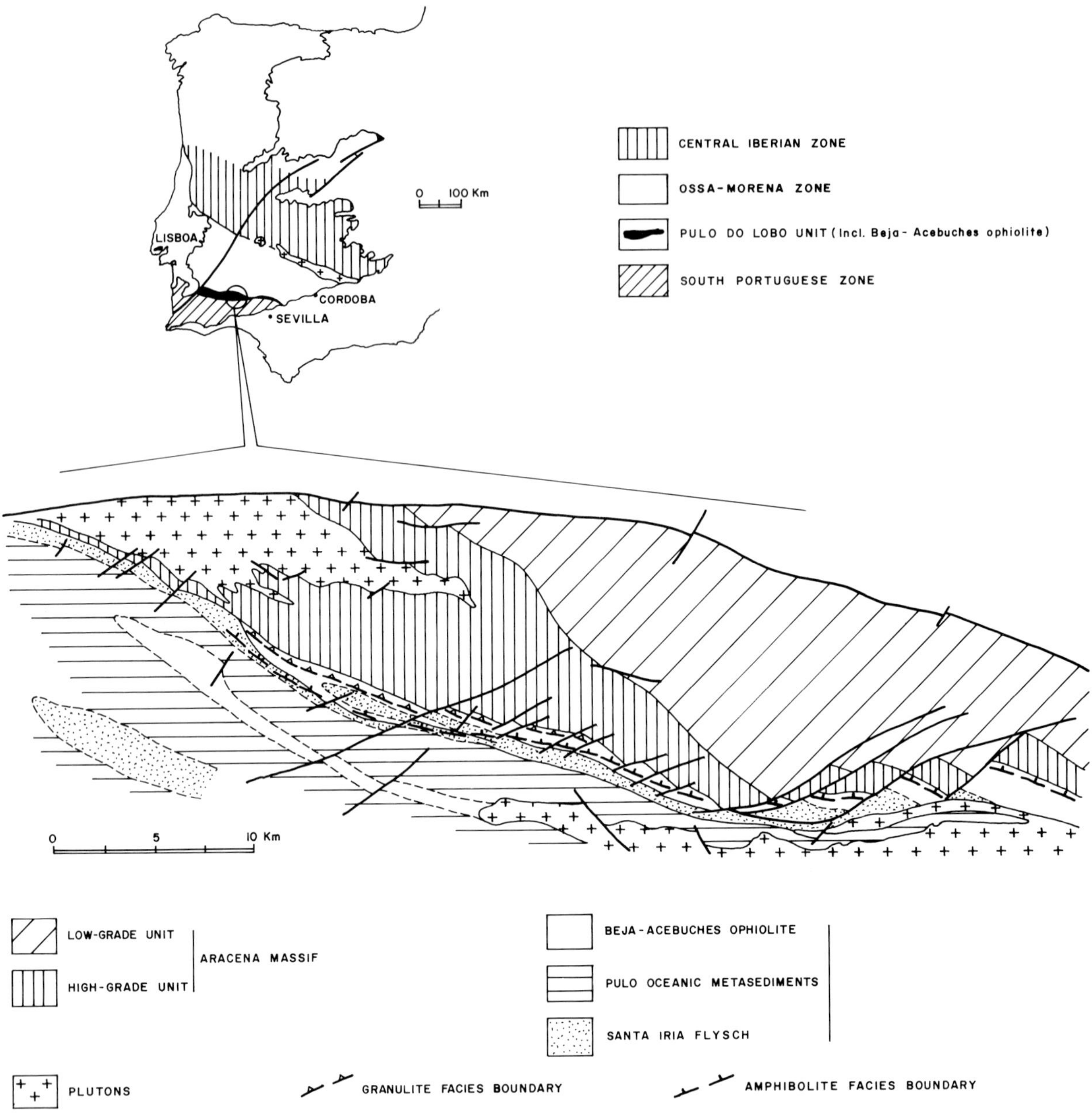

Fig. 4. Major structural units within the Aracena Massif with indication of their metamorphic grade

In close spatial association with the Aracena complex lies the so-called Acebuches amphibolites (Bard 1969), which have been recently interpreted as being part of the Beja-Acebuches ophiolite (Munhá et al. 1986). In a late, post-obduction event, the Beja-Acebuches ophiolite was disrupted by overthrusting which brought up the high-grade metamorphic rocks of the Aracena complex over the ophiolite sequence; related to this event, a high-grade tectonic mélange formed at the contact and a steep inverted metamorphic gradient developed in the ophiolite sequence, upgrading to the N from zeolite-prehnite-pumpellyite facies to the granulite facies (Castro et al. 1987).

5.3 Concluding Remarks

On the basis of the data presented above, the existence of two orogenic and consequent tectonothermal events, sufficiently separated in time to be recognized as independet, becomes clear to us.

Precambrian metamorphic events, although always more or less reactivated during Hercynian times, are still well recognized in several areas. Unfortunately, the available quantitative data are still very scarce and much more detailed work in specific areas is a first-order requirement. Nevertheless, we have been able to detect, at least in the Blastomylonitic Formation, evidence of an early high-pressure Precambrian metamorphic event. The P, T-time path displayed by these rocks, involving initial heating and subsequent almost adiabatic decompression, from eclogite through granulite to amphibolite facies, strongly suggests that collision tectonics was operative during Late Precambrian times in this terrain. The remaining Precambrian metamorphic complexes in the Ossa-Morena Zone correspond mostly to thermal doming and, at present, give little indication concerning tectonic setting.

On the other hand, the available data give no suggestion of high-pressure Hercynian metamorphic regimes in the Ossa-Morena Zone. The regional Hercynian metamorphic framework reflects the combined effects of thermal doming and uplifting of deep crustal levels. Thermal doming is most probably the result of geothermal uprisal related to plutonic rock emplacement. Simultaneously, magma underplating at the base of the crust should have kept relatively high heat flow values, resulting in low pressure granulite formation (cf. Aracena complex). Within the regional tectonic framework requiring northeasterly directed oblique subduction (Ribeiro et al., Part VIII), we interpret the uplifting event as a consequence of subsequent continental collision.

Acknowledgements. Comments by an anonymous reviewer are deeply appreciated. R. Fernández Toledo, for carefully typing various versions of the manuscript, and J. Vallejo, for drawing all the artwork, are also kindly acknowledged.

References

Apalategui O, Roldan J, Pérez F, Cueto LA, Quesada C (1985) Mapa Geológico de España. Escala 1:50000, Hoja n° 880: Espiel. IGME

Apalategui O, Contreras F, Jorquera A, Villalobos M, Eguiluz L, Cueto LA, Quesada C (1989). Mapa Geológico de España. Escala 1:50000, Hoja n° 804: Oliva de Mérida. IGME

Arriola A, Eguiluz L, Fernández J, Garrote A, Sánchez R (1984) Zonación metamórfica en el área Monesterio-Fuente de Cantos: criterios texturales y mineralógicos. Cuad Lab Xeol Laxe 8: 169-193

Bard JP (1969) Le métamorphisme régional progressif des Sierras d'Aracena en Andalousie occidentale (Espagne). Sa place dans le segment hercynien sub-ibérique. Thesis, Univ Montpellier, pp 1-309

Bladier Y (1974) Structure et pétrologie de la bande blastomy lonitique de Badajoz-Cordoue. Les roches cataclastiques-Classification-Interprétation. Thesis, Univ Montpellier, pp 1-105

Blatrix P, Burg JP (1981) ^{40}Ar-^{39}Ar dates from Sierra Morena (Southern Spain): Variscan metamorphism and Cadomian orogeny. Neues Jahrb Mineral Monatsh 1981/10: 470-478

Brun JP, Burg JP (1982) Combined thrusting and wrenching in the Ibero-Armorican Arc: a corner effect during continental collision. Earth Planet Sci Lett 61: 319-332

Burg JP, Iglesias M, Laurent P, Matte P, Ribeiro A (1981) Variscan intracontinental deformation: the Coimbra-Córdoba shear zone (SW Iberian Peninsula). Tectonophysics 78: 161-177

Castro P, Quesada C, Munha J (1987) Metamorphic regime in the Beja-Acebuches Variscan ophiolite (SW Iberia). Extended Abstract in Int Conf Tectonothermal evolution of the West African orogens and circum-Atlantic terrane linkages. IGCP Project 233. Nouakchott (Mauritania), December 1987

Chacón J (1974) Metamorfismo dinámico en la banda Badajoz-Córdoba: Los gneises milonitas de Higuera de Llerena (Provincia de Badajoz). Bol Geol Min 85/6: 700-712

Chacón J (1979) Estudio geológico del sector central del anticlinorio Portalegre-Badajoz-Córdoba (Macizo Ibérico Meridional). Thesis, Univ Granada, pp 1-728

Crespo A (1987) El macizo de Aracena (Macizo Ibérico Meridional): Propuesta de división sobre la base de nuevos datos estructurales y petrográficos. Bol Geol Min 98/4: 507-515

Dallmeyer RD, Quesada C (1989) Geochronological constraints to the structural development of the Badajoz-Cordoba Shear Belt (southwest Iberia). TERRA Abstracts 1 (1): pp 366

Delgado M, Liñán E, Pascual E, Pérez F (1977) Criterios para la diferenciación de dominios en Sierra Morena Central. Stud Geol Salmant 12: 75-90

Eguiluz L, Fernández J, Garrote A, Liñán E, Quesada C (1984) Sucesiones estratigráficas del anticlinorio de Olivenza-Monesterio en la transversal Montemolín-Arroyomolinos. Cuad Lab Xeol Laxe 8: 117-123

Eguiluz L, Quesada C (1980) La sucesión precámbrica de la transversal de Monesterio (Badajoz): Nota preliminar. Temas Geo Min 4: 123-141

Fernández J, Coullaut JL, Eguiluz L, Garrote A, Quesada C, Cueto LA (1983) Mapa Geológico de España. Escala 1: 50000. Hoja n° 897: Monesterio. IGME

García Casquero JL, Boelrijk NAIM, Chacón J, Priem HNA (1985) Rb-Sr evidence for the presence of Ordovician granites in the deformed basement of the Badajoz-Córdoba Belt, SW Spain. Geol Rundschau 74/2: 379-384

Garrote A (1976) Asociaciones minerales del núcleo metamórfico de Sierra Albarrana (Prov. de Córdoba). Sierra Morena Central. Mem Not Univ Coimbra 82: 17-39

Garrote A, Delgado M, Contreras MC (in press) Mapa Geológico de España. Escala 1:50000. Hoja n° 900: La Cardenchosa. IGME

Laurent P (1974) Structure et pétrologie de la bande blastomylonitique de Badajoz-Cordoue à l'est d'Azuaga. Thesis, Univ Montpellier, pp 1-120

Mata J, Munha J (1986) Geodynamic significance of high-grade metamorphic rocks from Degolados-Campo Maior (Tomar-Badajoz-Córdoba Shear Zone). Maleo (Bol Soc Geol Port) 2/13: 28

Matte P (1983) Two Geotraverses across the Ibero-Armorican Variscan Arc of western Europe. In: Rast N, Delany M (eds) Profiles of orogenic belts. AGU Geodynam Ser 10: 53-81

Matte P (1986) Tectonics and plate tectonics model for the Variscan Belt of Europe. Tectonophysics 126: 329-374

Munha J, Oliveira JT, Ribeiro A, Oliveira V, Quesada C, Kerrich R (1986) Beja-Acebuches ophiolite: characterization and geodynamic significance. Maleo 2 (13): 31

Muñoz M, Vegas R (1974) Paraneises y ortoneises de la banda metamórfica Badajoz-Córdoba. Bol Geol Min 85/4: 450-463

Pascual E (1981) Investigaciones geológicas en el sector Córdoba-Villaviciosa de Córdoba (Sector Central de Sierra Morena). Thesis, Univ Granada, pp 1-521

Pascual E (1985) Las rocas polimetamórficas del área de Cerro Muriano (Córdoba, España). Edad y consideraciones petrológicas. Temas Geol Min 8: 81-89

Pérez Lorente F (1978) Geología de la zona Ossa-Morena al Norte de Córdoba (Pozoblanco-Belmez-Villaviciosa de Córdoba). PhD Thesis, Univ Granada 281: 1-340

Priem HNA, Boelrijk NAIM, Verschure RH, Hebeda EH, Verdurmen EAT (1970) Dating events of acid plutonism through the Paleozoic of the Western Iberian Peninsula. Eclogae Geol Helv 63/1: 255-274

Quesada C (1975) Geología de un sector de la parte central del anticlinorio Olivenza-Monesterio. Alrededores de Monesterio (Badajoz). Thesis, Univ Granada, pp 1-128

Quesada C (1983) El Carbonífero de Sierra Morena. In: Martínez C (ed) Carbonífero y Pérmico de España. IGME, pp 243-278

Ribeiro A, Iglesias M, Ribeiro ML, Pereira E (1983) Modéle géodynamique des Hercynides Ibériques. Comun Serv Geol Port 69 (2): 291-293

6 Metallogeny

J. Locutura, F. Tornos, P. Florido and L. Baeza[1]

6.1 Introduction and General Setting

Ossa-Morena is one of the zones in which the Iberian Massif is currently divided according to its peculiar geologic features. It extends across the Portugal - Spain borderline, defining a 120×240 km^2 belt with NW-SE trend.

Besides its characteristic geologic features it differs from other zones of the Massif, both in the high density and in the great diversity of ore deposits. Moreover, it has played, in the past, an important role in the Spanish mining industry, and has been one of the most significant producers of Zn, Pb, Ag and Fe of Iberia. However the continued descent of metal prices in the last years has led to the closure of most of the mines.

This metallogenic significance of the area is tightly bound to its complex evolution. Several sucessive tectonosedimentary cycles may be recognized in the evolution of the Ossa-Morena (Quesada, Part V, Chap. 1), each one having its own associated mineralizations, and remobilizing the previous ones. The complex metallogenic history of the area and the consequent abundance and diversity of ore deposits explain the strong exploration activity which is being carried out at present in the Ossa-Morena Zone by state and private companies.

The salient characteristics of the geological and petrological setting of mineralizations have been described in previous chapters. Therefore, only metallogenically relevant geologic aspects will be referred to here.

Abundant literature exists concerning the local and partial analysis of metallogenic features and ore deposits, but no synthesizing paper has ever been published. General descriptions of the Ossa-Morena ore deposits can de found in the papers by Arribas (1979): Vázquez (1978) and Schermerhorn (1982).

This contribution is a first attempt to synthesize the available metallogenic information concerning the Ossa-Morena Zone, and to integrate it into the new interpretative tectonic models, described in detail in previous sections of this book.

6.2 Mineral Deposits

The various major tectonosedimentary events summarized by Quesada (Part V, Chap. 1, Fig. 2) to account for the evolution of the Ossa-Morena in Late Precambrian and Paleozoic times have been considered in order to classify and subdivide the ore deposits. It is worthwile mentioning that most deposits fit rather precisely with the theoretical characteristics expected for their respective tectonic environments, thus providing strong support for the evolution model.

Along the different tectonic events, two main kinds of ore deposits predominate: volcanosedimentary and plutonic related mineralizations. Excepting those associated with Late Hercynian felsic granites (Sn-W), they generally follow WNW-ESE trends, parallel to the Hercynian structural grain (Fig. 1 B). Main ore districts are shown in Fig. 1 B and their respective tectonosedimentary setting is described in Fig. 2. In the following, the numbers in parentheses correspond to those in Fig. 1 B.

6.2.1 Pre-Cadomian Mineralizations

Ophiolitic Cr Mineralizations (27)

Economically unimportant chromite and magnetite pods and lenses with 25%-35% Cr_2O_3 occur in serpentinites within the Malcocinado Fm. (Arriola et al. 1984). These serpentinite bodies have been interpreted as being olistoliths formed during the obduction of a dismembered ophiolite sheet (remnants of a pre-Vendian oceanic crust) onto the continental margin of Iberia in the course of the Cadomian orogeny (Quesada et al. 1987).

[1] Dirección de Recursos Minerales, Instituto Geológico y Minero de España, Rios Rosas, 23 28003 Madrid, Spain

R. D. Dallmeyer and E. Martínez García (Eds.)
Pre-Mesozoic Geology of Iberia

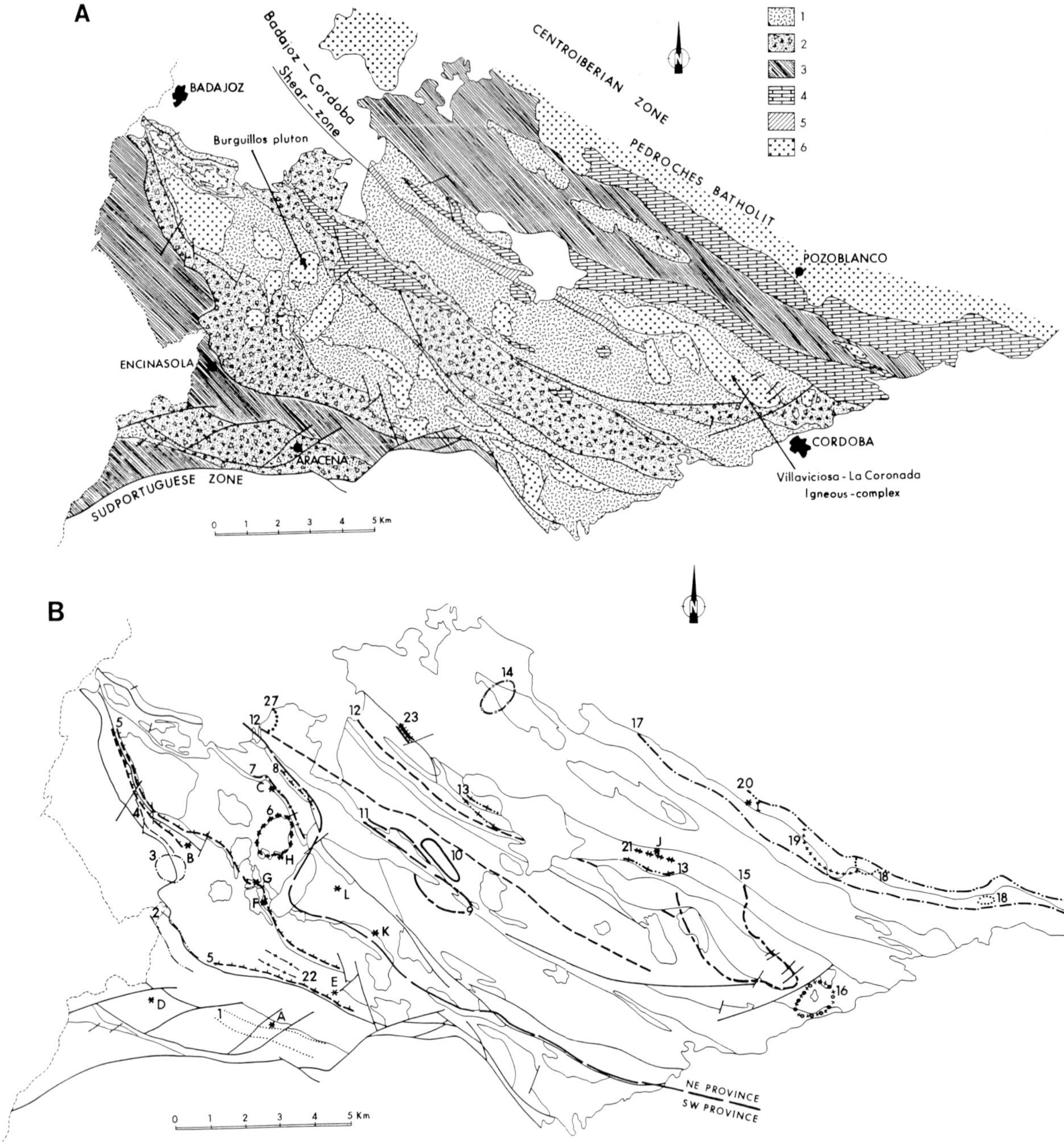

Fig. 1. Geological and metallogenic features of the Ossa-Morena Zone. **A** Geology: *1* Precambrian pre- and synorogenic rocks *2* Cambrian *3* Lower Paleozoic (Ordovician to Devonian) *4* Lower Carboniferous *5* Badajoz Córdoba Shear Zone *6* Plutonic rocks. **B** Metallogeny: Main ore districts *1* Pb-Zn Aracena *2* Cu-Fe Hinojales *3* W (Bi, Au) Oliva de la Frontera *4* Mn (Fe) Zahinos *5* Fe (Cu) Sw limb of the Olivenza Monesterio structure (San Guillermo, la Berrona, La Valera, La Bilbaina, La Bóveda, El Soldado, Bismark mines) *6* Fe (Cu, U) Burguillos pluton (Monchi mine) *7* Fe (Cu) NE flank Olivenza Monesterio structure (Alfredo mine) *8* Cu (Ni, Co) Burguillos *9* baryte Llerena *10* Cu Llerena *11* Hg (Pb-Ba-Sb-Au) Usagre *12* Pb-Zn Valencia de las Torres-Cerro Muriano shear zone *13* Cu-Pb-Zn Matachel *14* W (Bi) San Nicolás *15* baryte Los Ojuelos la Coronada *16* polymetallic (Cu, Zn, Pb, Sn, F, Ba) Cerro Muriano *17* Pb zone in Pedroches batholith *18* W (Zn, Bi) Pedroches batholith *19* Bi (Co, Ni) Pedroches batholith *20* Cu Pedroches batholith *21* Sn Oropesa *22* baryte Hinojales *23* Cu (Pb-Zn) Puebla de la Reina *24* Pb Santa Marta pluton. *A* Maria Luisa Mine (Cu, Pb, Zn); *B* La Bóveda Mine (Fe), *C* Alfredo Mine (Fe); *D* Santa Teresa Mine (Mn, Fe); *E* Cala Mine (Fe); *F* La Berrona Mine (Fe); *G* Colmenar Santa Justa Mine (Fe); *H* Monchi Mine (Fe); *I* El Soldado Mine (Pb, Zn); *J* Oropesa prospect (Sn); *K* Cabra Baja Mine (U); *L* Calzadilla prospect (Cr); *M* La Nava-Paredon prospect (Zn-Cu-Pb massive sulphides)

	NE Province Pb-Zn-(Cu-Ag)	SW Province Fe-(Cu-Mn)	Aracena massif Pb-Zn-Fe
* Hercynian - Tension fractures (Epithermal)	+ Pb-Zn-shear zones (12) + Hg-(Ba-Pb-Au) (11) + Ba-F-(Pb-Zn-Cu) veins (15, 16) + U-(Co-Ni-Cu) shear zones + Sn-(As-Cu-Pb) replacement (21)		
- Postcollisional granitoids - Leucogranites	+ Bi-(Co-Ni) lodes, shear zones (19) + W-(Bi-Au) lodes, greisens (3, 14, 18) + Scheelite disseminations + Sn-(Nb-Ta-REE) pegmatites (18) + Zn-(Pb-Sn-Cu) skarns (16) + Pb-Zn-(F-Ba) veins? (16)		
- Adamellites and granites	+ Cu (U) lodes (inner zones, 20) + Pb-Zn (Cu-Ag) lodes (external zones, 17, 27) + Sb lodes (peripheral)	+ Fe-(Cu-Au) skarns (5, 24)	+ Cu veins
- Pull-apart basins	+ Zn-Cu-Pb massive sulphides (13, 25) + Fe in volcanosedimentary, rocks (25) + Coal deposits		
- Active margin igneous rocks		+ Fe-(Cu-Au-Co-As-REE) skarns (16) + Cu-(As-Ni-Co) lodes and skarns (8)	
- Regional metamorphism and tectonics	+ + + + + + + + + + + + + + + REE-U (Cu) pegmatites and aplites? + + + + + + + + + + + + + + Pb-Zn veins	+ Cu-(Sb-Ba-Fe) veins and disseminations (2)	+ Cu-(Fe) skarnoids
* Pre-Hercynian - Stable platform		+ Fe, Mn in psammites + P in shales + Cu in quartzites (2)	+ Mn, Cu, Zn in black schists?
- Rifting		+ + + + + + + + + Fe-(Cu) in detritic rocks (7) + + + + + + + + + + + + + + + Fe-(Cu) in volcanosed rocks (5, 7) + + + Mn-(Fe) in volcanosedimentory rocks (4) + Fe in carbonates (5, 7) + Fe in volcanic chimneys	+ Pb-Zn in marbles (1) + Zn-(Fe-Ag-Ba) in volcanosedimentary rocks (1)
* Cadomian - Synorogenic squences - Flysch	+ Zn-Pb in detritic rocks (12)		
- Volcanosedimentary	+ Zn-Cu-Pb-(Sb-As-Ag) massive sulphides (23, 26) + Cu disseminations (10) + Ba-(Cu) stratiform (9, 22)		+ Zn-Pb in marbles (1) + Zn-(Fe-Ba) in volcanosedimentary rocks (1)
- Subvolcanic plutonism	+ Cu porphyries (10)		
* Pre-Cadomian - Stable platform and rifting?	+ Zn-Pb-(Cu) in black shales + Au in conglomerates + Chromite in serpentinites	+ Cu in black shales	

Fig. 2. Tectonic setting of mineralizations in the Ossa-Morena zone. The numbers of ore deposits and districts are those of Fig. 1 and referred in the text

Mineralizations Related to Black Shales and Quartzites
Sphalerite, galena, chalcopyrite, bornite and pyrite occur as disseminations in black schists and quartzites of the Serie Negra at several localities. Those disseminations are thought to be related to exhalative processes similar to the ores described by Gustafson and Williams (1980) for sediment-hosted Pb-Zn-Cu deposits. They do not have economic importance, but their remobilizations along fractures, locally with high Ag grade, have been exploited in the past (e. g., at the Badajoz Córdoba shear zone).

Recently, disseminated gold in conglomerates filling paleonchannels is being investigated in the Monesterio Antiform (Apalategui et al., this Vol.).

6.2.2 Cadomian Synorogenic Mineralizations (Upper Riphean-Lower Vendian)

The continental margin evolution during the Cadomian orogenesis involving subduction and volcanic arc construction produce some characteristic ore deposits of that tectonic environment, such as small porphyry-like and volcanosedimentary mineralizations.

Porphyry-Type Mineralizations (10)
Subvolcanic calc-alkaline plutons (i. e., Ahillones) associated with volcanic rocks show a strong hydrothermal alteration, with local development of phyllic, potassic and propilitic aureoles with sparse disseminations and veinlets carrying chalcopyrite, sphalerite, magnetite and rare molybdenite. No economic deposits of this group have been found yet.

Volcanic Hosted Polymetallic Mineralizations
In the volcanic rocks which appear in genetic association with the previously mentioned metagranites, two kinds of stratabound deposits occur.

1. Zn-Cu-Pb with minor Ag-As-Sb volcanosedimentary deposits comprise [Puebla de la Reina (23), IGME 1984; Mina María Luisa (26), Vázquez 1972] hectometric lenticular bodies of massive suphides interbedded with hydrothermally altered (kaolinitized, silicified and chloritized) mafic and salic flows and pyroclasts of Vendian age (Malcocinado Fm.). The most common rocks between the volcanics have an intermediate composition and they are, as a whole, of a calc-alkaline nature.
2. In the distal facies of the most acid rocks (Loma del Aire Fm.), baryte, with some chalcopyrite and galena in beds up to 4 m thick and several kilometers long [Llerena (9) and Hinojales (22) areas], appear within alternating shales, limestones, acid tuffs and cherts. These ore deposits are thought to be of a volcanic exhalative origin (distal facies) (Moro and Arribas 1982; Miras and al. 1987; Presur 1987) in an arid environment. At present, they are being exploited, and baryte reserves exceed 300000 metric tons.

Disseminated Zn-Pb Mineralizations in a Flysch Environment (12)
Ore disseminations have been described in flysch series (Azuaga Fm.), related to mafic and intermediate calc-alkaline volcanic rocks. In the SE part of the area, where volcanics appear widespread as layers or intercalations in the flysch, sequence mineralizations occur, disseminated in andesites and in the transition to tuffs and shales, whereas in the NW, they are associated with sedimentary materials. They are thought to be linked with sedimentary exhalative processes, probably being distal equivalents of the previously described Zn-Cu-Pb mineralizations.

6.2.3 Post-Cadomian mineralizations (Upper Vendian-Devonian)

Ore occurrences are very common in Upper Vendian and Cambrian rocks, related to a rift event, while they are less abundant in platform environments of Ordovician to Early Devonian age.

The Zn-Pb Belt (1)
In the southern part of the Ossa-Morena Zone there is a Pb-Zn belt (Portel-Aracena belt, Schermerhorn 1982) directly related to calcareous and volcanosedimentary rocks. At its east end (Aracena area) mineralizations appear near the top of the calcareous unit and below the alkaline bimodal volcanism. The ore is Pb- rich and Ba-Fe-poor in the carbonate-rich sequences, but in the laterally correlative series, in which volcanosedimentary rocks predominate, Zn is dominant and the ore is rich in magnetite, Ag and Ba (Fuenteheridos deposit, near Aracena).

Those deposits are interpreted as being sedimentary exhalative mineralizations related to bimodal volcanism associated with rifting.

Volcanosedimentary Fe (Cu, Mn) Deposits (5, 7)
Iron deposits mainly appear in the western Ossa-Morena, along two Cambrian belts which are parallel to the Olivenza-Monesterio antiformal structure.

Individual lenses can occur along 6–7 km long bands with thicknesses up to 7 m. They are directly related to rifting sedimentation and bimodal (alkaline to tholeiitic) volcanism (Doetsch et al. 1973; Ruiz 1976; Dupont 1978; Vázquez and Fernández Pompa 1976; Coullaut et al. 1980) of Vendian and Lower to Middle Cambrian age. Magnetite and haematite with minor chalcopyrite, pyrite, galena, bornite and pyrrothite occur as disseminations or as stockworks in diatremes and volcanic chimneys of acidic composition (La Valera), or interbedded with volcanic and volcanoclastic rocks (tuffs, agglomerates and flows, La Bilbaina, La Bóveda) or with detritic and carbonate rocks of Vendian (Aracena Massif, El Soldado and Bismark mines) or Cambrian age (Alfredo mine).

Possible distal mineralizations of these groups are the Mn-Fe ore deposits of Zahínos (Ruiz de Almodovar and Galán 1984) and Alconchel (4). In the latter, acid tuffs with disseminated chalcopyrite are covered by metacherts and basic volcanics with Fe and Mn mineralizations.

Such deposits are envisaged as possible correlatives of the Lahn-Dill type (Ziztmann and Neumann Redlin 1977). An important observation is that volcanosedimentary iron mineralizations in the rest of the European Hercynides are dominantly of Ordovician age, whereas those south of the Badajoz-Córdoba shear zone are mostly of Cambrian age.

Mineralizations of Ordovician – Devonian Age
Mineral occurrences of Ordovician-Devonian age are quite abundant and ubiquitous, but they do not have economic significance. All these mineralizations are in good agreement with ore models in shallow platforms (e. g. Mitchell and Garson 1981), which are the conditions of the area from Ordovician to Early Devonian times.

Ore occurrences are of the following types:

1. Thin Cu-rich layers with chalcopyrite and chalcocyte cementing quartz in the basal white quartzite of Silurian age of the Barrancos – Hinojos syncline.
2. Phosphate nodules in black shales of Silurian age, which have been described near Encinasola (Presur 1987).
3. Stratiform Fe, Mn mineralizations which are related to ferruginous psammites of the Sierra Colorada Fm. (Ordovician).
4. Some Cu, Fe and Mn mineralizations are commonly found as disseminations in the Ordovician and Silurian black schists and shales. Beltran (1976) relates them to a coeval volcanism.

In the Cubito Schists Fm., of uncertain age (Cambrian to Devonian), stratiform Mn oxides and Cu, Fe, Zn sulphides appear associated with cherts and acid volcanics (St. Teresa Mine). Recent investigations (Quesada, pers. comm.) show that the Cubito Schists could be metamorphic equivalents of oceanic sediments tectonically emplaced between different terranes of Hercynian age (Ossa-Morena and sub-Portuguese zones). Hence, probably such mineralizations are related to exhalative processes in oceanic crust.

6.2.4 Hercynian Mineralizations (Upper Devonian-Permian Age)

Previous stable platform conditions were disturbed by the Hercynian orogeny and, as in the whole Iberian Massif, major mineralizations in the Ossa-Morena Zone are related to this orogenic event. They occur in a wide spectrum of geologic settings, from early metamorphic events to late tensional basins.

Metamorphic and Metasomatic Mineralizations (1)
In the Aracena area, some uneconomic disseminations of chalcopyrite, pyrite, pyrrothite and magnetite appear related to calc-silicate rocks in medium- to high-grade metamorphic areas. Such disseminations are thought to be produced by diffusive metasomatic processes under metamorphic conditions (Einaudi et al. 1981).

In relationship with high-grade regional metamorphism (Sillimanite) in the Sierra Albarrana unit, some mineralized pegmatites occur. Mineralization, composed of pyrite, chalcopyrite, uraninite, brannerite and other complex REE minerals, is disseminated in such pegmatites (Garrote 1976). Paragenesis shows variations after the metamorphic grade.

Mineralizations Related to Thrust and Shear Zones (2)
Veins with chalcopyrite, pyrite, arsenopyrite with baryte and quartz appear in close relationship to stratiform ores of Ordovician or Silurian age. They are thought to be remobilizations of such ores or related to hydrothermal circulation along major Hercynian shear zones and thrusts (Cueto and Quesada, pers. comm.).

Volcanosedimentary Mineralizations (13, 25)
In close association with alkaline volcanism of Upper Tournaisien-Lower Visean age developed in pull-apart extensional basins, rather similar and roughly coeval to the Pyrite Belt basin in the South

Portuguese Zone (Gabaldón and Quesada 1986), there are some polymetallic Zn-Cu-Pb massive sulphides and Fe mineralizations (Baeza et al. 1978, 1982), in deltaic and lagoon depositional environments (Gabaldón et al. 1985).

Ore appears as lenses of massive ore or as stockwork in hydrothermally altered acid pyroclasts and lava flows. They are predominantly composed of pyrite, sphalerite, galena, chalcopyrite and minor tetrahedrite.

Granite Related Deposits

The geotectonic evolution during the Hercynian orogeny produced a complex sequence of igneous rocks, typical of an active continental margin, with development of an early "I" type basic to acid plutonism (Aparicio et al. 1977; Pons 1982), followed by postcollisional granites: peraluminous granodiorites to leucogranites, "H" granites (Pitcher 1982), which are similar to the mineralized granites of northern areas of the Hesperian Massif.

The first plutonism appears along the Zafra-Monesterio and Arroyomolinos units and is directly related to the Hercynian deformation (Pons 1982). The late magmatic activity is dominant in the eastern zone, where the large Pedroches batholith (200 km long) outcrops; in the western area small plutons appear in the extension of and with the same characteristics as the Pedroches batholith. The emplacement of theses late granites is probably conditioned by deep-seated shear zones.

Iron Copper Skarns (5, 6, 24). In the contact zone between Early (Burguillos) and Late (Cala, S. Guillermo) Hercynian plutonic rocks with marbles of Precambrian or Lower Cambrian age, calcic iron skarns occur (Velasco 1976; Casquet 1980) and are frequently developed. These skarns constitute the most important deposits in the Ossa-Morena Zone in terms of economic importance; total reserves largely exceed 100 million metric tons with 25%-35% Fe, almost exclusively in Cala (Casquet and Velasco 1978), La Berrona (Coullaut et al. 1980) and San Guillermo (Coullaut 1980) mines. Some of them have appreciable Cu grades (up to 0.4% Cu in Cala mine).

Skarns are mainly composed of andradite and hedembergite. Late aposkarn is made of actinolite, epidote and chlorite. Endoskarn with potassic feldspar, actinolite, magnetite and epidote is locally extensive on igneous rocks, as in the Berrona Mine (Coullaut et al., op. cit.). Mineralization is disseminated or massive in the early aposkarn and it is composed of major magnetite with chalcopyrite, hematite, pyrite, pirrothite and traces of gold, arsenopyrite, etc. These skarns are locally characterized by the presence of U and REE minerals such as uraninite and monazite (Arribas 1962a). At the Monchi Mine (Ruiz 1976; Vázquez 1978) the deposit is also characterized by the presence of complex Co-Bi minerals and the abundance of ilvaite and vonsenite. In proximal skarn areas (in the vicinity of the plutonic rock) gangue minerals are mostly Ca-Mg-Fe silicates, whereas, further away from the contact, mineralization (magnetite) is exclusively related to carbonates (distal skarns).

The characteristics of calcic iron skarns in the Ossa-Morena Zone fit very well with those of the Fe (Cu, Au) skarn type of Einaudi et al. (1981) and Einaudi and Burt (1982). Those skarns form a very well-defined group geotectonically related to early plutonism in an active continental margin.

Copper Deposits (8). As an external zone of the iron skarns in the Burguillos pluton, some small Cu-skarns and ore lodes exist. The latter appear along important fractures (up to 5 km long) of near N-S trend. Chalcopyrite is the main ore mineral with pyrite and As-Ni-Co minerals as accessories.

Cu-Zn-Pb Lodes. These mineralizations occur in direct relationship only with Late Hercynian intermediate to acid rocks (granodiorites to granites, 320-305 m.y. old). In the west area, only the Santa Marta Pb-Zn ore field (27) is related to a very small stock of granodiorites. In the Pedroches area (E), a well defined zonation appears (Ovtrach and Tamain 1973; IGME 1972; Delgado et al. 1978): Cu lodes with chalcopyrite, pyrite, bornite, marcasite and tetrahedrite with minor uraninite (20) appear (Arribas 1964) in the internal areas of the batholith, whereas Pb-Zn ores preferentially occur in the metamorphic aureole (17). Major trends of these mineralizations are 0-20° E and 60-70° E and lodes can extend to several kilometers. The most external zone of that periplutonic zonation is made of Sb lodes (Gumiel 1983). The only exception to the general scheme is the El Soldado Group, a Pb-Zn lode on a shear zone with NW-SE trend, which outcrops in the central part of the Pedroches batholith.

The Cerro Muriano metallogenic area is a relatively small zone, in which important skarn and vein mineralizations are associated with Los Arenales subvolcanic granite and associated dykes (16). Hydrothermal circulation related to this granite was active along a considerable time span; early Cu-Zn-Sn skarns were followed by Cu-Pb-Zn-F-Ba (with minor As-Sb-Ag) lodes and late fluorite-silica-veins, all of them extensively mined up to the present.

Finally, Cu veins are irregularly developed in association with intermediate intrusions all around the Ossa-Morena Zone (Castilblanco, Castillo de las Guardas, Salvatierra, etc. granites).

W-Sn (Bi, Au) Mineralizations. In close relationship with highly evolved epizonal leucogranites (320–290 m. y.) there are scheelite disseminations, W-(Bi) and Bi-(Co-Mi) lodes and scarce Sn-(Nb-Ta-REE) pegmatites (Ovtrach and Tawain 1973; Ruiz de Almodovar 1980). In the Pedroches batholith, leucogranites cut granodiorites and granites and intrude in the contact between culm slates and such previous granites (18). Two groups of mineralizations appear in the west area: the Oliva de la Frontera (3) (Ruiz de Almodovar 1980, Gumiel et al. 1983) and the San Nicolás (14) ore fields, but only W-(Bi) lodes and scheelite disseminations are found there. W-Bi mineralizations, as stockworks or individual lodes, are the most important and appear in tight connection with hydrothermal phenomena such as greisenization and tourmalinization. They are characterized by the widespread presence of wolframite and scheelite, with Cu-Bi-Au minerals in a gangue of quartz, fluorite, moscovite and some topaz and beryl.

Bi-(Co-Ni) Mineralizations. Bi-(Co-Ni) deposits of the Pedroches batholith are among the most characteristic mineralizations in the area (19, Espiel mines), in which ore is related to shear zones and veins near leucogranites.

Zn-(Pb-Sn) Skarns (16). Calcic and magnesian F-rich skarns are related to subvolcanic granites in the Cerro Muriano area (Delgado et al. 1978). Ore, composed of sphalerite, chalcopyrite, galena, magnetite and other minor minerals, appears in a gangue of clinopyroxene, Sn-rich garnet, epidotes, hastingsite and biotite.

Sn Deposits (21). The Oropesa prospect is a new Sn deposit discovered recently by IGME, and at the moment the ore reserves are estimated at more than 10 million metric tons, with 0.25% SnO_2. The ore is associated with conglomerates, psammites and lutites of Carboniferous age. It appears as a stockwork in extensively hydrothermally altered (silicified, chloritizid and carbonatized) country rocks. The paragenesis includes cassiterite, arsenopyrite, pyrite, chalcopyrite, bismuthinite, stannite, galena and minor sphalerite in quartz - chlorite - carbonate gangue (IGME 1987a). Although the deposit is still under study it seems to have many features similar to the Sn replacement deposits of the Renison Bell type.

Magmatic Mineralizations (Ni, Cr)

Anecdotic Ni (pentlandite - pyrrothite disseminations) and Cr (chromite - magnetite - talc) ore showings appear associated with serpentinites that outcrop related to Carboniferous shear zones (Crousilles et al. 1974) in the Guadalbarbo river valley (NW of Córdoba).

Epithermal Ore Deposits

Brittle Hercynian deformation related to late extensional and/or strike-slip motion along regional-scale fracture zones produced convective cells of hydrothermal circulation, in close association with which different kinds of ore deposits were formed.

Hg-(Pb, Sb, Au) Ore Deposits (11). The Usagre mine is an old Hg mine exploited from the sixteenth century to 1971. Either massive or disseminated ore appears in irregular bodies of baryte and quartz in direct relationship with Cambrian limestones. It has been interpreted as a stratiform (Vázquez and Fernández Pompa 1976; Arribas 1979; Schemerhorn 1982; Apalategui and Higneras 1983) or hydrothermal orebody (IGME 1987b). Now this mine seems to be a part of a group of ore showings with the same parageneses and the same metal association (Hg-Ba-Pb-Sb-Au), related to a major Late-Hercynian fracture zone with associated epizonal igneous rocks. Careful study of ore-host rock relationships shows mineralization is always related to epigenetic silicification of limestones, which makes this deposit comparable to the epithermal carbonate hosted ore deposits (Tornos and Locutura 1988).

Pb-Zn Lodes (12). Multiple Pb-Zn lodes appear inside or near the Badajoz-Córdoba shear zone, in two sets of fractures (0–45° E and 80–100° E) oblique to the shear zone strike. Mineralization is composed of galena (with high Ag content) and sphalerite with minor chalcopyrite, pyrite, tetrahedrite in a brecciated matrix of silicified host rock and quartz with carbonates, barite and fluorite. Deposits show little extension (less than 40 m) and great variations in grade. The veins are interpreted as tensional fractures devoloped as a consequence of late sinistral brittle movements of the shear zone. Ore formation processes are related to circulation of fluids of epithermal characteristics (Th = 200 °C, hyposaline solutions, no traces of CO_2, Chacón et al. 1984) that remobilize possible original stratiform ores such as the ones of the Azuaga Fm.

Ba-F-(Pb, Zn, Cu) Veins. Near the Villaviciosa - La Coronada magmatic belt (15) and other minor zones (preferentially Precambrian volcanic rocks), baryte-fluorite lodes with sparse chalcopyrite, pyrite, galena and sphalerite appear. Some of them can be related to Late Hercynian plutonism [such as the ones of Los Pedroches (17) and Cerro Muriano (16) area], but others clearly cut post-Hercynian rocks (Triassic beds). For these, a plutonic related origin is not clear and perhaps they are related to mixing of epithermal and surface origin fluids, such as other Ba-F deposits in the Hercynian Massif (Locutura and Tornos 1987) in which Ba is related to hydrothermal or meteoric alteration of igneous rocks.

Uranium Mineralizations. Spatially associated with black schists and quartzites of Precambrian age there are U-(Co-Ni-Cu) showings (Arribas 1962b) related to a possible low temperature Late Hercynian shear zone. From the main mine (Cabra Baja), now exhausted, 65000 metric tons with 1.5% U_3O_8 have been extracted.

Coal Deposits

Coal was formed in the three kinds of late postcollisional Hercynian basins; earlier pull-apart basins (Upper Tournaisian-Lower Visean), Westphalian intramontane and Late Stephanian basins. Main coal measures (up to 1-5 million m. t. reserves) are related to the second type (namely at the Peñarroya-Belmez coalfield) and are associated to fluviatile and lacustrine facies. Further detailed descriptions of these coal mineralizations can be found in Gabaldón and Quesada (1986), and Quesada and Garrote (1983).

6.3 Discussion and Conclusions

The most significant metallogenic feature of the Ossa-Morena Zone is the occurrence of two main metallogenic belts (heterochronous provinces in the sense of Routhier 1984, Fig. 1). Those belts are characterized by the occurrence of Pb-Zn (Cu) mineralizations at the NE of the Monesterio thrust zone, while at the SW, Fe (Cu) ore deposits predominate; the only exception are the ore showings of the Aracena Massif where Pb-Zn and Fe-Cu mineralizations appear, but original relationships of the Aracena Massif with other zones are unknown. This tectonic limit between the two belts is of Hercynian age, but metallogenic differences exist from Upper Precambrian to Late Hercynian rocks. In each domain, Fe or Pb-Zn mineralizations are successively recycled from at least Riphean (Zn-Pb) or Upper Vendian (Fe) to Carboniferous times (Fig. 2).

Hercynian mineralizations are characteristic of an active marginal setting (early Fe skarns, massive sulphides) with late collision (Cu-Pb-Zn lodes an skarns, W-Sn-Bi mineralizations) and final tensional tectonics with epithermal ore deposits, within igneous evolution similar to the one proposed by Harris et al. (1986). Previous mineralizations are less clear, because of unrecorded geological data and distorsion produced by the Hercynian event. The Cadomian orogeny is recorded by active margin mineralizations (untypical collision ores appear) and Cambrian to Early Devonian deposits are those typical of tensional (Cambrian) and stable passive margin environments. Such ore deposit models fit quite well in geotectonic models proposed by several authors (e. g. Quesada et al. 1987).

The Ossa-Morena Zone lies at the intervening area between the sub-Portuguese Zone with external basin and massive sulphide deposits, and the Central-Iberian Zone, with mainly collisional magmatism and associated mineralizations (Arribas 1978; Vázquez 1978; Schermerhorn 1982). Such situations permit both kinds of mineralizations along with precollision mineralizations and previous ones. Intensive thrust tectonics along with low-grade metamorphism and limited collisional plutonism allow the preservation of earlier ores, in contrast to northern zones. The importance of pre-Hercynian events with records of Cadomian events and Early Paleozoic evolution explain the complex spectrum of mineralizations.

Acknowledgements. This work is a resumé of metallogenic studies carried out by IGME in recent years in the Ossa-Morena Zone. We wish to thank A. Sanchez, C. Quesada and R. Urbano for their collaboration, discussions and improvement of the final manuscript.

References

Apalategui D, Higueras P (1983) Mapa Geológico de España, II serie, nº 855, Usagre IGME

Aparicio A, Barrero JL, Casquet C, Peinado M, Tineo JM (1977) Caracterización geoquímica del plutonismo postmetamórfico del Sw del macizo Hespérico. Stud Geol 12: 9-39

Arribas A (1962a) Mineralogía y metalogenia de los yacimientos españoles de uranio: Burguillos del Cerro (Badajoz). Estud Geol 18: 173-192

Arribas A (1962b) Mineralogía y metalogenia de los yacimientos de uranio: Monesterio (Badajoz). Not Comun IGME 70: 47-69

Arribas A (1962c) Mineralogía y metalogenia de los yacimientos españoles de uranio. Los indicios con davidita de Villanueva del Fresno (Badajoz). Estud Geol 18: 33-51

Arribas A (1964) Mineralogía y metalogenia de los yacimientos españoles de uranio: Cardeña (Córdoba). Not Comun IGME 76: 45-78

Arribas A (1979) Mineral paragénesis in the Variscan metallogeny of Spain. Stud Geol 14: 223-260

Arriola A, Cueto LA, Fernández J, Garrote A (1984) Serpentinitas y mineralizaciones de cromo asociadas, en el Proterozóico superior de Ossa-Morena. Cuad Lab Geol Laxe 88: 137-145

Baeza L, Ruiz C, Ruiz Montes M (1978) Presencia de formaciones volcanosedimentarias y mineralizaciones de hierro asociados en el eje magmático de La Coronada-Villaviciosa (Córdoba). Bol Geol Min vol 89-5: 431-437

Baeza L, Ruiz C, Ruiz Montes M, Sánchez A (1982) Mineralizaciones exhalativo sedimentarias de sulfuros polimetálicos en la Sierra Morena (Córdoba). Bol Geol Min 92-3: 203-216

Beltrán FJ (1976) Geología de un sector comprendido entre Encinasola y Aroche (Huelva). Thesis, Univ Granada

Casquet C (1980) Fenómenos de endomorfismo, metamorfismo y metasomatismo de contacto en los mármoles de Rivera de Cala (Huelva). Thesis, Univ Complutense de Madrid, 295 pp

Casquet C, Velasco F (1978) Contribución a la geología de los skarns cálcicos en torno a Santa Olalla de Cala (Huelva-Badajoz). Estud Geol 43: 399-405

Chacón J, Herrero JM, Velasco F (1984) Las mineralizaciones Zn-Pb de Llera (Badajoz). I Congr Esp Geol 2: 447-455

Coullaut J (1980) Geología y metalogenia del criadero de San Guillermo-Colmenar, Jerez de los Caballeros (Badajoz). l Curso Rosso de Luna, IGME

Coullaut JL, Babiano F, Fernández Carrasco J (1980) Mineralizaciones de hierro del Suroeste de España. Mina La Valera. J Min Metal Huelva, pp 97-108

Crousilles M, Dixaut C, Henry B, Tamain G (1974) L'alignement basique-ultrabasique de Varas-Guadalbarbo (Córdoba, Espagne). C R Acad Sci Paris Ser D 283: 141-143

Delgado M, Pascual E, Fenoll P (1978) A geological and metallogenic study of some occurrences of magnetite and sulphides in Sierra Morena (NW of Córdoba, Spain). Estud Geol 34: 461-474

Doetsch J, Rivas G, Locutura J, Arana M, Calvo B, Cueto L (1973) Estudio actual de la investigación de criaderos del SW de España. Bol Geol Min 94-2: 105-129

Dupont J (1978) Cadre geologique et metallogenese des gisements de fer du sud de la province de Badajoz (Sierra Morena, Espagne). Thesis, Univ Polytechnique Lorraine

Einaudi MT, Burt DM (1982) Introduction, terminology, classification and composition of skarn deposits. Econ Geol 77-4: 745-754

Einaudi M, Meinert L, Newberry R (1981) Skarn deposits. Econ Geol (anniversary volume) 75: 317-391

Gabaldón V, Quesada C (1986) Exemplex de bassins houillers liminiques du Sud-Quest de la Peninsula Ibérique: evolution sédimentaire et controle structural. Mem Soc Geol Fr 149: 29-38

Gabaldón V, Garrote A, Quesada C (1985) El Carbonifero Inferior del Novite de la zona de Ossa-Morena (SE de España). X Cong Geol Carbonifero, pp 173-185

Garrote A (1976) Asociaciones minerales del núcleo metamórfico de Sierra Albarrana (provincia de Córdoba), Sierra Morena Central. Mem Not Publ Mus Lab Geol Univ Coimbra 82: 17-39

Gumiel P (1983) Metalogenia de los yacimientos de antimonio de la Península Ibérica. Tecniterrae 53: 120

Gumiel P, Vindel E, Lunar R, Sierra J, López A (1983) Contribución al conocimiento de las mineralizaciones de W, bi del distrito de Oliva de la Frontera (Badajoz). Mater Procesos Geol 1: 229-248

Gustafson L, Williams N (1980) Sediment hosted stratiform deposist of Cu, Pb, Zn. Econ Geol (anniversary volume) 75: 139-178

Harris NBW, Pearce JA, Tindle AG (1986) Geochemical characteristics of collision zone magmatism. In: Coward R (eds) Collision tectonics. Geol Soc Spec Publ 19: 67-81

IGME (1972) Mapa Metalogenetico de España 1/200000 Hoja n° 69 (Pozoblanco)

IGME (1984) Exploración minera en las zonas de Oliva de Mérida y Puebla de la Reina (Badajoz). Inst Geol Min Esp Int Rep

IGME (1987a): Investigación minera en la zona de Oropesa. Inst Geol Min Esp Int Rep

IGME (1987b) Mapa Metalogénico de España escala 1/200000 n° 67-68. Inst Geol Min Esp Int Rep

Locutura J, Tornos F (1987) Aspectos genéticos de las mineralizaciones de F (Ba-Pb) del área de Colmenar de Arroyo (Sistema Central Español). Bol Geol Min 98-5: 680-694

Miras A, Hernando MJ, Galán E (1987) Estudio de algunas mineralizaciones de barita al Sur de Llerena (Badajoz). Bol Soc Esp Mineral 10: 55-56

Mitchell A, Garson M (1981) Mineral deposits and global tectonics setting. Academic Press, London, 410 pp

Moro C, Arribas A (1982) Los yacimientos de barita de España. Tecniterrae 264: 18-45

Ovtracht A, Tamain G (1973) Tectonique, migration des 'centres chauds' et minéralizations dans le Sud de la Meseta Ibérique (Espagne). In: Raguin E (ed) Les roches plutoniques dand leurs rapports avec les gites mineraux. Masson, Paris, pp 191-211

Pitcher WS (1982) Granite type and tectonic environment. In: Hsu (ed) Mountain building processes. London, pp 19-40

Pons J (1982) Un modéle d'evolution de complexes plutoniques: Gabbros et granitoides de la Sierra Morena Occidentale (Espagne). PhD Thesis, Pauel Sabatier-Toulouse

Presur (1987) Investigacion Minera en la zona de la Masaguera. Int Rep

Quesada C (1983) El Carbonífero de Sierra Morena. In: Martínez E (ed) Carbonífero y Pérmico en España. IGME, pp 243-278

Quesada C, Garrote A (1983) Carboniferous of the Sierra Morena. X Int Cong Geol Carboniferous, Guidebook of field trip D, 104 pp

Quesada C, Florido P, Gumiel P et al. (1987) Mapa geológico-minero de Extremadura. Junta de Extremadura, 131 pp

Routhier E (1984) Ou sont les metaux pour l'avenir? Mem BRGM 105: 410

Ruiz C (1976) Génesis de los depósitos de hierro del SW de la provincia de Badajoz: mina Monchi. Bol Geol Min 87-1: 15-31

Ruiz de Almodovar G (1980) Nota sobre las mineralizaciones de wolframio de Oliva de la Frontera, suroeste de Badajoz. Cuad Geol 11: 167-173

Ruiz de Almodovar G, Galán E (1984) Mineralizaciones de Manganeso-hierro del suroeste de Badajoz (zona de Ossa-Morena). Congr Esp Geol 2: 643-657

Schermerhorn LJG (1982) Framework and evolution of Hercynian mineralization in the Iberian Meseta. Comun Serv Geol Port 68 (1): 91–140

Tornos J, Locutura J (1988) Mineralizaciones epitermales de Hg (Pb, Au) en Ossa Morena (Usagre, Badajoz). Bol Soc Esp Min 11 (2): 124–127

Vázquez F (1972) Génesis de la Mina María Luisa, La Nava (Huelva, España): Una mineralización zonada. Bol Geol Min 83 (4): 377–386

Vázquez F (1978) Depósitos minerales de España. Temas Geol Min IGME, Tema II

Vázquez Guzman F, Fernández Pompa F (1976) Contribución al conocimiento geológico del suroeste de España en relación con la prospección de depósitos de magnetita. Mem IGME 80: 130

Velasco F (1976) Mineralogía y metalogenia de las skarns de Santa Olalla (Huelva). Thesis, Univ Lejona

Zitzmann A, Neumann Redlin C (1977) The iron ore deposits of Spain. In: Zitmann A (ed) Iron ore deposits of Europe and adjacent areas. Hannover

Part VI South Portuguese Zone

1 Introduction

J. T. OLIVEIRA[1]

The South Portuguese Zone (SPZ) represents the southern branch of the Iberian Variscides. For a long time, geological studies in this zone were focused mostly on its mid-part, the so-called Iberian Pyrite Belt, which is the locus of impressive reserves in polymetallic massive sulphides. It was only during the last two decades that the areas north and south of the Pyrite Belt attracted more attention. The existing knowledge resulted from persistent work carried out by geologists from private companies and State institutions.

The matters dealt with in this chapter cover the classical domains of the geology, i. e. stratigraphy/paleontology, structure, metamorphism and metallogeny. Furthermore, geochemical, sedimentological and geophysical data constitute information fundamental to the understanding of the geology of the SPZ, and these are referred to in the following chapters.

In south Portugal the most complete geotraverse of the SPZ crops out. The geology of this region has always been the subject of leading work, mainly in the fields of stratigraphy, structure, geochemistry and metallogeny. Therefore, it should not be surprising that, in all of the following reports, emphasis is put on this important segment of the SPZ. The works reported by the different authors constitute excellent syntheses concerning the subjects and, as a whole, this chapter represents the state of the art regarding the geology of the SPZ.

[1] Servicos Geológicos de Portugal, Rua de Academia des Ciencias, 1200 Lisbon, Portugal

In spite of the good global level of scientific knowledge, the reader will detect some areas of research in which problems remain to be solved. This is particularly true for the age control of important parts of the stratigraphic sequence, namely the units of the Pulo do Lobo Antiform and, to a certain extent, of the Pyrite Belt. This situation introduces constraints in time-stratigraphic correlations and lead, directly or indirectly, to some controversies and unavoidable speculations; for example, is it reasonable to infer from the available data that the stratigraphic sequence north of the Pyrite Belt forms an antiform (Oliveira, Part VI, Chap. 2)? Have the MORB-type basalts, interbedded in the Pulo do Lobo Formation, any genetic relationship with the Beja-Acebuches Ophiolite? If so, do they indicate an oceanic basin between the Pyrite Belt and the Ossa-Morena Zone (Silva et al., this Vol.)? Or is the Pulo do Lobo sequence a suspected terrane, as suggested by Ribeiro et al. (Chap. 8, this Vol.)?

These and other questions are now raised, due to the scientific advances achieved recently and which are incorporated in the following reports. It is hoped that research in progress will soon find adequate answers for such questions, in order to clarify the role of the SPZ within the context of the Variscan Orogen.

References

Carvalho D, Correia H, Inverno C (1976) Contribuição para o conhecimento geológico do Grupo de Ferreira-Ficalho. Suas relações com a Faixa Piritosa e o Grupo do Pulo do Lobo. Mem Not Mus Lab Min Fac Ciênc Coimbra 82: 145-169

R. D. Dallmeyer and E. Martínez García (Eds.)
Pre-Mesozoic Geology of Iberia

2 Stratigraphy and Synsedimentary Tectonism

J. T. Oliveira[1]

2.1 Introduction

The stratigraphy of the South Portuguese Zone (SPZ) has been described for many selected parts; however, regional studies are scarce (Schermerhorn 1971; Routhier et al. 1980; Oliveira 1983), and an overview of this palaeogeographic zone has not been published. This synthesis provides an interpretation of the geological environment based on sedimentary processes, volcanism and synsedimentary tectonism. A strong bias is put on the Portuguese part of the SPZ, where the stratigraphic sequence is the most complete and with which the author is more familiar. However, references are made to the fundamental work made in the Spanish part of the SPZ during the last two decades.

Within the SPZ, five main geological domains (Fig. 1) are recognized in this work; namely, Beja-Acebuches Ophiolite, Pulo do Lobo Antiform, Pyrite Belt, Baixo Alentejo (BA) Flysch Group and Southwest Portugal (Bordeira and Aljezur Antiforms). In detail, each of these domains have many unresolved stratigraphic problems, particularly the Pulo do Lobo Antiform and the Pyrite Belt. Despite these problems, however, clear geological relationships can be established between the Pyrite Belt, the BA Flysch and the SW Portugal domains. Relationships of the Beja-Acebuches Ophiolite and the Pulo do Lobo Antiform to the other geological domains are uncertain.

2.2 Beja-Acebuches Ophiolite

The Beja-Acebuches Ophiolite (Andrade 1977; Munhá et al. 1986) crops out exactly at the boundary between the Ossa-Morena Zone and the Pulo do Lobo Antiform. The ophiolite belt is less than 1500 m wide, and is more than 100 km long. Petrological evidence shows that the belt has a clear internal organization characteristic of ophiolites – that is, from base to top: metagabbros (granulites, flaser gabbros and amphibolites s. l.), with scattered serpentinite bodies, a sheeted dyke complex and metabasalts with intercalated cherts. Geochemical and geological data (Bard 1977; Dupuy et al. 1979; Munhá et al. 1986) suggest that this ophiolite was generated in a small oceanic basin. It has a steep metamorphic gradient and a tectonic inversion that are related to thrusting of the infracrustal metamorphic rocks of the Ossa-Morena Zone, over the ophiolite suite (Castro et al. 1987, unpublished work). The south boundary is sheared by left lateral thrust faulting. The ophiolite was dismembered by Late-Hercynian faults.

In spite of tectonic complications, field data indicate that the ophiolite stratigraphically underlies the Horta da Torre Formation, the upper unit of the Ferreira-Ficalho Group (see below). This formation has recently been dated in Portugal as Early to Middle Famennian (Oliveira et al. 1986a). Consequently, the ophiolite generation and emplacement are older than the Late Famennian to Middle Visean volcanosedimentary assemblage of the Pyrite Belt.

The locus of the ophiolite generation is uncertain. Available data suggest tectonostratigraphic and geochemical links between the Beja-Acebuches Ophiolite and the Pulo do Lobo Formation, as discussed below.

2.3 Pulo Do Lobo Antiform

The stratigraphic units of the Pulo do Lobo Antiform are schematically shown in Table 1. For the geological mapping of these units see Oliveira et al. (1986a).

Pulo do Lobo Formation. This unit is composed of highly sheared phyllites and quartzites, minor intercalations of felsic volcanics and some deformed fel-

[1] Serviços Geológicos de Portugal, Rua de Academia das Ciencias 1200 Lisbon, Portugal

R. D. Dallmeyer and E. Martínez García (Eds.)
Pre-Mesozoic Geology of Iberia

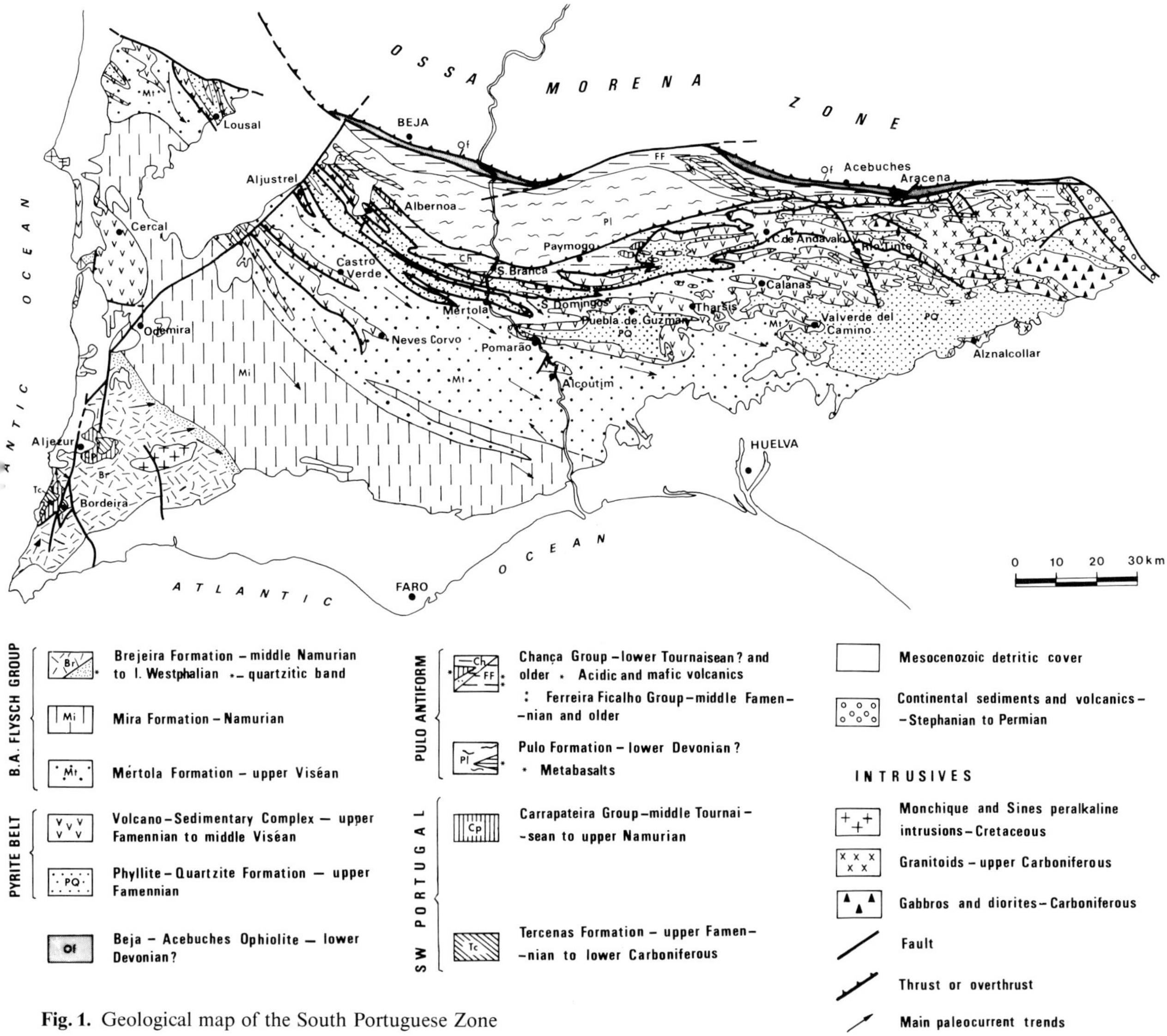

Fig. 1. Geological map of the South Portuguese Zone

sic dykes (Pfefferkorn 1968). Towards the base are metabasalts of MORB-type geochemical affinity (Munhá 1983). A typical feature of this unit is the widespread occurrence of exsudation quartz veinlets, which are related to strong tectonic deformation and associated cleavages that resulted from three episodes of folding.

Ferreira-Ficalho Group. This group comprises the following units, from base to top (Pfefferkorn 1968; Carvalho et al. 1976; Oliveira et al. 1986a; IGME map sheets 915, 916, 917): Ribeira de Limas Formation, made up of phyllites, metagreywackes and minor tuffites; Santa Iria Formation, a succession of shales, siltstones and greywackes that form a flysch-like succession; and Horta da Torre Formation, composed of shales, impure sandstones and siltstones, and quartzite beds towards the top. The Ribeira de Limas Formation grades downwards to the Pulo do Lobo Formation and is deformed by the same three episodes of folding. The contact between the Santa Iria and Horta da Torre Formations appears gradational. These formations are less deformed than the underlying units, and they show only one main NW to SE trending episode of folding with associated clearage. Tectonic thickening complicates the determination of stratigraphic thickness but it is probably about 500 m for the whole succession. From the units of the Ferreira-Ficalho Group only the Horta da Torre Formation has yielded spores and acritarchs of Early to Middle Famennian age (Oliveira et al. 1986a). In Spain

Table 1. Main lithostratigraphic units of the Pulo do Lobo Antiform and Pyrite Belt

	PULO DO LOBO ANTIFORM		PYRITE BELT
	North limb	South limb	
Middle Visean to lower Tournaisean			VS Volcano-Sedimentary Complex
Upper Famennian	FERREIRA FICALHO GROUP	CHANÇA GROUP: Rp / Gf — Rp-Represa Fm, Gf-Gafo Fm	PQ / BH — PQ-Phyllite-Quartzite Fm, BH-Barranco do Homem Fm
Middle to lower Famennian	FERREIRA FICALHO GROUP: HT / SI — HT-Horta da Torre Fm, SI-Santa Iria Fm	CHANÇA GROUP: At — Atalaia Fm	
Frasnian and older	FERREIRA FICALHO GROUP: RL — Ribeira de Limas Fm		
Lower Devonian ?	Pl — Pulo Fm		

greywackes and shales ascribed to the Santa Iria Formation yielded spores whose age was contradictorily determined as Upper Famennian (Giese et al. 1988) or Upper Frasnian (O. Lake, pers. comm.).

Chança Group. This group, here informally defined, also includes three main units (Pfefferkorn 1968; Carvalho 1976, unpublished work; Carvalho et al. 1976; Salpeteur 1976; Oliveira et al. 1986; Silva et al., this Vol.): Atalaia Formation composed mostly of phyllites and sandstones; Gafo Formation, a thick (650 m) pile of greywackes, siltstones and shales, with intercalations of acidic and mafic volcanics (Trindade volcanics) and intrusions of basic and acidic dykes; Represa Formation, a local variation of the Gafo Formation in the Mina de São Domingos area, which is composed of siliceous siltstones, shales and quartzwackes. The group thickness is estimated at 1100 m.

The age of the Chança Group units is doubtful, due to the lack of guide fossils. The Represa Formation provided recently a poor assemblage of Early Upper Famennian spores and acritarchs (Cunha, pers. comm.). This suggests that all the Chança Group units are older than Upper Famennian, and should be considered time-stratigraphic equivalents of the units of the Ferreira-Ficalho Group, as previously pointed out by Schermerhorn (1971) and Carvalho et al. (1976).

The location of the original depocentre for the units of the Pulo do Lobo Antiform is open to debate. The oceanic character of the metabasalts interbedded in the Pulo do Lobo Formation suggests that these basalts and the Beja-Acebuches Ophiolite were originated in close depositional settings, possibly in a small oceanic basin. This inference seems supported by the occurrence of a well-marked positive gravimetry anomaly over the Pulo do Lobo Formation. Although not proven, this anomaly may be due to the occurrence of basic rocks below the Pulo do Lobo Formation. The overlying units of the Ferreira-Ficalho and Chança Groups represent the terrigenous infilling of this basin. Tectonostratigraphic relationship of this basin, with respect to the Ossa-Morena and Pyrite Belt terranes is uncertain. Two possibilities for further research are: (1) the small oceanic basin originated between the Pyrite Belt and the Ossa-Morena Zone, and which is at or close to its original position (Munhá et al. 1986) and (2) the whole Pulo do Lobo Antiform/Beja-Acebuches Ophiolite, which now forms an allochthonous terrane whose original position is unknown.

2.4 Pyrite Belt

The Pyrite Belt is divided into two main branches. The first, to the south, consists of rooted (parautochthonous) structures; the second, to the north, forms a predominantly allochthonous belt (Fig. 1; see also Fig. 1 in Silva et al., this Vol.).

2.4.1 South Branch

This branch (Fig. 1) includes the major anticlines of Valverde del Camino, Puebla de Guzman, Castro Verde-Ourique, Cercal and the less impressive structures of Alcoutim and São Francisco da Serra.

Phyllite-Quartzite Formation (PQ). This unit, with variable local names, forms the detrital basal formation of the Pyrite Belt. It consists of alternations of sandstone packets, which are metres to tens of metres thick, and shales. The sandstone lithologies include mostly quartzites and quartzwackes, and rare conglomerates. These rocks are highly deformed and no reliable stratigraphic logging is possible. However, the combination of sedimentary structures (graded and massive bedding, parallel lamination, current and wave ripples, soft sediment deformation and bioturbation) indicates that this unit was deposited in a shallow water environment, possibly an epicontinental sea.

At the top, the sandstones are overlain by a very continuous shaly horizon (30 m) that displays lenses and nodules of bioclastic limestones. This horizon is here interpreted as a transitory mud-carbonate shelf. The limestones provided minor faunas of trilobites, clymenids and brachyopods (Pruvost 1912/13) and abundant faunas of conodonts (Boogaard 1963; Boogaard and Schermerhorn 1980, 1981). The latter range from the lower *marginifera* zone (Forno da Cal, SE of Castro Verde) through the *velifer* and *styriacus* zones (Mértola, Pomarão, Puebla de Guzman Anticline) to the upper *costatus* zone (near Nerva). Therefore, the shaly/limestone horizon has a Middle to Late Famennian age, and the quartzites below may be older.

Volcanosedimentary Complex (VS). The interfingering of volcanic and sedimentary rocks is a common feature of this complex (Fig. 2b, c, d, e, f, g). Palaeontological data (Boogaard 1963; Boogaard and Schermerhorn 1980; Oliveira 1983) show clearly that it has a Late Famennian to Early Late Viséan age. This age range is based on the assumption that the VS complex conformably overlies the PQ, and also on the occurrence of conodont-bearing limestones of Late Tournaisian *anchoralis* zone, near Sotiel (Boogaard and Schermerhorn 1975), and of Early Late Viséan *billineatus* zone, east of Castro Verde (Oliveira 1983).

In the central area of the Pyrite Belt (Puebla de Guzmán Anticline and the west part of the Valverde Anticline), three main volcanic acidic episodes are recognized (Boogaard 1967; Routhier et al. 1980; IGME map sheets, Fig. 2c, d, e): the first episode (V_1) comprises fine and coarse pyroclastics and minor lavas of rhyolitic composition; the second (V_2), also contains pyroclastics and subordinate rhyolitic to rhyodacitic lavas; the third (V_3), is mostly composed of reworked tuffs and siliceous shales.

Mafic lavas, sometimes with pillowed structures, are usually interbedded between the V_1 and V_3 volcanic units.

Immediately below - the V_3 tuffs crops out a horizon of purple and green shales, and the pair V_3/purple shales forms an excellent stratigraphic marker of regional significance. Below this marker, the volcanics interfinger with a large variety of sedimentary rocks, which include black (sometimes chloritic) shales, siltstones and fine impure sandstones, jaspers (exhalites), nodules and lenses of Fe/Mn oxides and carbonates, and minor limestones. These sediments constitute the intermediate series, so called by the Spanish workers.

The lower half of the stratigraphic sequence is commonly intruded by sills and dykes of diabase, which are considered to be feeders of the interbedded post-V_1 mafic lavas. Locally, as in the Pomarão region, andesites may also occur (Munhá 1983).

Outside the central area these three volcanic episodes cannot be positively recognized, and the lack of stratigraphic information precludes any reliable correlation among the different volcanic cycles.

The thickness of the VS assemblage ranges between 100 and 600 m, depending on the proximity to volcanic centres.

2.4.2 North Branch

This branch is bounded, to the north and south, by major thrust faults. Two distinct geotectonic settings occur: parautochthonous volcanosedimentary successions and allochthonous assemblages. What follows is based mostly on the Portuguese part of this branch, with which the author is most familiar.

Parautochthonous Volcanosedimentary Complex. This succession is well represented in the Albernoa,

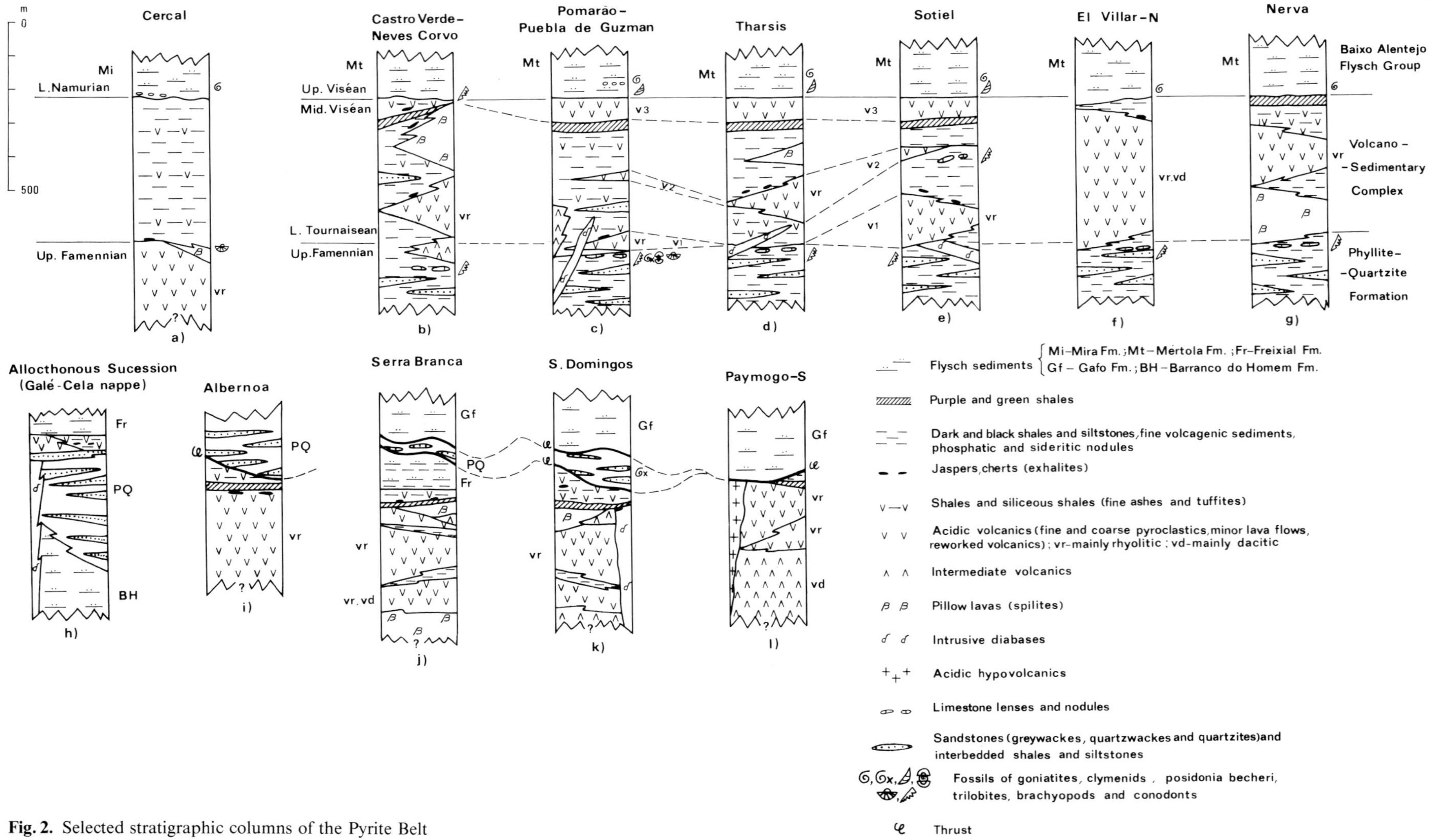

Fig. 2. Selected stratigraphic columns of the Pyrite Belt

Serra Branca and Mina de São Domingos Anticlines (Portugal) and probably in the Paymogo and Cerro de Andevalo Anticlines (Spain). The selected stratigraphic columns (Fig. 2i, j, k, l) show the variability of the lithostratigraphic sequence. They illustrate one main acidic and pyroclastic episode in Albernoa, as compared to one acidic event (with some lavas), plus two distinct episodes of mafic lava deposition and minor andesites, at Serra Branca and São Domingos. The Serra Branca assemblage is overlain by the Freixial Formation, a flyschoid sequence, about 200 m thick, which has thin intercalations of tuffite layers. This unit, whose type section is exposed along the road cut on the west side of the Guadiana river near Rocha da Galé, marks the beginning of turbidite deposition.

The stratigraphic thickness of these successions is not known, because the base is never exposed and the top is truncated by overthrusts.

A review of the published work (Routhier et al. 1980; IGME sheet maps) suggests that the Paymogo and Cerro de Andevalo volcanic belts in Spain are in close continuation with the parautochthonous VS Complex of Portugal. In Spain, the belts are also dominated by one main acidic volcanic episode; rhyodacitic lavas and ignimbrites are relatively common (Routhier et al. 1980). Acidic hypovolcanics, as illustrated in the Paymogo column (Fig. 2, *1*) are also represented. Basic volcanics are, however, much rarer than in Portugal.

Allochthonous Assemblages. Over large areas of the north branch, PQ-type sedimentary rocks appear "mixed" with scattered intercalations of fine-grained acidic volcanics, rare mafic lavas and diabase intrusions. This assemblage of lithologies commonly overlie autochthonous successions of the VS Complex.

The results of detailed work in the Mértola area (Oliveira and Silva, in press; Oliveira 1988, unpublished work; Silva 1985) and of a regional survey from Mértola to Aljustrel, show that the lithologies of the upper assemblage form allochthonous thrust sequences which are organized in two distinct successions. These successions form the Galé-Cela and Mértola Nappes (Silva et al., Fig. 1, this Vol.) and have a consistent stratigraphy.

Galé-Cela Nappe. This nappe includes lithologies of the Barranco do Homem and PQ Formations, the VS Complex and Freixial Formation (Fig. 2h). The Barranco do Homem Formation is composed of quartzwackes, siltstones and shales, and forms a terrigenous sequence of regional significance from Mértola to Aljustrel. In Aljustrel, it corresponds to what has been described as the Gomes Quartzite Formation (Schermerhorn and Stanton 1969). The unit is highly deformed and its boundary with the PQ Formation is commonly tectonized. Structural data suggest that it occupies the cores of large recumbent folds (Silva et al., this Vol.; Oliveira and Silva, in press). The best exposed section, about 300 m thick, crops out along the Barranco do Homen stream, 2 km SW of the village of Corte da Velha (Mértola).

The Barranco do Homem Formation has many lithological similarities with the Represa Formation of the Chança Group. The lack of fossils in the Barranco do Homem Formation and the impressive tectonism are major difficulties that prevent their correlation.

The PQ Formation has the same sedimentological characteristics as in the Puebla de Guzman Anticline, but locally the quartzites are thicker bedded and form continuous ridges hundred of metres in length. As in the Puebla de Guzman Anticline, the sandstones appear frequently as isolated blocks surrounded by shales. This geological setting may result from the combination of sedimentary features (rapid pinch out of sedimentary beds, soft sediment deformation and gravity slides) and superimposed tectonic shearing.

The VS Complex lithologies comprise volcanogenic sediments (dark siliceous shales and cherts) that interfinger with tuffites. The VS stratigraphic succession is about 50 m thick.

The Freixial Formation is composed of greywackes, shales and siltstones everywhere.

Mértola Nappe. The internal stratigraphy of this nappe comprises minor quartzites and shales of the PQ Formation, distal volcanics of the VS Complex and greywackes, siltstones and shales of the Freixial Formation. The latter shows, in places, a very low sand/shale ratio.

From early work (Carvalho 1979; Routhier et al. 1980; IGME map sheets), it was suggested that a stratigraphic continuity exists between the autochthonous Upper Famennian to Middle Visean volcanosedimentary suite and the allochthonous sequences. Limestone nodules, quartzites and shales of the Galé-Cela Nappe, in the Mina de São Domingos and Mértola areas, have yielded conodonts (*Striacus* zone), clymenids (*Platyclymenia* sp., *Cymaclymenia* sp.) and spores of Upper Famennian age, and conodonts (*anchoralis* zone) of late Tournaisean age (Fantinet et al. 1976; Oliveira 1983; Oliveira et al. 1986b, Cunha, pers. comm.). Therefore, these rocks are partially coeval to the PQ Formation and VS Complex of the Puebla de Guzman Anticline and are now in an allochthonous position (Silva et al., this Vol.). The stratigraphic sequence of the Mértola Nappe is also considered a time equivalent of that of the Puebla de Guzman Anticline.

2.4.3 Volcanism, Sedimentation and Tectonics

In previous work, volcanism, mineralizations and sedimentation of the Pyrite Belt were related to E-W and NE-SW (Carvalho 1974; Routhier et al. 1980) or NW-SE (Rambault 1978) tectonic lineaments. Furthermore, Carvalho et al. (1976) and Routhier et al. (1980) have suggested a northward migration of the volcanism. These interpretations were based on the distribution of the volcanic centres and on the assumption that the stratigraphic sequence is predominantly autochthonous. Stratigraphic control of the different volcanic events is not fully understood and, at the present stage of our knowledge, any reconstruction is always somewhat speculative. Despite these difficulties, the examination of available data permits a different interpretation.

From the overall distribution of the volcanism, and taking tectonic transport into consideration, it appears that the stratigraphic sequence of the North Branch is dominated by the occurrence of one main acidic event (with lavas and ignimbrites) and minor mafic volcanics at the lower and upper levels. Over large areas of the South Branch, three acidic episodes and interbedded mafic lavas occur. The acidic volcanics are commonly reworked. Thus, one can infer volcanic eruptions from fissures in the northern area, probalby linked to an early rifting stage, and a progressive development of the volcanism to the south. On the other hand, mafic lavas display a westward migration, as shown by their common occurrence below the main acid-volcanic event, in Nerva and surrounding areas (IGME map sheet No. 938), and by their predominance towards the top of the stratigraphic sequence, in Castro Verde and Lousal regions (Leca et al. 1983; Oliveira 1984). This migration was possibly related to the westward development of transverse faults.

Details of the initial stress regime are poorly known. However, the stratigraphic record (compare, for instance, columns b, c, d, e, f with column h, Fig. 2) of the above inferred transverse faults and the intracontinental character of the volcanism (Schermerhorn 1975; Routhier et al. 1980; Munhá 1983) strongly suggest that the Pyrite Belt depositional area may have been generated in southward prograding and late disrupted graben and half grabens. This extensional regime was operative after an initial pre-Famennian collision (coupled with left lateral shearing?) between the Pulo do Lobo Antiform and the Ossa-Morena Zone.

Tectonostratigraphic events critical to the evolution of the Pyrite Belt are as follow:

1. The rupture of the shallow water detritic platform and the transitory mud-carbonate shelf, during the Upper Famennian to Middle Viséan time (Fig. 3). After an initial stage of rifting, the depositional area was progressively divided into half grabens, which became the main marine depocentres and the locus of renewed volcanism. Terrigenous and volcanogenic sediments and remmants of the older mud-carbonate shelf were accumulated in the down-going troughs. Over the uplifted blocks only fine sediments and minor volcanics were preserved.

As pointed out above, tectonostratigraphic relationships between the south limb of the Pulo do Lobo Antiform and the Pyrite Belt are uncertain. The uncertainties are due to the poor information for various units of the Pulo do Lobo Antiform, and the scarcity of geochemical data to characterize the intrusive and extrusive volcanics that occur in the Gafo and Represa Formations. The units of the Chança Group show three episodes of folding and associated cleavages, while the units of the Pyrite Belt show only two episodes with associated cleavages. The Chança Group units were probably under compression during the deposition of the VS Complex of the Pyrite Belt.

2. Following renewed Lower Carboniferous oblique (?) subduction and continental collision between the SPZ and the Ossa-Morena Zone (Santos et al., in press; Silva et al., this Vol.), the former extensional regime gradually turned to a compressional one (tectonic inversion) which gave rise, Upper Visean onwards, to the development of the synorogenic flysch sedimentation (Fig. 4).

2.5 Baixo Alentejo Flysch Group

This group comprises mostly gravity flow sediments that form a continuous and southward prograding detritic cover (Fig. 1). Sedimentological and palaeontological data (Oliveira et al. 1979; Oliveira 1983; Oliveira and Wagner Genthis 1983) indicate three

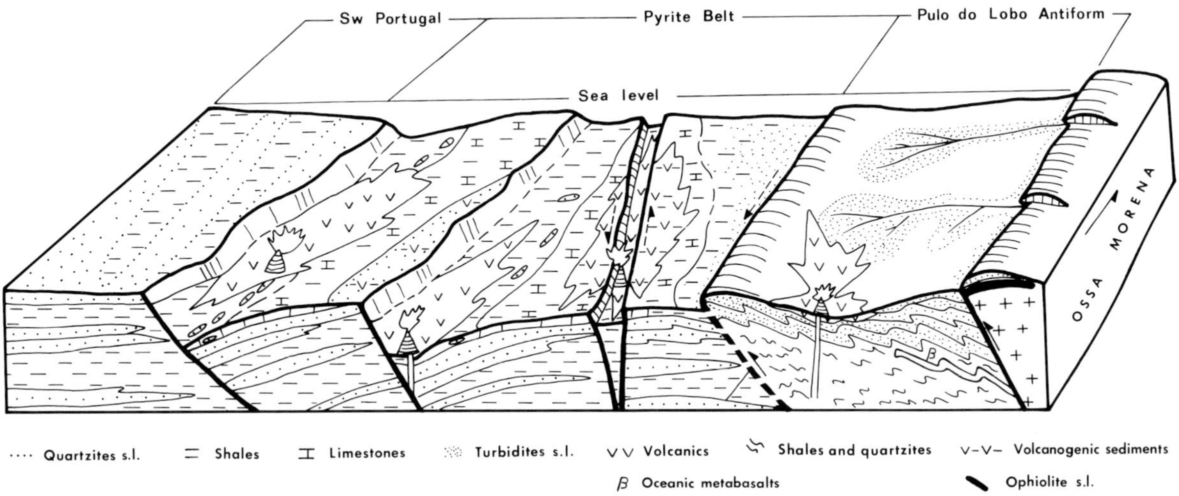

Fig. 3. Block diagram (schematic) of the South Portuguese Zone during the Upper Famennian to Middle Viséan times

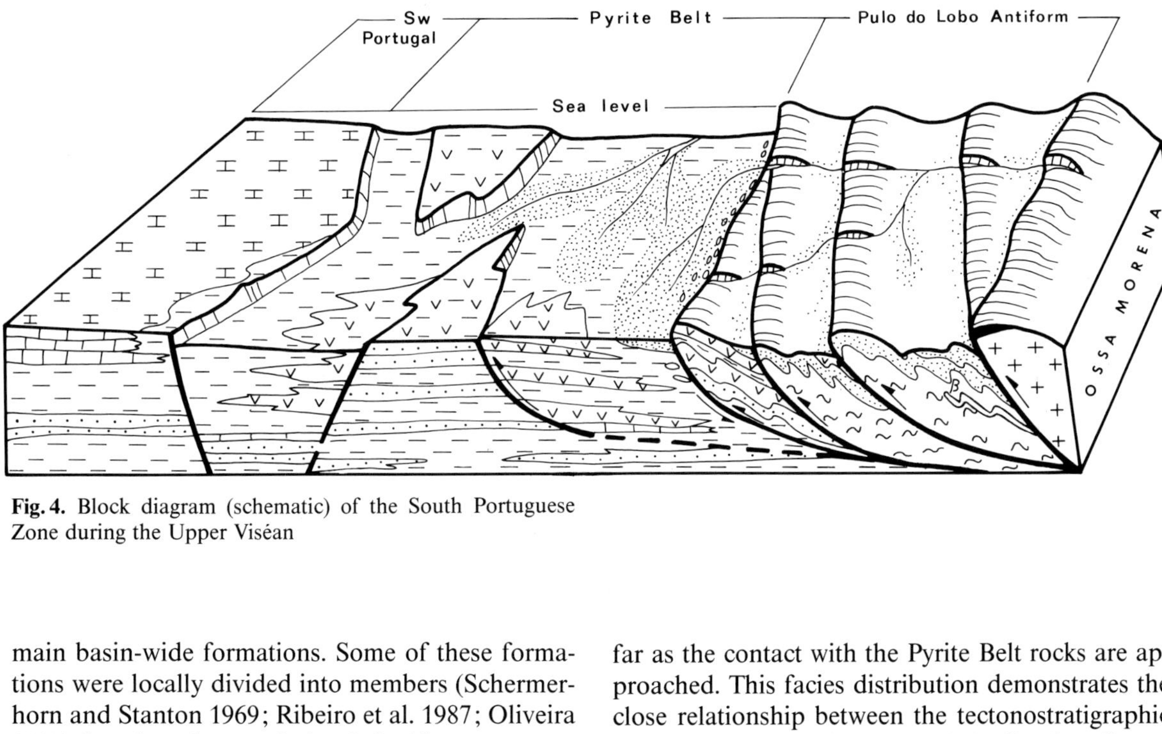

Fig. 4. Block diagram (schematic) of the South Portuguese Zone during the Upper Viséan

main basin-wide formations. Some of these formations were locally divided into members (Schermerhorn and Stanton 1969; Ribeiro et al. 1987; Oliveira 1988), but these have only local significance.

Mértola Formation. It is built up by sequences of thick- and thin-bedded sandstones, and alternating bands of shales and thin-bedded siltstones. The sandstone/shale ratio is high. The sandstones display many of the sedimentary features usually described for gravity flow sediments. Massive and amalgamated sandstones, fluidized sandstones, slumps, conglomerates, sand dykes, mud flows and isolated sheeted sandstones are more common as far as the contact with the Pyrite Belt rocks are approached. This facies distribution demonstrates the close relationship between the tectonostratigraphic evolution of the Pyrite Belt and the flysch sedimentation (Oliveira 1983). The Mértola Formation has yielded goniatites of the Upper Viséan $G_0\alpha$, $G_0\beta$ and $G_0\gamma$ zones, and abundant specimens of *Posidonia becheri* (Oliveira 1983; Oliveira 1988).

In the Mértola region, this unit has been divided into five members separated by five shaly horizons, taken as markers beds (Oliveira 1988). The members correspond to sedimentary lobes and the shaly horizons to interlobe sediments. The sedimentary lobes prograded to SSW.

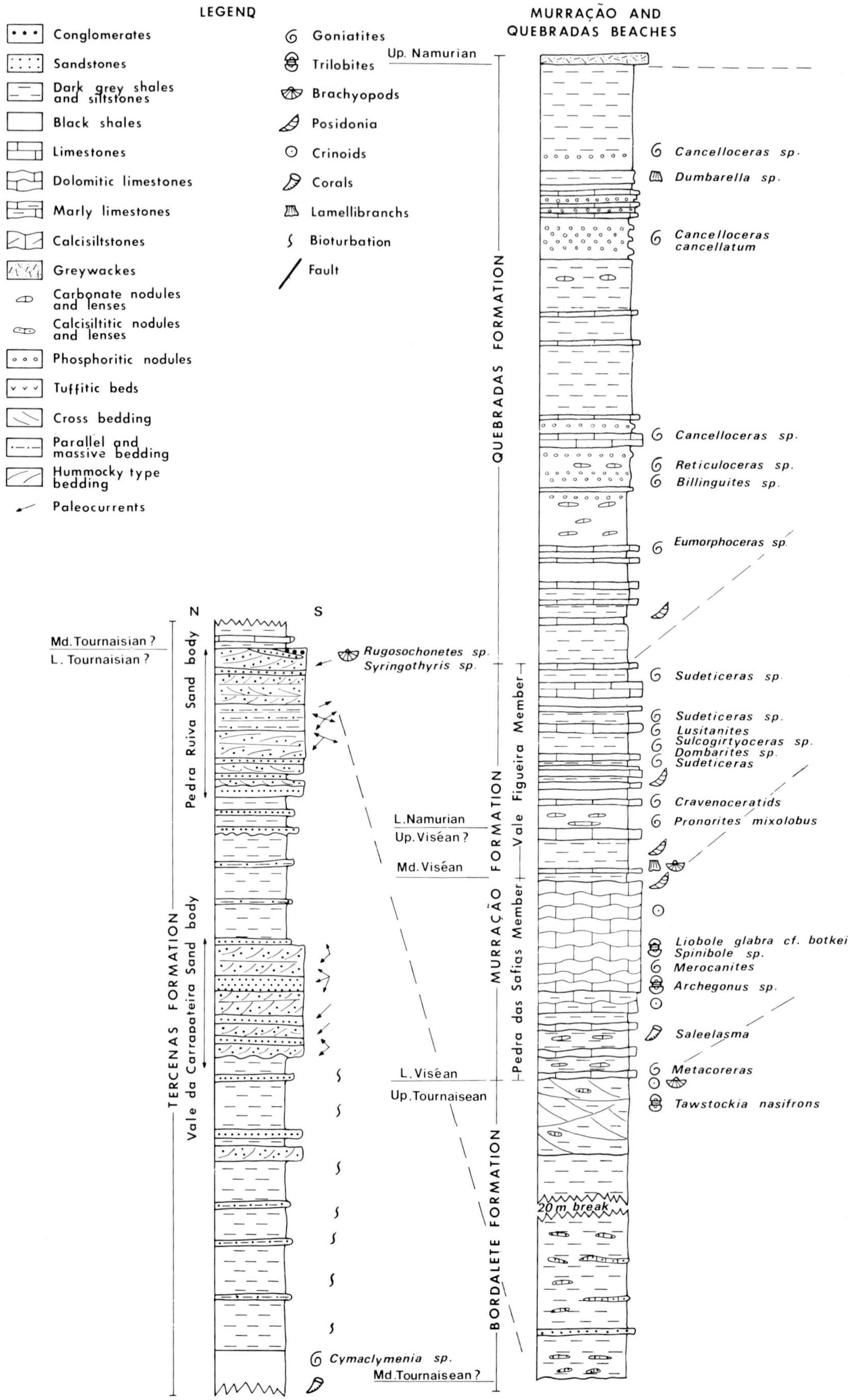
LEGEND
Conglomerates
Sandstones
Dark grey shales and siltstones
Black shales
Limestones
Dolomitic limestones
Marly limestones
Calcisiltstones
Greywackes
Carbonate nodules and lenses
Calcisiltitic nodules and lenses
Phosphoritic nodules
Tuffitic beds
Cross bedding
Parallel and massive bedding
Hummocky type bedding
Paleocurrents
Goniatites
Trilobites
Brachyopods
Posidonia
Crinoids
Corals
Lamellibranchs
Bioturbation
Fault
MURRAÇÃO AND QUEBRADAS BEACHES
Up. Namurian
QUEBRADAS FORMATION
Cancelloceras sp.
Dumbarella sp.
Cancelloceras cancellatum
Cancelloceras sp.
Reticuloceras sp.
Billinguites sp.
Eumorphoceras sp.
Sudeticeras sp.
Sudeticeras sp.
Lusitanites
Sulcogirtyoceras sp.
Dombarites sp.
Sudeticeras
Cravenoceratids
Pronorites mixolobus
Liobole glabra cf. botkei
Spinibole sp.
Merocanites
Archegonus sp.
Saleelasma
Metacoreras
Tawstockia nasifrons
MURRAÇÃO FORMATION
Vale Figueira Member
Pedra das Safias Member
L. Namurian
Up. Viséan ?
Md. Viséan
L. Viséan
Up. Tournaisean
20 m break
BORDALETE FORMATION
Md. Tournaisean ?
N
S
Md. Tournaisian ?
L. Tournaisian ?
Rugosochonetes sp.
Syringothyris sp.
Pedra Ruiva Sand body
Vale da Carrapateira Sand body
TERCENAS FORMATION
Cymaclymenia sp.

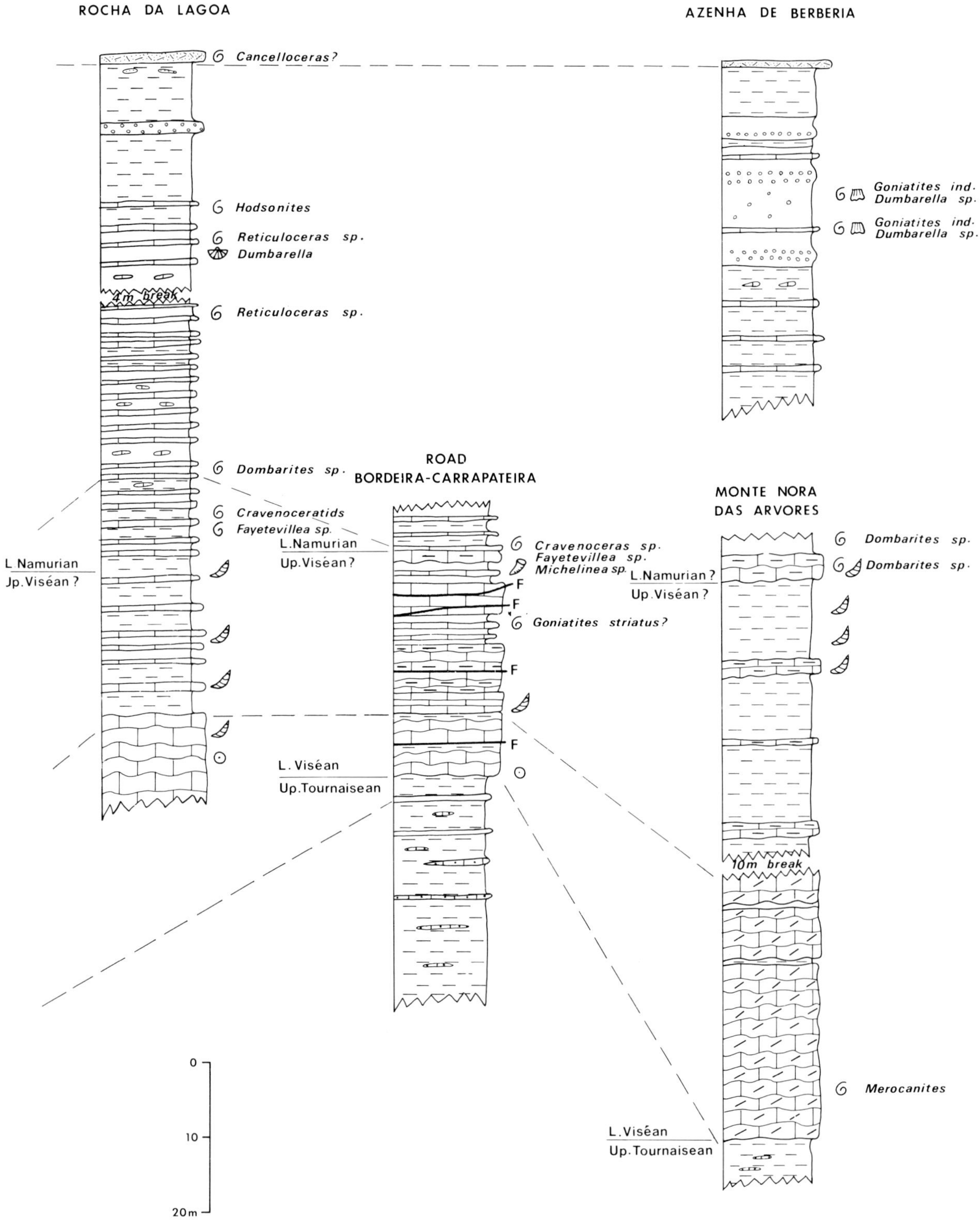

Fig. 5. (pp. 342 and 343) Stratigraphic logs from selected areas of the Aljezur and Bordeira Antiforms (modified from Oliveira et al. 1985)

Mira Formation. The turbidites of this formation are commonly thin-bedded and show a sandstone/shale ratio <1. An exception is the southeastern area were they have a sandstone/shale ratio similar to that of the Mértola Formation. The contact with the Mértola Formation is marked by a continuous shaly horizon, 50 to 100 m thick, which yielded goniatites of the $G_0\gamma$ zone (Oliveira and Wagner Genthis 1983). Scattered goniatites of the El, H and R zones indicate that the unit is of Late Viséan to Early Late Namurian age (Oliveira et al. 1979).

Brejeira Formation. This unit is broadly divided into two distinct sedimentological belts: one, 5 to 10 km wide, is composed of packets of relatively mature sandstones, several metres thick, and alternating shales; the other, to the southwest, crops out over a large area and is mostly made up of dark immature greywackes and shales. In the Aljezur and Bordeira Antiforms, the lower part of the Brejeira Formation is locally composed of thick-bedded and coarse-grained greywackes. In this case, an erosional contact with the underlying rocks of the Carrapateira Group is discernible. Goniatites of H, R and G zones indicate a Middle Namurian to Lower Westphalian age for this unit.

The southward decrease of volcanic fragments in the greywackes indicates a northerly source area for the flysch sediments. Palaeocurrents show a predominant NNW to SSE flow direction, but in the southern areas they are more variable (Fig. 1). This northern confinement is probably due to well-defined submarine reliefs. Note that while the currents flowed to SSE the flysch prograded to SW. The sedimentary progradation was thus governed by southwest directed compressive tectonic instability, which led to successive discharges of terrigenous sediments to the formerly confined basins.

2.6 Southwest Portugal (Bardeira and Aljezur Antiforms)

The stratigraphic succession of these antiforms is, in many ways, different from that in the northern areas of the SPZ. A full description of this sequence has been given in Oliveira et al. (1985); Oliveira et al. (1986b); Ribeiro et al. (1987).

Tercenas Formation. Like the PQ Formation of the Pyrite Belt, this unit also represents the detrital basal formation of these antiforms. The unit is composed essentially of two coarsening upward sequences of sandstones and alternating heterolithic sandstone/shale beds and sheeted sandstones. The top of the upper sandstone beds carry, in places, a metric intercalation of conglomerates and breccias, and local beds of calcarenites. The exposed thickness of the Tercenas Formation is about 100 m (base not exposed). Faunas of clymenids and corals (at the lower shaly horizon) and of brachyopods and chonetacea (at the top conglomerates) indicate a Late Famennian to Early Carboniferous age for this formation (Fig. 5).

The sedimentary structures (large-scale cross-bedding, herringbone cross-bedding, parallel and low-angle cross-lamination, large hummocks, synsedimentary folds), the vertical organization of the sandstone sequences and their lateral correlation indicate deposition in a tidal environment. Thin-bedded sandstones, lenses and flaser structures, sheeted sandstones and the intense bioturbation of the heterolithic horizons are congruent with sedimentation in a subtidal environment. The two main sandstone bodies may coalesce laterally, as suggested by the occurrence of a single thick (100 m) sandstone packet 1 km NE of the Bordeira village. The conglomerates and the Fe- and Mn-rich calcarenites show that the sediments were reworked in shallow water, possibly due to a local marine regression. Paleocurrents have variable trends but northward orientations are most common.

Carrapateira Group. The mud/carbonate character of this group make it clearly distinct from the underlying Tercenas Formation sandstones and shales. Selected logs of the stratigraphic succession are schematically shown in Fig. 5. The Bordalete Formation contains shales, thin-bedded siltstones, and interbedded nodules and lenses of calcisiltstones. The Murração Formation consists of alternating beds of bioclastic dolostones, marls and dark pyritic shales. The Quebradas Formation is composed of alternations of bioclastic dolostones, black shales and conspicuous horizons of phosphoritic nodules at the upper part. A 5-cm-thick bed of evaporites in the Murração Formation and layers of tuffites in the Murração and Quebradas Formations are distinctive lithologies of these formations.

Faunas of corals, trilobites, bivalves and goniatites (Fig. 5) show that the Bordalete, Murração and Quebradas Formations have Middle to Late Tournaisian, Lower Visean to Lower Namurian and Early to Late Namurian ages respectively.

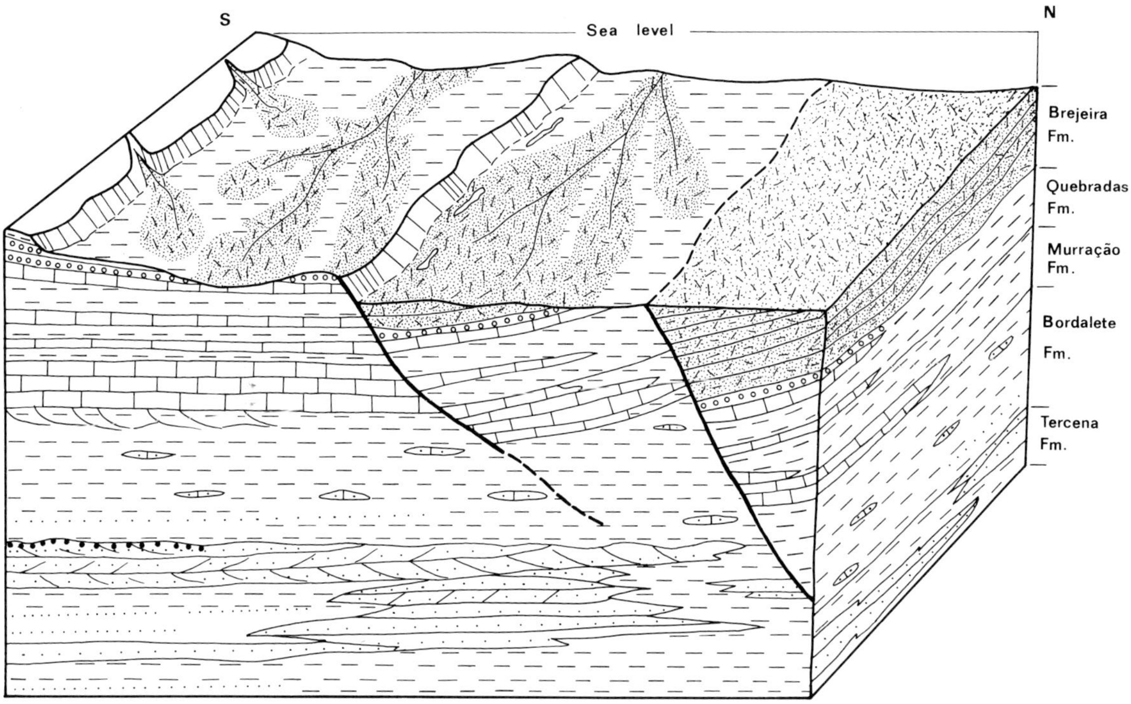

Fig. 6. Sedimentary environments and synsedimentary faults in the Aljezur and Bordeira Antiforms during the Upper Namurian to Lower Westphalian times. Symbols as in Fig. 5

The deposition of a thick succession of mudstones and siltstones over the shallow water sandstones of the Tercenas Formation points to a general transgression during the Middle Tournaisian. This was followed by a regressive tendency during the Upper Tournaisian as suggested by the large scale cross-bedded siltstones (subtidal channels?) at the top of the Bordalete Formation. The bioclastic dolostones (with algae remnants), marls and pyritic shales of the Murração Formation, and the interbedded evaporitic layer, all appear consistent with deposition in a subtidal carbonate platform. This platform may have graded northward to a deeper depositonal environment, as indicated by the increase in the amount of shale in the Aljezur Antiform. Limestones and black shales of the Quebradas Formation were deposited in a more restricted environment, whose depositional meaning is, however, not clear. Phosphates at the upper part of this unit may be indicative of upwelling at the edge of the carbonate platform. Their occurrence in conglomeratic nodules reflect synsedimentary reworking related to tectonic instability, during the Upper Namurian.

This facies interpretation (Fig. 6) is reinforced by the dolostone's lithogeochemistry, which shows an increase in organic matter from the base of the Murração Formation to the top of the Quebradas Formation (Sweenen, written comm.). This increase in organic matter is also reflected by the predominance of black carbonaceous shales in the Quebradas Formation. The Bordeira and Aljezur areas were therefore sites of progressive shallowing during the Visean, and deepening Lower Namurian onwards.

The Brejeira Formation turbidites rest conformably on shales and limestones of the Quebradas Formation. However, in some places, the turbidites are in direct stratigraphic contact with the Murração Formation lithologies and even with the shales of the Bordalete Formation. This disconformity is due to local submarine erosion coupled with block tilting ascribed to the southward directed tectonism (Fig. 6).

2.7 Conclusions

The SPZ is divided into five main tectonostratigraphic domains; that is, Beja-Acebuches Ophiolite, Pulo do Lobo Antiform, Pyrite Belt, Baixo Alentejo Flysch Group and Southwest Portugal. Each of these domains has a distinctive stratigraphy.

The Beja-Acebuches Ophiolite and the terrigenous succession of the Pulo Antiform were probably related to the same sedimentary basin whose original locus is uncertain. Despite their distinct stratigraphic sequences, the Pyrite Belt and Southwest Portugal areas shared a similar style of tectonostratigraphic evolution. Both depositional areas have a shallow-water detritic basement suggestive of a large epicontinental sea in Upper Devonian time. During the Late Famennian to Middle Viséan, this detritic platform was affected by an extensional regime which led to the development of southward migrating marine basins (half grabens) filled with fine terrigenous sediments and bimodal volcanics. Upper Viséan onwards this extensional regime turned to a south-directed compressional one and gave rise to concomitant development of a thick succession of gravity flow sediments. Paleoflows were mostly from NNW to SSE, while the flysch progradated southwestward.

In southwest Portugal, where the stratigraphic succession is less tectonically disturbed, it is possible to detect regressive (Upper Famennian and Viséan to Namurian) and transgressive (Tournaisean) pulses which were the responses to regional tectonics.

Faunal assemblages are scattered and impoverished, except for the period between the Upper Viséan and the Lower Namurian, which contains continuous faunal successions. The faunas, mainly goniatites, always had a cosmopolitan character, with Eurasia, North America and North Africa affinities (Kullmann 1985).

The South Portuguese Zone is a key area within the context of the Variscan orogeny. Details of its structural and palaeogeographic evolution are given in Silva et al. (this Vol).

Acknowledgements. This work benefited from fruitful discussions with A. Ribeiro and J. B. Silva, from the Department of Geology, Faculty of Sciences, Lisbon. I want to express my sincere thanks to them.

References

Andrade AS (1977) The Beja Ophiolitic Complex in southern Portugal: a preliminary description. Mem Not Publ Mus Lab Min Geol Univ Coimbra 84: 49-52

Bard JP (1977) Signification tectonique des metatholeites d'affinité abysal de la ceinture métamorphigue de basse pression d'Aracena (Huelva, Espagne). Bull Soc Geol Fr 17 (2): 385-393

Boogaard MV Den (1963) Conodonts of upper Devonian and lower Carboniferous age from southern Portugal. Geol Mijnbouw 42 (8): 248-259

Boogaard MV Den (1967) Geology of the Pomarão region (southern Portugal). Thesis, Graf Cent Deltro Rotterdam

Boogaard MV Den, Schermerhorn LJG (1975) Conodont faunas from Portugal and southwestern Spain. Part 3 - Carboniferous conodonts from Sotiel Coronada. Scr Geol 29: 37-43

Boogaard MV Den, Schermerhorn LJG (1980) Conodont faunas from Portugal and southwestern Spain. Part 4 - A Famennian conodont fauna near Nerva (Rio Tinto). Scr Geol 56: 1-14

Boogaard MV Den, Schermerhorn LJG (1981) Conodont faunas from Portugal and southwestern Spain. Part 6 - A lower Famennian conodont fauna at Monte Forno da Cal (South Portugal). Scr Geol 63: 1-16

Carvalho D (1974) Lineament patterns and hypogene mineralization in Portugal. Symp IAGOD. Problems of ore deposition 2: 444-453

Carvalho D (1976a) Considerações sobre o vulcanismo de Cercal-Odemira. Suas relações com a Faixa Piritosa. Comun Serv Geol Port 60: 215-238

Carvalho D (1979) Geologia, metalogenia e metodologia da investigação de sulfuretos polimetálicos do Sul de Portugal. Comun Serv Geol Port 65: 169-191

Carvalho D, Correia M, Inverno C (1976) Contribuição para o conhecimento geológico do Grupo de Ferreira-Ficalho. Suas relações com a Faixa Piritosa e o Grupo do Pulo do Lobo. Mem Not Mus Lab Min Fac Ciênc Coimbra 82: 145-169

Dupuy C, Dostal J, Bard JP (1979) Trace element geochemistry of paleozoic amphibolites from SW Spain. Tschermaks Mineral Petrol Mitt 26: 87-93

Fantinet D, Dreesen R, Dusar M, Termier G (1976) Faunes Famenniennes de certains horizons calcaires dans la formation quartzitophylladique aux environs de Mértola (Portugal méridional). Comun Serv Geol Port 60: 121-138

Giese V, Reitz E, Walter R (1988) Contributions to the stratigraphy of the Pulo de Lobo succession in southwest Spain. Comun Serv Geol Port 74: 79-84

IGME Geological map sheets, scale 1:50000, numbers: 915 - Rosal de la Frontera; 916 - Aroche; 917 - Aracena; 919 - Almaden de la Plata; 936 - Paymogo; 937 - El Cerro de Andevalo; 938 - Nerva; 939 - Castillo de las Guardas; 940 - Castilblanco de los Arroyos; 958 - Puebla de Guzman; 959 - Calanas; 960 - Valverde del Camino; 961 - Aznalcollar; 962 - Alcala del Rio; 980 - San Silvestre de Guzman; 981 - Gibraleon; 982 - La Palma del Condado

Kullmann J (1985) Relaciones faunisticas de los ammonoideos carboniferos de Espanña-Portugal (Peninsula Ibérica). X Congr Int Strat Géol Carbonifère Madr (1983) C R 2: 509-512

Leca X, Ribeiro A, Oliveira JT, Silva JB, Albouy L, Carvalho D, Merino F (1983) Cadre géologique des mineralizations de Neves-Corvo (Baixo-Alentejo, Portugal). Lihostratigraphie, paléogeographie et tectonique. Mem Bur Rech Géol Min 121: 80 pp

Munhá J (1983) Hercynian magmatism in the Iberian Pyrite Belt. In: Sousa MJL, Oliveira JT (Eds) The Carboniferous of Portugal. Mem Serv Geol Port 29: 39-81

Munhá J, Oliveira JT, Ribeiro A, Oliveira V, Quesada C, Kerrich R (1986) Beja-Acebuches Ophiolite: characterization and geodynamic significance. Maleo (Abstr) 2 (13): 31

Oliveira JT (1983) The marine Carboniferous of South Portugal: a stratigraphic and sedimentological approach. In: Sousa MJL, Oliveira JT (Eds) The Carboniferous of Portugal. Mem Serv Geol Port 29: 3-37

Oliveira JT (Coordenador) (1984) Carta Geológica de Portu-

gal à escala 1/200000. Serv Geol Port, Notícia Explicativa da Folha 7: 77 pp
Oliveira JT (in press) Estratigrafia, sedimentologia e estrutura do flysch da Formação de Mértola, na região de Mértola. Comun Serv Geol Port 74
Oliveira JT, Silva JB (1988) Carta Geológica de Portugal à escala 1/50000. Notícia Explicativa da folha 46-D Mértola. Serv Geol Port
Oliveira JT, Wagner Genthis C (1983) The Mértola and Mira formational boundary between Dogueno and Almada do Ouro, marine Carboniferous of South Portugal. In: Lemos Sousa MT (ed) Contributions to the Carboniferous Geology and Palaeontology of the Iberian Peninsula. Fac Cienc Porto pp 1-39
Oliveira JT, Horn M, Paproth E (1979) Preliminary note on the stratigraphy of the Baixo Alentejo Flysch Group, Carboniferous of Portugal, and on the palaeogeographic development compared to corresponding units in northwest Germany. Comun Serv Geol Port 65: 151-168
Oliveira JT, Horn M, Kullman J, Paproth E (1985) Stratigraphy of the upper Devonian and Carboniferous sediments of southwestern Portugal. X Congr Int Strat Géol Carbonifére Madr (1983) Co Re 1: 107-120
Oliveira JT, Cunha T, Streel M, Vanguertaine M (1986a) Dating the Horta da Torre Formation, a new lithostratigraphic unit of the Ferreira-Ficalho Group, South Portuguese Zone: geological consequences. Comun Serv Geol Port 72: 1987
Oliveira JT, García Alcalde JL, Liñán E, Truyóls J (1986b) The Famennian of the Iberian Peninsula. Ann Soc Géol Belg 109: 159-174
Pfefferkorn H (1968) Geologie des Gebietes zwischen Serpa und Mértola (Baixo Alentejo, Portugal). Münster Forsch Geol Paläontol 9: 943
Pruvost P (1912/13) Sur la presence de fossiles d'age Dévonien superieur dans les schistes à Nereites de São Domingos. Comun Serv Geol Port 9: 58-68
Rambault F (1978) Distribución de focos volcánicos y yacimientos en la banda pirítica de Huelva. Bol Geol Min LXXXIX-III: 223-233
Ribeiro A, Oliveira JT, Ramalho M, Ribeiro L, Silva C (1987) Carta Geológica de Portugal à escala 1/50000. Notícia Explicativa da folha 48-D Bordeira. Serv Geol Port, 30 pp
Routhier P, Aye F, Boyer C, Lécolle M, Molière P, Picot P, Roger G (1980) La ceinture sud-ibérique à ames sulfurés dans ce partie espagnole. Tableau géologique et métallogenique. Synthese sur le type ames sulfurés volcano-sedimentaires. Mem Bur Rech Géol Minière (26e Cong Int Géol, Paris, 1980) 94 (1): 265 pp
Salpeteur I (1976) Étude structurale et petrographique de la Zone de Paymogo (Nord de la province de Huelva). Métallogenèse des amas sulfurés. Thesis, Nancy
Santos JFP, Mata J, Gonçalves F, Munhá J (1987) Contribuição para o conhecimento geológico-petrológico da região de Santa Suzana: o Complexo vulcano-sedimentar de Toca de Moura. Comun Serv Geol Port 73 (42): 29-40
Schermerhorn LJG (1971) An outline stratigraphy of the Iberian Pyrite Belt. Bol Geol Min 82 (3/4): 239-268
Schermerhorn LJG, Stanton WI (1969) Folded overthrusts at Aljustrel (South Portugal). Geol Mag 106 (2): 130-141
Silva JB (1985) Structure of the Pyrite Belt: the Mértola Traverse Papers on the Carboniferous of the Iberian Pyrite Belt. An Fac Cienc Porto Suppl to 64: 41-47 (1983), (Spec Vol)

3 Structural Outline

J. B. Silva[1], J. T. Oliveira[2] and A. Ribeiro[1]

3.1 Introduction

In the southwest branch of the Iberian Variscan Arc, the South Portuguese Zone (SPZ) is an arcuate thrust belt (Fig. 1 in Part IV, Chap. 2, p. 148). The dominant structures change from a north-south direction on the Atlantic seaboard, to an east-west direction near the Spanish border.

The following main structural domains are recognized: (1) Beja-Acebuches Ophiolite, (2) Pulo do Lobo Antiform, (3) Pyrite Belt, (4) Baixo Alentejo Flysch Group and (5) Southwest Portugal. These domains have been considered as a succession of simple anticlines and synclines (Carvalho et al. 1971; Schermerhorn 1971). Feio and Lombard (1958) were the first to admit thrust structures and Schermerhorn and Stanton (1969) described pre-cleavage overthrusts. More recently Ribeiro et al. (1983 a) ascribed a thin-skinned tectonic model for the SPZ.

From a geotectonic point of view, the SPZ has been variously interpreted as: (1) intracontinental orogen (Schermerhorn 1975; Routhier et al. 1980), (2) an accretionary prism related to a north-dipping subduction below the SPZ (Bard 1971; Carvalho 1972; Bard et al. 1973; Vegas and Muñoz 1976; Monteiro and Carvalho 1987), (3) a back-arc basin (Munhá 1983 b; Ribeiro et al. 1983 b) and (4) as an oblique-slip mobile belt (Badham 1982; Andrews 1983).

This work provides interpretation of the SPZ based essentially on its more complete geotraverse; that is, the Portuguese part. Details of each of the structural domains are described separately, and the thin-skinned model of tectonic deformation is emphasized for the Pyrite Belt and Southwest Portugal. This stratigraphic and structural data, coupled with the occurrence of a pre-Famennian ophiolite at the SPZ – Ossa-Morena Zone (OMZ) boundary (Munhá et al. 1986), the evidence of a Lower Carboniferous calc-alkaline suite in the northwestern part of the OMZ (Santos et al. 1987) suggesting subduction and collision, and the intracontinental character of the Upper Famennian to Middle Visean volcanism of the Pyrite Belt, allow a distinct geodynamic reconstruction.

3.2 Structural Divisions of the South Portuguese Zone

The stratigraphy of the five structural domains of the SPZ is described in Oliveira (Part VI, Chap. 2). The tectonic episodes of deformation within each structural domain is shown in Table 1.

Table 1. Timing of folding episodes (F) and associated cleavage (S)

	Pulo do Lobo Antiform	Pyrite Belt	Baixo Alentejo Flysch Group	Southwest Portugal
Middle Westphalian	F_3 (S_3)	F_{1b} (S_{1b})	late kinks late kinks F_{1b} (S_{1b})	F_3 F_2 (S_2) F_1 (S_1)
Namurian to Low. Westphalian	?			
Upper Visean	F_2 (S_2)	F_{1a} (S_{1a})		
pre-Lower Tournaisian	? F_1 (S_1)			

[1] Departamento de Geologia da Faculdade de Ciências, Universidade de Lisboa, 1700 Lisboa, Portugal
[2] Serviços Geológicos de Portugal Rua de Academia das Ciencias, 1200 Lisboa, Portugal

R. D. Dallmeyer and E. Martínez García (Eds.)
Pre-Mesozoic Geology of Iberia

3.2.1 Beja - Acebuches Ophiolite

This ophiolite is exposed along the boundary of the SPZ and the OMZ. The rocks of this ophiolite, particularly the amphibolites and metabasalts (Munhá et al. 1986) show two stretching lineations (L1 and L2), each associated with ductile shearing. These lineations vary in attitude from north-northwest/south-southeast (L1) to west-northwest/east-southeast (L2) directions. The latter is related to a penetrative mylonitization. The stretching lineations together with the sense of shear indicate transport towards north-northwest (L1) and west-northwest (L2) and probably developed during structural emplacement of the ophiolite on the Iberian margin.

The ophiolite is bounded to the north by a thrust which is locally (Acebuches village, Spain) marked by a tectonic mélange (Fig. 1). This mélange is composed of a gneissic matrix in which blocks of distinct lithologies are included. The southern boundary of the ophiolite and the neighbouring Horta da Torre Formation are disrupted by a thrust with south vergence and associated sinistral shearing. The ophiolite was gently folded following the Famennian.

3.2.2 Pulo do Lobo Antiform

The Pulo do Lobo Formation is the oldest unit of the Pulo do Lobo Antiform (Fig. 1), and has been deformed by three episodes of folding (F1, F2 and F3), and associated cleavages (S1, S2 and S3; see Table 1). F1 episode has a north-south orientation, and the others trend northwestward. This unit is comprised of sheared phyllites and quartzites, in which acidic and basic volcanics are interbedded. In the north limb of this antiform, the Pulo do Lobo Formation is stratigraphically overlain by sedimentary rocks of the Ferreira-Ficalho Group (Ribeira de Limas, Santa Iria and Horta da Torre Formations). All of these units are older than Upper Famennian and have a terrigenous character (Oliveira et al. 1986). The Santa Iria and Horta da Torre Formations exhibit one main cleavage, with a northwest orientation and steep northward dips.

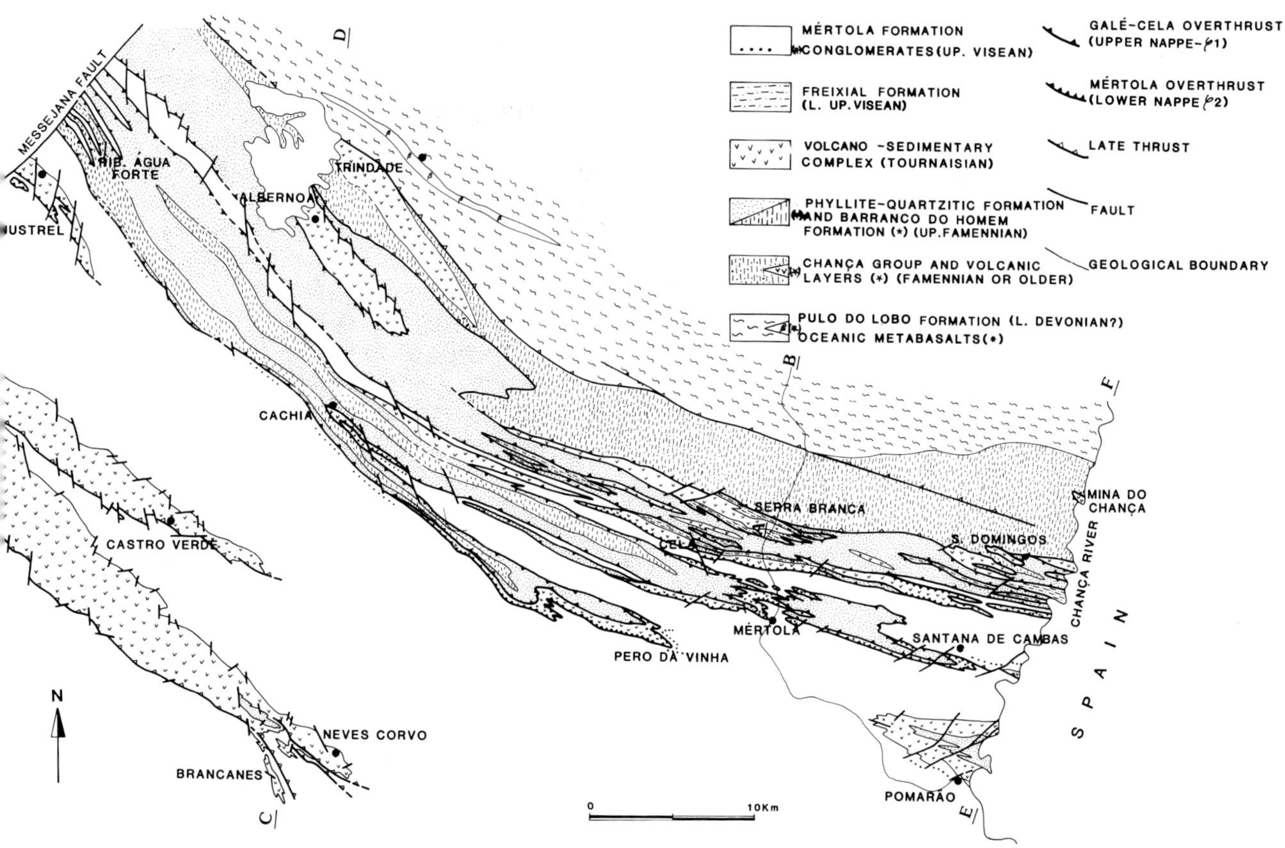

Fig. 2. Structural map of part of the Portuguese Pyrite Belt, showing relationships between the main units

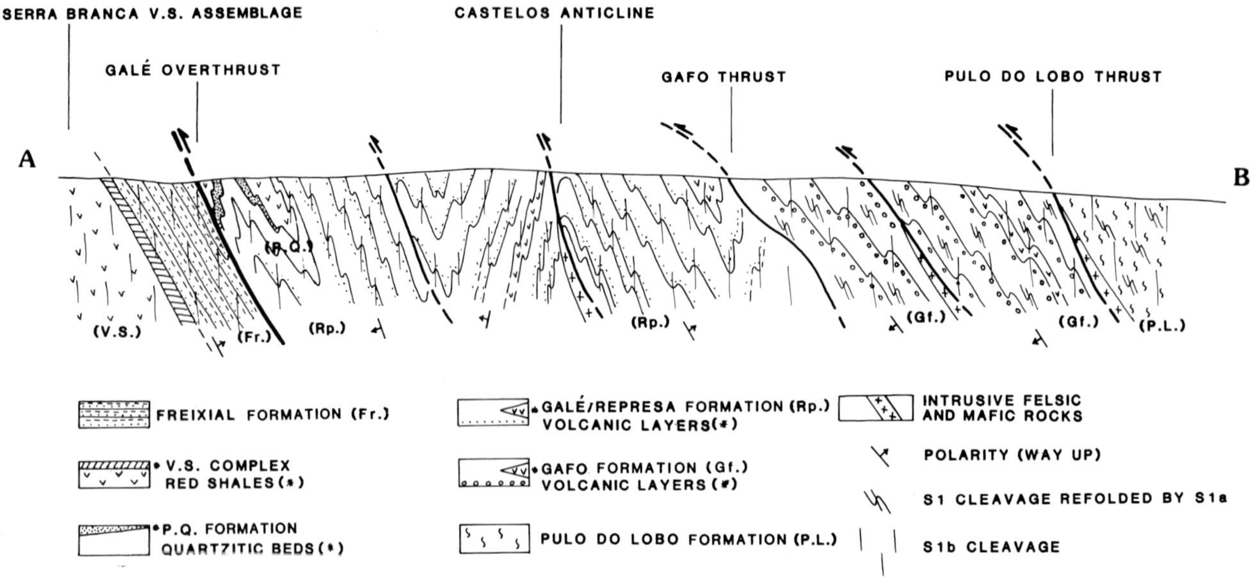

Fig. 3. Schematic cross-section *A–B* along the Guadiana Valley, from the Pulo do Lobo Formation to the Serra Branca Anticline (After Silva 1986)

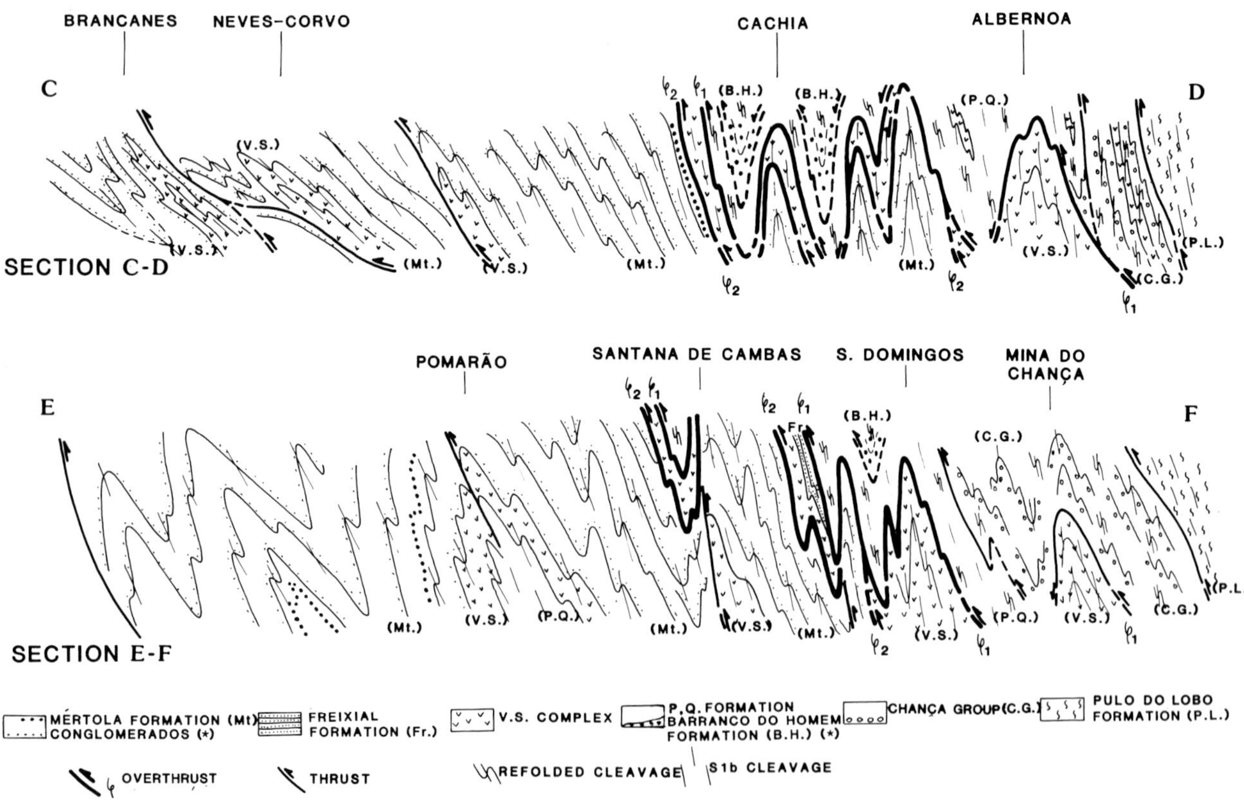

Fig. 4. Schematic cross-sections along selected areas of the Pyrite Belt. *C–D* between Neves-Corvo and Albernoa; *E–F* between SW of Pomarão and Mina de Chança

The underlying Ribeira de Limas Formation shows a tectonic deformation similar to that of the Pulo do Lobo Formation. The deformational contrast between basement (the Ribeira de Limas and Pulo do Lobo Formations) and its cover (the Santa Iria and Horta da Torre Formations) suggests the existence of an unconformity.

Along the south limb of the Pulo do Lobo Antiform (Fig. 2), the Pulo do Lobo Formation is bounded by the units of the Chança Group (in ascending order these include the Atalaia, Gafo and Represa Formations). The uppermost two formations have interbedded acidic and basic volcanics. A detailed traverse along the Guadian Valley (Silva 1986; Silva 1989) shows that the Gafo Formation is overturned (Fig. 3). Here the Atalaia Formation is absent and the contact with the Pulo do Lobo Formation is marked by a north-dipping fault. The Gafo Formation structurally overlies the Galé Formation (a local name for the Represa Formation) and this in turn is also in allochthonous position over the volcanosedimentary assemblage of the the Serra Branca Anticline (Oliveira, Part VI, Chap. 2) (Figs. 2, 3). The units of the Chança Group show the same three episodes of folding that affected the Pulo do Lobo Formation. Acidic and basic veins were intruded along post-Sl thrust faults (Fig. 3). These intrusives show petrographic affinities with the bimodal Pyrite Belt volcanics (Munhá, pers. comm.) and are deformed and cleaved by the same episodes of folding that affected the Pyrite Belt. This suggests that the intrusives may have acted as volcanic feeders.

In the regions of Albernoa - Neves Corvo (Fig. 4 C-D) and São Domingos - Pomarão (Fig. 4 E-F), the Chança Group has similar structural relationships with the Pyrite Belt succession. This structural evidence, together with the Upper Devonian palynomorphs recently recovered from the Represa Formation (Cunha and Oliveira, in press), indicate that the units of the Chança Group are older than the volcanosedimentary assemblage of the Pyrite Belt.

The stratigraphic units of the Pulo do Lobo Antiform are affected by a progressive metamorphism which changes from greenschist facies in the Pulo do Lobo Formation to a very low-grade metamorphism in the upper units. The relationship between metamorphism and folding has not been studied in detail.

3.2.3 Pyrite Belt

In the Pyrite Belt, two main northwest-southeast episodes of folding and cogenetic cleavages occur. The first one (F1a) is linked to thrust tectonics, and is equivalent to the F2 episode of folding in the Pulo do Lobo Antiform (see Table 1). The second episode (F1b) is equivalent to the F3 episode of the Pulo do Lobo Antiform, and is related to regional cleavage (S1b).

The stratigraphic succession of the Pyrite Belt includes (from base to top) the Barranco do Homem and Phyllite Quartzite (PQ) Formations of Famennian age and the Volcanosedimentary Complex (VS) of Upper Famennian to Low-Upper Viséan age (Oliveira, Part VI, Chap. 2).

The Pyrite Belt is dominated by thrust tectonics. Structures in footwall and hangingwall positions are identified (Figs. 2, 4).

3.2.3.1 Footwall Structures (Parautochthonous)

These structures occur both in the internal and the external zones of the Pyrite Belt (respectively the al-

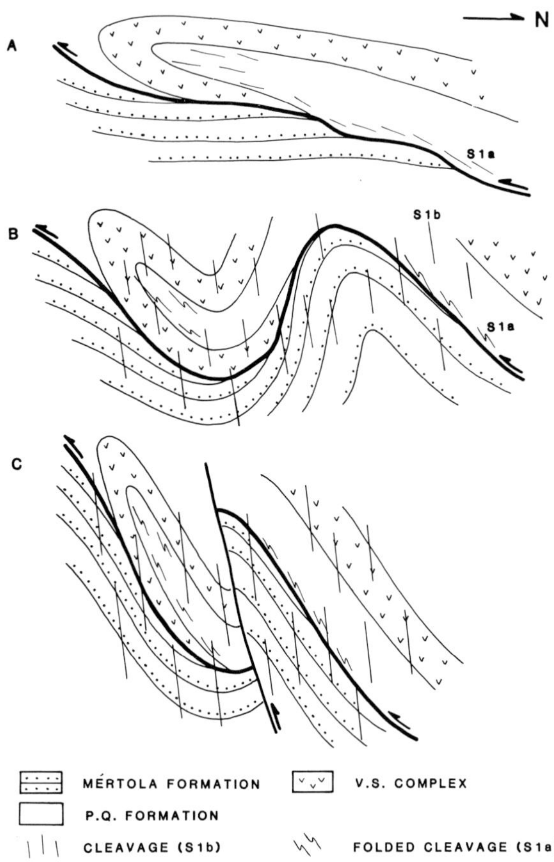

Fig. 5. Interpretative model of the Mértola nappe emplacement (After Silva 1983): *A* Recumbent fold of the first folding phase (F_{1a}-S_{1a}), in the fronta, part of the nappe; *B* refolding of the nappe surface during the second tectonic phase (F_{1a}-S_{1b}): *C* reversed limb lamination and thrusting of the allochthon by the autochthon during the F_{1b} phase

lochthonous and autochthonous branches of Oliveira (Part VI, Chap. 2). Rocks exposed in the internal zones are represented by anticlines of the VS Complex over which an allochthon involving mostly lithologies of the Barranco do Homem and PQ Formations was emplaced (Fig. 4). A common feature of these structures is that the hangingwall is usually more deformed than the underlying footwall.

In the external zones of the Pyrite Belt only the footwall is exposed. This is represented by the PQ Formation, the VS Complex and the Mértola Formation turbidites (in normal stratigraphic order), and this succession is well-exposed in the anticlines of Valverde del Camino, Puebla de Guzman, Pomarão, Castro Verde, Cercal and the less impressive structures of Alcoutim and São Francisco da Serra (Figs. 2, 3). These anticlines were generated during the second regional episode of folding (F1b), after thrust imbrication of F1a episode.

3.2.3.2 Hangingwall Structures (Allochthonous)

Galé-Cela Nappe. This nappe is exposed over extensive areas of the Pyrite Belt and mostly involves lithologies of the Barranco do Homem and PQ Formations and minor assemblages of VS Complex and Freixial Formation. Nappe emplacement was related to a first phase (F1a) of recumbent folding and concomitant thrusting that affected the reverse limbs of the folds (Figs. 4, 5, 6). During thrusting, both hangingwall and footwall structures were involved in tectonic cutting. In some cases, tectonic mélanges including lithologic elements of the hangingwall and footwall were formed.

The Galé-Cela Nappe was folded during a second (F1b) episode of deformation. The Barranco do Homem Formation occupies the core of large F1b synforms (Fig. 4). Structural position of the Barranco do Homem Formation can be ascribed to the F1a recumbent episode of folding, or alternatively to thrust emplacement over the PQ Formation.

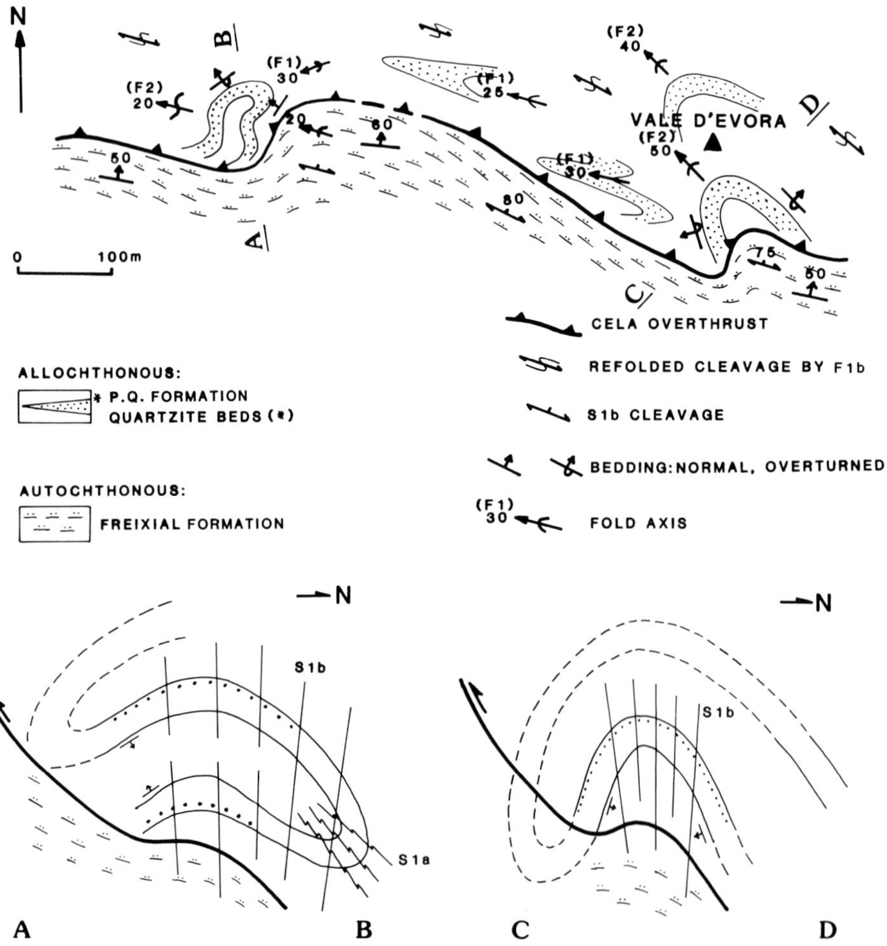

Fig. 6. Detailed map of Cela nappe near Vale d'Evora Hill, 4 km N Mértola. The schematic cross-sections *A–B* and *C–D* illustrate F1a recumbent folds in front of the overthrust, refolded by the second folding phase (After Silva 1983)

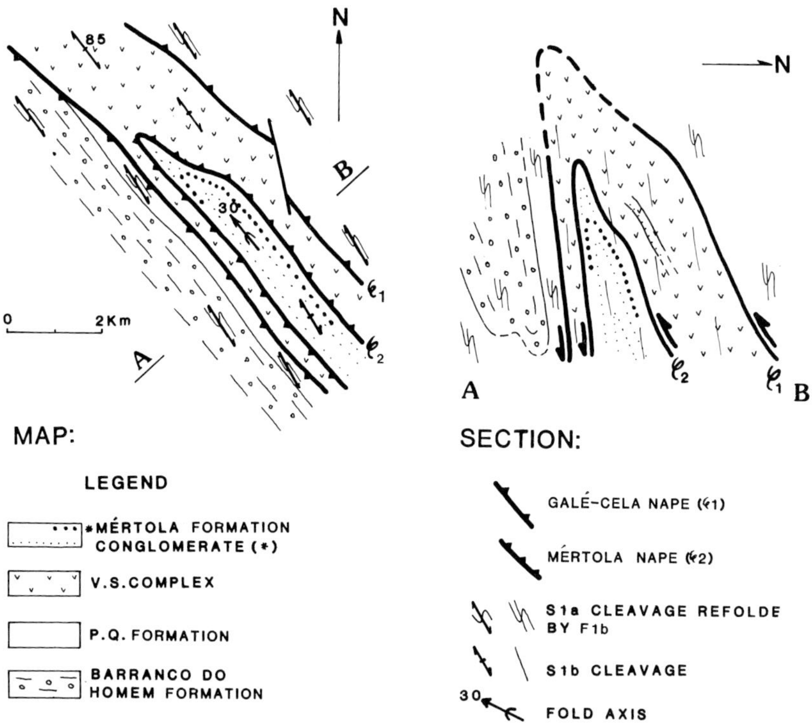

Fig. 7. Map *(left)* and schematic cross-section *(right)*, showing superposition of the Galé-Cela and Mértola nappes in the Cachia region (see Fig. 2)

Mértola Nappe. This nappe, including mostly lithologies of the VS Complex and Freixial Formation, underlies the Galé-Cela Nappe and in turn structurally overlies the turbidites of the Mértola Formation. An example of this superposition can be seen in Monte da Cachia area (Fig. 7) where turbidites of upper parts of the Mértola Formation have intercalations of breccias and conglomerates with pebbles and blocks of the VS Complex and the PQ Formation. These conglomerates and breccias are structurally overlain by the Mértola Nappe, which contains lithologies of the VS Complex (shales, purple shales, tuffites) in an upright position. Above the Mértola Nappe occurs an allochthon composed of quartzites and shales of the PQ Formation, which is ascribed to the Galé-Cela Nappe. The Mértola Formation turbidites and conglomerates, and rocks of the VS Complex of Mértola Nappe, show a single cleavage (S1b); the shales and quartzites above display two cleavages (S1a + S1b).

This tectonic, stratigraphic and sedimentological relationship can be observed in other places; for instance, near Mértola (Fig. 8) and in the Pero da Vinha (Oliveira 1983) and Moreanes regions (Oliveira and Silva, in press). The Mértola Nappe was emplaced before development of the regional cleavage (S1b) and it is interpreted as a synsedimentary thrust. Southward movement of this thrust induced gravitysliding which generated, by itself, conglomerates, breccias and slumps that occur interbedded in the Mértola Formation turbidites.

3.2.3.3 Overall Structural Interpretation of the Pyrite Belt

Tectonostratigraphic relationships between units of the Pyrite Belt during F1a episode of folding is exemplified in Fig. 9.

The Galé-Cela Nappe was emplaced first. Within the allochthon, cleavage (S1a) was mainly associated with the hinge zones of the folds. The Mértola Nappe emplacement was contemporaneous with deposition of the Mértola Formation turbidites. Progression of these nappes, with uncertain magnitude, is explained by the thin-skinned model, and the tangential arrangement of thrusts shows a forward propagation (piggy-back sequence). The frontal parts of the nappes were sites of important secondary thrusting.

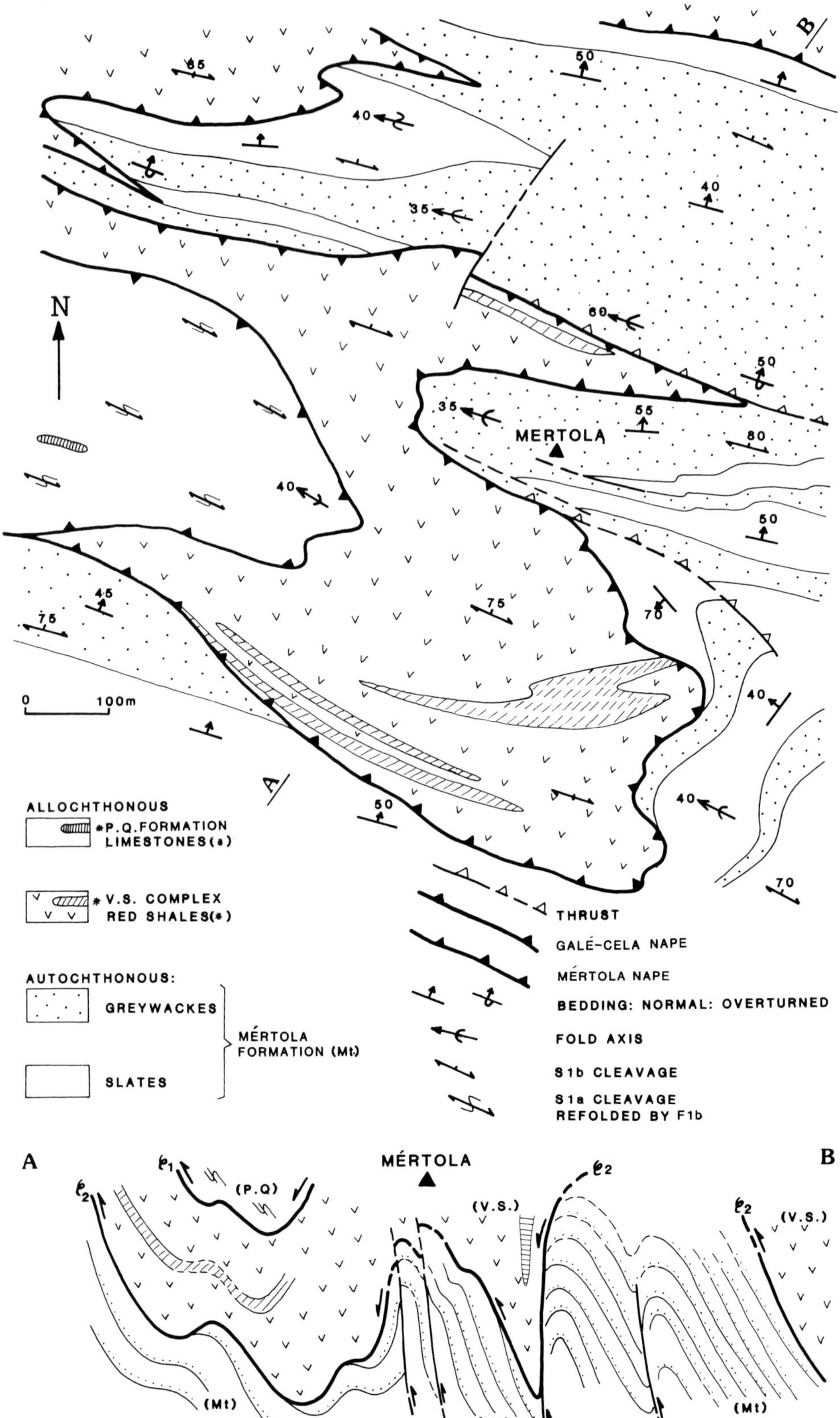

Fig. 8. Detailed map of Mértola area (see Fig. 2), showing the geological relationship between Galé-Cela and Mértola nappes *(A)* and illustrative cross-section *(B)* (After Silva 1983)

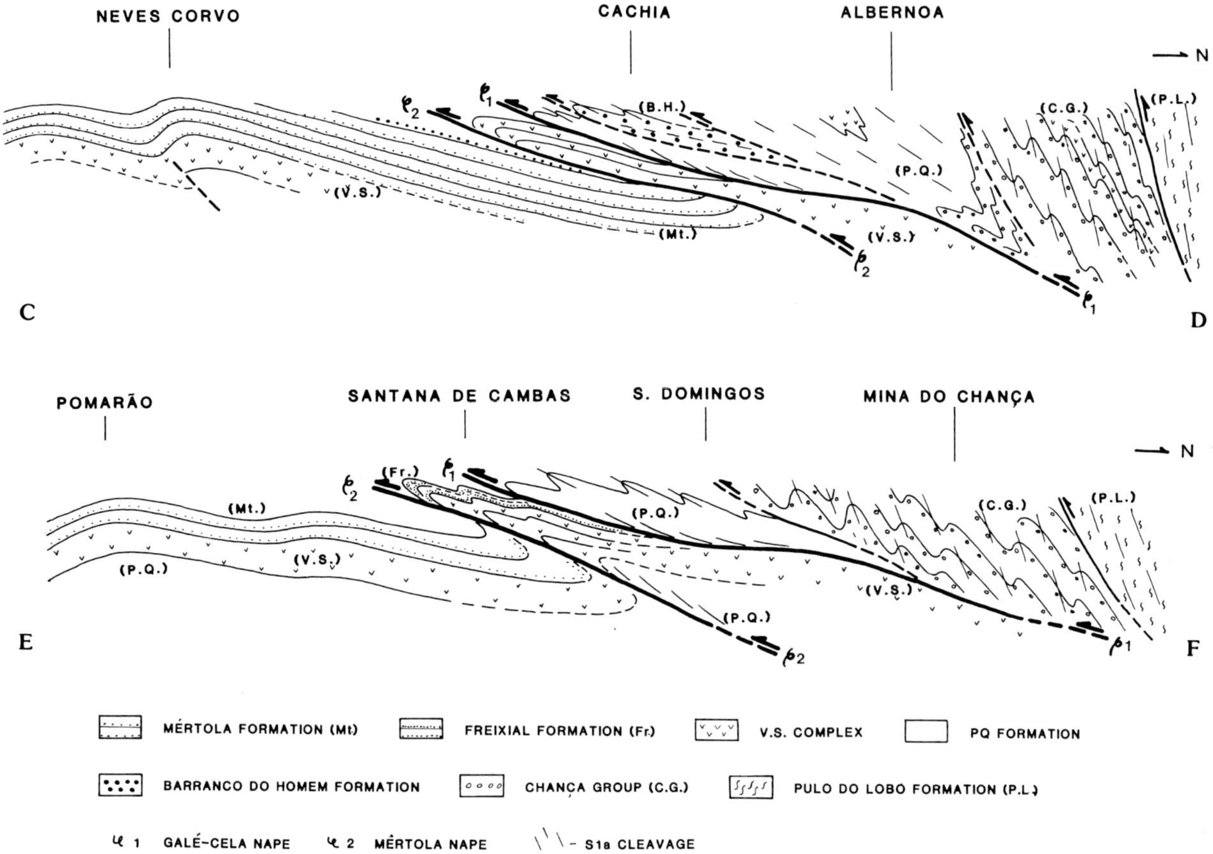

Fig. 9. Tectonostratigraphic relationships between units of the Pyrite Belt, based on cross-sections *C–D* and *E–F* of Fig. 4, before the regional folding (F1b)

During the second regional stage of deformation (F1b), both footwall and hangingwall were folded with generation of the regional cleavage (S1b). At deeper structural levels, S1a and S1b can be differentiated in the hinge zones of the F1b folds only. Usually S1a and S1b are approximately coplanar, thus indicating that S1a was transposed by S1b in the limbs of similar folds. Within intermediate structural levels, F1a flexural folds were flattened with development of S1b cleavage. In upper structural levels, folds and associated cleavage were generated for the first time. The footwall exhibits a tectonic deformation similar to that of the upper structural levels of the hangingwall (Fig. 10). Formation of both cleavages in the allochthon can be explained by a superposition of two homogeneous strains, the result of which is a total homogeneous strain that can be expressed by a single strain ellipse (Ramsay 1967, Chaps. 3–11; Silva 1981, 1983, 1985).

Reverse limbs of the F1b folds were locally affected by thrust faults contemporaneous to S1b cleavage. Both thrusts and cleavage become gradually steeper towards the internal zones of the Pyrite Belt. The syn-F1b thrusts led to tectonic complications that include, among others, thrusting of the allochthon by the footwall (Fig. 4 E–F, near Santana de Cambas village; Figs. 5C, 8). Tectonic slides, contemporaneous to S1b cleavage generation, are identified in mesoscopic and macroscopic structures. They are marked by thickening of hinge zones and stretching of limbs, generally without rupture.

North/northwest-south/southeast and north/northeast-southsouthwest postmetamorphic crenulations are locally observed. They are related to chevron folds and conjugate kink bands that occur in high anisotrophic lithologies. Northeast-southwest left lateral and north/northwest-south/southeast dextral faults disrupted the previous structures.

In the Pyrite Belt, and also in the Baixo Alentejo Flysch Group (Figs. 1, 2) the macroscopic folds display an *en échelon* arrangement, which is shown by variable plunges of the axis of the major structures. This *en échelon* arrangement is due to the effect of a

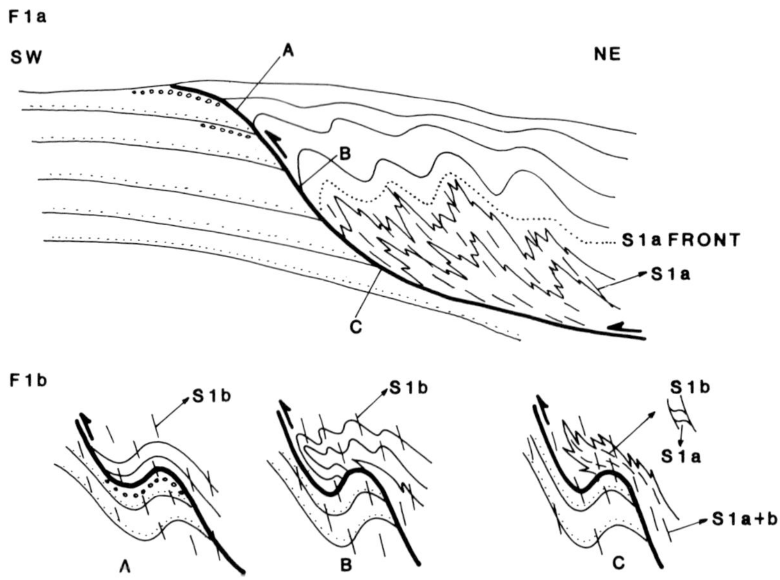

Fig. 10. Thrust and folding relationships in the Pyrite Belt (After Ribeiro et al. 1983a). During F1a: Development of S1a cleavage in the lower structural level *(C)*; Flexural folding without cleavage in the intermediate structural level *(B)*; no deformation in the upper structural level *(A)*. During F1b: Refolding during F1b, with development of regional cleavage *(S1b)*; *S1a* and *S1b* are differentiated in the hinge zones (S1a+S1b) and are transposed in the limbs (S1a+b)

left lateral shear component concomitant with the main compression normal to the axial planes of the folds. This non-coaxial deformation caused the transection of the folds (Ribeiro et al. 1979; Ribeiro and Silva 1983). Counterclockwise transection is prevailing, as shown by the northerly trend of cleavage with respect to the axial planes of the folds

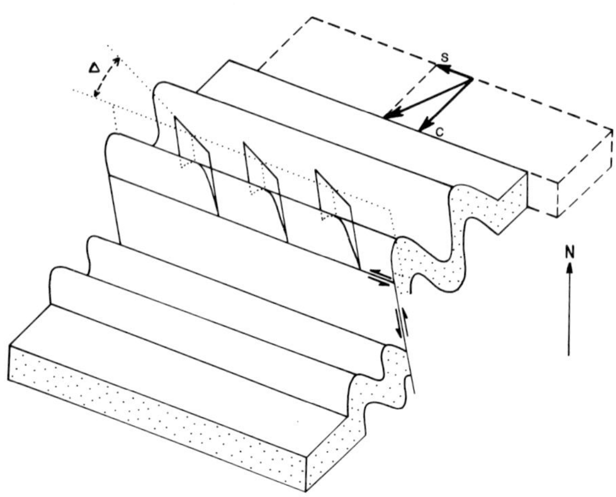

Fig. 11. The transection mechanism in the South Portuguese Zone (after Ribeiro et al. 1979). Counterclockwise transection (Δ - angle between axial plane and cleavage) is due to a sinistral shear component *(S)* added to the main compressive component *(C)*

(closer to the northwest-southeast direction). L1 (So ^ S1) intersection lineation around the folds (Figs. 11, 12) show northward plunges in the north limb (back limb) and southward plunges in the southern limb (fore-limb). This lineation arrangement confirms the counterclockwise transection. The non-coaxial regime, linked to the genesis of the arcuate belt (Fig. 16-C), is not well documented for the F1a stages of folding. However, it was certainly operative during F1b deformation.

Metamorphic regime in the Pyrite Belt belongs mainly to the prehnite-pumpellyite facies (Schermerhorn 1975; Munhá 1976, 1983a, b). Metamorphic assemblages are mostly contemporaneous to S1b cleavage, indicating that the regional metamorphism is associated with the second episode of folding (F1b). Growth of a metamorphic regime during F1a folding has not been demonstrated but cannot be excluded, particularly in deep structural levels where S1a cleavage was developed.

3.2.4 Baixo Alentejo Flysch Group

The Baixo Alentejo Flysch Group is composed of three main lithostratigraphic units (Mértola, Mira and Brejeira Formations, Oliveira et al. 1979). These units were deformed by F1b, southwest ver-

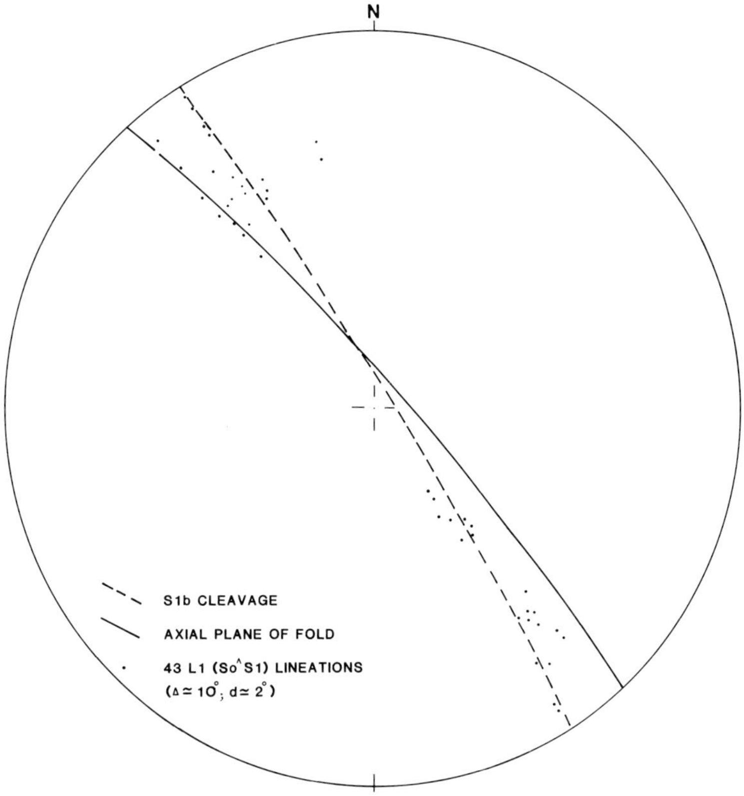

Fig. 12. Equal area projection of L1 (S0 ^ S1) lineations in an F1b transected fold, near Ribeira de Agua Forte, Aljustrel region (see Fig. 2). The angle between S1b cleavage and axial plane (see also Fig. 11), is due to counterclockwise transection

gent folding. In northern areas, vertical symmetric folds predominate over asymmetric ones. Towards the south, asymmetric folds, with long normal limbs and short reversed limbs, are dominant. Slaty cleavage has a northwest-southeast systematic orientation and is usually axial planar or fanning, and transection is also common. The dip of the cleavage changes from steep in general to more flat near thrust planes. This variation is related to south-directed thrusting, which affected mainly the reversed limbs of the folds. Cleavage and thrusts become less penetrative and more gently southward (Ribeiro 1981). When S1b cleavage is flat, as is the case of the Cercal Anticline, a post-metamorphic crenulation cleavage is developed (Ribeiro 1983). Northeast-southwest and north-south Late Hercynian faults are widespread. Some of these faults, as for instance the Messejana Fault, may result from reactivation of older fractures. These faults focused fluid flow which led to the formation of quartz veins, sometimes with minor amounts of barite and copper carbonates.

3.2.5 Southwest Portugal

Southwestern portions of the SPZ are dominated by the antiformal structures of Aljezur and Bordeira (Figs. 1, 13). Both antiforms expose similar lithostratigraphic sequences. From base to top, these include shales and quartzites of the Tercenas Formation, a shaly/carbonate succession (Carrapateira Group) and turbidites of the Brejeira Formation.

In this sector, F1 (see Table 1) is the most important episode of deformation (Ribeiro 1983). It is characterized by northwest-southeast trending folds and cogenetic axial planar or fanning slaty cleavage (S1). This cleavage changes from less penetrative and northeast steeply dipping, in the higher structural levels, to more intense and flat, near the Carrapateira thrust plane (Fig. 14). This change in the cleavage attitude shows that the Carrapateira thrust is a shear zone coeval to the F1 episode of folding. Detailed work in the Bordeira region (Ribeiro 1983; Ribeiro et al. 1987) demonstrates that the Carrapateira thrust plane was a ductile shear in the north

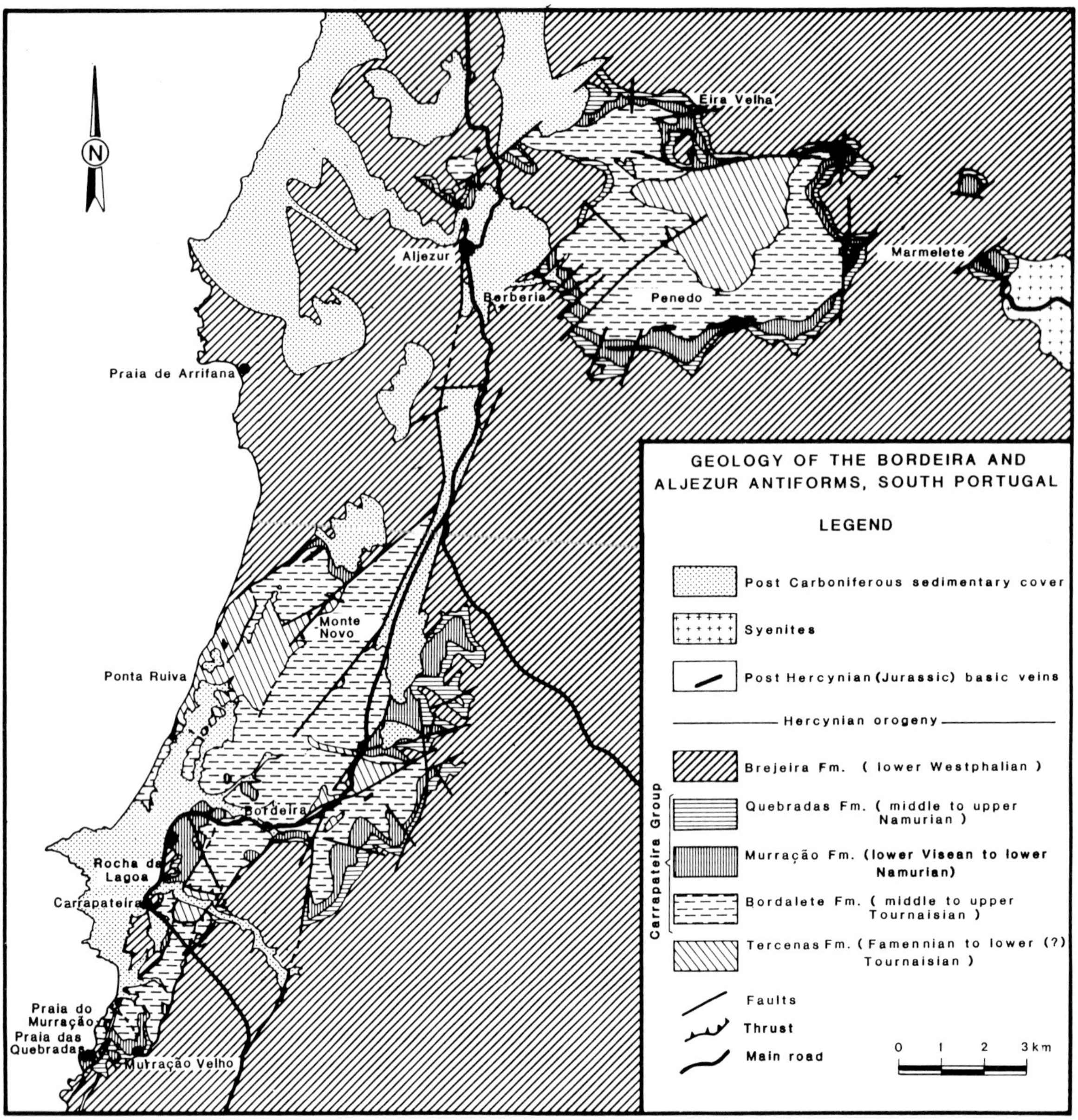

Fig. 13. Geological map of the Bordeira and Aljezur Antiforms (After Oliveira 1984)

(lower structural level) and a more brittle fault in the south (high structural level). Stretching lineations and sheath folds near the thrust plane indicate southwestward tectonic transport. The thrust dies out at the core of a recumbent fold, in Quebradas beach (Ribeiro 1983). In front of this large recumbent fold, chevron folds with subhorizontal axial planes were also developed.

F1 structures were refolded by two less intense stages of deformation (see Table 1). The second one (F2) generated northwest-southeast trending folds with subvertical axial planes and S2 crenulation cleavage restricted to small areas. This stage corresponds probably to anticlines developed above frontal ramps of the sole thrust. The third one (F3), with northeast-southwest orientation, is responsible for the box-fold profile (with limbs cut by normal faults) of the Bordeira Antiform.

West-northwest/east-southeast sinistral, and east-northeast/west-southwest dextral faults are related to the stress field of the third episode of deformation.

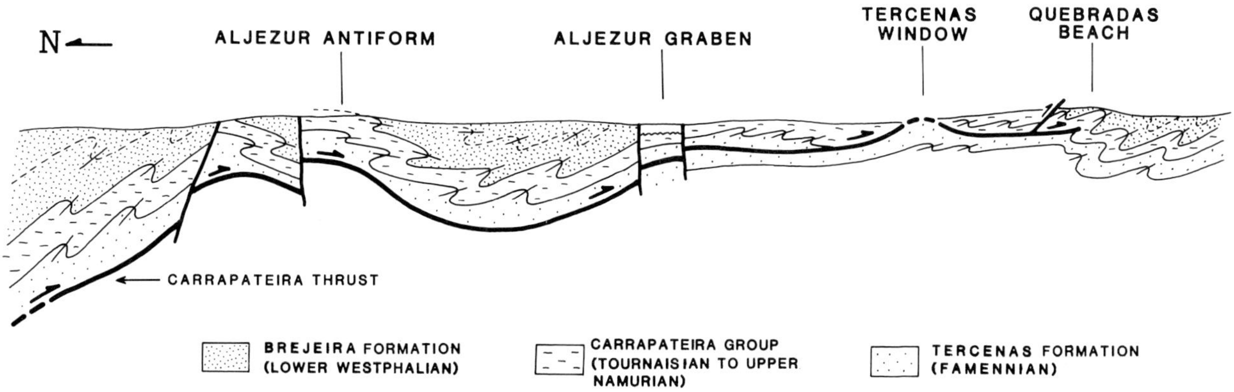

Fig. 14. Schematic section across the Aljezur and Bordeira Antiforms and regional interpretation of the Carrapateira Nappe (After Ribeiro 1981 and Oliveira 1984)

The structural style of the Bordeira and Aljezur Antiform is also of thin-skinned type.

All the stratigraphic sequence of Southwest Portugal is affected by a very low-grade metamorphism (Schermerhorn 1971).

3.3 Structural Synthesis

Structural development of the SPZ is characterized by southwestward vergent folding and thrust displacement, and by a general decrease in deformational intensity in that direction. Metamorphism grades from the greenschist facies in the northern region, through prenite-pumpellyite facies in the Pyrite Belt, to anchimetamorphic regime in the south (Schermerhorn 1975; Munhá 1983b). Because the age of formations and flysch deposition is also younger southwestward, a clear tectonometamorphic and palaeogeographic polarity is indicated.

Stratigraphic units of the Pulo do Lobo Antiform (except the Santa Iria and Horta da Torre Formations) show an earlier north-south trending episode of deformation. This tectonic episode, not studied in detail, may have developed during east-west compression or, alternatively, to sheath folds related to north-south compression.

Between the Upper Famennian and Middle Visean the SPZ was affected by an extensional deformation which led to the development of bimodal volcanism (Schermerhorn 1971; Munhá 1983a). Acidic and basic volcanic rocks are also interbedded in the Gafo and Represa Formations (Fig. 2). The geochemical signature of these volcanics has not been studied and their geodynamic significance is unknown.

From the Upper Visean a major tectonic inversion occurred (Oliveira, Part VI, Chap. 2). This led to a continuous process from sedimentation to hard-rock deformation, through slumping and soft-rock synsedimentary deformation (Ribeiro et al. 1983a). Folds and nappes were generated in a forward propagation, now forming the imbricate complex of the SPZ.

In the Pyrite Belt and in Southwest Portugal, cores of the main anticlines never expose formations older than Upper Devonian. This and relationships observed in deep seismic profile (Mueller et al. 1973; Prodehl et al. 1975) suggest a major decollement at the base of the imbricate complex (Ribeiro et al. 1983a; Fig. 15).

3.4 Geodynamic Model

Location of the Beja-Acebuches Ophiolite suite indicates a major geosuture between the SPZ and OMZ. The ophiolite and the Pulo do Lobo Formation are older than the Horta da Torre Formation of Middle to Upper Famennian age (Oliveira et al. 1986; Giese et al., in press). Available geochemical date (Munhá 1983b) show that the metabasalts interbedded in the Pulo do Lobo Formation have MORB affinities. Rare earth elements from the amphibolites of the Beja-Acebuches ophiolite suite indicate that these amphibolites were former basalts with oceanic to transitional affinities (Bard 1977; Dupuy et al. 1982; Munhá et al. 1986). All this suggests that the metabasalts of the Pulo do Lobo Formation and the Beja-Acebuches ophiolite suite were probably generated in the same pre-Famennian oceanic basin (Fig. 16A). The Horta da Torre

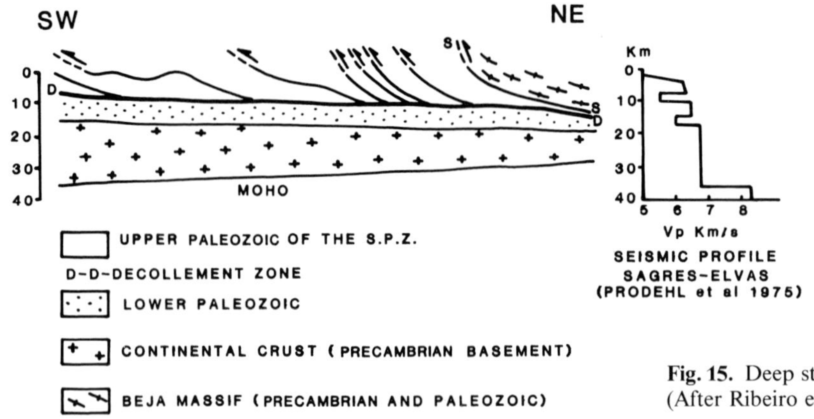

Fig. 15. Deep structure of the South Portuguese Zone (After Ribeiro et al. 1983a)

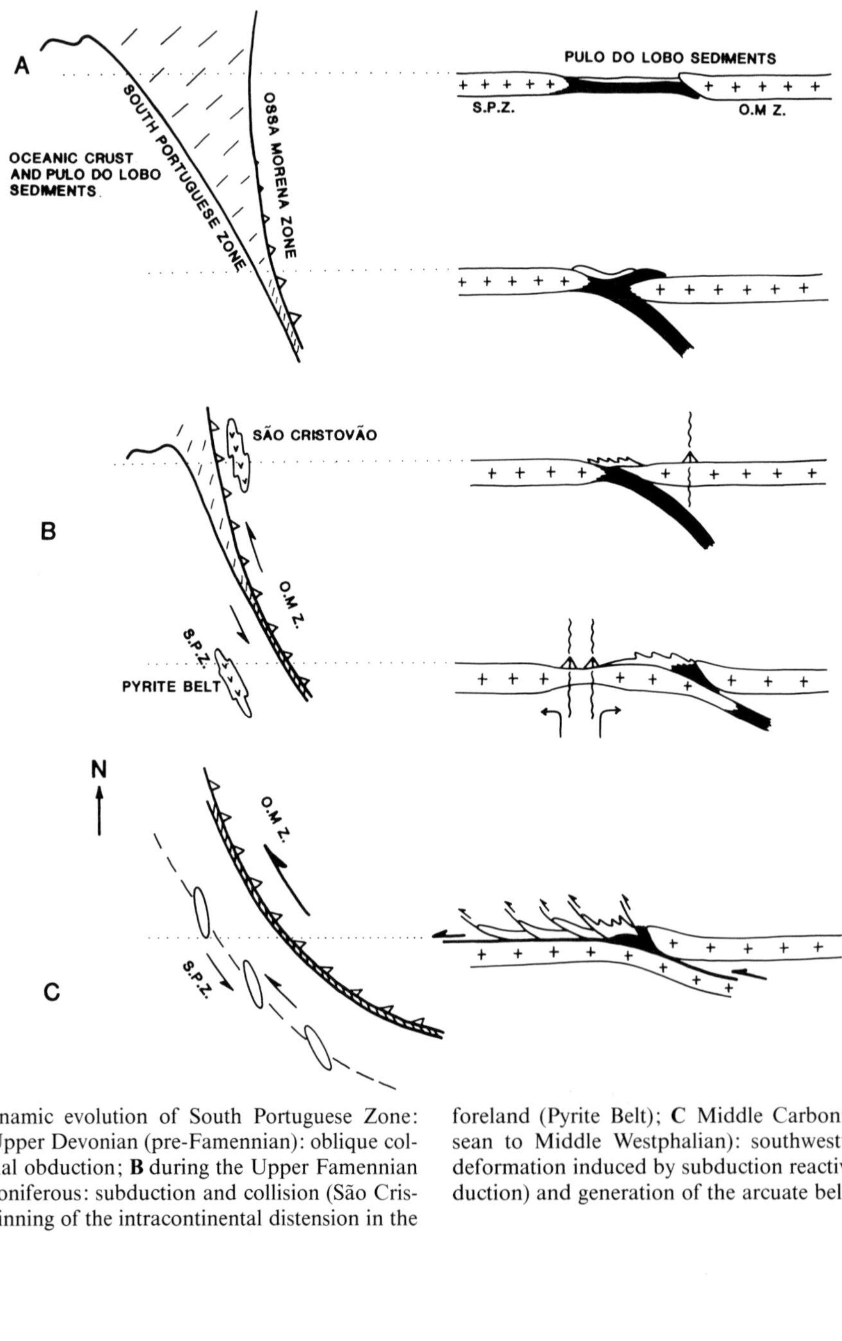

Fig. 16. Geodynamic evolution of South Portuguese Zone: **A** During the Upper Devonian (pre-Famennian): oblique collision and partial obduction; **B** during the Upper Famennian to Lower Carboniferous: subduction and collision (São Cristovão) and beginning of the intracontinental distension in the foreland (Pyrite Belt); **C** Middle Carboniferous (Upper Visean to Middle Westphalian): southwestward transpressive deformation induced by subduction reactivation (type A subduction) and generation of the arcuate belt

Formation is considered in a overstep position over the Beja-Acebuches ophiolite. This indicates that a first collisional event between the SPZ and OMZ may have occurred in pre-Famennian times.

The occurrence of an orogenic calc-alkaline suite in the western border of the OMZ (São Cristóvão and Santa Suzana regions, 90 km northwest of Beja) suggests a north dipping subduction in that area (Santos et al. 1987). Unfortunately, the age of that volcanic suite is not known. However, recent spore determinations on shales interbedded in the upper part of Santa Suzana volcanics indicate an Upper Tournaisian age (Cunha, pers. comm.). The Lower Carboniferous age of the São Cristóvão calc-alkaline suite gives rise to the possibility that the collision between the SPZ and OMZ was oblique, with a northward (present coordinates) propagation (Fig. 16 B).

Bimodal volcanics of the Pyrite Belt (Munhá 1983 a) are slightly younger than that oblique collision, suggesting that they were generated in an intracontinental environment affected by a transtensil regime induced by that collision.

The transpressive deformation of the SPZ, between the Upper Visean and the Lower Westphalian, is tentatively related to the reactivation of the former subduction plane, now acting as a kind of type A subduction. This reactivation may have been responsible for the major decollement below the SPZ and the associated thin-skinned deformation (Fig. 16 C). The arcuate geometry, the *en échelon* arrangement of the main structures and the transection of the folds were due to a left lateral shear component, probably induced by that type A subduction. Together, these characteristics indicate that the SPZ deformation can be ascribed to a transpressive orogen. Previous authors (Badham 1982; Andrews 1983) have also suggested this type of tectonic deformation. We emphazise that the thin-skinned deformation is a main component of this orogen.

Acknowledgements. The authors are indebted to D. Dallmeyer for his criticism, which improved the first version of the manuscript. The text was typed by Fernanda Oliveira from Serviços Geológicos de Portugal.

References

Andrews JR (1983) The Iberian Pyrite Belt, an oblique-slip mobile zone. Notes Mém Serv Géol Maroc Rabat (1986) 335

Badham JPN (1982) Strike-slip orogens – an explanation for the Hercynides. J Geol Soc (Lond) 139: 493–504

Bard JP (1971) Sur l'alternance des zones métamorpgiques et granitiques dans de segment hercynien sud-ibérique; comparaison de la variabilité des caractéres géotectoniques de ces zones avec les orogènes "orthotectoniques". Bol Geol Min Madr 82 (3/4): 324–345 (Reun. Geólogos del SO de la Peninsula Ibérica – la – Badajoz (España); Elvas (Portugal) – 1970

Bard JP (1977) Signification tectoniques des metatholeites d'affinité abyssale de la ceinture métamorphique de basse pression d'Aracena (Huelva, Espagne). Bull Soc Géol Fr (Paris) 7e Ser 19 (2): 375–383

Bard JP Capdevilla R Matte P, Ribeiro A (1973) Geotectonic model for the Iberian Variscan Orogen. Nature 241 (107): 50–52

Carvalho D (1972) The metallogenetic consequences of plate tectonics and the upper Paleozoic evolution of southern Portugal. Estud Not Trab Serv Fomento Min (Porto) 20 (3/4): 297–320

Carvalho D, Goinhas J, Oliveira V, Ribeiro A (1971) Observações sobre a geologia do Sul de Portugal e consequências metalogenéticas. Estud Not Trab Serv Fomento Min (Porto) 20 (1/2): 153–199

Cunha T, Oliveira JT (in press) Upper Devonian palynomorphs from the Represa and Phyllite-Quartzite Formations, Mina de São Domingos Region, southeast Portugal: tectono-stratigraphic implications

Dupuy C, Bostal J, Marcelot G, Bougaultt H, Toron JL, Treuil M (1982) Geochemistry of basalts from central and southern New Hebrides Arc: implications for their source rock composition. Earth Planet Sci Lett 60 (2): 207–225

Feio M, Lombard A (1958) La série de Murração. In: Lombard A Observations lithologiques et tectoniques dans le Sud du Portugal. III. Comun Serv Geol Port 42: 123–128

Giese E, Reitz E, Walter R (in press) Contributions to the stratigraphy of the Pulo do Lobo succession in southwest Spain. Comunicacões dos Sewicos Geológicos de Portugal, vol 74

Monteiro JH, Carvalho D (1987) Sea floor volcanism and polymetallic sulfide deposits in ancient active margins: the case of the Iberian Pyrite Belt. In: Teleki DG, Robson MR, Moore JR, Stockelberg von (eds) Marine minerals. Advances in Research and Resource Assessment. Reichel, Dordrecht

Mueller S, Prodehl C, Mendes AS, Moreira VS (1973) Crustal structure in the southwestern part of the Iberian Peninsula. Tectonophysics 20: 307–318

Munhá JM (1976) Nota preliminar sobre o metamorfismo na Faixa Piritosa portuguesa. Comun Serv Geol Port 60: 151–161

Munhá JM (1983 a) Hercynian magmatism in the Iberian Pyrite Belt. In: Sousa MJL, Oliveira JT (Eds) The Carboniferous of Portugal. Mem Serv Geol Port 29: 39–81

Munhá JM (1983 b) Low grade regional metamorphism in the Iberian Pyrite Belt. Comun Serv Geol Port 69 (1): 3–35

Munhá JM, Oliveira JT, Ribeiro A, Oliveira V, Quezada C, Kerrich R (1986) Beja-Acebuches ophiolite: characterization and geodynamic significance. Maleo (Abstract) (13) 31 pp

Oliveira JT (1983) The marine Carboniferous of South Portugal: a stratigraphic and sedimentological approach. In: Sousa MJL, Oliveira JT (eds) The Carboniferous of Portugal. Mem Serv Geol Port 29: 3–37

Oliveria JT (Coord inador) (1984) Carta Geológica de Portugal à escala 1:200.000. Noticia explicativa da folha 7. Serv Geol Port, 77 pp

Oliveira JT, Silva JB (in press) Carta Geológica de Portugal à escala 1:50.000. Noticia explicativa da folha 46-D Mértola. Serv Geol Port

Oliveria JT, Horn M, Parroth E (1979) Preliminary note on the stratigraphy of the Baixo Alentejo Flysch Group, Carboniferous of Portugal, and on the palaeogeographic development compared to corresponding units in northwest Germany. Comun Serv Geol Port 65: 151-168
Oliveira JT, Cunha T, Streel M, Vanguestaine M (1986) Dating the Horta da Torre Formation, a new lithostratigraphic unit of the Ferreira-Ficalho Group, South Portuguese Zone: geological consequences. Comun Serv Geol Port (1987) 72: 129-137
Prodehl C, Moreira VS, Mueller S, Mendes AS (1975) Deep-seismic sounding experiments in central and southern Portugal. General Assembly of the European Seismological Commission (14th - Berlin). DDR National-Komitee für Geodäsie und Geophysik Berlin, pp 261-266
Ramsay JG (1967) Folding and fracturing of rocks. Mc Graw-Hill, New York, 568 pp
Ribeiro A (1981) A geotraverse through the Variscan Fold Belt in Portugal. In: Zwart HJ, Dornsiefen UF (eds) The Variscan orogen in Europe. Geol Mijnbow 60 (1): 41-44
Ribeiro A (1983) Structure of the Carrapateira Nappe in the Bordeira area, SW Portugal. In: Sousa MJL, Oliveira JT (Eds) The Carboniferous of Portugal. Mem Serv Geol Port 29: 91-97
Ribeiro A, Silva JB (1983) Structure of the South Portuguese Zone. In: Sousa MJL, Oliveira JT (Eds) The Carboniferous of Portugal. Mem Serv Geol Port 29: 83-89
Ribeiro A, Antunes MT, Ferreira MP, Rocha RB, Soares AF, Zbyszewski G, Almeida FM, Carvalho D, Monteiro JH (1979) Intruduction à la géologie générale du Portugal, 114 pp., Serv Geol Port, 11 + pp. 26e Cong Int Géol, Paris, 1980
Ribeiro A, Oliveira JT, Silva JB (1983a) La estructura de la Zona Sur Portuguesa. In: Comba JA (Coordinador) Geología de España. Inst Geol Min Esp Madr 1: 504-511
Ribeiro A., Iglesias M, Ribeiro ML, Pereira E (1983b) Modéle géodynamique des Hercynides Ibériques. Comun Serv Geol Port 69 (2): 291-293
Ribeiro A, Oliveira JT, Ramalho M, Ribeiro L (1987) Carta Geológica de Portugal à escala 1/50.000. Noticia explicativa da folha 48-D Bordeira. Serv Geol Port, 30 pp
Routhier P, Aye F, Boyer C, Lecolle M, Moliére P, Picot P, Roger G (1980) La ceinture sud-ibèrique à amas sulfurés dans sa partie espagnole. Tableau géologique et métallogenique. Synthese sur le type amas sulfurés volcano-sedimentaires. Mem Bur Rech Géol Min (26e Cong Int Géol, Paris, 1980) 94 (1): 265 pp
Santos JF, Mata J, Gonçalves F, Munhá JM (1987) Contribuição para o conhecimento geológico-petrológico da região de Santa Suzana: o Complexo Vulcano-Sedimentar de Toca da Moura. Comun Serv Geol Port 73 (1/2): 29-48
Schermerhorn LJG (1971) An outline stratigraphy of the Iberian Pyrite Belt. Bol Geol Min Madr 82 (3/4): 239-268
Schermerhorn LJG (1975) Spilites, regional metamorphism and subduction in the Iberian Pyrite Belt: some comments. Geol Minjbouw 54 (1): 23-35
Schermerhorn LJG, Stanton WI (1969) Folded overthrusts at Aljustrel (South Portugal). Geol Mag 106: 130-141
Silva JB (1981) Estudo estrutural dum sector da Faixa Piritosa (Região de Mértola). Comun 2° Encontro Geociênc Coimbra
Silva JB (1983) Estrutura da Faixa Piritosa: o estado actual dos conhecimentos com base na cartografia estrutural de uma área na zona de Mértola. Thesis, Univ Lisboa, Lisboa, 70 pp
Silva JB (1985) Structure of the Pyrite Belt: the Mértola Traverse. Papers on the Carboniferous of the Iberian Pyrite Belt. An Fac Ciênc Porto, Suppl to 64 (1983): 41-47 (Spec Vol)
Silva JB (1986) Relações tectonoestratigráficas entre unidades da Faixa Piritosa a Sul da Formação do Pulo do Lobo (Sul de Portugal, região de Mértola). Maleo (Abstract) 2 (13): 42 pp
Silva JB (1989) Estrutura de uma Geotransversal da Faixa Piritosa: Zona do Vale do Guadiana. Estudo da tectónica pelicular em Regime de: Deformação não coaxial. PhD Thesis, Univ Lisboa, Lisboa, 294 pp
Vegas R, Muñoz M (1976) El Contacto entre las zonas Surportuguesa y Ossa-Morena en el SW de España. Una nueva interpretation. Comun Serv Geol Port 60: 31-51

4 Metamorphic Evolution of the South Portuguese/Pulo Do Lobo Zone

J. MUNHA[1]

4.1 Introduction

The South Portuguese Zone (see Fig. 1) of the Iberian Variscan chain consists of an Upper Paleozoic sedimentary series with intercalations of felsic to mafic volcanics which were metamorphosed under low-grade conditions during Hercynian times. The metamorphic phenomena is complex and reflects the superposition of a regional (dynamothermal) event onto an early hydrothermal alteration stage, contemporaneous with massive sulphide ore formation and cooling of the Pyrite Belt volcanic piles.

This paper summarizes current knowledge on the metamorphic petrology of the South Portuguese/Pulo do Lobo zone.

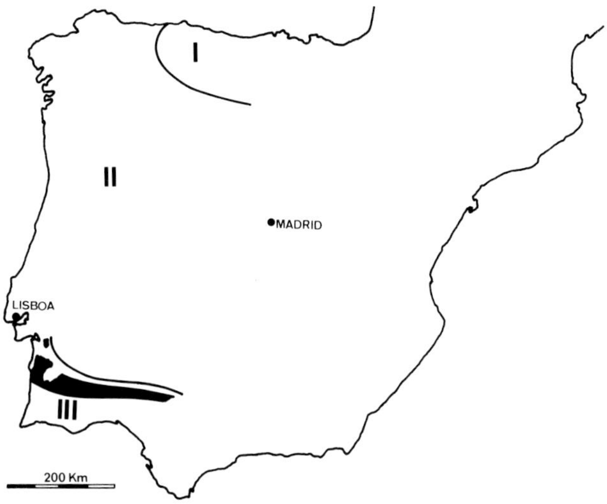

Fig. 1. Location of the main geotectonic zones in the Iberian Variscan Belt. *I* Cantabrian Zone; *II* Centro Iberian Zone; *III* South Portuguese/Pulo do Lobo Zone (*black*, Pyrite Belt)

4.2 Hydrothermal Metamorphism of the Pyrite Belt Volcanic Rocks

Spilitic and quartz-keratophyric rocks are widespread in the Volcanosedimentary Complex of the Iberian Pyrite Belt (Strauss 1970; Schermerhorn 1970, 1975a). They correspond to a bimodal magmatic suite (see Munhá 1983a) which represents the igneous stage in the development of the South Portuguese Zone from Late Devonian to Lower Visean times (Schermerhorn 1971; Carvalho 1976). Volcanism took place in a submarine environment (Williams 1966), and detailed textural and mineralogical observations (Munhá 1979, 1981, unpublished work) indicate that the volcanic rocks acquired their spilitic character prior to the onset of tectonic deformation and associated regional metamorphism.

Recent studies on the petrogenesis of spilitic and quartz-keratophyric rocks from the Pyrite Belt (Munhá 1979; Munhá and Kerrich 1980; Munhá et al. 1980; Barriga and Kerrich 1984) strongly contradict earlier views considering this as primary igneous rocks (cf. Rambaud Pérez 1969; Soler 1973; Schermerhorn 1975a), and rather support the hypothesis that they represent the products of hydrothermal metamorphism of basaltic (rifting tholeiites to within plate alkaline basalts) and crustal rhyolitic or dacitic percursors. It is now firmly established that the hydrothermal fluid was a seawater type solution, as evidenced by oxygen, hydrogen and sulphur isotopic compositions, variations in the oxidation states of iron and hydrothermal mineral assemblages (Munhá and Kerrich 1980; Barriga and Kerrich 1984; Munhá et al. 1986).

The hydrothermal metamorphic grade changes abruptly (on a decametric scale) and ranges from zeolite facies, through chlorite-epidote-bearing assemblages, to the greenschist facies characterized in mafic rocks by the occurrence of chlorite-epidote-actinolite (locally with hornblende). Also, in some gabbroic cumulate rocks the hydrothermal metamorphism produced mineral assemblages characterized by grossular-tremolite-diopside. These features, combined with pertinent experimental data

[1] Departamento de Geologia, Faculdade de Ciencias, Rua da Escola Politecnica 58, 1294 Lisboa Codex, Portugal

R. D. Dallmeyer and E. Martínez García (Eds.)
Pre-Mesozoic Geology of Iberia

(cf. Liou 1971; Maruyama et al. 1983), suggest that high temperature gradients were operative and that the hydrothermal metamorphism experienced by the Pyrite Belt volcanic rocks took place under temperatures ranging from 75 ± 25 °C to 375 ± 25 °C.

Hydrothermal metamorphism was markedly non-isochemical and produced a wide spectrum of compositional variation out of the percursor volcanic rocks. This process involved extensive Na ± K exchange, hydration, oxidation and carbonatization (cf. Strauss 1970; Munhá and Kerrich 1980; Munhá et al. 1980) clearly requiring introduction of large amounts of fluid from a reservoir external to the volcanic rocks. Indeed, calculated hydrothermal fluid $\delta^{18}O$ (mostly within the range, 0- +3‰) and δD (−10- +18‰) values close to that of seawater (cf. Munhá and Kerrich 1980; Munhá et al. 1986) indicate that hydrothermal alteration must have proceeded under conditions of generally high water: rock ratios; sometimes, however, positive shifts in $\delta^{18}O_{fluid}$ (up to +6‰) at constant δD suggest that the system evolved into low water: rock values (Barriga and Kerrich 1981, 1984; Munhá and Kerrich 1981; Munhá et al. 1986), reflecting hotter hydrothermal streamlines characterized by lowered rock permeabilities. Mineral and bulk rock chemical data (Munhá and Kerrich 1980; Barriga 1983 unpublished work) also suggest that the hydrothermal fluids became progressively reduced, Ca- and Si-rich and transition metal bearing by interaction with the Pyrite Belt volcanic rocks, which is similar to those geochemical trends found in both natural and experiemental seawater-rock interaction systems (cf. Tomasson and Kristmanndottir 1972; Bischoff and Dickinson 1975; Motll and Holand 1978). Further evidence for large-scale seawater and rock interaction in the Pyrite Belt is provided by the rodingitic gabbros occurring near the Tharsis mine (Huelva, Spain) which are similar to rodingites formed by rapid hydrothermal discharge through basalts and gabbros found at modern ocean ridges (Honnorez and Kirst 1975) and ophiolites (Barriga et al. 1985).

As seen from the previous discussion, it seems clear that the Pyrite Belt hydrothermal metamorphic pattern is analogous to that found in ophiolites (cf. Spooner and Fyfe 1973; Spooner et al. 1974), which is well explained as a consequence of seawater circulation through the volcanic pile. Fluid motion might have been related to thermally driven convective circulation as a result of heat released due to contemporaneous igneous activity (Lister 1972; Wolery and Sleep 1976; Norton and Knight 1977). Whether or not convective flow in a porous media is possible depends on thermal gradient and permeability; preliminary calculations using boundary conditions appropriate for hydrothermal activity in the Pyrite Belt (Munhá and Kerrich 1980) suggest that not only convective circulation of fluids was possible but also that it must have proceeded in a vigorous manner. As the process results from flow of seawater from cold input zones to zones of hot discharge, the observed/estimated highly variable rock and fluid chemistry is by no means unexpected. Convective circulation of seawater through the volcanic rocks eventually modified seawater into a hot, reduced and mineralized brine; a focused return flow and appropriate geological conditions at the site of discharge (see Barriga and Fyfe 1988) would have led to precipitation of metals generating the massive sulphide deposits and associated metalliferous sediments which characterize the Iberian Pyrite Belt (metallogenesis in the Iberian Pyrite Belt is reviewed by Barriga, Part VI, Chap. 5).

4.3 Regional Metamorphism

4.3.1 Deformation - Recrystallization Relationship

In the South Portuguese Zone, Variscan tectonics produced a typical thrust belt (Ribeiro and Silva 1983) with dominant structures verging to SW and showing increasing deformation towards the NE.

The Pulo do Lobo subzone is an antiformal structure located at the boundary between the Ossa-Morena and the South Portuguese Zones. Its NE limb is formed by the Santa Iria Formation (Carvalho et al. 1976), which was affected by a single episode of folding with generation of axial plane cleavage and very low to low-grade regional metamorphism. However, the underlying formations (Ribeira de Limas and Pulo do Lobo) show a much more complex and intense deformation (see Silva et al., this Vol.), with slaty cleavage (S_1) transposed by a dominant crenulation cleavage (S_2), later refolded by a weaker (S_3) crenulation. Recrystallization of metamorphic minerals took place essentially during the development of S_2, being only insipiently retrograded by F_3. The age of this metamorphic episode is still unknown.

Within the Iberian Pyrite Belt, regional metamorphism was contemporaneous with, and to some extent postdated, the first (main) folding phase of pre-Westphalian D age (Schermerhorn 1975a,b; Munhá 1976). New-grown metamorphic minerals are observed, both oriented parallel to and growing across S_1 being deformed later by S_2, which is a typical post-metamorphic crenulation cleavage. In the southernmost areas of the South Portuguese Zone (e.g. Castro Marim, South Portugal) almost unmetamorphosed culm type sediments were folded but not cleaved.

4.3.2 Metamorphic Zoning

Since Boogard's (1967) first description of prehnite-pumpellyite facies rocks from Pomarão (South Portugal) the metamorphic petrology of the South Portuguese Zone has been the subject of several studies (cf. Schermerhorn 1975b; Lecolle and Roger 1976; Munhá 1976, 1979, 1983b; Simancas 1983, unpublished work).

The metamorphic zonation in the South Portuguese/Pulo do Lobo Zone is shown in Fig. 2; however, the general pattern of the metamorphic zones must be considered as schematic, due to the scarcity and discontinuous nature of the mafic metavolcanic rock outcrops which carry the typomorphic Ca-aluminosilicates.

Four metamorphic zones were distinguished, both on the basis of the observed mineral assemblages in mafic metavolcanics and (limited) illite crystallinity data in the metasedimentary rocks. The metamorphic grade was low and, generally, it increases northward from zeolite facies in the extremel south (zone 1), through prehnite-pumpellyite / greenschist facies in the Pyrite Belt (zones 2 and 3), to high-rank greenschist facies on the Pulo do Lobo Formation (zone 4).

In zone 1, laumontite (or any other zeolite), definitive of the zeolite facies was absent because of the rock chemistry (only sedimentary rocks of culm type are known to occur in this zone), but the metamorphic grade was determined by the occurrence of mixed-layer phyllosilicates (mostly chlorite – illite ± smectites) and illite-crystallinite data (mostly within the range 0.35 to 0.7 2θ° on Kubler's (1968) scale), which suggest that the upper limit of the an-

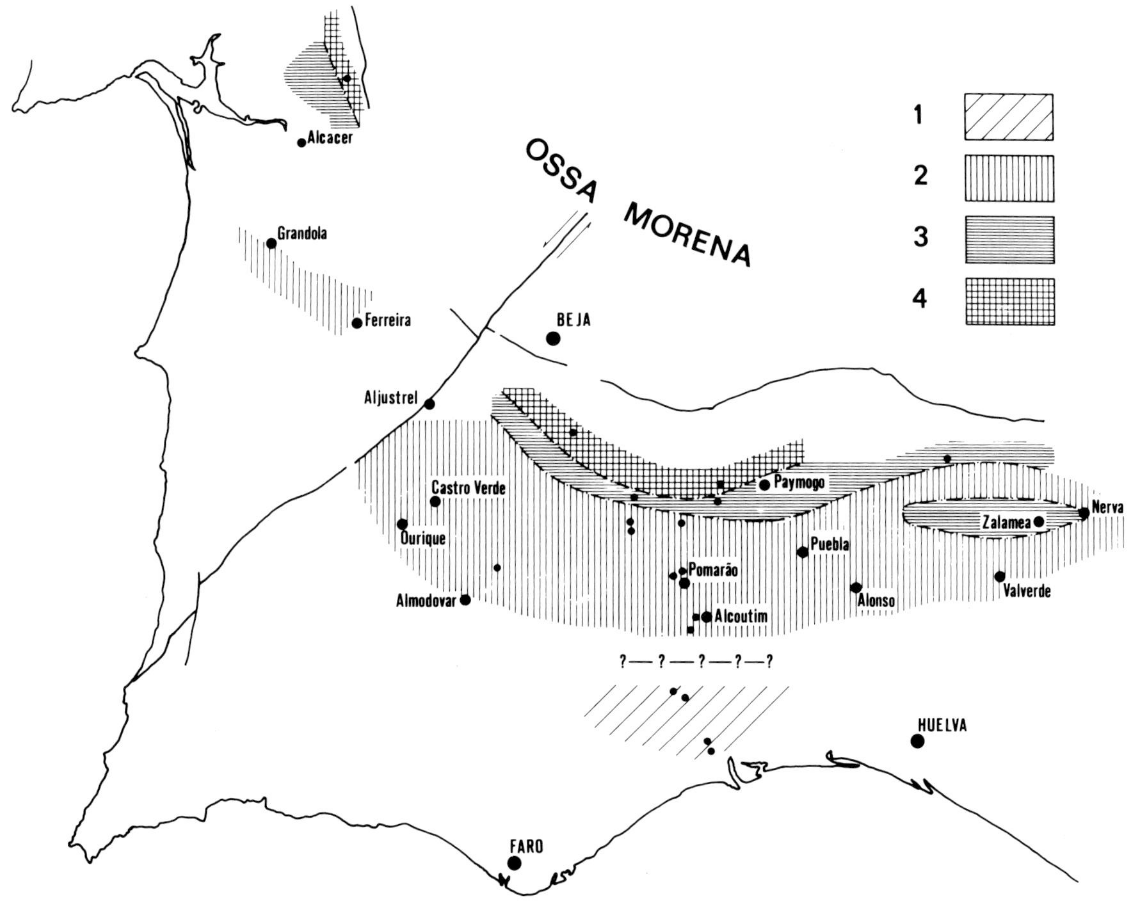

Fig. 2. Metamorphic zonation in the South Portuguese/Pulo do Lobo Zone

chizone (=very low-grade metamorphism) has only been barely reached in zone 1.

Zone 2 dominates the Iberian Pyrite Belt. The critical paragenesis for this zone is albite - pumpellyte - prehnite - chlorite, but actinolite may also occur as a replacement product of igneous clinopyroxene; however, actinolite was observed giving way to chlorite + prehnite (Munhá 1981, unpublished work), and, consequently, it is not considered to be stable under the prehnite - pumpellyte facies conditions which prevailed during regional metamorphism in zone 2. Both the actinolite and some epidote - calcite-bearing assemblages which predate the formation of pumpellyite - prehnite were interpreted as meta-stable relicts developed during the earlier hydrothermal metamorphic event (see Munhá 1979). Illite + chlorite dominate the phyllosilicate assemblage in metasedimentary rocks.

Zone 3 is transitional between prehnite - pumpellyite and greenschist facies; pumpellyite and (particularly) prehnite became less frequent and actinolite appears as a stable metamorphic phase in mafic rocks. Remarkable features of zone 3 are the mineral associations actinolite - epidote and actinolite - pumpellyite, the latter being critical to the pumpellyite - actinolite facies as defined by Hashimoto (1966); nevertheless, in the South Portuguese Zone, actinolite - pumpellyite occurs together with dominant pumpellyite-free assemblages and prehnite may also be present, indicating metamorphic conditions close to the boundary between pumpellyite - actinolite facies and prehnite - actinolite facies (see Liou et al. 1985).

Zone 4 corresponds to the greenschist facies. In contrast with the previous zones at this metamorphic grade the original textures are only very rarely preserved and no relict minerals are left in the mafic metavolcanics; the rocks show penetrative deformation and develop typical nematoblastic texture. Pumpellyite and prehnite are absent, chlorite shows a significant decrease in modal abundance (relative to the previous zones) and hornblende occurs sporadically as narrow rims on actinolite. Typical metamorphic assemblages are: albite, actinolite (± hornblende), epidote, chlorite, quartz, sphene, and muscovite - chlorite - quartz ± albite - tourmaline - iron oxides, in mafic metavolcanics and metasedimentary rocks respectively. Crespo Blanc (1987) reports oligoclase - actinolite - epidote - chlorite - sphene and andaluzite - biotite - muscovite - quartz - oligoclase - tourmaline metamorphic assemblages in Pulo Formation mafic metavolcanic and metasedimentary rocks from near Aroche (SW Spain); these mineral assemblages indicate (local) metamorphic conditions transitional to the lower amphibolite facies, being (most probably) related to thermal doming due to plutonic igneous activity in the area.

4.3.3 P/T Conditions of Regional Metamorphism

Estimates of the temperature stability based on mixed-layer phyllosilicates and illite-chlorite assemblages, provided by Velde (1972), and coupled with the low stability limit of epidote, discussed by Liou (1979) suggest a temperature of approximately 200 °C for the zone ½ transition. A temperature of 290 °C was obtained by Munhá (1983b) using oxygen isotope geothermometry from coexisting albite - hematite ($\delta^{18}O_{alb}$ = +13.82‰; $\delta^{18}O_{hem}$ = −0.16‰) extracted from a meta-basalt outcropping near the zone ⅔ boundary; use of the P-T-X grid developed by Liou et al., (1985), with the constraints imposed by the observed compositional range in epidotes [$X(Fe^{3+})$=0.23-0.27] from buffered assemblages (pumpellyite - prehnite - epidote - actinolite - chlorite - albite - quartz - sphene), indicates similar temperature conditions (290-300 °C) for the zone ⅔ transition. On the basis of the compositional variation of epidotes [$X(Fe^{3+})$=0.27-0.17; Munhá 1983b] coexisting with pumpellyite - chlorite - actinolite (see Nakajima et al. 1977), it was also shown that pumpellyite-bearing assemblages in zone 3 represent equilibration over a temperature range of about 50 °C. The highest metamorphic grade observed in the South Portuguese/Pulo do Lobo Zone is in the greenschist side close to the transitional assemblage to the amphibolite facies determined by Maruyama et al. (1983). According to these data, a maximum temperature in the range of 400° to 450 °C is suggested by the appearance of hornblende and decreasing chlorite modal content in albite - epidote meta-basaltic rocks of zone 4.

Relatively low metamorphic pressures are suggested by geological evidence (indicative of restricted subsidence; see Schermerhorn 1971; Oliveira et al. 1979), which is also substantiated by the absence of any typical high-pressure mineral in the studied samples. The P-T-X grid of Liou et al. (1985), with the constraints detailed above, indicates a pressure of approximately 1.8 kb at the entrance of zone 3. White mica b_0 values from zone 4 (average 8.994 Å; Munhá 1983b) and from the Pyrite Belt (average 8.981 Å; Simancas 1983, unpublished work) pelitic rocks (see Sassi and Scolari 1974) and Na-M4 content of Ca-amphiboles (≤0.1;

Munhá 1979, 1983b) in buffered assemblages from mafic rocks of zone 3 (see Brown 1977) also suggest a low-pressure metamorphic regime and that pressure should not have exceeded about 2 kb at the entrance of greenschist facies. If, as previously discussed, it is accepted that the temperature range within zone 3 was 300° to 350 °C, geothermal gradients of 40° to 50 °C/km can be assumed.

As discussed by Schermerhorn (1975b), it should further be noted that the observed northward increase in metamorphic grade across the same stratigraphic horizon (Volcaniosedimentary Complex of the Pyrite Belt) implies a regional increase in heat flow towards the north; the metamorphism was therefore not simply related to burial, but to orogenic metamorphism, probably associated with thermal doming.

Acknowledgements. The manuscript was much improved by the comments of Prof. D. Santallier and two anonymous reviewers. This is a contribution to the INIC research project Regimes Magmatico e Metamorfico na Cadeia Varisca Iberica.

References

Barriga FJAS, Fyfe WS (1988) Giant pyritic base-metal deposits: the example of Feitais (Aljustrel, Portugal). Chem Geol 69-331-343

Barriga FJAS, Kerrich R (1981) High $\delta^{18}O$ fluids, circulation regimes and mineralization at Aljustrel Iberian Pyrite Belt. Geol Soc Am Abstr Prog 13: 403-404

Barriga FJAS, Kerrich R (1984) Extreme ^{18}O-enriched volcanics and ^{18}O-evolved marine water, Aljustrel, Iberian Pyrite Belt: transition from high to low Rayleigh number convective regimes. Geochim Cosmochim Acta 48: 1021-1031

Barriga FJAS, Munhá J, Fyfe WS, Vibeti J (1985) Extreme hydrothermal alteration in the intrusive layers of the Troodos ophiolite (Cyprus). EOS 66: 1128

Bischoff JL, Dickson FW (1975) Sea water-basalt interaction at 200 °C and 500 bars; implications for origin of sea-floor heavy metal deposits and regulation of sea water chemistry. Earth Planet Sci Lett 25: 385-397

Boogaard M van den (1967) Geology of the Pomarao region (southern Portugal). PhD Thesis, Amsterdam, Univ Amsterdam 110 p

Brown EH (1977) The crossite content of Ca-amphibole as a guide to pressure during metamorphism. J Petrol 18 (1): 58-72

Carvalho D (1976) Considerações sobre o vulcanismo na região de Cercal-Odemira. Suas relações com a Faixa Piritosa. Comun Serv Geol Port 60: 215-238

Carvalho D, Correia HAC, Inverno CMC (1976) Contribuição para o conhecimento geologico do Grupo de Ferreira Ficalho. Suas relações com o Grupo do Pulo do Lobo e com a Faixa Piritosa. Mem Not (Univ Coimbra) 82: 145-169

Cresp Blanc A (1987) El macizo de Aracena (macizo Iberico meridional): Propuesta de división sobre la base de nuevos datos estructurales y petrograficos. Bol Geol Min Esp 98 (4): 507-515

Hashimoto M (1966) On the prehnite-pumpellyite metagreywacke facies. J Geol Soc Jpn 72: 253-265

Honnorez J, Kirst P (1975) Petrology of rodingites from the equatorial mid-Atlantic fracture zones and their geotectonic significance. Contrib Mineral Petrol 49: 233-257

Kubler B (1968) Evaluation quantitative du metamorphisme par la cristallinite de l'illite. Eta des progres realises ces dernieres annes. Bull Centre Rech Pau-SNPA 2: 385-397

Lecolle M, Roger G (1976) Metamorphism regional Hercynien de faible degre dans la province pyrito-cuprifere de Huelva (Espagne). Bull Soc Geol Fr (7), XVIII, (6): 1687-1698

Liou JG (1971) Synthesis and stability relations of prehnite, $Ca_2Al_2Si_3O_{10}(OH)_2$. Am Mineral 56: 507-531

Liou JG (1979) Zeolite facies metamorphism of basaltic rocks from the East Taiwan ophiolite. Am Mineral 64: 1-14

Liou JG, Maruyama S, Cho M (1985) Phase equilibria and mineral parageneses of metabasites in low-grade metamorphism. Mineral Mag 49: 321-333

Lister CRB (1972) On the thermal balance of a mid-ocean ridge. Geophys J R Astronomic Soc 26: 515-535

Maruyama S, Suzuki K, Liou JG (1983) Greenschist-amphibolite facies equilibria at low pressures. J Petrol 24 (4): 583-604

Motll M, Holland HD (1978) Chemical exchange during hydrothermal alteration of basalt by sea water - I. Experimental results for major and minor components of sea water. Geochim Cosmochim Acta 42: 1103-1115

Munhá J (1976) Nota preliminar sobre o metamorfismo na Faixa Piritosa Portuguesa. Comun Serv Geol Port 60: 151-161

Munhá J (1979) Blue amphiboles, metamorphic regime and plate tectonic modelling in the Iberian Pyrite Belt. Contrib Mineral Petrol 69: 279-289

Munhá J (1983a) Hercynian magmatism in the Iberian Pyrite Belt. Mem Serv Geol Port 29: 39-81

Munhá J (1983b) Low-grade regional metamorphism in the Iberian Pyrite Belt. Comun Serv Geol Port 69: 3-35

Munhá J, Kerrich R (1980) Sea water basalt interaction in spilites from the Iberian Pyrite Belt. Contrib Mineral Petrol 73: 191-200

Munhá J, Kerrich R (1981) High temperature (>400 °C) equilibrium of marine water with ocean floor: evidence from ^{18}O enriched fluids of the Iberian Pyrite Belt. EOS 63: 142

Munhá J, Fyfe WS, Kerrich R (1980) Adularia, the characteristic mineral of felsic spilites. Contrib Mineral Petrol 75: 15-19

Munhá J, Barriga FJAS, Kerrich R (1986) High ^{18}O ore-forming fluids in volcanic hosted base metal massive sulfide deposits: geologic, $^{18}O/^{16}O$, and D/H evidence from the Iberian Pyrite Belt; Crandon, Wisconsin; and Blue Hill, Maine. Econ Geol 81: 530-552

Nakajima T, Banno S, Suzuki T (1977) Reactions leading to the disappearance of pumpellyite in low grade metamorphic rocks of the Sanbagawa metamorphic belt in central Shikoku, Japan. J Petrol 18: 263-284

Norton D, Knight J (1977) Transport phenomena in hydrothermal systems: cooling plutons. Am J Sci 277: 937-981

Oliveira JT, Horn M, Paproth E (1979) Preliminary note on the stratigraphy of the Baixo Alentejo Flysch Group, Carboniferous of Portugal, and on the paleogeographic development compared to corresponding units in northwest Germany. Comun Serv Geol Port 65: 151-168

Rambaud Perez F (1969) El sinclinal Carbonifero de Rio Tinto y sus mineralizaciones asociadas. Mem Inst Geol Min Esp 71: 229

Ribeiro A, Silva JB (1983) Structure of the South Portuguese Zone. Mem Serv Geol Port 29: 83-89

Sassi FP, Scolari A (1974) The b_0 value of the potassic white micas as a barometric indicator in low-grade metamorphism of pelitic schist. Contrib Mineral Petrol 45: 143-152

Schermerhorn LJG (1971) An outline stratigraphy of the Iberian Pyrite Belt. Bol Geol Min Esp 80(3-4): 239-268

Schermerhorn LJG (1975a) Spilites, regional metamorphism and subduction in the Iberian Pyrite Belt. Geol Mijnbouw 54 (1): 23-35

Schermerhorn LJG (1975b) Pumpellyite facies metamorphism in the Spanish Pyrite Belt. Petrologie 1(1): 71-86

Soler E (1973) L'association spilites - quartz keratophyres du Sud Ouest de la Peninsule Iberique. Geol Mijnbouw 52: 277-288

Spooner ETC, Fyfe WS (1973) Sub-sea floor metamorphism, heat and mass transfer. Contrib Mineral Petrol 42: 287-304

Spooner ETC, Beckinsale RD, Fyfe WS, Smewing JD (1974) ^{18}O-enriched ophiolitic metabasaltic rocks from Liguria (Italy), Pindos (Greece), and Troodos (Cyprus). Contrib Mineral Petrol 47: 41-62

Strauss GK (1970) Sobre la geología de la provincia piritifera del Suroeste de la Peninsula Ibérica y de sus yacimentos, en especial sobre la mina de Lousal (Portugal). Mem Inst Geol Min Esp Tomo 77: 266 pp

Tomasson J, Kristmannwsdottir H (1972) High temperature alteration minerals and thermal brines, Reykjanes, Iceland. Contrib Mineral Petrol 36: 123-134

Velde R (1972) Phase equilibria for dioctahedral expandable phases in sediments and sedimentary rocks. Proc Int Clay Conf (Madr), pp 235-248

Williams D (1966) Volcanism and ore genesis. Freiberger Forsch C210: 93-111

Wolery TJ, Sleep NH (1976) Hydrothermal circulation and geochemical flux at mid-ocean ridges. J Geol 84: 249-275

5 Metallogenesis in the Iberian Pyrite Belt

F. J. A. S. BARRIGA[1]

5.1 Introduction

One of the truly outstanding characteristics of the South Portuguese Zone (SPZ) is the abundance and dimensions of the massive sulphide deposits contained herein. The part of the SPZ where these ores exist or are likely to be found is well known in the literature as the Iberian Pyrite Belt (IPB). Also abundant, although modest in size and in economic significance, are manganese deposits, which usually are found not far from the massive sulphide deposits.

In this brief summary it is not possible to adequately review the literally hundreds of important contributions that have been published concerning mineral deposits within the IPB. The reader is therefore referred to previous reviews such as those by Carvalho et al. (1976); Strauss et al. (1977); Carvalho (1979); Routhier et al. (1980); and Barriga and Carvalho (1983).

The Iberian Pyrite Belt massive sulphide deposits have been known and exploited since pre-Roman times, mainly as gold, silver and copper ore from the supergene enrichment zones of exposed deposits. It is worth noting that the only recently discovered deposit not near those of the Roman or pre-Roman mines is the Salgadinho deposit (Carvalho 1976), intersected by drilling in 1976.

About 60 mines have operated during the last 100 years, and about 280 million metric tonnes of massive polymetallic sulphides were extracted in this period (Strauss and Madel 1974), mainly for sulphur (average grade 46%), copper (0.7%) and sometimes for iron (40%), although ore nearly always contains appreciable amounts of zinc and lead (combined average 4%, Carvalho et al. 1976), precious metals (0.8 ppm Au, 30 ppm Ag, ibid) and a host of alloy metals such as Sn, Cd, Co, Hg, Bi, Se and many others in concentrations ranging from tens to hundreds of ppm. Many of these metals (including Pb and Zn) are often not recovered because of the fine-grained nature of most ores, which in turn is responsible for the expression Iberian *Pyrite* Belt (IPB) and for frequent incorrect literature descriptions of these deposits as barren pyrite and pyrite orebodies. However, Pyrites Alentejanas SARL (Aljustrel Mines) will start industrial production of Cu, Pb and Zn concentrates in the near future. The discovery of the outstanding Neves-Corvo deposits (Beja district, Portugal) generated renewed interest in the economic potential of the Iberian Pyrite Belt, as they include $\approx$ 40 metric t of high grade copper ore (av. Cu $\approx$ 8.5%), and a comparable amount of high grade Zn-Pb ore ($\approx$ 6% Zn, 1.2% Pb).

The Neves-Corvo deposits were intersected by drilling in 1977, by a syndicate of French and Portuguese companies (Albouy et al. 1981). As early as 1972 the Neves-Corvo area was already a highly promising target, based on work carried out by Serviço de Fomento Mineiro (the Portuguese government exploration agency). Full scale ore production started in 1988.

Reserves of massive sulphide ore in the IPB presently amount to about 700 Mt (megatonnes), essentially concentrated at seven major locations (Fig. 1): (Aljustrel (250 Mt) and Neves-Corvo (130 Mt) in Portugal, and Tharsis (130 Mt), La Zarza (60 Mt), Rio Tinto (55 Mt), Aznalcóllar (50 Mt) and Sotiel (40 Mt) in Spain. Other mines, exhausted or nearly exhausted, which were important in the past, include S. Domingos (30 Mt) and Lousal (20 Mt?). A very large part of these reserves (which do not include stockwork-type ores) were discovered after the early 1960s, as a consequence of widespread acceptance of the exhalative-sedimentary theory of ore genesis (Oftedhal 1958), originally advocated by Klockman (1894), eloquently illustrating the importance of correct geological understanding and models in mineral exploration (Strauss et al. 1977; Carvalho 1979).

[1] Departamento de Geologia, Faculdade de Ciências de Lisboa Rua da Escola Politécnica, 58 1294 Lisboa Codex, Portugal

R. D. Dallmeyer and E. Martínez García (Eds.)
Pre-Mesozoic Geology of Iberia

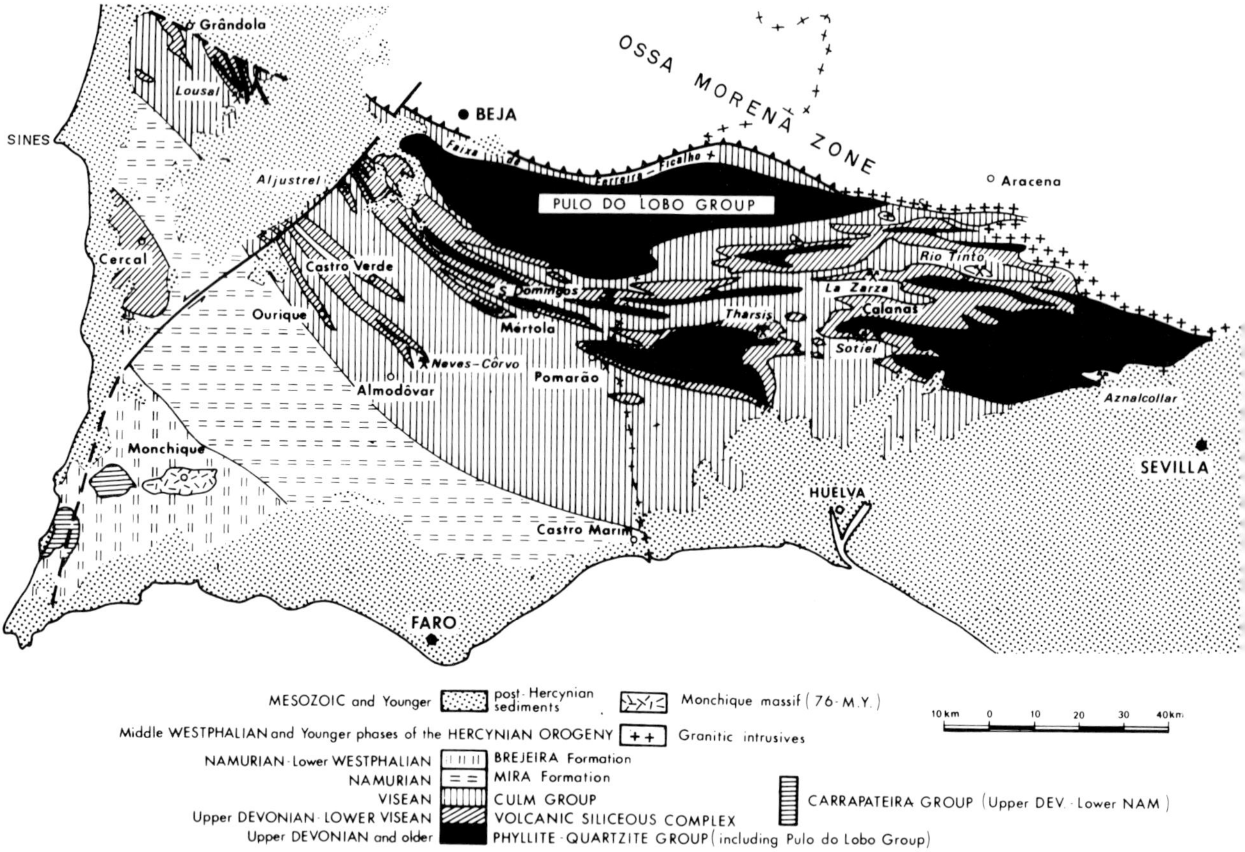

Fig. 1. Location map of the mines and mineral deposits cited in the text

5.2 Ore Geology

As a result of many detailed geologic studies during the last two decades, and in the light of the exhalative-sedimentary theory, several important characteristics bearing on the origin of the sulphide orebodies have been established. They are: (1) syngenetic *(sensu lato),* as evidenced by the well-defined horizons in which they occur, the common sedimentary features they depict and also because they experienced the same tectonic and metamorphic history of the rocks that host them; (2) volcanogenic, in the sense that they formed in close association with the waning stages of (felsic) volcanism; (3) exhalative, i. e. resulting from metalliferous aqueous solutions permeating through footwall rocks, as shown by chemical zonation, the presence of underlying stockwork zones and prominent hydrothermal alteration of the host lithologies in the vicinity of autochthonous orebodies; and (4) submarine, from the submarine lithologies and facies that host them (mostly submarine felsic tuffs, cherts, jaspers) and from the subaqueous textures of the ores themselves.

The IPB massive sulphide deposits invariably occur in statigraphic horizons corresponding to waning stages of felsic volcanism at each of various volcanic centres, or, if stacked felsic volcanic cycles occur, towards the end of one or more of the volcanic periods. The orebodies occur not only at or near the top of felsic volcanic piles, but also anywhere along the strike, in laterally equivalent positions, both on more distal volcanic facies or even mainly on sedimentary sequences (shales, black shales), again laterally equivalent to the volcanic rocks (Fig. 2). As a consequence of this range of host lithologies, the orebodies can be found anywhere from the top to near the base of the ore-hosting Volcanic-Siliceous (or Sedimentary) Complex (VS, see Oliveira, Part VI, Chap. 2), as they are hosted in thick felsic pyroclastic piles or in sediments respectively.

Sulphide orebodies lying on extensive pyroclastic sequences are invariably underlain by stockworks of stringer and disseminated mineralization

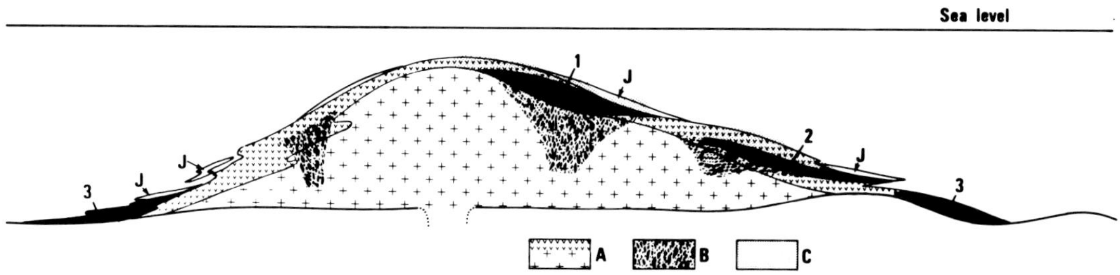

Fig. 2. Schematic representation of the relations between sulphide ores, siliceous sediments, ore zone alteration and lithostratigraphy in the Iberian Pyrite Belt (after Carvalho 1979; and Barriga 1983). *1* Autochthonous type sulphide ore; *2* transitional type ore; *3* allochthonous type ore; *J* jasper; *A* felsic volcanic rocks; *B* predominantly chloritic alteration and cupriferous stockwork; *C* predominantly sericitic alteration

in highly altered host lithologies. Such stockworks are generally not found under orebodies hosted in sediments or on distal volcanic facies. Also, soft sediment deformation structures such as slumping, scour and fill, synsedimentary breccias and even turbidite textures are ubiquitous in orebodies not associated with stockworks and are rare or absent in orebodies rooted in stockworks.

These characteristics suggest that massive sulphide deposits in the IPB formed from metalliferous aqueous solutions rising through footwall rocks, and precipitating sulphides at or near the coeval sea floor (Strunian to Viséan). Slope instability and thixotropic behaviour of the sulphide muds precipitated on the hydrothermal vents may, in some cases, have induced down-slope movements along volcano flanks, with redeposition in deeper waters, among distal volcanic beds or even within sediments (Schermerhorn 1971). Clear-cut examples of autochthonous orebodies are Rio Tinto and Perrunal-La Zarza (Spain) and Feitais-Estação (Aljustrel, Portugal). Good examples of allochthonous orebodies are Filón Norte and S. Guillermo (Tharsis, Spain) and Lousal (Portugal, Carvalho 1979). The Salgadinho disseminated deposit, hosted in felsic tuffs depicting stockwork-type alteration (Carvalho 1976; Plimer and Carvalho 1982) may represent the roots of an ore-forming system, and massive mineralization (not yet found in the area) may have moved downslope from its top (Barriga and Carvalho 1983). The relations between the various types of sulphide deposits, lithostratigraphy, siliceous sediments and ore zone alteration are schematically illustrated in Fig. 2.

Key references for detailed descriptions of individual mining areas are summarized below:

1. *Aljustrel* contains the largest massive sulphide reserves in the IPB, and constitutes one of the largest concentrations of volcanogenic massive sulphides in the world. It is one of the more extensively studies areas in the whole IPB. Schermerhorn and Stanton (1969); Freire d'Andrade and Schermerhorn (1971); and Carvalho et al. (1976) described the general geology. The geological map produced by Freire d'Andrade and Schermerhorn is still used today, with some additional geological observations (e. g. Ribeiro et al. 1983; Barriga 1983). Barriga and Fyfe (1988) have recently summarized the geology of the area.
2. *The Neves Corvo* geology and mineral deposits were described by Albouy et al. (1981) and Leca et al. (1983, 1985).
3. *Lousal* (Strauss 1965) and *S. Domingos* (Webb 1958) were important Portuguese mines (now exhausted).
4. *Tharsis* descriptions can be found in Strauss and Madel (1974) and Carvalho et al. (1976).
5. Strauss et al. (1981) published a geological and geochemical study of the La Zarza deposits.
6. *The Rio Tinto* ores were the first to gain international fame, and are believed to represent the largest known concentration of volcanogenic massive sulphides ever discovered (> 500 Mt), 55 Mt still remaining after erosion and mining. Another outstanding feature of Rio Tinto is the size and economic significance of the Cerro Colorado stockwork ($\approx$200 Mt of stringer and disseminated Cu ore). Rambaud Pérez (1969) and Garciá Palomero (1977, 1980) studied the area in detail (see also Williams et al. 1975).
7. Specific accounts on *Aznalcóllar* are those by Sicilia et al. (1975) and Sierra (1984).

5.3 Manganese Deposits and Other Metalliferous Sediments

Hundreds of small manganese deposits are known from within the Volcanosedimentary VS Complex of the IPB, and have been the object of minor mining operations. They occur within various lithologies, namely tuffites, siliceous slates (often red or purple, hematitic) and cherts and jaspers. The ensemble of the Mn concentrations and their immediate host lithologies constitute mappable units, and lie invariably on volcanic rocks, both mafic and felsic, or on their lateral equivalents. They constitute good field evidence of a late- to post-volcanic environment, but cannot be considered true marker horizons in exploration for massive sulphide deposits, because they occur associated with both productive and non-productive volcanic episodes (at least as far as is known). Despite this limitation, the occurrence of Fe-Mn formations can be an indication of ore proximity, as some such Fe-Mn rich horizons are indeed related to massive sulphide deposition (Strauss and Madel 1974; Carvalho et al. 1976). Jaspers are particularly significant, and are perhaps the most typical rocks of the IPB (Carvalho 1979). Jaspers usually occur in the field as lenticular bodies of various sizes (up to kilometres), and not as continuous strata. However, this may result largely from deformation, either synsedimentary or tectonic. In several localities (Pomarão, Aljustrel, etc.) it can be seen that the "lenses" are in fact boudins.

Barriga and Oliveira (1986) presented a geochemical study of 16 occurrences of cherts, jaspers and manganese ores related to felsic volcanism throughout the IPB, aimed at elucidating the origin of these siliceous and metalliferous sediments. They concluded that these rocks are almost exclusively chemical hydrothermal sea-floor precipitates. These may have accumulated as silica gels with Fe and Mn oxides or carbonates, in relatively short time spans, given the general lack of detrital material. Low abundances of transition trace metals and REE and negative Ce anomalies were interpreted as indicative of relatively rapid precipitation under hydrothermal conditions. V-shaped REE patterns suggest precipitation from sea water or sea water derived hydrothermal fluids, or both. Variations in Eu systematics (negative, zero, or positive anomalies) were interpreted as reflecting various degrees of Eu enrichment in hydrothermal fluids in the course of hydrothermal metamorphism. Barriga and Oliveira reported only one occurrence of hydrogenous-type metalliferous sediments, which may have precipitated more slowly, esssentially in equilibrium with cold, unmodified marine water.

A similar but much more detailed study of the Aljustrel siliceous and metalliferous sediments (Barriga 1983; Barriga and Fyfe 1988) indicates a hydrothermal origin for these rocks, and unveils an intimate relationship with the genesis of the associated sulphide ores. The jaspers are an early product of ore-related hydrothermal activity, later reworked and altered by subsequent stages of the same global hydrothermal process (see below).

5.4 Hydrothermal Alteration

Hydrothermal alteration is a prominent feature of several lithologies of the VS Complex, notably igneous rocks in general and essentially all host rocks to autochthonous massive sulphide deposits.

5.4.1 Regional Alteration

Most, if not all volcanic and hypabissal rocks in the IPB have been affected by (sea water) hydrothermal metamorphism (Munhá 1979; Munhá and Kerrich 1980; Barriga and Kerrich 1981; Barriga 1983). This has produced marked chemical changes in the rocks. Despite undebatable textural evidence for metasomatism, this fact has not been adequately realized in the past. Mafic igneous rocks were originally basaltic, tholeiitic transitional to arc tholeiites at the base of VS, and become progressively enriched in incompatible elements upwards, where they are typical "within plate" alkaline basalts (Munhá 1983), deeply spilitized (Schermerhorn 1975; Munhá 1979; Munhá and Kerrich 1980). Most felsic rocks in the area were originally rhyolitic, often high-silica rhyolites (Munhá 1981; Barriga 1983). Their present composition is mostly quartz-keratophyric to quartz-kalikeratophyric (Schermerhorn 1970, 1973; Salpeteur 1976; Barriga 1983). These cumbersome terms obscure the fact that they are simply hydrothermally altered rhyolites (felsic spilites according to Munhá et al. 1980). It should be emphasized that conclusions on the original chemical nature of the rocks require careful screening of all metasomatic effects and consideration of primary features only: relic igneous minerals (often minute or absent) and the less mobile elements.

Alteration of the igneous rocks on a regional scale has importance in metallogenetic studies and in mineral exploration. In some instances it has been possible to document metal leaching from volcanic rocks, implying that mineralized solutions may have originated from such processes. For example, the analyses of a doleritic sill intersected by drilling near Garrochal show that its margins, more susceptible to sea water access, are preferentially hydrated, oxidized and enriched in alkali metals, and depleted in Co, Cu and Zn (Munhá and Kerrich 1980). At Aljustrel it has been possible to show that regional alteration is stratigraphically controlled, with generation of an alteration zonation essentially parallel to bedding (Barriga 1983, 1986). The uppermost tens of metres of the Aljustrel volcanic rocks are markedly oxidized, hydrated and enriched in Mg (up to $\approx 7\%$ MgO in rhyolites), whereas stratigraphically deeper levels are reduced and markedly depleted in transition metals, including Fe, Cu and Zn. It is thus possible to estimate the stratigraphic position of a given outcrop or drilling intersection, even with complicated tectonic situations. Here the alteration pattern cannot be explained but through pervasive sea water circulation shortly after deposition of the volcanic rocks. It should be mentioned that this is one of the few studies where the ability of sea water to leach metals from rocks has been documented in an ore-bearing area (see Franklin et al. 1981, p. 573).

5.4.2 Ore Zone Hydrothermal Alteration

During the early twentieth century, ore zone hydrothermal alteration was an important aspect of both metallogenic and exploration studies in the IPB (e.g. Collins 1922; Edge 1928; Bateman 1927; Williams 1934; Webb 1958). With the renaissance of the exhalative sedimentary theory in the 1960s, sedimentary features were given priority at the expense of hydrothermal aspects. Hydrothermal alteration studies have recovered their status since Carvalho (1976) successfully used alteration studies as ore guides, in the discovery of the Salgadinho deposit (Fig. 1; see Barriga and Carvalho 1983). At the present time there is unanimous acceptance of their importance.

Ore zone hydrothermal alteration is the result of interaction of mineralizing fluids, wall rocks and sea water. This produced several alteration mineral assemblages, subordinated to a more or less well-zoned, concentric pattern centered in the main domain, or domains, of hydrothermal fluid supply on each mineralized system. Thus the more prominent and better recognized type is *stockwork alteration*. Stockwork ores are composed of sulphide (chalcopyrite, pyrite, minor pyrrhotite, sphalerite) veinlets and disseminations in a silicate ($\pm$carbonates) matrix essentially composed of quartz + chlorite and/or sericite. Barriga and Carvalho (1983) proposed the name *stockwork rock* for this lithotype. Stockworks are often zoned. At Feitais (Aljustrel) a prominent stockwork exists (Barriga 1983), composed of a quartz - chlorite - sulphide core surrounded by a peripheral halo of quartz - sericite - sulphides that grades progressively into regionally altered footwall rocks (Aljustrel volcanics). The presence of relics of quartz-keratophyre in the stockwork indicates that regional hydrothermal alteration predated sulphide deposition. The zonation of the stockworks can also be seen around individual sulphide veins. From the periphery to the core of the stockwork, original volcanic textures are progressively obliterated by corrosion of the original minerals and concomitant precipitation of alteration phases. Stockwork alteration extends laterally beyond the limits of the sulphide orebodies, and thus can be used as an ore guide in exploration.

In some instances, *hanging wall alteration* has been described, such as above the Salgadinho deposit (Plimer and Carvalho 1982), and above Feitais (Barriga 1983; Barriga and Fyfe 1988). The presence of this type of alteration indicates that hydrothermal activity persisted after deposition of the immediate hanging wall rocks. These may include not only volcanic rocks but also shales, phyllites, tuffites, cherts and jaspers. Hanging wall alteration may be mineralogically similar to stockwork alteration (especially in volcanic rocks), but can also consist of manganese enrichment (oxides or carbonates: rhodocrosite, Mn-calcite, kutnahorite).

Ore zone alteration is invariably reducing. In instances where originally oxidized rocks are affected the results can be striking. This is the case of the hanging wall of the Feitais orebody, where jasper (hematitic red chert) is the predominant lithotype. In the immediate vicinity and especially above the orebody the jasper is partially or totally altered into bluish-gray, pyritic chert, with reduction of hematite to magnetite and pyrite. Bright red jasper above massive sulphide ore is restricted to relics of various sizes, up to a few tens of centimetres, with gradational contacts to chert of various shades of gray, containing magnetite, pyrite and even chlorite in replacement of hematite. Vein-controlled reducing alteration can often be seen, from the vein wall gray chert to unaltered red jasper. In other cases a network or a subparallel set of closely spaced veins leaves no unaltered relics, transforming the jasper into completely reduced pyritic and/or chloritic chert. The pretectonic nature of the alteration of the jasper is proven by the fact that alteration features are often deformed, probably by synsedimentary deformation. Minerals in these veins can include (singly or in combination): quartz, carbonates, chlorite, pyrite, magnetite, chalcopyrite, sphalerite, barite and cobaltite (Barriga and Fyfe 1988). These transformations have genetic significance. It should be noted that bluish gray chert (altered jasper) is known for long as an empirical highly favourable exploration guide in the IPB (Carvalho 1979).

5.5 Isotope Geochemistry

5.5.1 Sulphur Isotopes

Munhá and Kerrich (1980) summarized sulphur isotope data reported for the IPB by Rambaud Pérez (1969); Williams et al. (1975); Lécolle (1977); and Routhier et al. (1980): $\delta^{34}S$ in pyrites from stockwork mineralizations, +5 to +10 per mil, average +8.5; in one vein, +4.3 per mil; massive sulphide ores (1) cores, +6 to 0 per mil, av. +3; (2) periphery, −4 to +5 per mil, av. +2; and from disseminations in cherts, −8 to −32 per mil, av. −24 per mil. Munhá and Kerrich concluded that the more likely interpretation of the isotope data was that reduced sea water sulphur was the main source of sulphur in the pyritic ores, whereas the markedly negative $\delta^{34}S$ values of pyrite disseminated in cherts could be explained by bacterial reduction of sea water sulphate (based on the dependence of sulphur isotope fractionations among solid species *and* several possible oxidized and reduced sulphur species in solution, and their dependence on factors such as fO_2 and pH, and, of course, temperature and isotopic composition of the solution). Arnold et al. (1977) reported analyses on five Spanish deposits, and found values similar to those above in autochthonous deposits (San Miguel, La Zarza), but far lower values in allochthonous deposits (New Sotiel, hosted in carbonaceous shales: −34 to −25 per mil $\delta^{34}S$), and intermediate values in transitional deposits (San Platón, Concepción), where whole rock (rich in framboidal pyrite) determinations are markedly negative (−25 to −19 per mil), and pyrite separates (from San Platón) vary between −4 and +8.5 per mil $\delta^{34}S$ (these differences are puzzling and remain to be explained). Arnold et al. interpreted their data to indicate that sulphur in autochthonous deposits resulted from leaching of host rocks, whereas allochthonous ores were progressively contaminated with sulphur derived from marine sulphate. More recently, Mitsuno et al. (1988) reported ≈300 sulphur isotope measurements of pyrite, chalcopyrite, sphalerite and barite separated from ore specimens from 11 mines throughout the IPB. Values range between −10 and +10 per mil $\delta^{34}S$. The Concepción and San Telmo samples (see Fig. 1) are all positive, all others (except Tharsis, which is exclusively negative) include both positive and negative values (Lousal, Aljustrel, Neves-Corvo, S. Domingos, La Zarza, Sotiel, Rio Tinto and Aznalcóllar). Stockwork samples are always positive. Mitsuno et al. noted that the predominance of negative values is anomalous, and interpreted their data as reflecting various degrees of incorporation of bacteriogenic sulphur, maximal in fine-grained banded pyrite.

5.5.2 Oxygen and Hydrogen Isotopes

Oxygen and hydrogen isotopic compositions are perhaps the best tracers of fluid reservoir signatures, not only because these elements are the essential components of aqueous fluids, but also in view of the simplicity of their fractionation mechanisms, which depend only on temperature and composition of solid phases (e. g. Hoefs 1980). Given that fluid inclusions in pretectonic hydrothermal minerals from the IPB ore deposits are far too small to enable study, oxygen isotope compositions also constitute the best geothermometers available.

$^{18}O/^{16}O$ and D/H fractionation data on whole rock samples and mineral separates from various mine areas have been reported by Barriga and Kerrich (1981, 1984), Munhá and Kerrich (1981), and Munhá et al. (1986). The Aljustrel data is by far the more extensive, enabling interpretations more detailed than elsewhere. The minerals analyzed are quartz, chlorite, sericite, manganese, oxides, rhodonite and albite, mostly in pairs. The preservation of a systematic disparity between the oxygen isotope composition of quartz in stockwork and massive ore, coupled with the retention of a high ^{18}O signature for islands of less altered volcanics in stockworks (see above, and Barriga and Kerrich 1984), collectively suggest an absence of significant isotopic homogenization during waning hydrothermal activity and/or the ensuing low-grade regional metamorphism. Furthermore, the minerals analyzed are believed to have crystallized during hydrothermal activity, on the basis of paragenetic evidence (Barriga 1983), as opposed to being earlier phases isotopically modified by hydrothermal activity.

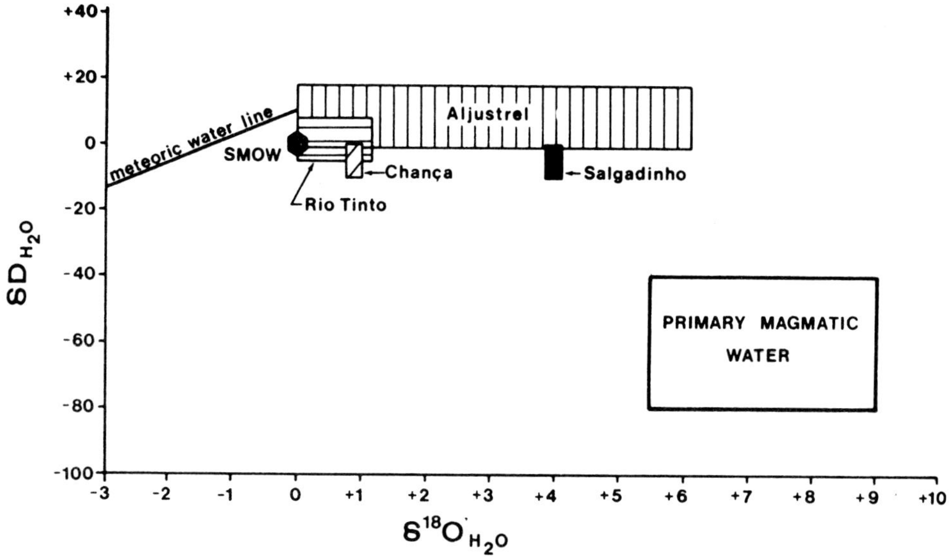

Fig. 3. The δD-$\delta^{18}O$ values calculated for ore-forming fluids in various massive sulphide deposits of the Iberian Pyrite Belt (After Munhá et al. 1986)

Thus, there is good reason to believe that the observed variations in oxygen compositions reflect real variations of temperature in the hydrothermal systems.

Oxygen isotope geothermometry in IPB mineral deposits indicates that stockworks formed at $\approx 200°$-$270\,°C$ (Aljustrel, Rio Tinto, Salgadinho), massive sulphide ores at $\approx 200\,°C$ (Aljustrel), and hanging wall cherts and jaspers up to $\approx 120\,°C$ (Aljustrel, Lagoas do Paço, Mértola). $\delta^{18}O$ and δD values calculated for hydrothermal fluids are represented in Fig. 3. $\delta^{18}O$ varies considerably between 0 and 6 per mil, whereas δD is remarkably constant (-10 to $+15$ per mil). These data are interpreted as meaning that mineralizing fluids in the IPB were mostly derived from sea water, perhaps with small contributions of fluids derived from other reservoirs (magmatic, metamorphic), although these are not required by the isotope data.

The large variation in $\delta^{18}O_{fluid}$ calculated for the Feitais-Estação orebody of Aljustrel can only be interpreted in terms or interaction of marine water with high ^{18}O rocks, under conditions of low water/rock ratios (see Fig. 4). Under these constraints, production of large quantities of isotopically heavy hydrothermal fluid requires multi-pass, essentially closed-system convective circulation through permeable rocks, under an impermeable cover. Other mechanisms that have been invoked to account for high ^{18}O ore fluids in volcanogenic environments include: (1) participation of magmatic or metamorphic fluid along with sea water, (2) derivation by expulsion of isotopically evolved connate fluids of marine origin, (3) isotopic enrichment by shale ultrafiltration, (4) prior rapid evaporation of seawater in a closed basin and/or (5) hydrothermal boiling (see Munhá et al. 1986, for a discussion on these mechanisms). All these would induce far larger δD shifts than observed (decreases in the case of magmatic or metamorphic fluids, positive for the remaining hypotheses), as depicted in Fig. 4., and it is therefore unlikely that any of these processes are actually reflected by the isotope data.

At Aljustrel oxygen isotope data includes volcanic rocks unaffected by ore zone alteration, either remote from sulphide ores or from relics of quartz-keratophyre preserved within the stockworks (see above; Barriga and Kerrich 1984). $\delta^{18}O$ values are relatively uniform at 15.7 to 18.1 per mil (average 16.7 per mil). These are some of the highest values ever reported for volcanic rocks (e.g. Taylor 1978), and reflect the intense hydrothermal alteration experienced by the Aljustrel volcanic rocks. This is in good agreement with mineralogy and textures (Barriga 1983). $\delta^{18}O$ of whole rock stockwork samples are very constant at ≈ 12 per mil (± 0.5), and recognizable quartz-keratophyres affected by weak stockwork alteration depict intermediate values ≈ 14 per mil. On the basis of these data, Barriga and Kerrich (1984) concluded that ore formation postdated the early hydrothermal event that enriched the Aljustrel volcanics in ^{18}O, as described above. The present

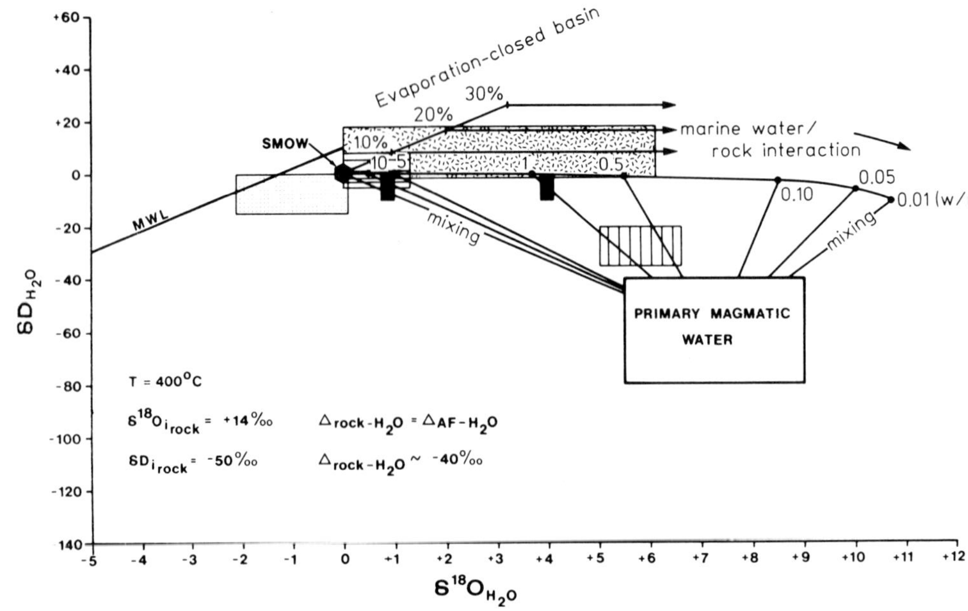

Fig. 4. Schematic representation of the δD and δ^{18}O values expected in ore fluids derived from various processes and fluid reservoirs (see text). *Boxes,* Isotope compositions of ore fluids. *Dotted,* Crandon, Wisconsin; *vertical lines,* Blue Hill, Maine; *all others,* Iberian Pyrite Belt (see Fig. 3). Compare Blue Hill, with a likely contribution of magmatic fluid, with the IPB fluids, derived from sea water via interaction with rock (After Munhá et al. 1986)

δ^{18}O values of 12 per mil of the stockwork rocks are therefore believed to have been reached through two successive stages of hydrothermal alteration: early ^{18}O enrichment to values near 17 per mil (regional alteration), followed by ^{18}O loss to around 12 per mil (stockwork alteration).

5.6 Genetic Models

5.6.1 General Model

There is little doubt that the IPB massive sulphide deposits precipitated at or near the coeval sea floor, either during the waning stages of local volcanic activity and/or shortly thereafter, from hot, reducing, slightly acidic metal-bearing solutions largely of sea-water derivation. These were formed during cooling of the highly permeable felsic pyroclastic rocks that invariably constitute the footwall of autochthonous deposits, via convective circulation of sea water. The cherts, jaspers and manganiferous ores that almost invariably accompany the sulphide orebodies, as lateral equivalents and cap rocks, constitute another product of the same overall hydrothermal process. They have the genetic significance of the exhalites, key tuffites and cherty iron formations of the North American literature (Ridler 1971; Hutchinson 1973; Sangster and Scott 1976) or the ferruginous beds of the Kuroko (Japan) deposits (Lambert and Sato 1974; Kalogeropoulos and Scott 1983). Barriga and Carvalho (1983) summarized the more salient aspects of ore genesis, and compared the IPB data with that from similar metallogenic provinces elsewhere, including present day analogues. They concluded that there is firm evidence to support the applicability of the general sea water convective model on the genesis of the IPB ores. Metals were probably leached from host rocks (mostly volcanic rocks). Most sulphur is derived from sea water, probably upon mixture of metalliferous fluids with cold unmodified sea water.

5.6.2 Giant Massive Sulphides

The discovery of "black smokers" (e. g. Corliss et al. 1979; Francheteau et al. 1979) convinced many that massive sulphide deposits might have formed in the same way. Massive sulphide deposits in ophiolites are indeed often comparable to black smokers, but the IPB sulphide ores are typically much larger that these. Also, the autochthonous orebodies do not usually contain intercalated oxidic facies, or oxidized sulphides, are not diluted by detrital material and do not appear to be explained by this model. Abundant venting of mineralized fluids onto the sea floor may not always generate large mineral deposits. The conditions at the site of ore deposition are of prime importance in massive sulphide generation.

Barriga and Fyfe (1988) have addressed this question and pointed out that the ideal set of variables for the production of giant massive sulphides must have been met at various locations in the IPB. Among these variables stand out the presence of felsic volcanics at the marine interface and the possibility of large plutonic energy sources (to drive the hydrothermal systems during significant time spans). These do not suffice, however. The conditions at the site of ore deposition are of prime importance as well, and a means of preventing dispersion (as ineffective black smoke) must have been present. It is interesting to note that the IPB ores do not appear to have accumulated in stagnant brine pools such as those of the Red Sea, because the mineralizing fluids had to be less dense than sea water, otherwise they would not have risen through the highly porous footwall submarine pyroclastic rocks. Instead, they would have accumulated at the base of the permeable layer (see Zierenberg 1984).

Other systems where dispersion is reduced are the sulphides within sediments of the Guaymas basin (Von Damm et al. 1985), and the deeper Salton Sea (California) brines (see McKibben and Elders 1985). In these situations sulphides precipitate within sediments and few ore components are lost as black smoke.

Barriga (1983) and Barriga and Fyfe (1988) interpreted the Aljustrel data to indicate that hydrothermal activity started as soon as the volcanic rocks were deposited, with drifting cellular convection that altered the rocks under conditions of high water/rock ratio and down to temperatures in the vicinity of 100 °C. Unfocused discharge of the modified fluid may have generated the jaspers, as an initially unconsolidated silica gel. Hydrothermal activity likely continued, and the essentially impermeable siliceous and metalliferous sediments induced a major change in the convective regime through the volcanic rocks, to closed-system, multi-pass circulation, and associated reheating, as heat loss became conductive through the permeability barrier. Massive sulphide precipitation subsequently occurred under the cap of unconsolidated jasper precursors, as a consequence of the new hydrodynamic, chemical and thermal conditions. Barriga and Fyfe (1988) envisaged two mechanisms of creating space for sulphide precipitation under the jasper. One involves precipitation within the base of the unconsolidated jaspers, and accumulation on top of the volcanics through some sort of zone-refining process. Alternatively (or concomitantly), the buoyancy of the ore-forming fluid may have floated the silica gel (specific density ≈ 1.3). Hanging wall alteration in the jasper would then represent minor leakage of ore fluids through the otherwise impermeable blanket. The effect of a cap rock is two-fold: (1) the available heat is used with great efficiency, as heat loss is minimal (via conduction across the impermeable barrier) and (2) sulphide precipitation through mixing of M^{2+} -rich fluids with S^{2-} under an impermeable cover will facilitate essentially quantitative metal precipitation, with minimal loss as black smoke. Additional research is needed to test the applicability of this model to other situations in the IPB.

5.7 Exploration Implications

Since the early 1960s, exploration in the IPB has been quite successful, once detailed geophysical surveys were performed on target areas selected on the basis of sound geological knowledge (see Strauss et al. 1977; Carvalho 1979). But the current exploration practice has two drawbacks: (1) in several cases, very promising targets have been abandoned after drilling that neither intersects ore nor explains the favourable data (geophysical anomalies, etc.) and (2) no known tools capable of distinguishing low- and high-grade ore exist. There is great need of means to overcome both problems.

There is research in progress aimed at solving both of the above issues, through establishment of new criteria to be used in addition to the current exploration methodologies (Barriga 1986). Factors requiring detailed elucidation include: (1) total amount and rate of supply of mineralized fluids, (2) composition and physical characteristics of such fluids (e. g. density, temperature, Eh, pH), (3) fluid pathways, (4) chemical reactions leading to ore mineral precipitation, (5) wall rock influence on this, (6) barriers to dispersion of potential ore components (including conditions as diverse as palaeotopography and structure, presence of physical, chemical and/or biological barriers to dispersion) and (7) post-depositional evolution.

5.8 Conclusion

The economic and scientific importance of the ore deposits of the Iberian Pyrite Belt constitutes one of its more salient features. Our knowledge of metallogenesis in the IPB is, despite all limitations, quite considerable, and involves many aspects, from detailed field geology to the use of sophisticated methods of understanding natural processes. The key to this success is the treatment of metallogenesis as the result of an array of normal geological processes, requiring contributions from all the different fields of geology.

Acknowledgements. The ideas expressed in this chapter result in part from discussions with many colleagues and friends. Mining and exploration companies operating in the IPB, Instituto Geologico y Minero de España, Serviço de Fomento Mineiro, Serviços Geológicos de Portugal and Departamento de Geologia da Faculdade de Ciências de Lisboa have repeatedly organized fruitful field trips and mine visits, and provided other forms of support as well. They are all gratefully acknowledged.

This is a contribution to EEC project MA1M - 0029 - P.

References

Albouy L, Conde LN, Foglerini F, Leca X, Morikis A, Callier L, Carvalho P, Songy JC (1981) Le gisement de sulfures massifs polymétalliques de Neves-Corvo (Baixo Alentejo, Sud Portugal). Chron Rech Minière 460: 5-27

Arnold M, Bernard AJ, Soler E (1977) Premier apport de la géochimie des isotopes du soufre à la comprehension de la genese des minéralisations pyriteuses de la province de Huelva (Espagne). Mineral Dep 12: 197-218

Barriga FJAS (1983) Hydrothermal metamorphism and ore genesis at Aljustrel, Portugal. Ph D Thesis, Univ Western Ontario, Canada, 386 p

Barriga FJAS (1986) Lithogeochemistry and petrography as exploration tools in the Iberian Pyrite Belt: the Aljustrel example. Maleo Soc Geol Port 2(13): 11

Barriga FJAS, Carvalho D (1983) Carboniferous volcanogenic sulphide mineralizations in South Portugal (Iberian Pyrite Belt). In: Lemos de Sousa MJ, Oliveira JT (eds) The Carboniferous of Portugal. Mem Serv Geol Port 29: 99-113

Barriga FJAS, Fyfe WS (1988) Giant pyritic base-metal deposits: the example of Feitais (Aljustrel, Portugal). Chem Geol 69: 331-343

Barriga FJAS, Kerrich R (1981) High ^{18}O fluids, circulation regimes and mineralization at Aljustrel, Iberian Pyrite Belt. Geol Soc Am Abstr Progr 13(7): 403-404

Barriga FJAS, Kerrich R (1984) Extreme ^{18}O-enriched and ^{18}O-evolved marine water, Aljustrel, Iberian Pyrite Belt: transition from high to low Rayleigh number convective regimes. Geochim Cosmochim Acta 48: 1021-1031

Barriga FJAS, Oliveira JT (1986) Geochemical study of cherts, jaspers and manganese ores from the Iberian Pyrite Belt. Maleo Soc Geol Port 2(13): 11

Bateman AM (1927) Ore deposits of the Rio Tinto district. Econ Geol 22: 569-614

Carvalho D (1976) Considerações sobre o vulcanismo da região de Cercal-Odemira. Suas relações com a Faixa Piritosa. Comun Serv Geol Port 60: 215-238

Carvalho D (1979) Geologia, metalogenia e metodologia da investigação de sulfuretos polimetálicos do sul de Portugal. Comun Serv Geol Port 65: 169-191

Carvalho D, Conde L, Hernández Henrile J, Oliveira V, Schermerhorn LJG, García Palomero F, Lopera Caballero E, Madel J, Martin E (1976) Livro-Guia das excursões geológicas na Faixa Piritosa Ibérica (24-27 de Junho de 1975). Comun Serv Geol Port 6: 271-315

Collins HF (1922) The igneous rocks of the province of Huelva and genesis of the pyritic orebodies. Trans Inst Mining Metall 31: 61-169

Corliss JB, Dymond J, Gordon LI, Edmond JM, von Herzen RP, Ballard RD, Green K, Williams D, Bainbridge A, Crane K, van Andel TH (1979) Submarine thermal springs on the Galápagos rift. Science 203: 1073-1083

Damm KL von, Edmond JM, Measures C, Grant B (1985) Chemistry of hydrothermal solutions at Guaymas basin, Gulf of California. Geochim Cosmochim Acta 49: 2221-2237

Edge AB (1928) Observations on the pyritic orebodies of southern Spain and Portugal. C R XIV Cong Géol Int (Madr) Fasc Suget VIII, Théor Mod Métall 3: 1027-1233

Francheteau J, Choukroune P, Juteau T, Seguret M, Ballard RD, Fox PJ, Normark W, Carranza A, Cordoba D, Guerrero J, Rangin C, Bougault H, Cambon P, Hékinian R (1979) Massive deep sea sulphide ore deposits discovered on the East Pacific Rise. Nature 277: 523-528

Franklin JM, Lydon JW, Sangster DF (1981) Volcanic-associated massive sulfide deposits. Econ Geol 75th Anniversary Vol, pp 485-627

Freire d'Andrade R, Schermerhorn LJG (1971) Aljustrel and Gavião. In: Carvalho D, Goinhas DAC (eds) Principais jazigos minerais do Sul de Portugal. Livro-guia excursão nº 4, I Congr Hispano Luso Americano Geol Econ, pp 32-59

García Palomero F (1977) Caracteres geológicos y relaciones morfológicas y genéticas de las mineralizaciones del anticlinal de Rio Tinto. Acta Salmant, Salamanca, 22 p

García Palomero F (1980) Caracteres geológicos y relaciones morfológicas y genéticas de las mineralizaciones del anticlinal de Rio Tinto. Arquivos de la provincia de Huelva, Huelva

Hoefs J (1980) Stable isotope geochemistry, 2nd edn. Springer, Berlin Heidelberg New York, 208 p

Hutchinson RW (1973) Volcanogenic sulfide deposits and their metallogenic significance. Econ Geol 68: 1223-1246

Kalogeropoulos SI, Scott SD (1983) Mineralogy and geochemistry of tuffaceous exhalites (tetsusekiei) of the Fukasawa mine, Hokuroku district, Japan. Econ Geol Monogr 5: 412-432

Klockman F (1894) Über die lagerartige Natur der Kiesvorkommen des südlichen Spaniens und Portugals. Sitzungsber Preussisch Akad Wiss Berlin 46: 1173-1181

Lambert IB, Sato T (1974) The Kuroko and associated ore deposits of Japan: a review of their features and metallogenesis. Econ Geol 69: 1215-1236

Leca X, Ribeiro A, Oliveira JT, Silva JB, Albouy L, Carvalho P, Merino H (1983) Cadre géologique des minéralisations de Neves-Corvo, Baixo Alentejo, Portugal. Mém Bur Rech Géol Min 121, 79 p

Leca X, Albouy L, Aye F, Picot P (1985) Caractéristiques principales du gisement de Neves-Corvo (Portugal). Chron Rech Min 481: 53-58

Lécolle M (1977) La ceinture sud-ibérique: un exemple de

province à amas sulfurés volcano-sédimentaires (tectonique, métamorphisme, stratigraphie, volcanisme, paléogéographie et métallogénie). Thesis, Univ P et M Curie, Paris

McKibben MA, Elders WA (1985) Fe-Zn-Cu-Pb mineralization in the Salton Sea geothermal system, Imperial Valley, California. Econ Geol 80: 539-559

Mitsuno C, Nakamura T, Kanehira K, Sugita M, Kase K, Thadeu D, Carvalho D, Arribas A (1988) Geological studies of the "Iberian Pyrite Belt" - with special reference to its genetical correlation of the Yanahara ore deposit and others in the inner zone of south-west Japan. Univ Okayama, Japan, 300 p

Munhá J (1979) Blue amphiboles, metamorphic regime and plate tectonic modelling in the Iberian Pyrite Belt. Contrib Mineral Petrol 69: 279-289

Munhá J (1981) Igneous and metamorphic petrology of the Iberian Pyrite Belt volcanic rocks. Ph D Thesis, Univ Western Ontario, Canada, 711 p

Munhá J (1983) Hercynian magmatism in the Iberian Pyrite Belt. In: Lemos de Sousa MJ, Oliveira JT (eds) The Carboniferous of Portugal. Mem Serv Geol Port 29: 39-81

Munhá J, Kerrich R (1980) Sea water-basalt interaction in spilites from the Iberian Pyrite Belt. Contrib Mineral Petrol 73: 191-200

Munhá J, Kerrich R (1981) High temperature (>400 °C) equilibrium of marine water with ocean floor: evidence from ^{18}O enriched fluids of the Iberian Belt. Eos 62: 1056

Munhá J, Fyfe WS, Kerrich R (1980) Adularia, the characteristic mineral of felsic spilites. Contrib Mineral Petrol 75: 15-19

Munhá J, Barriga FJAS, Kerrich R (1986) High ^{18}O ore-forming fluids in volcanic-hosted base metal massive sulfide deposits: geologic, $^{18}O/^{16}O$, and D/H evidence from the Iberian Pyrite Belt; Crandon, Wisconsin; and Blue Hill, Maine. Econ Geol 81: 530-552

Oftedhal C (1958) A theory of exhalative-sedimentary ores. Geol Foren I Stockh Forh 80: 1-19

Plimer IR, Carvalho D (1982) The geochemistry of hydrothermal alteration at the Salgadinho copper deposit, Portugal. Mineral Dep 17: 193-211

Rambaud Pérez F (1969) El sinclinal Carbonífero de Rio Tinto y sus mineralizaciones associadas. Mem Inst Geol Min Esp 71, 229 p

Ribeiro A, Oliveira JT, Silva JB (1983) La estructura de la Zona Surportuguesa. In: Libro Jubilar de J. M. Rios (Geologia de España). Inst Geol Min Esp, pp 504-511

Ridler RH (1971) Relationship of mineralization to stratigraphy in the Archean Rankin Inlet-Ennadai Belt. Can Mining J 92(4): 50-53

Routhier P, Aye F, Boyer C, Lécolle M, Molière P, Picot P, Roger G (1980) La Ceinture Sud-Ibérique à amas sulfurés dans sa partie espagnole mediane. Mém Bur Rech Géol Minières 94, 265 p

Salpeteur I (1976) Étude structural et petrographique de la zone de Paymogo (Nord de la province d'Huelva). Metallogenèse des amas sulfurées associés. Thesis, Univ Nancy I, 234 p

Sangster DF, Scott SD (1976) Precambrian strata-bound, massive Cu-Zn-Pb sulfide ores of North America. In: Wolf KH (ed) Handbook of strata-bound and stratiform ore deposits, Elsevier, Amsterdam, pp 129-222

Schermerhorn LJG (1970) The deposition of volcanics and pyritite in the Iberian Pyrite Belt. Mineral Dep 5: 273-279

Schermerhorn LJG (1971) Pyritite emplacement by gravity flow. Bol Geol Min 82(III-IV): 304-308

Schermerhorn LJG (1973) What is a keratophyre? Lithos 6: 1-11

Schermerhorn LJG (1975) Spilites, regional metamorphism and subduction in the Iberian Pyrite Belt: some comments. Geol Mijnbouw 54(1): 23-35

Schermerhorn LJG, Stanton WI (1969) Folded overthrusts at Aljustrel (South Portugal). Geol Mag 106: 130-141

Sicilia JLCS, Sampere MS, Plaza JAA (1975) Investigacion de piritas y sulfuros complejos en la zona de Aznalcóllar. Jomadas Min Metal Bilbao II: 155-178

Sierra J (1984) Geología, mineralogía y metalogenía del yacimiento de Aznalcóllar. Bol Geol Min 95: 553-568

Strauss GK (1965) Zur Geologie der SW-Iberischen Kiesprovinz und ihre Lagerstätten, mit besonderer Berücksichtigung der Pyritgrube Lousal/Portugal. Thesis, Univ Munich, 152 pp

Strauss GK, Madel J (1974) Geology of massive sulphide deposits in the Spanish-Portuguese Pyrite Belt. Geol Rundschau 63(1): 191-211

Strauss GK, Madel J, Fernández Alonso F (1977) Exploration practice for strata-bound volcanogenic sulphide deposits in the Spanish-Portuguese Pyrite Belt. In: Klemm DD, Schneider H (eds) Time and strata-bound ore deposits. Springer, Berlin Heidelberg New York, pp 55-93

Strauss GK, Roger G, Lécolle M, Lopera E (1981) Geochemical and geological study of the volcano-sedimentary sulfide orebody of La Zarza, Huelva Province, Spain. Econ Geol 76: 1975-2000

Taylor HP Jr (1978) Oxygen and hydrogen isotope studies of plutonic granitic rocks. Earth Planet Sci Lett 38: 177-210

Webb JS (1958) Observations on the geology and origin of the S. Domingos pyrite deposits, Portugal. Comun Serv Geol Port 42: 129-143

Williams D (1934) The geology of the Rio Tinto mines, Spain. Trans Inst Mining Metall 43: 593-640

Williams D, Stanton RL, Rambaud F (1975) The Planes-San Antonio pyritic deposits of Rio Tinto, Spain: its nature, environment and genesis. Trans Inst Mining Metall B84: B73-B82

Zierenberg RA (1984) The Mattagami Lake mine Archean Zn-Cu sulfide deposit, Quebec: hydrothermal co-precipitation of talc and sulfides in a sea floor brine pool - evidence from geochemistry, $^{18}O/^{16}O$, and mineral chemistry - a discussion. Econ Geol 79: 1951-1952

Part VII Sedimentary and Faunal Domains in the Iberian Peninsula During Lower Paleozoic Times

Sedimentary and Faunal Domains in the Iberian Peninsula During Lower Paleozoic Times

M. Robardet[1] and J. C. Gutierrez Marco[2]

1 Introduction

The purpose of this chapter is to define, characterize and illustrate the distinct domains that composed the Iberian Penisula during the Lower Paleozoic (from Cambrian-Ordovician to Devonian), and to discuss and evaluate the affinities or the differences that existed between these domains and to compare them with other regions of southwestern Europe. The Upper Devonian and Carboniferous periods have not been considered here: in those times the different regions were on the verge of assembly in the Hercynian Belt, and their distinctive sedimentary and faunal characteristics were most probably partially or totally obliterated. Consequently, the South Portuguese Zone, which consists entirely of Upper Devonian-Carboniferous, is outside the scope of the present chapter.

During the Lower Paleozoic the global paleogeographical pattern was characterized by several distinct faunal "realms" or "provinces". These paleobiogeographical entities have been defined on the basis of the distribution of fossil faunas and sediments. These data can give useful indications about the paleolatitude and provide constraints on the relative position and the distances separating the different provinces. On the other hand, paleomagnetic studies have also contributed significantly to our paleogeographical knowledge in yielding quantitative information about paleolatitudes and the past continental configuration.

The global reconstructions that have been proposed for the Ordovician period indicate that the Iberian Peninsula was part of the Mediterranean Province or Selenopeltis Province (Spjeldnaes 1961, 1967; Dean 1967, 1976; Whittington and Hughes 1972; Cocks and Fortey 1982; Gutiérrez Marco and Rabano 1987) which extended over southern and central Europe and broadly corresponded to the high paleomagnetic inclination domain (Fig. 1) or the "Armorica plate" (Van Der Voo 1979, 1982, 1988; Perroud et al. 1984). On the other hand, no general agreement has been reached as to the geographical position (Fig. 2) and the evolution of the

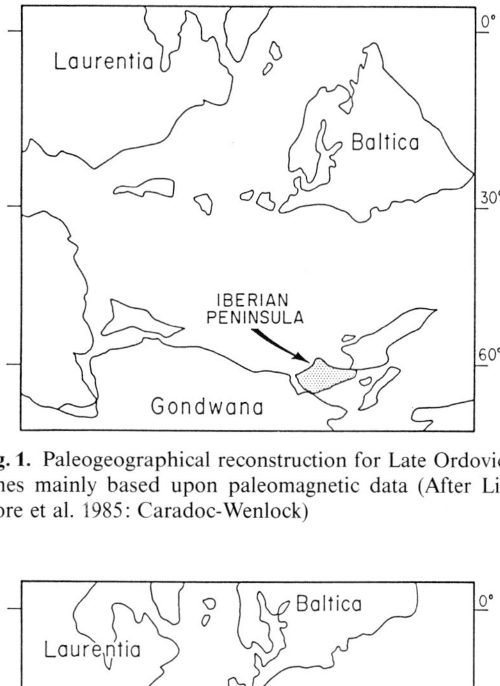

Fig. 1. Paleogeographical reconstruction for Late Ordovician times mainly based upon paleomagnetic data (After Livermore et al. 1985: Caradoc-Wenlock)

Fig. 2. Paleogeographical reconstruction for Early Devonian times mainly based upon paleomagnetic data (after Livermore et al. 1985: Ludlow-Emsian). This reconstruction is one of various alternatives

[1] Laboratoire de Paléontologie et de Stratigraphie, Université de Rennes I, 35042 Rennes Cedex, France. U. R. A. 1364 "Paléozoique marin gondwanien" associée au CNRS
[2] U. E. I. de Paleontología, Instituto de Geología Económica (C. S. I. C.-U. C. M.), Facultad de Ciencias Geológicas, 28040 Madrid, Spain

R. D. Dallmeyer and E. Martínez García (Eds.)
Pre-Mesozoic Geology of Iberia

Fig. 3. Subdivision of the Iberian Hercynian Belt after Lotze (1945) (outlined areas: pre-Mesozoic rocks). *KZ* Kantabrische Zone (Cantabrian Zone); *WLZ* Westasturisch-Leonische Zone (West-Asturian-Leonese Zone); *GKZ* Galizisch-Kastilische Zone (Galaico-Castillan Zone); *OAZ* Ostlusitanisch-Alcudische Zone (Luso-Alcudian Zone); *OMZ* Ossa-Morena Zone; *SPZ* Sudportugiesische Zone (South Portuguese Zone)

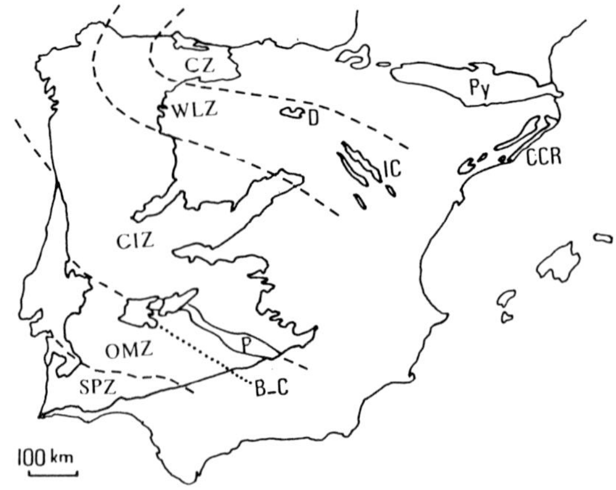

Fig. 4. Subdivison of the Iberian Hercynian Belt after Julivert et al. (1974). *CZ* Cantabrian Zone; *WLZ* West-Asturian-Leonese Zone; *CIZ* Central-Iberian Zone; *OMZ* Ossa-Morena Zone; *SPZ* South Portuguese Zone; *Py* Pyrenees; *D* Sierra de la Demanda; *IC* Iberian Chains (Celtiberia); *CCR* Cataloni-an Coastal Ranges; *P* Los Pedroches batholith; *B-C* Badajoz-Cordoba shear zone

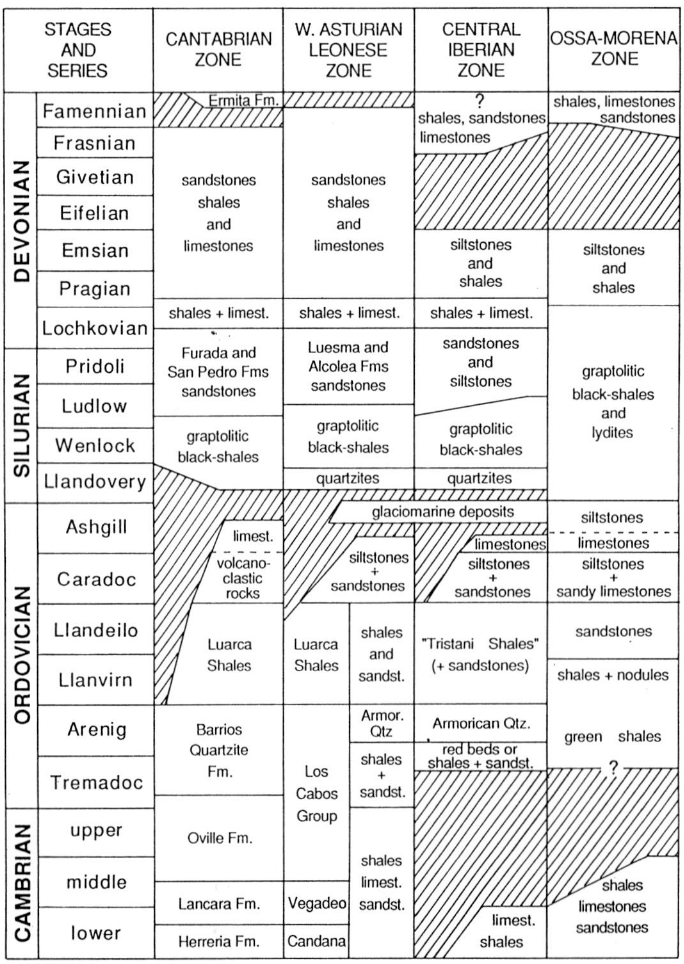

Fig. 5. Major characteristics of the pre-Carboniferous successions in the different zones of the Iberian Peninsula (*hatched* erosional or stratigraphical gaps)

different continental plates during the Silurian and the Devonian (see Perroud et al. 1984; Kent et al. 1984; Scotese et al. 1985; Livermore et al. 1985; Van Der Voo 1988).

As mentioned above, all the regions which compose the Hercynian Belt in Europe present sedimentary and faunal similarities and are therefore included in the Mediterranean Province. However, detailed studies of the lithological successions, the rock facies and the benthic faunas (compared at the species level) have indicated that this province can be divided into several domains, each one being characterized by a distinctive pattern of sedimentation and distinctive benthic faunas. Such conclusions were proposed for the southwest of Europe and more precisely for the Iberian Peninsula (Robardet 1976; Paris and Robardet 1977) and are supported by recent studies in the Ossa-Morena Zone (Gutiérrez Marco et al. 1984; Racheboeuf and Robardet 1986; Robardet 1986).

The paleobiogeographical data are presented here using sections dealing with the Cambrian, Ordovician, Silurian and Devonian periods, so that it is possible to compare the paleogeographical pattern in the successive periods and to discern a possible evolution of the different units.

The Iberian part of the Hercynian Belt has been divided into different zones (Lotze 1945), each being characterized by distinctive stratigraphical and structural features. These subdivisions (Fig. 3) have subsequently been slightly modified (Julivert et al. 1974) and the present chapter refers to this amended zonal scheme (Fig. 4). However, it must be pointed out that the limit between the Central-Iberian Zone and the Ossa-Morena Zone defined by Julivert et al. (1974) along the Los Pedroches batholith (Fig. 4, P) has not been adopted unanimously; several authors (Robardet 1976; Chacon and Pascual 1979, etc.) rather favour the Badajoz-Córdoba shear zone (Fig. 4 B–C) as discussed below. The major characteristics of the pre-Carboniferous succession in the different zones are summarized in Fig. 5.

2 Cambrian

Cambrian rocks are known throughout the whole Iberian Peninsula, except the South Portuguese Zone. Numerous studies (listed in Zamarreno 1983 and Perejón 1984) have led to biostratigraphical subdivisions and correlations, mainly based upon trilobite and archeocyath faunas (Lotze 1961; Sdzuy 1961; Linan 1978 unpublished work; Perejon 1984).

The northern regions (Cantabrian Zone and West-Asturian-Leonese Zone, including Sierra de la Demanda and Celtiberia; see Fig. 4) comprise a complete Cambrian sequence with a continuous passage to the overlying Ordovician. In the south, Lower and Middle Cambrian rocks occur in the Ossa-Morena Zone. In the central regions, only Lower Cambrian rocks are known, and this may be partly related to erosional processes preceding the Ordovician transgression.

All the paleogeographical reconstructions that have been proposed refer to that of Lotze (1961), who considered the Iberian Peninsula as a wide marine area comprising a northern trough (Kantabrisch-Iberischer Trog), a central uplifted area with incomplete successions (Mittel-schwelle, z. T. Heraushebungen), and a southern trough (Andalusischer Trog). This reconstruction has been regarded as oversimplified, but present knowledge of the Cambrian in the Iberian Peninsula does not permit a precise reconstruction (Julivert 1978; Vegas 1978; Zamarreno 1983, pp. 186–188), as the distribution of the different lithofacies in time and space do not clearly define an unequivocal sedimentary pattern.

The paleontological data have rarely been considered from a paleobiogeographical point of view. In any case they do not clearly indicate the existence of distinct faunal domains in the Iberian Peninsula. The variations in the faunal assemblages seem mainly due to lithological changes (carbonate facies faunas, shale facies faunas) and do not clearly support evidence for an important biogeographical differentiation (Gil Cid et al. 1987).

3 Ordovician

The Iberian Peninsula contains extensive areas of Ordovician rocks which crop out in the different zones of the Hercynian Meseta (except the South Portuguese Zone) and in regions which were later included in the Alpine belt (Celtiberia, Catalonia, Pyrenees). The lithological and faunal major characteristics of the Iberian Ordovician are those of the Mediterranean Province (= *Selenopeltis* Province). The lithological succession is almost completely terrigenous and consists mainly of dark grey-green siltstones and shales with alternating sandstone formations; limestones only occur in the uppermost part of the succession (Ashgill). These Ordovician rocks yield rich benthic faunas composed of trilobites, brachiopods, molluscs, echinoderms, ostracods, etc., whose main characteristics and affinities have been defined and illustrated in numerous publications (see references in Hammann et al. 1982; Romano 1982; Julivert and Truyóls 1983; Rabano 1984; Cocks and Fortey 1982). They are regarded as cold-water faunas closely related to those of North Africa. This agrees perfectly with a north-Gondwanan position of all southern Europe supported by both paleomagnetic and sedimentological data (Robardet and Doré 1988).

Considering the different types of rocks, the sedimentary development and the fossil faunas, numerous regional units have been distinguished in Spain and Portugal (Hammann et al. 1982). However, most of these units only differ in characteristics which most probably do not have any major paleogeographical significance and result rather from minor differences in environment and ecology. The main paleogeographical units can only be defined on the basis of first-order characteristics, such as those concerning the relations between the Ordovician and the underlying rocks, the main types of rock facies and benthic faunas.

3.1 Central-Iberian Domain

The best understanding of the Ordovician in the Iberian Peninsula comes from south of the central regions (Ostlusitanisch-Alcudische Zone, *sensu* Lotze 1945) where the succession is the most complete, with extensive outcrops and rich fossiliferous formations in numerous localities (cf. Hammann et al. 1982; Gutiérrez Marco et al. 1984; Rabano 1984). These types of successions and faunas may be regarded as a standard and a reference for comparison with other regions of the Peninsula.

The basal arenaceous formations of the Ordovician sequence transgressively and discordantly overlie Lower Cambrian or Precambrian rocks. The Armorican Quartzite Formation with trace fossils of the *Cruziana-Skolithos* ichnofacies (Arenig) is locally underlain by nonfossiliferous conglomerates, red beds or shale-sandstone alternations, regarded either as Tremadoc or Arenig in age. The dark shales with oolithic ferruginous levels (the so-called *Neseuretus tristani* Shales) which overlie the Armorican Quartzite Formation correspond to the Llanvirn-Llandeilo time interval. They yield graptolites (Gutiérrez Marco 1986, unpublished work) and very rich benthic faunas of trilobites, brachiopods, molluscs, echinoderms, ostracods, etc., and a wealth of paleontological information has been published for more than 100 years (see references in Julivert and Truyols 1983). In the later Ordovician, the lithology is less uniform, with Caradoc sandstone, siltstone and shale units, locally overlain either by Early Ashgillian limestones, by glacio-marine deposits (most probably of Hirnantian age) or directly by Silurian rocks. This rather complicated stratigraphical setting results both from local sedimentary variations and from pre-Silurian erosional episodes in response to the Late Ordovician African glaciation.

The lithological and stratigraphical succession and the benthic faunas of the Luso-Alcudian Zone show close similarities with those of the Armorican Massif (North and Central Armorican Domain, *sensu* Paris and Robardet 1977). Trilobites (Hammann and Henry 1978; Rabano 1984 and listed references), ostracods (Vannier 1986a, b) and in general the whole benthic fauna of the *Neseuretus* community (Fortey and Owens 1978) are strongly similar to those of the Armorican Massif. A large number of species are common to both domains, indicating similar ecological conditions and the absence of any paleogeographical barrier and of any reproduction isolation mechanism that could have separated the inner shelf marine populations. The occurrence in both regions of the same three subdomains (Fig. 6) with slightly differing characteristics (Clarkson and Henry 1969; Hammann and Henry 1978) and the astonishing identity (Fig. 7) of the whole Ordovician succession and fauna (Henry et al. 1974, 1976) in Crozon (NW Brittany) and Buçaco (central Portugal) strongly argue for geographical continuity between the North-Central Armorican and the Central-Iberian Domains in Ordovician times.

Towards the south, this Central-Iberian Domain includes the regions located immediately to the south of the Los Pedroches (Fig. 4, P) batholith (i. e. the northern part of the Ossa-Morena Zone as defined in the present volume = Portalegre-Montoro subzone of Hammann et al. 1982) where the southernmost outcrops of the Armorican Quartzite Fm. and *Neseuretus* dark shales occur (Pérez Lorente 1979; Hammann et al. 1982; Hammann 1983, pp. 64–85; Gutiérrez Marco 1986 unpublished work pp. 134–137; Gutiérrez Marco et al. 1988). Thus, the present southern limit of the Paleozoic Central-Iberian Domain coincides with the Badajoz-Córdoba shear zone (Fig. 4 B–C).

3.2 Northern Iberian Domain?

Towards the north, the contrast is not as strong and the northern limit of the Central-Iberian Domain is indistinct.

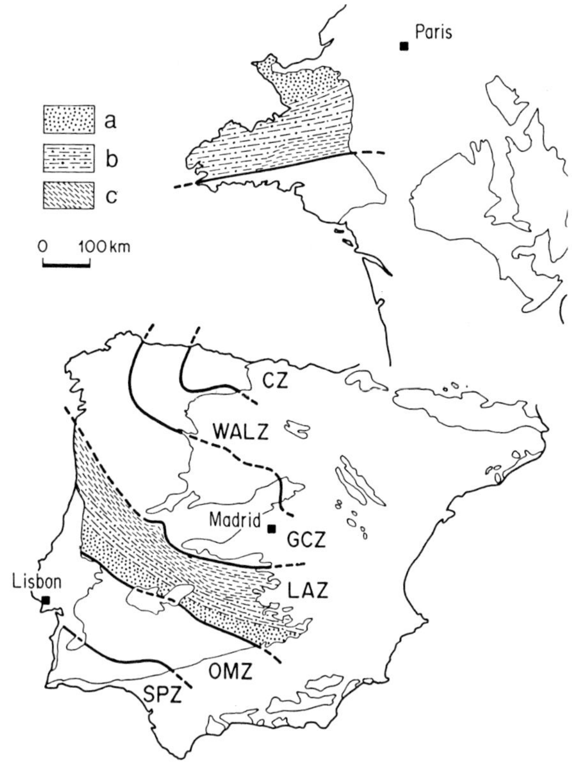

Fig. 6. Subdivision of the North Central Armorican-Central-Iberian Domain during Llandeilo times (after Hammann and Henry 1978). The subdivision into three subdomains is based upon lithofacies: *a* sandstones; *b* siltstones and sandstones; *c* siltstones and shales; and trilobite assemblages. Zonal subdivision of Iberia after Lotze (1945): *CZ* Cantabrian Zone: *WALZ* West-Asturian-Leonese Zone; *GCZ* Galaico-Castillan Zone; *LAZ* Luso-Alcudian Zone; *OMZ* Ossa-Morena Zone: *SPZ* South Portuguese Zone (see Fig. 3)

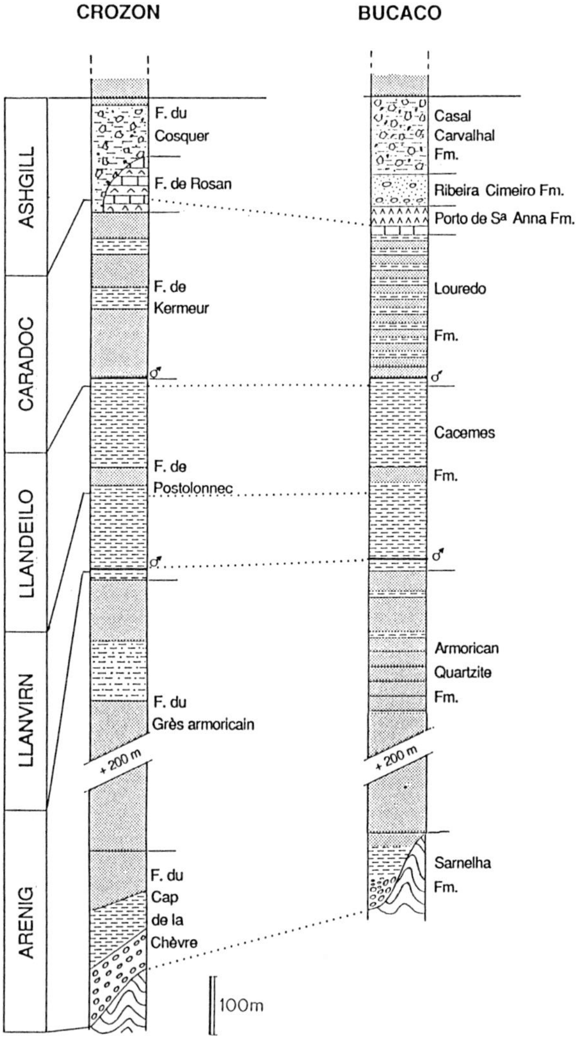

Fig. 7. The Ordovician successions from Crozon (western Brittany) and Buçaco (central Portugal)

The most important differences concern the lowermost Ordovician and the relationship with the Cambrian. In the West-Asturian-Leonese Zone and in the Iberian Chains (Celtiberia, NE Spain), a thick Tremadoc Series conformably overlies the Cambrian. This criterion has been used to distinguish a north-eastern area (Julivert 1978). This area probably includes Catalonia (Julivert et al. 1986) and Pyrenees where the understanding of the Ordovician succession is incomplete due to metamorphism and complex structure and to the scarcity of fossils. Lithofacies and some fossils suggest affinities to the Montagne Noire (S France).

However, the lithological succession is not completely different from that of the Central-Iberian Domain and the equivalents of the Armorican Quartzite Fm. and of the *Tristani* Shales (here Luarca Shales) are known in the West-Asturian-Leonese and Cantabrian Zones.

In the Llanvirn-Llandeilo time interval the benthic faunas do not strongly differ: there are some variations in the relative abundance of the trilobite genera but several species are common to all the different areas, from central Iberia to Cantabria and Celtiberia (Hammann 1983; Rabano 1984, pp. 276–278). Nevertheless, the scarcity of Llanvirnian benthic fossils in the West-Asturian-Leonese region and the presence of mesopelagic graptolites, such as *Pterograptus,* both in Leon and Celtiberia (unknown from the other regions of the Iberian Peninsula) suggest a certain paleogeographical individuality for these regions. On the other hand, the Llandeilo and Caradoc brachiopods of Asturias, the Iberian Chains and Catalonia show close affinities to those of the Armorican Massif and central Iberia (Villas 1980, 1983, 1985; Villas et al. 1987).

Thus, it is not possible to delineate precisely the northern limit of the Central-Iberian Domain, and the assumption of a distinct Northern Domain would principally be based upon the existence of a continuous Cambrian-Ordovician sequence, a more argillaceous sedimentation in Early Ordovician times and some barely investigated but apparently significant faunal particularities.

3.3 South Iberian Domain

As mentioned above, the Badajoz-Córdoba shear zone corresponds to the southern limit of the Ordovician outcrops of central Iberian type.

In the Ossa-Morena Zone paleontological data are much sparser than in the central Iberian regions, and the Ordovician succession is not yet completely elucidated. However, fossiliferous sequences are known from Portugal (Barrancos) and from Spain, to the north of Aracena (Venta del Ciervo) and in the northern part of the Sevilla province (Valle and Cerrón del Hornillo synclines).

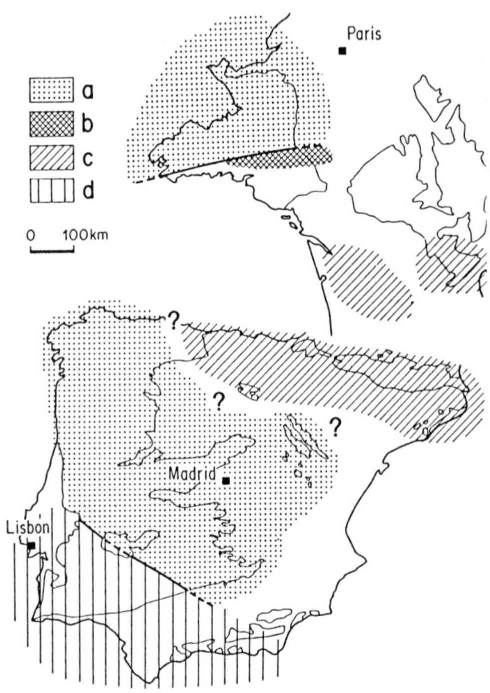

Fig. 8. Sedimentary and faunal domains in the Iberian and Armorican regions during Ordovician times. *a* North-Central Armorican-Central Iberian Domain; *b* Ligerian Domain; *c* North Iberian-Aquitanian Domain; *d* South Iberian Domain

The Armorican Quartzite Fm. and the *Neseuretus* dark shales that are so typical of the Central-Iberian Domain are totally absent. The lower part of the Ordovician sequence is composed of grey-green shales with graptolites of Hunnebergian age (Gutiérrez Marco et al. 1984; Gutiérrez Marco unpublished work 1986; Erdtmann et al. 1987). The distinctness of the Ossa-Morena Zone is also illustrated in the benthic faunas (Gutiérrez Marco et al. 1984b). Moreover, the limestones of the Ashgill (a period of global homogenization of the faunas) yield bryozoan, trilobite and conodont faunas, the composition of which perceptibly differs from those of the other regions of the Iberian Peninsula (Hafenrichter 1979, 1980). The Caradocian chitinozoan assemblages also show some peculiarities (Paris 1981, p. 325). Thus, the sedimentary development and the faunas illustrate the special characteristics (Robardet 1976) of the Ossa-Morena Zone during the Ordovician Period (Fig. 8).

4 Silurian

4.1 Faunas

During the Silurian Period both southern Europe and northern Gondwana were characterized by euxinic black shales and black limestones with pelagic or epipelagic faunas (mainly graptolites). The planktonic graptolite faunas everywhere have the same gross composition, even in widely separated areas, most species being of intercontinental or even global distribution. In detail slight differences may be apparent but it is not clear to what extent the observed differences are real or due to incomplete knowledge in the different areas (Jaeger and Robardet 1979, p. 695). Concerning the graptolite faunas, special mention must be made of the so-called Sardinian faunas, which are mainly of Upper Wenlock age and have been regarded as "giant faunas". Such Sardinian graptolites have been described from the Iberian Peninsula (Romariz 1962, 1969; Romariz et al. 1971). It has already been stated that these Sardinian giant faunas are composed of strongly deformed graptolites (Jaeger 1976; Jaeger and Robardet 1979; Hollard and Willefert 1985, p. 314). The concept of a Sardinian graptolite province has therefore proved to be ill-conceived and essentially due to negligence of the tectonic deformation of the fossils.

4.2 Sedimentary Evolution of the Iberian Peninsula

In several regions of the Iberian Peninsula the Silurian overlies Ordovician rocks of different ages, locally as old as Arenig quartzites. The Ordovician-Silurian transition is thus marked by a hiatus of varying importance, resulting both from non-deposition and from erosion. The precise time limits of these processes cannot be defined, even in the most complete succession, as nonfossiliferous quartzitic levels generally underlie the lowest Silurian graptolites. However, when fossiliferous beds directly overlie the Ordovician they indicate the middle part of the Llandovery Series (Aeronian Stage, zones 19–21). The Ossa-Morena Zone seems to be the only region where there is no hiatus (or a very small one), as the earliest Llandovery graptolite Zone (*Acuminatus* Zone) is known at the base of the Silurian sequence.

The most characteristic lithofacies of the Silurian are graptolitic black shales, and the supposed uniformity of facies partly results from the attention that has been paid to these richly fossiliferous rocks. However, other types of rocks also occur in the Silurian sequences and their occurrence and relative importance introduce differences in the type of succession (Figs. 5, 9).

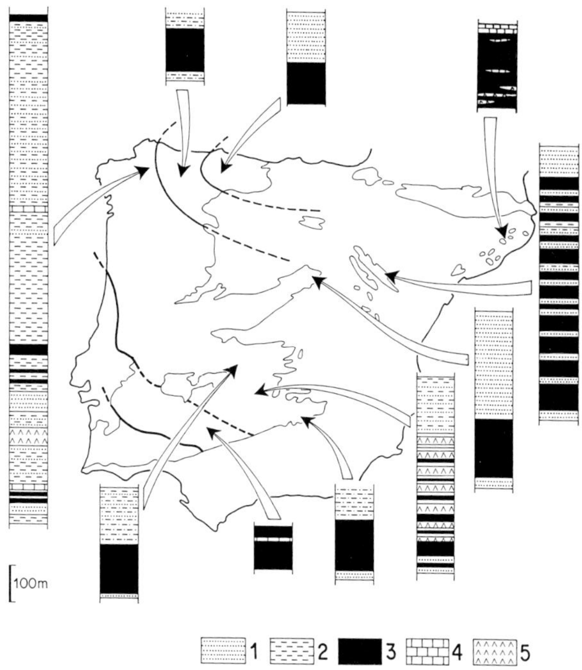

Fig. 9. Different types of Silurian successions in the Iberian Peninsula (after Iglesias and Robardet 1980). *1* Sandstones; *2* siltstones; *3* black shales (ampelites); *4* limestones; *5* volcanics; for details see text

Volcanic rocks are known in the Silurian in three areas: (1) in the Catalonian Coastal Ranges where intermediate to basic volcanic rocks occur locally in the Llandovery (Julivert et al. 1986); (2) in central Galicia where the thick Silurian succession also includes various rock facies such as quartzites, green and purple shales and siltstones (Iglesias and Robardet 1980); and (3) in the Almaden area (southern Central-Iberian Zone) where basaltic lavas and volcanoclastic rocks are interbedded in the Silurian black shales (Saupé 1973).

However, the occurrence of these volcanics does not permit delineation of major paleogeographical units.

On the other hand, different areas can be distinguished, based upon the whole Silurian succession and sedimentary evolution.

In numerous regions of northern and central Iberia, the lowermost part of the Silurian consists of nonfossiliferous quartzite or sandstone formations ("Valent Quartzit" or "Llandoverium Quartzit" of the German geologists; Criadero, Castellar Quartzites), somewhat similar to those which occur in several regions of the Armorican Massif ("Grès culminant", La Lande Murée Fm.).

In the same regions of the Iberian Peninsula, the upper part of the Silurian succession includes either thick sandstone formations, such as the Furada-San Pedro Formations in the Cantabrian Zone and Formations in Celtiberia and Guadarrama, or siltstone-sandstone alternations in the Central-Iberian Zone. These clastic facies, which appeared as early as Ludlow or Pridoli, extend into the lowermost Devonian, and the Silurian-Devonian boundary cannot be precisely located in these poorly fossiliferous sequences (chitinozoans, acritarchs, spores and brachiopods; see Carls 1977).

The Ossa-Morena Zone Silurian succession strongly differs from those of northern and central Iberia. In the Ossa-Morena Zone (Jaeger and Robardet 1979) the whole Silurian is composed of a rather condensed sequence of euxinic graptolitic black shales extending from the true base of the Llandovery Series up through the Upper Pridoli and into the lowermost part of the Devonian (Lochkovian). The sandstone formations which occur in the central and northern regions of the Iberian Peninsula do not exist in the Ossa-Morena Zone. A similar type of euxinic succession is known in the Catalonian Coastal Ranges (Julivert et al. 1986) and might also occur in the Pyrenees (Degardin 1984).

Additional details can be pointed out (cf. Jaeger and Robardet 1979, p. 696). In the Ossa-Morena Zone, no indications have been noted of some typical Saharian forms of graptolites such as *Diplograptus modestus fezzanensis* Desio and *Paraclimacograptus innotatus brasiliensis* (Ruedemann). The latter species, common in the Sahara through much of the Llandovery, has not been found in the abundant and diversified faunas of the Ossa-Morena Zone but it occurs in the Upper Llandoverian assemblages of Almaden in the southern part of the Central-Iberian Domain (Jaeger in Blachère 1978, unpublished work pp. 41–42) and in Celtiberia (Wehner 1984, unpublished work). This might possibly be important to estimate the relative longitudinal position of the Central-Iberian and the South Iberian Domains but more data of that type would be necessary to fully discuss that problem.

5 Devonian

To establish the Devonian paleogeography of the Iberian Peninsula is a rather difficult problem, owing to the uneven geographical distribution of the Devonian rocks and the different state of understanding of the succession in the different areas. The Devonian System is abundantly represented in the northern part of the Peninsula (Cantabrian and West Asturian-Leonese Zones, Iberian Chains, Catalonia and Pyrenees) and a wealth of detailed paleontological and stratigraphical information has been published (see references in Julivert et al. 1983). The central regions are almost devoid of Devonian rocks. Devonian outcrops and successions occur in the southern part of the Central-Iberian Zone and in the Ossa-Morena Zone (Pardo-Alonso and Garcia-Alcalde 1984; Racheboeuf and Robardet 1986) with Lower and Upper Devonian sequences; the absence of Middle Devonian most probably corresponds to an important stratigraphical gap.

Considering the foregoing, comparisons between different regions can only concern the Early Devonian period.

The geographical distribution of the Lower Devonian lithofacies (cf. Puschmann 1970; Julivert et al. 1983, p. 298) permits the distinction of four facies types:

1. The Cantabrian region with dominant marl, limestone and dolomite formations succeeding the Early Gedinnian sandstones; the western Pyrenees succession may roughly be considered of that type despite local variations;
2. Catalonia and eastern and central Pyrenees with a rather condensed sequence of black to grey limestones and shales with pelagic faunas (Hercynian "magnafacies");
3. The central Iberian regions with terrigenous deposits, mainly sandstones with subordinate siltstones and shales;
4. The Ossa-Morena Zone where dominant shale facies succeed the graptolitic black shales of the Lochkovian.

Considering the faunal characteristics, a sharp contrast appears between the Central-Iberian and the Ossa-Morena Zones. In the southern Central-Iberian Zone (Almaden) the Lower Devonian sequence (Pardo-Alonso and García Alcalde 1984) is mainly quartzitic and the fauna of Rhenish type includes many brachiopods among which spiriferids dominate.

In the Ossa-Morena Zone, the Devonian sequence (Valle syncline) is mainly argillaceous: black shales of Lochkovian age are overlain by Pragian-Emsian shales and siltstones. The brachiopod faunas of dominant Hercynian type strongly differ from those of Almaden and have pronounced affinities to those of the Moroccan Meseta (Racheboeuf and Robardet 1986; Brice et al. 1984).

All those who have studied the Devonian benthic faunas in the Iberian Peninsula have emphasized the similarities between Armorican and Celtiberian lithological successions and fossil assemblages. Close affinities exist for brachiopods (Garcia-Alcalde and Racheboeuf 1975; Racheboeuf 1981), trilobites (Gandl 1972; Carls 1979; Morzadec 1983), and crinoids (Le Menn 1985): in these groups many forms are closely related or even conspecific. Similar affinities are known for the whole benthic fauna. Such affinities also exist between the Armorican Massif and some regions of the Central-Iberian Zone in Portugal, but they are not so well documented.

On the other hand, paleontological studies have also emphasized the relative distinctiveness of the Cantabrian area. This region has some species (with a large geographical extension) in common with the Armorican Massif, and this indicates that marine connections existed between the two regions, but most of the species are distinctive and some trilobite genera are endemic to the Cantabrian area (Arbizu 1979; García Alcalde et al. 1979). Thus, it does not seem possible to consider that the Cantabrian and the Armorican regions were included in the same paleogeographical domain (Morzadec 1983, p. 292).

Finally, the Devonian succession of the Palentine domain (or Pisuerga-Carrion Unit), located SE of the Cantabrian Zone, has pronounced lithological and faunal Hercynian characteristics (= Palentine facies) from Upper Emsian up to the end of the Devonian (Jahnke et al. 1983). This region apparently represents a tectonically allochthonous unit, displaced from SW to NE (possibly from the West-Asturian-Leonese Zone) during the Namurian (Marcos 1979; Marquínez and Marcos 1984).

6 Discussion

Considering the whole Ordovician-Devonian time interval (Fig. 10), the most prominent fact is the distinctiveness of a South Iberian Domain (to the south of the Badajoz-Córdoba shear zone). The Ordovician, Silurian and Lower Devonian successions and faunas strongly differ from those which occur immediately to the north. Therefore, it has been proposed to consider the Badajoz-Cordoba shear zone as the real limit between the Central-Iberian and the Ossa-Morena Zones during the Paleozoic (Robardet 1976; Chacon and Pascual 1979; Florido and Quesada 1984).

The precise extent of this South Iberian Domain with Moroccan affinities cannot easily be fixed, as no Lower Paleozoic rocks are known from the South Portuguese Zone, which entirely consists of Upper Devonian and Carboniferous. The occurrence of a metamorphic ophiolitic suite along the boundary between the Ossa-Morena and the South Portuguese Zones (Oliveira 1986) may indicate a rather complex or composite domain. Nevertheless, this South Iberian Domain is clearly distinct from the Central-Iberian Domain.

The Central-Iberian Domain shows close and clearly established affinities to the Armorican Massif. During the Ordovician Period, these affinities are numerous and precise enough to argue strongly for geographical continuity between these areas. During the Lower Devonian the best demonstrated similarities are those between the Armorican Massif and the Iberian Chains (and, to some degree, the Central-Iberian Zone in Portugal). The northward extent of the Central-Iberian Domain is not precisely determined and the northern limit cannot be easily recognized. In Ordovician times the northern regions of Spain (Cantabrian and West-Asturian-Leonese Zones) display some local characteristics (Cambrian-Ordovician succession, Lower Ordovician lithofacies, faunal affinities to south France) but do not totally contrast with the Central-Iberian regions. In a rather similar way the Cantabrian regions, Catalonia and Pyrenees appreciably differ from the other regions of the Ibero-Armorican arc during the Early Devonian.

The existence and the respective affinities of the South Iberian and the Central-Iberian Domains are clearly established. Considering the arguments mentioned above, a North Iberian Domain is much more speculative. However, it must be noted that additionnal data come from the stratigraphical study of the basement of the Aquitaine Basin in France (Paris and Robardet 1985; Paris et al. 1987). Similarities between the Lower Paleozoic Series in the Aquitaine, Montagne Noire, Pyrenees and possibly part of the Cantabrian region may argue in favour of the existence of a North Iberian (or Ebro-Aquitanian) Domain; a somewhat similar point of view had previously been postulated on geodynamical grounds (North-Spain microplate of Riding 1974).

Returning to the relations between the South Iberian and the Central-Iberian Domains, paleomagnetic data must be mentioned. Paleomagnetic studies in southern Portugal (Perroud et al. 1985) have concluded that North-Central and South Iberia (on opposite sides of the Badajoz-Cordoba shear zone) were separated (about 15° in latitude) in the Late Devonian, South Iberia being latitudinally close to Africa and North-Central Iberia (and the other regions of the Armorica plate) being already assembled with the northern continents (Baltica and Laurentia). These propositions are in accordance with paleobiogeographical arguments (Robardet 1976, 1986) that suggest the independence of the South Iberian Domain. However, they also imply an important separation between the Armorica plate (including North-Central Iberia) and Gondwana (and the satellite areas such as South Iberia), which is in total discordance with paleobiogeographical data, indicating that the affinities between northern Africa and southern Europe persisted during the whole Paleozoic. That point of the discussion leads up to the more general problem of the possible existence of a Paleotethys ocean separating Armorica and Gondwana and this is beyond the scope of the present chapter (see Robardet et al. 1990).

To sum up, the paleogeographical pattern of the Iberian Peninsula during Lower Paleozoic times consists of three parts (Fig. 10):

1. A South Iberian Domain, with Moroccan affinities, that has no equivalent in the northern branch of the Ibero-Armorican arc;
2. A Central-Iberian Domain which represents the Iberian part of a larger paleogeographical entity including the North-Central Armorican Massif;

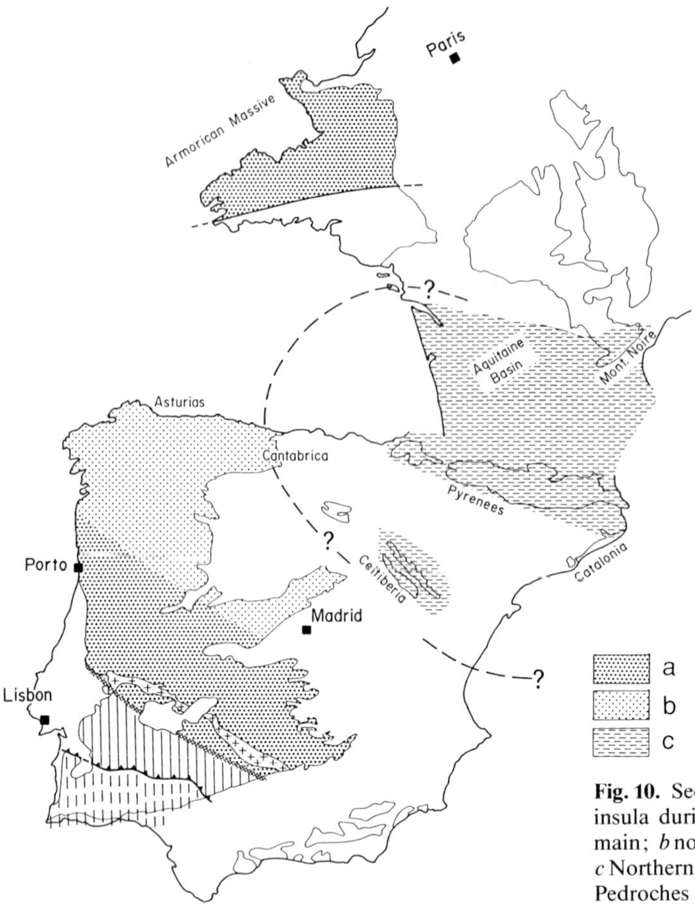

Fig. 10. Sedimentary and faunal domains in the Iberian Peninsula during Lower Paleozoic times. *a* Central-Iberian Domain; *b* northern extension of the Central-Iberian Domain; *c* Northern Iberian Domain; *d* South Iberian Domain; *e* Los Pedroches granitic batholith; *f* Badajoz-Cordoba shear zone

3. Possibly a North Iberian Domain (whose limits are not clearly established) in relation with southern France.

The distribution of lithofacies and benthic faunas can be regarded as important lines of evidence leading to characterization and identification of the diverse terranes that were accreted to compose the Hercynian Belt and to reconstruct past geographical configurations. The characterization and identification of the distinct domains that compose the Iberian Peninsula impose constraints upon any geodynamical model of the Hercynian Belt. In this way it would be impossible to agree upon reconstructions that ignore the close relationships between North Armorican and Central-Iberian regions. In the same way it would not be possible to accept any location for oceanic sutures.

The paleogeographical pattern of the Iberian Peninsula outlined in the present chapter certainly needs improvements and requires further investigations for a better understanding of the South European Hercynian Belt and of the global paleogeographical evolution during the Paleozoic Era.

Acknowledgements. We thank Drs. D. L. Bruton and R. B. Neuman for reviewing the present work and offering constructive comments that allowed us to improve the text. Daniele Bernard and Michel Lautram prepared the figures and Monique Le Moigne typed the manuscript.

References

Arbizu M (1979) Asteropyginae (Trilobita) du Dévonien des Montagnes cantabriques (Espagne). Bull Soc Géol Min Bretagne, 1977 (C) 9 (2): 59–102

Brice D, Charrière A, Drot J, Régnault S (1984) Mise en évidence, par les faunes de Brachiopodes de l'extension des formations dévoniennes dans la Boutonnière d'Immouzer du Kandar (Sud de Fès, Maroc). Ann Soc Géol Nord 103: 445–458

Carls P (1977) The Silurian-Devonian boundary in northeastern and central Spain. In: The Silurian-Devonian Boundary. IUGS, Ser A, 5: 143–158

Carls P (1979) Emsian-Eifelian stratigraphy of the eastern Iberian Chains. In: Garcia-Alcalde et al. (eds) Meeting International Subcommission Devonian stratigraphy, Guidebook of the field trip. Ser Publ Univ Oviedo, pp 35–39

Chacón J, Pascual E (1979) El anticlinorio Portalegre. Badajoz-Cordoba divisoria entre las zonas Centro-Ibérica y Ossa-Morena (Sector SW del Macizo Iberico). Cuad Geol Univ Granada 8–9: 18–31

Clarkson ENK, Henry JL (1969) Sur une nouvelle espèce du genre Crozonaspis (Trilobite) découverte dans l'Ordovicien de la Mayenne. Bull Soc Géol Fr (7) 11: 116-123

Cocks LRM, Fortey RA (1982) Faunal evidence for oceanic separations in the Palaeozoic of Britain. Geol Soc (Lond) 139: 465-478

Dean WT (1967) The distribution of Ordovician shelly faunas in the Tethyan region. In: Adams CG, Ager DV (eds) Aspects of Tethyan biogeography. Syst Assoc Publ 7: 11-44

Dean WT (1977) Some aspects of Ordovician correlation and trilobite distribution in the Canadian Appalachians. In: Bassett MG (ed) The Ordovician System. University of Wales Press, pp 227-250

Degardin JM (1984) Graptolites du Silurien de la région de Camptodon, Province de Gérone, Pyrénées orientales espagnoles. Ann Soc Géol Nord 103: 55-74

Erdtmann BD, Maletz J, Gutiérrez Marco JC (1987) The new Early Ordovician (Hunneberg Stage) graptolite genus *Paradelograptus* (Kinnegraptidae), its phylogeny and biostratigraphy. Palaontol Z 61 (1-2): 109-131

Florido P, Quesada C (1984) Estado actual de conocimientos sobre el Macizo de Aracena. Introducción a la excursión en la parte española. III Reunion Grupo Ossa-Morena. Cuad Lab Xeol Laxe 8: 257-277

Fortey RA, Owens RM (1978) Early Ordovician (Arenig) stratigraphy and faunas of the Carmarthen district, southwest Wales. Bull Br Mus (Nat Hist) Geol 30 (3): 225-294

Gandl J (1972) Die Acastavinae und Asteropyginae (Trilobita) Keltiberiens (NE Spanien). Abh Senckenb Naturforsh Ges 530: 1-184

García Alcalde JL, Arbizu M, García Lopez S, Méndez Bedia I (eds) (1979) Meeting of the International Subcommission on Devonian stratigraphy. Guidebook of the field trip, Spain 1979. Serv Publ Univ Oviedo, pp 1-41

García Alcalde JL, Racheboeuf P (1975) Données paléobiologiques et paléobiogéographiques sur quelques Strophochonetinae du Dévonien d'Espagne et du Massif Armoricain. Lethaia 8 (4): 329-338

Gil Cid MD, Gutiérrez Marco JC, Rabano I (1987) Comments on the Cambrian biogeography of Spain. Geogaceta 3: 27-29

Gutiérrez Marco JC, Apalategui O, Rabano I (1988) Fossiles ordovicicos del N.O. de Adamuz (Cordoba): aportación paleobiogeográfica a la concepción del límite entre las Zonas Centroibérica y Ossa Morena durante el Paleozoico. Geogaceta 3: 22-24

Gutiérrez Marco JC, Rabano I (1987) Paleobiogeographical aspects of the Ordovician Mediterranean faunas. Geogaceta 2: 24-26

Gutiérrez Marco JC, Rabano I, Prieto M, Martín J (1984) Estudio bioestratigráfico del Llanvirn y Llandeilo (Dobrotiviense) en la parte meridional de la Zona Centroiberica (Espana). Cuad Geol Ibérica 9: 287-319

Gutiérrez Marco JC, Rabano I, Robardet M (1984) Estudio bioestratigrafico del Ordovícico en el Sinclinal del Valle (Provincia de Sevilla, S.O. de Espana). Mem Not Coimbra 97: 12-37

Hafenrichter M (1979) Palaontologisch-okologische und lithofazielle Untersuchungen des "Ashgill-Kalkes" (Jungordovizium) in Spanien. Arb Palaontol Inst Wurzburg 3: 1-139

Hafenrichter M (1980) The lower and upper boundary of the Ordovician system of some selected regions (Celtiberia, eastern Sierra Morena) in Spain. Part II: The Ordovician-Silurian boundary in Spain. Neues Jahrb Geol Palaontol Abh 160 (1): 138-148

Hammann W (1983) Calymenacea (Trilobita) aus dem Ördovizium von Spanien: ihre Biostratigraphie, Okologie and Systematik. Abh Senckenb Naturforsch Ges 542: 1-177

Hammann W, Henry JL (1978) Quelques espèces de Calymenella, Eohomalonotus et Kerfornella (Trilobita, Ptychopariida) de l'Ordovicien du Massif Armoricain et de la Péninsule ibérique. Senckenb Lethaea 59: 401-429

Hammann W, Robardet M, Romano M (1982) The Ordovician System in southwestern Europe (France, Spain and Portugal). IUGS Publ 11: 1-47

Henry JL, Mélou M, Nion J, Paris F, Robardet M, Skevington D, Thadeu D (1976) L'apport de Graptolites de la zone à *G. teretiusculus* dans la datation de faunes benthiques lusitano-armoricaines. Ann Soc Géol Nord 96 (4): 275-281

Henry JL, Nion J, Paris F, Thadeu D (1974) Chitinozoaires, Ostracodes et Trilobites de l'Ordovicien du Portugal (Serra de Buçaco) et du Massif Armoricain: essai de comparaison et signification paléogéographique. Comun Serv Geol Port (1973-1974) 57: 303-345

Hollard H, Willefert S (1985) Silurian. In: Destombes J, Hollard H, Willefert S, Holland CH (eds) Lower Palaeozoic rocks of Morocco. Lower Palaeozoic of north-western and west central Africa. Wiley, New York, pp 238-336

Iglesias M, Robardet M (1980) El Silurico de Galicia Media (Central), su importancia en la paleogeografía varisca. Cuad Lab Xeol Laxe 1: 99-115

Jaeger H (1976) Das Silur und Unterdevon vom thuringischen Typ in Sardinien und seine regionalgeologische Bedeutung. In: Franz Kosmat Symposion. Nova Acta Leopoldina, Halle (Saale) 224 (45): 263-299

Jaeger H, Robardet M (1979) Le Silurien et le Dévonien basal dans le Nord de la Province de Séville (Espagne). Géobios 12 (5): 687-714

Jahnke H, Henn A, Mader H, Schweineberg J (1983) Silur und Devon im Arauz-Gebiet (Prov. Palencia, N-Spanien). Newsl Stratigr 13 (1): 40-66

Julivert M (1978) Algunas bases para una correlación paleogeográfica entre los Macizos Hercinianos del Occidente de Europa. In Geología de la parte norte del Macizo Ibérico. Cuad Semin Estud Ceram Sargadelos 27: 159-191

Julivert M, Duran H, García Lopez S, Truyóls Massoni M, Villas E (1986) The Paleozoic pre-Carboniferous sequence in the Catalonian Coastal Ranges (N. E. Spain) and its faunas. IGCP Projet n° 5 "Correlation of Prevariscan and Variscan Events in the Alpine Mediterranean Mountain Belts". Final Meeting, Sardinia, May 1986. Abstr, pp 31-37

Julivert M; Fontboté J, Ribeiro A, Conde L (1974) Memoria explicativa del Mapa Tectónico de la Peninsula Ibérica y Baleares E. 1:1000000. Inst Geol Min Esp, pp 1-113

Julivert M, Truyóls J (1983) El Ordovícico en el Macizo Ibérico. In: Geología de España, Libro Jubilar J.M. Rios, Inst Geol Min Esp 1: 192-246

Julivert M; Truyóls J, Verges J (1983) El Devónico en el Macizo Ibérico. In: Geologia de España, Libro jubilar JM Rios, Inst Geol Min Esp 1: 265-311

Kent DV, Dia O, Sougy JMA (1984) Paleomagnetism of Lower-Middle Devonian and Upper-Proterozoic-Cambrian (?) rocks from Mejeria (Mauritania, West Africa). In: Van Der Voo R, Scotese CR, Bonhommet N (eds) Plate reconstruction from Paleozoic paleomagnetism. Geodynamics Ser 12: 99-115 (Am Geophys Union)

Livermore RA, Smith AG, Briden JC (1985) Paleomagnetic constraints on the distribution of continents in the late Silurian and early Devonian. In: Chaloner WG, Lawson JC (eds) Evolution and environment in the late Silurian and early Devonian. Philos Trans R Soc London B Biol Sci 309: 29-56

Lotze F (1945) Zur Gliederung der Varisziden der Iberischen Meseta. Geotekt Forsch 6: 78-92

Lotze F (1961) Das Kambrium Spaniens. I: Stratigraphie. Akad Wiss Lit 6: 285-498

Marcos A (1979) Facies differentiation caused by wrench deformation along a deep-seated fault system (Leon Line), Cantabrian Mountains, North Spain: discussion. Tectonophysics 60: 303-309

Marquínez J, Marcos A (1984) La estructura de la Unidad del Gildar-Monto (Cordillera Cantabrica). Trab Geol 14: 53-64

Le Menn J (1985) Les Crinoïdes du Dévonien inférieur et moyen du Massif Armoricain. Systématique, Paléobiologie, Evolution, Paléoécologie, Biostratigraphie. Mém Soc Géol Min Bretagne 30: 1-268

Morzadec P (1983) Le Dévonien (Emsien-Famennien) de la rade de Brest (Massif armoricain). Lithologie, cartographie, stratigraphie, paléogéographie. Géol Fr (2) 4: 269-310

Oliveira JT (1986) The South-Portuguese terrane: tectonostratigraphic evolution. Int Conf Iberian terranes and their regional correlation. IGCP 233, Oviedo September 1986. Abstracts, pp 28-29

Pardo Alonso M, Garcia-Alcalde JL (1984) Bioestratigrafia del Devónico de la región de Almaden (Ciudad Real, Espana). Trab Geol 14: 79-120

Paris F (1981) Les Chitinozoaires dans le Paléozoïque du Sud-Ouest de l'Europe (Cadre géologique - Etude systématique - Biostratigraphie). Mém Soc Géol Min Bretagne 26: 1-412

Paris F, Le Pochat G, Pelhate A, Holtzapffel T (1987) Synthèse des connaissances sur le Paléozoïque Nord-Aquitain. In Géologie Profonde de la France, thème 7. Doc Bur Rech Géol Min 144: 1-57

Paris F, Robardet M (1977) Paléogéographie et relations ibéroarmoricaines au Paléozoïque anté-carbonifère. Bull Soc Géol Fr (7) 19: 1121-1126

Paris F, Robardet M (1985) Evaluation des affinités entre le Paléozoïque caché sous l'Aquitaine et les formations armoricaines contemporaines. In: Géologie Profonde de la France, Thème 7. Doc Bur Rech Géol Min 95-7: 11-36

Perejón A (1984) Bioestratigrafía de los Arqueociatos en España. Cuad Geol Ibérica 9: 213-265

Pérez Lorente F (1979) Geología de la Zona Ossa-Morena al Norte dè Córdoba (Pozoblanco-Belmez-Villaviciosa de Cordoba). Thesis, Univ Granada 281: 1-340

Perroud H, Bonhommet N, Ribeiro A (1985) Paleomagnetism of Late Paleozoic igneous rocks from southern Portugal. Geophys Res Lett 12 (1): 45-48

Perroud H, Voo R van der, Bonhommet N (1984) Paleozoic evolution of the Armorica plate on the basis of paleomagnetic data. Geology 12: 579-582

Puschmann H (1970) Eine Palaogeographie des Devons auf der Iberischen Halbinsel. Z Geol Ges 120: 107-118

Rabano I (1984) Trilobites ordovicicos del Macizo Hesperico espanol: una vision bioestratigrafica. Cuad Geol Ibérica 9: 267-287

Racheboeuf P (1981) Chonetacés (Brachiopodes) siluriens et dévoniens du Sud-Ouest de l'Europe (Systématique, Phylogénie, Biostratigraphie, Paléobiogéographie). Mém Soc Géol Min Bretagne 27: 1-294

Racheboeuf P, Robardet M (1986) Le Pridoli et le Dévonien inférieur de la Zone d'Ossa Morena (Sud-Ouest de la Péninsule Ibérique). Etude des Brachiopodes. Geol Paleontol 20: 11-37

Riding R (1974) Model of the Hercynian Foldbelt. Earth Planet Sci Lett 24: 125-135

Robardet M (1976) L'originalité du segment hercynien sudibérique au Paléozoïque inférieur: Ordovicien, Silurien et Dévonien dans le Nord de la Province de Séville, Espagne. C R Acad Sci Paris D 283: 999-1002

Robardet M (1986) The Lower Paleozoic Series of the Ossa-Morena terrane, stratigraphy and paleogeography. Int Conf Iberian terranes and their regional correlation, IGCP 233, Oviedo September 1986. Abstr, p 47

Robardet M, Doré F (1988) The Late Ordovician diamictic formations from southwestern Europe: north-Gondwana glaciomarine deposits. Palaeogeogr Palaeoclimatol Palaeoecol 66: 19-31

Robardet M, Paris F, Racheboeuf PR (1990) Palaeogeographic evolution of southwestern Europe during early Palaeozoic times. In: McKerrow WS, Scotese CR (eds) Palaeozoic

Romano M (1982) The Ordovician biostratigraphy of Portugal. A review with new data and re-appraisal. Geol J 17: 89-110

Romariz C (1962) Graptolitos do Silurico Portugues. Rev Fac Cienc Lisboa 2e Ser, C, 10 (2): 115-312

Romariz C (1969) Graptolitos siluricos do noroeste peninsular. Comun Serv Geol Port 53: 107-155

Romariz C, Arche A, Barba A, Gutiérrez-Elorza M, Vegas R (1971) The Mediterranean graptolitic fauna of the Wenlockian in the Iberian Peninsula. Bol Soc Geol Port 18: 57-61

Saupé F (1973) La géologie du gisement de mercure d'Almaden (Province de Ciudad Real, Espagne). Sci Terre Mém 29: 1-342

Scotese CR, Voo R Van der, Barrett SF (1985) Silurian and Devonian base maps. In: Chaloner WG, Lawson JD (eds) Evolution and environment in the late Silurian and early Devonian. Philos Trans R Soc Lond B Biol Sci 309: 57-77

Sdzuy K (1961) Das Kambrium Spaniens. II: Trilobiten. Akad Wiss Lit 7-8: 499-690

Spjeldnaes N (1961) Ordovician climatic zones. Norsk Geol Tidsskr 41: 45-77

Spjeldnaes N (1967) The palaeogeography of the Tethyan region during the Ordovician. In: Adams CG, Ager DV (eds) Aspects of Tethyan biogeography. Syst Assoc Publ 7: 45-57

Vannier J (1986a) Ostracodes Binodicopa de l'Ordovicien (Arenig-Caradoc) ibéro-armoricain. Palaeontographica (A) 193: 77-143

Vannier J (1986b) Ostracodes Palaeocopa de l'Ordovicien (Arenig-Caradoc) ibéro-armoricain. Palaeontographica (A) 193: 145-218, 74 fig, 13 pl

Vegas R (1978) Sedimentation and tectonism in the Iberian massif prior to the Hercynian deformation (Late Precambrian to Silurian times). In: Geologia de la parte norte del Macizo Iberico. Cuad Semin Estud Ceram Sargadelos 27: 269-286

Villas E (1980) El Ordovicico de los alrededores de Alpatir (Cadenas Ibericas Orientales: NE de España). Bull Soc Géol Min Bretagne (C) 12: 77-91

Villas E (1983) Las formaciones del Ordovicico medio y superior de las Cadenas Ibericas y su fauna de braquiopodos. Estud Geol 39: 359-377

Villas E (1985) Brachiopodos del Ordovicico medio y superior de las Cadenas Ibericas Orientales. Mem Mus Paleontol Univ Zaragoza 1: 1-153, 34 pl

Villas E, Duran H, Julivert M (1987) The Upper Ordovician clastic sequence of the Catalonian Coastal Ranges and its brachiopod fauna. Neues Jahrb Geol Paläontol Abh 174 (1): 55-74

Voo R van der (1979) Paleozoic assembly of Pangea: a new plate tectonic model for the Taconic, Caledonian and Hercynian orogenies. EOS, Trans Am Geophys Union 60, 241 (Abstract)

Voo R van der (1982) Pre-Mesozoic paleomagnetism and plate tectonics. Ann Rev Earth Planet Sci 10: 191-220

Voo R van der (1988) Paleozoic paleogeography of North America, Gondwana and intervening displaced terranes: comparisons of paleomagnetism with paleoclimatology and biogeographical patterns. Geol Soc Am Bull 100: 311-324

Wehner G (1984) Graptolithen aus der Badenas Formation N. E. Spaniens. Diplomarbeit, Univ Wurzburg, pp 1-52 (unpubl)

Whittington HB, Hugues CP (1972) Ordovician geography and faunal provinces deduced from trilobite distribution. Philos Trans R Soc (Lond) B 263 (850): 235-278

Zamarreno I (1983) El Cambrico en el Macizo Iberico. In: Geologia de España. Libro Jubilar JM Rios. Inst Geol Min Esp 1: 117-191

Part VIII Geodynamic Evolution of the Iberian Massif

Geodynamic Evolution of the Iberian Massif

A. Ribeiro[1], C. Quesada[2] and R. D. Dallmeyer[3]

1 Introduction

The Iberian Massif represents the largest continuous exposure of pre-Permian rocks within the Iberian Peninsula (Fig. 1). It was first described in terms of a coherent tectonic unit by Staub (1926) who considered it to be comprised of Archean basement sequences that had been successively reworked during Caledonian and Herycnian orogenic events. Stille (1924, 1929) contested these ideas and argued, instead, that the Iberian Massif largely records the effects of Hercynian tectonothermal events. This was later documented in more detail by Lotze (1945), who divided the massif into six zones on the basis of contrasting lithologic, structural and/or metamorphic characteristics (Fig. 1). With the advent of plate tectonic concepts, a new view of the Iberian Massif emerged, and many different geodynamic plate models were proposed. Van der Voo (1982) suggested that the Iberian Massif constituted the southwesternmost and largest piece of the so-called Armorican plate, a continental block which appears to have rifted from Gondwana in the early Paleozoic. During the Lower Paleozoic it was separated from Laurentia by a proto-Atlantic ocean, from Baltica by the Tornquist Sea, and from Gondwana by the Theic Ocean (Fig. 2). Subsequent closing of the intervening oceans and the sequential reassembly of the various continental plates resulted in multiphase Middle and Late Paleozoic orogenesis within the circum-Atlantic realm. Many conflicting plate tectonic models have been proposed to explain the evolution of the whole or various parts of the Iberian Variscan Belt (e. g., Bard 1971; Dewey and Burke 1973; Autran and Cogné 1980; Lefort and Ribeiro 1980; Matte 1983, 1986; Van der Voo 1982, 1983; Ziegler 1984; Lorenz and Nichols 1984). These show a confusing array of models, correlations and "fits", and have been mainly based on paleomagnetic or geophysical, and, more rarely, biostratigraphic information. These models should be considered in terms of two time periods: those published before or after the significant paper by Robardet (1976), who first applied the concepts of faunal provincialism into geological studies within the Iberian Massif. Prior to this paper, tectonic models had mostly considered the Iberian Peninsula to have been a single continental unit throughout the Paleozoic, and had concluded that all tectonothermal activity was of Hercynian (Carboniferous) age (e. g., Bard 1971; Carvalho 1972; Bard et al. 1973; Schermerhorn 1975). Robardet (1976) suggested different paleobiographic positions for the Ossa-Morena and Central-Iberian Zones during the Paleozoic on the basis of their faunal characteristics. This, together with recognition of the Badajoz-Córdoba shear zone as a major structural feature, resulted in a set of new interpretations in which the Iberian Peninsula was considered to consist of two continental blocks: a northern one with Armorican affinities and connected with the rest of the European Hercynian, and a southern one with Gondwana affinities (e. g., Brun and Burg 1982; Matte 1983, 1986). The intervening suture was though to be of Carboniferous age, and to have been located within the Badajoz-Córdoba shear zone (now more or less a cryptic feature because of Late Hercynian tangential tectonic overprinting: Burg et al. 1981).

The potential structural complexity of the world's orogenic belts has been more clearly defined over the last decade, and it is now apparent that their evolutionary development may be much more complicated than suggested by initial plate tectonic models (e. g., Coney et al. 1980; Blake et al. 1982; Howell and Jones 1984; Howell et al. 1985). Much recent work suggests that many orogenic belts are composed of fault-bounded structural units (termed tectonostratigraphic terranes) which record mutually contrasting stratigraphic, structural, and/or metamorphic histories. These terranes appear to have originated in palinspastic settings

[1] Departamento Geologia, Faculdade Ciências, 1294 Lisbon, Portugal
[2] I. T. G. E., Rios Rosas 23, 28003 Madrid, Spain
[3] Department of Geology, University of Georgia, 30602 Athens, Georgia, USA

R. D. Dallmeyer and E. Martínez García (Eds.)
Pre-Mesozoic Geology of Iberia

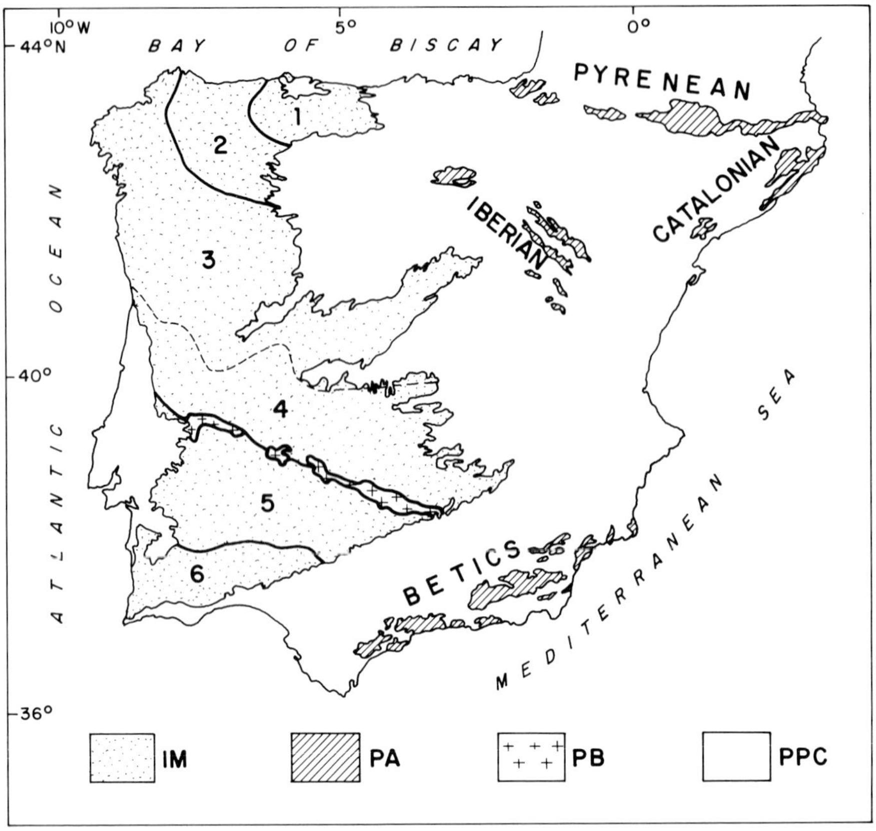

Fig. 1. Paleozoic and Precambrian exposures in the Iberian Peninsula. *IM* Iberian Massif (zonal subdivision after Lotze 1945 and Julivert et al. 1974): *1* Cantabrian Zone; *2* West Asturian-Leonese Zone; *3* and *4* Central-Iberian Zone of Julivert et al. 1974, which includes Lotze's Galicial-Castilian (3), and Alcudian (4) Zones; *5* Ossa-Morena Zone; *6* South Portuguese Zone. *PA* Paleozoic and Precambrian exposures in Alpine ranges. *PB* Pedroches Batholith. *PPC* post-Paleozoic cover

removed from the continent on which they are presently situated, and have been interpreted to represent fragments of various crustal elements (oceanic lithosphere, intraoceanic arcs, continental arcs, etc.) which have been tectonically juxtaposed during plate movements. Detailed terrane analyses have been reported for the generally well-exposed Mesozoic-Cenozoic orogens in the circum-Pacific realm (e.g., Coney et al. 1980; Jones et al. 1983a, b; Howell and Jones 1984). Application of concepts learned in these settings in the older, circum-Atlantic orogens is complicated by many factors, including: (1) removal of supracrustal levels through extensive erosion; (2) pervasive metamorphism which precludes establishment of paleontologic provenance and/or paleomagnetic determination of paleolatitudes; and, (3) disruption or removal of critical overstepping sequences which inhibits determination of amalgamation and/or accretionary relationships. In spite of these limiting constraints, the terrane concept has been successively applied in many of the Paleozoic circum-Atlantic orogenic belts (e.g., Dallmeyer 1989). We feel that a generalized terrane perspective is also appropriate for the Iberian Massif, and have outlined four major terrane elements (Ribeiro et al. 1987b). These include (Fig. 3): (1) an extensive, reference autochthonous terrane (Iberian Terrane) composed of both internally imbricated and relatively parautochthonous elements, which include the Cantabrian, West Asturian-Leonese and Ossa-Morena Zones as well as most of the Central-Iberian Zone (Julivert et al. 1974); (2) various oceanic terranes which are exotic relative to the Iberian Terrane and occur as basal allochthonous sheets in the northern Central-Iberian zone (Cabo Ortegal, Ordenes, Bragança and Morais nappe complexes) as well as south of the Iberian Terrane, where they constitute the Pulo do Lobo Zone (which includes the Beja-Acebuches ophiolite, Munhá et al. 1986) (3) a Northern Continental Exotic Terrane represented by upper allochthonous nappes in the Cabo Ortegal, Ordenes,

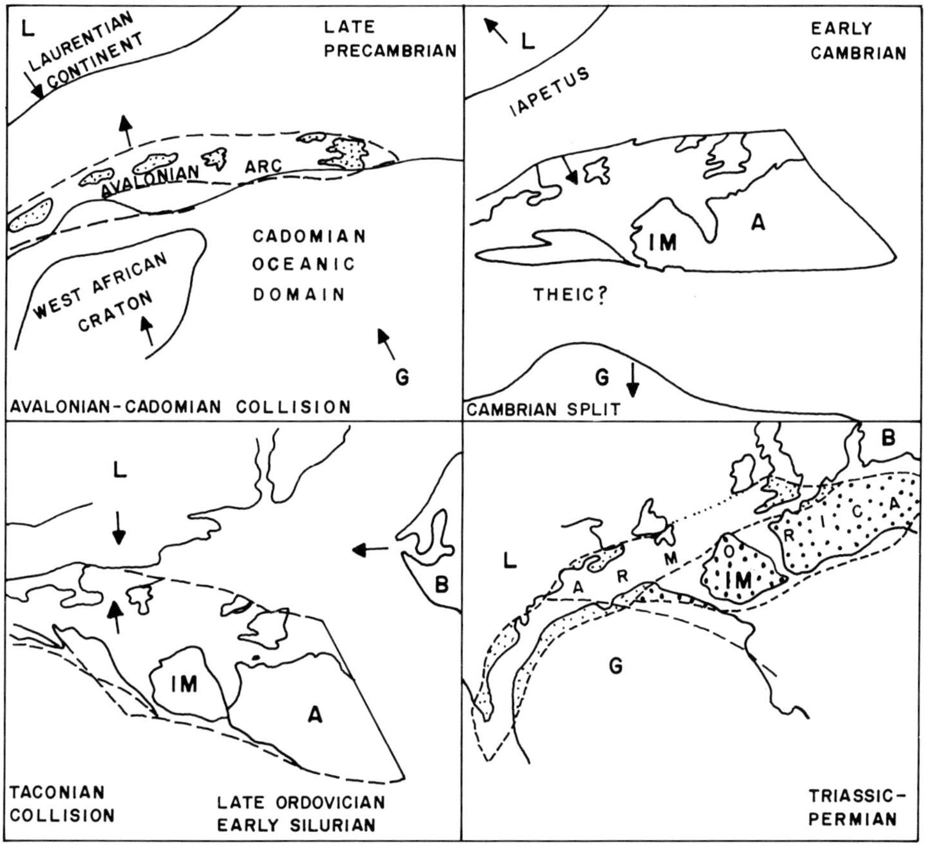

Fig. 2. Speculative reconstruction of the development of Armorica (after Van der Voo 1982, 1983). *A* Armorica; *B* Baltica; *G* Gondwana; *IM* Iberian Massif; *L* Laurentia

Bragança and Morais complexes of the northern Central-Iberian Zone; and, (4) the suspect South Portuguese Terrane which is separated from the Iberian Terrane by the Pulo do Lobo oceanic unit. In this closing chapter we attempt to compile and focus the geologic relationships described in preceding chapters, in an effort to produce an internally consistent and geologically viable model for the overall geodynamic evolution of the Iberian Massif.

2 Precambrian Evolution of Iberian Terranes

With the obvious exception of ophiolitic units, representing remnants of newly formed Paleozoic oceanic crust, Precambrian rocks constitute the basement of the other, continental, Iberian terranes, and has been variably (generally penetratively) reactivated during the Late Paleozoic Variscan orogeny. In the South Portuguese suspect terrane, approximately equivalent to the South Portuguese zone, situated south of the Pulo do Lobo ophiolitic terrane, no rocks older than Upper Devonian are exposed (Oliveira, Part VI, Chap. 2). Nevertheless, the continental nature of this unit is reflected by: (1) the platform characteristics of most of the sedimentary record (Oliveira, Part VI, Chap. 2); (2) the crustal origin of felsic volcanics which constitute most of the Iberian Pyrite Belt (Munhá 1983); and, (3) the existence of crust-derived S-type granitoids which intruded South Portuguese Zone cover rocks in the Upper Carboniferous (Simancas 1984).

The lack of basement exposure in this terrane, along with the absence of regional geophysical studies (namely magnetolluric sounding) prevents any correlation between the South Portuguese terrane and other continental suspect terranes which occur in NW Iberia. These, conversely, are mostly composed of basement rocks (Ribeiro et al., Part IV, Chap. 3.1). Their evolution is outlined below.

2.1 Precambrian Evolution of Suspect Exotic Terranes in Northwestern Iberia

As it has been described elsewhere (Ribeiro et al., Part IV, Chap. 3.1), allochthonous units which occur in the northwest of the Central-Iberian Zone constitute a composite nappe complex, largely removed from respective original settings, which was thrust eastward onto parautochthonous sequences of to the Iberian Terrane. A lower allochthonous sheet of ophiolites and an upper continental nappe complex are distinguished in this region. Both are interpreted as remnants of exotic terranes (Ribeiro et al. 1987b). The latter, which occurs at the Cabo Ortegal, Ordenes, Bragança and Morais massifs, consists of both Precambrian basement and Lower Paleozoic cover sequences.

A telescoped section of continental lithospheric block is reflected by the Morais Massif (Ribeiro 1983; Ribeiro et al. 1987a), in which a basal garnet lherzolite sheet, interpreted as a subcontinental mantle slice, is overlain by a mafic granulite thrust sheet (lower crust), in turn succeeded by a granitic augengneiss (upper crust) covered unconformably by a micaschist succession, including Upper Cambrian to Lower Ordovician fossils (Timofeev, in Chacón 1979), which is interpreted as a lower Paleozoic cover sequence.

All this sequence underwent a penetrative tectonothermal overprint by the Variscan orogeny, during which it was obducted onto the Iberian miogeocline. This overprint precludes reliable recognition of Precambrian events in these sequences. Evidence suggesting they represent a basement may be correct are: (1) existence of a foliation in augengneisses that predates emplacement of mafic dykes and sills that also intrude overlying Lower Paleozoic micaschists (the mafic dykes were subsequently affected by the Variscan deformation and transformed into amphibolites); and, (2) mafic granulite and metaperidotite sheets show at least two phases of penetrative deformation, accompanied by granulite facies metamorphism, prior to formation of dilation veins in which hornblende crystals grew with long axis parallel to Variscan stretching lineations.

An alternative interpretation of the mafic granulites and metaperidotites has been recently set forth by Peucat et al. (1990) and Gil Ibarguchi et al. (1990) on the basis of geochemical and geochronological results from similar rocks exposed in the Cabo Ortegal complex. They suggested they formed in an arc environment in the Lower Ordovician were metamorphosed under eclogitic/granulitic conditions soon after formation (in the 490–470 Ma interval) in a subduction tectonic setting. The apparent contradiction between these data and former interpretations (Ribeiro et al. 1987a) clearly shows the need of further research. Two aspects with important connotations for the palinspastic origin of these exotic terranes must be emphasized: (1) the lack of pan-African ages; and, (2) the frequent reference to ca. 2–2.5 Ga zircon ages (Kuijper 1979; Kuijper et al. 1982; Peucat et al., 1990), interpreted as the age of the continental source for some of the rocks in these terranes, which is similar to the source age of many Late-Precambrian and Paleozoic rocks within the Iberian Terrane (Schäfer et al. 1988, 1989; Nägler et al. 1988, 1989).

2.2 Precambrian Evolution of the Iberian Terrane

The Iberian Terrane is comprised of the clasically defined Palentian, Cantabrian, West Asturian-Leonese and Ossa-Morena Zones, together with autochthonous/parautochthonous sequences exposed in the Central-Iberian Zone (cf. Figs. 1, 3). Apart from local, possible relics of older basement that have been suggested to occur in the Ossa-Morena Zone (Pérez Lorente 1979; Pascual 1981), the Precambrian history recorded within the Iberian Terrane initiated in the Upper Proterozoic and continued through a complete Wilson cycle, terminating in tectonothermal activity variably termed the Cadomian or pan-African orogeny. Throughout most of the Iberian Terrane the Precambrian record is fragmentary because of penetrative tectonothermal reworking during the Paleozoic. However, in portions of the Central-Iberian and Ossa-Morena zones this overprinting was less penetrative, and a clearer picture of the Precambrian evolution may be composed. Here, it appears that an initial Late Proterozoic rifting eventually resulted in development of oceanic lithosphere of uncertain extent. Quesada et al. (1987) have suggested that this history is, in part, recorded by a bimodal volcanic suite (Blastomylonitic Formation) and serpentinites (also, see Quesada et al., Part V, Chap. 2.1). Older basement components may be present within the Blastomylonitic Formation, however, this has not been clearly demonstrated.

A passive continental margin developed during Late Proterozoic drifting. Sequences deposited during this Middle-Upper Riphean interval are represented by lower portions of the Serie Negra Group (Montemolin succession of Eguiluz 1987). Upper sedimentary units within the Serie Negra Group

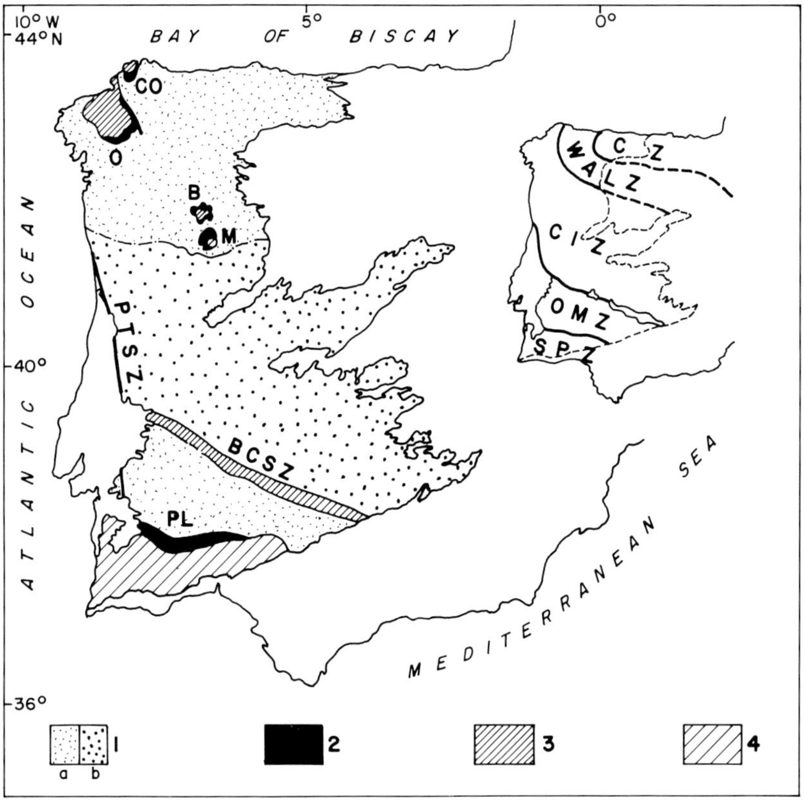

Fig. 3. Simplified preliminary terrane map of the Iberian Massif. *1* Iberian terrane, *a* imbricated margins, *b* parautochthonous core; *2* oceanic exotic terranes: *PL* Pulo do Lobo ophiolitic terrane; *CO* Cabo Ortegal, *O* Ordenes, *B* Bragança, *M* Morais complexes; *3* Northwestern Continental Exotic terranes; *4* South Portuguese Suspect Terrane. *BCSZ* Badajoz-Cordoba Shear Zone; *PTSZ* Porto-Tomar Shear Zone. *Inset* Julivert et al. (1974) zonal division for comparison

(Tentudia succession of Eguiluz 1987) constitute a coarsening-upward marine sequence which contains an increasingly significant abundance of calc-alkaline volcanic detritus. Quesada (1989) suggested this material represents a back-arc or foreland basin fill which marked initiation of outboard tectonic activity. Deformation appears to have propagated continentward, and by the Upper Riphean affected the Serie Negra Group where it is recorded as initial phases of the Cadomian orogeny (Fig. 4). The subhorizontal attitude of associated structural elements and the local development of high-pressure metamorphic assemblages may indicate that this deformation was of collisional character. An extensive suite of calc-alkaline volcanic rocks unconformably overlies deformed Serie Negra units, and appears to indicate that this collision involved an arc terrane and the older miogeocline (Quesada 1989). Subsequent subduction after arc accretion is stratigraphically recorded in the Upper Riphean-Lower Vendian interval (see, Quesada et al., this Vol.). This terminated in a second phase of Cadomian tectonothermal activity in the Middle-Late Vendian (Fig. 4). Associated synorogenic volcanic rocks are relatively enriched in K_2O towards the south (Sánchez Carretero et al., this Vol.), perhaps indicating that the associated subduction zone was inclined in that direction. A-type subduction of the northern continental domain during final closure of the intervening oceanic tract probably ultimately led to formation of a foreland basin which filled with synorogenic flysch sequences, represented by Late-Precambrian parts of the so-called "Schist and Greywacke Complex" of the Central-Iberian Zone and related sequences in other zones. Subsequent deformation appears to have migrated northward (Quesada 1989) from an area approximately coincident with the present location of the Badajoz-Córdoba shear zone (Fig. 3). This shear zone is locally marked by development of high-pressure metamorphic rocks derived from mafic and ultramafic protoliths (Mata and Munhá 1986). It may, therefore, represent a cryptic suture which was structurally overprinted during subsequent Paleozoic tectonothermal activity.

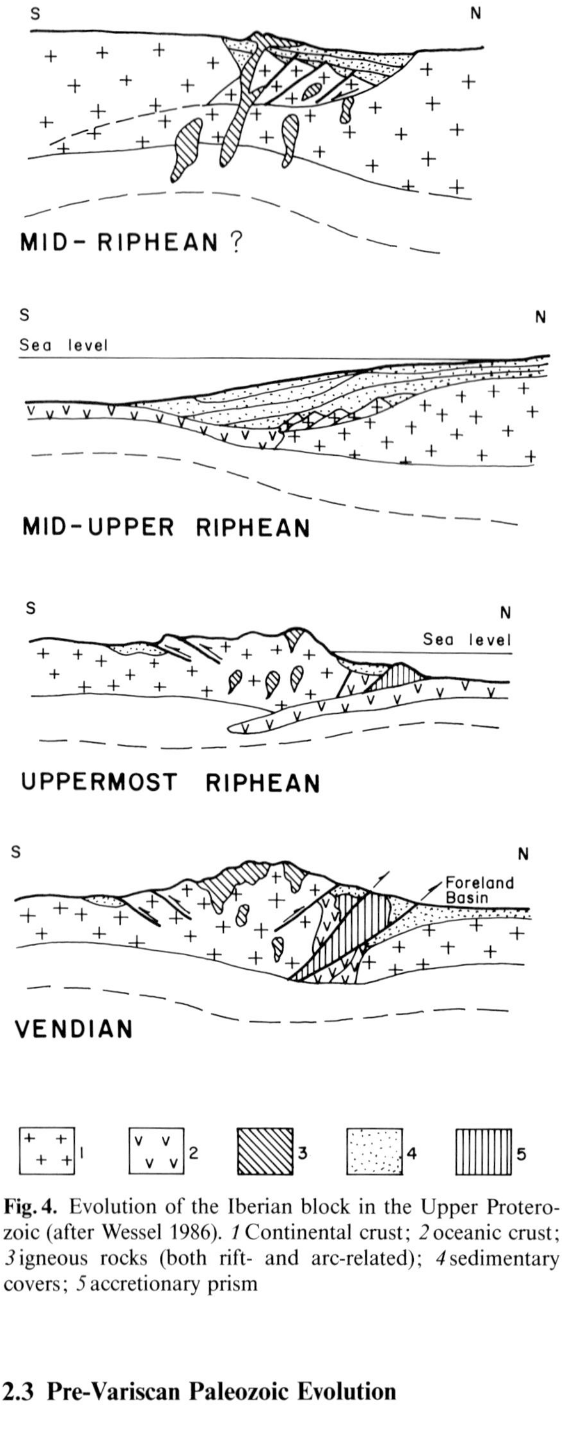

Fig. 4. Evolution of the Iberian block in the Upper Proterozoic (after Wessel 1986). *1* Continental crust; *2* oceanic crust; *3* igneous rocks (both rift- and arc-related); *4* sedimentary covers; *5* accretionary prism

2.3 Pre-Variscan Paleozoic Evolution

Initiation of a Paleozoic Wilson cycle is recorded in Lower Paleozoic units exposed in all tectonic elements of the Iberian Terrane. The pre-accretionary history which they collectively define initiated with a prolonged phase of intracontinental rifting which led to eventual separation and formation of oceanic lithosphere. This was followed by a period of seafloor spreading during which miogeoclinal successions accumulated along the passive continental margin of the Iberian Terrane (Fig. 5). Although this evolution is suggested by geologic relationships within all elements of the Iberian Terrane, there exists no evidence to suggest that any of the continental units exposed within exotic Iberian terranes correspond to the rifted continental counterparts. Conversely, there is ample evidence of strike-slip, tangential faulting during both Lower Paleozoic extension and subsequent Paleozoic convergence.

2.3.1 Pre-Variscan Evolution of Continental Suspect Terranes

The continental, suspect terranes exposed within Iberia exhibit incomplete Paleozoic successions; therefore, there is little direct evidence of their pre-Late Paleozoic evolution. A rifting event is likely recorded in the northwestern Upper Allochthon where alkaline mafic dikes cut basement units and cover sequences (Lagoa micaschist) in the Morais massif (Ribeiro et al., Part IV, Chap. 3.1). By contrast, passive margin conditions are reflected in Upper Devonian shallow-water sedimentary units exposed in the South Portuguese suspect terrane (Oliveira, Part VI, Chap. 2). However, the lack of an older record precludes determination of the prior tectonothermal evolution.

The recent interpretation proposed by Gil Ibarguchi et al. (1990) and Peucat et al. (1990) (considering that a part of the rocks forming the exotic terranes in the northern Central-Iberian Zone originated in subduction-related arc setting, in which both newly-formed and basement rocks underwent a complex tectonothermal history involving eclogitic and granulitic stages in the Ordovician-Silurian), largely contrasts with the situation in the Iberian Terrane during the same time span, characterized by miogeocline development in passive margin conditions followed by continental rifting in the Upper Ordovician-Silurian (see below). This is an important difference, providing strong support to the exotic consideration of the northwestern Upper Allochthons.

2.3.2 Pre-Variscan Paleozoic Evolution of the Iberian Terrane

Throughout the Paleozoic, miogeoclinal successions accumulated within the Iberian Terrane, with deposition of shallow-water, platform type sequences in northwestern areas (Cantabrian Zone).

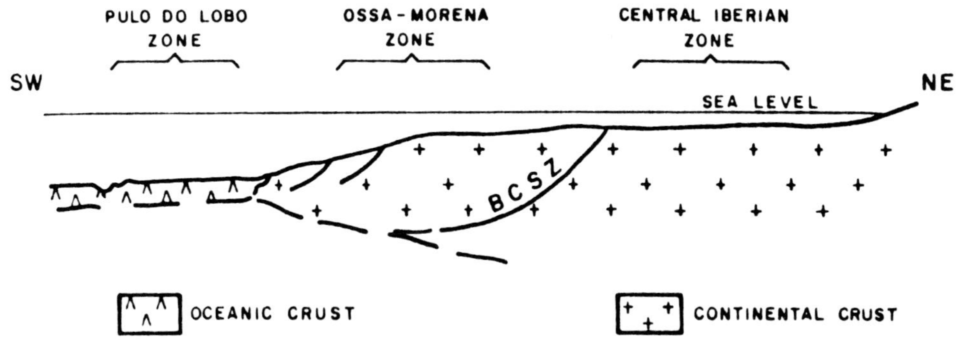

Fig. 5. Palinspastic section of the southwestern margin of the Iberian Terrane in the Lower Paleozoic. *BCSZ* Badajoz-Córdoba Shear Zone

Progressively thicker sequences accumulated southwestward, and these display deep-water characteristics. The miogeocline has a complex internal geometry and records a prolonged evolution. Initial continental rifting in the Cambrian resulted in development of a central axial trough within the Central-Iberian Zone in which flyschoid units accumulated (Cambrian, upper part of the Slate and Graywacke Complex). Shallow-water assemblages were deposited in flanking areas to the northeast and southwest. In the Central-Iberian and Ossa-Morena Zones the end of the Cambrian is marked by the Sardic deformational phase, which is not recorded northeastward (Cantabrian or West Asturian-Leonese zones). In the Central-Iberian Zone, Sardic deformation produced folds at high angles to Cambrian isofacial lines. Synsedimentary thrusts also developed at this time as a result of gravity sliding into the trough axis. Lefort and Ribeiro (1980) suggested that this deformation likely developed in response to weak transpressive stresses associated with dextral motion along the boundary between the Ossa-Morena and Central-Iberian zones (e. g., reactivation of the Badajoz-Córdoba shear zone). In the Ossa-Morena Zone, Sardic deformation was less penetrative and only resulted in regional tilting of pre-Ordovician sequences. We suggest this developed in response to final continental breakup and initial drifting of the Iberian Terrane from Gondwana, following the prolonged Cambrian phase of rifting.

Sedimentation rates diminished over the Iberian Terrane during the Ordovician, and there was widespread deposition of shallow-water, platform facies. During the Lower Ordovician, the widespread Armorican quartzite was deposited, indicating proximity of a large continental sediment supply (Gondwana?). Local records of Upper Ordovician glaciation in Iberia (e. g., Perroud and Bonhommet 1981; Robardet and Doré 1988) indicate a polar latitude, and are consistent with the high latitudinal position also suggested for Gondwana at this time (e. g., Bonhommet and Perroud 1986). These relationships suggest that although the Iberian Terrane drifted from Gondwana, it was likely not far removed; a conclusion also suggested by faunal relationships (see Robardet and Gutiérrez Marco, this Vol.).

The miogeocline was subjected to a new accelerated tension episode during the Upper Ordovician. The axis of maximum subsidence migrated to the SW and become located in the SW border of the Ossa-Morena Zone.

An intracontinental rift system developed along the Badajoz-Córdoba shear zone and its northern prolongation; this suggests the reactivation of an older Precambrian zone of weakness leading to crustal thinning; this thinned belt was tectonically inverted during the Variscan compression, producing a deep flower structure. The magmatic suite characteristic of this tectonic setting contains members evolving from an alkaline-transitional bimodal sequence (peralkaline metarhyolite end members), to a sequence of oceanic tholeiitic basalts; this Silurian assemblage is well preserved in the hanging-wall simplex of Macedo de Cavaleiros (Ribeiro 1986, 1987) on top of the Lower Allochthonous thrust complex of the northern Central-Iberian Zone.

Thus, an incipient oceanic rift developed in the Silurian, but the age of maximum width of the ocean is not known precisely (Upper Silurian or Lower Devonian). It is perhaps not a coincidence that fragmentation of the Armorican Plate occurred during the period of maximum velocity of drift, from an Upper Ordovician south-pole position to a near-equator Devonian position. The newly-formed ocean between the Armorican Plate and Gondwana is now headed reflected by the Beja-Acebuches Ophiolite Terrane (Munhá et al. 1986) and associated Pulo do Lobo sequence of oceanic affinities. This ocean was connected in some way with the wide Rheic ocean, indicated by a difference in paleolatitude for Lower Paleozoic rocks in Armorica and Baltica, and was the origin of the oceanic exotic terranes in the Iberian Massif.

2.4 The Variscan Orogeny

It is not known at exactly what time initial subduction of oceanic lithosphere located southwest and west of the Iberian Terrane began. However, commencement of related orogenic activity is clearly marked by an unconformity developed at the base of Frasnian sequences exposed in southwestern sectors of the Ossa-Morena Zone. This unconformity becomes progressively younger northeast and eastward, and is opposite to the vergence of first-phase Variscan structures in this area (a backward propagating structural sequence). These geometric and chronologic relationships, shown by both stratigraphy (Quesada et al., Part V, Chap. 2.4) and geochronology (Dallmeyer and Quesada, 1990), together with the typical I-type magmatism which accompanied initial Variscan deformation, suggests an east or northeast subduction polarity (Fig. 6). Variscan tectonic activity appears to have commenced with obduction of oceanic sequences (Beja-Acebuches ophiolite and Pulo do Lobo units) onto the southern Iberian Terrane (Ossa-Morena Zone). Obduction polarity appears to have been northeastward, similar to that of the subduction polarity. This indicates development of a flake geometry (Fig. 6), which may have been induced by an anomalous oceanic zone which was too buoyant to allow subduction. Therefore, it was tectonically stripped from its hot oceanic lithosphere during convergence.

Less certain are the palinspastic derivation and obduction mechanics of the oceanic exotic allochthons exposed in northern segments of the Iberian Terrane. We favour, however, their correlation with the Pulo do Lobo ophiolite, provided the similarities of the Upper Ordovician-Silurian rifting processes are recorded in both southern (Ossa-Morena Zone) and northern (lowermost northwestern Allochthon) sectors of the Iberian Terrane. 40Ar/39Ar hornblende cooling ages of ca. 385 Ma, reported from the Morais ophiolite by Dallmeyer and Gil Ibarguchi (1990), were interpreted as dating cooling after a tectonothermal event associated with the obduction of the ophiolite sheet. This may be taken as a good estimate for the age of ophiolite obduction onto the Iberian Terrane.

Within southern sectors of the Iberian Terrane, flake development continued in the Late Devonian and ultimately resulted in delamination when uppermost portions of the Paleozoic cover of the Ossa-Morena Zone were thrust northeastward (Fig. 6), while lower crustal elements (basement and basal cover units) were underthrust below the Central-Iberian Zone. This mechanical behaviour may have been favoured because of a prior thinning of the continental crust within this zone during the Lower Paleozoic rifting process. By the Middle-Upper Devonian, continued closure of the southwestern ocean ultimately resulted in accretion of a continental terrane now concealed beneath the South Portuguese Zone. Continent-continent collision was probably nearly completed by the Lower Carboniferous in the corner of the Cantabrian arc, which appears to have acted as a promontory during continental impingement (Brun and Burg 1982). However, southwest of the Iberian Terrane the oceanic domain was not yet completely consumed, as demonstrated by the calc-alkaline character of Lower Carboniferous magmatism within the Ossa-Morena Zone. Associated subduction polarity appears to have remained northeast, and we therefore favour a model of continued flake development (Fig. 6), with the western continental element undergoing delamination and simultaneous overthrusting to the southeast. This led to detachment of continental elements which are now represented in the exotic continental terranes which structurally overlie the Iberian Terrane. The obduction of these exotic units is better constrained than that of the underlying ophiolites; their progressive cooling through ca. 500 °C is dated at ca. 380 Ma; through ca. 400 °C at ca. 370 Ma; and finally, through ca. 300 °C at ca. 350 Ma (Peucat et al. 1990).

During the Lower Carboniferous a continent-continent collisional regime gradually developed along the southwestern margin of the Iberian Terrane, and a suite of pull-apart basins formed. These sequentially overstepped contacts between the Ossa-Morena Zone and Pulo do Lobo successions. Because of thin-skinned tectonic development within the South Portuguese Terrane (Ribeiro and Silva 1983), the exact location of the suture between the Iberian Terrane and the western continental element (of uncertain basement affinity) is not known. However, this collision is marked by deposition of increasingly shallow-water sequences southwest-

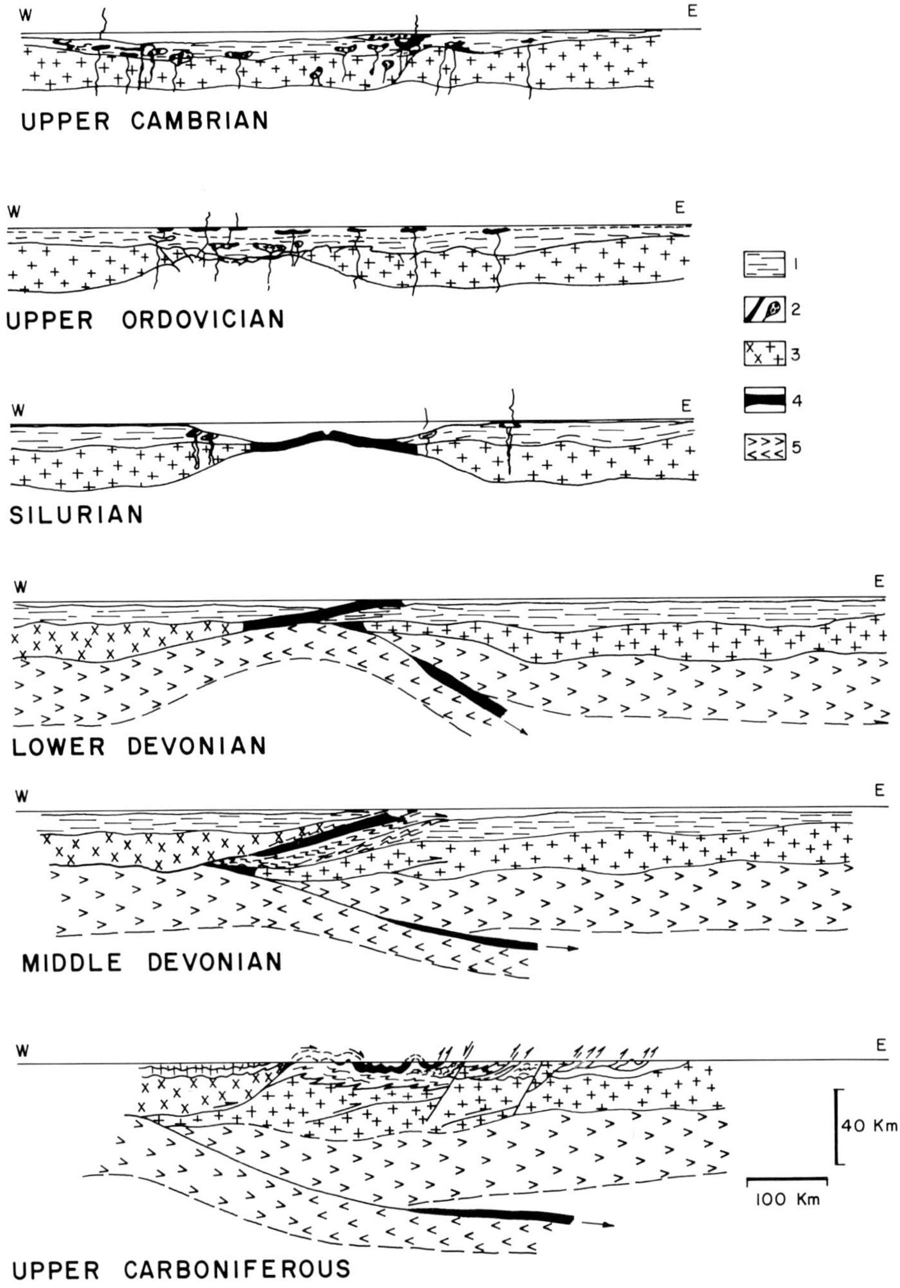

Fig. 6. Schematic geodynamic evolution of the Iberian Massif in the Paleozoic (after Iglesias et al. 1983). *1* Paleozoic cover sequences; *2* igneous rocks (mafic and acid); *3* continental crust; *4* oceanic crust; *5* upper mantle

ward across the South Portuguese Terrane. The margin of this southwestward continent was likely extended, allowing emplacement of bimodal volcanic sequences now exposed in the Pyrite belt of the South Portuguese Terrane (Silva et al., this Vol.). These and host sedimentary units were subsequently metamorphosed to a low grade and deformed in a sinistral, transpressional regime during the Visean-Westphalian. At the same time, in northern sectors of the Iberian Terrane, a continent-continent collision proceeded through successive footwall collapse (Bastida et al. 1987), with development of typical thin-skinned structures (in the Cantabrian Zone). This extended to deeper crustal levels westward and involved basement elements in the West Asturian-Leonese Zone and portions of the Central-Iberian Zone. The age of deformation in northern portions of the Iberian Terrane becomes younger in the sense of structural vergence, so that, in a general sense, a forwardly propagating structural sequence is represented, but, in detail, this was complicated by out-of-sequence thrust which led to formation of antiformal stacks over major crustal discontinuities. Associated metamorphism in northwestern portions of the Iberian Terrane progressively intensified with eventual generation of the S-type granites which are widespread in the Central-Iberian and West Asturian-Leonese Zones. However, these intrusives are often associated with mixed and even I-type magmatic units that we interpret to have originated above an east-dipping subduction zone (Fig. 6).

As a result of superposed Variscan deformation the Ibero-Armorican arc was progressively tightened. The genesis of the arc can be explained by kinematic analysis of the structures (Matte and Ribeiro 1975; Ries and Shackleton 1976) and by interpretation of paleomagnetic data (Perroud 1985). Both approaches point to a secondary origin for the arc, with progressive arcuation, but some disagreement persists on the precise mechanism of arcuation and its relation to the sense of movement of opposed forelands.

Termination of the Variscan orogeny was marked by development of brittle fault systems. The Porto-Tomar ductile fault (Fig. 3) was reactivated at this time as a dip-slip, high-angle reverse structure, downthrown to the east. This indicates an essentially east-west maximum compressive direction. At approximately the same time, conjugate strike-slip faults were generated. These indicate a north-south maximum compresive stress direction. This constrictional regime likely reflected a corner effect within a second-order Iberian block, which was compressed within a diffuse, large-scale, right-lateral shear zone extending between the Ural and southern Appalachian orogens (e. g.) (Arthaud and Matte 1975). This brittle movement reflects the final manifestation of an earlier ductile transpressive regime that had been active with the same vergence throughout most of the Variscan convergence onto the Iberian Terrane.

References

Arthaud F, Matte P (1975) Les décrochements tardihercyniens du sud-ouest de l'Europe. Géometrie et éssai de réconstitution des conditions de la déformation. Tectonophysics 25: 139–171

Autran A, Cogné J (1980) La zone interne de l'orogène varisque dans l'Ouest de la France et sa place dans le dévelopement de la chaîne hercynienne. In: Cogné J, Slanski M (eds) Geology of Europe. Mém BRGM 108: 191–202

Bard JP (1971) Sur l'alternance des zones métamorphiques dans le segment hercynien sud-ibérique; comparaison de la variabilité des caractères géotectoniques de ces zones avec les orogènes 'orthotectoniques'. Bol Geol Min 82 (3–4): 324–345

Bard JP, Capdevila R, Matte P, Ribeiro A (1973) Geotectonic model for the Iberian Variscan Orogen. Nature 241: 50–52

Bastida F, Martínez-Catalan JR, Perez-Estaun A (1987) Crustal thickening and deformation sequence in the footwall to the suture of the Hercynian Belt of NW Spain. Conf on Deformation and plate tectonics, Gijón (Spain), Abstr vol, pp 14

Blake MC Jr, Howell DG, Jones DL (1982) Preliminary tectonostratigraphic terrane map of California. US Geol Surv Open-File Rep 82–593, scale 1:500000

Bonhommet N, Perroud H (1986) Apport du paléomagnétisme à la compréhension de l'orogènese hercynienne en Europe occidentale. Bull Soc Géol Fr 8: 2: 35–42

Brun JP, Burg JP (1982) Combined thrusting and wrenching in the Ibero-Armorican Arc: a corner effect during continental collision. Earth Planet Sci Lett 61: 319–332

Burg JP, Iglesias M, Laurent PH, Matte P, Ribeiro A (1981) Variscan intracontinental deformation: the Coimbra-Córdoba Shear Zone (SW Iberian Peninsula). Tectonophysics 78: 161–177

Carvalho D (1972) The metallogenetic consequences of plate tectonics and the upper Paleozoic evolution of southern Portugal. Est Not Trab SFM 20 (3–4): 297–320

Chacón J (1979) Estudio geológico del sector central del anticlinorio Portalegre-Badajoz-Córdoba (Macizo Ibérico Meridional). Thesis, Univ Granada, pp 1–728

Coney P, Jones DL, Monger JWH (1980) Cordilleran suspect terranes. Nature 288: 329–333

Dallmeyer RD (ed) (1989) Terranes in the Circum-Atlantic Paleozoic Orogens. GSA Spec Pap, n° 230

Dallmeyer RD, Gil Ibarguchi JI (1990) Age of amphibolitic metamorphism in the ophiolitic unit of the Morais Allochthon (Portugal): implications for early Hercynian orogenesis in the Iberian Massif. J Geol Soc Lond

Dewey JF, Burke K (1973) Tibetan, Variscan and Precambrian basement reactivation, products of continental collision. J Geol 81: 683–692

Eguiluz L (1987) Petrogénesis de rocas ígneas y metamórficas en el antiforme Burguillos-Monesterio. Macizo Ibérico Meridional. PhD Thesis, Univ Bilbao, pp 1–694

Gil Ibarguchi JI, Mendia M, Girardeau J, Peucat JJ (1990) Eclogites and granulites in the Cabo Ortegal Complex (northwestern Spain). Lithos

Howell DG, Jones DL (1984) Tectonostratigraphic terrane analysis and some terrane vernacular. In: Howell DG, Jones DL, Cox A, Nur A (eds) Proc Circum-Pacific Terrane Conference. Stanford Univ Publ Geol Sci 18: 6–9

Howell DG, Jones DL, Schemer ER (1985) Tectonostratigraphic terranes in the Circum-Pacific region. In: Howell DG (ed) Tectonostratigraphic terranes of the Circum-Pacific region. Houston, Circum-Pacific Council for Energy and Mineral Resources. Earth Sci Ser 1: 3–30

Iglesias M, Ribeiro ML, Ribeiro A (1983) La interpretación aloctonista de la estructura del noroeste peninsular. In: Geología de España, Libro Jubilar J M Ríos IGME 1: 459–466

Jones DL, Howell DG, Coney PJ, Monger JWH (1983a) Recognition, character and analysis of tectonostratigraphic terranes in western North America. In: Hashimoto M, Uyeda S (eds) Advances in earth and planetary sciences. Terra Sci, Tokyo, pp 31–35

Jones DL, Howell DG, Coney PJ, Monger JWH (1983b) Recognition, character and analysis of tectonostratigraphic terranes in western North America. J Geol Educ 31: 295-303

Julivert M, Fontboté JM, Ribeiro A, Conde L (1974) Mapa Tectónico de la Península Ibérica y Baleares. Escala 1: 1000000. IGME, pp 1-101

Kuijper RP (1979) U-Pb systematics and the petrogenetic evolution of infracrustal rocks in the Paleozoic basement of western Galicia, NW Spain. Vehr ZWO Lab Isotopen Geol (Amst) 5: 1-101

Kuijper RP, Priem HNA, Tex E Den (1982) Late Archean-Early Proterozoic source areas of zircons in rocks from the Paleozoic orogen in western Galicia. Precambrian Res 19 (1): 1-29

Lefort JP, Ribeiro A (1980) La faille Porto-Badajoz-Cordoue a-t-elle contrôlé l'evolution de l'océan paléozoique sud-armoricain?. Bull Soc Géol Fr (7) 22 (3): 455-462

Lorenz V, Nicholls JA (1984) Plate and intraplate processes of Hercynian Europe during the late Paleozoic. Tectonophysics 107: 25-26

Lotze F (1945) Zur Gliederung der Varisciden der Iberischen Meseta. Geoteckt Forsch 6: 78-92

Mata J, Munhá J (1986) Geodynamic significance of high-grade metamorphic rocks from Degolados-Campo Maior (Tomar-Badajoz-Córdoba Shear Zone). Maleo 2 (13): 28

Matte P (1983) Two Geotraverses across the Ibero-Armorican Variscan Arc of western Europe. In: Rast N, Delany M (eds) Profiles of orogenic belts. AGU Geodyn Ser 10: 53-81

Matte P (1986) Tectonics and plate tectonics models for the Variscan Belt of Europe. Tecnonophysics 126: 329-374

Matte P, Ribeiro A (1975) Forme et orientation de l'ellipsoïde de déformation dans la virgation hercynienne de Galice. Rélations avec le plissement et hypothèses sur la genèse de l'arc ibéro-armoricain. C R Acad Sci Paris D 280: 2825-2828

Munhá J (1983) Hercynian magmatism in the Iberian Pyrite Belt. In: Lemos MJ, Oliveira JT (eds) The Carboniferous of Portugal. Mem Serv Geol Port 29: 39-81

Munhá J, Oliveira JT, Ribeiro A, Oliveira V, Quesada C, Kerrich R (1986) Beja-Acebuches Ophiolite: characterization and geodynamic significance. Maleo 2 (13): 31

Nägler TF, Gebauer D, Schäfer HJ, Quadt A Von (1988) Sm-Nd and Pb- isotope geochemistry of Paleozoic sediments of the Almadén and Guadalmez synclines (Central Iberian Zone, Spain). In: Cinturones Orogénicos, II Cong Geol Esp, Symp vol, pp 45-50

Nägler TF, Gebauer D, Schäfer HJ (1989) Nd-, Sr- and Pb-isotope geochemistry of two pre-Permian sedimentary profiles from the Spanish Meseta: evidence of both crustal recycling and crustal growth. Terra Abstr 1 (1): 336

Pascual E (1981) Investigaciones geológicas en el sector Córdoba-Villaviciosa de Córdoba (Sector Central de Sierra Morena). Thesis, Univ Granada, pp 1-521

Pérez Lorente F (1979) Geología de la zona Ossa-Morena al Norte de Córdoba (Pozoblanco-Belmez-Villaviciosa de Córdoba). Thesis, Univ Granada 281: 1-340

Perroud H (1985) Synthèse des résultats paléomagnétiques sur le Massif Armoricain. Hercynica 1 (1): 65-71

Perroud H, Bonhommet N (1981) Paleomagnetism of the Ibero-Armorican arc and the Hercynian orogeny in Western Europe. Nature 292: 445-448

Peucat JJ, Bernard Griffiths J, Gil Ibarguchi JI, Dallmeyer RD, Menot RP, Cornichet J, Iglesias M (1990) Geochemical and geochronological cross section of the deep Variscan crust: the Cabo Ortegal high-pressure nappe (NW Spain). Tectonophysics 177: 263-292

Quesada C (1989) Precambrian successions in SW Iberia: their relationship to Cadomian orogenic events. J Geol Soc (Lond) Spec Issue: The Cadomian Orogeny (in litt.)

Quesada C, Larrea FJ, Florido P et al. (1987) Mapa Geológico Minero de Extremadura. Dir Gen Ind Energ Min Junta Extremadura, pp 1-131

Ribeiro A (1983) Los complejos de Bragança y Morais. In: Geologia de España, Libro Jubilar J M Ríos. IGME 1: 450-454

Ribeiro A, Silva JB (1983) Structure of the South Portuguese Zone. In: Sousa MJL, Oliveira JT (eds) The Carboniferous of Portugal. Mem Serv Geol Port 29: 83-89

Ribeiro A, Dias R, Pereira E, Merino H, Sodre Borges F, Noronha F, Marques M (1987a) Guide-book for the Miranda do Douro-Porto Excursion. Conf Deformation and plate tectonics, Gijón (Spain), pp 1-50

Ribeiro A, Quesada C, Dallmeyer RD (1987b) Tectonostrátigraphic terranes and the geodynamic evolution of the Iberian Variscan Fold Belt. Conf Deformation and plate tectonics, Gijón (Spain), Abstr vol, pp 60-61

Ribeiro ML (1986) Geología e petrologia da regiao a SW de Macedo de Cavaleiros (Tràs-os-Montes oriental). PhD Thesis, Univ Lisbon, pp 1-202

Ribeiro ML (1987) Petrogenesis of early Paleozoic peralkaline rhyolites from the Macedo de Cavaleiros region (NE Portugal). Geol Rundschau 76 (1): 147-168

Ries AC, Shackleton RM (1976) Pattern of strain variation in arcuate fold belts. Philos Trans R Soc Lond A Math Phys Sci 283: 281-288

Robardet M (1976) L'originalité du segment hercynien sud-ibérique au Paléozoique inférieur: Ordovicien, Silurien et Dévonien dans le Nord de la province de Seville (Espagne). C R Acad Sci Paris D 283: 999-1002

Robardet M, Doré F (1988) The late Ordovician diamictic formations from southwestern Europe: north-Gondwana glaciomarine deposits. Palaeogeogr Palaeoclimatol Palaeoecol 66: 19-31

Schäfer HJ, Gebauer D, Nägler ThF (1989) Pan-African and Caledonian ages in the Ossa-Morena Zone (southwest Spain): a U-Pb and Sm-Nd study. Terra Abstr 1 (1): 350-351

Schäfer HJ, Gebauer D, Nägler ThF, Von Quadt A (1988) U-Pb zircon and Sm-Nd studies of various rock types of the Ossa-Morena Zone (southwest Spain): In: Cinturones Orogénicos, II Cong Geol Esp, Granada. Symp vol, pp 51-57

Simancas JF (1984) Geología de la extremidad oriental de la Zona Sud-Portuguesa. Thesis, Univ Granada, pp 1-447

Staub R (1926) Gedanken zur Tektonik Spaniens. Vierteljs Naturforsch Ges Zürich 71: 196-260

Stille H (1924) Grundfragen der vergleichenden Tektonik. Brontraeger pp 1-443

Stille H (1929) Die subvarizische Vortiefe. Z Dtsch Geol Ges 81: 339-354

Dér Voo R Van (1982) Pre-Mesozoic paleomagnetism and plate tectonics. Ann Rev Earth Planet Sci 10: 191-220

Dér Voo R Van (1983) Paleomagnetic constraints on the assembly of the Old Red Continent. Tectonophysics 91: 271-283

Wessel GR (1986) The geology of plate margins. GSA Map and Chart Series, MC-59

Ziegler PA (1984) Caledonian and Hercynian consolidation of western and central Europe, a working hypothesis. Geol Mijnbouw 63: 93-108

Subject Index